The
Paramyxoviruses

THE VIRUSES

Series Editors
HEINZ FRAENKEL-CONRAT, *University of California*
Berkeley, California

ROBERT R. WAGNER, *University of Virginia School of Medicine*
Charlottesville, Virginia

THE VIRUSES: Catalogue, Characterization, and Classification
Heinz Fraenkel-Conrat

THE ADENOVIRUSES
Edited by Harold S. Ginsberg

THE BACTERIOPHAGES
Volumes 1 and 2 • Edited by Richard Calendar

THE HERPESVIRUSES
Volumes 1–3 • Edited by Bernard Roizman
Volume 4 • Edited by Bernard Roizman and Carlos Lopez

THE INFLUENZA VIRUSES
Edited by Robert M. Krug

THE PAPOVAVIRIDAE
Volume 1 • Edited by Norman P. Salzman
Volume 2 • Edited by Norman P. Salzman and Peter M. Howley

THE PARAMYXOVIRUSES
Edited by David W. Kingsbury

THE PARVOVIRUSES
Edited by Kenneth I. Berns

THE PLANT VIRUSES
Volume 1 • Edited by R. I. B. Francki
Volume 2 • Edited by M. H. V. Van Regenmortel and Heinz Fraenkel-Conrat
Volume 3 • Edited by Renate Koenig
Volume 4 • Edited by R. G. Milne

THE REOVIRIDAE
Edited by Wolfgang K. Joklik

THE RHABDOVIRUSES
Edited by Robert R. Wagner

THE TOGAVIRIDAE AND FLAVIVIRIDAE
Edited by Sondra Schlesinger and Milton J. Schlesinger

THE VIROIDS
Edited by T. O. Diener

The
Paramyxoviruses

Edited by
DAVID W. KINGSBURY
Howard Hughes Medical Institute
Bethesda, Maryland

PLENUM PRESS • NEW YORK AND LONDON

Library of Congress Cataloging-in-Publication Data

The Paramyxoviruses / edited by David W. Kingsbury.
     p.   cm. -- (The Viruses)
  Includes bibliographical references and index.
  ISBN 0-306-43553-5
  1. Paramyxoviruses. I. Kingsbury, David W. II. Series.
  [DNLM: 1. Paramyxoviridae--physiology. QW 168.5.P2 P222]
 QR404.15.P37 1991
 576'.6484--dc20
 DNLM/DLC
 for Library of Congress   90-14321
                                      CIP

ISBN 0-306-43553-5

Printed in the United States of America

Contributors

Thomas Barrett, AFRC Institute for Animal Health, Pirbright Laboratory, Woking, Surrey GU24 0NF, United Kingdom

Graham J. Belsham, AFRC Institute for Animal Health, Pirbright Laboratory, Woking, Surrey GU24 0NF, United Kingdom

Martin A. Billeter, Institute for Molecular Biology I, University of Zürich, CH 8093 Zürich, Switzerland

Francis L. Black, Department of Epidemiology and Public Health, Yale University School of Medicine, New Haven, Connecticut 06510

Benjamin M. Blumberg, Departments of Neurology and Microbiology and Immunology, University of Rochester Medical Center, Rochester, New York 14642

Roberto Cattaneo, Institute for Molecular Biology I, University of Zürich, CH 8093 Zürich, Switzerland

John Chan, University of Medicine and Dentistry New Jersey—New Jersey Medical School, Newark, New Jersey 07103

Peter L. Collins, Laboratory of Infectious Diseases, National Institute of Allergy and Infectious Diseases, National Institutes of Health, Bethesda, Maryland 20892

Richard W. Compans, Department of Microbiology, University of Alabama at Birmingham, Birmingham, Alabama 35294

Joseph Curran, Department of Microbiology, University of Geneva Medical School, C.M.U., 1211 Geneva 4, Switzerland

Mark S. Galinski, Department of Molecular Biology, Cleveland Clinic Foundation, Cleveland, Ohio 44195

Sandra M. Horikami, Department of Immunology and Medical Microbiology, University of Florida College of Medicine, Gainesville, Florida 32610

Daniel Kolakofsky, Department of Microbiology, University of Geneva Medical School, C.M.U., 1211 Geneva 4, Switzerland

Robert A. Lamb, Department of Biochemistry, Molecular Biology, and Cell Biology, Northwestern University, Evanston, Illinois 60208-3500

Brian W. J. Mahy, AFRC Institute for Animal Health, Pirbright Laboratory, Woking, Surrey GU24 0NF, United Kingdom; *present address:* Division of Viral and Rickettsial Diseases, Centers for Disease Control, Atlanta, Georgia 30333

Mary Ann K. Markwell, Section on Molecular Pharmacology, Clinical Neuroscience Branch, National Institute of Mental Health, Bethesda, Maryland 20892

Exeen M. Morgan, Biotechnology Services Division, Microbiological Associates, Inc., Rockville, Maryland 20850

Trudy Morrison, Department of Molecular Genetics and Microbiology, University of Massachusetts Medical School, Worcester, Massachusetts 01655

Sue A. Moyer, Department of Immunology and Medical Microbiology, University of Florida College of Medicine, Gainesville, Florida 32610

Erling Norrby, Department of Virology, Karolinska Institute, School of Medicine, S-105 21 Stockholm, Sweden

Reay G. Paterson, Department of Biochemistry, Molecular Biology, and Cell Biology, Northwestern University, Evanston, Illinois 60208-3500

Mark E. Peeples, Department of Immunology/Microbiology, Rush Medical College, Chicago, Illinois 60612

Allen Portner, Department of Virology and Molecular Biology, St. Jude Children's Research Hospital, Memphis, Tennessee 38101-0318

Craig R. Pringle, Department of Biological Sciences, University of Warwick, Coventry CV4 7AL, United Kingdom

R. E. Randall, Department of Biochemistry and Microbiology, University of St. Andrews, St. Andrews, Fife KY16 9AL, Scotland, United Kingdom

Ranjit Ray, Secretech, Inc., Birmingham, Alabama 35205

Gian Guido Re, Pediatric Leukemia Research Program, Division of Pediatrics, University of Texas M. D. Anderson Cancer Center, Houston, Texas 77030

Laurent Roux, Department of Microbiology, University of Geneva Medical School, C.M.U., 1211 Geneva 4, Switzerland

W. C. Russell, Department of Biochemistry and Microbiology, University of St. Andrews, St. Andrews, Fife KY16 9AL, Scotland, United Kingdom

Shaila M. Subbarao, AFRC Institute for Animal Health, Pirbright Laboratory, Woking, Surrey GU24 0NF, United Kingdom; *permanent address:* Department of Microbiology and Cell Biology, Indian Institute of Science, Bangalore 560012, India

Wayne M. Sullender, Department of Microbiology, University of Alabama School of Medicine, Birmingham, Alabama 35294

Stephen A. Udem, University of Medicine and Dentistry New Jersey—New Jersey Medical School, Newark, New Jersey 07103

Sylvia Vidal, Department of Microbiology, University of Geneva Medical School, C.M.U., 1211 Geneva 4, Switzerland

Steven L. Wechsler, Department of Virology Research, Cedars-Sinai Medical Center, Los Angeles, California 90048

Gail W. Wertz, Department of Microbiology, University of Alabama School of Medicine, Birmingham, Alabama 35294

Preface

What justifies the size of this compendium of reviews on the paramyxoviruses? As intracellular parasites that reproduce with almost complete indifference to nuclear activities, paramyxoviruses have not been providing insights about genes that regulate cellular activities and development, topics that account for much of the excitement in modern biology. For contributions of virus research to those topics, we must look to the retroviruses, which have the propensity to steal developmentally important genes and subvert them to malignant purposes, and to the nuclear DNA viruses, whose gene expression depends heavily upon cellular transcription machinery, making them exceptionally useful tools for identifying and characterizing components of that machinery. From this perspective, it may appear that purely lytic viruses like the paramyxoviruses are sitting on the sidelines of contemporary biology.

But there is plenty of action on the sidelines. Paramyxoviruses remain unconquered, devastating agents of disease. Human deaths attributable to paramyxoviruses worldwide, especially in children, are numbered in the millions annually. There are many pathogenic paramyxoviruses and too few effective vaccines, and those vaccines (against measles and mumps) are affordable only by relatively affluent nations. Moreover, the paramyxoviruses are intrinsically interesting organisms, presenting the challenge of understanding the self-replication of RNA and many other challenges peculiar to the structures and functions of their proteins, not only as individual entities, but also as they act in concert during virus reproduction and interact with vital functions of the cells they infect and often (but not always) destroy.

Consider also the sheer volume of information that has been accumulating about this virus family since the advent of cloning and sequencing. The last comprehensive review of the paramyxoviruses that appeared in the predecessor of the current series, *Comprehensive Virology*, was written by P. W. Choppin and R. Compans and published in 1975, well before this avalanche of new knowledge. Who could have anticipated that we would now have in hand the complete sequences of the entire genomes of members of each of the currently recognized paramyxovirus genera? The task of presenting and digesting this kind of information fills much of the present volume.

These and all the other activities in this field called for the participation of a large number of authors who are actively engaged in the research. I was surprised and delighted to receive so many positive responses to my invitations. Although the emphasis is on molecular biology, topics with a more biological slant are well represented, with chapters on receptors, immunology, persistence, and epidemiology.

Admittedly, the size of the book could have been reduced by removal of some redundancies. But, as one contributor pointed out to me, people do not read this kind of book from cover to cover; they use it as a reference work according to their interests. Furthermore, one can hope that the variety of interpretations of new or controversial topics by different authors will provide perspectives that readers will find instructive.

David W. Kingsbury

Contents

Chapter 3

The Molecular Biology of the Morbilliviruses

Thomas Barrett, Shaila M. Subbarao, Graham J. Belsham,
and Brian W. J. Mahy

Chapter 4

The Molecular Biology of Human Respiratory Syncytial Virus (RSV) of the Genus *Pneumovirus*

Peter L. Collins

Chapter 5

Evolutionary Relationships of Paramyxovirus Nucleocapsid-Associated Proteins

Exeen M. Morgan

Chapter 6

The Nonstructural Proteins of Paramyxoviruses

Robert A. Lamb and Reay G. Paterson

Chapter 7

Paramyxovirus RNA Synthesis and _P_ Gene Expression

Daniel Kolakofsky, Sylvia Vidal, and Joseph Curran

Chapter 8

Function of Paramyxovirus 3′ and 5′ End Sequences: In Theory and Practice

Benjamin M. Blumberg, John Chan, and Stephen A. Udem

Chapter 9

The Role of Viral and Host Cell Proteins in Paramyxovirus Transcription and Replication

Sue A. Moyer and Sandra M. Horikami

Chapter 10

Deletion Mutants of Paramyxoviruses

Gian Guido Re

Chapter 11

Paramyxovirus Persistence: Consequences for Host and Virus

R. E. Randall and W. C. Russell

Chapter 12

Molecular Biology of Defective Measles Viruses Persisting in the Human Central Nervous System

Martin A. Billeter and Roberto Cattaneo

Chapter 13

Structure, Function, and Intracellular Processing of the Glycoproteins of Paramyxoviridae

Trudy Morrison and Allen Portner

Chapter 14

The Unusual Attachment Glycoprotein of the Respiratory Syncytial Viruses: Structure, Maturation, and Role in Immunity

Wayne M. Sullender and Gail W. Wertz

Chapter 15

New Frontiers Opened by the Exploration of Host Cell Receptors

Mary Ann K. Markwell

Chapter 16

Paramyxovirus M Proteins: Pulling It All Together and Taking It on the Road

Mark E. Peeples

Chapter 17

Intracellular Targeting and Assembly of Paramyxovirus Proteins

Ranjit Ray, Laurent Roux, and Richard W. Compans

Chapter 18

Immunobiology of Paramyxoviruses

Erling Norrby

Chapter 19

Epidemiology of *Paramyxoviridae*

Francis L. Black

Appendix

Annotated Nucleotide and Protein Sequences for Selected *Paramyxoviridae*

Mark S. Galinski

CHAPTER 1

The Genetics of Paramyxoviruses

Craig R. Pringle

I. INTRODUCTION: THE GENOME STRATEGY OF THE PARAMYXOVIRUSES

The replication of RNA viruses, though sometimes requiring integrity of the cell nucleus, with a single exception takes place in the cytoplasm of the host cell independently of DNA replication. According to taxonomic type, either the positive strand or the negative strand or both may become encapsidated in the virion. The negative-strand viruses include the nuclear hepatitis delta virus and the members of the families *Arenaviridae, Bunyaviridae, Filoviridae, Orthomyxoviridae, Paramyxoviridae,* and *Rhabdoviridae.* Hepatitis delta virus is exceptional in having a genome consisting of a single-stranded, covalently-closed, circular molecule, and although replication competent in some cells, it is dependent on hepatitis B virus for encapsidation and transmission (Wang *et al.,* 1986; Taylor *et al.,* 1987). The linear negative-strand RNA viruses are a diverse collection of viruses replicating in the cytoplasm. They exhibit a genome strategy whereby the noncoding RNA strand is sequestered in the extracellular virion and the positive-strand template RNA and mRNAs are generated by a virion-associated transcriptase/replicase from the complementary template strand when the infectious process is initiated.

The six families of linear negative-strand RNA viruses comprise viruses with circular or linear helical nucleocapsid core structures invested by an envelope composed of viral proteins associated with the host-cell plasma membrane, or an internal membrane in the case of the bunyaviruses. The viruses in these families have relatively large particles, but relatively small genomes ranging between 11,100 and 15,900 nucleotides. The six families are to some extent differentiated by morphology and fall into two distinct catego-

CRAIG R. PRINGLE • Department of Biological Sciences, University of Warwick, Coventry CV4 7AL, United Kingdom.

ries in terms of genome structure. The arenaviruses, bunyaviruses, and orthomyxoviruses have segmented genomes containing two, three, and seven or eight subunits, respectively, whereas the filoviruses, paramyxoviruses and rhabdoviruses have linear undivided genomes. Segmentation of the genome may be a device to uncouple the transcription of individual viral genes, or an adaptation to compensate for the low fidelity of transcription of RNA polymerases, allowing error correction by reassortment of subunits.

This dichotomy of the negative-strand viruses, whatever its functional significance, has profound genetic consequences. Mutation appears to be the sole means of generation of variation among those viruses with nonsegemented genomes. Unlike the positive-strand viruses, intramolecular recombination has not been detected in viruses with linear negative-strand genomes (see Section III.A). The segmented-genome viruses, on the other hand, can acquire variation by reassortment of genome subunits as well as by mutation, and therefore appear to be inherently more adaptable. In bunyaviruses and orthomyxoviruses, the gene pool may be very large. The propensity of the paramyxoviruses in particular to become persistent may reflect their reduced capacity for variation and their consequent reliance on alternate mechanisms to combat host immune responses. Certain paramyxoviruses associated with recurrent epidemic disease in humans appear remarkably stable. For example, the HN surface glycoprotein gene of parainfluenza virus type 3 and the F protein gene of respiratory syncytial (RS) virus appear to be highly conserved, showing little sequence variation during the course of 20 years since first isolation (Coelingh et al., 1986; Baybutt and Pringle, 1987). However, other genes are less conserved and the apparent stability of the surface glycoproteins may merely reflect the existence of consistent selective forces.

The negative-strand unsegmented genome RNA viruses have many features in common. They are similar in terms of genetic organization (Pringle, 1987), with genes of homologous function transcribed in the same order from a single 3' terminus-adjacent promoter site. A block of core protein genes is followed by a block of envelope protein genes with the polymerase protein gene located farthest from the promoter and accounting for approximately half the coding capacity of the genome (Fig. 1). The genome of the *Vesiculovirus* genus of the *Rhabdoviridae* is the least complex, comprising five linearly arrayed genes with minimal di- or trinucleotide intergenic junctions and highly conserved gene start and gene stop signals (Fig. 2). The NS (or P) phosphoprotein gene encodes in-frame a second gene product, a feature shared with the homologous P protein gene of some paramyxoviruses. Other rhabdoviruses have genomes of greater complexity, resembling the paramyxovirus pattern. The *G–L* intergenic region of rabies virus has the structure of a pseudogene, and a transcribed gene of unknown function is present between the *G* and *L* genes of the fish rhabdovirus, infectious hematopoietic necrosis virus. The paramyxoviruses and rhabdoviruses differ most with regard to particle morphology and general biology. The paramyxoviruses are pleomorphic, in contrast to the uniform morphology of the rhabdoviruses. The paramyxoviruses are restricted in host range and confined to vertebrates, whereas the rhabdoviruses have a broad host range, infecting vertebrates, invertebrates, and plants. The limited host range and routes of transmission are probably responsible for the strong asso-

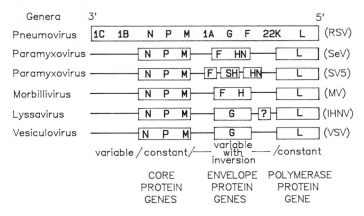

FIGURE 1. Comparison of the linear order of gene functions in the different genera of paramyxoviruses and rhabdoviruses. Negative-stranded (genomic) polarity is depicted. Key: RSV, human respiratory syncytial virus; SeV, Sendai virus; SV5, simian virus 5 (and mumps virus); MV, measles virus; IHNV, infectious hematopoeitic necrosis virus; VSV, vesicular stomatitis virus. 1C and 1B, genes of unknown function; N, nucleoprotein gene; P, phosphoprotein gene; M, nonglycosylated membrane protein gene; 1A, glycosylated membrane protein gene; G, H, and HN, attachment protein gene; F, fusion protein gene; SH, small hydrophobic protein gene; ?, transcribed gene of unknown function (pseudogene in rabies virus); L, polymerase protein gene.

ciation of paramyxoviruses with respiratory tract infection and disease, or vice versa.

The family *Paramyxoviridae* as presently constituted (Kingsbury *et al.*, 1978) is by no means homogeneous. Two of the three genera, *Morbillivirus* and *Paramyxovirus*, are clearly related viruses by criteria such as nucleotide sequence similarities, gene number and order, and detailed morphology. The third genus, *Pneumovirus*, is distinct. Indeed, the differences distinguishing the pneumoviruses from the other members of the family are greater than those differentiating the rhabdoviruses from the other paramyxoviruses. Evi-

FIGURE 2. Gene start, end, and intergenic consensus sequences. Consensus sequences are given for the start and end regions of the genes of viruses where all or most of the genome has been sequenced. The intergenic junctions are given as consensus sequences or as numbers of intervening nucleotides. Key: N, variable nucleotide; NDV, Newcastle disease virus; PIV3, parainfluenza virus type 3; RSV, human respiratory syncytial virus; PVM, pneumonia virus of mice; VSV, vesicular stomatitis virus. Compiled from data published up to the end of 1988 (except in the case of PVM).

GENE ENDS/ JUNCTIONS /GENE STARTS

Family Paramyxoviridae

Genus Paramyxovirus:

ANUCUUUUUUU/	1–48	/UGCCCAUCNUN	*NDV*
UNAUUCU/	GNN	/UCCCANUUUC	*Sendai*
UUNAAUNNUUUUU (UU) /	GAA	/UCCNNUUC	*PIV3*
ANNUNUUUUUU (U) /	1–7	/UNCNNNNNU	*Mumps*

Genus Morbillivirus:

NNNUUUU (UU) /	GNA	/UCCNNNNUNCN	*Measles*

Genus Pneumovirus:

UCANNNNNNUU (UU) /	1–52	/CCCNNUUU	*RSV*
UANUUANNN/			*PVM*

Family Rhabdoviridae

Genus Vesiculovirus:

AUACU/	NA	/UUGUCNNUAG	*VSV*

dence for the apparent uniqueness of the pneumoviruses has come solely from characterization of the bovine and human respiratory syncytial (RS) viruses, although current work on pneumonia virus of mice (PVM), until recently the only other recognized member of the genus, indicates that the structure and function of its genome are virtually identical to those of the RS viruses (Chambers *et al.*, 1990 and unpublished data; Ling and Pringle, 1989a,b). The identification of turkey rhinotracheitis virus (TRTV), a new avian respiratory pathogen, which has similar molecular characteristics to the RS viruses and PVM (Ling and Pringle, 1988; Cavanagh and Barrett, 1988), has increased the biological diversity of the pneumoviruses and strengthened the case for regarding them as a distinct group. Ultimately, the pneumoviruses may be accorded separate taxonomic status; however, for the present they are still ranked as members of the family *Paramyxoviridae*.

Figure 1 compares diagrammatically the structures of the genomes of the rhabdoviruses and the paramyxoviruses. Comparisons of nucleotide and deduced amino acid sequences have confirmed the distinctiveness of the pneumoviruses among the other paramyxoviruses. Alignment of protein sequences suggests possible conservation of cysteine residues and hydrophobic regions in the fusion (F) protein, but little more (Morrison, 1988; Rima, 1989). On the contrary, clear relationships are discernible among all members of the genera *Paramyxovirus* and *Morbillivirus* which have been compared so far.

It is more informative to align and compare amino acid sequences, because the inherent variability of RNA genomes is undoubtedly constrained by the selective forces which maintain protein structure and function. From such comparisons, it has been concluded that the paramyxoviruses can be separated into two distinct groups (Rima, 1989). The first group is fairly homogeneous and comprises the bovine and human strains of parainfluenza virus type 3 and Sendai virus (parainfluenza virus type 1), whereas the second is more heterogeneous and includes Newcastle disease virus (avian parainfluenza virus type 1), simian virus 5 (SV5), mumps virus, and possibly parainfluenza virus type 4 (Komada *et al.*, 1989). This subdivision is reflected in the differing organization of the genome, and supported by immunoprecipitation studies (Ito *et al.*, 1987). The former group is more closely related to the morbilliviruses than is the latter.

In the first group the *P* gene encodes a nonstructural protein in an alternate reading frame. In the second group the internal nonstructural protein gene is encoded in the same reading frame within the *P* gene, and some of these viruses have an additional gene coding for a small hydrophobic (SH) protein inserted between the F and HN envelope glycoprotein genes. Which genome represents the higher form cannot be deduced from such data. On present evidence loss of genes by paramyxoviruses with progression toward the basic pattern of the vesiculoviruses, or acquisition of genes with progression toward the greater complexity exhibited by the pneumoviruses (Fig. 1), may equally be the direction of evolution in the family *Paramyxoviridae*.

Within the morbillivirus group, protein sequence comparisons and monoclonal antibody studies (Sheshbaradaran *et al.*, 1986) suggest that measles virus and rinderpest virus are more closely related than measles virus and canine distemper virus. Norrby *et al.* (1985) have proposed on the basis of monoclonal antibody studies that rinderpest virus represents the archetypal morbillivirus from which canine distemper virus diverged at an earlier time in evolution than measles virus.

II. GENOME ORGANIZATION

A. Genome Structure and Function

Details of the structure and function of the genome of paramyxoviruses are discussed elsewhere in this volume and are summarized in Fig. 1. The features of relevance to this chapter are that the paramyxoviruses and morbilliviruses possess six or seven genes and the pneumoviruses possess ten genes. The RS virus genome is unique among paramyxoviruses in that the normally 3'-terminal N-protein gene is preceded by two small nonstructural protein genes, 1C and 1B, of unknown function. In general, gene start and stop signals correspond to consensus sequences which are different for the different genera. Intergenic junctions show more variability among individual viruses and a genus-specific pattern is less evident (Fig. 2). There does not appear to be any temporal control of transcription in paramyxoviruses, but there is a progressive attenuation of transcription with increasing distance from the promoter, so that the 3'-terminus-adjacent *NP* or *N* mRNA is the most abundant species and the L-protein mRNA the least. These attributes are shared in common with the rhabdoviruses, with the exception that in the pneumoviruses the messages transcribed from the 3'-terminal enigmatic 1C and 1B nonstructural protein genes are the most abundant species of viral RNA in the infected cell. Positive-sense and negative-sense leader RNAs have not been unequivocally identified in measles virus-infected cells (Crowley *et al.*, 1988).

The genera are also differentiated by the nature of their surface glycoproteins, as described elsewhere in this volume. All three genera possess a fusion (F) protein gene, and infectivity is probably dependent on proteolytic cleavage of the gene product, although this has not been demonstrated in every case. The functional activities of the attachment protein vary among genera and include hemagglutinin and neuraminidase (HN) activities in the paramyxoviruses, hemagglutinin (H) alone in the morbilliviruses, and neither in RS virus. The pattern in the pneumoviruses is not uniform, however, since PVM can hemagglutinate mouse cells, and this activity has been localized to the putative attachment (G) protein (Ling and Pringle, 1989b). The H protein of morbilliviruses is structurally and in part functionally analogous to the HN protein of parainfluenzaviruses. The G protein of pneumoviruses is structurally different, being a small polypeptide heavily modified by attachment of predominantly O-linked carbohydrate chains. It does resemble the HN attachment proteins of Sendai virus, Newcastle disease virus, and SV5 by its insertion into the lipid bilayer, with the carboxy terminus located externally. The attachment (G) protein and fusion (F) protein genes of RS virus are inverted in the gene order relative to other paramyxoviruses (Fig. 1). This inversion is unlikely to represent a recombinational event, since the block of envelope protein genes in RS virus has been expanded from the three [M, F, H(N)] of other paramyxoviruses to five [M, 1A, G, F, 22K(M2)]. The pneumoviruses are unique among enveloped RNA viruses in having two additional membrane-associated proteins. The precise function of these additional membrane proteins is unknown, but one of them, the glycosylated 1A protein, appears to contain two overlapping T cell-stimulating sites (Nicholas *et al.*, 1989). Although the genome of RS virus contains more genes than the genomes of the other paramyxoviruses characterized so far, the individual genes are in general smaller. In fact, the sequence of the *L* gene of RS virus has now been determined (P. L. Collins, personal communication), but the size of the complete genome is similar to that of other paramyxoviruses (see p. 7).

The descending order of overall conservation of the structural proteins of paramyxoviruses appears to be L > M > F > N > H(N) > P (Rima, 1989). The *L* genes and proteins of Sendai virus and measles virus exhibit very close similarity. The L proteins of NDV and measles virus are less similar, but there is only a distant resemblance between the L proteins of VSV and measles virus or Sendai virus. However, the L proteins of NDV and VSV do show considerable similarity (Yusoff *et al.*, 1987), and Blumberg *et al.* (1988) have proposed that this represents an evolutionary pathway leading from VSV via NDV to Sendai virus and ultimately to measles virus. Furthermore, they suggest that this pathway is consistent with the progressive narrowing of the host range of these four viruses and that the evolution of two separate glycoproteins of the paramyxoviruses from the *G* gene region of VSV may be part of this process. The L proteins of paramyxoviruses exhibit more extensive similarities, containing conserved sequences common to a number of viral RNA-dependent RNA polymerases, including the Gly-Asp-Asp motif which is considered to represent the nucleic acid recognition or active site of an ancestral RNA polymerase (Kamer and Argos, 1984).

The matrix proteins of paramyxoviruses are hydrophobic in nature, but show little conservation of sequence. Nonetheless, the similarity of the hydropathy plots for the M proteins of measles virus and Sendai virus, and the dissimilarity of those of NDV and mumps virus, reaffirms the close relationship of the morbilliviruses and the Sendai/parainfluenzavirus type 3 subdivision of the paramyxoviruses referred to previously. The overall conservation of cysteine residues, helix-breaking residues, and paired basic residues suggests that a common tertiary structure is maintained (Morrison, 1988; Rima, 1989).

Likewise, among the fusion proteins of paramyxoviruses there is overall conservation of those residues likely to be important determinants of tertiary structure, and the same functional regions (signal region, cleavage site, and membrane anchor region) are preserved throughout the family (Morrison, 1988). Glycosylation sites are not rigidly conserved; in the morbilliviruses there is a relative preponderance of glycosylation sites in the F_1 fragment, whereas in the paramyxoviruses and pneumoviruses the inverse is the case.

The N proteins of parainfluenza viruses appear to possess three distinct domains which may correspond to an amino-terminal RNA-binding region, a central region involved in transcription, and a carboxy-terminal, negatively-charged region possibly exposed on the outside of the nucleocapsid (Buckland *et al.*, 1988). The central region shows the most conservation and the carboxy-terminal region the least.

The attachment proteins of the paramyxoviruses differ in their functional properties: the HN of the paramyxoviruses has hemagglutinating and neuraminidase activities, the H of the morbilliviruses has hemagglutinating activity only, and the G of pneumoviruses has neither in the RS viruses and hemagglutinating activity only in PVM (Ling and Pringle, 1989b). The sequence similarities of the attachment proteins follows the taxonomic subdivision of the family. Within genera there is considerable resemblance, whereas between genera there is little, or none in the case of the pneumoviruses. The cysteine residues are well conserved in H(N) molecules, indicating possibly structural similarity. The G protein of RS virus is exceptional in its rich endowment in O-glycosylation sites.

The P proteins exhibit lesser sequence similarity than the other proteins of paramyxoviruses. They resemble the NS proteins of rhabdoviruses in that the amino-terminal region is hypervariable, conservation being limited to the carboxy-terminal region (Banerjee, 1987).

B. Coding Potential

The size of paramyxovirus genomes is remarkably uniform. The four genomes which have been sequenced completely lie within a few hundred nucleotides of each other; 15,892 nucleotides for measles virus (Crowley et al., 1988), 15,222 for human RS virus (P. L. Collins, personal communication), 15,285 for Sendai virus (Shioda et al., 1986), 15,156 for NDV (Yusoff et al., 1987), and 15,463 for parainfluenza virus type 3 (see Chapter 2 and Appendix to this book). The genomes of the paramyxoviruses are larger than those of the other negative-strand genomes sequenced so far; e.g., the genome of an influenza A virus has been determined as 13,588 nucleotides, that of a bunyavirus as 12,278 nucleotides, and that of the Indiana serotype of VSV as 11,162 nucleotides. By contrast, the circular, negative-stranded hepatitis delta virus has the smallest known RNA genome at 1678 nucleotides.

The coding potential of negative-strand RNA viruses may be extended by a variety of devices, including obligate mRNA splicing, alternate mRNA splicing, and translation from overlapping reading frames. In the paramyxoviruses, the P genes of Sendai virus, parainfluence type 3 virus, and measles virus encode a second gene product in an overlapping reading frame which is translated from a functionally bicistronic mRNA. In Sendai virus, a third gene product (C') of the P gene has been identified, which is translated from an anomalous ACG initiation codon upstream from the AUG codons initiating translation of the P and C polypeptides (Curran and Kolakofsky, 1988).

In other paramyxoviruses the P gene products appear to be encoded in the same reading frame. The SV5 genome encodes a P protein, which is a virion protein and a component of the transcription complex, and a V protein of unknown function. The P and V proteins are amino-coterminal proteins with 164 amino acids in common, but with different carboxy termini. Thomas et al. (1988) have demonstrated that these proteins are translated from two separate mRNAs that differ by the presence or absence of two nontemplated G residues. The presence of these two G residues converts the two overlapping reading frames into a single reading frame encoding a sequence of 392 amino acids. The mechanism of addition of the two nontemplated nucleotides is unknown, but could be accounted for by reiterative copying at a point where the template has a run of four C residues, and indeed a sequence resembling the polyadenylation signal is located just upstream from the four C residues. The unique C terminus of the V protein has a cysteine-rich region which is present as a highly conserved element of the open reading frame of the P gene of other paramyxoviruses, suggesting that it is functionally important. SV5 appears to be unique among paramyxoviruses in having two mRNAs associated with the P gene.

In summary, despite the overall similarity of genetic organization of viruses in the paramyxovirus group, there is diversity in the strategy of transcription and in posttranscriptional processing. The overlap of the 22K- and L-protein genes of respiratory syncytial virus may be another unique and as yet unresolved mechanism of transcriptional control (Collins et al., 1987).

III. GENETIC INTERACTIONS

A. Absence of Genetic Recombination

Genetic recombination is now a well-established phenomenon in picornaviruses (King et al., 1982). In laboratory experiments, both intra- and interserotypic recombinants of both poliovirus and of foot-and-mouth disease virus have been isolated in progeny from mixed infections, and recombinant polioviruses are generated in the human intestine during natural infection. It has also been shown that recombination occurs at high frequency in coronaviruses (Lai et al., 1985). However, recombination is not universal in positive-strand RNA viruses, and reports of recombination in alphaviruses have not been confirmed. Recombination appears to be restricted to viruses which lack nucleoprotein, and presumably the investment of the genome by nucleoprotein prevents intramolecular exchange.

There are no confirmed reports of genetic recombination involving negative-strand RNA viruses, other than the reassortment of genome subunits which is universal among the segmented genome viruses. Granoff (1959a,b, 1961a,b, 1962) and Dahlberg and Simon (1968, 1969a,b) demonstrated that presumptive non-ts recombinants among the progeny virus from mixed infections of ts mutants of Newcastle disease virus (NDV) were in fact complementing heterozygotes. UV irradiation and density gradient electrophoresis were used to investigate the physical structure of these heterozygotes, but the data were conflicting and open to several interpretations (Kingsbury and Granoff, 1970). The pleomorphic morphology of paramyxoviruses suggests that multiploid particles are abundant components of the progeny of all paramyxoviruses. Complementation of ts mutants is a special case of the more general phenomenon of phenotypic mixing, and phenotypic mixing involving serotype and thermal stability has been recorded also in paramyxoviruses.

Kirkegaard and Baltimore (1986) have shown that recombination in poliovirus involves the negative template strand generated during replication, therefore there would appear to be no inherent barrier to recombination in negative-strand RNA viruses. It must be presumed that the rapid association of the nucleoprotein with the negative and positive RNA strands which is characteristic of the replication of all negative-strand RNA viruses prevents interactions between RNA molecules which might result in recombination. There is also no record of rescue into infectious virus of genetic information from the defective-interfering viruses which are present in laboratory stocks of most paramyxoviruses propagated by high-multiplicity cycles of infection.

Drake (1962) and Kirvatis and Simon (1965) observed that high multiplicity of infection with UV-irradiated Newcastle disease virus resulted in a greater number of infected cells or yield of virus than expected from the surviving infectivity. Multiplicity reactivation by a recombinational mechanism cannot account for this phenomenon, since recombinants were not identified. Likewise, production of complementing heterozygotes by UV irradiation is not a sustainable explanation, since a UV-induced lesion anywhere in the genome would abort transcription of the genome. Peeples and Bratt (1982a) showed that at high multiplicity of infection, input virions could supply a UV-resistant function, probably a virion protein, which could rescue other *ts* mutants by complementation. In the paramyxoviruses, therefore, multiplicity reactivation may be another extreme manifestation of phenotypic mixing.

Recently Sakaguchi *et al.* (1989) and Toyoda *et al.* (1989) analyzed the sequences of the hemagglutinin-neuramindase genes of 13 strains of NDV isolated over a period of 50 years and the sequences of the fusion gene of 11 of these strains by a method involving comparison of synonymous nucleotide substitutions in the coding regions. The rate of synonymous substitution can serve as an indicator of evolutionary progression, since it is approximately constant for different genes and greater than the number of amino acid replacements (Miyata and Yasunaga, 1980). Three distinct lineages could be defined which appeared to have cocirculated for a considerable period. Virulent strains were associated with one lineage and avirulent strains with another, whereas the third lineage consisted solely of North American isolates, but contained both virulent and avirulent strains. It was concluded that the different strains appeared to have evolved by accumulation of point mutations and that no gene exchange by recombination had occurred in generation of the three lineages.

Sakaguchi *et al.* (1989) concluded that the evolutionary pattern of Newcastle disease virus with its multiple lineages and absence of progressive antigenic drift more resembled that of the influenza B and C viruses than that of the influenza A viruses, which are characterized by linear evolution and progressive antigenic change.

Comparison of the frequencies of occurrence of monoclonal antibody escape mutants indicated that the inherent genetic variabilities of a paramyxovirus (Sendai virus), a rhabdovirus (vesicular stomatitis virus), and an orthomyxovirus (influenza virus type A) were not markedly different (Portner *et al.*, 1980). Therefore, the occurrence of progressive antigenic drift in influenza virus and its apparent absence in the paramyxoviruses is due to factors other than differences in basic mutation rate.

B. Complementation Analysis with Conditional Lethal Mutants

Genetic analysis in negative-strand RNA viruses with unsegmented genomes is limited to the use of complementation analysis to identify and define the functions of individual genes of the viral genome. Suppressor-mediated chain-terminating mutations are rare in mammalian viruses, and among negative-strand RNA viruses have only been identified unequivocally in the case of VSV (White and McGeoch, 1987). In most viruses, including the paramyxoviruses, the conditional-lethal mutants employed in genetic analysis are exclusively of the temperature-sensitive type. In both the rhabdoviruses and the paramyxoviruses, collections of temperature-sensitive (ts) mutants have been classified into groups containing mutants which do not complement inter se, but which complement mutants in other groups. With some minor exceptions, the pattern of complementation is nonoverlapping. In favorable circumstances, the underlying assumption that the complementation groups correspond to the component genes of the viral genome has been substantiated by characterization of representative mutants and identification of the mutational lesion. For example, the 5/6 complementation groups of vesicular stomatitis virus (VSV) have been equated with a fair degree of certainty with individual genes [reviewed in Pringle (1988)].

In the paramyxoviruses and the other rhabdoviruses the assignment of complementation groups is less advanced and is complicated by inability to distinguish between intergenic and intragenic complementation. Where intragenic complementation is prevalent, the number of complementation groups will be overestimated. Intragenic complementation is a consequence of the multifunctional nature of some viral gene products. In the rhabdoviruses the multifunctional role of the L protein appears to be responsible for most intragenic complementation (Gadkari and Pringle, 1980), whereas in paramyxoviruses, as discussed below, the multifunctional properties of the hemagglutinin/neuraminidase (attachment) protein appear to be responsible for most intragenic complementation.

Temperature-sensitive mutants of Newcastle disease virus and Sendai virus (representing the paramyxoviruses), measles virus (representing the morbilliviruses), and respiratory syncytial virus (representing the pneumoviruses) have been obtained by chemical mutagenesis and classified into complementation groups. Five groups have been identified in Newcastle disease virus, seven groups in Sendai virus and measles virus, and eight in human respiratory syncytial virus. The data are summarized in the following four subsections and in Tables I–IV.

1. The Temperature-Sensitive Mutants of Newcastle Disease Virus

a. Complementation Groups

Newcastle disease virus exhibits great variation in its biological and disease-producing properties and it has a long history of genetic analysis, the early stages of which have been reviewed comprehensively by Bratt and Hightower (1977). The most definitive analysis of temperature-sensitive mutants has been carried out by Tsipis and Bratt (1976). Forty-nine mutants were derived from the Australia-Victoria strain of Newcastle disease virus by chemical mutagenesis. Twenty-eight were obtained by nitrosoguanidine treatment, 15 by nitrous acid treatment, and three by growth in the presence of 5-fluorouracil.

TABLE I. The Complementation Groups of Newcastle Disease Virus[a]

Group	RNA phenotype at restrictive temperature (41.8°C)	Gene assignment (tentative)	Nature of evidence
A	−	(L)	Largest group (47% of mutants); defective in primary transcription; no rescue by L-deficient nc mutants; UV transcription mapping
B/C/BC	+	HN	Defective in HN-related functions
D	+	M	Electrophoretic migration difference
E	−	P	Defective in secondary transcription; complement L-deficient nc mutants; UV transcription mapping

[a]For references, see text.

However, five spontaneous mutants were also isolated and the frequency of ts mutants in mutagenized stocks was only 3–5% compared with 2% in untreated material. Nine of the mutants were regular temperature-sensitive mutants, whereas the other six mutants plated with equal efficiency at the permissive (36°C) and restrictive (41.8°C) temperatures, but produced only minute plaques at the restrictive temperature. These 15 mutants were classified into five (initially six) complementation groups. Group A contained five mutants with RNA-negative phenotypes; group B, four RNA-positive mutants; group C, one RNA-negative mutant; group D, two RNA-positive mutants; and group E, one RNA-positive mutant. A low level of complementation was observed between some mutants in groups B and C, and a sixth group, BC, containing two mutants was tentatively defined. Subsequent detailed analysis of the phenotypic properties of these and additional mutants added to groups B, C, and BC confirmed one of the several interpretations of these data, namely that the complementation observed between mutants in these groups represented intragenic complementation. Therefore, only four distinct complementation groups are represented in this collection of mutants, i.e., groups A, B/C/BC, D, and E (Table I). The B/C/BC complex clearly represents mutations affecting the HN-protein gene, and the A, D, and E groups have been tentatively assigned to represent the L, M, and P genes, respectively (see below).

b. Assignments

The RNA-negative mutants of group A have been assigned to the L-protein gene (Madansky and Bratt, 1981a,b; Peeples and Bratt, 1982a; Peeples et al., 1982). Four of the seven mutants in group A which were studied in detail were partially or completely defective in primary transcription, as defined by RNA synthesis in the presence of cycloheximide and actinomycin D. Following release of the cycloheximide block, there was no restoration of RNA synthesis during incubation at the restrictive temperature. Similarly, no RNA synthesis

was observed following shift from permissive to restrictive temperature during
the first 3 hr of incubation. Except in the case of mutant tsA1, which was
rapidly inactivated at nonpermissive temperature, shift-up at later times allowed synthesis, suggesting that secondary transcription from progeny genomes was not deficient. Non-ts revertants of all the mutants had normal
transcriptive ability restored. The group A mutants were assigned to the L-
protein gene on the basis of their predominance (47%) and the fact that the L-
protein gene accounts for approximately half of the genome, and the observation that the L-protein-deficient, noncytopathic mutants (Madansky and Bratt,
1981a,b), described in Section C.1, complemented the group E RNA-negative
mutant, but not the group A mutants. Furthermore, UV inactivation of the
ability of mutant tsE1 to complement tsA1 suggested that the tsA1 defect was
located furthest from the promoter site, as is the L-protein gene (Collins et al.,
1980; Peeples and Bratt, 1982a).

The balance of evidence suggests that the other RNA-negative group, complementation group E defined by the solitary mutant tsE1, represents a lesion
in the P-protein gene. Complementation of mutant tsE1 by tsA1 was inactivated by UV irradiation at the same rate as the P gene, and, as noted above, all
of the noncytopathic putative L-protein mutants complemented tsE1. Furthermore, UV-irradiated mutant tsB1, which has a non-ts mutation affecting the
electrophoretic mobility of P, complemented both tsE1 and tsA1, but did not
amplify the expression of P in multiply-infected cells, which is consistent with
failure of the irradiated virus to replicate despite the presence of components
necessary for the replication of tsE1. Mutant tsE1 was able to sustain primary
transcription at 50% of the wild-type level; however, cycloheximide release
and temperature shift-up before 4 hr postinfection suggested that secondary
transcription was defective. RNA synthesis was observed on shift-up after 4 hr,
suggesting that RNA transcription from progeny genomes was unimpaired. A
non-ts revertant of tsE1 exhibited normal RNA synthesis at the restrictive
temperature of 41.8°C.

The assignment of the B, C, and BC groups to the HN-protein gene was
confirmed by study of the thermolability of four virion functions associated
with the HN protein: hemagglutination, neuraminidase activity, hemolysis
and infectivity (Peeples et al., 1983). Some mutants in groups B, C, and BC were
much less stable in all four functions than wild-type virus or RNA-positive
mutants belonging to group D. Three of four non-ts revertants of the most
thermolabile mutant exhibited normal HN functions. A large proportion of the
non-ts revertants of the HN-protein gene (mutants in groups B, C, and BC)
appeared to be pseudorevertants. Peeples et al. (1988) explored the basis for the
decreased incorporation of HN into virions and the temperature sensitivity of
the 11 B/C/BC group mutants. The HN of two of the mutants had altered
electrophoretic migration rates. In all cases, as much HN protein was synthesized in mutant-infected cells as in cells infected by wild-type virus, but the
HN protein of six mutants was rapidly degraded. The HN (and F) protein of
NDV normally undergoes antigenic maturation early in infection from a form
unable to react with antibodies generated against mature HN to a reactive
form. All the mutants, including the six with metabolically unstable HN, were
deficient in conversion of the HN protein to the mature antigenically reactive

form, indicating an early block in processing. With the exception of one mutant (BC2), the neuraminidase activity of infected cells was temperature sensitive, but the hemadsorbing properties of infected cells were not. The HN protein of mutant BC2 was extremely thermolabile and both activities were temperature sensitive. These results further substantiate the hypothesis that mutants classified in group B/C/BC have defective HN proteins, and that the complementation observed between these mutants represents intragenic complementation.

The group D mutants exhibited normal hemagglutinating and neuraminidase activities, but had lower specific infectivities and hemolytic activities, indicating mutation in a gene for a protein required for membrane fusion and infectivity (Peeples and Bratt, 1982a,b). Unlike the thermostability of the hemagglutinating and neuraminidase activities, the thermostabilities of infectivity and hemolytic activity were salt dependent. Two of the three group D mutants exhibited increased thermolability in low-salt buffer, which is consistent with the assignment of this group to a gene other than the HN-protein gene. These lower activities were correlated with decreased amounts of the cleaved fusion $(F_1 + F_2)$ glycoprotein in virions of group D ts mutants of Newcastle disease virus propagated in eggs. In fact, incorporation of $F_1 + F_2$ into virions was diminished at both permissive and restrictive temperatures when the group D mutants were propagated in cultured chick embryo cells, yet infectivity was correlated with the amount of $F_1 + F_2$ in virions. One of the group D mutants, tsD1, produced an M protein which migrated faster in SDS-polyacrylamide gel, and the M protein of three of four non-ts revertants of this mutant exhibited the electrophoretic mobility of the wild-type M protein. In the case of these three mutants, and also three non-ts revertants of both tsD2 and tsD3, there was coreversion of the low-specific-infectivity and reduced-hemolytic-activity phenotypes to near normal levels and normal incorporation of $F_1 + F_2$ into virions. These observations indicate that the group D mutants represent lesions of the M protein gene and that a defective M protein is responsible for the $F_1 + F_2$ related phenotypes (Peeples and Bratt, 1984). The group D mutants imply a specific interaction between the M and F proteins of Newcastle disease virus which has not been detected by other means. At least seven of the ten revertants of tsD mutants examined were considered to be pseudorevertants, since their heat stability was significantly different from both their parents and the wild-type virus.

c. Other Temperature-Sensitive Mutants of Newcastle Disease Virus

Dahlberg and Simon (1968) obtained 48 temperature-sensitive mutants of the Beaudette-C strain of Newcastle disease virus using a restrictive temperature 42.5°C and nitrous acid mutagenesis. These mutants were classified into complementation groups; five groups were described by Dahlberg and Simon (1968) and nine by Dahlberg (1968). These mutants were employed subsequently in the study of multiploid particles (Dahlberg and Simon, 1969a,b). The properties of six of these mutants from different complementation groups and non-ts revertants of each mutant were studied by Sampson et al. (1981). Two-dimensional electrophoresis was employed to discriminate changes in size and/or isoelectric point of individual proteins which might occur as a result of missense mutation. One of the six mutants appeared to be a double mutant with a temperature-sensitive lesion affecting the P protein and a non-temperature-sensitive lesion involving the HN protein. This mutant had an RNA-negative phenotype at the restrictive temperature, suggesting involvement of the P protein in RNA polymerase activity. Further characterization of these mutants has not been reported.

Preble and Youngner (1973a) isolated four spontaneous and four nitrous acid-induced *ts* mutants of the Herts strain of NDV using a restrictive temperature of 43°C. All were RNA-negative. Preble and Youngner (1972, 1973b, 1975) also described RNA-negative temperature-sensitive variants isolated from a persistently infected line of L cells which had been initiated by infection with the same strain of Newcastle disease virus. Two types of variant were defined, one which continued to synthesize RNA for some time after shift-up from permissive to restrictive temperature and one which did not. Three other temperature-sensitive variants isolated from these cells possessed virion transcriptases which were less stable than the wild-type enzyme when assayed *in vitro* at 42°C (Stanwick and Hallum, 1976).

These mutants were used by Kowal and Youngner (1978) to analyze the functions required for induction of interferon by Newcastle disease virus. Newcastle disease virus is a poor inducer of interferon in unprimed chick embryo cells. However, when infectivity is inactivated by UV irradiation, Newcastle disease virus can become a potent inducer of interferon in unprimed cells. From their analysis they concluded that no single virus function appeared to be responsible for interferon induction. At the restrictive temperature, the interferon-inducing capacity of some UV-irradiated mutants was dependent on prior exposure of the cells to interferon, and one mutant (*ts*100) was defective at permissive temperature as well as restrictive temperature. The interferon-inducing ability of this mutant could be partially restored by coinfection with heavily-irradiated wild-type virus. The molecular basis of interferon induction has not been resolved.

A variety of other variants of Newcastle disease virus have been recovered from persistent infection of cultured cells and this work has been comprehensively reviewed by Bratt and Hightower (1977).

2. The Temperature-Sensitive Mutants of Sendai Virus

a. Complementation Groups

Ten temperature-sensitive mutants of the Enders strain of Sendai virus have been described by Portner *et al.* (1974). These mutants were isolated following exposure to *N*-methyl-*N*-nitro-*N*-nitrosoguanidine or growth in the presence of 200 μg/ml 5-fluorouracil. The parental and mutant viruses were grown in chick embryo lung (CEL) fibroblasts at 30°C with a restrictive temperature of 38°C, some 5–6 deg below that employed with NDV. The ten mutants were placed in seven nonoverlapping complementation groups, five of the groups (A, D, E, F, and G) being represented by single mutants and groups B and C by two and three mutants, respectively (Table II). Since the mutants in

TABLE II. The Complementation Groups of Sendai Virus[a]

Group	RNA phenotype at restrictive temperature (38°C)	Gene assignment (tentative)	Nature of evidence
A	−	None	—
B	−	None	—
C	−	None	—
D	−	(N) or (L) or (P) defect	Nucleocapsid assembly
E	−	None	—
F	−	None	—
G	+	HN	Production of HN-deficient particles; enhanced transcription (due to absence of HN-mediated inhibition)

[a]For references, see text.

six of the seven groups had RNA-negative phenotypes and individual complementation values ranged between 2 and 30,000, it is likely that the number of discrete complementation groups has been overestimated due to inability to discriminate intergenic and intragenic complementation.

b. Assignments

Biochemical characterization of the ten mutants distinguished at least four phenotypic groups. Temperature-shift experiments revealed three phenotypes among the RNA-negative mutants. Mutants in groups A, B and C appeared to be defective in an early function which was not required throughout the growth cycle, whereas the group D mutant was defective in RNA synthesis throughout the entire cycle, with the mutants in groups E and F intermediate in behaviour.

Mutant ts271, the single RNA-positive mutant representing group G, was defective in hemagglutinating activity. Portner et al. (1975) subsequently showed that noninfectious particles which lacked both the M_r 70,000 virion envelope protein and hemagglutinin and neuraminidase activities were released from ts271-infected CEL cells maintained at the restrictive temperature. These M_r 70,000 protein-deficient particles failed to attach to susceptible cells, thereby confirming that the HN protein was the attachment protein. The HN polypeptide was synthesized in infected cells at the restrictive temperature and was presumed to be unable to adopt a functional configuration. This conclusion was supported by the fact that mutant virions released from cells maintained at the permissive temperature exhibited a reversible ability to hemagglutinate erythrocytes on temperature shift to 38°C and back to 30°C. The neuraminidase activity of the protein was unaffected by temperature shift, indicating that the hemagglutinating and neuraminidase activities of the molecule were functionally separate. The behavior of mutant ts271 confirmed that morphogenesis of Sendai virus particles did not depend on maintenance of the structural and functional integrity of the HN protein, and that neuraminidase played no role in any virus-specified event between eclipse and release. Because of the absence of cell killing at the restrictive temperature, it was concluded that the native HN protein was a determinant of cytopathogenicity. The HN-deficient particles exhibited sevenfold enhanced *in vitro* transcriptase activity, which correlated with previous observation of an inhibitory effect of HN protein on *in vitro* transcriptase activity (Marx *et al.,* 1974). No revertants of *ts*271 could be isolated and comparison of the nucleotide sequence of the HN genes of *ts*271 and wild type revealed three amino acid replacements, two in close proximity at positions 262 and 264 and one distant at position 461. The two adjacent replacements are close to antigenic site I as defined by neutralization escape mutants (Fig. 2) and may be responsible for the temperature-sensitive phenotype (Thompson and Portner, 1987).

Only one of the RNA-negative, temperature-sensitive mutants of Sendai virus, mutant *ts*105 of group D, has been examined in detail (Portner, 1977). The *ts*105 mutation appeared to involve a nucleocapsid protein, possibly P or NP, which is involved in the synthesis or stability of viral RNA. There was no evidence of any defect in viral mRNA translation, but on shift to restrictive temperature there was a time-dependent loss of ability of the NP protein to assemble into nucleocapsids. NP synthesized at restrictive temperature, on the other hand, could assemble into nucleocapsids on temperature down-shift. Association of the P protein with nucleocapsids was unaffected by the temperature of incubation. However, since the behavior of the L protein was not determined in these experiments and revertants were not studied, the temperature-sensitive phenotype could not be associated with any specific core polypeptide. Interestingly, replicative RNA synthesis was favored over transcription on resumption of RNA synthesis on shift-down to permissive temperature.

3. The Temperature-Sensitive Mutants of Measles Virus

a. *Complementation Groups*

A total of five (possibly seven) complementation groups have been defined by analysis of three independently isolated series of temperature-sensitive mutants. Temperature-sensitive mutants of the Rapp strain were isolated by Yamazi and Black (1972), of the Edmonston strain by Bergholz *et al.* (1975) and Ju *et al.* (1980), and of the Schwarz vaccine strain by Haspel *et al.* (1975). In each series a temperature of 39°C was used as the restrictive temperature.

Yamazi and Black (1972) obtained seven *ts* mutants following mutagenesis with proflavine or 5-fluorouracil. Complementation was observed with some combinations of mutants and the results were interpreted as indicative of the existence of three groups. Temperature-shift experiments were carried out which identified mutants with early and late functional defects (Yamazi *et al.,* 1975), but further characterization of these mutants has not been reported.

Bergholz *et al.* (1975) obtained nine mutants from an unattenuated Edmonston virus grown in the presence of 100 μg/ml 5-fluorouracil, and these mutants were classified into three complementation groups. Group A (*ts*1) and group B (ts2, ts3, ts4, ts5, ts6, and ts7) contained RNA-negative mutants, and group C (ts9) contained a single RNA-positive mutant. Mutant ts8 complemented the two mutants representing groups A and C and accordingly was classified in group B, but it was distinguishable from the other mutants of group B by its intermediate phenotype and its propensity to interfere with their multiplication. These ts mutants were attenuated in their ability to induce fatal illness following intracerebral inoculation of hamsters, and animals surviving infection exhibited no sequelae for up to 12 months following recovery.

Haspel et al. (1975) obtained 24 mutants after mutagenesis with 5-fluorouracil, 5-azacytidine, and proflavine from a derivative of the Schwarz vaccine strain. A wild-type stock of this strain (designated CC) able to grow at 39°C was established by cocultivation of persistently infected hamster cells and uninfected BS-C-1 cells. Three complementation groups were defined: Group I contained 21 mutants of RNA-negative phenotype, group II contained 2 mutants of RNA-positive phenotype, and group III contained a single mutant of RNA-negative phenotype. Twenty-three of the 24 mutants were derived from the CC wild type and one (tsA) was derived directly from the Schwarz vaccine strain.

Breschkin et al. (1977) carried out cross-complementation experiments to establish the homologies of the groups. Groups I and B of the two series did not cross-complement and were considered to be homologous, whereas the other groups of each series did cross-complement and therefore appeared to be unique. Consequently, five complementation groups of measles virus were identified and the groups were redesignated as A, B, C, D, and E (Table III). Group B contained 28 of the 33 ts mutants isolated.

Two more complementation groups were identified tentatively by Ju et al. (1980), and the phenotypic properties of these mutants are given in Table III. Ju et al. (1978) obtained their ts mutants without the use of mutagen from human lymphocytes persistently infected with the Edmonston strain. Eighty percent of the clones isolated from these cells were temperature sensitive in Vero cells. Twenty-one of the mutants were classified into the four groups A, B, C, and D by complementation with the standard groups defined by Breschkin et al.

TABLE III. The Complementation Groups of Measles Virus[a]

Group[b]	RNA phenotype at restrictive temperature (39°C)	Gene assignment (tentative)	Nature of evidence
A (A)	−	None	—
B (B or I)	−	(L)	Largest group (85% of mutants)
C (III)	−	None	—
D (C)	+	(F)	Hemolysin defect
E (II)	+	None	—
F	+	None	—
G	+	None	—

[a]For references, see text.
[b]Earlier designation in parentheses.

(1977). Group E could not be determined directly in this way, due to leakiness of the group E prototype stock, but two mutants complemented groups A, B, C, and D were presumed to represent group E, since they had phenotypic properties in common with other group E mutants. Two additional groups (F and G) were identified tentatively on the basis of their RNA-positive phenotypes. However, many (6/21) of the isolates obtained from persistently infected lymphocytes appeared to be multiple mutants, with mutation of the gene corresponding to group B the most common (17/22).

b. Miscellaneous Phenotypic Properties

The interferon-inducing properties of the mutants isolated by Haspel et al. (1975) and Bergholz et al. (1975) were investigated by McKimm and Rapp (1977). Unlike the three wild types, all the mutants tested and a revertant of one (tsG3), were incapable of inducing interferon at either permissive or restrictive temperature. Ability to induce interferon was not restored by complementation of different mutants. Coinfection of a noninducing mutant and wild type had no inhibitory affect on interferon induction. The most likely explanation of the failure of all the mutants to induce interferon was that the three wild-type stocks were heterogeneous for this property, and the mutants were derived from a predominant noninducing type.

Mutant tsG3 exhibited reduced neurovirulence in intracerebrally inoculated newborn hamsters and induced a high frequency of hydrocephalus (Breschkin et al. 1976; Haspel and Rapp, 1975; Haspel et al. 1975). Significantly, however, lower titers of infectious virus and less viral antigen were detected in the brain of animals infected with mutant tsG3. Antigen was predominantly localized to the meninges, and hydrocephalus was presumed to be a consequence of occlusion of the subarachnoid space and closure of the foramena of the fourth ventricle, rather than as a result of infection of ependymal cells. The pattern of immunofluorescence suggested that measles virus spread perivascularly and via the meningeal surface (Woyciechowska et al., 1977).

Chui et al. (1986) described a temperature-sensitive mutant (ts38) of the LEC strain of measles with an unusual property. This mutant had been derived by 5-fluorouracil mutagenesis and possessed an RNA-negative phenotype at a restrictive temperature of 39°C. NP protein presynthesized in ts38-infected cells at 32°C was transported into the nucleus when the temperature was raised to 39°C. Nuclear accumulation of the NP protein is usually associated with isolates of measles virus obtained from SSPE patients by biopsy or from postmortem material, although it can be observed in both lytic and persistent infections. Robbins (1983) proposed that the extent of viral nuclear invasion was a reliable indicator of morbidity. Consequently, mutant ts38 may be useful in elucidating the role of nuclear invasion in measles virus pathogenesis.

Mutant ts38 has also proved a useful experimental tool in analysis of measles virus-induced immunosuppression. Measles virus can infect and multiply in stimulated human peripheral blood mononuclear cells, but at 37°C no infectious virus was released from ts38-infected cells, although virus protein synthesis was detected by immunofluorescence and RNA synthesis by means of an N gene-specific probe detecting positive-sense RNA (Vydelingum et al., 1989). The viability of ts38-infected cells was unaffected, but their response to mitogens and antigens was severely diminished. This experiment suggests that the immunosuppression observed during measles virus infection is not a consequence of destruction of immune cells. Antibodies to alpha-interferon partially reversed the virus-associated inhibition of lymphocyte mitogenesis, indicating that alpha-interferon may play a role in immunosuppression.

c. Persistent Infection

Haspel et al. (1973) and Armen et al. (1977) reported that a high proportion of the virus released from persistently infected cells was temperature sensitive, although in the former case Fisher and Rapp (1979a,b) found that defective virus rather than temperature-sensitive virus was released at later passages. This particular persistent infection had been initiated by infection of hamster embryo fibroblasts with a derivative of the Schwarz vaccine strain and during the initial phase of propagation did not spontaneously release infectious virus, although virus could be recovered by cocultivation with susceptible cells (Knight et al., 1972). Subsequently, Fisher (1983) initiated persistent infections of Vero cell with the Schwarz vaccine strain, the Halle SSPE strain, the Edmonston strain, and mutant ts841 derived from the Edmonston strain. She found that only the cells infected with ts841 released temperature-sensitive virus. Wild and Dugre (1978), on the other hand, could find no involvement of either temperature-sensitive virus or defective virus in a persistent infection of BCG cells initiated by infection with the Halle strain. They concluded that a host-cell factor played the major role in restricting virus replication.

4. The Temperature-Sensitive Mutants of Human Respiratory Syncytial Virus

a. Complementation Groups

Induced temperature-sensitive mutants of the A2 (Gharpure et al., 1969), the RSN-2 (Faulkner et al., 1976), and the RSS-2 (McKay et al., 1988) strains of human respiratory syncytial virus have been isolated from wild-type viruses propagated in secondary bovine kidney cells, BS-C-1 cells, and MRC-5 human diploid cells, respectively. The mutagens employed were 5-fluorouracil and 5-fluorouridine for the A2 strain, and 5-fluorouracil for the RSN-2 and RSS-2 strains. In the case of the A2 and RSS-2 strains, second-stage mutations were induced using nitrosoguanidine (L. S. Richardson et al., 1977) or the acridinelike compounds ICR340 and ICR372 (McKay et al., 1988), respectively. The restrictive temperature was 37°C for the A2 strain and 39°C for the other two strains.

Seven temperature-sensitive mutants of the A2 strain were classified into three complementation groups and the RSN-2 strain mutants into six groups. Although it was determined subsequently that the A2 strain belonged to the A and the RSN-2 strain to the B antigenic subgroups of respiratory syncytial virus, interstrain complementation was observed and the homology of the

TABLE IV. The Complementation Groups of Human Respiratory Syncytial Virus[a]

Group	RNA phenotype at restrictive temperature (37°C, 39°C)	Gene assignment (tentative)	Nature of evidence
A	+	(F)	F_0 mobility difference
B	(+)	G	G processing defect[b]
B'	−	(L)	Majority (45%) group
C	+	None	—
D	+	M	M instability and apparent electrophoretic mobility difference[c]
E	−	P	Loss of monoclonal antibody reactivity: Serine (AGU) present at position 172 in this mutant, and glycine (GGU) in the wild type and a non-ts revertant
F	(+)	None	—
G	(+)	None	—

[a]Compiled from Pringle et al. (1981) and unpublished data of C. Caravokyri.
[b]This mutant also exhibits a non-ts defect involving the F protein.
[c]The P protein of this mutant has a non-ts mutation at position 217 which affects its electrophoretic mobility. Asparagine (AAU) in the wild type is replaced by aspartic acid (GAU) in the mutant.

groups could be established. The conditions for successful complementation were critical and strain dependent (Gimenez and Pringle, 1978). Seven (Gimenez and Pringle, 1978) and subsequently eight (Pringle et al., 1981) distinct complementation groups were defined. Complementation group A was common to both strains, groups B and C were unique to the A2 strain, and groups B', D, E, F, and G were unique to the RSN-2 strain (Table IV). Six of the eight groups are represented by single mutants; hence it is not possible in these cases to discriminate between intergenic and intragenic complementation. However, since the genome of the A2 strain of respiratory syncytial virus is composed of at least ten genes (Collins et al., 1984), it is likely that the complementation groups correspond to individual genes, since two of the single mutant groups have been successfully assigned (see below).

b. Assignments

The temperature-sensitive lesions in mutants ts1 and ts19 of the RSN-2 strain representing groups D and E involve the M and P proteins, respectively. The group D mutant has an M protein which is synthesized, but rapidly degraded, at restrictive temperature. This mutant can be complemented by a vaccinia virus recombinant expressing the RS virus M protein (C. R. Pringle, unpublished data). The group E mutant does not react with a unique anti-P monoclonal antibody, and non-ts revertants regain reactivity.

Mutant ts2 (A2 strain) of group A appears to have a temperature-sensitive mutation affecting the G protein and a non-temperature-sensitive mutation affecting the F protein (C. Caravokyri and C. R. Pringle, unpublished data). The dual lesion in ts2 correlates with the complex phenotype of this mutant, which appears to have a temperature-sensitive defect in attachment/adsorption and a non-temperature-sensitive lesion of the fusion function (Belshe et al., 1977).

c. Vaccine Potential

Temperature-sensitive mutants of the A2 strain have been extensively evaluated as potential live vaccines [reviewed in Chanock (1982) and McIntosh and Chanock (1985)]. These mutants are restricted at 37°C and it was anticipated that they would replicate sufficiently in the upper respiratory tract where the temperature is in the range 32–34°C to induce a local immune response, but not penetrate into the lower respiratory tract. Mutant *ts*1 (complementation group A) was administered intranasally to seropositive adults and seropositive and seronegative children. No disease occurred, nor was there any indication of reversion of mutant *ts*1 during multiplication *in vitro* in cell culture, or *in vivo* in hamsters and adults. A significant immune response was observed in previously seronegative children, but sporadic mild disease was observed and there was evidence of loss of temperature sensitivity during replication in fully susceptible individuals. An attempt was made to obtain greater genetic stability and reduction of residual virulence by remutagenization with a different mutagen (NTG) and selection of more temperature-sensitive virus (L. S. Richardson *et al.*, 1977). However, the result was overattenuation (Belshe *et al.*, 1978) and the development of a *ts*1-based vaccine was abandoned despite its early promise.

Mutant *ts*2 (complementation group B) of the A2 strain was found to be avirulent in primates and defective in adsorption/penetration in cultured cells at 37°C. However, this mutant proved to be poorly infectious both in adult volunteers and in seronegative children, and consequently unsuitable for vaccine development.

The A2 strain mutants were isolated and propagated in bovine secondary cells and hence were vulnerable to contamination with extraneous agents. More recently, single-stage *ts* mutants restricted at 39°C and two-stage *ts* mutants restricted at 38°C have been derived from the RSS-2 strain entirely in human diploid cell culture. One of the two-stage mutants (*ts*1B) was tested in adult volunteers and proved to be almost as immunogenic as wild-type virus, but with greatly reduced disease-producing potential (McKay *et al.*, 1988; Watt *et al.*, 1990).

These independent vaccine trials, together with other studies of cold-temperature-adapted virus, establish that a substantial immune response can be generated by RS virus restricted to multiplication in the upper respiratory tract. The real value of these experiments is that they demonstrate that vaccination can be achieved without the exacerbation of disease encountered in the earlier trials of formalin-inactivated vaccine (McIntosh and Fishaut, 1980).

C. Other Mutants

1. The Noncytopathic Mutants of Newcastle Disease Virus

Noncytopathic (*nc*) mutants of the Australia-Victoria strain of NDV do not induce plaque formation in chick embryo fibroblasts, although near normal amounts of infectious virus are produced by infected cells. The *nc* mutants were detected because red blood cells adsorb to the surface of infected cells.

These mutants can be cloned and titrated by scoring hemadsorbing foci on infected chick embryo fibroblast monolayers. All the *nc* mutants were associated with extended mean embryo death times, and the mutants with the smallest foci exhibited the greatest prolongation of mean embryo death time. These observations suggested that the *nc* mutants have reduced virulence for chickens (Madansky and Bratt, 1978, 1981a,b).

Viral RNA and protein synthesis were reduced in *nc*-infected cells, and there was a disproportionate reduction in the amount of L protein in infected cells, though not in virions. Inhibition of host macromolecule synthesis was also diminished. Plaque-forming revertants of *nc* mutants exhibited coreversion, regaining plaque-forming ability, normal viral RNA and protein synthesis, and normal host protein synthesis inhibition. A new protein (X) identified as an altered form of the P protein was present in cells infected with two of the mutants, *nc*4 and *nc*16, and also in released virions. A subclone of *nc*4 which produced larger hemadsorbing foci had the normal form of the P protein restored (Madansky and Bratt, 1981a,b). Thus, the lesion in the P protein did not appear to be a primary determinant of cytopathogenicity. Similarly, a lesion in the F protein affecting its cleavability by trypsin was associated with mutant *nc*7, but did not appear to be a determinant of cytopathogenicity, since a plaque-forming revertant of *nc*7 still exhibited the F protein defect of the mutant. Thus, the P and F protein lesions in mutants *nc*4, *nc*7, and *nc*16 were genetically distinct from the noncytopathogenic phenotype. Nonetheless, the P protein and F protein lesions caused an extension of the mean embryo death time, and similar phenotypes appear in naturally avirulent strains. Consequently, the determination of virulence *in vivo* is likely to be polygenic in nature.

2. The Plaque Morphology Mutants of Newcastle Disease Virus

Although there is a positive correlation between plaque size and virulence of Newcastle disease virus in chickens (Schloer and Hanson, 1968), the ability to spread in one cell type *in vitro* cannot be assumed to correspond to replicative ability in the tissues of the natural host. Indeed, Granoff (1961a) reported that certain mutagen-induced small-plaque mutants isolated in chick embryo cell cultures exhibited enhanced neurovirulence in mice. Estupinan and Hanson (1971a) distinguished six plaque morphology mutants present in the Delaware-Hickman strain of Newcastle disease virus, which differed in their virulence for chickens. A large, clear, plaque-forming virus was the most virulent, and a small, red plaque-former the least virulent. The spectrum of chicken virulence was paralleled by the mean embryo death time and rapidity of cytopathic effects in chick embryo fibroblast monolayers. Minor antigenic differences were observed between the mutants as measured by neutralization test and resistance to challenge. These different mutants also differed in immunogenicity as measured by their inability to induce good immunity to themselves. In general, the most virulent mutants were the best antigens, even after inactivation. Virulence decreased with diminishing plaque size in the case of both clear and red plaques. Ability of mutants to induce interferon was not related to virulence in chick embryos, nor were there any correlations among virulence, rate of release of virus, or neuraminidase or hemagglutinin activities. Virulence was correlated, however, with a shorter lag period during single-cycle growth (Schloer and Hanson, 1971).

The rate of mutation of red to clear plaque type was high in this and other strains of Newcastle disease virus (Granoff, 1961a, 1964; Thiry, 1964), whereas the reverse was rare (Estupinan and Hanson, 1971b). Schleor and Hanson (1971) estimated by the fluctuation test that the mutation rate from red to clear plaque was 3.3×10^{-6} and that from large to small plaque was 1×10^{-7} per particle per generation.

Estupinan and Hanson (1971a,b) observed that a heterogeneous population could be restablished within a few passages from any of the red plaque types present in the original Delaware-Hickman isolate, whereas the clear mutants appeared to be genetically stable, and it was apparent that components of the original isolate were soon lost during propagation in cultured cells. Consequently, the apparent genetic homogeneity of the common laboratory stocks of Newcastle disease virus may not represent the true genetic potential of this virus, and can complicate the analysis of virulence and pathogenesis. The genetic heterogeneity of field isolates of Newcastle disease virus, on the other hand, is particularly obvious.

Lomniczi (1975) characterized plaque morphology mutants of the mesogenic Herts vaccine strain and discounted the possibility of strain heterogeneity in the original isolate. Small plaque-forming mutants isolated from this strain lacked neurovirulence, contrary to previous reports that lentogenic viruses did not form plaques on chick embryo cells.

The general conclusion from these and similar studies is that there are multiple determinants for both plaque size and virulence.

3. Monoclonal Antibody-Resistant Mutants of Newcastle Disease Virus and Sendai Virus

Monoclonal antibodies have been used extensively to map antigenic sites and epitopes in Newcastle disease virus and Sendai virus. Usually, neutralization-resistant ("escape") mutants have been used for this purpose (e.g., Abenes et al., 1986; Nishikawa et al., 1983; Portner, 1981, 1984; Portner et al., 1987a,b; Iorio and Bratt, 1983; Toyoda et al., 1987). Iorio and Bratt (1985) described the isolation of unique nonneutralizable mutants with lesions located in the hemagglutinin-neuraminidase protein gene of Newcastle disease virus. A high proportion of anti-HN protein monoclonal antibodies do not neutralize the infectivity of Newcastle disease virus, although they bind to virions. Nevertheless, Iori and Bratt (1984a) were able to define four antigenic sites by competitive binding and additive neutralization assays. Antibodies to all four sites were required for maximum neutralization (Iorio and Bratt, 1984b). Persistent nonneutralizable fractions and enhanced neutralization following addition of rabbit anti-mouse IgG were further indications of the prevalence of nonneutralizing monoclonal antibodies. Iorio and Bratt (1985) developed a protocol which enhanced the recovery of neutralizing monoclonal antibodies. Pretreatment of monoclonal antibody-exposed virus with rabbit anti-mouse immunoglobulin prior to plating reduced the amount of infectious virus with bound nonneutralizing monoclonal antibodies. Two types of mutant were isolated from the residual infectious virus; typical "escape" mutants resistant to neutralization by the monoclonal antibody used in their selection, and unique mutants binding the selecting antibody without loss of infectivity. Both types of mutant have value in the fine mapping of epitopes and the methodology is applicable to other viruses.

Three antigenic determinants have been identified on the F protein of Newcastle disease virus (Abenes et al., 1986; Toyoda et al., 1987). All the monoclonal antibodies used in defining these sites had neutralizing, hemolysis-inhibiting, and fusion-inhibiting activities, but none could recognize either the nascent polypeptide or the denatured F protein, suggesting that the three antigenic determinants, though functionally related, were dependent on protein folding. Sequencing of the F genes of a series of escape mutants confirmed that the three sites were located at a distance from the fusion-inducing N terminus of the F_1 subunit. Sites I and III were located in the F_1 subunit and site II in the F_2 subunit. Therefore, either the three antigenic sites are located adjacent to the fusion-inducing domain in the folded protein and inhibition is by steric interference, or more than one domain in the F protein is involved in membrane fusion.

Monoclonal antibodies directed against the HN protein of Newcastle disease virus have been classified into three groups according to their biological activities (Nishikawa et al., 1983). Data on the frequency of isolation of escape mutants and competitive binding assays indicated that two of the three sites were topologically distinct and nonoverlapping.

Only one epitope has been defined on the F protein of Sendai virus (Portner et al., 1987b). The sequences of the F protein gene of three escape mutants have been determined. In all three mutants there was a single amino acid substitution (Pro → Glu) at a site (residue 399) remote from the putative fusion-inducing F_1 N terminus. The loss of a proline could affect protein folding and result in loss of antigenic recognition.

By contrast, four antigenic sites have been topographically mapped on the HN protein of Sendai virus by competitive binding assays (Portner et al., 1987a). Antibodies to the different sites had different biological activities. Antibodies to sites I, III, and IV inhibited hemagglutination and neuraminidase, hemagglutination alone, or neuraminidase alone, respectively, whereas antibodies to site II inhibited hemolysis, an F-protein function, suggesting either steric interference or a direct involvement of the HN protein in membrane fusion. The combined result of the mapping neutralization escape mutants and ts mutants of Sendai virus is illustrated in Fig. 3.

Shioda et al. (1988) sequenced the M, F, and HN genes of three laboratory strains derived from a wild strain of bovine parainfluenza virus type 3 which differ in their syncytium-inducing activity. No amino acid substitutions were identified in the deduced F proteins. A syncytium-formation-deficient variant (MR) was isolated from the strain with the most pronounced syncytium-inducing activity (M). The single amino acid change identified was located not in the F protein, but in the HN protein, where phenylalanine at residue 193 in the M strain was replaced by leucine in the MR variant. These observations also suggest that the HN protein may contribute to syncytium formation in addi-

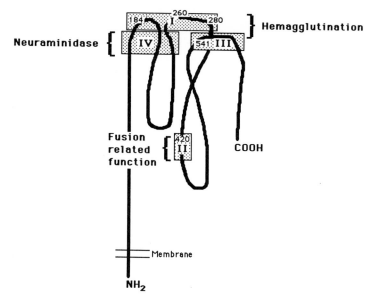

FIGURE 3. Hypothetical model of the HN glycoprotein of Sendai virus. [Reproduced with permission from Thompson and Portner (1987).] The boxes represent antigenic sites on the HN protein. Sites I, II, and III have been located by mapping of antigenic and temperature-sensitive mutants. Site IV has not yet been located. The function associated with each antigenic site is indicated.

tion to the F protein. An alternative explanation, however, is that any mutation affecting maturation may simply enhance syncytium formation by increasing the amount of F protein in the plasma membrane.

4. A Neuraminidase Activity-Deficient Variant of Mumps Virus

Adaptation of mumps virus to growth in cultured cells is usually accompanied by development of a syncytial cytopathic effect. Waxman and Wolinsky (1986) have shown that the neuraminidase protein plays a role in cell-to-cell fusion, and that fusion is not simply (as in Sendai virus) a consequence of mutation affecting proteolytic processing of the fusion (F) protein. By propagation of the nonfusing O'Take strain of mumps virus in CV-1 cells in the presence of the neuraminidase inhibitor 2-deoxy-2,3-dehydro-N-acetylneuraminic acid, a syncytium-forming variant was obtained. This variant had no detectable neuraminidase activity. Hemagglutinin activity was unaffected and as a consequence the syncytium-forming variant could agglutinate red blood cells, but, unlike wild-type virus, was unable to elute once adsorbed. The isolation of this variant demonstrates that the viral neuraminidase modulates the cytopathogenicity of mumps virus, and may be a determinant of neurovirulence, since a correlation has been observed between fusogenic ability in cell cultures and neurotropism in newborn hamsters.

Löve et al. (1985) isolated four mutants of the neurotropic Kilham strain of mumps virus which have enhanced neuraminidase activity. These mutants were obtained by passage in the presence of neutralizing anti-HN monoclonal antibody. The four mutants had markedly different biological properties in addition to alteration of the target epitope. One of the four mutants had reduced neurovirulence, again implicating the surface glycoproteins of mumps virus as major determinants of pathogenicity and virulence.

5. Plaque Morphology Mutants of Canine Distemper Virus

Cosby et al. (1981) described a stable, small-plaque-forming variant isolated from the large-plaque-forming Onderstepoort vaccine strain of canine distemper virus, which had altered neurovirulence for weanling hamsters. The large-plaque-former induced acute neurological illness and little neutralizing antibody, whereas the small-plaque-former induced high titers of neutralizing antibody, but no disease. By altering the levels of antibody early in infection, the outcome could be modified, showing that the major factor in determination of virulence was the differing immunogenicity of these viruses (Cosby et al., 1983). Other small-plaque mutants were isolated from a culture of Vero cells persistently infected with the large-plaque form of canine distemper virus and shown to have phenotypic properties similar to the original small-plaque isolate; i.e., enhanced immunogenicity and decreased neurovirulence.

Plaque morphology mutants induced by 5-fluorouracil treatment, however, exhibited a spectrum of plaque types and disease-producing potential. The mutagen-induced variants did not show the same stability on passage as the small-plaque type derived previously. It was concluded that the small-plaque type was probably generated during high-multiplicity passage and was probably the result of multiple mutations (Cosby et al., 1985).

6. The Protease Activation Mutants of Sendai Virus

The orthomyxoviruses and the paramyxoviruses proper adsorb to neuraminic acid-containing receptors. Since neuraminic acid-containing glycoproteins and glycolipids are ubiquitous components of the cell membrane of vertebrates, the availability of surface receptors may play a lesser role in determining host range and tissue tropism than it does in other enveloped viruses, such as the retroviruses.

In the case of the paramyxoviruses, and perhaps also the morbilliviruses and pneumoviruses, the second surface glycoprotein which mediates fusion and hemolysis is a major determinant of host range and tissue tropism. Cleavage of the F protein is essential for activation of infectivity and membrane fusion, and this specific cleavage is mediated by a host protease (Homma and Ohuchi, 1973; Scheid and Choppin, 1974). Consequently, the host range, tissue tropism, and pathogenicity of paramyxoviruses are dependent on the availability of the appropriate cellular protease.

Some unique genetic studies by Scheid and Choppin (1976) provided the essential experimental confirmation of the predominant role of cellular proteases in activation of infectivity. The F_0 protein of wild-type Sendai virus is cleaved by trypsin and in the presence of this protease, plaques are formed on monolayers of the otherwise nonpermissive MDBK line of bovine kidney cells.

Other proteases, such as chymotrypsin and elastase, did not promote plaque formation. However, when MDBK cells were infected in the presence of chymotrypsin with Sendai virus mutagenized by treatment with 1 M sodium nitrite at pH 4.4, a cytopathic effect was observed in a small proportion of cultures. Two protease activation mutants, pa-c1 and pa-c2, were isolated from these cultures, which were able to form plaques on MDBK cell monolayers in the presence of chymotrypsin. Similarly, a series of elastase-activated mutants, pa-e1 to pa-e8, were obtained from mutagenized virus by incubation of infected cells in the presence of elastase. Mutants activated by plasmin or thermolysin were also obtained. Some of these pa mutants retained the propensity to be activated by trypsin, whereas others did not. Analysis of the polypeptide composition of pa mutants grown in different host systems confirmed the association of F protein cleavage and activation of infectivity.

The importance of these protease activation specificities in determining host range in the whole organism was demonstrated by injecting pa mutants into the allantoic sac of the chick embryo. Mutant pa-c1 was unable to undergo multiple cycles of replication in the allantoic sac unless chymotrypsin was injected simultaneously, and similarly, mutant pa-1e required elastase for activation in the allantoic sac. In acquiring their new specificity, both of these mutants had become resistant to cleavage by the proteases normally present in the allantoic sac of the chick embryo. These mutants of Sendai virus provide sensitive probes for identifying the proteases present in tissues.

An important implication of these results is that when a paramyxovirus is isolated from a living organism by inoculation into cultured cells, the virus which grows out may be one which is susceptible to cleavage by the protease present in the cell membrane of the detector system. Consequently, the isolation procedure may introduce a bias by selecting a minor component or rare mutant which is not typical of the virus predominating in the organism. This may be one reason why virus propagated in culture rarely has the disease-producing potential of the original isolate. Similarly, the rare progression of disease by invasion of a normally inviolate tissue may be due to the appearance of a novel pa mutant. So far, however, protease activation mutants have only been described for Sendai virus, and Sendai virus is the only paramyxovirus known which produces noninfectious virions containing uncleaved F_0 protein.

The sequence of 60–70 amino acids around the cleavage site has been derived from the mRNA sequence for five of the pa mutants, revealing in each case one or two amino acid replacements near or at the cleavage sites (Hsu et al., 1987). Figure 4 shows that in three instances the change in protease specificity appeared to be due to a replacement at the cleavage site, whereas in the remaining two it was more remote. Itoh and Homma (1988) confirmed independently that the replacement of Arg by Ile at position 116 is responsible both for the loss of trypsin sensitivity and gain of chymotrypsin sensitivity, although amino acid sequence analysis showed that chymotrypsin cleaved between residues 114 and 115, whereas trypsin cleaved between residues 116 and 117. In the work of Itoh et al. (1987), the complete F gene nucleotide sequence was determined and amino acid changes at more remote sites in the F protein could be excluded. Protease activation mutants derived by passage in the presence of chymotrypsin independently in different laboratories had identical

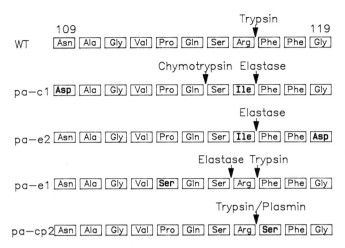

FIGURE 4. Amino acid sequences around the cleavage activation site of the F protein of wild-type and four protease activation (*pa*) mutants of Sendai virus. The cleavage site of each protease is indicated by an arrow. The amino acids are numbered from the amino terminus of the F protein. Mutant *pa*-c1 was selected with chymotrypsin, mutants *pa*-e1 and *pa*-e2 with elastase, and mutant *pa*-cp2 with chicken plasmin (see text).

mutational changes at positions 109 (Asn → Asp) and 116 (Arg → Ile). The Asp at 109 may be essential for chymotrypsin sensitivity, or alternatively Asn may be dispensable, but neither is sufficient, since Asn was unchanged in a chymotrypsin-resistant revertant TSrev-58.

Although the chick embryo is not the natural host of Sendai virus, Tashiro and Homma (1983), using the trypsin-sensitive wild type and a chymotrypsin-sensitive, trypsin-resistant mutant (TR-2), showed that a trypsinlike activity in the bronchial epithelium of the mouse was responsible for activation and multiplication of Sendai virus in the mouse lung. A trypsin-sensitive revertant (TS-rev58) of trypsin-resistant mutant TR-5 had the pneumopathogenicity of the wild-type virus restored (Mochizuki *et al.*, 1988). The amino acid sequences around the cleavage sites of TR-2, TR-5, and TR-rev58 were predicted from the nucleotide sequence. In mutants TR-2 and TR-5 there are two replacements, one at the cleavage site (Arg → Ile at 116) and another at position 109 (Asn → Asp). In the revertant, there was a single reversion of Ile → Arg at position 116, restoring the wild-type amino acid, leaving the Asp at 109 unchanged (Itoh *et al.*, 1987; Itoh and Homma, 1988). It is clear that a single nucleotide change in the F protein gene in certain circumstances can alter the specificity of protease activation and thereby affect the host range and tissue tropism of the virus.

The biological importance of the specific cleavage of the F protein of paramyxoviruses is emphasized on the one hand by the conservation of sequence of the N termini of the F_1 proteins and on the other hand by the inhibitory activity of oligopeptides which structurally mimic this region (Choppin and Scheid, 1980). C. D. Richardson *et al.* (1980) established that the most inhibitory oligopeptide was a heptapeptide with the same sequence as the N terminus of Sendai virus. The presence of a carboxybenzoxy group on the

N terminus enhanced activity, but the correct peptide sequence was more important. This was verified by isolation of a mutant of measles virus resistant to the inhibitory action of carbobenzoxy-phenylalanine-phenylalanine–nitro-arginine [Z-Phe-Phe–L(NO2)Arg] by growth in its presence, which remained sensitive to the oligopeptide Z-Phe-Phe-Gly. Sequencing of the *F* gene of this mutant revealed that mutations were present at residues 338, 352, and 460 in the deduced F_0 protein sequence and not in the cleavage site region (Hull *et al.*, 1987). It is likely, therefore, that conformation of the F protein is important in functionally orientating the fusion sequence.

IV. ANALYSIS OF GENE FUNCTION

Although genetic analysis of paramyxoviruses is limited by the inability to achieve successful marker rescue (see Section III), the methodology of recombinant DNA technology has a role in analysis of gene function.

A. Transport and Glycosylation of the G Glycoprotein of Respiratory Syncytial Virus

Three C-terminally-deleted mutants of the G glycoprotein of the A2 strain of respiratory syncytial virus have been described (Vijaya *et al.*, 1988; Olmsted *et al.*, 1989). Plasmids expressing truncated products containing the N-terminal 71, 180, and 230 amino acids of the 298-amino acid G glycoprotein were constructed by mutagenesis which introduced a restriction enzyme site followed by an in-frame termination codon and another restriction enzyme site after proline residues 71, 180, and 230. The 71-amino acid product reacted poorly with specific antiserum and a reporter sequence was added which consisted of a cysteine and the 12 C-terminal amino acids of the 1A glycoprotein of RS virus. This chimera could be detected with a 1A-specific C-terminal peptide antiserum. The truncated proteins expressed by all three constructs contained O-linked carbohydrates like the complete G protein, although interpretation of data for the 71-amino acid protein was compromised by the addition of the reporter sequence. The 180- and 230-amino acid proteins also contained N-linked carbohydrates and had M_r values estimated from electrophoretic mobility that were approximately twice the values calculated from their composition. Experiments with the inhibitor monensin suggested that this discrepancy was due to posttranslational addition of carbohydrates or dimerization in or beyond the Golgi apparatus, or both. All three truncated proteins were transported to the cell surface and had their C termini located extracellularly, like the complete G protein. Thus, the N-terminal 71-amino acid sequence contained all the structural information required for transport and membrane insertion.

Cotton rats were immunized with vaccinia recombinant viruses expressing these truncated proteins. The 230-amino acid truncated protein induced levels of protection and specific antibodies approaching those induced by the complete G protein, whereas the two shorter proteins failed to induce either response. The positive results with the 230-amino acid fragment indicate that the C-terminal 68-amino acid region of the 236-amino acid ectodomain does not contain major epitopes responsible for induction of protective immunity. The 230-amino acid product is the only one of the three which contains the highly conserved cysteine-rich region, indicating that this region of the molecule plays a key role in the antigenicity and function of the protein.

The tolerance of the G protein to drastic mutational trauma contrasts with the common experience that the introduction of mutations into integral membrane proteins adversely affects their transport to the plasma membrane. The efficient transport of truncated G proteins suggests that the conformational state of the G protein is not critical. The considerable antigenic and sequence variability of the G protein of respiratory syncytial virus (Johnson et al., 1987) is another indication of the tolerance of the ectodomain of this protein to mutational change.

This tolerance of the G protein to genetic manipulation has been utilized to design a carrier to facilitate transport of a foreign peptide sequence to the plasma membrane (Vijaya et al., 1988). The rationale was based on the assumption that protein folding begins during synthesis. If so, conformational disturbance should be minimized by attaching a foreign peptide to the C terminus of a protein with an uncleaved signal-anchor domain. The truncated G-protein genes described previously were flanked by vaccinia virus DNA to permit cloning and expression in a vaccinia virus vector. A sequence encoding four copies of the four-amino acid immunodominant repeating epitope of the circumsporozoite of the malaria parasite *Plasmodium falciparum* was attached to the truncated G gene and recombined into the vaccinia virus genome. A chimeric protein was expressed which was transported to the plasma membrane and exposed on the external surface of the cell.

B. Membrane Interactions of the F_1 Polypeptide of SV5

The hydrophobic N termini of the F_1 proteins of paramyxoviruses are implicated in the process of membrane fusion. The inactive F_0 precursor, however, is translocated across the cell membrane and subsequently is activated by proteolytic cleavage to disulfide bond-linked F_1 and F_2 polypeptides with loss of an interconnecting peptide of one to 5 basic amino acids. The exposed hydrophobic N terminus of the F_1 polypeptide is highly conserved in all paramyxoviruses. Paterson and Lamb (1987) examined the effect of the position of this hydrophobic region on its function by *in vitro* mutagenesis. Plasmids were constructed which expressed hybrid proteins comprising the hydrophobic N terminus of the F_1 protein, with or without the F_1/F_2 interconnecting peptide, translocated to the HA protein of influenza A virus in place of its own membrane anchorage region. The HA–F_1 hybrid protein without the cleavage activation region behaved as an integral membrane protein, but was not transported beyond the Golgi apparatus. The HA–F_1 hybrid protein which did contain the interconnecting peptide of five basic residues did not behave as an integral membrane protein and was exported from the cell. These observations were interpreted in terms of differential hydrophobicity thresholds according to location; when located internally in its normal location, the N terminus of F_1 did not exceed the threshold of hydrophobicity to function as an anchor sequence, whereas, when located terminally, the critical threshold was exceeded. Since the five-residue interconnecting peptide was found not to be necessary for transport of the F_0 protein of SV5, it was concluded that its presence in the HA–F_1 hybrid protein lowered its hydrophobicity below the threshold.

C. Gene-Specific Hypermutation in Measles Virus

It is well established that under conditions of persistent infection, mutations accumulate in the genomes of negative-strand RNA viruses primarily as a consequence of both the inherent low fidelity of transcription of RNA polymerases consequent upon the lack of a proofreading mechanism, and an apparent relaxation of selective pressure in nonlytic infection (Domingo et al., 1978, Holland et al., 1979; Rowlands et al., 1980; Steinhauer and Holland, 1987). Although genomic variability is intrinsically high, many paramyxoviruses exhibit stable biological properties and in general there is no progressive antigenic variation in response to immunological pressure. It is likely that a stable consensus sequence is established in RNA virus populations maintained under constant selective pressure, masking the inherent variability of the genome.

Cattaneo et al. (1988) attempted to estimate the mutational load accumulated by measles virus in the course of persistent infection of human neural tissue by cloning full-length transcripts of measles virus genes directly from affected tissue obtained at autopsy. Approximately 2% of the nucleotides of the genome had undergone mutation, with 35% of these changes resulting in amino acid replacements. cDNAs were obtained from the brains of individuals with subacute sclerosing panencephalitis (SSPE) and a single individual with measles inclusion body encephalitis (MIBE), as well as from the lytic Edmonston strain. Since the original infecting viruses were undefinable and the Edmonston strain inappropriate because of its complicated passage history, a consensus sequence was derived to identify the mutations accumulated in these viruses during persistent infection of neural tissue. The consensus sequence represented the most frequent nucleotides in nine independently determined sequences (from three lytic and six defective viruses), and deviations from this consensus were designated mutations. Remarkably, mutations were not distributed randomly, but were more abundant in the M protein gene from the MIBE case. Moreover, the mutations were predominantly U → C transitions. In fact, 132 of the 266 U residues in the consensus sequence were changed to C residues in this gene. The number of U → C transitions in the M gene from the MIBE case was 20-fold greater than that of other mutations, whereas in the corresponding genes from the SSPE cases the U → C transitions were not more frequent than other mutations. The high frequency of U → C transitions was confined within the boundaries of the M protein gene. These mutational changes introduced a frameshift which created a termination codon resulting in a truncated gene product.

A significant feature of this phenomenon was that A → G changes corresponding to the U → C mutations in the other genomic strand were not increased, indicating that the U → C transitions were introduced exclusively into one strand. This suggests either a singular hypermutational event or repetitive cycles of strand-specific restricted mutation. Cattaneo *et al.* (1988) concluded from the apparent conservation of changes in different clones from the same source that a singular mutational event was the more acceptable explanation, and they inclined to the view that the phenomenon was mediated by an aberrant RNA polymerase. (This gene-specific hypermutation is discussed in greater detail in Chapter 12).

Lamb and Dreyfuss (1989) put forward an explanation of the hypermutation phenomenon in the M protein that is based on the finding of an ATP-dependent unwinding activity which covalently modifies double-stranded RNA molecules by converting adenosine residues to inosine residues (Baas and Weintraub, 1988). If, during transcription of the *M* gene of the SSPE virus, the mRNA transcript fails to separate from the genomic template, the dsRNA unwinding/modifying activity might change A residues to I residues in the transcript. In the next round of transcription of mRNA, these I residues would be converted to C residues. There is little evidence that double-helical RNA structures are normal intermediates in the replication of paramyxoviruses, and hypermutation in the *M* gene may represent a rare failure of separation of template and progeny strands. Indeed, the absence of double-stranded RNA structures in the replication cycle of single-stranded RNA viruses may be a requirement for protection from potentially lethal hypermutational events (Weissmann, 1989).

V. PROSPECTS

Full application of genetic approaches to the study of paramyxoviruses awaits development of techniques which will enable genetic information manipulated *in vitro* to be rescued into infectious virus. The lack of this experimental facility is a barrier to progress. It is a general problem for genetic analysis of all negative-strand RNA viruses, and represents one of the remaining major technical limitations of recombinant DNA technology.

Interim expedients may be the use of virus vectors carrying individual paramyxovirus genes to confirm and extend the assignment of temperature-sensitive mutations, and refinement of mapping by the RNase A mismatch cleavage procedure (Lopez-Galindez *et al.*, 1988). In the immediate future existing methods of genetic analysis will have particular application in the elucidation of the functions and homologies of nonstructural proteins and in the unraveling of the determinants of virulence.

Note added in proof: The first significant steps towards this end have now been taken. Enami *et al.* (*Proc. Natl. Acad. Sci. USA* **87**:3902–3905, 1990) have succeeded in rescuing infectious influenza virus from cells transfected with RNAs derived from specific recombinant DNAs following addition of purified influenza virus polymerase complex. Ballart *et al.* (*EMBO J.* **9**:379–384, 1990) have been able to generate the infectious measles virus from cloned measles virus cDNA by the microinjection of committed transcription complexes into the cytoplasm of helper cells which supply the appropriate proteins required for encapsidation and transcription/replication. These developments make the site-specific mutagenesis of negative strand RNA virus a reality, though not yet a universally applicable technique.

VI. REFERENCES

Abenes, G., Kida, H., and Yamagawa, R., 1986, Antigenic mapping and functional analysis of the F protein of Newcastle disease virus using monoclonal antibodies, *Arch. Virol.* **90:**97–110.

Armen, R. C., Evermann, J. F., Truant, A. L., Laughlin, C. A., and Hallum, J. V., 1977, Temperature-sensitive mutants of measles virus produced from persistently infected HeLa cells, *Arch. Virol.* **53:**121–132.

Baas, B. L., and Weintraub, H., 1988, An unwinding activity that covalently modifies its double-stranded RNA substrate, *Cell* **55:**1089–1098.

Banerjee, A., 1987, The transcription complex of vesicular stomatitis virus, *Cell* **48:**363–364.

Baybutt, H. N., and Pringle, C. R., 1987, Molecular cloning and sequencing of the F and 22K membrane protein genes of the RSS-2 strain of respiratory syncytial virus, *J. Gen. Virol.* **68:**2789–2796.

Belshe, R. B., Richardson, L. S., Schnitzer, T. J., Prevar, D. A., Camargo, E., and Chanock, R. M., 1977, Further characterization of the complementation group B temperature-sensitive mutant of respiratory syncytial virus, *J. Virol.* **24:**8–12.

Belshe, R. B., Richardson, L. S., London, W. T., Sly, D. L., Camargo, E., Prevar, D. A., and Chanock, R. M., 1978, Evaluation of five temperature-sensitive mutants of respiratory syncytial virus in primates. II. Genetic analysis of virus recovered during infection, *J. Med. Virol.* **3:**101–110.

Bergholz, C. M., Kiley, M. P., and Payne, F. E., 1975, Isolation and characterization of temperature-sensitive mutants of measles virus, *J. Virol.* **16:**192–202.

Blumberg, B. M., Crowley, J. C., Silverman, J. I., Menonna, J., Cook, S., and Dowling, P. C., 1988, Measles virus L protein evidences elements of ancestral RNA polymerase, *Virology* **164:**487–497.

Bratt, M. A., and Hightower, L. E., 1977, Genetics and paragenetic phenomena of paramyxoviruses, *in* "Comprehensive Virology" (H. Fraenkel-Conrat and R. R. Wagner, eds.), Vol. 9, pp. 457–533, Plenum Press, New York.

Breschkin, A. M., Haspel, M. V., and Rapp, F., 1976, Neurovirulence and induction of hydrocephalus with parental, mutant, and revertant strains of measles virus, *J. Virol.* **18:**809–811.

Breschkin, A. M., Rapp, F., and Payne, F. E., 1977, Complementation analysis of measles virus temperature-sensitive mutants, *J. Virol.* **21:**439–441.

Buckland, R., Giraudon, P., and Wild, T. F., 1988, Antigenic variation of the internal proteins of measles virus: Identification and expression of the individual epitopes in bacteria, *Virus Res.* (Suppl.) **2:**46.

Cattaneo, R., Schmid, A., Eschle, D., Baczko, K., ter Meulen, V., and Billeter, M., 1988, Biased hypermutation and other genetic changes in defective measles viruses in human brain infections, *Cell* **55:**255–265.

Cavanagh, D., and Barrett, T., 1988, Pneumovirus-like characteristics of the mRNA and proteins of turkey rhinotracheitis virus, *Virus Res.* **11:**241–256.

Chambers, P., Barr, J., Pringle, C. R., and Easton, A. J. 1990, Molecular cloning of pneumonia virus of mice, *J. Virol.* **64:**1869–1872.

Chanock, R. M., 1982, Respiratory syncytial virus, *in* "Virus Infections of Humans; Epidemiology and Control" (A. S. Evans, ed.), pp. 471–488, Plenum Press, New York.

Choppin, P. W., and Scheid, A., 1980, The role of viral glycoproteins in adsorption, penetration, and pathogenicity of viruses, *Rev. Infect. Dis.* **2:**40–61.

Chui, L. W.-l., Vainionpaa, R., Marusyk, R., Salmi, A., and Norrby, E., 1986, Nuclear accumulation of measles virus nucleoprotein associated with a temperature-sensitive mutant, *J. Gen. Virol.* **67:**2153–2162.

Coelingh, K. J., Winter, C. C., Murphy, B. R., Rice, J. M., Kimball, P. C., Olmsted, R. A., and Collins, P. L., 1986, Conserved epitopes on the hemagglutinin-neuraminidase proteins of human and bovine parainfluenza type 3 viruses: Nucleotide sequence analysis of variants selected with monoclonal antibodies, *J. Virol.* **60**:90–96.

Collins, P. L., Hightower, L. E., and Ball, L. A., 1980, Transcriptional map for Newcastle disease virus, *J. Virol.* **35**:682–693.

Collins, P. L., Huang, Y. T., and Wertz, G. W., 1984, Identification of a tenth mRNA of respiratory syncytial virus and assignment of polypeptides to the 10 viral genes, *J. Virol.* **49**:572–578.

Collins, P. L., Olmsted, R. A., Spriggs, M. K., Johnson, P. R., and Buckler-White, A. J., 1987, Gene overlap and site-specific attenuation of transcription of the viral polymerase L gene of human respiratory syncytial virus, *Proc. Natl. Acad. Sci USA* **84**:5134–5138.

Cosby, S. L., Lyons, C., Fitzgerald, S. P., Martin, S. J., Pressdee, S., and Allen, I. V., 1981, The isolation of large and small plaque canine distemper viruses which differ in their neurovirulence for hamsters, *J. Gen. Virol.* **52**:345–353.

Cosby, S. L., Morrison, J., Rima, B. K., and Martin, S. J., 1983, An immunological study of infection of hamsters with large and small plaque canine distemper viruses, *Arch. Virol.* **76**:201–210.

Cosby, S. L., Lyons, C., Rima, B. K., and Martin, S. J., 1985, The generation of small-plaque mutants during undiluted passage of canine distemper virus, *Intervirology* **23**:157–166.

Crowley, J. C., Dowling, P. C., Menonna, J., Silverman, J. I., Shuback, D., Cook, S. D., and Blumberg, B. M., 1988, Sequence variability and function of measles virus 3′ and 5′ ends and intercistronic regions, *Virology* **164**:498–506.

Curran, J., and Kolakofsky, D., 1988, Ribosomal initiation from an ACG codon in the Sendai virus P/C mRNA, *EMBO J.* **7**:245–251.

Dahlberg, J. E., 1968, Ph. D. Thesis, Purdue University, West Lafayette, Indiana.

Dahlberg, J. E., and Simon, E. H., 1968, Complementation in Newcastle disease virus, *Bacteriol. Proc.* **1968**:162.

Dahlberg, J. E., and Simon, E. H., 1969a, Recombination in Newcastle disease virus (NDV): The problem of complementing heterozygotes, *Virology* **38**:490–493.

Dahlberg, J. E., and Simon, E. H., 1969b, Physical and genetic studies of Newcastle disease virus: Evidence for multiploid particles, *Virology* **38**:666–678.

Domingo, E., Sabo, D., Taniguchi, T., and Weissmann, C., 1978, Nucleotide sequence heterogeneity of an RNA phage population, *Cell* **13**:735–744.

Drake, J. W., 1962, Multiplicity reactivation of Newcastle disease virus, *J. Bacteriol.* **84**:352–356.

Estupinan, J., and Hanson, R. P., 1971a, Methods of isolating six mutant classes from the Hickman strain of Newcastle disease virus, *Avian Dis.* **15**:798–804.

Estupinan, J., and Hanson, R. P., 1971b, Mutation frequency of red and clear plaque types of the Hickman strain of Newcastle disease virus, *Avian Dis.* **15**:805–808.

Faulkner, G. P., Follett, E. A. C., Shirodaria, P. V., and Pringle, C. R., 1976, Respiratory syncytial virus *ts* mutants and nuclear immunofluorescence, *J. Virol.* **20**:487–500.

Fisher, L. E., 1983, Characterization of four cell lines persistently infected with measles virus, *Arch. Virol.* **77**:51–60.

Fisher, L. E., and Rapp, F., 1979a, Role of virus variants and cells in maintenance of persistent infection by measles virus, *J. Virol.* **30**:64–68.

Fisher, L. E., and Rapp, F., 1979b, Temperature-dependent expression of measles virus structural proteins in persistently infected cells, *Virology* **94**:55–60.

Gadkari, D. A., and Pringle, C. R., 1980, Temperature-sensitive mutants of Chandipura virus. I. Inter- and intra-group complementation, *J. Virol.* **33**:100–114.

Gharpure, M. A., Wright, P. F., and Chanock, R. M., 1969, Temperature-sensitive mutants of respiratory syncytial virus, *J. Virol.* **3**:414–421.

Gimenez, H. B., and Pringle, C. R., 1978, Seven complementation groups of respiratory syncytial virus temperature-sensitive mutants, *J. Virol.* **27**:459–464.

Granoff, A., 1959a, Studies on mixed infection with Newcastle disease virus. I. Isolation of Newcastle disease virus mutants and test for genetic recombination between them, *Virology* **9**:636–648.

Granoff, A., 1959b, Studies on mixed infection with Newcastle disease virus. II. The occurrence of Newcastle disease virus heterozygotes and the study of phenotypic mixing involving serotypes and thermal stability, *Virology* **9**:649–670.

Granoff, A., 1961a, Induction of Newcastle disease virus mutants with nitrous acid, *Virology* **13**:402–408.

Granoff, A., 1961b, Studies on mixed infection with Newcastle disease virus. III. Activation of nonplaque-forming virus by plaque-forming virus, *Virology* **14**:143–144.

Granoff, A., 1962, Heterozygosis and phenotypic mixing with Newcastle disease virus, *Cold Spring Harbor Symp. Quant. Biol.* **27**:319–326.

Granoff, A., 1964, Nature of Newcastle disease virus population, in "Newcastle Disease Virus, an Evolving Pathogen" (R. P. Hanson, ed.), pp. 107–118, University of Wisconsin Press, Madison, Wisconsin.

Haspel, M. V., and Rapp, F., 1975, Measles virus: An unwanted variant causing hydrocephalus, *Science* **187**:450–451.

Haspel, M. V., Knight, P. R., Duff, R. G., and Rapp, F., 1973, Activation of a latent measles virus infection in hamster cells, *J. Virol.* **12**:690–695.

Haspel, M. V., Duff, R., and Rapp, F., 1975, Isolation and preliminary characterization of mutants of measles virus, *J. Virol.* **16**:1000–1009.

Holland, J. J., Grabau, E. A., Jones, C. L., and Semler, B. L., 1979, Evolution of multiple genome mutations during long-term persistent infections by vesicular stomatitis virus, *Cell* **16**:495–504.

Homma, M., and Ohuchi M., 1973, Trypsin action on the growth of Sendai virus in tissue culture cells, *J. Virol.* **12**:1457–1465.

Hsu, M.-C., Scheid, A., and Choppin, P. W., 1987, Protease activation mutants of Sendai virus: Sequence analysis of the mRNA of the fusion protein (F) gene and direct identification of the cleavage-activation site, *Virology* **156**:84–90.

Hull, J., Krah, D., and Choppin, P., 1987, Resistance of a measles virus mutant to fusion inhibiting oligopeptides is not associated with mutations in the fusion peptide, *Virology* **159**:368–372.

Iorio, R. M., and Bratt, M. A., 1983, Monoclonal antibodies to Newcastle disease virus: Delineation of four epitopes on the HN glycoprotein, *J. Virol.* **48**:440–450.

Iorio, R. M., and Bratt, M. A., 1984a, Monoclonal antibodies as functional probes of the HN glycoprotein of Newcastle disease virus: Antigenic separation of the hemagglutinating and neuraminidase sites, *J. Immunol.* **133**:2215–2219.

Iorio, R. M., and Bratt, M. A., 1984b, Neutralization of Newcastle disease virus by monoclonal antibodies to the hemagglutinin-neuraminidase glycoprotein: Requirement for antibodies to four sites for complete neutralization, *J. Virol.* **51**:445–451.

Iorio, R. M., and Bratt, M. A., 1985, Selection of unique antigenic variants of Newcastle disease virus with neutralizing monoclonal antibodies and anti-immunoglobulin, *Proc. Natl. Acad. Sci. USA* 7106–7110.

Ito, Y., Tsurudome, M., and Hishiyama, M., 1987, Immunological relationships among human and non-human paramyxoviruses revealed by immunoprecipitation, *J. Gen. Virol.* **68**:1289–1297.

Itoh, M., and Homma, H., 1988, Single amino acid change at the cleavage site of the fusion protein is responsible for both enhanced chymotrypsin sensitivity and trypsin resistance of a Sendai virus mutant TR-5, *J. Gen. Virol.* **69**:2907–2911.

Itoh, M., Shibuta, H., and Homma, M., 1987, Single amino acid substitution of Sendai virus at the cleavage site of the protein confers trypsin resistance, *J. Gen. Virol.* **68**:2939–2944.

Johnson, P. R., Spriggs, M. K., Olmsted, R. A., and Collins, P. L., 1987, The G glycoprotein of human respiratory syncytial virus of subgroups A and B: Extensive sequence divergence between antigenically related proteins, *Proc. Natl. Acad. Sci. USA* **84**:5625–5629.

Ju, G., Udem, S., Rager-Zisman, B., and Bloom, B. R., 1978, Isolation of a heterogeneous population of temperature-sensitive mutants of measles virus from persistently infected human lymphoblastoid cell lines, *J. Exp. Med.* **147**:1637–1652.

Ju, G., Birrer, M., Udem, S., and Bloom, B., 1980, Complementation analysis of measles virus mutants isolated from persistently infected lymphoblastoid cell lines, *J. Virol.* **33**:1004–1012.

Kamer, G., and Argos, P., 1984, Primary structural comparison of RNA-dependent polymerases from plant, animal and bacterial viruses, *Nucleic Acids Res.* **12**:7269–7282.

King, A. M. Q., McCahon, D., Slade, W. R., and Newman, J. W., 1982, Recombination in RNA, *Cell* **29**:921–928.

Kingsbury, D. W., and Granoff, A., 1970, Studies on mixed infection with Newcastle disease virus. IV. On the structure of heterozygotes, *Virology* **42**:262–265.

Kingsbury, D. W., Bratt, M. A., Choppin, P. W., Hanson, R. P., Hosaka, Y., ter Meulen, V., Norrby, E., Plowright, W., Rott, R., and Wunner, W. H., 1978, *Paramyxoviridae, Intervirology* **10**:137–152.

Kirkegaard, A., and Baltimore, D., 1986, The mechanism of RNA recombination in poliovirus, *Cell* **47**:433–443.

Kirvatis, J., and Simon, E. H., 1965, A radiobiological study of the development of Newcastle disease virus, *Virology* **26**:545–553.

Knight, P. R., Duff, R., and Rapp, F., 1972, Latency of human measles virus in hamster cells, *J. Virol.* **10**:995–1001.

Komada, H., Tsurudome, M., Bando, H., Nishio, M., Yamada, A., Hishiyama, M., and Ito, Y., 1989, Virus-specific polypeptides of human parainfluenza virus type 4 and their synthesis in infected cells, *Virology* **171**:254–259.

Kowal, K. J., and Youngner, J. S., 1978, Induction of interferon by temperature-sensitive mutants of Newcastle disease virus, *Virology* **90**:90–102.

Lai, M. M. C., Baric, R. S., Makino, S., Keck, J. G., Egbert, J., Leibowitz, J. L., and Stohlman, S. A., 1985, Recombination between nonsegmented RNA genomes of murine coronaviruses, *J. Virol.* **56**:449–456.

Lamb, R. A., and Dreyfuss, G., 1989, RNA structure. Unwinding with a vengeance, *Nature* **337**:19–20.

Ling, R., and Pringle, C. R., 1988, Turkey rhinotracheitis virus: *in vivo* and *in vitro* polypeptide synthesis, *J. Gen. Virol.* **69**:917–923.

Ling, R., and Pringle, C. R., 1989a, Polypeptides of pneumonia virus of mice I: Immunological crossreactions and post-translational modifications, *J. Gen. Virol.* **70**:1427–1440.

Ling, R., and Pringle, C. R., 1989b, Polypeptides of pneumonia virus of mice II: Characterization of the glycoproteins, *J. Gen. Virol.* **70**:1441–1452.

Lomniczi, B., 1975, Properties of non-neurovirulent plaque-forming mutants of Newcastle disease virus, *Avian Dis.* **20**:126–134.

Lopez-Galindez, C., Lopez, J. A., Melero, J. A., de la Fuente, L., Martinez, C., Ortin, J., and Perucho, M., 1988, Analysis of genetic variability and mapping of point mutations in influenza virus by the RNase A mismatch cleavage method, *Proc. Acad. Natl. Sci. USA* **85**:3522–3526.

Löve, A., Rydbeck, R., Kristensson, K., Örvell, C., and Norrby, E., 1985, Hemagglutinin-neuraminidase glycoprotein as a determinant of pathogenicity in mumps virus hamster encephalitis: Analysis of mutants selected with monoclonal antibodies, *J. Virol.* **53**:67–74.

Madansky, C. H., and Bratt, M., 1978, Noncytopathic mutants of Newcastle disease virus, *J. Virol.* **26**:724–729.

Madansky, C. H., and Bratt, M., 1981a, Noncytopathic mutants of Newcastle disease virus are defective in virus specific RNA synthesis, *J. Virol.* **37**:317–327.

Madansky, C. H., and Bratt, M., 1981b, Relationships among virus spread, cytopathogenicity, and virulence as revealed by the noncytopathic mutants of Newcastle disease virus, *J. Virol.* **40**:691–702.

Marx, P. A., Portner, A., and Kingsbury, D. W., 1974, Sendai virion transcriptase complex: Polypeptide composition and inhibition by virion envelope proteins, *J. Virol.* **13**:107–112.

McIntosh, K., and Chanock, R. M., 1985, Respiratory syncytial virus, in "Virology" (B. N. Fields *et al.*, eds.), pp. 1285–1304, Raven Press, New York.

McIntosh, K., and Fishaut, J. M., 1980, Immunopathologic mechanisms in lower respiratory tract disease of infants due to respiratory syncytial virus, *Prog. Med. Virol.* **26**:94–118.

McKay, E., Higgins, P., Tyrrell, D., and Pringle, C. R., 1988, Immunogenicity and pathogenicity of temperature-sensitive modified respiratory syncytial virus in adult volunteers, *J. Med. Virol.* **25**:411–421.

McKimm, J., and Rapp, F., 1977, Stability of measles virus temperature-sensitive virus mutants to induce interferon, *Virology* **76**:409–415.

Miyata, T., and Yasunaga, T., 1980, Molecular evolution of mRNA: A method for estimating evolutionary rates of synonymous and amino acid substitutions from homologous nucleotide sequences and its application, *J. Mol. Evol.* **16**:23–36.

Mochizuki, Y., Tashiro, M., and Homma, M., 1988, Pneumopathogenicity in mice of a Sendai virus mutant, TSrev-58, is accompanied by *in vitro* activation with trypsin, *J. Virol.* **62**:3040–3042.

Morrison, T. G., 1988, Structure, function, and intracellular processing of paramyxovirus membrane proteins, *Virus Res.* **10**:113–136.

Nicholas, J. A., Levely, M. E., Mitchell, M. A., and Smith, C. W., 1989, A 16-amino acid peptide of respiratory syncytial virus 1A protein contains two overlapping T-cell stimulating sites distinguishable by clan II MHC restriction elements, *J. Immunol.* **143**:2790–2796.

Nishikawa, K., Isomura, S., Suzuki, S., Watanabe, E., Hamaguchi, M., Yoshida, T., and Nagai, Y., 1983, Monoclonal antibodies to the HN glycoprotein of Newcastle disease virus. Biological characterization and use for strain comparisons, *Virology* **130**:318–330.

Norrby, E., Sheshberadaran, H., McCullough, K. C., Carpenter, W. C., and Örvell, C., 1985, Is rinderpest virus the archevirus of the *Morbillivirus* genus?, *Intervirology* **23**:228–232.

Olmsted, R. A., Murphy, B. R., Lawrence, L. A., Elango, N., Moss, B., and Collins, P., 1989, Processing, surface expression, and immunogenicity of carboxy-terminally truncated mutants of G protein of human respiratory syncytial virus, *J. Virol.* **63**:411–420.

Paterson, R. G., and Lamb, R. A., 1987, Ability of the hydrophobic fusion-related external domain of a paramyxovirus F protein to act as a membrane anchor domain, *Cell* **48**:441–452.

Peeples, M. E., and Bratt, M. A., 1982a, UV irradiation analysis of complementation between, and replication of, RNA-negative temperature-sensitive mutants of Newcastle disease virus, *J. Virol.* **41**:965–973.

Peeples, M. E., and Bratt, M. A., 1982b, Virion functions of RNA +ve temperature-senitive mutants of Newcastle disease virus, *J. Virol.* **42**:440–446.

Peeples, M. E., and Bratt, M. A., 1984, Mutation in the matrix protein of Newcastle disease virus can result in decreased fusion glycoprotein into particles and decreased infectivity, *J. Virol.* **51**:81–90.

Peeples, M. E., Rasenas, L. L., and Bratt, M. A., 1982, RNA synthesis by Newcastle disease virus temperature-sensitive mutants in two RNA-negative complementation groups, *J. Virol.* **42**:996–1006.

Peeples, M. E., Glickman, R. L., and Bratt, M. A., 1983, Thermostabilities of virion activities of Newcastle disease virus: Evidence that the temperature-sensitive mutants in complementation groups B, BC, and C have altered HN protein, *J. Virol.* **45**:18–26.

Peeples, M. E., Glickman, R. L., Gallagher, J. P., and Bratt, M. A., 1988, Temperature-sensitive mutants of Newcastle disease virus altered in HN glycoprotein size, stability or antigenic maturity, *Virology* **164**:284–289.

Portner, A., 1977, Association of nucleocapsid polypeptides with defective RNA synthesis in a temperature-sensitive mutant of Sendai virus, *Virology* **77**:481–489.

Portner, A., 1981, The HN glycoprotein of Sendai virus: Analysis of site(s) involved in hemagglutinating and neuraminidase activities, *Virology* **115**:375–384.

Portner, A., 1984, Monoclonal antibodies as probes of the antigenic structure and functions of Sendai virus glycoproteins, in: *Non-segmented Negative Strand Viruses* (D. H. L. Bishop and R. W. Compans, eds.), pp. 345–350. Academic Press, Orlando, FL.

Portner, A., Marx, P. A., and Kingsbury, D. W., 1974, Isolation and characterization of Sendai virus temperature-sensitive mutants, *J. Virol.* **13**:298–304.

Portner, A., Scroggs, R. A., Marx, P. A., and Kingsbury, D. W., 1975, A temperature-sensitive mutant of Sendai virus with an altered hemagglutinin-neuraminidase polypeptide: Consequences for virus assembly and cytopathology, *Virology* **67**:179–187.

Portner, A., Webster, R. G., and Bean, W. J., 1980, Similar frequencies of antigenic variants in Sendai, vesicular stomatitis and influenza A virus, *Virology* **104**:235–238.

Portner, A., Scroggs, R. A., and Metzger, D. W., 1987a, Distinct functions of antigenic sites of the HN glycoprotein of Sendai virus, *Virology* **198**:61–68.

Portner, A., Scroggs, R. A., and Naeve, C. W., 1987b, The fusion glycoprotein of Sendai virus: Sequence analysis of an epitope involved in fusion and virus neutralization, *Virology* **157**:556–559.

Preble, O. T., and Youngner, J. S., 1972, Temperature-sensitive mutants isolated from L cells persistently infected with Newcastle disease virus, *J. Virol.* **9**:200–206.

Preble, O. T., and Youngner, J. S., 1973a, Temperature-sensitive defect of mutants isolated from L cells persistently infected with Newcastle disease virus, *J. Virol.* **12**:472–480.

Preble, O. T., and Youngner, J. S., 1973b, Selection of temperature-sensitive mutants during persistent infection: Role in maintenance of persistent Newcastle disease virus of L cells, *J. Virol.* **12**:481–491.

Preble, O. T., and Youngner, J. S., 1975, Temperature-sensitive viruses and the etiology of chronic and inapparent infections, *J. Infect. Dis.* **131**:467–473.

Pringle, C. R., 1987, Paramyxoviruses and disease, *in* "Molecular Basis of Virus Disease" (W. C. Russell and J. W. Almond, eds.), pp. 51–90, Cambridge University Press, Cambridge.

Pringle, C. R., 1988, Rhabdovirus genetics, *in* "The Rhabdoviruses" (R. R. Wagner, ed.), pp. 167–243, Plenum Press, New York.

Pringle, C. R., Shirodaria, P. V., Gimenez, H. B., and Levine, S., 1981, Antigen and polypeptide synthesis by temperature-sensitive mutants of respiratory syncytial virus, *J. Gen. Virol.* **54**:173–183.

Richardson, C. D., Scheid, A., and Choppin, P. W., 1980, Specific inhibition of paramyxovirus and myxovirus replication by oligopeptides with amino acid sequences similar to those at the N-termini of the F1 or HA2 viral polypeptides, *Virology* **105**:205–222.

Richardson, L. S., Schnitzed, T. J., Belshe, R. B., Prevar, D. A., and Chanock, R. M., 1977, Isolation and characterization of further defective clones of a temperature-sensitive mutant (*ts*1) of respiratory syncytial virus, *Arch. Ges. Virusforsch.* **54**:53–60.

Rima, B. K., 1989, Comparison of amino acid sequences of the major structural proteins of the paramyxo- and morbilliviruses, in Genetics and Pathogenicity of Negative Strand Viruses (D. Kolakofsky and B. M. J. Mahy, eds.), pp. 254–263. Elsevier, Amsterdam.

Robbins, S. J., 1983, Progressive invasion of cell nuclei by measles virus in persistently infected human cells, *J. Gen. Virol.* **64**:2335–2338.

Rowlands, D., Grabau, E., Spindler, K., Jones, C., Semler, B., and Holland, J., 1980, Virus protein changes and RNA termini alterations evolving during persistent infection, *Cell* **19**:871–880.

Sakaguchi, T., Toyoda, T., Gotch, B., Inocencio, N. M., Kuma, K., Miyata, T., and Nagai, Y., 1989, Newcastle disease virus evolution. I. Multiple lineages defined by sequence variability of the haemagglutinin-neuraminidase gene, *Virology* **169**:260–272.

Samson, A. C. R., Chambers, P., Lee, C. M., and Simon, E., 1981, Temperature-sensitive mutant of Newcastle disease virus which has an altered nucleocapsid-associated protein, *J. Gen. Virol.* **54**:197–201.

Scheid, A., and Choppin, P. W., 1974, Identification of the biological activities of paramyxovirus glycoproteins. Activation of cell fusion, hemolysis, and infectivity by proteolytic cleavage of an inactive precursor protein of Sendai virus, *Virology* **57**:475–490.

Scheid, A., and Choppin, P. W., 1976, Protease activation mutants of Sendai virus. Activation of biological properties by specific proteases, *Virology* **69**:265–277.

Schloer, G. M., and Hanson, R. P., 1968, Relationship of plaque size and virulence for chickens of 14 representative Newcastle disease virus strains, *J. Virol.* **2**:40–47.

Schloer, G. M., and Hanson, R. P., 1971, Virulence and *in vitro* characteristics of four mutants of Newcastle disease virus, *J. Infect. Dis.* **124**:289–295.

Sheshbaradaran, H., Norrby, E., McCullough, K. C., Carpenter, W., and Örvell, C., 1986, The antigenic relationship between measles, canine distemper and rinderpest viruses studied with monoclonal antibodies, *J. Gen. Virol.* **67**:1381–1392.

Shioda, T., Iwasaki, K., and Shibuta, H., 1986, Determination of the complete nucleotide sequence of the Sendai virus genome RNA and the predicted amino acid sequences of the F, HN and L proteins, *Nucleic Acids Res.* **4**:1545–1563.

Shioda, T., Wakao, S., Suzo, S., and Shibuta, H., 1988, Differences in bovine parainfluenza 3 virus variants studied by sequencing of the genes of viral envelope proteins, *Virology* **162**:388–396.

Stanwick, T. L., and Hallum, J. V., 1976, Comparison of RNA polymerase associated with Newcastle disease virus and a temperature-sensitive mutant of Newcastle disease virus isolated from persistently infected L cells, *J. Virol.* **17**:68–73.

Steinhauer, D. A., and Holland, J. J., 1986, Direct method for quantitation of extreme polymerase error frequencies at selected single base sites in viral RNA, *J. Virol.* **57**:219–228.

Tashiro, M., and Homma, M., 1983, Pneumotropism of Sendai virus in relation to protease-mediated activation in mouse lungs, *Infect. Immun.* **39**:879–888.

Taylor, J., Mason, W., Summers, J., Goldberg, J., Aldrich, C., Coates, L., Gerin, J., and Gowans, E., 1987, Replication of human hepatitis delta virus in primary cultures of woodchuck hepatocytes, *J. Virol.* **61**:2891–2895.

Thiry, L., 1964, Some properties of chemically induced small-plaque mutants of Newcastle disease virus, *Virology* **24**:6–15.

Thomas, S. M., Lamb, R. A., and Paterson, R. G., 1988, Two mRNAs that differ by two nontemplated nucleotides encode the amino coterminal proteins P and V of the paramyxovirus SV5, *Cell* **54**:891–902.

Thompson, S. D., and Portner, A., 1987, Location of functional sites on the hemagglutinin-neuraminidase glycoprotein of Sendai virus by sequence analysis of antigenic and temperature-sensitive mutants, *Virology* **160**:1–8.

Toyoda, T., Sakaguchi, T., Imai, K., Inocencio, N. M., Gotch, B., Hamaguchi, M., and Nagai, M., 1987, Structural comparison of the cleavage-activation site of the fusion glycoprotein between virulent and avirulent strains of Newcastle disease virus, *Virology* **158**:242–247.

Toyoda, T., Sakaguchi, T., Hirota, H., Gotch, B., Kuma, K., Miyata, T., and Nagai, Y., 1989, Newcastle disease virus evolution. II. Lack of genetic recombination in generating virulent and avirulent strains, *Virology* **169**:273–282.

Tsipis, J. E., and Bratt, M., 1976, Isolation and preliminary characterization of temperature-sensitive mutants of Newcastle disease virus, *J. Virol.* **18**:848–855.

Vijaya, S., Elango, N., Zavala, F., and Moss, B., 1988, Transport to the cell surface of a peptide sequence attached to the truncated C terminus of an N-terminally anchored integral membrane protein, *Mol. Cell. Biol.* **8**:1709–1714.

Vydelingum, S., Ilonen, J., Salonen, R., Marusyk, R., and Salmi, A., 1989, Infection of human peripheral blood mononuclear cells with a temperature-sensitive mutant of measles virus, *J. Virol.* **63**:689–695.

Wang, K.-S., Choo, K.-L., Weiner, A. J., Ou, H.-J., Najarian, J. C., Thayer, R. M., Mullenbach, J. T., Denniston, K. J., Gerin, J. L., and Houghton, M., 1986, Structure, sequence and expression of the HDV genome, *Nature* **323**:508–514.

Watt, P. J., Robinson, B. S., Pringle, C. R., and Tyrrell, D. A. J., 1990, Determinants of susceptibility to challenge and the antibody response of adult volunteers given experimental RS virus vaccines, *Vaccine* **8**:231–236.

Waxam, M. N., and Wolinsky, J. S., 1986, A fusing mumps virus variant selected from a nonfusing parent with the neuraminidase inhibitor 2-deoxy-2,3-dehydro-N-acteylneuraminic acid, *Virology* **151**:286–295.

Weissmann, C., 1989, Single-strand RNA, *Nature* **337**:415–416.

White, B. T., and McGeoch, D. J., 1987, Isolation and characterization of conditional lethal amber nonesense mutants of vesicular stomatitis virus, *J. Gen. Virol.* **68**:3033–3044.

Wild, T. F., and Dugre, R., 1978, Establishment and characterization of a subacute sclerosing panencephalitis (measles) virus persistent infection in BGM cells, *J. Gen. Virol.* **39**:113–124.

Woyciechowska, J., Breschkin, A. M., and Rapp, F., 1977, Measles virus meningoencephalitis. Immunofluorescence study of brains infected with virus mutants, *Lab. Invest.* **36**:233–236.

Yamazi, Y., and Black, F. L., 1972, Isolation of temperature-sensitive mutants of measles virus, *Med. Biol.* (Jpn) **84**:47–51.

Yamazi, Y., and Black, F. L., Honda, H., Todome, Y., Suganuma, M., Watari, E., Iwaguchi, H., and Nagashima, M., 1975, Characterization of temperature-sensitive mutants of measles virus: Temperature-shift experiment, *Jpn. J. Med. Sci. Biol.* **28**:223–229.

Yusoff, K., Millar, N. S., Chambers, P., and Emmerson, P. T., 1987, Nucleotide sequence analysis of the *L* gene of Newcastle disease virus: Homologies with Sendai and vesicular stomatitis viruses, *Nucleic Acids Res.* **15**:3961–3976.

CHAPTER 2

The Molecular Biology of the *Paramyxovirus* Genus

MARK S. GALINSKI AND STEVEN L. WECHSLER

I. INTRODUCTION

A. History

The *Paramyxovirus* genus, one of the three genera classified in the family *Paramyxoviridae*, contains a number of important viruses causing disease in humans and other animals (Table I). Historically, parainfluenza, mumps, and Newcastle disease viruses were assigned to the *Myxoviridae* group based upon common biological properties shared with the influenza viruses (Andrewes *et al.*, 1955; Chanock, 1956; Chanock *et al.*, 1958). However, later studies indicated sufficient biochemical differences, in particular, the molecular organization of the genomes (segmented versus nonsegmented) and the melding of neuraminidase and hemagglutinating activities into a single glycoprotein (paramyxoviruses) versus separate glycoproteins (influenza virus), to reclassify both groups of viruses into the separate taxons *Paramyxoviridae* and *Orthomyxoviridae* (Waterson, 1962; Wildy, 1971). A number of viruses, including measles, canine distemper, respiratory syncytial, and mouse pneumonia viruses, do not share cross-reactive antigens with other *Paramyxoviridae*. However, because of their morphological and biochemical characteristics, they were classified in the *Paramyxoviridae* family as separate genera (Kingsbury *et al.*, 1978a).

The etymological origin of the family name is from the Greek *para*, meaning akin or closely related (indicating their morphological and biological relatedness with myxoviruses) and the root *myxo* (Greek *myxa*), meaning mucus, reflecting their affinity for mucoprotein receptors on erythrocytes from a

MARK S. GALINSKI • Department of Molecular Biology, Cleveland Clinic Foundation, Cleveland, Ohio 44195.   STEVEN L. WECHSLER • Department of Virology Research, Cedars-Sinai Medical Center, Los Angeles, California 90048.

41

TABLE I. *Paramyxoviridae* Family

Paramyxovirus Genus
    Human parainfluenza virus types 1, 2, 3, and 4
    Mumps virus
    Newcastle disease virus
    Simian virus 5 (canine parainfluenza 2)
    Sendai virus (murine parainfluenza 1)
    Bovine parainfluenza viruses
    Avian parainfluenza viruses
Morbillivirus genus
    Measles virus
    Canine distemper virus
    Rinderpest virus
    Peste des petits ruminants
Pneumovirus genus
    Human respiratory syncytial virus
    Bovine respiratory syncytial virus
    Pneumonia virus of mice

---

number of animal species (hemagglutinating abilities) (Andrews *et al.*, 1955). Further significance can be attached to the family name in that all *Paramyxoviridae* must first infect the respiratory tract upon entry into their hosts. In order to avoid confusion between the genus and family, throughout this chapter paramyxovirus will be used as the vernacular term to indicate a member of the genus, while the taxonomic term *Paramyxoviridae* will be used to indicate members of the family.

B. General Properties

The *Paramyxoviridae* are enveloped viruses whose major biological properties include hemagglutinating and neuraminidase activities, cell fusion (syncytia formation and hemolysis), and the ability to readily establish persistence upon passage in cell culture. The family contains three genera, *Paramyxovirus*, *Morbillivirus*, and *Pneumovirus* (Table I) (Kingsbury *et al.*, 1978a; Matthews, 1982). The major biological properties of these viruses are effected through the two surface glycoproteins, a receptor-binding protein and a fusion protein (F) (Choppin and Scheid, 1980; Matsumoto, 1982). Differences in the biological and antigenic properties of the glycoproteins have provided useful "descriptors" for classifying these viruses into the three genera (Matthews, 1982). Thus, members of the *Paramyxovirus* genus share cross-reacting antigenic sites that are unique from those present in the *Morbillivirus* or *Pneumovirus* genera. Another distinguishing characteristic separating paramyxoviruses from the other two genera is the presence of both hemagglutinating and neuraminidase activities in the receptor-binding glycoprotein HN. The receptor-binding proteins, designated H (hemagglutinating glycoprotein) in the *Morbillivirus* genus and G (glycoprotein) in the *Pneumovirus* genus, do not have neuraminidase activity (Choppin and Scheid, 1980; Matthews, 1982).

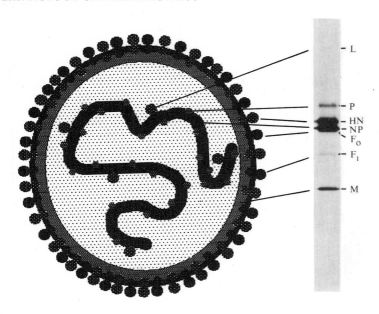

FIGURE 1. Schematic representation of the human parainfluenza virus type 3 virion and its associated proteins. The various virion-associated proteins (L, large protein; P, phosphoprotein; HN, hemagglutinin-neuraminidase glycoprotein; NP, nucleocapsid protein; F_0, uncleaved fusion protein; F_1, large-subunit polypeptide of the cleaved fusion protein; and M, matrix protein) were resolved by SDS–PAGE on a 7.5–15% polyacrylamide gradient gel. The locations of these proteins in the virion are indicated in the diagram.

An illustration of a typical *Paramyxoviridae* virion is shown in Fig. 1 using the human parainfluenza virus type 3 and its virion-associated proteins as a model. The genome consists of a nonsegmented, single-stranded RNA molecule of negative polarity with an approximate relative molecular mass (M_r) of 5.6×10^6 (approximately 15,000 nucleotides) (Matthews, 1982). The viral genomes contain six to ten genes. Although some genes are polycistronic and encode multiple proteins, there are at least six major structural proteins common to all *Paramyxoviridae*. The genomic RNA is never found as a free molecule, but is tightly associated with nucleocapsid proteins (NP) to form a structure termed the nucleocapsid core. Two ancillary proteins, closely associated with the nucleocapsid core, are a phosphoprotein (P) and a large protein (L). Together, these four components comprise the nucleocapsid. During virion assembly, the nucleocapsid is incorporated into budding particles at regions of the cell surface containing the viral matrix protein (M) and the two surface glycoproteins, HN, H or G, and F (Choppin and Compans, 1975; Choppin and Scheid, 1980).

A number of reviews have been published concerning the natural history, structure, function, and assembly of *Paramyxoviridae* (Chanock *et al.*, 1961; Chanock and Parott, 1965; Choppin and Compans, 1975; Bratt and Hightower, 1977; Chanock, 1979; Choppin and Scheid, 1980; Matsumoto, 1982; Welliver *et al.*, 1982; Morrison, 1988). Much of our knowledge about paramyxoviruses has been obtained from studies using Newcastle disease virus (NDV), Sendai virus, and simian virus 5 (SV5). Recently, recombinant cloning and sequence analysis of these viruses, as well as the human and bovine parainfluenza viruses type 3 (HPF3 virus and BPF3, respectively) and mumps virus, have provided a large body of information concerning the molecular organization and gene expression of these pathogens. Indeed, the entire nucleotide sequence for the genomes of Sendai virus, NDV, and HPF3 virus have been determined. The recent expansion of information derived from the molecular analysis of these various viruses will form the basis of discussion for this chapter.

II. VIRUS STRUCTURE

The fine structure of *Paramyxoviridae* has been determined largely by electron microscopy. In conjunction with biochemical and electrophoretic mobility analyses of the virion components, a reasonable understanding of the virion structure has been obtained. The following description will extend the basic features listed above for the paramyxoviruses.

A. Morphology

Paramyxoviruses are pleomorphic in size, ranging from 150 to 300 nm in diameter. Although the are generally spherical, filamentous forms are commonly observed (Hosaka *et al.*, 1966). Purified virions band at a buoyant density of 1.18–1.20 g/cm^3 in sucrose. The outer layer of the lipid membrane contains a number of morphologically homogeneous spikes 8–12 nm in length spaced 8–10 nm apart. These represent the two viral glycoproteins, F and HN. Although the precise location of the M protein is unclear in *Paramyxoviridae*, by analogy with the *Orthomyxoviridae* M protein (membrane protein), this protein is believed to be associated with the inner leaflet of the viral membrane. The absence of an electron-dense region below the inner leaflet, as has been observed for *Orthomyxoviridae* (Compans *et al.*, 1966), suggests that the paramyxovirus M protein may be evenly distributed throughout the virion.

In addition to the M protein, the membrane encloses the nucleocapsid, which has a left-handed helical symmetry with an outer diameter of 14–17 nm, an inner diameter of 5 nm, and an average length of 1 μm (Hosaka *et al.*, 1966; Compans and Choppin, 1967). The nucleocapsid consists of a nonsegmented, single-stranded genomic RNA (approximately 15,000 nucleotides with a sedimentation value of 50S) encapsidated with 2200–2600 NP protein molecules. Together, these two components form the nucleocapsid core, which has a buoyant density of 1.30–1.31 g/cm^3 in cesium chloride. In close association with the nucleocapsid core are 250–300 P-protein molecules and 20–30 L-protein molecules, which are believed to be the subunits of the RNA-dependent RNA polymerase (Choppin and Compans, 1975; Lamb *et al.*, 1976; Matthews, 1982).

Monoclonal and monospecific antibodies to NP, P, or L have been used as probes for the localization of these proteins on nucleocapsids by immunogold labeling (Portner and Murti, 1986; Portner *et al.*, 1988). The distribution of NP on nucleocapsids was uniform whether the nucleocapsids were isolated from virions or cells. This is in agreement with the assignment of NP as the primary structural component of nucleocapsid cores. In contrast, the P proteins were distributed in discrete clusters in nucleocapsids prepared from cells, while they were uniformly distributed in nucleocapsids prepared from virions. The variance in the distribution of the P proteins may reflect the different transcriptional activities of cellular versus virion nucleocapsids. This notion has further support from analyses using double-labeling techniques, where it was shown that clusters of P and L colocalized in cellular nucleocapsids, again suggesting an interaction of these proteins in transcription.

Independent clusters of P, devoid of L, were also observed at random locations on the cellular nucleocapsids. Whether these P proteins reflect a pool of unrecruited polypeptides to be used in transcription or serve some auxiliary function is unclear. The uniform distribution of P in virion nucleocapsids may reflect the transcriptional inactivity of these complexes. The results described above show that in quiescent virion nucleocapsids the various proteins are randomly distributed. Thus, some coordinated recruitment of the various proteins appears essential for transcription. These results also suggest that upon entry into the host cell the switch to primary transcription may involve a fundamental rearrangement of the nucleocapsid proteins into a transcriptionally active complex.

Studies on the localization of the M protein using monoclonal antibodies showed that M was present on both cellular and virion nucleocapsids (Portner and Murti, 1986). Although very little is known about the function and location of this protein in virions, its localization on nucleocapsids provides evidence for the direct interaction of this protein with the nucleocapsid complex. This also supports the putative role of M in modulating transcription and in mediating the alignment of nucleocapsids with the plasma membrane (Shimizu and Ishida, 1975; Marx et al., 1974; Peeples and Brat, 1984).

B. Virion Envelope and Envelope-Associated Proteins

1. Envelope

The viral envelope is derived from the host cell in which the virion was assembled. In a number of studies on SV5 by Klenk and Choppin (1969a, 1969b, 1970) it was demonstrated that the fatty acid composition of the viral envelope is strictly dependent upon the cell type used for virus growth. Thus, the major fatty acids incorporated into the mature virion are identical to those of the host cell in which they are grown, even if the host-cell membranes have significantly different lipid compositions. Although the bulk of the plasma membrane remains unaltered by virus infection, there are two subordinate means through which paramyxoviruses can affect some lipid components: fatty acid acylation of the surface glycoproteins and modification of gangliosides.

Fatty acid acylation of glycoprotein, during transit to the cell surface, provides a limited means by which paramyxoviruses can directly interact with a fatty acid component of the plasma membrane. The attachment of myristic acid (myristoylation) to glycine and palmitic or stearic acids (palmitoylation) to cysteine residues have been postulated as ancillary means of anchoring glycoproteins to the plasma membrane (Sefton and Buss, 1987). Recently, fatty acid acylation of the F and HN glycoproteins of NDV, SV5, and Sendai viruses has been investigated by Veit et al. (1989). The results indicate that, dependent on the paramyxovirus type, the various glycoproteins are acylated differently. The F glycoprotein in NDV and the HN glycoprotein in SV5 were exclusively modified with hydroxylamine-labile fatty acids (palmitoylation); however, in Sendai virus, neither of these glycoproteins was modified. These differences could not be directly accounted for by the lack of a cysteine residue as an attachment site, since both the SV5 F and Sendai virus HN proteins have cysteines within the transmembranal anchoring domain (Fig. 2). The biological significance of fatty acid acylation and the functional importance of the differences in glycoprotein acylation are currently unknown.

Gangliosides, which are sialic acid-containing glycolipids found primarily in the outer leaflet of the cell membrane, are believed to make important contributions to cell surface properties. Further, the gangliosides GD1a, GT1b, and GQ1b have been implicated as potential receptors for Sendai virus attachment (Markwell et al., 1981, 1984). Examination of the carbohydrate composition of the gangliosides in Sendai virions has shown that they contain primarily galactose and glucose, and that sialic acids constitute less than 0.001% of the virion (Kohama et al., 1978). The lack of sialic acids on the virion surface reflects the presence of neuraminidase activity with the receptor binding protein HN. Thus, in addition to the fatty acid acylation of the glycoproteins, the modification of gangliosides provides another means whereby these viruses can directly interact with a lipid component of the plasma membrane.

2. Glycoproteins

The membrane-associated glycoproteins F and HN are intimately involved in the attachment and penetration of paramyxoviruses into susceptible cells. The synthesis and assembly of these proteins require both cotranslational and posttranslational processing, including proteolytic cleavage, addition of core carbohydrate and successive modifications (trimming and addition of complex carbohydrates), attachment of fatty acids, and folding and homo-oligermerization of the proteins into their functional conformations (see Chapters 13 and 17). Together, these processes participate in the production of the biologically functional glycoproteins. The fine structure of the paramyxovirus glycoproteins has not been examined in as much detail as have the analogous proteins of the orthomyxoviruses. Consequently, much of our understanding of the paramyxovirus proteins has been derived by analogy to the influenza glycoproteins.

Although much information has been gathered from the direct biochemical analysis of the F and HN proteins, recent molecular cloning and sequence analysis of the genes/mRNAs has provided more insight into their structure and function. The *Paramyxoviridae* glycoprotein genes which have been molecularly cloned include those from respiratory syncytial [F (Collins *et al.*, 1984); G (Satake *et al.*, 1985; Wertz *et al.*, 1985)], measles [H (Alkhatib and Briedis, 1986); F (Richardson *et al.*, 1986)], canine distemper [F (Barret *et al.*, 1987)], rinderpest [F (Tsukiyama *et al.*, 1988)], Sendai [HN and F (Blumberg *et al.*, 1985a,b)], SV5 [HN (Hiebert *et al.*, 1985a); F (Paterson *et al.*, 1984b)], NDV [F (Chambers *et al.*, 1986a; McGinnes and Morrison, 1986); HN (Jorgensen *et al.*, 1987; McGinnes *et al.*, 1987)], mumps [F and HN (Waxham *et al.*, 1987, 1988; Kövamees *et al.*, 1989)], BPF3 [F and HN (Suzu *et al.*, 1987)], and HPF3 [HN (Elango *et al.*, 1986); F (Spriggs *et al.*, 1986)] viruses.

a. Fusion Glycoprotein

All paramyxoviruses contain a fusion glycoprotein (F) which can directly fuse virion and cell plasma membranes following adsorption of the virus to the cell surface. This process occurs at neutral pH and allows direct entry of an infectious virion into cells. In addition to viral penetration, expression of F at the infected cell surface induces fusion of contiguous uninfected cells, resulting in the formation of syncytia. This cytopathology allows virus to spread from cell to cell without export to the extracellular environment and is pathognomonic of paramyxovirus infection in tissue culture (Bratt and Gallaher, 1969; Choppin and Compans, 1975; Choppin and Scheid, 1980).

The fusion glycoprotein is synthesized as an inactive precursor F_0, which requires a cellular protease for activation (Scheid and Choppin, 1974). Cleavage of F_0 results in the generation of two subunit polypeptides F_1 and F_2 held together by a disulfide bond $F_{1,2}$ (Scheid and Choppin, 1977). Cleavage of F_0 is dependent upon the availability of cellular proteases, since cells lacking a protease capable of activating F_0 are unable to support replication of the virus. Thus, proteolytic activation is a determinant of the host range, tissue tropism, and pathogenicity of the virus (Choppin and Scheid, 1980). The cleavage of F_0 unmasks a highly conserved hydrophobic amino terminus on F_1 and generates a new carboxyl terminus on F_2 (Scheid *et al.*, 1978; Richardson *et al.*, 1980). The hydrophobic amino terminus is believed to participate directly in the fusion event based upon the following: (1) its necessity to be unmasked by proteolytic cleavage, (2) its hydrophobic physical properties, (3) its high conservation among the paramyxoviruses, and (4) the ability of synthetic oligopeptides resembling the amino acid sequence to inhibit hemolysis and replication of a number of paramyxoviruses (Richardson and Choppin, 1983; Richardson *et al.*, 1980). Inhibition of virus replication by oligopeptides appears to be mediated through a direct competition for cell surface receptors and not by inhibition of viral HN/H receptor binding. These results suggest that there are specific sites on the cell surface, in addition to HN binding sites, which interact specifically with F and are involved with viral fusion and penetration.

Molecular cloning and sequence analysis of nine different *Paramyxoviridae* F genes/mRNAs, including those of Sendai (Blumberg *et al.*, 1985a), SV5 (Paterson *et al.*, 1984b), NDV (Chambers *et al.*, 1986a; Jorgensen *et al.*, 1987), mumps (Waxham *et al.*, 1987); BPF3 (Suzu *et al.*, 1987); HPF3 (Spriggs *et al.*, 1986; Galinski *et al.*, 1987b), measles (Richardson *et al.*, 1986), CDV (Barret

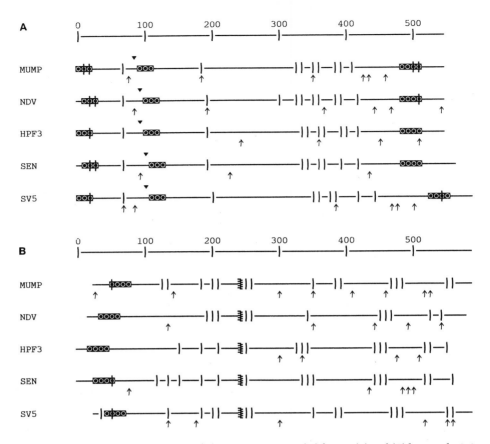

FIGURE 2. Schematic representation of the paramyxovirus (A) fusion (F) and (B) hemagglutinin-neuraminidase (HN) glycoproteins. The positions of hydrophobic domains (boxed circles), potential sites for N-linked glycosylation (↑), and cysteine residues (||) are indicated for each protein. (A) The cleavage activation of F_0 to produce $F_{1,2}$ is indicated ▼. (B) The position of the putative neuraminidase active site in HN is indicated by the zipper-like symbol. The viruses depicted include HPF3, BPF3, Sendai (SEN), mumps, SV5, and NDV. See text for more details.

et al., 1987), rinderpest (Tsukiyama et al., 1988), and RS (Collins et al., 1984; Wertz et al., 1985) viruses, has provided further information about these genes and their encoded proteins.

Some basic features of the paramyxovirus F proteins are shown in Fig. 2A. These proteins contain 529–565 amino acids. There are three distinct hydrophobic domains. They are located at the N termini, at amino acid 100 (approximate), and at the C termini. The N-terminal domain represents the cleaved signal peptide that guides the protein for cotranslational insertion into the endoplasmic reticulum for further glycosylation. The C-terminal domain, located between residues 480 and 520, functions as the transmembranal anchoring domain. The internal hydrophobic domain, approximately 100 residues from the N terminus, represents the fusion domain. This sequence is unmasked (becoming the N terminus of F_1) following proteolytic activation of F_0 to $F_{1,2}$.

Although the amino acid sequences of the paramyxovirus F glycoproteins are conserved (Fig. 3), the distribution of similarities across the proteins is not symmetrical. The F_1 subunit polypeptide contains the regions of greatest similarity, while the F_2 subunit polypeptide diverges significantly. Analysis of the F protein similarity suggests division of these viruses into two related groups; first, Sendai, BPF3, and HPF3 viruses, and second, mumps virus, SV5, and NDV. Members of each group are more closely related to one another than they are to members of the other group.

Amino acids which participate in protein folding (Cys, Gly, Pro) are more highly conserved in the F protein than other residues. Thus, in the F_2 subunit polypeptide, which contains two to three cysteines, the relative positions of these residues appear to be highly conserved. Further inspection of Fig. 2 shows that this is also the case in F_1. The relative abundance of these residues together with their conservation supports the notion that the folding of F into a specific configuration is an important structural feature of this protein.

The sequences surrounding the cleavage activation site of F are shown in Fig. 4. The high degree of amino acid sequence conservation observed here suggests that this domain plays an important role in fusion (Scheid et al., 1978; Richardson et al., 1980). Interestingly, the amino acids in the C terminus of F_2 are not as highly conserved except for the presence of basic residues. The availability of dibasic residues proximal to the cleavage site in F_2 plays a critical role in the proteolytic processing of F_0 (Toyoda et al., 1987; Glickman et al., 1988; Paterson et al., 1989). The importance of these residues has been characterized in certain strains of NDV which differ in their pathogenic characteristics. Toyoda et al. (1987) and Glickman et al. (1988) have shown that the pathogenic differences attributable to the proteolytic processing of F are correlated with the presence or absence of dibasic residues in F_2. Thus, in avirulent strains, which lack the dibasic residues, F is poorly cleaved, resulting in attenuation of pathogenicity.

In addition to the fusion domain, there appears to be another domain in F_1 which may participate in the fusion process. This is based upon the analysis of antigenic variants selected using an F monoclonal antibody which inhibits fusion and infectivity of Sendai virus (Portner et al., 1987). Sequence analysis of the mutant genes revealed that the alteration in the variant proteins mapped not to the fusion domain, but at some distance in the primary sequence from this domain (residue 399). These results suggest that this auxiliary domain may be topologically positioned near the N terminus of F_1. These results also support the notion that the folding of F into a specific conformation is an important structural feature of this protein.

There are three to six potential asparagine-linked (N-linked) glycosylation sites (Fig. 2A), most of which are clustered within the C-terminal one-third of F_1. All of the F_2 subunit polypeptides, except for that of HPF3 virus, contain a potential site for glycosylation. Which of these potential sites are used for carbohydrate attachment and the functional importance of glycosylation are unknown.

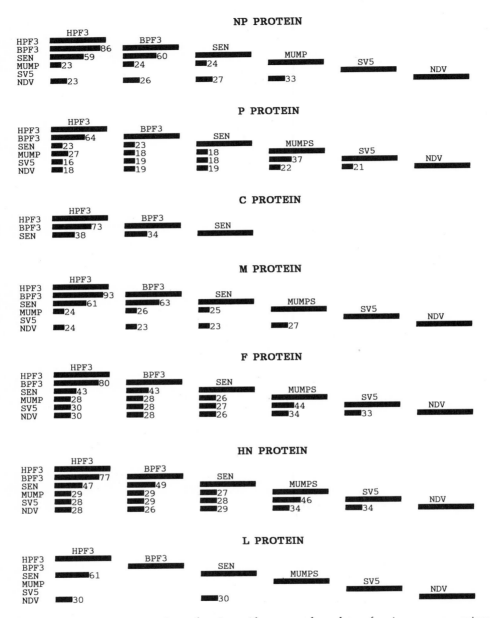

FIGURE 3. A two-way comparison of amino acid sequence homology of various paramyxovirus proteins. Comparisons of the reported amino acid sequences of the NP, P, C, M, F, HN, and L proteins from HPF3, BPF3, Sendai (SEN), mumps, SV5, and NDV viruses are shown. The bars are proportional in length to the percentage of amino acid sequence homology (indicated at the end of each bar) in pairwise comparisons. Gaps in the figure represent missing data, since not all of the genes of every virus have been determined. The *P* genes of mumps virus, SV5, and NDV do not encode unique C proteins as do BPF3, HPF3, and Sendai viruses.

```
          F2           ↓                    F1
HPF3    QESNENTDPRTKR  FFGGVIGTIALGVATSAQITAAVALVEAKQAR
BPF3    HETNNNTNSRTKR  ***EI******I*I******************K    84%
SEN     NDTTQNAGVPQSR  ***A***************GI**A**RE*K        78%
MUMP    IASPSPGSRRHKR  *A*IA**IA*******A**V****S**Q*QTNA     60%
SV5     IRNQLIPTRRRRR  *A*V***LA******A**V*******K*NENA      66%
NDV     ESVTTSGGRRQKR  *I*AI**GV******A*******A**IQ***NA     65%
```

FIGURE 4. Amino acid sequence conservation around the cleavage-activation site of F_0. The amino acid sequence of the C terminus of F_2 and the C terminus of F_1 are shown for the viruses HPF3, BPF3, Sendai (SEN), mumps, SV5, and NDV. The percentage of amino acid sequence identity of the fusion domains relative to HPF3 is indicated on the right. Identical amino acids are indicated by asterisks.

The role of F in directly mediating fusion has further support from the results of recombinant protein expression experiments. By the use of an SV40 expression system, the F protein of SV5 was shown to cause cell fusion in the absence of other viral proteins (Paterson *et al.*, 1985, 1989). This protein is cleaved and found abundantly on the cell surface. It is of interest that the degree of fusion observed was not as great as that observed during an acute infection. Whether this is the result of quantitative differences in expression or whether this reflects the absence of an auxiliary viral protein(s) is not clear. In contrast with these results, the expression of the HPF3 virus F protein in insect cells using a baculovirus vector did not result in cell fusion (Ray *et al.*, 1989). Although the protein was found on the cell surface, it was not cleaved. These results are consistent with loss of the F protein's fusion property due to a lack of posttranslational cleavage. Since the host cell supplies the protease necessary for activation, it would be of interest to determine whether the insect cells expressing the recombinant F would fuse following the addition of trypsin.

b. Hemagglutinin-Neuraminidase Glycoprotein

Although the distribution of F and HN over the surface of the virion appears homogeneous at the level of the electron microscope, higher orders of structure have been inferred from biochemical analysis. Two-dimensional polyacrylamide gel electrophoresis under nonreducing (first dimension) and reducing (second dimension) conditions has demonstrated that HN generally migrates not as a unit polypeptide, but rather as homo-oligomers (dimers and higher orders) for NDV, Sendai, and mumps viruses (Markwell *et al.*, 1981, 1984; Herrler and Compans, 1982; Waxham *et al.*, 1986). This indicates that much of the HN protein in these viruses interacts by disulfide bonding and that the biologically active component may not be the unit polypeptide. Although homo-oligomerization of HN appears to be an important feature for these three viruses, it has not been observed in HPF3 virus (Wechsler *et al.*, 1985b). This is surprising, since Sendai and HPF3 viruses appear to be closely related based on amino acid sequence homology.

Another important feature of the HN protein which is restricted to some strains of NDV is the proteolytic processing of a precursor, HN_0, to an active HN by removal of a small 90-amino acid peptide from the carboxyl terminus (Homma and Ohuchi, 1973; Scheid and Choppin, 1974). In avirulent NDV strains, HN_0 is not cleaved, suggesting that proteolytic processing is related to viral virulence (Nagai et al., 1976; Nagai and Klenk, 1977).

The molecular cloning of five different HN genes, including those of Sendai virus (Blumberg et al., 1985b), SV5 (Hiebert et al. 1985a), NDV (Jorgensen et al., 1987; McGinnes et al., 1987), mumps virus (Waxham et al., 1988; Kövamees et al., 1989), BPF3 virus (Suzu et al., 1987) and HPF3 virus (Elango et al., 1986), has been reported. The results of the molecular cloning of these genes (and a larger number of additional serotypes and variants) are schematically summarized in Fig. 2B. Some of the basic features shown in this figure include the transmembranal anchoring domains, potential sites for N-linked glycosylation, putative neuraminidase sites, and positions of cysteine residues. The HN proteins contain 565–582 amino acids and a single hydrophobic domain of sufficient length (22–28 amino acids) to span the plasma membrane. This domain begins approximately 17–32 amino acids from the amino terminus and presumably functions both as a transmembranal anchor and an uncleaved signal peptide for directing the protein for further processing (glycosylation). The absence of a cleavable signal peptide and the utilization of an N-terminal anchoring domain for HN are common structural features reported for all *Paramyxoviridae* receptor-binding proteins examined. Curiously, this protein orientation is identical to the analogous neuraminidase glycoprotein (NA) of the orthomyxovirus influenza virus (Fields et al., 1981; Blok et al., 1982). As described above, this orientation is opposite to that observed for the F glycoproteins, which are all anchored by C-terminal hydrophobic domains. There are a number (ranging from four to eight) of potential N-linked glycosylation sites predicted from the HN protein sequences. Although the relative locations are not highly conserved, most of the sites are similarly located in the carboxyl half of the proteins.

All the sequenced HN proteins from the *Paramyxovirus* genus share some amino acid sequence similarities, particularly in regions containing the amino acids which participate in protein folding (cysteine, glycine, and proline). This is seen for the cysteine residues depicted in Fig. 2. The distribution of related sequences across the HN proteins is asymmetric, in that the carboxyl two-thirds of the protein is more conserved than the amino one-third of the protein. Since the cytoplasmic tail and transmembranal anchor reside in the N-terminal third of the protein, it would appear that these sequences have significantly reduced constraints on alteration and the maintenance of function compared with the rest of the protein. The degree of sequence conservation across these proteins resembles that seen for the F protein. Together, these results support the notion that the paramyxoviruses share a common evolutionary history.

The possible role of the paramyxovirus neuraminidase activity in the fusion process (see below) and its location within the HN protein are undefined. Recently, amino acid sequence comparison of a number of paramyxovirus HN proteins has revealed only a single region (Asn-Arg-Lys-Ser-Cys-Ser; located between residues 223 and 259) which is completely conserved in six different viruses: Sendai virus (Blumberg et al., 1985b); BPF3 virus (Suzu et al., 1987); HPF3 virus (Elango et al., 1986); NDV (Jorgensen et al., 1987); mumps virus (Waxham et al., 1988); and SV5 (Hiebert et al., 1985a). Interestingly, most of the amino acid residues flanking this conserved domain are themselves not highly conserved. However, the relative positions of the flanking cysteine residues are highly conserved. This suggests that the ability to fold into a distinct conformation may be more important than the specific flanking residues. Blumberg *et al.* (1985b) predicted that the neuraminidase active site in Sendai virus would reside between residues 163 and 382 based on structural similarities with the analogous influenza virus protein. Jorgensen *et al.* (1987), who determined the sequence of the NDV HN protein, suggested that the six conserved amino acids may be part of, or close to, the sialic acid binding site for neuraminidase activity. This conclusion was also based on the potential folding of the HN protein into beta-sheet structures similar to that reported for the active site of influenza virus NA. Further support for the localization of the neuraminidase active site to this region of the protein was obtained using monoclonal anti-HN antibodies. Monoclonal antibodies that inhibit neuraminidase activity were used for the selection of Sendai and PF3 virus variants (Coelingh *et al.*, 1986, 1987; Thompson and Portner, 1987; Van Wyke Coelingh *et al.*, 1985, 1987). Sequence analysis of several antigenic variants showed that the amino acid alterations mapped close to the putative neuraminidase active site. In addition, another epitope was found near the C terminus in Sendai virus (residue 541). Whether these sites are topologically related or represent unique binding domains is uncertain.

Although the receptor-binding, hemagglutinating, and neuraminidase activities of HN are major properties of this protein, there is also some evidence that HN may play an intimate role in the expression of the fusion process. The basis for this includes the following: (1) An increase in syncytium formation has been reported for some mumps and PF3 virus HN variants (Merz and Wolinsky, 1981; Waxham and Wolinsky, 1986; Shibuta *et al.*, 1981, 1983). These variants did not appear to have any defects in the processing of the F protein, i.e., F_0 was not abundant and was readily cleaved to the $F_{1,2}$ active protein. Variants with increased fusion activity had decreased neuraminidase activity, while variants with low fusion activity had increased neuraminidase activity (Merz *et al.*, 1983). Furthermore, the latter variants with high neuraminidase activity (low fusion activity) could be made fusigenically active by proteolytic inactivation of the HN protein (Merz and Wolinsky, 1983). (2) Both mumps virus- and Sendai virus-induced fusion have been shown to be blocked by anti-HN monoclonal antibodies (mAbs) (Miura *et al.*, 1982; Merz *et al.*, 1981; Portner, 1984; Tsurudome *et al.*, 1986). These mAbs do not inhibit either hemagglutination (receptor binding) or neuraminidase activity. (3) A more direct interaction has been reported by Ozawa *et al.* (1979), Citovsky *et al.* (1986), and Gitman and Loyter (1984), who have worked with reconstituted viral envelopes and artificial membranes containing Sendai virus glycoproteins. Membranes containing only F glycoprotein are incapable of causing fusion, while membranes containing both F and HN can fuse. The ability of other ligands (i.e., anti-membrane antibody, polypeptide hormones, concanavalin A) to substitute as receptor-binding analogs of HN was also examined. Although the ligands were quite effective in binding, they were significantly less effective in mediating fusion than the authentic HN protein. Analogous results have been reported for NDV, where co-reconstitution of F and influenza glycoproteins NA or HA showed that fusion was dependent upon the availability both of the neuraminidase activity of NA and the receptor-binding activity of HA (Huang *et al.*, 1980). (4) A more intimate interaction between F and HN is suggested by circular dichroism measurements (Citovsky *et al.*, 1986). The conformation of individually reconstituted HN or F proteins differed from that of co-reconstituted HN and F proteins. Together these results support the notion of an intimate interaction between F and HN in the fusion process.

3. Matrix Protein

The matrix (M) proteins appear to play a critical role in the assembly of mature progeny during replication. Virus budding at the cell surface involves the association of nucleocapsids with regions of the membrane containing viral surface glycoproteins (Yoshida *et al.*, 1976; Roux *et al.*, 1984; Tuffereau and Roux, 1988). This nucleocapsid–membrane association is believed to be directly mediated by the matrix protein. This protein is thought to interact both with the plasma membrane, through hydrophobic interactions, and with the nucleocapsid core, through ionic interactions (Heggeness *et al.*, 1982; Shimizu and Ishida, 1975; Yoshida *et al.*, 1976). Biochemically, the M proteins are hydrophobic, as evidenced by their extraction into acidified chloroform–methanol (Giuffre *et al.*, 1982). Whether the interaction of the M protein is directly with the plasma membrane or coordinately with the membrane and the surface glycoproteins is unclear. The latter interaction has been proposed based on the isolation of NDV *M* gene mutants which have a defect in their ability to incorporate the F glycoprotein into infectious progeny (Peeples and Bratt, 1984). In addition, Heggeness *et al.* (1982) have reported immune electron microscopy studies which suggest that the association of the M protein with nucleocapsids is important for the correct positioning of the surface glycoproteins in Sendai virus. The association of M with nucleocapsids suggests that, in addition to participating in virion assembly, M may have a role in regulating transcription. Indeed, there is some evidence that the M protein shuts down transcription during assembly (Marx *et al.*, 1974).

Molecular cloning and sequence analyses of several paramyxovirus *M* genes/mRNAs, including those of BPF3 virus (Sakai *et al.*, 1987), HPF3 virus (Luk *et al.*, 1987; Galinski *et al.*, 1987b; Spriggs *et al.*, 1987), Sendai virus (Blumberg *et al.*, 1984; Hidaka *et al.*, 1984), mumps virus (Elango, 1989b), and NDV (McGinnes and Morrison, 1987), have provided some insight into the structure of this protein. The M proteins vary in size from 341 to 375 amino acids (approximate M_r 39,500–41,500). The amino acid compositions indicate that all of these proteins are highly basic (net charges at neutral pH ranging from +14 to +17) and contain an abundance of hydrophobic residues (ranging from 32 to 38%). The relative abundance of basic residues may reflect their importance in ionic interactions with NP proteins, which are acidic in all paramyxoviruses. The abundance of hydrophobic residues is consistent with interactions between M and the plasma membrane.

The M proteins of BPF3, HPF3, and Sendai viruses are highly conserved (Fig. 3). Indeed, the conservation of amino acid sequences (35% of the residues) between two genera of the *Paramyxoviridae* family (*Morbillivirus*—measles virus and CDV; and *Paramyxovirus*—BPF3, HPF3, and Sendai viruses) would indicate the importance of this protein in viral maturation (Bellini *et al.*, 1986; Galinski *et al.*, 1987a; Luk *et al.*, 1987; Sakai *et al.*, 1987). Surprisingly, despite the high degree of conservation between these virus groups, NDV and mumps virus have a significantly reduced sequence similarity in comparison with the other paramyxoviruses. Only 25% of the amino acids of mumps virus and NDV are identical to BPF3, HPF3, and Sendai viruses. Why the M protein has diverged in these viruses is unclear. The results of the sequence comparisons reported here for the M protein are similar to those observed for the F and HN proteins, in that Sendai, BPF3, and HPF3 viruses appear to be more closely related to one another than to the mumps virus, SV5, and NDV group.

The association of a number of paramyxoviruses with several chronic diseases in humans, including the slow neurological disease subacute sclerosing panencephalitis (SSPE) (Wechsler and Meissner, 1981; Morgan and Rapp, 1977), multiple sclerosis (Morgan and Rapp, 1977), and Paget's disease of the bone (Goswami *et al.*, 1984; Basle *et al.*, 1985), and the relative ease with which these viruses establish persistent infections in cell culture is a biological characteristic of much interest and consequence (Wechsler *et al.*, 1987). Altered synthesis and processing of M has been reported for both measles virus (Hall and Choppin, 1979; Wechsler *et al.*, 1979) and Sendai virus (Roux and Waldvogel, 1982) during persistent infections. It has been suggested that the reduction of M protein expression may abolish viral budding and induce the persistent state. In support of this notion is the recent finding that the defective measles viruses associated with human chronic central system infections (SSPE and measles inclusion body encephalitis) have a disproportionately greater number of mutations in the M gene (Cattaneo *et al.*, 1988). The role of M in modulating persistent infection, while not firmly established, is suggesting enough to warrant continued investigation (see Chapters 11, 12, and 16).

C. Internal Virion and Nonstructural Proteins

1. Phosphoprotein and C Protein

The nucleotide sequences of a number of paramyxovirus *P* genes/mRNAs have been determined, including those of Sendai virus (Giorgi *et al.*, 1983; Shioda *et al.*, 1983), BPF3 virus (Sakai *et al.*, 1987), HPF3 virus (Galinski *et al.*, 1986b; Luk *et al.*, 1986; Spriggs and Collins, 1986b), mumps virus (Takeuchi *et al.*, 1988), and NDV (McGinnes *et al.*, 1988). In some paramyxoviruses, the *P* gene is unusual in that it contains multiple cistrons (see Chapters 6 and 7). In Sendai virus (Giorgi *et al.*, 1983; Shioda *et al.*, 1983; Gupta and Kingsbury, 1985b), BPF3 virus (Sakai *et al.*, 1987), and HPF3 virus (Galinski *et al.*, 1986b; Luk *et al.*, 1986), the *P* gene encodes both the major structural phosphoprotein and a smaller, nonstructural C protein in overlapping cistrons (Fig. 5). The utilization of overlapping reading frames for the encoding of separate cistrons is a feature also seen in the *Morbilliviruses* measles (Bellini *et al.*, 1985) and canine distemper virus (Rozenblatt *et al.*, 1985). In Sendai virus a number of additional nonstructural proteins also appear to be encoded in the *P* gene, including an alternate form of the C protein, C' (Giorgi *et al.*, 1983; Shioda *et*

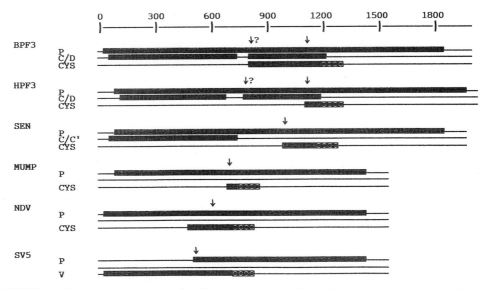

FIGURE 5. Cistron usage in the phosphoprotein genes of several paramyxoviruses. The three alternate reading frames for the phosphoprotein genes of BPF3 virus, HPF3 virus, Sendai virus, mumps virus, NDV, and SV5 are shown scaled to the length of each gene. Putative cistrons within each reading frame are indicated by black bars for P, C, D, and V proteins. In SV5, the accession of a full-length P-protein cistron is through the production of an mRNA containing two additional nontemplated nucleotides incorporated at an aberrant transcription-termination sequence (arrow). The position of a cysteine-rich ORF (boxed circles) in the various *P* genes, analogous to the C-terminal domain of the SV5 V protein, is indicated. The positions of transcription-termination sequences similar to those found in SV5 which putatively function in the accession of these domains are indicated with arrows. A second set of aberrant transcription-termination sequences present in BPF3 and HPF3 is indicated with arrows and question marks. See the text for further discussion.

al., 1983), Y_1 and Y_2 proteins (Curran and Kolakofsky, 1987; Gupta and Patwardhan, 1988), and an X protein (Curran *et al.*, 1986).

In SV5, the *P* gene also encodes two proteins, P and V, in partially overlapping cistrons (Fig. 5) (Thomas *et al.*, 1988). The V cistron resides at the 3' end of the gene, while the P cistron is contained in two discontinuous regions, accessed through a transcriptional fusion (frameshift) process involving the addition of two nontemplated nucleotides at an aberrant transcription termination sequence (for a detailed discussion of transcription in these viruses see below). Thus, the P protein of SV5 contains 164 amino acids, which are coterminal with the V protein. However, the V protein (222 amino acids) contains 58 amino acids at the C terminus which are unique to this protein. These residues are most notable for the relative abundance of cysteines (Fig. 6). This latter feature resembles the nonrepetitive cysteine-rich zinc fingers characteristic of a number of DNA-dependent transcriptional regulators, including adenovirus E1A protein (Moran and Mathews, 1987) and the steroid hormone receptor superfamily (Evans, 1988). Together with the assumed role of P in transcription, these results suggest that the cysteine-rich V protein may have a regulatory function in RNA transcription.

```
SV5    GFHRREYSIGWVGDEVKVTE WCNPSCSPITAAARRFECTCHQCPVTCSECERDT
MUMPS  |G||||W|LS||QG||R|F|  |*||I*||||||||FHS*K*GN*|AK*DQ*|||Y     63%
NDV    |  ||||H||S|TMGG|TTIS  |*|||*||VR|EP|QYS*I*GS*|A|*RL*AG|D     50%
SEN    |  ||||HI|YER |GYI|D|S|*||V*|R|RVIP||EL*V*KT*|KV*KL*RD|I      46%
BPF3   |  ||||H||YRE||YII ||S|*||I*|K|RPVP|QES*V*GE*|KQ*GY*IE|R      48%
HPF3   |  Y|||Q||YRE|NYSI A|S|*|SIYIKTRFIS|QTS*M*SK*TKQ*RY*IK|R      33%

MV     |  ||AQI|LI|N||R|FIDR |*||M*|KV|LGTI|AR*||*GE*|RV*EQ*RT||     46%
CDV    |  ||||V|LT|N||SCWIDK |*||I*TQVNWGII|AK*F*GE*|P|*N|*KD|P      44%
```

FIGURE 6. Comparison of the C-terminal 54 amino acids of the SV5 V protein with putative polypeptides encoded in the *P* genes of several paramyxoviruses. Amino acids 169–222 of the SV5 V protein are aligned for maximal similarity with subsets of the cysteine-rich polypeptides found in mumps, NDV, Sendai (SEN), BPF3, and HPF3 viruses. Identical residues are indicated with vertical lines, while the conserved cysteines are indicated with asterisks. The percentage identity of each subset with the SV5 sequence is given on the right.

A search for amino acid sequence similarities in the cysteine-rich domain (58 amino acids) of the SV5 V protein with other paramyxovirus proteins showed no conservation. However, a more detailed analysis of several paramyxovirus *P* genes revealed an intriguing feature (Fig. 5). Comparison of all the potential amino acid sequences encoded in the three alternate reading frames of the P mRNA revealed that each *P* gene contained, in an overlapping reading frame, a cysteine-rich peptide that shared extensive amino acid sequence similarity with the V protein of SV5 (Fig. 6). The sequence homology ranged from 32 to 65%, which is significantly greater than that observed for the analogous P proteins encoded with the same nucleotides in an alternate reading frame (Fig. 3). Further, proximal to the 5' end of the coding sequences for the cysteine-rich peptide are sequences which resemble the aberrant transcription termination sequence of SV5 (Fig. 5, arrows). Accession of each of these open reading frames (ORFs) is only possible through a frameshift, since there are no methionine codons within these domains, with the exception of NDV and HPF3 virus.

More recently, a similar coding strategy has been reported for the *Morbillivirus* measles virus (Cattaneo *et al.*, 1989). Together, these findings strongly support the biological significance of this form of gene expression as a general feature of paramyxoviruses.

The arrangement of the BPF3 and HPF3 virus *P* genes differs from the paradigm described above in that they contain two internal sets of sequences which resemble aberrant transcription termination signals (Fig. 5). One set is upstream of the cysteine-rich domain, while the second set is found upstream of another overlapping ORF previously described for the D protein in HPF3 virus (Galinski *et al.*, 1986b). Curiously, the second sequence set more closely resembles the homologous transcription termination sequences than does the first set. Although accession of either ORF through aberrant transcription is uncertain, three independent studies have reported divergent numbers of G residues following the second sequence set (Galinski *et al.*, 1986b; Luk *et al.*, 1986; Spriggs and Collins, 1986b).

The paramyxovirus P proteins are quite divergent in terms of physical structure, ranging in size from 391 to 603 amino acids (approximate M_r 41,000–67,600) and with net charges ranging from −15 to +3. Reflecting these physical differences, the amino acid sequences are also significantly divergent, as evidenced by the lack of sequence similarities between paramyxoviruses and morbilliviruses. More importantly, there is a reduced or absence of detectable similarity among the various paramyxovirus P proteins. Thus, the degree of amino acid sequence conservation among the P proteins of the closely related Sendai, BPF3, and HPF3 viruses is less than that of any other viral protein. Furthermore, there is little similarity among the P proteins of Sendai virus (and, by extension, BPF3 and HPF3 viruses), mumps virus, SV5, and NDV.

An additional small-subunit polypeptide (X protein), derived from P, has been identified in Sendai virus-infected cell lysates (Curran and Kolakofsky, 1987). This protein accumulates to significant levels in infected cells and appears to be almost as abundant as the P protein based upon its intracellular molar concentration. Antibody analysis has revealed that the X protein represents the C-terminal 95 amino acids of P. The lack of any precursor–product relationship with P suggests that synthesis of the X protein is initiated independently of P.

In NDV (McGinnes et al., 1988) and mumps virus (Takeuchi et al., 1988), the analogous C proteins appear to be encoded in the same cistron as the P protein. Synthesis of these "C-like" proteins is by initiation at internal methionine codons in the P ORF, resulting in the production of a subunit polypeptide of P. The biological significance of these proteins as well as that of the X protein of Sendai is unclear.

In Sendai virus, much of the P protein can be found as homo-oligomeric trimers (Markwell and Fox, 1980), although this property has not been observed for the closely related HPF3 virus (Wechsler et al., 1985a). In addition to oligomerization, P appears to interact with the vRNA and the NP protein, since all of these nucleocapsid subunits can be chemically cross-linked to P (Raghow and Kingsbury, 1979; Raghow et al., 1979). Analysis of the functional domains of P using limited proteolysis has shown that the C-terminal half of the protein contains the binding site(s) for nucleocapsids and the site(s) which participate in transcription (Chinchar and Portner, 1981). A more detailed analysis of the binding site using in vitro-expressed recombinant P protein localized this domain to the 224 carboxyl-terminal residues (Ryan and Kingsbury, 1988). These studies examined the ability of a recombinant P protein and a series of carboxyl-terminal and internally deleted proteins to bind to nucleocapsids derived from infected cells. Deletion of the 30 C-terminal residues abrogated all binding to nucleocapsids. Further, a series of proteins containing internal deletions mapped the binding site to 195 residues upstream and overlapping (by a single amino acid) the 30-residue C-terminal binding domain. Since this domain functions in a direct protein–protein interaction, these results suggest that the C terminus is accessible at the protein surface. In further support of this notion is the clustering of at least three epitopic sites (recognized by a number of monoclonal antibodies) in this portion of the protein (Deshepande and Portner, 1985; Vidal et al., 1988). In addition to acting as a nucleocapsid-binding domain, this region may also participate in transcription, since several of the monoclonal antibodies were able to inhibit this activity.

Although NP and M are to some extent phosphorylated, P is the major protein site for the addition of phosphate to the virion components (Lamb and Choppin, 1977a,b). Thus, the addition of phosphate is the primary biochemical modification during the posttranslational processing of P. The phosphorylation sites are believed to occur at serine and threonine residues. These amino acids are quite abundant in P and account for approximately 25% of the total residues. Limited mapping of the phosphorylation sites in Sendai virus by partial V8 protease digestion indicated that in *in vivo*-synthesized P, the bulk of the phosphates are contained in the N-terminal one-quarter of the protein (Hsu and Kingsbury, 1982). In contrast, in *in vitro*-synthesized P, the bulk of the phosphates are contained in the second one-quarter of the protein (Vidal *et al.*, 1988). Although the results are not in agreement, the variations may reflect differences in the source of the phosphorylation complexes.

The paramyxovirus C protein is a nonstructural protein whose precise function is unknown, but may be involved in the induction of interferon (Taira *et al.*, 1987). The "unique" C proteins of Sendai, BPF3, and HPF3 viruses range in size from 199 to 205 amino acids (approximate M_r 23,300–24,000) and are highly positively charged (+7 to +10). The net positive charge contrasts with the highly negative charge of the P proteins. The abundance of charged residues and the deficit of hydrophobic residues suggest that these proteins may mediate their function through ionic interactions.

The Sendai C cistron differs from the BPF3 and HPF3 cistrons in that it encodes two proteins, C and C' (Giorgi *et al.*, 1983; Shioda *et al.*, 1983). These proteins are initiated at alternate codons, and thus contain different N termini (Gupta and Kingsbury, 1985b; Curran *et al.*, 1986). The ORFs encoding these various proteins are all accessed during translation by initiation at alternate codons. Interestingly, deletion and site-directed mutational analysis of the P/C mRNA of Sendai virus revealed that the C' protein does not initiate from a methionine codon, but rather uses a nonstandard ACG codon (Patwardhan and Gupta, 1988; Gupta and Patwardhan, 1988). Whether this unusual initiation strategy for protein synthesis plays a regulatory role in the expression of the C' protein is not clear. Another unusual feature of the C cistron is the accession of two additional subunit polypeptides of C, namely Y_1 and Y_2, by internal initiation of protein synthesis (Curran and Kolakofsky, 1987; Gupta and Patwardhan, 1988).

The unique C proteins are all encoded in cistrons with ORFs which are +1 nucleotide out of phase in relation to the P ORFs. This molecular organization is a conserved feature for all those *Paramyxoviridae* which contain overlapping P/C cistrons. This curious organization results in overlapping codons in which the third position of the C codons use the same nucleotide present in the first position of the overlapping P codons. Because of the degeneracy of the genetic code, mutations occurring at the third position of a codon are less likely to change the encoded amino acid than are mutations occurring in the first position. Thus, the *Paramyxoviridae* have adopted a strategy which allows two different proteins encoded by the same nucleotide sequence to diverge at different rates. Indeed, this has been observed for the P and C proteins, which show markedly different degrees of amino acid sequence conservation between related viruses (Galinski *et al.*, 1986b). The various P proteins are the least conserved of all the *Paramyxoviridae* proteins. In contrast, the C proteins, encoded from the same nucleotide sequence, show substantially greater amino acid sequence conservation (Fig. 3).

2. Nucleocapsid Protein

The nucleocapsid protein (NP) is closely associated with the genomic RNA and functions to maintain the structural integrity of the viral RNA. This protein also interacts with the P and L proteins in the nucleocapsid complex (see Chapters 5 and 9). In Sendai virus, and by analogy in other paramyxoviruses, NP is found at a copy number of 2600 molecules per viral RNA molecule (Lamb *et al.*, 1976) and therefore represents the major protein component of the nucleocapsid.

The paramyxovirus *NP* genes/mRNAs which have been molecularly cloned and sequenced include those of BPF3 virus (Sakai *et al.*, 1987), HPF3 virus (Galinski *et al.*, 1986a; Jambou *et al.*, 1986; Sanchez *et al.*, 1986), mumps virus (Elango, 1989a), Sendai virus (Shioda *et al.*, 1983; Blumberg *et al.*, 1984; Morgan and Kingsbury, 1984), and NDV (Ishida *et al.*, 1986). Based on the predicted primary structures of the various NP proteins, the NP proteins range from 489 to 553 amino acids (M_r 53,167–57,896). Generally, the proteins have net acidic charges ranging from -7 to -12, with the exception of mumps virus ($+2$). The hydropathic profiles of these proteins indicate that the N- and C-terminal ends of the proteins are highly hydrophilic and are therefore most likely to be found near the surface of the molecule.

Comparison of the amino acid sequences among the various paramyxovirus NP proteins shows that they are closely related (Fig. 3). In particular, the NP proteins of Sendai, HPF3, and BPF3 viruses appear to be closely related to one another (approximately 60–86% similarity), while each of these viruses is more distantly related to NDV and mumps virus (approximately 25% similarity). These results are similar to those observed for the matrix proteins. The distribution of identical sequences is not uniform. The N-terminal residues 1–133 and 160–386 appear to be more highly conserved than other regions (Galinski *et al.*, 1986a; Morrison, 1988; Elango, 1989a). Interestingly, comparison of the amino acid sequences with the morbillivirus NP protein sequences deduced for measles and CDV showed that these same regions are highly conserved (Galinski *et al.*, 1986a).

The binding of NP to the genomic RNA is believed to be mediated through specific interactions with sequences at the 5′ end of the genomic RNA or the antigenomic RNA. This is based on three observations: (1) the various mRNAs are not incorporated into nucleocapsids, presumably because they lack the proper leader sequence; (2) there are significant sequence similarities between the 5′ ends of both viral and replicative RNAs; and (3) unencapsidated genome-length RNA is absent during infection. Further support for this notion comes from the sequence analysis of defective-interfering particles (DIs). These aberrant RNAs arise from internal deletions of the various protein coding sequences, but all contain either direct copies or copy-back sequences of the genome 3′- and 5′-end sequences (Amesse *et al.*, 1982; Re *et al.*, 1983a,b; Re and Kingsbury, 1986).

Monoclonal antibodies have been used to probe functional domains of Sendai virus NP (Deshpande and Portner, 1985). By immune (Western) blot analysis, the NP protein contains at least two topologically distinct epitopes. The configuration of the tertiary folding of the NP protein is believed to expose these sites at the surface of the protein. These sites are also required for transcription, since a number of purified monoclonal antibodies were able to inhibit *in vitro* transcription to the same extent.

The role of NP in transcription is not well understood; however, its assumed binding to the plus and minus leader RNA species make NP a likely candidate to be the decisive factor for switching from transcription to replication. In addition, the ability of monoclonal antibodies to inhibit transcription would suggest that another role of this protein is to interact with the P and L proteins coordinately. It would be of interest to determine whether the nucleocapsid binding site on P (Ryan and Kingsbury, 1988) interacts with the same NP sites that are involved in transcription.

3. L Protein

The large (L) protein of *Paramyxoviridae* is believed to be multifunctional, with the capacity to act as an RNA-dependent RNA polymerase for the production of mRNAs, replicative intermediates, and progeny viral RNAs. In addition, L is believed to participate in the synthesis of both positive- and negative-strand leader sequences, and the capping, methylation, and polyadenylation of the various mRNAs (Choppin and Compans, 1975; Matsumoto, 1982). The L proteins are all very large (M_r greater than 250,000), presumably because of their need to carry out multiple functions.

Due to the length (approximately 6900 nucleotides) and location (5' end of the genome) of the *L* gene, the L mRNA is the least abundant viral transcript. This presumably reflects a need for only catalytic amounts of this protein. Since limited amounts of this protein have been available for study, most of our understanding concerning its biochemical structure has come from recombinant cloning and sequence analysis of *L* genes. The entire *L* genes from Sendai virus (Shioda *et al.*, 1986; Morgan and Rakestraw, 1986), NDV (Yusoff *et al.*, 1987), and HPF3 virus (Galinski *et al.*, 1988) have been molecularly cloned and sequenced. These three paramyxovirus L proteins are all very similar in size, 2204–2233 amino acids (range of M_r 248,854–255,812) and composition. Although the amino acid compositions are similar, differences in the numbers of charged residues result in net charges which are disparate (HPF3, 0; Sendai, +4.5; and NDV, +27). Despite the charge differences, the various L proteins share extensive amino acid sequence similarities, with 62% amino acid conservation between HPF3 and Sendai viruses, and 28% amino acid conservation between HPF3 virus and NDV (Fig. 3). Similar to the observations made with NP and M proteins, amino acid sequence conservation of the L proteins extends beyond the three paramyxovirus L proteins and includes the *Morbillivirus* measles virus, which shares 39% of its L-protein amino acids with that of HPF3 virus (Blumberg *et al.*, 1988; M. S. Galinski, unpublished observations). These results support the notion that *Paramyxovirus* and *Morbillivirus* share a common evolutionary history.

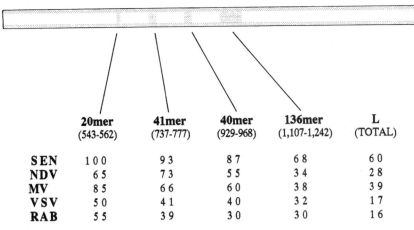

	20mer (543-562)	41mer (737-777)	40mer (929-968)	136mer (1,107-1,242)	L (TOTAL)
SEN	1 0 0	9 3	8 7	6 8	6 0
NDV	6 5	7 3	5 5	3 4	2 8
MV	8 5	6 6	6 0	3 8	3 9
VSV	5 0	4 1	4 0	3 2	1 7
RAB	5 5	3 9	3 0	3 0	1 6

FIGURE 7. Schematic representation of the conserved amino acid domains between the *Paramyxoviridae* and *Rhabdoviridae* L proteins. The HPF3 L protein is shown at the top and the areas showing greatest homology are shaded. Tabulated below are the length, position, and percentage amino acid identity with the HPF3 virus L protein for Sendai (SEN), NDV, measles (MV), VSV, and rabies (RAB) viruses. The total represents the percentage of amino acid identity spanning the length of the protein.

In addition to these intrafamilial relationships, there are at least four smaller peptide domains (Fig. 7) which appear to be conserved between the four *Paramyxoviridae* L proteins and the analogous proteins from the *Rhabdoviridae* VSV (Schubert *et al.*, 1984) and rabies virus (Tordo *et al.*, 1988). These domains range in length from 20 to 136 residues and have percentages of shared amino acids from 30 to 55% between the two families. The longest invariant peptide shared by all six L proteins is Gln-Gly-Asp-Asn-Gln (approximately 750 amino acids from the N termini). This peptide and the flanking hydrophobic residues partially resemble the consensus sequence identified as an active or recognition sequence common to a number of bacterial, plant, and animal RNA transcriptases (Kamer and Argos, 1984). This consensus sequence contains Gly-Asp-Asp and flanking hydrophobic residues as the central motif.

Although interfamilial conservation (*Paramyxoviridae* and *Rhabdoviridae*) of other structural proteins is not apparent, the molecular organization of the genomes of members of these two taxa and the length and placement of intergenic and extracistronic transcriptional control sequences support the notion that *Paramyxoviridae* and *Rhabdoviridae* share a common evolutionary history. Further circumstantial evidence for a common ancestry comes from the lack of any significant amino acid sequence conservation with the *Orthomyxoviridae* (influenza virus) polymerase proteins or the *Arenaridae* [lymphocytic choriomeningitis virus (Salvato *et al.*, 1989) and Tacaribe virus (Iapalucci *et al.*, (1989)] L proteins (M. S. Galinski, unpublished data). Since all these viruses have negative-strand RNA genomes, this observation would argue against the apparent relatedness of the *Paramyxoviridae* and *Rhabdoviridae* proteins based on convergent evolution due to similarity of function. Additionally, it is noteworthy that the segmented negative-strand viral polymerase proteins appear to be significantly different from their non-segmented counterparts.

4. SH Protein

The SV5 and mumps virus genomes contain an additional gene between the *F* and *HN* genes which encodes a small hydrophobic protein, SH (Hiebert *et al.*, 1985b, 1988; Elango *et al.*, 1989). No analogous counterparts are present in the genomes of Sendai, HPF3, or BPF3 viruses or NDV. In SV5, the SH protein consists of 44 amino acids (M_r 5012) and has the biochemical properties of an integral membrane protein (Hiebert *et al.*, 1988). The predicted protein has a C-terminal domain of 25 hydrophobic amino acids (45% of the total amino acids), a length which is sufficient to span the plasma membrane. Indeed, Hiebert *et al.* (1988) demonstrated that the SH protein is associated with the cell membrane. Their analysis indicates that the protein is anchored by the hydrophobic domain with approximately five amino acids of the C terminus exposed to the extracellular environment. The hydrophilic N-terminal domain resides in the cell cytoplasm. The functional role of SH in the replicative cycle of SV5 is unknown. Its biological significance in viral replication is further obscured by the observation that SH is not incorporated into mature virions.

In mumps virus, the SH protein consists of 57 amino acids (M_r 6719), and, like the analogous SV5 protein, contains a hydrophobic domain of 25 amino acids. However, this domain resides in the N-terminal portion of the protein, in contrast with the C-terminal location found in SV5. Interestingly, there is no amino acid sequence conservation between these two proteins. Although SH mRNA transcripts were demonstrated in mumps virus-infected Vero cells, the protein was not found in these cells.

III. VIRAL REPLICATION

A. Adsorption, Penetration, and Uncoating

The initial steps in paramyxovirus infection of a susceptible cell include *adsorption*, binding of the virion to the cell surface; *penetration*, internationalization of the virion; and *uncoating*, the removal of the virion envelope and activation of transcriptive processes. Although many details of these initial events are cloaked in mystery, together they are critical determinants of whether a productive infection will ensue.

The receptor-binding protein of paramyxoviruses, HN, contains both hemagglutinating and neuraminidase activities. It is generally believed that hemagglutination is mediated through receptor binding of HN to glycophorin, the major sialic acid-containing glycoprotein of erythrocytes (Steck, 1974; Paulson *et al.*, 1979; however, see Chapter 15). Although paramyxoviruses can bind to and hemolyze erythrocytes, infection does not occur. This suggests that on the surface of a susceptible cell other receptors may be present which are involved in the binding process. The precise cell surface receptors for paramyxoviruses remain undefined; however, the one feature most consistently

reported is the requirement for sialic acid in the viral receptor. All the paramyxoviruses have neuraminidase activity, which presumably allows viral elu-
tion from the cell surface. Thus, the neuraminidase activity of the HN protein
could provide a further means of modulating viral adsorption and/or
penetration.

It is generally believed that viral neuraminidase activity may alter either
the primary receptor or a secondary receptor, which can then interact with the
F protein, allowing direct membrane fusion. The proposal of a secondary receptor for the F protein is based upon the observed inhibition of infectivity and
hemolysis using synthetic oligopeptides which resemble the amino terminus
of F_1 (Richardson et al., 1980). This inhibition appears to be mediated through a
direct competition for cell surface receptors and not through inhibition of viral
HN receptor binding (Richardson and Choppin, 1983). These results suggest
that there are specific sites on the cell surface, in addition to the HN binding
site, which interact specifically with F and which are involved with viral
fusion and penetration.

Certain gangliosides (sialic acid-containing glycolipids) have been shown
to be important receptors for Sendai virus attachment (Haywood, 1974, 1975).
These gangliosides are also substrates for the neuraminidase activity (Suzuki
et al., 1980). Further studies (Holmgren et al., 1980) have shown that attachment occurs specifically with GD_{1a}, GT_{1b}, and GQ_{1b}. Whether these gangliosides are the primary or secondary receptors for Sendai is currently uncertain (but see Chapter 15).

Recently, an alternate route of infection was described for Sendai virus
using a temperature-sensitive mutant which lacks HN at the nonpermissive
temperature (Markwell et al., 1985). In this experimental system, virus binding
bypassed the normal sialic acid-containing receptors of the cell. Thus, this
HN^- mutant, which was incapable of entering cells containing natural receptors for the virus, could infect and cause syncytium formation in a cell line
(Hep G2) that expresses an asialoglycoprotein receptor (lectin). This lectin is
able to bind proteins containing terminal galactose or N-acetylgalactosamine.
Entry into these cells was presumably through direct interaction of the F
glycoprotein with the lectin (see Chapter 15 for further discussion).

The course of events during penetration and uncoating of paramyxoviruses
following adsorption to a susceptible cell is not well understood. Membrane
fusion, allowing penetration and uncoating of the virus, is an infectious entry
process shared by a number of RNA virus families (White et al., 1983). There
are two known pathways used for entering the host cell by fusion. First, the
majority of these viruses (Togaviridae, Orthomyxoviridae, Rhabdoviridae, and
Retroviridae) enter the cell through endocytic vesicles. The fusigenic proteins
are then activated by a drop to pH 5–6 following the fusion of the endocytic
vesicles with prelysosomal vacuoles. Following activation, the viral membrane
fuses with the endosomal membrane and discharges its genome into the
cytoplasm. The second known pathway for entry is utilized by the Paramyxoviridae. After attachment to cellular receptors, the F protein directly interacts
with the cell membrane and catalyzes fusion of the viral–cell membranes,
releasing the viral genome into the cell (Ishida and Homma, 1978; Choppin
and Scheid, 1980). In contrast to the first pathway, fusion in this latter pathway
occurs at a neutral pH.

Although there is good evidence for the participation of HN in the cell-to-cell fusion observed in infected tissue culture monolayers, the role of HN in membrane fusion during penetration and uncoating is less certain. In support of the participation of HN in viral entry, fusion induced by both mumps and Sendai virus has been shown to be blocked by anti-HN monoclonal antibodies which have no inhibitory effects toward either hemagglutination or neuraminidase activity (Miura *et al.*, 1982; Merz *et al.*, 1981; Tsurudome *et al.*, 1986). Although providing supportive evidence, these findings do not exclude the possibility that the receptor-binding domain of HN is distinct from the hemagglutinating domain.

B. Molecular Organization of the Genome

Recent molecular cloning and sequence analysis has provided, either directly or indirectly, the complete nucleotide sequence for the genomes of Sendai virus (Gupta and Kingsbury, 1984; Shioda *et al.*, 1983, 1986; Hidaka *et al.*, 1984), NDV (Kurilla *et al.*, 1985; Ishida *et al.*, 1986; Millar *et al.*, 1986; Chambers *et al.*, 1986a,b; Yusoff *et al.*, 1987; McGinnes *et al.*, 1988), and HPF3 virus (Côté *et al.*, 1987; Dimock *et al.*, 1986; Elango *et al.*, 1986; Galinski *et al.*, 1986a,b, 1987a,b, 1988; Jambou *et al.*, 1986; Luk *et al.*, 1986, 1987; Prinoski *et al.*, 1987; Sanchez *et al.*, 1986; Spriggs and Collins, 1986a,b; Spriggs *et al.*, 1986, 1987; Storey *et al.*, 1987). Together with the sequence analysis of most of the structural genes/mRNAs of BPF3 virus (Sakai *et al.*, 1987; Suzu *et al.*, 1987), SV5 (Paterson *et al.*, 1984a,b; Hiebert *et al.*, 1985a,b; Thomas *et al.*, 1988), and mumps virus (Waxham *et al.*, 1987, 1988; Elango *et al.*, 1988, 1989; Takeuchi *et al.*, 1988; Elango, 1989a,b), there is abundant nucleotide sequence data.

The molecular organization of the genomes follows the general pattern seen in other nonsegmented negative-stranded RNA viruses. There are two extracistronic regions found at the 3' and 5' ends of the genome. By analogy with other nonsegmented negative-stranded RNA viruses, these regions are believed to function as the templates for negative- and positive-strand leader sequences (see Chapter 8).

The 3'-end extracistronic sequences reported for HPF3 virus (Dimock *et al.*, 1986; Galinski *et al.*, 1988), BPF3 virus (Sakai *et al.*, 1987), Sendai virus (Shioda *et al.*, 1983, 1986), NDV (Kurilla *et al.*, 1985; Yusoff *et al.*, 1987), and mumps virus (Elango *et al.*, 1988) are shown in Fig. 8 for both the viral minus (−3') and complementary plus (+3') RNAs. It is apparent that the viral 3'-end proximal 20 nucleotides are highly conserved among these viruses. These sequences diverge as they approach their 5'-end junctions with the *NP* genes. Similarly, the viral complementary plus 3'-end sequences appear to be conserved between HPF3 and Sendai viruses, but not between these viruses and NDV. The apparent lack of conservation may indicate that the sequences proximal to the viral complementary 3' end of the NDV genome may not be accurate. Since the sequences in this region were determined from a single recombinant clone, the precise end of the genome may not have been accurately determined (Yusoff *et al.*, 1987). Another important feature of these regions is the apparent conservation of sequences at the 3' ends of the plus and

GENOME RNA

```
                    10        20        30        40        50
HPF3        UGGUUUGUUCUCUUCUUUGAACAAACCUUUAUAUU UAAAUUUAAUUUUAAUUGAA-----------
BPF3        ***************C*****G****C*****A* ***G*C**U***********-----------
SEN         **************U****U***U*** C****CA *U*C**C***A*GUCC*A**A---------
MUMP        N****CCCCUCU*****CU*C*CU*UAACC**C** GUUUA*C*CA**CUU*GUC*-----------
NDV         ******* ******AGGCAUU***UG*  ***U**CCGCU*CCUCG**A*U*CA*CG---------
```

ANTIGENOME RNA

```
                    10        20        30        40
HPF3        UGGUUUGUUCUC   UUCUUGAGACAAACCAUAUAUAUAUAAUGUAC--------------------
SEN         ****C********AAA****CU*U***U*GG* *A**U** **A*A**AGAAGAACAUUC-------
NDV            ****GU*A*U****UCA*GC*UA*GCUC*AGAUUCCUCAGCCUCAAGUU---
```

FIGURE 8. Comparison of the nucleotide sequences of the 3' ends of the genome and antigenome RNAs for several paramyxoviruses. The genome (−3') and antigenome (+3') sequences of BPF3, Sendai (SEN), mumps, and NDV are aligned with that of HPF3 virus. The sequences are aligned from the 3' to 5' direction. Where necessary, gaps were introduced for maximal alignment of identical sequences. Nucleotides identical to the HPF3 virus sequence are indicated with asterisks.

minus strands. This is particularly evident in HPF3 virus, where 33 of the first 39 nucleotides are identical (Galinski et al., 1988). These sequences presumably function in transcription initiation of the leader RNA and mRNAs, and in replication of full-length viral plus and progeny minus strands. The utilization of a common sequence for recognition of the 3' ends of either strand may indicate parsimonious control of transcription initiation. Another consequence of the use of similar sequences at the 3' ends would be the intramolecular complementarity of the 3' and 5' ends of the plus or minus strands. Although complementarity of the ends of the genomic RNA has been reported to allow the formation of circular structures for other negative-stranded RNA viruses (Obijeski et al., 1976; Pettersson and von Bonsdorff, 1975), the formation of similar structures in paramyxoviruses is not clear. The biological significance of this complementarity is unknown, but may simply reflect the use of similar sequences at the 3' ends of the genomic RNAs for transcriptional control and/or the NP protein binding site.

The genome contains six major structural genes (NP, P, M, F, HN, and L), a feature common to all Paramyxoviridae. An additional gene (SH) follows the F gene in SV5 and mumps viruses (Hiebert et al., 1985b; Elango et al., 1989). Further, the phosphoprotein gene is unique in some Paramyxoviridae in that it contains multiple cistrons encoding different proteins (P, C, and V) either discontinuously, as in SV5 (Thomas et al., 1988) and in the Morbillivirus measles virus (Cattaneo et al., 1989), or in overlapping reading frames, as in Sendai virus (Giorgi et al., 1983) and HPF3 virus (Galinski et al., 1986b; Luk et al., 1986). The genes have been unambiguously ordered in the genome, based upon nucleotide sequence analysis, as

3' NP-P(C/V)-M-F-(SH)-HN-L 5'

At the beginning and end of each gene are conserved transcriptional control sequences, which are copied into the mRNAs. The transcriptional control

```
          SENDAI                                              PF3

   UCCCAGUUUC    NP    AUUCUUUUU            UCCUAAUUUC    NP    UUUAUUCUUUUU
   **********    P     *********            **********    P     ***********
   *****C****    M     *********            **********    M     ******UCUAUUAG*****
   AU**CUA**U    F     *********            ****GU****    F     **A**AU*****
   *****C****    HN    *********            ****C*****    HN    *****AU****
   *****C**A*    L     *********            ****CG****    L     **C*********


           NDV                                            MUMPS

   UGCCCAUCUUC   NP    AAUCUUUUUU           UUCGGUCCUUCACC  NP  AAUUCUUUUUU
   ***********   P     *U********           *C***G****UCUU  P   *U**A******
   **********A   M     **********           ****UG*U*GUGUU  M   *UA********
   **********    F     **********           *****AU****CUA  F   **A********U
   **G***A*A**   HN    *U********           ***UUA*U*AG*GG  SH  *U*********
   *G******C*G   L     **********U          ******U*GUCUG   HN  ***********
                                            *C*****U*A*CG*  L   -----------


           SV5                                            BPF3

   ----------    NP    -----------------    UCCUAAUUUC    NP    UUCAUUCUUUUU
   ACCGGCCUGCC   P     ACUCCCAAAAUCUUUUU    *******G*    P      C*A*********
   ----------    M     -----------------    *****U****    M     **UU*AG*****
   UUCGUGCUUUC   F     AUUAAAAAAAGUUUUUUU   ****G*****    F     **U**GU*****
   UCCUGGCUUGG   SH    AUAAAAUUUGUUUUUU     ****UG****    HN    **A**GU*****
   UCCAUUAGCUC   HN    ACCAAAAUUCUUUUUUUU   ****CC****    L     -----------
   ----------    L     -----------------
```

FIGURE 9. Nucleotide sequences found at the boundaries of the *NP, P, M, F, HN, SH,* and *L* genes. Nucleotide sequences are aligned for maximal similarity with their respective viral *NP* genes (except for SV5). The sequences are oriented in the 3' to 5' direction. Nucleotides which are identical to the *NP* transcription start/stop sequences are indicated with an asterisk. The *M*-gene transcriptional stop sequence for the HPF3 virus is aberrant and contains an additional eight intervening nucleotides (UCUAUUAG). The gene end sequences have not been reported for the SV5 *NP, M,* and *L* genes and the mumps virus *L* gene (- - -). See the text for further discussion.

sequences found at the gene-end boundaries for Sendai virus (Gupta and Kingsbury, 1984), HPF3 virus (Spriggs and Collins, 1986a), BPF3 virus (Sakai *et al.,* 1987; Suzu *et al.,* 1987), NDV (Kurilla *et al.,* 1985; Ishida *et al.,* 1986; Millar *et al.,* 1986; Chambers *et al.,* 1986a,b; Yusoff *et al.,* 1987; McGinnes *et al.,* 1988), and mumps virus (Elango *et al.,* 1988) are shown in Fig. 9. In this figure, the sequences of contiguous genes are aligned at the 3' and 5' ends demarcating the coding sequences for the mRNA. Located between adjacent genes are the intercistronic sequences described below. Although the gene-end boundaries have been divided into three separate domains (transcription termination, intergenic sequences, and transcription initiation), the coordinate interaction of the three domains in transcriptional control is unknown.. Alterations in the intergenic sequences of Sendai and PF3 virus variants have been shown to allow readthrough transcription into the next gene (Hsu *et al.,* 1985; Gupta and Kingsbury, 1985a; Galinski *et al.,* 1987a), suggesting that these sequences are involved with transcription termination. The conservation of nucleotide sequence among these viruses, in particular that of the transcription termination sequences, is quite striking, and underscores the biological importance of these domains.

Between the gene-end boundaries are intergenic sequences which are not present in the monocistronic mRNAs. The intergenic sequences in HPF3 virus are 3'GAA and in Sendai virus either 3'GAA or 3'GGG. In contrast to the conservation of the intergenic trinucleotides of Sendai and HPF3 viruses, mumps virus displays a variability in the intergenic sequences with 3'AA (NP-P), 3'A (P-M-F), 3'GAUUUUA (F-SH), 3'CG (SH-HN), and 3'G (HN-L) being found. Similarly, NDV contains variable intergenic sequences, with 3'CA (NP-P), 3'A (P-M), 3'G (M-F), and 31 and 48 nucleotides between the F-HN and HN-L genes, respectively.

C. Transcription

The early events following release of the virion-associated contents into infected cells are not well understood. The precise form of the nucleocapsid complex which is released into the cell and the changes required for initiating transcription are unknown. Much of our understanding of primary transcription and replication have been gleaned from *in vitro* systems using NDV and Sendai virus. The transcriptase activity associated with purified Sendai virions residues in the nucleocapsid, which is a complex structure consisting of the genomic RNA encapsidated with NP and two additional proteins, P and L (Stone *et al.*, 1972; Marx *et al.*, 1974; Emerson and Yu, 1975; Lamb and Choppin, 1978).

During negative-strand virus replication, it is generally thought that synthesis of the 3'-end extracistronic leader sequence (approximately 55 nucleotides) initiates the primary transcriptive phase (Leppert *et al.*, 1979; Dethlefsen and Kolakofsky, 1983). Synthesis of the leader sequence terminates at the boundary of the NP gene with reinitiation and synthesis of the NP mRNA. The 5' ends of the mRNAs are capped and methylated (Colnno and Stone, 1975, 1976). The mechanism which allows successive termination of leader synthesis and initiation of mRNA synthesis is unknown. Inspection of Fig. 8 (5'-end nucleotides) shows that the sequences at the leader–NP junction in HPF3 and BPF3 viruses closely resemble the consensus transcription termination sequences and intergenic trinucleotides of these viruses. This feature suggests that leader termination in these two viruses might employ a mechanism analogous to mRNA termination. In contrast with BPF3 and HPF3 viruses, the leader–NP junctions in Sendai virus, mumps virus, and NDV do not contain sequences which resemble transcription termination sequences, although in Sendai virus the trinucleotide AAA is present at the leader–NP junction. This sequence is similar to the GAA intergenic trinucleotides found in this virus (Gupta and Kingsbury, 1984).

Both initiation and termination of mRNA synthesis occur at conserved transcriptional control sequences present at the gene-end boundaries of the NP gene (Gupta and Kingsbury, 1984; Spriggs and Collins, 1986a; Sakai *et al.*, 1987; Kurilla *et al.*, 1985; Ishida *et al.*, 1986; Elango *et al.*, 1988). The consensus termination sequence AUUCUUUUU found at the end of the NP gene and of all the other genes is believed to function as the template for addition of the polyadenylated tail onto the 3' end of the mRNA through some form of transcriptional chattering. Following the addition of the polyadenylated tail onto the NP mRNA, it is believed that the transcription complex passes over the intergenic sequences and reinitiates transcription of the next gene, P. Thus, the intergenic sequences do not ordinarily act as templates for mRNA synthesis.

The mechanism which allows for chattering and polyadenylation, passing over the intergenic sequences, and precise reinitiation of transcription (including capping and methylation of the new mRNA) at the start site of the next gene is unknown. The complicated nature of this process probably plays a role in the attenuation of reinitiation at successive genes due to the release of the transcription complex. The net result of attenuation is that the relative abundance of each mRNA decreases from the 3' to the 5' end of the genomic RNA (Glazier *et al.*, 1977; Carlsen *et al.*, 1985; Moyer *et al.*, 1986). Accordingly, the first gene, *NP*, is transcribed at the highest levels, while the final cistron in the genome, the *L* gene, is transcribed at the lowest. Thus, the characteristic molecular organization of the paramyxovirus genome may function as another level of control for gene expression. Recent studies on the expression of the VSV L protein have provided some evidence that attenuation may play an important role in the replication of this rhabdovirus. Meier *et al.* (1987) have shown that VSV L protein expressed from recombinant cDNA can complement temperature-sensitive polymerase mutants at the restrictive temperature. However, only cells expressing low levels of L protein were able to rescue the defective virus, while cells expressing high levels were not. This result suggests that the normal attenuation of L expression may have a biological function in VSV replication, and by analogy in the replication of all negative-strand viruses.

Polycistronic transcripts are made following aberrant termination and continued transcription into the next gene (Kolakofsky *et al.*, 1974b; Varich *et al.*, 1979; Collins *et al.*, 1982). This form of transcription may be an inherent property of the transcriptive process or may result from the alteration of nucleotide sequences within the gene-end boundaries (Hsu *et al.*, 1985; Spriggs and Collins, 1986a; Galinski *et al.*, 1987a). During this form of transcription the intergenic nucleotides are incorporated into the polycistronic mRNA, and, in addition, the polyadenylate sequences normally introduced by chattering during transcription termination are not present between the contiguous gene sequences (Wilde and Morrison, 1984; Gupta and Kingsbury, 1985a). These results suggest that polyadenylation and transcription termination are coordinated events. The biological significance of this form of transcription, if any, is unknown; it may reflect the inherent low fidelity of paramyxovirus transcription or may provide further means of transcription/translation attenuation.

The sum result of transcription is the production of mRNAs representing each of the viral genes, which then serve as templates for the synthesis of the various viral proteins. The mRNAs produced are generally monocistronic, except for the readthrough polytranscripts described above and the P mRNA in certain *Paramyxoviridae*, which are functionally bicistronic, as discussed above.

There is further complexity in the expression of the *P* genes of SV5 (Thomas *et al.*, 1988), measles virus (Cattaneo *et al.*, 1989), and perhaps generally for all *Paramyxoviridae P* genes. As described above, the *P* gene of SV5 also encodes two proteins, a M_r 44,000 P protein and a M_r 24,000 V protein (Thomas *et al.*, 1988). The two proteins contain identical amino termini, based upon the finding of common tryptic peptides, and are encoded by the same mRNA, as determined by hybrid arrest of translation. Direct dideoxy sequence analysis of the viral gene revealed two overlapping ORFs (encoding proteins of predicted sizes of 222 and 250 amino acids), neither of which was of sufficient length to encode a protein with a molecular mass as great as the P protein (Thomas *et al.*, 1988). However, sequence analysis of a number of recombinant cDNA clones derived from mRNA revealed two types. One set contained a faithful copy of the gene, while the second set contained an additional two nucleotides not present in the viral gene template, which had been introduced into the transcript in a purine-rich region (5′ UUUAAGAGGGGCACC 3′; additional nucleotides underlined). The additional nucleotides allow accession of the second ORF by introducing a frameshift. This functionally fuses the two cistrons and permits synthesis of a "full-length" P protein (392 amino acids).

The addition of two nucleotides at a precise position in the mRNA suggests that transcription may be involved in the generation of the novel mRNA. Support for this notion comes from the gene sequences upstream of the chatter site for the addition of the nontemplated nucleotides (3′ AAAAUUCUCCCC 5′; underlined sequences), which resemble the transcription termination sequence of the HN gene (3′ AAAAUUCUUUUUUUU 5′) of SV5 (Hiebert *et al.*, 1985a; Thomas *et al.*, 1988). Further, this sequence resembles the more general consensus sequence for paramyxovirus transcription termination, AUUCUUUUUU. Together, these results support the notion that aberrant transcription termination and chattering on the four C residues is the mechanism for the introduction of the additional nontemplated residues (see also Chapters 6 and 12).

The use of overlapping reading frames and transcriptional chattering to encode multiple proteins in a single gene reveals a unique robustness to the gene expression of *Paramyxovirus*. Why encode multiple proteins in a single gene? Condensation of the genomic sequences to encode more proteins seems unlikely, since the proteins accessed through the alternate reading frames are relatively small, and viral genomes of greater than 15,000 nucleotides could surely incorporate a few hundred additional nucleotides. More likely, this organization evolved in response to the necessity for the coexpression of these proteins during the replicative cycle. The coexpression of proteins would be useful if they interact or function at a common step in replication. One could easily imagine that the cysteine-rich proteins function in transcriptional regulation of mRNA synthesis (perhaps specifically regulating *P* gene transcription or transcription of some distal gene) or the switching from transcription to replication.

Molecular cloning and sequence analyses have provided the complete molecular structures of the Sendai virus, HPF3 virus, and NDV genomic RNAs. There are no other long open reading frames spanning contiguous genes, and no genes are encoded in the 3′ end of the viral complementary RNA as has been observed in the ambisense genomes described for *Arenavirus* and *Bunyavirus* (Auperin *et al.*, 1984; Ihara *et al.*, 1984).

D. Genome Replication

The switching from transcription to replication in *Paramyxoviridae* is a process about which even the most fundamental aspects are not well understood. A number of studies have investigated these events using the rhabdovirus VSV as a model for nonsegmented negative-strand RNA replication [for reviews see Ball and Werz (1981), Emerson (1985), and Banerjee (1987)]. Although similar studies have addressed paramyxovirus replication, the individual proteins and their role in replication have not been as well characterized as have the analogous VSV proteins.

The shift to replication of the genomic RNA can be viewed as a modification of the normal transcriptive process (see also Chapter 8). Replication requires suppression of the normal termination of leader synthesis at the junction of the *NP* gene, and continuation of transcription across the gene-end boundaries at each successive gene. Antitermination at each of these junctions permits "readthrough" synthesis of a full-length viral complementary or plus RNA (antigenome) from the viral (minus) RNA template. The antigenomic RNA is encapsidated with NP protein and presumably resembles the vRNA nucleocapsid. The antigenomic nucleocapsid can then serve as a complex for the production of progeny vRNA.

Although the vcRNA is of the same polarity as mRNA, the virion-associated and cellular nucleocapsids contain exclusively genomic 50S RNA. (Portner and Kingsbury, 1970; Robinson, 1970; Kolakofsky *et al.*, 1974b; Kingsbury, 1974). This indicates that the encapsidation of the 50S RNA by NP protein requires the presence of the extracistronic leader sequences in the 5' ends of the antigenomes and genomes. Further support for this notion comes from *in vitro* Sendai virus replication studies where incomplete genome synthesis occurs (Portner, 1982; Carlsen *et al.*, 1985). The genomic RNA in these studies appeared as a ladder of micrococcal nuclease-resistant fragments which were smaller than the parental 50S RNA. These results indicate that the RNAs were associated with NP protein during their synthesis and before their completion.

Amplification of the progeny vRNA and subsequent assembly into nucleocapsids serves two functions. First, amplification produces genomic nucleocapsids for incorporation into mature virions, and second, the accumulation of these nucleocapsids provides additional transcriptionally active complexes for further production of mRNAs and subsequent amplification of the proteins required for virion assembly (Choppin and Compans, 1975; Choppin and Scheid, 1980).

During viral replication, viral RNA and viral complementary RNA are never found free of NP protein. Although transcription does not require *de novo* protein synthesis, replication does (Robinson, 1971; Carlsen *et al.*, 1985). These findings indicate that primary viral transcription is dependent upon the availability of preformed proteins, ostensibly of viral origin; however, a role for cellular proteins cannot be ruled out (Hamaguchi *et al.*, 1983; Carlsen *et al.*, 1985; Moyer *et al.*, 1986). Since replication is dependent upon continued protein synthesis, expression of viral genes is a necessary condition of replication.

The best studied *in vitro* systems for paramyxovirus replication have used Sendai virus (Portner, 1982; Carlsen *et al.*, 1985; Moyer *et al.*, 1986) and NDV (Huang *et al.*, 1971; Hamaguchi *et al.*, 1983, 1985). Although these systems can support both transcription and replication, the precise roles of the various proteins are not understood. This has been primarily due to the lack of a simple system which can utilize individually purified proteins. Reconstitution experiments using either purified proteins or proteins produced from recombinant cDNA, similar to studies done with VSV, have yet to be exploited in paramyxoviruses. A further hindrance to studying paramyxovirus transcription may be the need for some infected host cell factor(s) not present, or limiting, in the virion (Hamaguchi et al., 1983; Carlsen et al., 1985).

The minimum complex required for transcription and presumably for replication consists of the genomic viral RNA and the three nucleocapsid-associated viral proteins, NP, P, and L. Fractionation of the nucleocapsids into nucleocapsid cores and purification of P and L by phosphocellulose chromatography provides the three components which when reconstituted will function in replication, albeit at low levels (Hamaguchi et al., 1983). In vitro replication systems employing infected cell lysates generally function well in transcription. In contrast, purified virion nucleocapsids are not as efficient unless the nucleocapsids are supplemented with cell extracts. These results imply that a cellular component(s) may be required for appropriate transcription. Recently, tubulin has been shown to act as a positive transcriptional factor in vitro for both VSV and Sendai virus (Moyer et al., 1986). In addition, anti-tubulin antibody completely inhibited in vitro transcription and replication using extracts prepared from either cell lysates or disrupted virions (Moyer et al., 1986).

Examination of the infected cell-associated transcriptive complexes has shown that NP, P, and probably L become associated with the cytoskeleton soon after they are synthesized (Hamaguchi et al., 1985). Once associated, they remain with the cytoskeleton during assembly into nucleocapsids, subsequent transcription/replication, and incorporation into mature virions (Kingsbury et al., 1978; Hamaguchi et al., 1983, 1985). Since the cytoskeletal microfilaments and microtubules are composed of repeating polymers of G-actin and tubulin, the intimate association of the nucleocapsid complexes with the cytoskeleton supports the positive transcriptional role of tubulin described above for Sendai virus (Moyer et al., 1986).

IV. REFERENCES

Alkhatib, G., and Briedis, D. J., 1986, The predicted primary structure of the measles virus hemagglutinin, Virology 150:479–490.

Amesse, L. A., Pridgen, C. L., and Kingsbury, D. W., 1982, Sendai virus DI RNA species with conserved virus genome termini and extensive internal deletions, Virology 118:17–27.

Andrewes, C. H., Bang, F. B., and Burnet, F. M., 1955, A short description of the myxovirus group (influenza and related viruses), Virology 1:176–184.

Auperin, D. D., Romonowski, V., Galinski, M., and Bishop, D. H. L., 1984, Sequencing studies of Pichinde arenavirus S RNA indicate a novel coding strategy, an ambisense viral S RNA, J. Virol. 52:897–904.

Ball, L. A., and Wertz, G. W., 1981, VSV RNA synthesis: How can you be positive?, Cell 26:143–144.

Banerjee, A. K., 1987, Transcription and replication of rhabdoviruses, Microbiol. Rev. 51:66–87.

Barret, T., Clarke, D. K., Evans, S. A., and Rima, B. K., 1987, The nucleotide sequence of the gene encoding the F protein of canine distemper virus: A comparison of the deduced amino acid sequence with other paramyxoviruses, Virus Res. 8:373–386.

Basle, M. F., Russell, W. C., Goswami, K. K. A., Rebel, A., Giraudon, P., Wild, F., and Filmon, R.,

1985, Paramyxovirus antigens in osteoclasts from Paget's bone tissue detected by monoclonal antibodies, *J. Gen. Virol.* **66:**2103–2110.

Bellini, W. J., Englund, G., Rozenblatt, S., Arnheiter, H., and Richardson, C. D., 1985, Measles virus *P* gene codes for two proteins, *J. Virol.* **53:**908–919.

Bellini, W. J., Englund, G., Richardson, C. D., Rozenblatt, S., and Lazzarini, R. A., 1986, Matrix genes of measles virus and canine distemper virus: Cloning, nucleotide sequences, and deduced amino acid sequences, *J. Virol.* **58:**408–416.

Blok, J., Air, G. M., Laver, W. G., Ward, C. W., Lilley, G. G., Woods, E. F., Roxburgh, C. M., and Inlis, A. S., 1982, Studies on the size, chemical composition and partial sequence of the neuraminidase from type A influenza viruses show that the N-terminal region of NA is not processed and serves to anchor the NA in the viral membrane, *Virology* **119:**109–121.

Blumberg, B. M., Rose, K., Simona, M. G., Roux, L., Giorgi, C., and Kolakofsky, D., 1984, Analysis of the Sendai virus *M* gene and protein, *J. Virol.* **52:**656–663.

Blumberg, B. M., Giorgi, C., Rose, K., and Kolakofsky, D., 1985a, Sequence determination of the Sendai virus fusion protein gene, *J. Gen. Virol.* **66:**37–331.

Blumberg, B., Giorgi, C., Roux, L., Raju, R., Dowling, P, Chollet, A., and Kolakofsky, D., 1985b, Sequence determination of the Sendai virus *HN* gene and its comparison to the influenza virus glycoproteins, *Cell* **41:**269–278.

Blumberg, B. M., Crowley, J. C., Silverman, J. L., Menonna, J., Cook, S. D., and Dowling, P. E., 1988, Measles virus L protein evidences elements of ancestral RNA polymerase, *Virology* **164:**487–497.

Bratt, M. A., and Gallaher, W. R., 1969, Preliminary analysis of the requirements for fusion from within and fusion from without by Newcastle disease virus, *Proc. Natl. Acad. Sci. USA,* **64:**536–543.

Bratt, M. A., and Hightower, L., 1977, Genetics and paragenetic phenomena of paramyxovirus, *in* "Comprehensive Virology," Vol. 9 (H. Fraenkel-Conrat and R. R. Wagner, eds.), pp. 457–533, Plenum Press, New York.

Carlsen, S. R., Peluso, R. W., and Moyer, S. E., 1985, *In vitro* replication of Sendai virus wild-type and defective interfering particle genome RNAs, *J. Virol.* **54:**493–500.

Cattaneo, R., Schmid, A., Eschle, D., Backzo, K., ter Mulen, V., and Billiter, M., 1988, Biased hypermutation and other genetic changes in defective measles viruses in human brain infections, *Cell* **55:**255–265.

Cattaneo, R., Kaelin, K., Baczko, K., and Billeter, M. A., 1989, Measles virus editing provides an additional cysteine-rich protein, *Cell,* **56:**759–764.

Chambers, P., Millar, N. S., and Emmerson, P. T., 1986a, Nucleotide sequence of the gene encoding the fusion glycoprotein of Newcastle disease virus, *J. Gen. Virol.* **67:**2685–2694.

Chambers, P., Millar, N. S., Platt, S. G., and Emmerson, P. T., 1986b, Nucleotide sequence of the gene encoding the matrix protein of Newcastle disease virus, *Nucleic Acids Res.* **14:**9051–9061.

Chanock, R. M., 1956, Association of a new type of cytopathogenic myxovirus with infantile croup, *J. Exp. Med.* **104:**555–576.

Chanock, R. M., 1979, Parainfluenza viruses, *in* "Diagnostic Procedures for Viral, Rickettsial and Chlamydial Infections," 5th ed. (E. H. Lennette and N. I. Schmidt, eds.), pp. 611–632, American Public Health Association, Washington, D.C.

Chanock, R. M., and Parott, R. H., 1965, Acute respiratory disease in infancy and childhood: Present understanding and prospects for prevention, *Pediatrics* **36:**21–39.

Chanock, R. M., Parrott, R. H., Cook, M. K., Andrews, B. E., Bell, J. A., Reichelderfer, T., Kapikian, A. Z., Mastrota, F. M., and Huebner, R. J., 1958, Newly recognized myxoviruses from children with respiratory disease, *N. Eng. J. Med.* **258:**207–213.

Chanock, R. M., Bell, J. A., and Parrott, R. H., 1961, Natural history of parainfluenza infection, *in* "Perspectives in Virology," Vol. 2 (M. Pollard, ed.), pp. 126–139, Burgess Pub. Co., Minneapolis, Minnesota.

Chinchar, V. G., and Portner, A., 1981, Functions of Sendai virus nucleocapsid polypeptides: Enzymatic activities in nucleocapsids following cleavage of polypeptide P by *Staphylococcus aureus* protease V8, *Virology* **109:**59–71.

Choppin, P., and Compans, R., 1975, Reproduction of paramyxoviruses, *in* "Comprehensive Virology," Vol. 4 (H. Fraenkal-Conrat and R. R. Wagner, eds.), pp. 95–178, Plenum Press, New York.

Choppin, P. W., and Scheid, A., 1980, The role of viral glycoproteins in adsorption, penetration and pathogenicity of viruses, *Rev. Infect. Dis.* **2**:40–61.

Citovsky, V., Yanai, P., and Loyter, A., 1986, The use of circular dichroism to study conformational changes induced in Sendai virus envelope glycoproteins: A correlation with the viral fusogenic activity, *J. Biol. Chem.* **261**:2235–2239.

Coelingh, K. J., Winter, C. C., Murphy, B. R., Rice, J. M., Kimball, P. C., Olmstead, R. A., and Collins, P. L., 1986, Conserved epitopes on the hemagglutinin-neuraminidase proteins of human and bovine parainfluenza type 3 viruses: Nucleotide sequence analysis of variants selected with monoclonal antibodies, *J. Virol.* **60**:90–96.

Coelingh, K. L., Winter, C. C., Jorgensen, E. D., and Murphy, B. R., 1987, Antigenic and structural properties of the human parainfluenza type 3 virus: Sequence analysis of variants selected with monoclonal antibodies which inhibit infectivity, hemagglutination, and neuraminidase activities, *J. Virol.* **61**:1473–1477.

Collins, P. L., Wertz, G. W., Ball, L. A., and Hightower, L. E., 1982, Coding assignments of the five smaller mRNAs of Newcastle disease virus, *J. Virol.* **43**:1024–1031.

Collins, P. L., Huang, Y. T., and Wertz, G. W., 1984, Nucleotide sequence of the gene encoding the fusion (F) glycoprotein of human respiratory syncytial virus, *Proc. Natl. Acad. Sci. USA* **81**:7683–7687.

Colonno, R. J., and Stone, H. O., 1975, Methylation of messenger RNA of Newcastle disease virus *in vitro* by a virion-associated enzyme, *Proc. Natl. Acad. Sci. USA* **72**:2611–2615.

Colonno, R. J., and Stone, H. O., 1976, Newcastle disease virus mRNA lacks 2-O-methylated nucleotides, *Nature* **261**:611–614.

Compans, R. W., and Choppin, P. W., 1967, The length of the helical nucleocapsid of Newcastle disease virus, *Virology* **33**:344–346.

Compans, R. W., Holmes, K. V., Dales, S., and Choppin, P. W., 1966, An electron microscopic study of single-cycle infection of chick embryo fibroblasts by influenza virus, *Virology* **30**:411–426.

Côté, M.-J., Storey, D. G., Kang, C. Y., and Dimock, K., 1987, Nucleotide sequence of the coding and flanking regions of the human parainfluenza virus type 3 fusion glycoprotein gene, *J. Gen. Virol.* **68**:1003–1010.

Curran, J. A., Richardson, C., and Kolakofsky, D., 1986, Ribosomal initiation at alternate AUGs on the Sendai virus P/C mRNA, *J. Virol.* **57**:684–687.

Curran, J. A., and Kolakofsky, D., 1987, Identification of an additional Sendai virus non-structural protein encoded by the P/C mRNA, *J. Gen. Virol.* **68**:2515–2519.

Deshpande, K. L., and Portner, A., 1985, Monoclonal antibodies to the P protein of Sendai virus define its structure and role in transcription, *Virology* **140**:125–134.

Dethlefsen, L., and Kolakofsky, D., 1983, *In vitro* synthesis of the nonstructural C protein of Sendai virus, *J. Virol.* **46**:321–324.

Dimock, K., Rud, E. W., and Kang, C. Y., 1986, 3' Terminal sequence of human parainfluenza virus 3 genomic RNA, *Nucleic Acids Res.* **14**:4694.

Elango, N., 1989a, The mumps virus nucleocapsid mRNA sequence and homology among the *Paramyxoviridae* proteins, *Virus Res.* **12**:77–86.

Elango, N., 1989b, Complete nucleotide sequence of the matrix protein mRNA of mumps virus, *Virology* **168**:426–428.

Elango, N., Coligan, J. E., Jambou, R. C., and Venkatesan, S., 1986, Human parainfluenza type 3 virus hemagglutinin-neuraminidase glycoproteins: Nucleotide sequence of mRNA and limited amino acid sequence of the purified protein, *J. Virol.* **57**:481–489.

Elango, N., Varsanyi, T. M., Kovamees, J., and Norrby, E., 1988, Molecular cloning and characterization of six genes, determination of gene order and intergenic sequences and leader sequence of mumps virus, *J. Gen. Virol.* **69**:2893–2900.

Elango, N., Kovamees, J., Varsanyi, T. M., and Norrby, E., 1989, mRNA Sequence and deduced amino acid sequence of the mumps virus small hydrophobic protein gene, *J. Virol.* **63**:1413–1415.

Emerson, S. U., 1985, Rhabdoviruses, in "Virology" (B. N. Fields, ed.), pp. 1119–1132, Raven Press, New York.

Emerson, S. U., and Yu, Y.-H., 1975, Both NS and L proteins are required for in vitro RNA synthesis by vesicular stomatitis virus, *J. Virol.* **15**:1348–1356.

Evans, R. A., 1988, The steroid and thyroid hormone receptor superfamily, *Science* **240**:889–895.

Fields, S., Winter, G., and Brownlee, G. G., 1981, Structure of the neuraminidase gene in human influenza virus A/PR/8/34, *Nature* **290**:213–217.

Galinski, M. S., Mink, M. A., Lambert, D. M., Wechsler, S. L., and Pons, M. W., 1986a, Molecular cloning and sequence analysis of the human parainfluenza 3 virus RNA encoding the nucleocapsid protein, *Virology* **149**:139–151.

Galinski, M. S., Mink, M. A., Lambert, D. M., Wechsler, S. L., and Pons, M. W., 1986b, Molecular cloning and sequence analysis of the human parainfluenza 3 virus mRNA encoding the P and C proteins, *Virology* **154**:46–60.

Galinski, M. S., Mink, M. A., Lambert, D. M., Wechsler, S. L., and Pons, M. W., 1987a, Molecular cloning and sequence analysis of the human parainfluenza 3 virus gene encoding the matrix protein, *Virology* **152**:24–30.

Galinski, M. S., Mink, M. A., Lambert, D. M., Wechsler, S. L., and Pons, M. W., 1987b, Molecular cloning and sequence analysis of the human parainfluenza 3 virus genes encoding the surface glycoproteins, F and HN, *Virus Res.* **12**:169–180.

Galinski, M. S., Mink, M. A., and Pons, M. W., 1988, Molecular cloning and sequence analysis of the human parainfluenza 3 virus gene encoding the L protein, *Virology* **165**:499–510.

Giorgi, C., Blumberg, B. M., and Kolakofsky, D., 1983, Sendai virus contains overlapping genes expressed from a single mRNA, *Cell* **35**:829–836.

Gitman, A. G., and Loyter, A., 1984, Construction of fusogenic vesicles bearing specific antibodies: Targeting of reconstituted Sendai virus envelopes towards neuraminidase-treated human erythrocytes, *J. Biol. Chem.* **259**:9813–9820.

Giuffre, R. M., Tovell, D. R., Kay, C. M., and Tyrrell, D. L. J., 1982, Evidence for an interaction between the membrane protein of a paramyxovirus and actin, *J. Virol.* **42**:963–968.

Glazier, K., Raghow, R., and Kingsbury, D. W., 1977, Regulation of Sendai virus transcription: Evidence for a single promoter *in vivo, J. Virol.* **21**:863–871.

Glickman, R., Syddal, R., Iorio, R., Sheehan, J., and Bratt, M., 1988, Quantitative basic residue requirements in the cleavage activation site of the fusion glycoprotein as a determinant of virulence of Newcastle disease virus, *J. Virol.* **62**:354–356.

Goswami, K. K. A., Cameron, K. R., Russel, W. C., Lange, L. S., and Mitchel, D. N., 1984, Evidence for the persistence of paramyxoviruses in human bone marrows, *J. Gen. Virol.* **65**:1881–1888.

Gupta, K. C., and Kingsbury, D. W., 1984, Complete sequences of the intergenic and mRNA start signals in the Sendai virus genome: Homologies with the genome of vesicular stomatitis virus, *Nucleic Acids Res.* **12**:3829–3841.

Gupta, K. C., and Kingsbury, D. W., 1985a, Polytranscripts of Sendai virus do not contain intervening polyadenylate sequences, *Virology* **141**:102–109.

Gupta, K. C., and Kingsbury, D. W., 1985b, Translational modulation *in vitro* of a eucaryotic viral mRNA encoding overlapping genes: Ribosome scanning and potential roles of conformational changes in the P/C mRNA of Sendai virus, *Biochem. Biophy. Res. Commun.* **131**:91–97.

Gupta, K. C., and Patwardhan, S., 1988, ACG, the initiator codon for a Sendai virus protein, *J. Biol. Chem.* **263**:8553–8556.

Hall, W. W., and Choppin, P. W., 1979, Evidence for lack of synthesis of the M polypeptide of measles virus in brain cells in subacute sclerosing panencephalitis, *Virology* **99**:443–447.

Hamaguchi, M., Yoshida, T., Nishikawa, K., Naruse, H., and Nagai, Y., 1983, Transcriptive complex of Newcastle disease virus. I. Both L and P proteins are required to constitute an active complex, *Virology* **128**:105–117.

Hamaguchi, M., Nishikawa, K., Toyoda, T., Yoshida, T., Hanaichi, T., and Nagai, Y., 1985, Transcriptive complex of Newcastle disease virus. II. Structural and functional assembly associated with the cytoskeletal framework, *Virology* **147**:295–308.

Haywood, A. M., 1974, Characteristics of Sendai virus receptors in a model membrane, *J. Mol. Biol.* **83**:427–436.

Haywood, A. M., 1975, Phagocytosis of Sendai virus by model membranes, *J. Gen. Virol.* **29**:63–68.

Heggeness, M. H., Smith, P. R., and Choppin, P. W., 1982, *In vitro* assembly of the nonglycosylated membrane protein (M) of Sendai virus, *Proc. Natl. Acad. Sci. USA* **79**:443–447.

Herrler, A., and Compans, R., 1982, Synthesis of mumps virus polypeptides in infected Vero cells, *Virology* **119**:430–438.

Hidaka, Y., Kanda, T., Iwasaki, K., Nomoto, A., Shioda, T., and Shibuta, H., 1984, Nucleotide sequence of a Sendai virus genome region covering the entire M gene and the 3' proximal 1013 nucleotides of the F gene, *Nucleic Acids Res.* **12**:7965–7973.

Hiebert, S. W., Paterson, R. G., and Lamb, R. B., 1985a, Hemagglutinin-neuraminidase protein of the paramyxovirus simian virus 5: Nucleotide sequence of the mRNA predicts an N-terminal membrane anchor, *J. Virol.* **54**:1–6.

Hiebert, S. W., Paterson, R. G., and Lamb, R. B., 1985b, Identification and predicted sequence of a previously unrecognized small hydrophobic protein, SH, of the paramyxovirus simian virus 5, *J. Virol.* **55**:744–751.

Hiebert, S. W., Richardson, C. D., and Lamb, R. A., 1988, Cell surface expression and orientation in membranes of the 44-amino-acid SH protein of simian virus 5, *J. Virol.* **62**:2347–2357.

Holmgren, J., Svennerholm, L., Elwing, H., Fredman, P., and Strannegard, O., 1980, Sendai virus receptor: Proposed recognition structure based on binding to plastic-adsorbed gangliosides, *Proc. Natl. Acad. Sci. USA* **77**:1947.

Homma, M., and Ohuchi, M., 1973, Trypsin action on the growth of Sendai virus in tissue culture cells. III. Structural difference of Sendai virus grown in eggs and tissue culture cells, *J. Virol.* **12**:1457–1465.

Hosaka, Y., Kitano, H., and Ikeguchi, S., 1966, Studies on the pleomorphism of HVJ virions, *Virology* **29**:205–221.

Hsu, C.-H., and Kingsbury, D. W., 1982, Topography of phosphate residues in Sendai virus proteins, *Virology* **120**:225–234.

Hsu, C. H., Re, G., Gupta, K. C., Portner, A., and Kingsbury, D. W., 1985, Expression of Sendai virus defective-interfering genomes with internal deletions, *Virology* **146**:38–49.

Huang, A. S., Baltimore, D., and Bratt, M. A., 1971, Ribonucleic acid polymerase in virions of Newcastle disease virus: Comparison with the vesicular stomatitis virus polymerase, *J. Virol.* **7**:389–394.

Huang, R. T. C., Rott, R., Wahn, K., Klenk, H.-D., and Kohama, T., 1980, The function of the neuraminidase in membrane fusion induced by myxoviruses, *Virology* **107**:313–319.

Iapalucci, S., Lopez, R., Rey, O., Lopez, N., Franze-Fernandez, M. T., Cohen, G. N., Lucero, M., Ochoa, A., and Zakin, M. M., 1989, Tacaribe virus L gene encodes a protein of 2210 amino acid residues, *Virology* **170**:40–47.

Ihara, T., Akashi, H., and Bishop, D. H. L., 1984, Novel coding strategy (ambisense genomic RNA) revealed by sequence analysis of Punta Toro phlebovirus S RNA, *Virology* **136**:293–306.

Ishida, N., and Homma, M., 1978, Sendai virus, *in* "Advances in Virus Research," Vol. 123 (K. Maramorosch, F. A. Murphy, and A. J. Shatkin, eds.), pp. 349–383, Academic Press, New York.

Ishida, N., Taira, H., Omata, T., Mizumoto, K., Hattori, S., Iwasaki, K., and Kawakita, M., 1986, Sequence of 2,617 nucleotides from the 3' end of Newcastle disease virus genome RNA and the predicted amino acid sequence of viral NP protein, *Nucleic Acids Res.* **14**:6551–6564.

Jambou, R. C., Narayanasamy, E., Sundararajan, V., and Collins, P. L., 1986, Complete sequence of the major nucleocapsid protein gene of human parainfluenza type 3 virus: Comparison with other negative strand viruses, *J. Gen. Virol.* **67**:2543–2548.

Jorgensen, E. D., Collins, P. L., and Lomedico, P. T., 1987, Cloning and nucleotide sequence of Newcastle disease virus hemagglutinin-neuraminidase mRNA: Identification of a putative sialic acid binding site, *Virology* **156**:12–24.

Kamer, G., and Argos, P., 1984, Primary structural comparison of RNA-dependent polymerases from plant, animal and bacterial viruses, *Nucleic Acids Res.* **12**:7269–7282.

Kingsbury, D. W., 1974, The molecular biology of paramyxoviruses, *Med. Microbiol. Immunol.* **160**:73–83.

Kingsbury, D. W., Brat, M. A., Choppin, P. W., Hanson, R. P., Hosaka, Y., ter Meulen, V., Norbby, E., Plowright, W., Rott, R., and Wunner, W. H., 1978a, *Paramyxoviridae, Intervirology* **10**:137–152.

Kingsbury, D. W., Hsu, H., and Murti, K. G., 1978b, Intracellular metabolism of Sendai virus nucleocapsids, *Virology* **91**:86–94.

Klenk, H.-D., and Choppin, P. W., 1969a, Chemical composition of the parainfluenza virus SV5, *Virology* **37**:155–157.

Klenk, H.-D., and Choppin, P. W., 1969b, Lipids of plasma membranes of monkey and hamster kidney cells and of parainfluenza virions grown in these cells, *Virology* **38**:255–268.

Klenk, H.-D., and Choppin, P. W., 1970, Plasma membrane lipids and parainfluenza virus assembly, *Virology* **40**:939–947.

Kohama, T., Shimizu, K., and Ishida, N., 1978, Carbohydrate composition of the envelope glycoproteins of Sendai virus, *Virology* **90**:226–234.

Kolakofsky, D. E., de la Tour, E. B., and Delius, H., 1974a, Molecular weight determination of Sendai and Newcastle disease virus RNA, *J. Virol.* **13**:261–268.

Kolakofsky, D., de la Tour, E. B., and Brushi, A., 1974b, Self-annealing of Sendai virus RNA, *J. Virol.* **14**:33–39.

Kövamees, J., Norrby, E., and Elango, N., 1989, Complete nucleotide sequence of the hemagglutinin-neuraminidase (HN) mRNA of mumps virus and comparison of paramyxovirus HN proteins, *Virus Res.* **12**:87–96.

Kurilla, M. G., Stone, H. O., and Keene, J. D., 1985, RNA sequence and transcriptional properties of the 3' end of the Newcastle disease virus genome, *Virology* **145**:203–212.

Lamb, R. A., and Choppin, P. W., 1977a, The synthesis of Sendai virus polypeptides in infected cells. II. Intracellular distribution of polypeptides, *Virology* **81**:371–381.

Lamb, R. A., and Choppin, P. W., 1977b, The synthesis of Sendai virus polypeptides in infected cells. III. Phosphorylation of polypeptides, *Virology* **81**:382–397.

Lamb, R. W., and Choppin, P. W., 1978, Determination by peptide mapping of the unique polypeptides in Sendai virions and infected cells, *Virology* **84**:469–478.

Lamb, R. A., Mahy, B. W. J., and Choppin, P. W., 1976, The synthesis of Sendai virus polypeptides in infected cells, *Virology* **69**:116–131.

Leppert, M., Rittenhouse, L., Perrault, J., Summers, D. F., and Kolakofsky, D., 1979, Plus and minus strand leader RNAs in negative strand virus-infected cells, *Cell* **18**:735–747.

Luk, D., Sanchez, A., and Banerjee, A. K., 1986, Messenger RNA encoding the phosphoprotein (P) of human parainfluenza virus 3 is bicistronic, *Virology* **153**:318–325.

Luk, D., Masters, P. S., Sanchez, A., and Banerjee, A. K., 1987, Complete nucleotide sequence of the matrix protein mRNA and three intergenic junctions of human parainfluenza virus type 3, *Virology* **156**:189–192.

Markwell, M. A. K., and Fox, C. E., 1980, Protein–protein interactions within paramyxoviruses identified by native disulfide bonding or reversible chemical crosslinking, *J. Virol.* **33**:152–166.

Markwell, M. A. K., Svennerholm, L., and Paulson, J. C., 1981, Specific gangliosides function as receptors for Sendai virus, *Proc. Natl. Acad. Sci. USA* **78**:5406–5410.

Markwell, M. A. K., Fredman, P., and Svennerholm, L., 1984, Receptor ganglioside content of three hosts for Sendai virus. MDBK, HeLa, and MDCK cells, *Biochim. Biophys. Acta* **775**:7–16.

Markwell, M. A. K., Portner, A., and Schwartz, A. C., 1985, An alternate route of infection for viruses: Entry by means of the asialoglycoprotein receptor of a Sendai virus mutant lacking its attachment protein, *Proc. Natl. Acad. Sci. USA* **82**:978–982.

Marx, P. A., Portner, A., and Kingsbury, D. W., 1974, Sendai virion transcriptase complex: Polypeptide composition and inhibition by virion envelope proteins, *J. Virol.* **13**:107–112.

Matsumoto, T., 1982, Assembly of paramyxoviruses, *Microbiol. Immunol.* **26**:285–320.

Matthews, R. E. F., 1982, Classification and nomenclature of viruses, *Intervirology* **17**:104–105.

McGinnes, L., and Morrison, T., 1986, Nucleotide sequence of the gene encoding the Newcastle disease virus fusion protein and comparisons of paramyxovirus fusion protein sequence, *Virus Res.* **5**:343–356.

McGinnes, L., and Morrison, T., 1987, The nucleotide sequence of the gene encoding the Newcastle disease virus membrane protein and comparison of membrane protein sequences, *Virology* **156**:221–228.

McGinnes, L., Wilde, A., and Morrison, T., 1987, Nucleotide sequence of the gene encoding the Newcastle disease virus hemagglutinin-neuraminidase protein and comparisons of paramyxovirus hemagglutinin-neuraminidase protein sequences, *Virus Res.* **7**:187–202.

McGinnes, L., McQuain, C., and Morrison, T., 1988, The P protein and the nonstructural 38K and 29K proteins of Newcastle disease virus are derived from the same open reading frame, *Virology* **164**:256–264.

Meier, E., Harmison, G. G., and Schubert, M., 1987, Homotypic and heterotypic exclusion of vesicular stomatitis virus replication by high levels of recombinant polymerase protein L, *J. Virol.* **61**:3133–3142.

Merz, D. C., and Wolinsky, J. S., 1981, Biochemical features of mumps virus neuraminidases and their relationship with pathogenicity, *Virology* **114**:218–227.

Merz, D. C., and Wolinsky, J. S., 1983, Conversion of nonfusing mumps virus infections to fusing infections by selective proteolysis of the HN glycoprotein, *Virology* **131**:328–340.

Merz, D. C., Scheid, A., and Choppin, P. W., 1980, Importance of antibodies to the fusion glycoprotein of paramyxoviruses in the prevention and spread of infection, *J. Exp. Med.* **151**:275–288.

Merz, D. C., Scheid, A., and Choppin, P. W., 1981, Immunological studies of the functions of paramyxovirus glycoproteins, *Virology* **109**:94–105.

Merz, D. C., Server, A. C., Waxham, M. N., and Wolinsky, J. S., 1983, Biosynthesis of mumps virus F glycoprotein: Non-fusing strains efficiently cleave the F glycoprotein precursor, *J. Gen. Virol.* **64**:1457–1467.

Millar, N., Chambers, P., and Emmerson, P., 1986, Nucleotide sequence analysis of the hemmagglutinin-neuraminidase gene of Newcastle disease virus, *J. Gen. Virol.* **67**:1917–1927.

Miura, N., Uchida, T., and Okada, Y., 1982, HVS (Sendai virus)-induced envelope fusion and cell fusion are blocked by monoclonal anti-HN protein antibody that does not inhibit hemagglutination activity of HVJ, *Exp. Cell Res.* **141**:409–420.

Moran, E., and Mathews, M. B., 1987, Multiple functional domains in the adenovirus E1A gene, *Cell* **48**:177–178.

Morgan, E. M., and Kingsbury, D. W., 1984, Complete sequence of the Sendai virus NP gene from a cloned insert, *Virology* **135**:279–287.

Morgan, E. M., and Rakestraw, K. M., 1986, Sequence of the Sendai virus L gene: Open reading frames upstream of the main coding region suggest that the gene may be polycistronic, *Virology* **154**:31–40.

Morgan, E. M., and Rapp, F., 1977, Measles virus and its associated diseases, *Bacteriol. Rev.* **41**:636–666.

Morrison, T. G., 1988, Structure, function, and intracellular processing of paramyxovirus membrane proteins, *Virus Res.* **10**:113–136.

Moyer, S. A., Baker, S. C., and Lessard, J. L., 1986, Tubulin: A factor necessary for the synthesis of both Sendai virus and vesicular stomatitis virus RNAs, *Proc. Natl. Acad. Sci. USA* **83**:5405–5409.

Nagai, Y., and Klenk, H.-D., 1977, Activation of precursors to both glycoproteins of Newcastle disease virus by proteolytic cleavage, *Virology* **77**:125–134.

Nagai, Y., Klenk, H.-D., and Rott, R., 1976, Proteolytic cleavage of the viral glycoproteins and its significance for the virulence of Newcastle disease virus, *Virology* **72**:494–508.

Obijeski, J. G., Bishop, D. H. L., Palmer, E. L., and Murphy, F. A., 1976, Segmented genome and nucleocapsid of La Crosse virus, *J. Virol.* **20**:664–675.

Ozawa, M., Asano, A., and Okada, Y., 1979, Biological activities of glycoproteins of HVJ (Sendai virus) studied by reconstitution of hybrid envelope and by concanavalin A-mediated binding: A new function of HANA protein and structural requirement for F protein in hemolysis, *Virology* **99**:197–202.

Paterson, R. G., Harris, T. J. R., and Lamb, R. A., 1984a, Analysis and gene assignment of mRNAs of a paramyxovirus, simian virus 5, *Virology* **138**:310–323.

Paterson, R. G., Harris, T. J. R., and Lamb, R. A., 1984b, Fusion protein of the paramyxovirus simian virus 5: Nucleotide sequence of mRNA predicts a highly hydrophobic glycoprotein, *Proc. Natl. Acad. Sci. USA* **81**:6706–6710.

Paterson, R. G., Hiebert, S. W., and Lamb, R. A., 1985, Expression at the cell surface of biologically active fusion and hemagglutinin/neuraminidase proteins of the paramyxovirus simian virus 5 from cloned cDNA, *Proc. Natl. Acad. Sci. USA* **82**:7520–7524.

Paterson, R. G., Shaughnessy, M. A., and Lamb, R. A., 1989, Analysis of the relationship between cleavability of a paramyxovirus fusion protein and length of the connecting peptide, *J. Virol.* **63**:1293–1301.

Patwardhan, S., and Gupta, K. C., 1988, Translation initiation potential of the 5' proximal AUGs of the polycistronic P/C mRNA of Sendai virus, *J. Biol. Chem.* **263**:4907–4913.

Paulson, J. C., Sadler, F. G., and Hill, R. L., 1979, Restoration of specific myxovirus receptors to asialoerythrocytes by incorporation of sialic acid with pure sialyltransferases, *J. Biol. Chem.* **254**:2120–2124.

Peeples, M. E., and Brat, M. A., 1984, Mutation in the matrix protein of Newcastle disease virus can result in decreased fusion glycoprotein incorporation into particles and decreased infectivity, *J. Virol.* **51**:81–90.

Pettersson, R., and von Bonsdorff, C. H., 1975, Ribonucleoproteins of Uukuniemi virus are circular, *J. Virol.* **15**:386–392.

Portner, A., 1982, Synthesis of message and genome RNAs *in vitro* by Sendai virus infected cell nucleocapsids, *J. Gen. Virol.* **60**:67–75.

Portner, A., 1984, Monoclonal antibodies as probes of the antigenic structure and functions of Sendai virus glycoproteins, in "Nonsegmented Negative Strand Viruses" (D. H. L. Bishop and R. W. Compans, eds.), pp. 345–350, Academic Press, Orlando, Florida.

Portner, A., and Kingsbury, D. W. 1970, Complementary RNAs in paramyxovirions and paramyxovirus-infected cells, *Nature* **228**:1196–1197.

Portner, A., and Murti, K. G., 1986, Localization of P, NP, and M proteins on Sendai virus nucleocapsid using immunogold labeling, *Virology* **150**:469–478.

Portner, A., Scroggs, R. A., and Naeve, C. W., 1987, The fusion glycoprotein of Sendai virus: Sequence analysis of an epitope involved in fusion and virus neutralization, *Virology* **157**:556–559.

Portner, A., Murti, K. G., Morgan, E., and Kingsbury, D. W., 1988, Antibodies against Sendai virus L protein: Distribution of the protein in nucleocapsids revealed by immunoelectron microscopy, *Virology* **163**:236–239.

Prinoski, K., Cote, M.-J., Kang, C. Y., and Dimock, K., 1987, Nucleotide sequence of the human parainfluenza virus 3 matrix protein gene, *Nucleic Acids Res.* **15**:3181.

Raghow, R., and Kingsbury, D. W., 1979, Protein–RNA contacts in Sendai virus nucleocapsids revealed by photo-crosslinking, *Virology* **98**:267–271.

Raghow, R., Kingsbury, D. W., Portner, A., and George, S., 1979, Topography of a flexible ribonucleoprotein helix: Protein–protein contacts in Sendai virus nucleocapsids, *J. Virol.* **30**:701–710.

Ray, R., Galinski, M. S., and Compans, R. W., 1989, Expression of the fusion glycoprotein of human parainfluenza type 3 virus in insect cells by a recombinant baculovirus and analysis of its immunogenic property, *Virus Res.* **12**:169–180.

Re, G. G., and Kingsbury, D. W., 1986, Nucleotide sequences that affect replicative and transcriptional efficiencies of Sendai virus deletion mutants, *J. Virol.* **58**:578–582.

Re, G. G., Gupta, K. C., and Kingsbury, D. W., 1983a, Genomic and copyback 3' termini in Sendai virus defective interfering RNA species, *J. Virol.* **45**:659–664.

Re, G. G., Gupta, K. C., and Kingsbury, D. W. 1983b, Sequence of the 5' end of the Sendai virus genome and its variable representation in complementary form at the 3' ends of copy-back defective interfering RNA species: Identification of the L gene terminus, *Virology* **130**:390–396.

Richardson, C., and Choppin, P. W., 1983, Oligopeptides that specifically inhibit membrane fusion by paramyxoviruses: Studies on the site of action, *Virology* **131**:518–532.

Richardson, C. D., Scheid, A., and Choppin, P. W., 1980, Specific inhibition of paramyxovirus and myxovirus replication by oligopeptides with amino acid sequences similar to those at the N-termini of the F_1 or HA_2 viral polypeptides, *Virology* **105**:205–222.

Richardson, C., Hull, D., Greer, P., Hasel, K., Berkovich, A., Englund, G., Bellini, B., Rima, B., and Lazzarini, R., 1986, The nucleotide sequence of the mRNA encoding the fusion protein of measles virus (Edmonston strain): A comparison of fusion proteins from several different paramyxoviruses, *Virology* **155**:508–523.

Robinson, W. S., 1970, Self-annealing of subgroup 2 myxovirus RNAs, *Nature* **225**:944–945.

Robinson, W. S., 1971, Sendai virus RNA synthesis and nucleocapsid formation in the presence of cyclohexamide, *Virology* **44**:494–502.

Roux, L., Beffy, P., and Portner, A., 1984, Restriction of cell surface expression of Sendai virus hemagglutinin-neuraminidase glycoprotein correlates with its higher instability in persistently and standard plus defective interfering virus infected BHK-21 cells, *Virology* **138**:118–128.

Roux, L., and Waldvogel, F. A., 1982, Instability of the viral M protein in BHK-21 cells persistently infected with Sendai virus, *Cell* **28**:293–302.

Rozenblatt, S., Eizenberg, O., Englund, G., and Bellini, W. J., 1985, Cloning and characterization of DNA complementary to the canine distemper virus mRNA encoding matrix, phosphoprotein, and nucleocapsid protein, *J. Virol.* **53**:691–694.

Ryan, K. W., and Kingsbury, D. W., 1988, Carboxyl-terminal region of Sendai virus P protein is required for binding to viral nucleocapsids, *Virology* **167**:106–112.

Sakai, Y., Suzu, S., Shioda, T., and Shibuta, H., 1987, Nucleotide sequence of the bovine parainfluenza 3 virus genome: Its 3' end and the genes of NP, P, C and M proteins, *Nucleic Acids Res.* **15**:2927–2945.

Salvato, M., Shimomaye, E., and Oldstone, M. B. A., 1989, The primary structure of the lymphocytic choriomenigitis virus *L* gene encodes a putative RNA polymerase, *Virology* **169**:377–384.

Sanchez, A., Banerjee, A. K., Furuichi, Y., and Richardson, M. A., 1986, Conserved structures among the nucleocapsid proteins of the *Paramyxoviridae:* Complete nucleotide sequence of the human parainfluenza virus type 3 NP mRNA, *Virology* **152**:171–180.

Satake, M., Coligan, J. E., Elango, N., Norrby, E., and Venkatesan, S., 1985, Respiratory syncytial virus envelope glycoprotein (G) has a novel structure, *Nucleic Acids Res.* **13**:7795–7812.

Scheid, A., and Choppin, P. W., 1974, Identification of biological activities of paramyxovirus glycoproteins. Activation of cell fusion, hemolysis and infectivity by proteolytic cleavage of an inactive precursor protein of Sendai virus, *Virology* **57**:475–490.

Scheid, A., and Choppin, P. W., 1977, Two disulfide-linked polypeptide chains constitute the active F protein of paramyxoviruses, *Virology* **80**:54–66.

Scheid, A., Graves, M. C., Silver, S. M., and Choppin, P. W., 1978, Studies on the structure and function of paramyxovirus glycoproteins, *in* "Negative Strand Viruses and the Host Cell" (B. W. J. Mahy and R. D. Barry, eds.), pp. 181–193, Academic Press, London.

Schubert, M., Harmison, G. G., and Meier, E., 1984, Primary structure of the vesicular stomatitis virus polymerase (*L*) gene: Evidence for a high frequency of mutations, *J. Virol.* **51**:505–514.

Sefton, B. M., and Buss, J. E., 1987, The covalent modification of eukaryotic proteins with lipid, *J. Cell Biol.* **104**:1449–1453.

Shibuta, H., Kanda, T., Hazama, A., Adachi, A., and Matumoto, M., 1981, Parainfluenza 3 virus: Plaque-type variants lacking neuraminidase activity, *Infect. Immunol.* **34**:262–267.

Shibuta, H., Nozawa, A., Shioda, T., and Kanda, T., 1983, Neuraminidase activity and syncytial formation of variants of parainfluenza 3 virus, *Infect. Immunol.* **41**:780–788.

Shimizu, K., and Ishida, N., 1975, The smallest protein of Sendai virus: Its candidate function of binding nucleocapsid to envelope, *Virology* **67**:427–436.

Shioda, T., Hidaka, Y., Kanda, T., Shibuta, H., Nomoto, A., and Iwasaki, K., 1983, Sequence of 3,687 nucleotides from the 3' end of Sendai virus genome RNA and the predicted amino acid sequences of viral NP, P, and C proteins, *Nucleic Acids Res.* **11**:7217–7330.

Shioda, T., Iwasaki, K., and Shibuta, H., 1986, Determination of the complete nucleotide sequence of the Sendai virus genome RNA and the predicted amino acid sequences of the F, HN and L proteins, *Nucleic Acids Res.* **14**:1545–1563.

Spriggs, M. K., and Collins, P. L., 1986a, Human parainfluenza virus type 3: Messenger RNAs, polypeptide coding assignments, intergenic sequences, and genetic map, *J. Virol.* **59**:646–654.

Spriggs, M. K., and Collins, P. L., 1986b, Sequence analysis of the P and C protein genes of human parainfluenza virus type 3: Patterns of amino acid sequence homology among paramyxovirus proteins, *J. Gen. Virol.* **67**:2705–2719.

Spriggs, M. K., Olmstead, R. A., Venkatesan, S., Coligan, J. E., and Collins, P. L. 1986, Fusion glycoprotein of human parainfluenza virus type 3: Nucleotide sequence of the gene, direct identification of the cleavage-activation site, and comparison with other paramyxoviruses, *Virology* **152**:241–251.

Spriggs, M. K., Johnson, P. R., and Collins, P. L., 1987, Sequence analysis of the matrix protein gene of human parainfluinza virus type 3: Extensive sequence homology among paramyxoviruses, *J. Gen. Virol.* **69**:1491–1497.

Steck, T. L., 1974, The organization of proteins in the human red blood cell membrane. A review, *J. Cell Biol.* **62**:1–19.

Stone, H. O., Kingsbury, D. W., and Darlington, R. W., 1972, Sendai virus-induced transcriptase from infected cells: Polypeptides in the transcriptive complex, *J. Virol.* **10**:1037–1043.

Storey, D. G., Côté, M.-J., Dimock, K., and Kang, C. Y., 1987, Nucleotide sequence of the coding and flanking regions of the human parainfluenza virus 3 hemagglutinin-neuraminidase gene: Comparison with other paramyxoviruses, *Intervirology* **27**:69–80 (1987).

Suzu, S., Sakai, Y., Shioda, T., and Shibuta, H., 1987, Nucleotide sequences of the bovine parainfluenza 3 virus genome: The genes of the F and HN glycoproteins, *Nucleic Acids Res.* **15**:2945–295.

Suzuki, Y., Morioka, T., and Matsumoto, M., 1980, Action of ortho- and paramyxovirus neuraminidase on gangliosides. Hydrolysis of ganglioside GM1 by Sendai virus neuraminidase, *Biochim. Biophys. Acta* **619**:632–639.

Taira, H., Kanda, T., Omata, T., Shibuta, H., Nomoto, A., and Iwasaki, K., 1987, Interferon induction by transfection of Sendai virus C gene cDNA, *J. Virol.* **61**:625–628.

Takeuchi, K., Hishiyama, M., Yamada, A., and Sugiura, A., 1988, Molecular cloning and sequence analysis of the mumps virus gene encoding the P protein: Mumps virus P gene is monocistronic, *J. Gen. Virol.* **69**:2043–2049.

Thomas, S., Lamb, R. A., and Paterson, R. G., 1988, Two mRNAs that differ by two nontemplated nucleotides encode the amino coterminal proteins P and V of the paramyxovirus SV5, *Cell* **54**:891–902.

Thompson, S. D., and Portner, A., 1987, Localization of functional sites on the hemagglutinin-neuraminidase glycoprotein of Sendai virus by sequence analysis of antigenic and temperature-sensitive mutants, *Virology* **160**:1–8.

Tordo, N., Poch, O., Ermine, A., Keith, G., and Rougeon, F., 1988, Completion of the rabies virus genome sequence determination: Highly conserved domains among the L (polymerase) proteins of unsegmented negative-strand RNA viruses, *Virology* **165**:565–576.

Toyoda, T., Sakaguchi, T., Imai, K., Inocencio, N. M., Hamaguchi, M., and Nagai, Y., 1987, Structural comparison of the cleavage-activation site of the fusion glycoprotein between virulent and avirulent strains of Newcastle disease virus, *Virology* **158**:242–247.

Tsukiyama, K., Yoshikawa, Y., and Yamanouchi, K., 1988, Fusion glycoprotein (F) of rinderpest virus: Entire nucleotide sequence of the F mRNA, and several features of the F protein, *Virology* **164**:523–530.

Tsurudome, M., Yamada, A., Hishiyama, M., and Ito, Y., 1986, Monoclonal antibodies against the glycoprotein of mumps virus: Fusion and inhibition by anti-HN monoclonal antibody, *J. Gen. Virol.* **67**:2259–2265.

Tuffereau, C., and Roux, L., 1988, Direct effects of Sendai virus DI particles on virus budding and on M protein fate and stability, *Virology* **162**:417–426.

Van Wyke Coelingh, K. L., Winter, C., and Murphy, B. R., 1985, Antigenic variation in the hemagglutinin-neuriminidase protein of human parainfluenza type 3 virus, *Virology* **143**:569–582.

Van Wyke Coelingh, K. L., Winter, C. C., Jorgensen, E. D., and Murphy, B., 1987, Antigenic and structural properties of the hemagglutinin-neuraminidase glycoprotein of human parainfluenza virus type 3: Sequence analysis of variants selected with monoclonal antibodies which inhibit infectivity, hemagglutination, and neuraminidase activities, *J. Virol.* **61**:1473–1477.

Varich, N. L., Lukashevich, I. S., and Kaverin, N. V., 1979, Newcastle disease virus-specific RNA: An analysis of 24S and 35S RNA transcripts, *Acta Virol.* **23**:273–283.

Veit, M., Schmidt, M. F. G., and Rott, R., 1989, Different palmitoylation of paramyxovirus glycoproteins, *Virology* **168**:173–176.

Vidal, S., Curran, J., Orvell, C., and Kolakofsky, D., 1988, Mapping of monoclonal antibodies to the Sendai virus P protein and the location of its phosphates, *J. Virol.* **62**:2200–2203.

Waterson, A. P., 1962, Two kinds of myxoviruses, *Nature* **193**:1163–1164.

Waxham, M. N., and Wolinsky, J. S., 1986, A fusing mumps virus variant selected from a nonfusing parent with neuraminidase inhibitor 2-deoxy-2,3-dehydro-*N*-acetylneuraminic acid, *Virology* **151**:286–295.

Waxham, M. N., Merz, D., and Wolinsky, J. S., 1986, Intracellular maturation of mumps virus hemagglutinin-neuraminidase glycoprotein: Conformation changes detected with monoclonal antibodies, *J. Virol.* **59**:392–400.

Waxham, N. M., Server, A. C., Goodman, H. M., and Wolinsky, J. S., 1987, Cloning and sequencing of the mumps virus fusion protein, *Virology* **159**:381–388.

Waxham, M. N., Aronowski, J., Server, A. C., Wolinsky, J. S., Smith, J. A., and Goodman, H. M., 1988, Sequence determination of the mumps virus HN gene, *Virology* **164**:318–325.

Wechsler, S. L., and Meissner, H. C., 1981, Measles and SSPE viruses: Similarities and differences, *Prog. Med. Virol.* **28:**64–95.

Wechsler, S. L., Rustigian, R., Stallcup, K. C., Byers, K. B., Winston, S. H., and Fields, B. N., 1979, Measles virus specified polypeptide synthesis in two persistently infected HeLa cell lines, *J. Virol.* **31:**677–684.

Wechsler, S. L., Lambert, D. M., Galinski, M. S., Heineke, B. E., and Pons, M. W., 1985a, Human parainfluenza virus 3: Purification and characterization of subviral components, viral proteins and viral RNA, *Virus Res.* **3:**339–351.

Wechsler, S. L., Lambert, D. M., Galinski, M. S., and Pons, M. W., 1985b, Intracellular synthesis of human parainfluenza type 3 virus-specified polypeptides, *J. Virol.* **54:**661–664.

Wechsler, S. L., Lambert, D. M., Galinski, M. S., Mink, M. A., Rochovansky, O., and Pons, M. W., 1987, Immediate persistent infection by human parainfluenza virus 3: Unique fusion properties of the persistently infected cells, *J. Gen. Virol.* **68:**1737–1748.

Welliver, R., Wong, D. T., Choi, T.-S., and Ogra, P. L., 1982, Natural history of parainfluenza virus infection in childhood, *Pediatrics* **101:**180–187.

Wertz, G., Collins, P., Huang, Y., Gruber, C. Levine, S., and Ball, L., 1985, Nucleotide sequence of the G protein gene of human respiratory syncytial virus reveals an unusual type of membrane protein, *Proc. Natl. Acad. Sci. USA* **82:**4075–4079.

White, J., Kielian, M., and Helenius, A., 1983, Membrane fusion proteins of enveloped animal viruses, *Q. Rev. Biophys.* **16:**151–195.

Wilde, A., and Morrison, T., 1984, Structural and functional characterization of Newcastle disease virus polycistronic RNA species, *J. Virol.* **51:**71–76.

Wildy, P., 1971, Classification and nomenclature of viruses (First Report of the ICNV), *Monogr. Virol.* **5:**181.

Yoshida, T., Nagai, Y., Yoshi, S., Maeno, K., Matsumoto, T., and Hoshino, M., 1976, Membrane (M) protein of HVJ (Sendai virus): Its role in virus assembly, *Virology* **71:**143–161.

Yusoff, K., Millar, N. S., Chambers, P., and Emmerson, P. T., 1987, Nucleotide sequence analysis of the L gene of Newcastle disease virus: Homologies with Sendai and vesicular stomatitis viruses, *Nucleic Acids Res.* **15:**3961–3976.

CHAPTER 3

The Molecular Biology of the Morbilliviruses

THOMAS BARRETT, SHAILA M. SUBBARAO, GRAHAM J. BELSHAM, AND BRIAN W. J. MAHY

I. INTRODUCTION

The *morbilliviruses* are a small, antigenically related genus within the family *Paramyxoviridae*. They are distinguished from the genus *Paramyxovirus* by their lack of neuraminidase activity. The group includes an important human virus, measles virus (MV), and three animal viruses, canine distemper virus (CDV), which causes disease in dogs and *Mustelidae*, rinderpest virus (RPV), causing disease in cattle and other large ruminants, and peste des petits ruminants virus (PPRV), causing disease in sheep, goats, and other small ruminants. Although closely related antigenically, the viruses can be distinguished quite easily by differential neutralization using homologous and heterologous viruses and sera. Their host range is restricted, as indicated above, but they can infect other hosts, although then they do not generally cause disease. Thus, rinderpest viruses cause disease in large ruminants, but cause a subclinical infection in small ruminants. The latter can, however, act as carriers of the disease in some areas of the world and this is of great epidemiological importance. Each virus is antigenically stable, but different strains can be isolated which show widely differing pathogenicity in the host species. In addition to the acute disease, two members of the group, measles and canine distemper viruses, are known to persist and cause chronic neurological disease in a small proportion of infected individuals (see Chapter 12). Within the last year, a new morbillivirus isolate has been described in European seals and porpoises and

THOMAS BARRETT, SHAILA M. SUBBARAO, GRAHAM J. BELSHAM, AND BRIAN W. J. MAHY • AFRC Institute for Animal Health, Pirbright Laboratory, Woking, Surrey GU24 0NF, United Kingdom. *Permanent address of S.M.S.:* Department of Microbiology and Cell Biology, Indian Institute of Science, Bangalore 560012, India. *Present address of B.W.J.M.:* Division of Viral and Rickettsial Diseases, Centers for Disease Control, Atlanta, Georgia 30333.

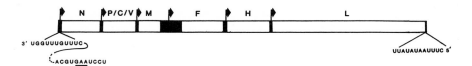

FIGURE 1. Diagrammatic representation of the structure of the MV genome. The relative sizes of the coding (white) and noncoding (black) regions are indicated and drawn to scale.

preliminary results suggest that it is a new virus of marine mammals. Analysis of the morbilliviruses at the molecular level has begun and this chapter is aimed at summarizing the results of this approach.

II. GENOME STRUCTURE AND REPLICATION STRATEGY

Because of its importance in human disease, measles virus (MV) is the most widely studied member of the group and the complete genome RNA of MV (Edmonston strain) has now been sequenced. It consists of a single negative-strand RNA 15,892 nucleotides long (Blumberg *et al.*, 1988). The genome structures of the other members of the group appear to be analogous in most respects to MV and the following features of MV are characteristic. There is a 56-base leader RNA template at the 3′ end of the genome, followed by the coding sequences for the structural proteins in the following order: the nucleocapsid (N) protein, the phosphoprotein (P), the matrix (M) protein, the fusion (F) protein, the hemagglutinin (H) protein, and the large (L) protein. The L protein is the RNA-dependent RNA polymerase protein and its coding sequence is followed by the 3′ antigenome leader RNA template. The gene order for MV is conserved in CDV and RPV (Rima *et al.*, 1986; T. Barrett, unpublished results). It is assumed that the leader RNAs contain the promoter sequences for transcription, replication, and RNA encapsidation. The intergenic region contains a triplet sequence GAA which is preceded by a semiconserved polyadenylation signal and followed by a semiconserved start signal for the next gene (Crowley *et al.*, 1988; also see Chapter 8). This is summarized in Fig. 1.

In the infected cell, the negative-sense RNA genome must first be transcribed into mRNAs to produce virus-specific proteins and there is a single transcription promoter at the 3′ end of the genome RNA. There is no evidence that the leader RNAs are transcribed as separate RNA species during the infectious cycle, but 3′ leader containing transcripts of the N gene (both monocistronic and bicistronic N–P transcripts) have been detected as well as leader-containing antigenome RNA; such leader-containing transcripts are sensitive to cycloheximide addition and so require the continued synthesis of a virus protein for their production (Castaneda and Wong, 1989). The latter authors have proposed a model for MV replication, different from that for VSV, where transcription always begins at the 3′ end, whereby transcription can initiate either at the beginning of the N gene to produce mRNAs or at the 3′ end of the genome RNA to initiate synthesis of the antigenome RNA template for virus replication. RNA replication has been shown to occur late in infection and its completion probably requires the accumulation of sufficient amounts of the N protein. The leader-containing N and N–P bicistronic transcripts may thus represent aborted attempts to synthesize full-length antigenome RNA.

Each gene, except the P gene, which in addition codes for two nonstructural proteins (see below), is transcribed into a monocistronic mRNA and the amount of each mRNA is controlled by the distance of its gene from the promoter, since at each intergenic junction there is a probability that the polymerase will detach from the template and reinitiate at the 3′ end. A steep gradient in the number of copies of each mRNA, from the most abundant 3′-proximal N-gene mRNA to the least abundant 5′-proximal L-gene mRNA, has been demonstrated in MV-infected cells and this gradient becomes more exaggerated in persistent infections of the virus (Cattaneo et al., 1987a,b; also see Chapter 12). During transcription there is sometimes a failure to stop at the intergenic junction before beginning a new mRNA, and readthrough to the next gene occurs, leading to the production of bi- and tricistronic mRNAs. These are found in all the paramyxoviruses and their function (if any) is unknown.

The order of genes in the MV virion RNA is the same as that found in other nonsegmented negative-strand viruses (the paramyxoviruses and rhabdoviruses); the genes are differentially transcribed and translated, probably reflecting the different amounts of each protein that are required for virus replication. In rhabdoviruses, such as vesicular stomatitis virus (VSV), complementation of ts mutants has been demonstrated, with the L protein expressed at low levels from an integrated DNA copy of the L gene when expressed, but not when higher levels of expression are induced. It was suggested that overproduction of L protein alters the ratio of L:NS protein required for shepherding L protein to the nucleocapsid, thereby inhibiting transcription (Meier et al., 1987). The only exception to this gene order is found in the Pneumovirus genus, where the two nonstructural protein genes do not overlap any other gene; these are located at the 3′ end of the genome and the receptor-binding protein (G protein) is located 3′ to the F protein (see Chapter 4).

The pattern of mRNAs synthesized by morbilliviruses in infected cells is highly conserved, the only difference being the length of the F mRNA of CDV, which is slightly shorter than the F mRNA of the other morbilliviruses (Barrett and Underwood, 1985).

III. GENETIC RELATIONSHIPS AMONG THE MORBILLIVIRUSES

All of the MV genome has been sequenced, as have the N, P, M, and F genes of CDV and the F and H genes of RPV. The individual protein genes of the morbilliviruses that have been sequenced show varying degrees of similarity, from the least conserved P and H genes to the highly conserved M and F genes. It is surprising that the region of the genome which is most highly used in terms of coding capacity, i.e., the P gene region, which encodes three virus proteins, is not strongly conserved in sequence. However, there are short, highly conserved sequence regions which enable the P gene to cross-hybridize

significantly with other members of the group (Barrett and Underwood, 1985).
These relationships are discussed in detail below.

A. Nucleocapsid Protein Gene

Nucleotide sequence comparisons of the morbillivirus N genes reveal regions
of high and low relatedness. There is a strongly conserved region in the middle
of the gene (residues 501–1215), a moderately conserved region at the beginning of the gene (residues 1–500), and a weakly conserved region (residues
1216–1625) at the end of the gene (Rozenblatt et al., 1985; T. Barrett and A.
Diallo, unpublished results). Monoclonal antibody analysis of measles virus
field strains has shown that the N protein varies in its reactivity with N-
specific monoclonal antibodies, whereas the surface glycoproteins are invariant in their reaction with a range of F- and H-specific monoclonal antibodies
(Giraudon et al., 1988). Mapping the locations of the epitopes for the N mono-
clonals revealed that the constant epitope (site I) is located nearer the N terminus (122–150) and the variable sites (sites II and III) are located toward the C
terminus (residues 457–476 and 519–525) of the protein (Buckland et al.,
1989).

The cDNA probes made from the N gene of these viruses can be used to
distinguish all the viruses in the group without the need for time-consuming
differential neutralization tests (Diallo et al., 1989). This is normally not
important clinically, since the diagnosis of measles is relatively simple, but it
can be important for veterinary epidemiological studies in regions where RPV
and PPRV coexist (see Section V on diagnosis).

B. The Phosphoprotein Gene

The P gene of the morbilliviruses is in many ways the most interesting.
Antigenically it is the least conserved (Sheshberadaran et al., 1986) and the
nucleotide sequence is one of the most variable, but it is the only cDNA which
cross-hybridizes between all members of the Morbillivirus genus. This is because, although overall sequence relatedness is low, there are interspersed
short regions of high sequence conservation (Barrett et al., 1985; Barrett and
Underwood, 1985). The analogous protein of the rhabdoviruses, the NS protein, is also very variable among serotypes and shows little or no sequence
conservation. However, the overall protein structure is conserved among
serotypes (Gill and Banerjee, 1985). In the paramyxoviruses the P protein often
migrates anomalously slowly on SDS–PAGE, indicating an approximate M_r of
70,000–75,000, whereas sequence data show that the true M_r is about 54,000–
55,000 (Barrett et al., 1985; Bellini et al., 1985). The most characteristic feature
of the P gene is the encoding of two virus nonstructural proteins in addition to
the main P protein product. The first nonstructural protein, termed the C
protein after an analogous protein in Sendai virus-infected cells (Lamb et al.,
1976), is encoded in an alternative reading frame beginning at an AUG some 20
bases downstream of the P protein AUG in both MV and CDV. Alkhatib et al.

(1988) found that equivalent amounts of C protein were synthesized in an adenovirus expression system regardless of the context of the upstream P protein initiator codon, suggesting that ribosomes may bind directly to the mRNA at an internal site at, or close to, the initiator codon for the C protein. Fluorescent antibody studies have shown that the C protein colocalizes with virus nucleocapsids in infected cells and may be involved in either nucleocapsid assembly or RNA transcription (Bellini et al., 1985).

The other nonstructural protein, originally known as protein X, now known as protein V after an analogous protein in SV5 virus-infected cells (Thomas et al., 1988), is encoded by an mRNA which is not a direct copy of the genome RNA but is edited by the addition of one G residue which changes the reading frame at position 751 in the P gene, which then continues in the −1 reading frame to produce a new protein with a cysteine-rich C terminus. This frameshifting phenomenon was originally discovered in SV5 virus (see Chapter 6). Sequence comparison of the MV and CDV P genes and other paramyxoviruses reveals a conserved consensus sequence for the presumed frameshift (Cattaneo et al., 1989). The consensus sequence for the presumed frameshift is identical in MV and RPV. We have found RPV P cDNA clones with an extra G residue inserted at the same position as in MV, and so editing of P gene transcripts is likely to be a common feature of all morbilliviruses (T. Barrett, unpublished results).

C. Matrix Protein Gene

The M genes of both MV (Edmonston) and CDV (Onderstepoort) have been sequenced and there is 67% identity at the nucleotide level and 76% at the amino acid level, with only chance identity in the nucleotide sequences of the 5′ and 3′ untranslated regions (Bellini et al., 1986; Greer et al., 1986). The M protein from a measles clinical isolate has also been sequenced and it shows a remarkably high sequence conservation when compared with the vaccine strain. In the coding region there were six nucleotide changes, two of which were silent mutations; no changes were seen in the 5′ noncoding region and only two in the long 3′ noncoding region (Curran and Rima, 1988). The protein is thought to mediate an interaction between the surface glycoproteins in the cell membrane and the nucleocapsid structures in the cytoplasm to form new virus particles. The C-terminal one-third of the protein is the most conserved across the paramyxovirus group and is hydrophobic in character, presumably to facilitate interaction with the lipid bilayer. The M protein is the smallest structural protein in the virus, consisting of 335 amino acids in both MV and CDV. Certain amino acids thought to be important in maintaining protein secondary and tertiary structural features, such as glycine, proline, and cysteine, are conserved between the M proteins of the two viruses (Bellini et al., 1986). The mRNA encoding the M protein is also the smallest; nevertheless, there is a long untranslated region of some 400 nucleotides at the 3′ end of the mRNA in both viruses. There are two small open reading frames (ORFs) (MX1 and MX2) immediately following the M stop codon in this region in MV, but they are not present in CDV and there is no evidence that they are translated into proteins *in vivo* (Bellini *et al.*, 1986; Wong *et al.*, 1987). Whether or not these ORFs are expressed in infected cells is unknown, but they would need to be accessed by internal initiation on the bicistronic mRNAs, since one of them spans the intergenic region. No sequence information is available for the other morbilliviruses. The long section of genome between the end of the M coding sequence and the beginning of the F coding region in the morbilliviruses is unique in the paramyxoviruses. It may act as a controlling signal or as a stabilizing feature of the genome RNA, but its function is unknown. This region is discussed further in Section IV.

D. Fusion Protein Gene

Two MV *F* gene sequences have been published—the Edmonston vaccine strain (Richardson *et al.*, 1986) and the SSPE-derived Hallé strain (Buckland *et al.*, 1987)—and they are virtually identical. One CDV *F* sequence [Onderstepoort vaccine strain (Barrett *et al.*, 1987)] and two *F* gene sequences of RPV have also been determined [the rabbit-adapted "lapinized" or L strain (Tsukiyama *et al.*, 1988) and one virulent isolate Kabete "O" (Hsu *et al.*, 1988)]. In both RPV sequences there is an open reading frame preceding the F protein sequence, but its position and the predicted amino acid sequence are not conserved, indicating that it may just be fortuitous. We have sequenced the *F* gene of the Plowright vaccine strain of RPV derived by multiple passage of the Kabete "O" strain in primary bovine kidney cells (Plowright and Ferris, 1962). This was found to be highly conserved when compared to the Kabete "O" parent strain, but both differed significantly from the L strain (T. Barrett and S. A. Evans, work in progress). Since the F protein is one of the most antigenically conserved (Sheshberadaran *et al.*, 1986) and the sequence data are available for three of the four known morbilliviruses, these data can be used to draw some conclusions regarding the evolution of these viruses. The two reported measles virus sequences differ by only four silent changes in the coding region, ten nucleotide changes in the long 5′ noncoding region, and two nucleotide dele-

TABLE I. Nucleotide Sequence Comparison
of Morbillivirus *F* Genes

Virus	5′ UTR (585)[b]	F_2	F_1	3′ UTR (134)[b]
RPV (Kabete "O")	97.5	99.3	99.5	99.2
RPV (L strain)	77.4	88.9	90.0	75.3
MV (Edmonston)	65.5 (260)	65.6	72.6	62.0 (62)
CDV (Onderstepoort)	65.0 (192)	59.4	65.9	70.0 (24)

[a]The sequences were compared to that of the RPV RBOK strain using the "Bestfit" program of the University of Wisconsin sequence analysis programs. Results are expressed as the percent of identical nucleotide residues. The 5′ and 3′ untranslated regions (UTRs) of RPV showed very little similarity to the equivalent sequences in MV and CDV. The lengths of continuous related sequences found are shown in parentheses after the percentage relationship.
[b]Actual lengths of the 5′ and 3′ UTRs in RPV.

tions in the 3' noncoding region (Buckland *et al.*, 1987). Similarly, the two related strains of rinderpest virus (Kabete "O" and the vaccine strain RBOK derived from it) are highly conserved, but they differ quite considerably from the rabbit-adapted L strain. This may reflect the changes needed for adaptation to a different host and a different geographic location for the original isolate, Asia versus Africa. Comparison of all the sequence data shows that MV and RPV are more closely related to each other than are MV and CDV or RPV and CDV (Table I and Fig. 2).

The surface glycoproteins of the morbilliviruses probably associate to form multimers at the cell surface before incorporation into virus particles, and a leucine zipper motif, which is thought to be involved in protein oligomerization, is found in the F proteins of many paramyxoviruses, including the morbilliviruses (Buckland and Wild, 1989).

E. The Hemagglutinin Protein Gene

The hemagglutinin protein is presumed to be responsible for the attachment of the virus to the cell receptor and so influences the cell tropism of the virus. Most neutralizing antibodies are directed against this protein, and monoclonal antibodies which inhibit hemagglutination by MV are protective in mice (Giraudon and Wild, 1985). It is a typical membrane-bound protein with several N-linked glycosylation sites and an N-terminal membrane anchor sequence. In this respect it resembles the neuraminidase of the orthomyxoviruses (Fields *et al.*, 1981). Nothing is known about the nature of the cell receptors for morbilliviruses, although the lack of neuraminidase activity in the virions would suggest that they lack sialic acid residues. Haemagglutinating activity is associated with MV virions, but only infrequently with CDV. So far only the sequences of the *HA* genes of two strains of MV and two strains of RPV have been published (Alkhatib and Briedis, 1986; Gerald *et al.*, 1986; Tsukiyama *et al.*, 1987; Yamanaka *et al.*, 1988). The HA proteins of the two MV strains are highly conserved, but vary quite significantly from the two strains of RPV, and this difference may in part account for their different host specificities. About 56% amino acid conservation is observed among the three viruses, but even the two RPV strains differ by about 12%. As with the great divergence between the *F* genes of the two RPV strains, this may reflect the extensive adaptation to the rabbit and subsequently to Vero cells of the L strain of RPV. However, the general structure of the HA molecules is conserved in that their glycosylation sites and cysteine residues are conserved (Fig. 3).

F. The L Protein Gene

Very little information is available for this protein in the morbilliviruses and only the *L* gene of MV has been sequenced. Comparisons of the sequence with the *L* genes of other paramyxoviruses have been published (Blumberg *et al.*, 1988) and the MV *L* gene was found to be highly conserved when compared to other paramyxoviruses, but less conserved when compared to VSV. In addi-

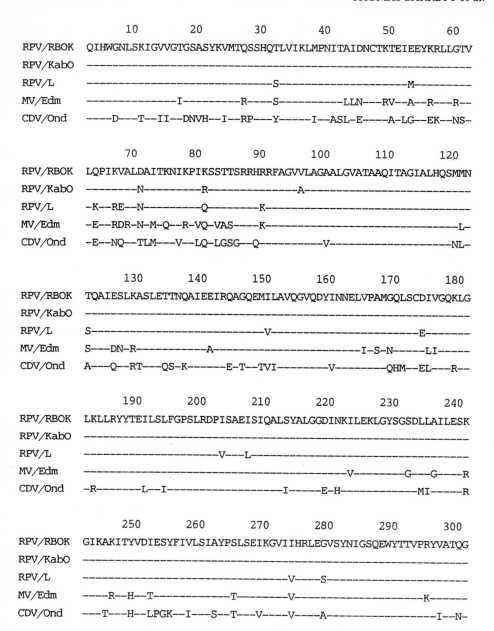

FIGURE 2. Amino acid sequence comparison of morbillivirus F proteins. A dash (–) indicates amino acid identity. RPV/RBOK, rinderpest virus Plowright vaccine strain; RPV/Kabo, rinderpest Kabete "O" strain, parental strain for Plowright vaccine; RPV/L, rinderpest lapinized strain; MV/Edm, measles virus Edmonston strain; CDV/Ond, CDV Onderstepoort vaccine strain.

tion, a Gly-Asp-Asp amino acid sequence motif thought to be characteristic of RNA-dependent RNA polymerases was found. Rima *et al.* (1986) have reported the isolation of a cDNA clone of the CDV *L* gene, but no sequence data are available for comparison.

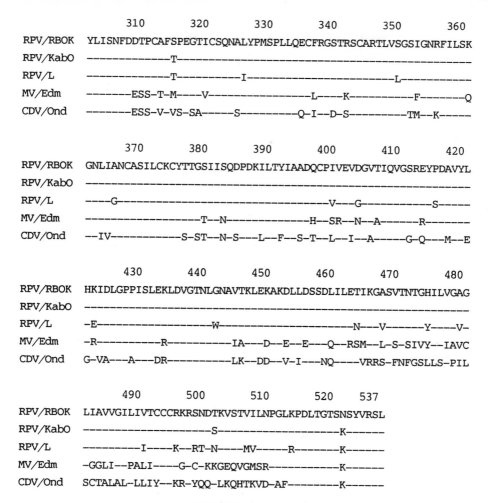

```
                    310        320        330        340        350        360
RPV/RBOK    YLISNFDDTPCAFSPEGTICSQNALYPMSPLLQECFRGSTRSCARTLVSGSIGNRFILSK
RPV/KabO    --------------T---------------------------------------------
RPV/L       --------------T-----------I-------------------------L--------
MV/Edm      --------ESS-T-M-----V---------------L-----K------------F-------Q
CDV/Ond     --------ESS-V-VS-SA------S----------Q-I--D-S----------TM--K-----

                    370        380        390        400        410        420
RPV/RBOK    GNLIANCASILCKCYTTGSIISQDPDKILTYIAADQCPIVEVDGVTIQVGSREYPDAVYL
RPV/KabO    ----------------------------------------------------------
RPV/L       ----G-------------------------------------V---G------------S-----
MV/Edm      ----------------------T--N--------------H--SR--N--A-------R--------
CDV/Ond     --IV-----------S-ST--N-S---L--F--S-T--L--I--A-----G-Q---M--E

                    430        440        450        460        470        480
RPV/RBOK    HKIDLGPPISLEKLDVGTNLGNAVTKLEKAKDLLDSSDLILETIKGASVTNTGHILVGAG
RPV/KabO    ----------------------------------------------------------
RPV/L       -E-----------------------W------------------------N---V------Y----V-
MV/Edm      -R-----------R-----------IA----D--E--E---Q--RSM--L-S-SIVY--IAVC
CDV/Ond     G-VA---A----DR-----------LK--DD--V-I----NQ----VRRS-FNFGSLLS-PIL

                    490        500        510        520    537
RPV/RBOK    LIAVVGILIVTCCCRKRSNDTKVSTVILNPGLKPDLTGTSNSYVRSL
RPV/KabO    ----------------------S--------------K-------
RPV/L       ----------I----K--RT-N----MV-----R-------K-------
MV/Edm      -GGLI---PALI-----G-C-KKGEQVGMSR-------------K-------
CDV/Ond     SCTALAL--LLIY--KR-YQQ-LKQHTKVD--AF---------K-------
```

FIGURE 2. (*continued*)

IV. FUNCTION OF THE 5′ AND 3′ UNTRANSLATED REGIONS

In addition to the semiconserved start signals, polyadenylation signals, and intergenic GAA triplet, each gene has variable lengths of 5′ and 3′ untranslated nucleotides. In the *M* genes of MV and CDV, the coding region stops some 400 nucleotides from the 3′ end of the mRNA and the first AUG in the F mRNA of MV is at position 574, although the AUG that is in the best Kozak context is the second at position 583 (Richardson *et al.*, 1986; Buckland *et al.*, 1987). Neither the sequence nor the arrangement of the start codons is conserved in the 5′ region of F genes of the other morbilliviruses. The only common feature is the high GC-rich nature of the sequence and the ability of the sequences to form stable secondary structures when analyzed by RNA folding

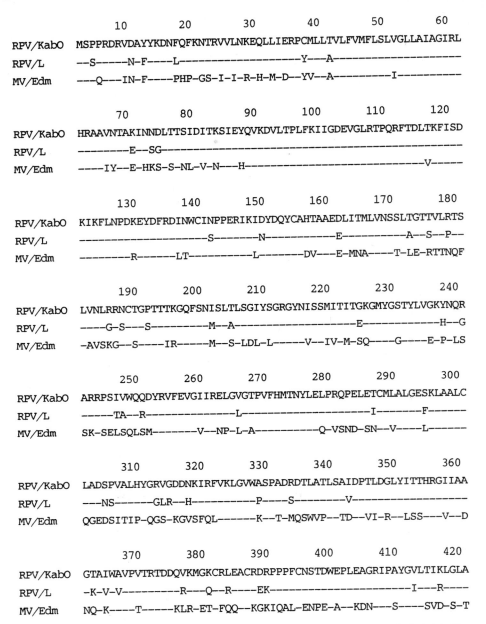

```
                  10        20        30        40        50        60
RPV/KabO  MSPPRDRVDAYYKDNFQFKNTRVVLNKEQLLIERPCMLLTVLFVMFLSLVGLLAIAGIRL
RPV/L     --S------N-F----L--------------------Y---A------------------
MV/Edm    ---Q---IN-F-----PHP-GS-I-I-R-H-M-D--YV--A----------I---------


                  70        80        90        100       110       120
RPV/KabO  HRAAVNTAKINNDLTTSIDITKSIEYQVKDVLTPLFKIIGDEVGLRTPQRFTDLTKFISD
RPV/L     ---------E--SG----------------------------------------------
MV/Edm    ----IY--E-HKS-S-NL-V-N---H---------------------------V------


                  130       140       150       160       170       180
RPV/KabO  KIKFLNPDKEYDFRDINWCINPPERIKIDYDQYCAHTAAEDLITMLVNSSLTGTTVLRTS
RPV/L     -----------------------S--------N------------E-----------A--S--P--
MV/Edm    ---------R-------LT-----------L--------DV---E-MNA----T-LE-RTTNQF


                  190       200       210       220       230       240
RPV/KabO  LVNLRRNCTGPTTTKGQFSNISLTLSGIYSGRGYNISSMITITGKGMYGSTYLVGKYNQR
RPV/L     ----G-S---S----------M--A--------------------E-------------H--G
MV/Edm    -AVSKG--S-----IR------M--S-LDL-L------V--IV-M-SQ----G----E-P-LS


                  250       260       270       280       290       300
RPV/KabO  ARRPSIVWQQDYRVFEVGIIRELGVGTPVFHMTNYLELPRQPELETCMLALGESKLAALC
RPV/L     ------TA--R---------------L--------------------I--------F-------
MV/Edm    SK-SELSQLSM--------V--NP-L-A----------Q-VSND-SN--V----L-------


                  310       320       330       340       350       360
RPV/KabO  LADSPVALHYGRVGDDNKIRFVKLGVWASPADRDTLATLSAIDPTLDGLYITTHRGIIAA
RPV/L     ---NS------GLR--H-----------P----S--------V-----------------
MV/Edm    QGEDSITIP-QGS-KGVSFQL-------K--T-MQSWVP--TD--VI-R--LSS---V--D


                  370       380       390       400       410       420
RPV/KabO  GTAIWAVPVTRTDDQVKMGKCRLEACRDRPPPFCNSTDWEPLEAGRIPAYGVLTIKLGLA
RPV/L     -K-V-V---------R---Q--R----EK--------------------I---R----
MV/Edm    NQ-K----T-----KLR-ET-FQQ--KGKIQAL-ENPE-A--KDN---S----SVD-S-T
```

FIGURE 3. Amino acid sequence comparison of morbillivirus H proteins. A dash (–) indicates amino acid identity. Abbreviations of viruses are as noted in Fig. 2.

computer programs. The only difference observed in the size of the mRNAs of the different morbilliviruses is a shorter *F* gene mRNA in CDV; all other mRNA species were identical (Barrett and Underwood, 1985). The difference in the size of the F protein mRNA of CDV is accounted for mainly by a shorter 5' region. The structures of the mRNAs for the F proteins of different morbilliviruses are shown in Fig. 4.

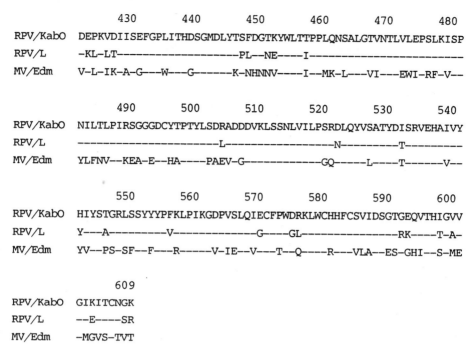

FIGURE 3. (continued)

In the case of CDV, there are three AUGs in frame and one out of frame in the first 500 nucleotides of the *F* gene and it is not certain which is the initiation codon. A protein starting at the fourth AUG, i.e., third in-frame (position 461), would be homologous in size and sequence with that of MV and we have suggested that this is the authentic initiator codon for the F protein of CDV (Barrett *et al.*, 1987). More recent work in our laboratory has shown that all other upstream AUGs can be deleted and this still allows authentic F protein to be produced, while F protein synthesis is abolished if the AUG codon at position 461 is deleted. However, in the transient expression system using a T7 promoter to direct transcription from plasmid constructs introduced into cells infected with a recombinant vaccinia virus expressing the T7 RNA polymerase (Fuerst *et al.*, 1986), we found that deleting the upstream AUGs, instead of increasing F protein production, reduced the amount of F protein expressed as each AUG in turn was deleted (Evans *et al.*, 1990). In contrast, RNA transcripts from the same constructs showed increased F protein synthesis in rabbit reticulocyte lysates as the upstream AUGs were deleted, as was found to be the case in MV where deleting the 5′ nontranslated sequences increased F protein synthesis *in vitro* (Hasel *et al.*, 1987).

In RPV, the *F* gene has a similar long sequence before the AUG initiation codon for the F protein, but in this case there is a second open reading frame in this 5′ region, which ends with a stop codon just before the start of the F protein. We found that deleting this upstream 5′ sequence reduced the efficiency of F protein production in cells infected with vaccinia virus recombinants containing the *F* gene (Fig. 5), but in cell-free translations of the full-length and

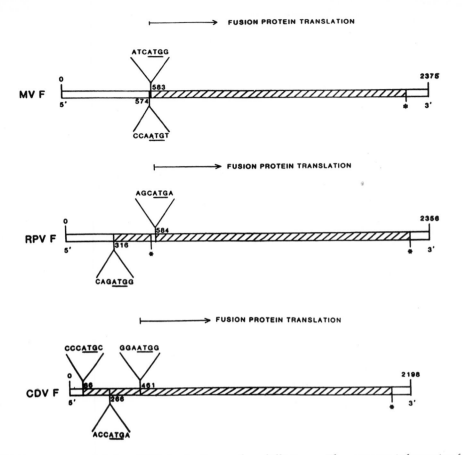

FIGURE 4. Structure of the mRNA for the *F* gene of morbilliviruses. The sequence information for MV, CDV, and RPV was taken from Richardson *et al.* (1986), Barrett *et al.* (1987), and T. Barrett and S. A. Evans (to be published). Open reading frames are shown by the hatched areas and the asterisk denotes a stop codon. The positions of the ATGs referred to in the text are indicated along with the adjacent nucleotide sequence.

truncated RNA made by transcription from plasmids containing the F gene, the loss of this sequence enhanced F protein synthesis. It thus appears that the presence of these sequences serves to direct efficient production of the F protein within cells, but in cell-free systems they are deleterious.

Such long untranslated 5' regions are not unique to the morbilliviruses. In picornaviruses, for example, similar long 5' noncoding regions with short ORFs are found which range from 650 to 1300 nucleotides. These 5' untranslated sequences serve to direct internal initiation by ribosomes at the correct initiator AUG in a cap-independent way and they serve to direct protein synthesis from the downstream gene in artificially constructed bicistronic mRNAs (e.g., Pelletier and Sonnenberg, 1988; Jang *et al.*, 1989). As already mentioned, bicistronic mRNAs occur naturally in paramyxovirus infections, and it is possible that these long untranslated regions between the end of the *M*

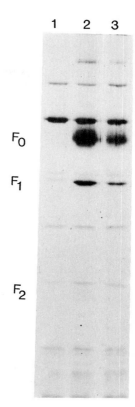

F_0

F_1

F_2

FIGURE 5. Immunoprecipitation of RPV F proteins from cells infected with recombinant vaccinia viruses containing the full-length RPV *F* (lane 2) or the truncated gene lacking the 5′ noncoding region (lane 3). The sample used in lane 1 was from cells infected with wild-type vaccinia virus (WR strain).

gene coding sequence and the beginning of the *F* gene coding sequence could act to direct internal initiation or to stabilize the RNAs. This has not been tested experimentally for *M–F* bicistronic mRNA, but in expression studies using an artificially produced *P–M* bicistronic mRNA, only the P protein could be detected, indicating that only the 5′-proximal gene was translated (Wong and Hirano, 1987). This construct, however, did not have a long untranslated region as is found between the *M* and *F* genes.

V. DIAGNOSIS USING MOLECULAR TECHNIQUES

In general, diagnosis of morbillivirus diseases, particularly measles, is a simple clinical one based on characteristic signs and the host from which the virus was isolated. However, mild cases of both RPV and PPRV can be confused with other diseases, such as mucosal disease. In addition, when a morbillivirus is isolated from an unusual host species, as in the recent seal distemper outbreak in Europe, diagnosis is not straightforward.

The two diseases of ruminants, RPV and PPRV, are known to have distinct epidemiological patterns in Africa, where the two viruses coexist independently. However, there are rare cases of RPV causing disease in small ruminants and PPRV has been shown to infect but not cause disease in large ruminants (Diallo

Radioactive Probes

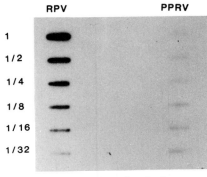

Non-radioactive Probes

Standard DNA

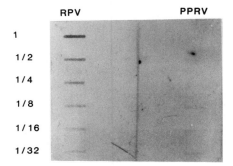

FIGURE 6. Use of cDNA probes to detect rinderpest sequences in the lymphocytes of a calf infected with rinderpest virus. Both radioactive and nonradioactive probes were compared for their sensitivity in detecting virus RNA. A cDNA probe derived from the *N* gene of the related PPR virus was used as a control for background hybridization. A sample of blood (50 ml) was taken during the febrile stage of the disease and RNA was extracted from the buffy coat. Starting with $\frac{1}{10}$ of the RNA, serial two-fold dilutions were made and the blot was probed with RPV- and PPRV-specific cDNA probes (Diallo *et al.*, 1989). The samples were analyzed using ^{32}P-labeled probes and digoxigenin-labeled nonradioactive probes (Boehringer Manheim, U.K., Ltd.)

et al., 1989). In India, on the other hand, PPRV has not been described, but a morbillivirus disease attributed to rinderpest has long been a problem in sheep and goats. In such cases it is important to know the nature of the virus causing the disease, since small ruminants could play a role as carriers of rinderpest. It is not feasible to use complicated hybridization techniques for routine field tests, but they can greatly facilitate diagnosis as tools in a more thorough epidemiological study. Using cDNA probes derived from the *N* gene of each virus, even those two very closely related viruses can be distinguished (Diallo *et al.*, 1989). It was thought that PPRV was not present in India and that small ruminants were susceptible to Indian strains of RPV. Recently, however, it has been shown using these cDNA probes that the two diseases cocirculate in India (Shaila *et al.*, 1989). In a similar way, we were able to show that the recent outbreak of rinderpestlike diseases in Sri Lanka was caused by RPV and not PPRV, although the source of infection was apparently healthy goats brought from the mainland of India during movement of troops (Anderson *et al.*, 1990).

Nucleic acid hybridization is a suitable method for analyzing blood and postmortem tissue and eliminates the need to isolate the virus for identification. During the acute stages of infection, virus RNA can be detected in lymphocytes using these cDNA probes. In Fig. 6, lymphocytes from a calf experimentally infected with spleen material obtained from an infected Sri Lankan animal were analyzed using *N* gene probes labeled either with ^{32}P or with a nonradioactive label and shown to react specifically with the RPV probe. Since these probes are nearly full-length, the stringency at which the washing is carried out is important. Low-level cross-hybridizations can be detected when the concentration of virus-specific RNA is high and so both probes must be used on identical RNA samples for unambiguous identification of the virus. Previous work with the N genes of CDV and MV had shown that these viruses could also be distinguished using the N gene cDNA probes (Rozenblatt et al., 1985). Thus, cDNA probes can be used to characterize the four known morbilliviruses.

The identification of the causative agent of the recent seal distemper epizootic in European seals as a morbillivirus led to the suggestion that it was CDV-like, since similar pathological signs were observed and serum from affected animals had high neutralization titers against CDV, but failed to inhibit the measles hemagglutination activity in a standard MHI test (Osterhaus and Vedder, 1988). Using a variety of nearly full-length hybridization probes to detect cross-hybridization with nonidentical morbilliviruses, we were able to confirm the presence of morbillivirus sequences in tissue samples from infected animals, but they were not identical to any of the four known viruses in the group (Mahy et al., 1988). Subsequent work showed that sera from infected animals were able to neutralize RPV efficiently and an ELISA test developed to detect RPV antibodies could be used to diagnose the presence of antibodies to the seal morbillivirus with a high degree of certainty. In addition, sera from infected seals precipitated the N, P, and F proteins of the other morbilliviruses (Bostock et al., 1990). In fact, high-titer seal serum was extremely efficient in precipitating the F protein from both the RPV F and CDV F vaccinia virus recombinants, which provide a good source of pure F protein, uncontaminated with other virus proteins (Fig. 7). This efficient precipitation is in agreement with Osterhaus et al. (1989), who reported the protection of seals from phocine (seal) distemper virus infection by vaccination with CDV F-containing subunit vaccines.

MV cDNA clones have been used extensively to look for persistent infections with the virus to establish its role as a possible etiological agent in a variety of diseases, including Paget's disease (Basle et al., 1986). Nucleic acid hybridization has also been used to study the sites of replication of MV in patients and it has been shown that measles replicates in cells not previously recognized to be involved. In addition to lymphoid organs, virus RNA was detected in epithelial cells of lung, gut, bile duct, bladder, and skin, while invasion of the brain parenchyma was found to be uncommon during acute measles virus infection (Moench et al., 1988).

VI. MORBILLIVIRUS VACCINES

It has proved relatively simple to develop live attenuated vaccine strains of each of the morbilliviruses by repeated passages in tissue culture (see Chapter 18). In addition, vaccine strains of rinderpest have been derived by passage in rabbits or in goats. The molecular basis of the attenuation is not known, and there has been no complete sequence comparison between a virulent and an

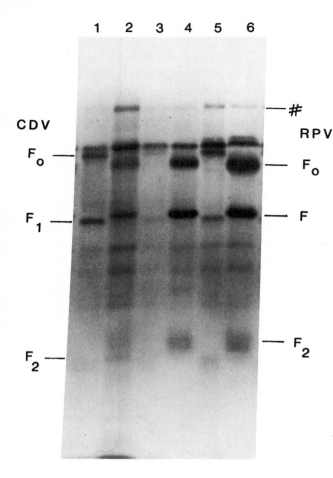

FIGURE 7. Immunoprecipitation of RPV F and CDV F proteins synthesized in cells infected with vaccinia recombinant viruses using rabbit anti-CDV antiserum (1,2), rabbit anti-RPV antiserum (3,4), and serum from a seal infected with the seal distemper virus (5,6). Samples 1, 3, and 5 are from cells infected with a CDV–F vaccinia virus recombinant and samples 2, 4, and 6 from cells infected with an RPV–F vaccinia virus recombinant. The positions of the F-related proteins are shown.

attenuated strain of any morbillivirus. It is known that avirulent strains of Newcastle disease virus have F proteins which are not readily cleavable by cellular proteases (Toyoda *et al.*, 1987). However, attenuated vaccine strains of all morbilliviruses have cleavable F proteins (Fig. 2) and it is probable that attenuation is associated with many point mutations introduced by multiple passage of the virus in tissue culture or in a nonnatural host. Recently, an analysis of several Edmonston-derived vaccine strains showed that the Edmonston-Zagreb vaccine was heavily contaminated with defective-interfering (DI) particles. It is not clear whether the presence of DI particles contributes to the attenuated phenotype, but their potentially harmful effect in helping to establish a persistent infection raises questions regarding vaccine safety (Calain and Roux, 1988). Research is now concentrated on developing subunit vaccines (see Chapter 18) and on live recombinant virus vaccines.

A vaccinia virus recombinant containing the G protein gene of rabies virus is undergoing field trials as a potential rabies vaccine in Europe (Pastoret *et al.*, 1988). Similar RPV–vaccinia recombinants could become useful, relatively heat-stable vaccines for use in areas of the world where it is difficult and expensive to provide proper refrigeration for live-virus vaccines. Such recombinants containing either the H or F gene of MV have been used successfully to protect mice against lethal challenge with intracerebral inoculation of MV (Drillien et al., 1988). We have shown that vaccinia virus recombinants containing the F protein gene of either CDV or RPV synthesize authentic F proteins which are cleaved to F_1 and F_2 and are glycosylated (Barrett et al., 1989; G. J. Belsham and T. Barrett, work in progress). Although the RPV and CDV F proteins are identical in size based on sequence data (527 amino acids, excluding the signal sequence), their migration in polyacrylamide gels is different, with the F_0 of CDV migrating appreciably slower than the F_0 of RPV and the F_1 slightly faster (see Fig. 6). Immunization with the F-protein vaccinia virus recombinant of RPV induces both an antivirus antibody as measured by ELISA and production of a neutralizing antibody against the virus and is protective in both a rabbit model system and in the normal cattle host (Barrett et al., 1989; Belsham et al., 1989). Similar protection of cattle using either the F or H gene of the Kabete "O" strain of RPV in vaccinia virus recombinants was reported by Yilma et al. (1988). Recombinant poxviruses other than vaccinia virus containing RPV genes are being investigated as possible field vaccines in areas of the world where the infectivity of vaccinia virus recombinants for humans might rule out their use. Other expression systems have been used to produce measles hemagglutinin protein. Alkhatib and Briedis (1988) have generated an adenovirus recombinant expressing high levels of biologically active protein, but its potential as a vaccine has not been investigated.

VII. CONCLUSIONS

The complete nucleotide sequence of the vaccine strain of MV has been determined and the genome organization is similar to other paramyxoviruses (excluding the pneumoviruses) and rhabdoviruses. However, there are unique features of the morbilliviruses, such as the absence of free 3' leader RNA and the long untranslated region between the end of the M gene and the start of the F gene, indicating differences in replication strategy and regulation of protein production which need to be investigated further. cDNA clones of individual virus genes are being used to diagnose virus infections and to study virus pathogenesis. Recombinant virus vaccines expressing individual morbillivirus glycoproteins have been shown to be protective both in animal model systems and against the disease in the natural host. Sequence analysis of the morbilliviruses has confirmed their common origin and supports the proposed evolutionary tree based on antigenic relationships that identifies RPV as the progenitor of the group.

ACKNOWLEDGMENTS. We thank S. Evans for making material available before publication. S. M. S. is the recipient of a Royal Society short-term fellowship and is also supported on a grant from the Wellcome Trust. Our molecular biology studies on rinderpest virus are supported by the Wellcome Trust (grant 14294/1.5).

VIII. REFERENCES

Alkhatib, G., and Briedis, D. J., 1986, The predicted primary structure of the measles virus hemagglutinin, *Virology* **150**:479–490.

Alkhatib, G., and Briedis, D. J., 1988, High level eucaryotic *in vivo* expression of biologically active measles virus hemagglutinin by using an adenovirus type 5 helper-free vector system, *J. Virol.* **62**:2718–2727.

Alkhatib, G., Massie, B., & Briedis, D. J., 1988, Expression of bicistronic measles virus P/C mRNA by using hybrid adenoviruses: Levels of C protein synthesized *in vivo* are unaffected by the presence or absence of the upstream P initiator codon, *J. Virol.* **62**:4059–4069.

Anderson, E. C., Hassan, A., Barrett, T., & Anderson, J., 1990, Observations on the pathogenicity and transmissibility for sheep and goats of the strain of virus isolated during the rinderpest outbreak in Sri Lanka in 1987, *Vet. Microbiol.* **21**:309–318.

Barrett, T., and Underwood, B., 1985, Comparison of messenger RNAs induced in cells infected with each member of the morbillivirus group, *Virology* **145**:195–199.

Barrett, T., Shrimpton, S. B., & Russell, S. E. H., 1985, Nucleotide sequence of the entire protein coding region of canine distemper virus polymerase-associated (P) protein mRNA, *Virus Res.* **3**:367–372.

Barrett, T., Clarke, D. K., Evans, S. A., and Rima, B. K., 1987, The nucleotide sequence of the gene encoding the F protein of canine distemper virus: A comparison of the deduced amino acid sequence with other paramyxoviruses, *Virus Res.* **8**:373–386.

Barrett, T., Belsham, G. J., Subbarao, S. M., and Evans, S. A., 1989, Immunization with a vaccinia recombinant expressing the F protein protects rabbits from challenge with a lethal dose of rinderpest virus, *Virology* **170**:11–18.

Basle, M. F., Fournier, J. G., Rozenblatt, S., Rebel, A., and Bouteille, M., 1986, Measles virus RNA detected in Paget's disease bone tissue by *in situ* hybridization, *J. Gen. Virol.* **67**:907–913.

Bellini, W. J., Englund, G., Rozenblatt, S., Arnheiter, H., and Richardson, C. D., 1985, Measles virus P gene codes for two proteins, *J. Virol.* **53**:908–919.

Bellini, W. J., Englund, G. Richardson, C. D., Rozenblatt, S., and Lazzarini, R. A., 1986, Matrix genes of measles virus and canine distemper virus: cloning, nucleotide sequence, and deduced amino acid sequences, *J. Virol.* **58**:408–416.

Belsham, G. J., Anderson, E. C., Murray, P. K., Anderson, J., and Barrett, T., 1989, Immune response and protection of cattle and pigs generated by a vaccinia virus recombinant expressing the F protein of rinderpest virus, *Vet. Rec.* **124**:655–658.

Blumberg, B. M., Crowley, J. C., Silverman, J. I., Menonna, J., Cook, S. D., and Dowling, P. C., 1988, Measles virus L protein evidences elements of ancestral RNA polymerase, *Virology* **164**:487–497.

Bostock, C. J., Barrett, T., and Crowther, J. R., 1990, Characterization of the European seal morbillivirus, *Vet. Microbiol.* (Suppl.), **23**:351–360.

Buckland, R., and Wild, F., 1989, Leucine zipper motif extends, *Nature* **338**:547.

Buckland, R., Gerald, C., Baker, R., and Wild, T. F., 1987, Fusion glycoprotein of measles virus: Nucleotide sequence of the gene and comparison with other paramyxoviruses, *J. Gen. Virol.* **68**:1695–1703.

Buckland, R., Giraudon, P., and Wild, F., 1989, Expression of measles virus nucleoprotein in *Escherichia coli:* Use of deletion mutants to locate the antigenic sites, *J. Gen. Virol.* **70**:435–441.

Calain, P., and Roux, L., 1988, Generation of measles virus defective interfering particles and their presence in a preparation of attenuated live-virus vaccine, *J. Virol.* **62**:2859–2866.

Castaneda, S. J., and Wong, T. C., 1989, Measles virus synthesizes both leaderless and leader-containing polyadenylated RNAs *in vivo*, *J. Virol.* **63**:2977–2986.

Cattaneo, R., Rebmann, G., Baczko, K., ter Meulen, V., and Billeter, M. A., 1987a, Altered ratios of measles virus transcripts in diseased human brains, *Virology* **160**:523–526.

Cattaneo, R., Rebmann, G., Schmid, A., Baczko, K., ter Meulen, V., and Billeter, M. A., 1987b, Altered transcription of a defective measles virus genome derived from a diseased human brain, *EMBO J.* **6**:681–688.

Cattaneo, R., Kaelin, K., Baczko, K., and Billeter, M. A., 1989, Measles virus editing provides an additional cysteine-rich protein, *Cell* **56:**759–764.

Crowley, J. C., Dowling, P. C., Menonna, J., Silverman, J. I., Schuback, D., Cook, S. D., and Blumberg, B. M., 1988, Sequence variability and function of measles virus 3′ and 5′ ends and intercistronic regions, *Virology* **164:**498–506.

Curran, M. D., and Rima, B. K., 1988, Nucleotide sequence of the gene encoding the matrix protein of a recent measles virus isolate, *J. Gen Virol.* **69:**2407–2411.

Diallo, A., Barrett, T., Barbron, M., Subbarao, S. M., and Taylor, W. P., 1989, Differentiation of rinderpest and peste des petits ruminants viruses using specific cDNA clones, *J. Virol. Meth.* **23:**127–136.

Drillien, R., Spehner, D., Kirn, A., Giraudon, P., Buckland, R., Wild, F., and Lecocq, J.-P., 1988, Protection of mice from fatal measles encephalitis by vaccination with vaccinia virus recombinants encoding either the haemagglutinin or the fusion protein, *Proc. Natl. Acad. Sci. USA* **85:**1252–1256.

Evans, S. A., Belsham, G. J., and Barrett, T., 1990, The role of the 5′ nontranslated regions of the fusion protein mRNAs of Canine Distemper virus and Rinderpest virus, *Virology* **177:**317–323.

Fields, S., Winter, G., and Brownlee, G. G., 1981, Structure of the neuraminidase gene in human influenza virus A/PR/8/34, *Nature* **290:**213–217.

Fuerst, T. R., Niles, E. G., Studier, F. W., and Moss, B., 1986, Eukaryotic transient expression system based on a recombinant vaccinia virus that synthesizes bacteriophage T7 RNA polymerase, *Proc. Natl. Acad. Sci. USA* **83:**8122–8126.

Gerald, C., Buckland, R., Barker, R., Freeman, G., and Wild, T. F., 1986, Measles virus haemagglutinin gene: Cloning, complete nucleotide sequence analysis and expression in COS cells, *J. Gen. Virol.* **67:**2695–2703.

Gill, D. S., and Banerjee, A. K., 1985, Vesicular stomatitis virus NS proteins: Structural similarity without extensive sequence homology, *J. Virol.* **55:**60–66.

Giraudon, P., and Wild, T. F., 1985, Correlation between epitopes on hemagglutinin of measles virus and biological activities: Passive protection by monoclonal antibodies is related to their hemagglutination inhibiting activity, *Viology* **144:**46–58.

Giraudon, P., Jacquier, M. F., and Wild, T. F., 1988, Antigenic analysis of African measles virus field isolates: Identification and localization of one conserved and two variable epitope sites on the NP protein, *Virus Res.* **10:**137–152.

Greer, P. A., Hasel, K. W., and Millward, S., 1986, Cloning and *in vitro* expression of the measles virus matrix gene, *Biochem. Cell Biol.* **65:**1038–1043.

Hasel, K. W., Day, S., Millward, S., Richardson, C. D., Bellini, W. J., and Greer, P. A., 1987, Characterization of cloned measles virus mRNAs by *in vitro* transcription, translation and immuno-precipitation, *Intervirology* **28:**26–39.

Hsu, D., Yamanaka, M., Miller, J., Dale, B., Grubman, M., and Yilma, T., 1988, Cloning of the fusion gene of rinderpest virus: comparative sequence analysis with other morbilliviruses, *Virology* **166:**149–153.

Jang, S. K., Davies, M. V., Kaufman, R. J., and Wimmer E., 1989, Initiation of protein synthesis by internal entry of ribosomes into the 5′ non-translated region of encephalomyocarditis virus RNA *in vivo, J. Virol.* **63:**1651–1660.

Lamb, R. A., Mahy, B. W. J., and Choppin, P. W., 1976, The synthesis of Sendai-virus polypeptides in infected cells, *Virology* **69:**116–131.

Mahy, B. W. J., Barrett, T., Evans, S., Anderson, E. C., and Bostock, C. J., 1988, Characterization of a seal morbillivirus, *Nature* **336:**115.

Meier, E., Harmison, G. G., and Schubert, M., 1987, Homotypic and heterotypic exclusion of VSV replication by high levels of recombinant protein L., *J. Virol.* **61:**3133–3142.

Moench, T. R., Griffin, D. E., Obriecht, C. R., Vaisberg, A. J., and Johnson, R. T., 1988, Acute measles in patients with and without neurological involvement: Distribution of measles virus antigen and RNA, *J. Inf. Dis.* **158:**433–442.

Osterhaus, A. D. M. E., and Vedder, E. J., 1988, Identification of virus causing recent seal deaths, *Nature* **335:**20.

Osterhaus, A. D. M. E., Utydehaag, F. G. C. M., Visser, I. K. G., Vedder, E. J., Reijnders, P. J. H., Kuiper, J., and Brugge, H. N., 1989, Seal vaccination success, *Nature* **337:**21.

Pastoret, P.-P., Brochier, B., Languet, B., Thomas, I., Paquot, A., Baudin, B., Kieny, M. P., Lecocq, J. P., Debruyn, J., Costy, F., Antoine, H., and Desmettre P., 1988, Field trial of fox vaccination against rabies using a vaccinia–rabies recombinant virus, *Vet. Rec.* **123:**481–483.

Pelletier, J., and Sonenberg, N., 1988, Internal initiation of translation of eukaryotic mRNA directed by a sequence derived from polio virus RNA, *Nature* **334:**320–325.

Plowright, W., and Ferris, R. D., 1962, Studies with rinderpest virus in tissue culture. The use of attenuated culture virus as a vaccine for cattle, *Res. Vet. Sci.* **3:**172–182.

Richardson, C., Hall, D., Greer, P., Hasel, K., Berkovich, A., Englund, G., Bellini, W. J., Rima, B., and Lazzarini, R., 1986, The nucleotide sequence of the mRNA encoding the fusion protein of measles virus (Edmonston strain): A comparison of fusion proteins from several different paramyxoviruses, *Virology* **155:**508–523.

Rima, B. K., Baczko, K., Clarke, D. K., Curran, M. D., Martin, S. J., Billeter, M. A., and ter Meulen, V., 1986, Characterization of clones for the sixth (L) gene and a transcriptional map for morbilliviruses, *J. Gen. Virol.* **67:**1971–1978.

Rozenblatt, S., Eizenberg, O., Ben-Levy, R., Lavie, V., and Bellini, W. J., 1985, Sequence homology within the morbilliviruses, *J. Virol.* **53:**684–690.

Shaila, M. S., Purushothaman, V., Bhavasar, D., Venugopal, K., and Venkatesan, R. A., 1989, Peste des petits ruminants of sheep in India, *Vet. Record* **125:**602.

Sheshberadaran, H., Norrby, E., McCullough, K. C., Carpenter, W. C., and Orvell, C., 1986, The antigenic relationship between measles, canine distemper and rinderpest viruses studied with monoclonal antibodies, *J. Gen. Virol.* **67:**1381–1392.

Thomas, S. M., Lamb, R. A., and Paterson, R., 1988, Two mRNAs that differ by two non-templated nucleotides encode the amino co-terminal proteins P and V of the paramyxovirus SV5, *Cell* **54:**891–902.

Toyoda, T., Sakaguchi, T., Imai, K., Inocencio, N. M., Gotoh, B., Hamaguchi, M., and Nagai, Y., 1987, Structural comparison of the cleavage activation site of the fusion glycoprotein between virulent and avirulent strains of Newcastle disease virus, *Virology* **158:**242–247.

Tsukiyama, K., Sugiyama, M., Yoshikawa, Y., and Yamanouchi, K., 1987, Molecular cloning and sequence analysis of the rinderpest virus mRNA encoding the hemagglutinin protein, *Virology* **160:**48–54.

Tsukiyama, K., Yoshikawa, Y., and Yamanouchi, K., 1988, Fusion glycoprotein (F) of rinderpest virus: Entire nucleotide sequence of the mRNA, and several features of the F protein, *Virology* **164:**523–530.

Wong, T. C., and Hirano, A., 1987, Structure and function of bicistronic RNA encoding the phosphoprotein and matrix protein of measles virus, *J. Virol.* **61:**584–589.

Wong, T. C., Wipf, G., and Hirano, A., 1987, The measles virus matrix gene and gene product defined by *in vitro* and *in vivo* expression, *Virology* **157:**497–508.

Yamanaka, M., Hsu, D., Crisp, T., Dale, B., Grubman, M., and Yilma, T., 1988, Cloning and sequence analysis of the hemagglutinin gene of the virulent strain of the rinderpest virus, *Virology* **166:**251–253.

Yilma, T., Hsu, D., Jones, L., Owens, S., Grubman, M., Mebus, C., Yamanaka, M., and Dale, B., 1988, Protection of cattle against rinderpest with vaccinia virus recombinants expressing the *HA* or *F* gene, *Science* **242:**1058–1061.

CHAPTER 4

The Molecular Biology of Human Respiratory Syncytial Virus (RSV) of the Genus *Pneumovirus*

PETER L. COLLINS

I. INTRODUCTION

The genus *Pneumovirus* of the family *Paramyxoviridae* contains pneumonia virus of mice (PVM), turkey rhinotracheitis virus (TRT), and the respiratory syncytial viruses of humans (RSV), cattle (BRSV), goats, and sheep (Kingsbury *et al.*, 1978; McIntosh and Chanock, 1985; Stott and Taylor, 1985; Spraker *et al.*, 1986; Collins and Gough, 1988). Like other members of the paramyxovirus family, pneumoviruses are pleomorphic, enveloped, cytoplasmic viruses that contain a single strand of negative-sense genomic RNA (vRNA).

The family *Paramyxoviridae* contains two other genera: *Morbillivirus*, which includes (naming only those viruses cited in this review) measles virus (MeV); and *Paramyxovirus*, which includes Sendai virus (SV), human parainfluenza virus types 1, 2, and 3 (PIV1, 2, and 3), mumps virus (MuV), simian virus type 5 (SV5), and Newcastle disease virus (NDV) (see Chapters 2 and 3).

The pneumoviruses are classified as a separate genus because of differences in the diameter of the nucleocapsid and the lack of detectable viral hemagglutination activity for RSV (Kingsbury *et al.*, 1978; Richman *et al.*, 1971), although PVM does have a hemagglutinin (Berthiaume *et al.*, 1974; Ling and Pringle, 1989b). The pneumoviruses, like the morbilliviruses, lack a neuraminidase. Additional support for the separate classification of the pneumoviruses is provided by molecular studies described below.

PETER L. COLLINS • Laboratory of Infectious Diseases, National Institute of Allergy and Infectious Diseases, National Institutes of Health, Bethesda, Maryland 20892.

RSV is the most thoroughly characterized pneumovirus because of its importance in human health: RSV is the single most important viral agent of pediatric respiratory tract disease worldwide (McIntosh and Chanock, 1985; Chanock *et al.*, 1988; Murphy *et al.*, 1988). Conventional approaches to vaccine development (such as the development of attenuated live strains for intranasal administration or of concentrated viral preparations for intramuscular injection) have been unsuccessful (McIntosh and Chanock, 1985; Stott and Taylor, 1985; Chanock *et al.*, 1988; Murphy *et al.*, 1988). Molecular studies have been of particular interest because new information and materials obtained with recombinant DNA techniques offer the basis for new approaches to vaccine development.

Preliminary molecular studies comparing other pneumoviruses to RSV indicate that they correspond closely by gene map and by the numbers and types of gene products (Cash *et al.*, 1977; Cavanagh and Barrett, 1988; Ling and Pringle, 1988, 1989a,b; Lerch *et al.*, 1989). Additionally, significant antigenic relatedness exists between RSV and BRSV and between RSV and PVM, and there is suggestive evidence for antigenic relatedness between PVM and TRT (Taylor *et al.*, 1984; Stott *et al.*, 1984; Gimenez *et al.*, 1984; Örvell *et al.*, 1987; Kennedy *et al.*, 1988; Ling and Pringle, 1988, 1989a; Beeler and Coelingh, 1989). Therefore, RSV appears to be representative of the genus.

This chapter summarizes recent information on the molecular biology of RSV as the prototype of the pneumoviruses, and is intended to complement published reviews and other chapters of this volume that discuss RSV epidemiology, immunobiology, pathogenesis, and vaccine development (McIntosh and Fishout, 1980; McIntosh and Chanock, 1985; Stott and Taylor, 1985; Murphy *et al.*, 1988; also see Chapters 14 and 18).

Significant impediments to studies of RSV are its poor growth in tissue culture, its biological and biochemical instability, the pleomorphism of the virion, and its tendency to remain closely associated with cellular material. Because of this, the discussions of some aspects of RSV molecular biology draw heavily from observations made with other paramyxoviruses and with vesicular stomatitis virus (VSV), which is a member of the other family of nonsegmented negative-strand viruses, *Rhabdoviridae* (Emerson, 1985). Because RSV has a number of significant differences in RNA and protein structure relative to these other viruses, these extrapolations are offered with caution.

II. STRUCTURES OF THE RSV VIRION, RNAs, AND PROTEINS

A. Virion Structure

RSV virions grown in tissue culture are heterogeneous in size and shape, consisting of pleomorphic spherical particles that are approximately 80–350 nm in diameter and filamentous particles that are approximately 60–100 nm in diameter and are up to 10 μm in length (Armstrong *et al.*, 1962; Norrby *et al.*, 1970; Kalica *et al.*, 1973; Bachi and Howe, 1973; Berthiaume *et al.*, 1974; Bachi, 1988).

The RSV virion consists of a ribonucleoprotein nucleocapsid contained

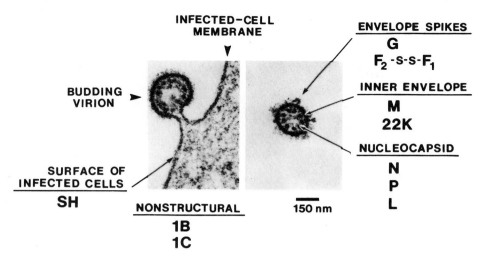

FIGURE 1. Electron photomicrographs of RSV strain A2 showing the locations of the ten major viral proteins. Left: A spherical virion budding from the plasma membrane of an RSV-infected HeLa cell. Right: A released spherical virion. The F protein is indicated as the two disulfide-linked subunits F_1 and F_2. The photomicrographs were adapted from published work (Kalica et al., 1973).

within a lipoprotein envelope. The nucleocapsid contains the single strand of negative-sense genomic vRNA (Huang and Wertz, 1982). Upon exposure to UV irradiation, RSV virions lose infectivity with single-hit kinetics, indicating that, despite their heterogeneity, most of the infectious particles contain a single functional vRNA molecule (Dickens et al., 1984; and our unpublished data).

Virions assemble at the plasma membrane of the infected cell and mature by budding, during which the intracellular nucleocapsid is packaged within a viral envelope that is derived directly from the host-cell plasma membrane (Fig. 1). The envelope contains spikelike projections that are approximately 11–20 nm in length and are spaced at intervals of 6–10 nm (Norrby et al., 1970; Bachi and Howe, 1973; Berthiaume et al., 1974; Bachi, 1988). Thus, the structure and maturation of RSV virions have general similarities to those of other paramyxoviruses. However, some differences exist: (1) the RSV nucleocapsid is narrower than those of other paramyxoviruses, (2) RSV has an additional structural protein (the 22K protein, described below), and (3) RSV virions are extremely labile, although infectivity can be stabilized by sucrose, magnesium sulfate, or other agents (Ueba, 1978; Fernie and Gerin, 1980).

Ten major RSV proteins have been identified, and most of these are structural components of the virion (Fig. 1, Table I). The G and F glycoproteins are the structural elements of the virion spikes and mediate viral attachment and penetration, respectively. Consistent with their surface localization, F and G can be removed with trypsin and are the two major antigens involved in in vitro neutralization of viral infectivity (Peeples and Levine, 1979; Taylor et al., 1984; Elango et al., 1986; Olmsted et al., 1986; Stott et al., 1986; Wertz et al., 1987; Walsh et al., 1987a,b; King et al., 1987; Johnson et al., 1987a; Routledge et al., 1988).

TABLE I. Genes and Proteins of RSV Strain A2

Gene	Gene length[a]	Protein length[b]	Protein M_r[c]	After glycosylation	Comments
Glycoproteins					
F	1903	574	63,453	68,000 to 70,000	Spike protein involved in viral penetration; precursor cleaved to disulfide-linked F_2 (1–130)-ss-F_1(130–574)
G	923	298	32,587	84,000 to 90,000	Spike protein involved in attachment; contains N- and O-linked carbohydrate; membrane-bound and secreted forms
SH (or 1A)	410	64	7,536	13,000, to 15,000, 21,000 to 30,000	Appears to be a structural component; multiple glycosylated and unglycosylated forms; an M_r 4800 form is made by initiation at second AUG
Membrane					
M	958	256	28,717		Basic, moderately hydrophobic; phosphorylated subpopulation
22K (or M2)	961	194	22,153		Basic, hydrophilic; phosphorylated and disulfide-linked subpopulations
Nucleocapsid					
N	1203	391	43,423		Tightly bound to vRNA
P	914	241	27,150		Acidic, phosphorylated
L	6578	2165	250,226		Basic, hydrophobic
Nonstructural					
1C (or NS1)	532	139	15,567		Slightly acidic; processing reduces M_r by 1500
1B (or NS2)	503	124	14,674		Basic

[a]Chain length in nucleotides.
[b]Chain length in amino acids.
[c]For the complete, unmodified polypeptide moiety, calculated from the predicted sequence.

In experiments in which RSV virions were dissociated with detergent and salt, the solubility characteristics of the M and 22K proteins suggested that both proteins are located on the inner face of the envelope (Peeples and Levine, 1979; Huang et al., 1985). Peeples and Levine (1979) noted that both proteins were solubilized more readily than are the M proteins of other paramyxoviruses and suggested that this might be one factor in the greater lability of RSV virions.

The viral nucleocapsid consists of vRNA and the N, P, and L proteins (Wunner and Pringle, 1976; Peeples and Levine, 1979; Huang et al., 1985). The association of N protein with vRNA is stable in nonionic detergent and high salt, as is characteristic of the nonsegmented negative-strand viruses (Wunner and Pringle, 1976; Peeples and Levine, 1979). The two other nucleocapsid proteins, P and L, are present in lower molar amounts and are less tightly bound to the nucleocapsid. By analogy to other paramyxoviruses, it is assumed that the RSV nucleocapsid contains an RNA-dependent RNA polymerase that directs the synthesis of RSV RNAs during transcription and replication.

The status of the remaining three RSV proteins, 1C, 1B, and SH (also called NS1, NS2, and 1A, respectively), as structural or nonstructural has been difficult to resolve because virion preparations characteristically are heavily contaminated by cellular debris and rapidly lose infectivity during purification (although this can now be remedied somewhat by the addition of magnesium sulfate as a stabilizing agent). The 1B and 1C proteins are considered to be nonstructural, whereas the SH protein appears to be a structural component of the virion (Huang et al., 1985; also see this chapter, Section II.F.6).

B. Overview: Identification of Genomic RNA (vRNA), mRNAs, and Proteins

1. Identification of vRNA

RSV virion preparations characteristically are heavily contaminated by intracellular viral and cellular RNAs (for example, Lambert et al., 1980). Identification of vRNA was achieved in metabolic labeling studies using actinomycin D, which strongly inhibits cellular transcription but does not affect RSV replication (Wunner et al., 1975; Lambert et al., 1980). Virions labeled under these conditions with [^{3}H]uridine contained a single major radiolabeled species of 50 S that was thereby identified as vRNA (Huang and Wertz, 1982). vRNA from highly purified virions was digested essentially to completion by the single-strand-specific nuclease RNase A, indicating that it was a single-stranded and of a single polarity (Huang and Wertz, 1982). It lacked polyadenylate as assayed by oligo(dT) chromatography (Huang and Wertz, 1982). Sequence analysis (described below) is nearly complete for vRNA of strain A2 and shows that it is a nonpolyadenylated strand that is approximately 15,222 nucleotides in length, similar in size to vRNA of SV (15,285 nucleotides), PIV3 (15,461), NDV (15,156), and MeV (15,892), and substantially larger than vRNA of the rhabdoviruses VSV (11,161) and rabies virus (11,863) (Schubert et al., 1984; Tordo et al., 1988; also see Chapters 2 and 3).

2. Demonstration of a Negative-Strand Coding Strategy

Intracellular RSV RNA radiolabeled in the presence of actinomycin D was separated by gel electrophoresis into a complex pattern (Lambert et al., 1980; Huang and Wertz, 1982; Collins and Wertz, 1983; Collins et al., 1984b; see this chapter, Section II.D.1 below). The largest gel band was concluded to contain 50S vRNA based on its comigration with virion-derived vRNA and its lack of polyadenylate. The other species ranged in estimated sizes from 500 to 7500 nucleotides and were polyadenylated (although small amounts of non-polyadenylated forms of some species also were observed). This showed that RSV RNA synthesis produces a number of subgenomic polyadenylated RNAs as well as progeny vRNA.

When added to *in vitro* translation systems, the poly(A)-containing infected-cell RNA directed the synthesis of several RSV proteins, including the N, P, and M proteins, an M_r 59,000 nonglycosylated form of the F protein, and several other proteins that appeared to be RSV-specific (Cash *et al.*, 1979; Huang and Wertz, 1983; Collins *et al.*, 1984b). This showed that the subgenomic poly(A)-containing RNAs are mRNAs.

Whereas virion-associated vRNA was highly sensitive to RNase A, hybridization of vRNA with infected-cell mRNA rendered it 93% resistant (Huang and Wertz, 1982). This showed that RSV vRNA is a single negative strand that is completely or almost completely transcribed into subgenomic, polyadenylated mRNAs.

3. Molecular Cloning of Strain A2. Identification of the Major mRNAs and Proteins

cDNA was synthesized using infected-cell mRNA or virion-associated vRNA as template and was cloned in the *Escherichia coli* plasmid pBR322 (Venkatesan *et al.*, 1983; Collins and Wertz, 1983). cDNAs were obtained for ten different viral genes, which account for most (97.3%) of RSV vRNA (Collins *et al.*, 1984b, 1986). The ten genes have been sequenced in their entirety (Table I) (Satake and Venkatesan, 1984; Satake *et al.*, 1984; Collins *et al.*, 1984a, 1985; Collins and Wertz, 1985a–c; Wertz *et al.*, 1985; Satake *et al.*, 1985; Elango *et al.*, 1985a,b) (for the *L* gene, David S. Stec and P. L. Collins, unpublished data).

Each cDNA was used to purify the corresponding mRNA by hybridization-selection, and these were identified by translation *in vitro*. Each mRNA encoded a single major polypeptide whose M_r was consistent with the length of the major translational open reading frame (ORF) (Venkatesan *et al.*, 1983; Collins *et al.*, 1984b; Satake and Venkatesan, 1984; Satake *et al.*, 1984; Elango *et al.*, 1985b). However, this analysis was not successful for the L mRNA and protein because of the difficulty of synthesizing a protein of M_r 250,000 *in vitro*.

When analyzed by polyacrylamide gel electrophoresis in the presence of sodium dodecyl sulfate (SDS–PAGE) in parallel with radiolabeled proteins from RSV-infected cells, the N, P, M, 22K, SH, and 1B proteins synthesized *in vitro* comigrated with their authentic counterparts from infected cells (Huang and Wertz, 1983; Venkatesan *et al.*, 1983; Collins *et al.*, 1984b; Satake and Venkatesan, 1984; Satake *et al.*, 1985; Elango *et al.*, 1985b). This suggested that these proteins are not processed extensively in the infected cell. For the 1C protein, a small (M_r 1500) reduction in size was observed in pulse-chase experiments, presumably due to posttranslational processing (Table I; see below) (Huang *et al.*, 1985; and our unpublished data). The identities of the *in vitro*-synthesized proteins were confirmed by peptide mapping (Huang and Wertz, 1983; Venkatesan *et al.*, 1983; Collins *et al.*, 1984b).

The G and F proteins were synthesized *in vitro* in their unglycosylated forms, which were identified by immunoprecipitation and peptide mapping (Wertz *et al.*, 1985; Huang, 1983). Amino acid sequencing of purified F and G proteins from infected cells provided partial sequences that were in agreement with those predicted by the nucleotide sequencing, confirming the identifications (Satake *et al.*, 1985; Elango *et al.*, 1985a; Hendricks *et al.*, 1988).

The L mRNA was identified by its large size, by the position of its gene in the gene map (described in the following section), and because its predicted protein has sequence relatedness with L proteins of other paramyxoviruses (our unpublished data, and Section II.F.1 below).

C. Genetic Map of Strain A2

1. Gene Order and Gene Junction Sequences

Intergenic and flanking gene regions in vRNA were analyzed by dideoxynucleotide sequencing of purified vRNA (Collins *et al.*, 1986). This determined the map order (Fig. 2) and provided sequences for the intergenic regions (Fig. 3).

Each gene starts with a conserved nine-nucleotide sequence, 3' CCC-CGUUUA, that is termed the gene-start signal. It encodes the 5' end of the corresponding mRNA and was conserved exactly among the nine smaller genes. Each gene-start signal is preceded by an A residue, suggesting that this nontranscribed residue also is part of the signal. This nontranscribed A residue also precedes the gene-start signal of five of six genes sequenced to date for RSV strain 18537 (Johnson and Collins, 1988b).

The downstream end of each gene terminates with a semiconserved 12- to 13-nucleotide sequence, termed the gene-end signal. In mRNA, the complement of this sequence constitutes the 3' end immediately preceding the poly(A) tail. The gene-end sequence exhibits considerable, but not exact, conservation among the different genes. It consists of the pentanucleotide 3' UCₒᵤU, followed by three or four variable nucleotides, followed by a run of four to seven U residues.

The nine smaller RSV genes are separated by short intergenic regions that by definition usually are not transcribed into mRNA. For strain A2, the intergenic regions vary in length from 1 to 52 nucleotides and do not contain any significant conserved sequence or obvious secondary structure except for the single conserved A residue noted above (Fig. 3).

Thus, whereas the conserved gene-start and gene-end sequences are likely to have functional significance, the diversity of the intergenic regions suggested that they might be nonspecific spacers.

2. Overlap between the *22K* and *L* Genes

Sequence analysis of RSV vRNA and cloned cDNAs of L mRNA showed that the *L* gene follows the *22K* gene in the 3' to 5' gene order (Fig. 2) (Collins *et al.*, 1987). Surprisingly, the gene-start signal for *L* is located within the *22K* gene, 68 nucleotides upstream of the end of the *22K* gene. Thus, the two genes

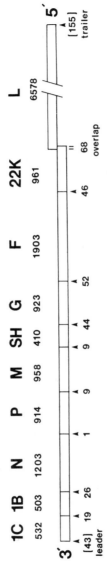

FIGURE 2. Genetic map of vRNA (3' to 5') of RSV strain A2. The nucleotide lengths of the ten genes are indicated over the map, and the lengths of intergenic, gene overlap, and proposed leader and trailer regions are indicated below the map. The gene lengths include the short oligo(U) tracts in vRNA that follow each gene and encode the poly(A) tail. The length of the leader region is in parentheses because its mapping and sequencing are incomplete. The genes are to scale except for L. The nucleotide length of vRNA is 15,222 (see Note added in proof).

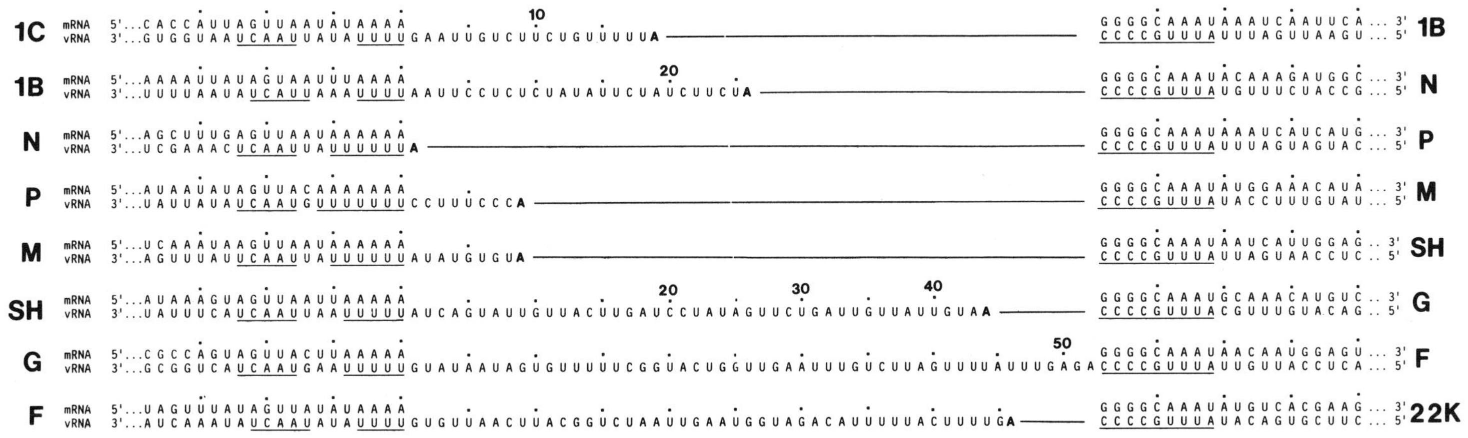

FIGURE 3. Intergenic regions of the nine smaller genes of RSV strain A2. Each sequence segment represents a gene junction, containing the downstream end of a gene and its encoded mRNA (left, double-stranded sequence), followed by the intergenic region (middle, single-stranded sequence) and the start of the adjacent downstream gene and its mRNA (right, double-stranded sequence). The 3' poly(A) tails of the mRNAs are not shown. Gene-start and gene-end sequences are underlined in the vRNA strand, and the intergenic regions are numbered. The junctions are listed in their 3' to 5' order in vRNA.

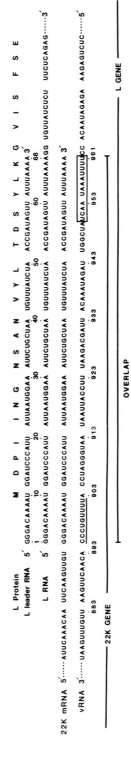

FIGURE 4. Overlap of the *22K* and *L̇* genes of RSV strain A2. The bottom sequence is vRNA (3' to 5') numbered according to the *22K* gene, which ends at position 961. The *L* gene-start (underlined) and *22K* gene-end (boxed) sequences are shown. Immediately above vRNA is the sequence of the downstream end of the 22K mRNA (5' to 3') without poly(A). Immediately above the 22K mRNA is the sequence of the upstream end of the L mRNA (5' to 3'). The top nucleotide sequence is of the L leader RNA without the poly(A) tail. The start of the L ORF is shown by means of the single-letter amino acid designations.

overlap by 68 nucleotides (Fig. 4). The two genes encode separate mRNAs that each contain the 68-nucleotide sequence encoded by the overlap. The overlap involves sequences that encode L protein, whereas the 22K mRNA overlap sequences are noncoding (Fig. 4). The *L* gene-start sequence contains two nucleotide differences from the others (3'CCCUGUUUU).

In general, the mRNAs of nonsegmented negative-strand viruses are non-overlapping. The overlap described here for RSV is an exception, but it is not unique: an analogous situation of overlapping mRNAs and genes exists for the filovirus (presently classified in *Rhabdoviridae*) Ebola virus (Sanchez *et al.*, 1987). The situation of "gene overlap" described here for RSV is different from the "gene overlap" that occurs for the P and C genes of SV, PIV3, and MeV (Chapters 2 and 3). For those viruses, the ORFs encoding the P and C proteins overlap, but are transcribed into an mRNA that encodes both proteins and that does not overlap with other mRNAs.

3. 3' and 5' Ends of vRNA

In other paramyxoviruses, the first gene in the 3' to 5' map, *NP*, is preceded by a 51- to 53-nucleotide leader region encoding a short leader RNA that initiates exactly at the 3' vRNA end (Chapter 8).

A partial sequence for the 3' end of RSV vRNA has been determined by direct chemical sequence analysis of 3' end-labeled vRNA and by analysis of a cDNA clone that was constructed using as template vRNA that had been polyadenylated at the 3' end with poly(A) polymerase and primed for cDNA synthesis with oligo(dT). These preliminary results indicated that the *1C* gene in vRNA is preceded by a leader region of 43–46 nucleotides. To date, 39 of the nucleotides have been identified (Fig. 5, bottom line). Additional work is required to determine the exact length of the leader region and to identify the 3'-terminal four to six nucleotides. [In Figs. 2 and 13, the value of 43 was used for the 3' leader region. (see Note Added in Proof)].

Primer extension on intracellular RSV mRNA identified a subclass of 1C mRNA that contained a 5'-terminal extension of 43 or 44 nucleotides (Collins and Wertz, 1985c; and our unpublished data) and is presumed to be a read-through transcript of the RSV leader region and the *1C* gene. Analogous leader-N readthrough mRNAs have been described for VSV and MeV (Herman and Lazzarini, 1981b; Castandea and Wong, 1989). A partial sequence for this 43- or

```
5'- NGG   AAAAAAAUGC GUACAACAAA CUUGCAUAAA CCAAAANNNN GGGGCAAAUA...
3'-(N)3-6 NUUUUUUACG CAUGUUGUUU GAACGUAUUU GGUUUUUUUA CCCCGUUUAU...
                                                         1C gene-start
```

FIGURE 5. Mapping and sequencing of the 3' end (leader region) of RSV vRNA of strain A2 and of a putative leader RNA. The upper line is a partial sequence (mRNA-sense) of a readthrough mRNA of the leader region and *1C* gene, determined from sequencing infected-cell mRNA. The lower line is a partial sequence (vRNA-sense) of the 3' end of vRNA, determined by sequence analysis of end-labeled vRNA and of a cloned cDNA of vRNA (M. A. Mink and P. L. Collins, unpublished data). The nine-nucleotide conserved gene-start sequence of the *1C* gene is underlined. N: unidentified nucleotide. (see Note added in proof).

44-nucleotide extension was determined by primer extension on intracellular mRNA under conditions of dideoxynucleotide sequencing and by chemical sequencing of the extended primer. As shown in Fig. 5 (top line), the available sequence was the exact complement of that determined from vRNA, confirming the identities of these sequences.

The putative RSV leader region is slightly shorter and relatively unrelated in sequence to the other paramyxovirus leader regions, which as a group have similar lengths and significant sequence relatedness. The sequence and mapping analysis shown in Fig. 5 raises the possibility that transcription of the RSV leader RNA might not initiate exactly at the 3' end of vRNA, but instead might initiate as many as three nucleotides from the end.

It is not known whether RSV-infected cells contain free leader RNA. Herman (1989) established an *in vitro* transcription reaction for RSV using lysolecithin-treated, RSV-infected cells and used gamma-thio-GTP to specifically label newly-initiated, uncapped transcripts. The products labeled under those conditions included a minor species of the appropriate size to be free leader RNA, but it was not identified directly.

Dideoxynucleotide sequencing of the 5' end of RSV vRNA together with sequence analysis of cloned cDNAs of vRNA showed that the *L* gene-end signal is followed by a 155-nucleotide "trailer" region (Fig. 6). Its length is substantially greater than those of the corresponding regions in SV, MeV, and PIV3 (54, 40, and 44 nucleotides, respectively) (Chapters 2 and 3).

Complementarity between the 3' and 5' ends of vRNA is characteristic of paramyxoviruses and rhabdoviruses (Emerson, 1985; Kingsbury, 1985, 1989; also see Chapters 2, 3, and 8). For example, the 25 nucleotides at the 3' and 5' termini of SV, MeV, and PIV3 have 75% sequence complementarity. Consistent with this, for RSV there is 81% sequence complementarity (with two single-nucleotide gaps and two single-nucleotide mismatches) for an alignment involving approximately 20 nucleotides from each terminus (not shown). The identification of an A residue as the 5'-terminal nucleotide of RSV vRNA (Fig. 6) suggests that the 3'-terminal nucleotide is a U residue, as is the case for several other paramyxoviruses.

```
     L Gene-End
... UCAAUAAUUUUU AAUUUUUAGU AUAUUAAAAA AUUUAUUGAA AAUCACUUGA

    UUAGGAUUUC AAAUAGUAAAA UUAGAACCUC CUUAUUUAAA UUUGGGAUUA

    GAUUAACCAA AUAUACACAU AAUUGAUUUA AUGCUCUAUA AUCAAAAACU

    GUGAAAAAAA GAGCA-5'
```

FIGURE 6. Sequence (vRNA-sense) of the 5' extracistronic (trailer) region of vRNA of strain A2. The gene-end sequence of the *L* gene is shown (underlined), followed by the 155-nucleotide extracistronic region determined by dideoxynucleotide sequencing of purified vRNA and from sequencing cloned cDNAs of vRNA (D. S. Stec and P. L. Collins, unpublished data).

D. Structures of the mRNAs

1. Major mRNAs

The gel pattern of metabolically labeled intracellular RSV mRNAs contained at least six major RNA gel bands and at least nine minor bands (Huang and Wertz, 1982; Collins and Wertz, 1983; Collins *et al.*, 1984b). In RNA blot hybridization experiments, cDNA for each gene hybridized to a single major RNA species of a size that was consistent with that predicted from the nucleotide sequence (Collins and Wertz, 1983; Collins *et al.*, 1984b). The gel pattern typically contains less than ten major RNA bands because some of the mRNAs comigrate due to close similarity in size (Table I).

The 5' end of each mRNA was mapped and sequenced by primer extension, and the 3' end was mapped by the location of terminal poly(A) tracts in the cDNA clones (Satake and Venkatesan, 1984; Collins *et al.*, 1984a; Satake *et al.*, 1984, 1985; Collins and Wertz, 1985a–c; Collins *et al.*, 1985; Wertz *et al.*, 1985; Elango *et al.*, 1985a,b; Collins *et al.*, 1987). The 5' end of each mRNA is the complement of the gene-start sequence; the 3' end is the complement of the gene-end signal followed by the poly(A) tail. The 5' ends of the mRNAs have not been further characterized, but, by analogy to other nonsegmented negative-strand viruses, might be capped and methylated by viral enzymatic activities associated with the nucleocapsid (Emerson, 1985; Kingsbury, 1985; also see Chapter 9).

For SV5, MeV, and SV, it was recently shown that "RNA editing" during transcription of the *P* gene gives rise to two forms of the P mRNA (Thomas *et al.*, 1988; Cattaneo *et al.*, 1989; S. Vidal and D. Kolakofsky, personal communication, cited in Cattaneo *et al.*, 1989). For example, for SV and MeV, one mRNA is an exact copy of vRNA and encodes the P and C proteins. A second species contains an insertion of nontemplated nucleotides at one specific site. This shifts the reading frame into an internal ORF, and the translation product of the resulting chimeric ORF is a chimeric protein, designated V, that consists of N-terminal residues of P fused to a polypeptide domain encoded by the internal ORF. It will be important to determine whether analogous structural microheterogeneity exists for the RSV mRNAs and might result in the synthesis of additional RSV proteins that have not yet been identified.

2. Polytranscripts

In RNA blot experiments described in the previous section, each of the cDNAs also hybridized with one or more of the nine minor RNA species (Collins and Wertz, 1983; Collins *et al.*, 1984b). These minor mRNAs were identified as polytranscripts, which are complete, linked transcripts of two or more adjacent genes. The leader–1C readthrough transcript described above and in Fig. 5 is another example of a polytranscript. Polytranscripts have been identified for every gene junction except for the one between the *SH* and *G* genes. Also, the polytranscript spanning the *G–F* junction was present in very low abundance. The reason for the absence or greatly reduced accumulation of certain polytranscripts is unknown; the others are present at 5–10% the level of the major mRNAs.

The structure of the gene junction region in several polytranscripts was determined by dideoxynucleotide sequencing of intracellular mRNA and by sequencing cDNAs of polytranscripts (Collins and Wertz, 1985a–c; Collins et al., 1985; Elango et al., 1985b). In all cases, the intergenic region in each polytranscript was the exact complement of the corresponding region on vRNA. This showed that the polytranscripts are generated by precise transcriptional readthrough across intergenic regions.

The synthesis of polytranscripts is a common feature of rhabdovirus and paramyxovirus transcription. Because the polytranscripts appear to contain complete gene transcripts, their 5'-proximal cistrons might function as mRNAs. Consistent with this expectation, polytranscripts of NDV have been shown to be associated with polysomes isolated from NDV-infected cells (Wilde and Morrison, 1984). It is likely that the polytranscripts of the nine smaller RSV genes are not essential features of RSV replication and are generated by polymerase error or by a mechanism similar to that responsible for the readthrough that occurs during vRNA replication.

3. L Leader RNA

Because transcription of the L gene initiates within the 22K gene, the polymerase must transcribe across the 22K gene-end signal in order to synthesize full-length L mRNA. The products of transcription of the L gene were analyzed, and the most abundant product was shown to be a 68-nucleotide transcript, called L leader RNA, that initiates at the L gene-start signal, terminates at the 22K gene-end signal, and contains a poly(A) tail (Fig. 4) (Collins et al., 1987). Full-length L mRNA was a much less abundant product and appeared to result from readthrough of the 22K gene-end signal, the same mechanism that generates the polytranscripts. Thus, the presence of the gene-end signal within the L gene results in a drastic attenuation of synthesis of full-length L mRNA (Collins et al., 1987).

L leader RNA contains the first 20 codons of the L ORF and therefore could encode a short protein (Fig. 4). The ORF in L leader RNA continues without termination into the poly(A) tail, which would be translated into a polylysine tail of approximately 30–70 residues, assuming a poly(A) length of 100–200 residues. The possible synthesis of this polypeptide has not been investigated.

4. Major ORFs

Each of the major RSV mRNAs contains a major ORF of the appropriate size to encode the corresponding polypeptide (Fig. 7). These are flanked by 5' and 3' noncoding regions that usually are short and lack noteworthy structural features.

For most of the mRNAs, the probable translational initiation codon is the first methionine codon in the sequence and is in a sequence context favored for efficient translational initiation (Kozak, 1987). One exception is the SH mRNA of strain A2 in which the initiation codon is in an unfavored sequence context (Collins and Wertz, 1985b). As described below (Section II.F.6), this allows some ribosomes to initiate translation at the second AUG in the ORF, resulting in the synthesis of a truncated form of the SH protein. The second exception is the G mRNA (of the three strains, A2, Long, and 18537, sequenced to date), in which the major ORF initiates with the second methionine codon in the sequence (Wertz et al., 1985; Satake et al., 1985; Johnson et al., 1987b). The first AUG in each G mRNA initiates a short ORF (15 codons for strain A2 and Long and 36 for strain 18537) that overlaps the translational start site of the major ORF and therefore might divert ribosomes from initiation at the major ORF. In general, these features of the SH and G mRNAs might have the effect of reducing the efficiency of translation of the major ORF, providing a modest amount of down-regulation of gene expression at the level of translation.

5. Additional ORFs

The M and 22K mRNAs of strain A2 each contain a second, internal ORF that initiates, respectively, at nucleotide 735 (reading frame 2) and 563 (reading frame 2) (Fig. 7). Each of these internal ORFs partially overlaps the upstream major ORF and could encode a polypeptide of 75 and 90 amino acids, respectively (Satake and Venkatesan, 1984; Collins and Wertz, 1985a; Elango et al., 1985b).

The second ORF in the 22K mRNA also occurs in RSV strain 18537 (our unpublished data). At the nucleotide level, the second ORF has 72% sequence identity between strains A2 and 18537, compared to 84% identity for the primary ORF and 34% identity for noncoding sequences in the 22K mRNA apart from the conserved termini. At the amino acid level, the predicted protein for the second ORF has 62% identity between the two strains, compared with the values of 53–96% sequence identity for the other proteins (Table II). The possible existence of this second polypeptide domain in infected cells can be investigated by immunoprecipitation using peptide-specific antibodies. The possible conservation of the second ORF of the M mRNA remains to be investigated.

Because the second ORFs in the M and 22K mRNAs are internal and are separated from the 5' end by many intervening AUG codons, they presumably would not be efficiently utilized by scanning ribosomes. However, there are situations where internal AUGs are utilized (Chapter 7). Alternatively, these ORFs might be accessed by ribosomes as they exit the primary ORF. Another possibility is that internal ORFs might be accessed by frameshifting due to the insertion of nontemplated nucleotides by RNA editing (Thomas et al., 1988; Cattaneo et al., 1989).

Numerous other short ORFs exist in the RSV mRNAs. For example, the N mRNA of strain A2 contains a second ORF that contains 43 codons and might be utilized because it initiates with the second AUG of the mRNA (Fig. 7) (Collins et al., 1985). But in the 18537 strain this ORF terminates after three codons (Johnson and Collins, 1989). Similarly, the F mRNA of strain A2 contains an ORF of 38 codons that initiates with the second AUG of the sequence (Fig. 7), but it was not conserved in the 18537 strain (Collins et al., 1984a;

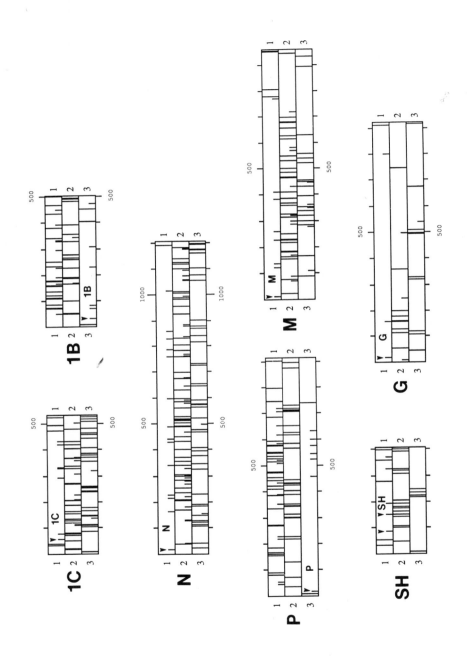

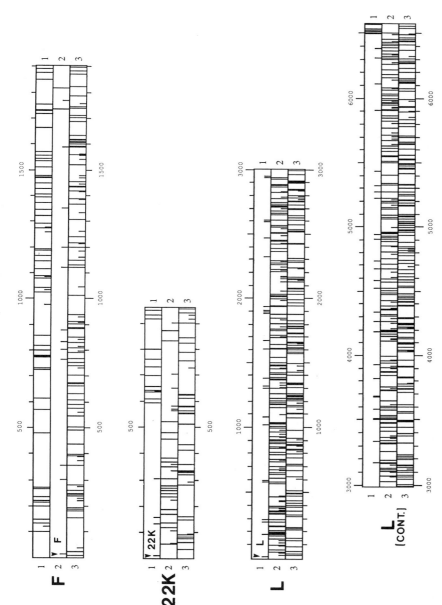

FIGURE 7. Open reading frame (ORF) analysis of the mRNAs of RSV strain A2. The vertical lines indicate methionine codons (half lines) and stop codons (full lines). ORFs utilized for the synthesis of known proteins are identified, and the probable initiating methionine codons are indicated with arrows. The horizontal scale indicates the nucleotide number. The maps are to scale except for L, which is half scale. Reading frames 1, 2, and 3 start with the first, second, and third nucleotides of the complete mRNA sequence.

TABLE II. Nucleotide and Amino Acid Sequence Diversity among Strains[a] of RSV Antigenic Subgroups A and B

| Gene | Strain | Diversity within subgroup A: percent sequence identity between indicated strain and A2 | | Diversity between subgroups A and B: percent identity with strains A2 and 18537 | |
		Amino acid	Nucleotide	Amino acid	Nucleotide
F	Long	98	98	91[b]	79 (−3)[c]
	SS2	97	97		
G	Long	94	96 (−1)[c]	53 (−6)[c]	67 (+3)[c]
SH	ND[d]			76 (+1)[c]	78 (+3)[c]
M	ND[d]			ND[d]	
22K	SS2	98	97	92	78 (−1)[c]
N	Long	100	99	96	86
P	Long	98	97	90	80
	Edinburgh	98	97		
L	ND[d]			ND[d]	
1C	ND[d]			87	78
1B	ND[d]			92	78
Intergenic regions	SS2	—	84[e] (+1)[c]	—	42[f] (+10)[c]
Gene overlap	ND[d]			100	92

[a]RSV strains from subgroup A: A2 (1961), Long (1956), SS2 (1976), and Edinburgh (1977). Strain 18537 (1962) represents subgroup B.
[b]Exclusive of the N-terminal 22 amino acids of the signal peptide region.
[c]Difference in amino acid or nucleotide sequence length relative to strain A2.
[d]Not determined.
[e]Intergenic region between F and 22K genes (44 nucleotides).
[f]Intergenic regions between 1C–1B, 1B–N, N–P, M–SH, G–F, and F–22K genes (165 nucleotides).

Johnson and Collins, 1988a). Any ORF that encodes a protein that is significant for RSV replication would be expected to be conserved among different strains. By that criterion, the second ORF in the 22K mRNA might encode an additional protein.

The nine smaller RSV mRNAs do not appear to contain an internal ORF for a cysteine-rich sequence that might be analogous to that of the V proteins of the other paramyxoviruses (Cattaneo et al., 1989).

E. Sequence Diversity among RSV Strains: RSV Antigenic Subgroups

1. Antigenic and Electrophoretic Differences

RSV is monotypic serologically. For example, postinfection sera neutralize heterologous strains in vitro with only modest (up to fourfold) differences in efficiency. Also, infection of experimental animals induces a high level of protective immunity against challenge with heterologous strains (Coates et al., 1966; Johnson et al., 1987a). However, these tests did detect a modest level of antigenic dimorphism, a finding that was confirmed and extended by antigenic analysis using polyclonal and monoclonal antibodies. RSV is now recognized as having at least two distinct antigenic subgroups, designated A and B (Anderson et al., 1985; Mufson et al., 1985; Gimenez et al., 1986; Hendry et al., 1986;

Johnson et al., 1987a; Walsh et al., 1987a; Morgan et al., 1987; Storch and Park, 1987; Tsutsumi et al., 1988; Akerlind et al., 1988; Garcia-Barreno et al., 1989). Commonly used laboratory strains representing the two subgroups include, for subgroup A: strains A2 (Australia, 1961), Long (Maryland, 1956), SS2 (United Kingdom, 1976), and Edinburgh (United Kingdom, 1977); and, for subgroup B: strains 18537 (Washington, D.C., 1962), RSN-2 (Newcastle, 1972), and 8-60 (Sweden, 1960).

Analysis of the two subgroups using monoclonal antibodies showed that several of the viral proteins, such as the N, F, P, M, and 22K proteins, were generally well-conserved antigenically but did contain some subgroup-specific differences (as well as a lower level of antigenic diversity within subgroups). The G protein was relatively well conserved within each subgroup, but, surprisingly, exhibited extensive antigenic divergence between subgroups. This was confirmed in binding assays with polyclonal postinfection sera, providing the estimate that G protein has 6% or less antigenic relatedness between subgroups, whereas the F proteins were well conserved antigenically within and between subgroups (Johnson et al., 1987a; Walsh et al., 1987a; Hendry et al., 1988).

Comparison of the proteins of the two subgroups by SDS–PAGE detected modest size differences in the P, F, G, and SH proteins of some strains. Within subgroup A, size differences were observed only occasionally, whereas strains of subgroup B commonly had intersubgroup and intrasubgroup differences in the gel mobilities of P and G (Norrby et al., 1986; Routledge et al., 1986; Morgan et al., 1987; Walsh et al., 1987a; Akerlind et al., 1988, our unpublished data). Sequence analysis of the P (our unpublished data), F (Johnson and Collins, 1988a), G (Johnson et al., 1987b), and SH (our unpublished data) genes for the A2 and 18537 strains (subgroups A and B, respectively) showed that the predicted proteins were similar in molecular weight between subgroups, indicating that the differences in gel mobility probably were due to differences in posttranslational processing (especially in the case of the glycosylated G protein) or were due to conformational or detergent-binding differences.

2. Sequence Diversity within Subgroup A

Sequence analysis has been initiated for several other strains of subgroup A, namely the Long, Edinburgh, and SS2 strains (Lambden, 1985; Johnson et al., 1987b; Baybutt and Pringle, 1987; López et al., 1988; Johnson and Collins, 1989). Comparison with strain A2 revealed a high degree of sequence conservation. For example: the N proteins of strains Long and A2 are predicted to be identical, the 22K proteins of the SS2 and A2 strains contain only three amino acid differences, the P proteins of the Edinburgh, Long, and A2 strains have only five differences in all, and the G protein, which has the most divergent sequence, exhibits only 20 amino acid differences between strains Long and A2 (Table II).

At the nucleotide level, there was 96% or greater sequence identity between genes. No differences were observed in the gene-start and gene-end signals, and intergenic sequences and gene sequences that did not encode protein were only slightly more divergent than ORF sequences.

Nucleotide sequence diversity among the G genes of subgroup A strains also has been documented using a ribonuclease protection assay (Storch *et al.*, 1989). This provides a rapid method for "fingerprinting" individual strains and for estimating rates of nucleotide substitution. It confirmed the existence of sequence heterogeneity among subgroup A strains, although the sensitivity of the assay remains to be determined. The assay has not yet been used for subgroup B strains.

In general, the level of sequence diversity within subgroup A is similar in magnitude to that observed among the *F* and *HN* genes of different strains of human PIV3 (Coelingh *et al.*, 1988; K. Coelingh, personal communication). Thus, each of the two subgroups appears to be relatively homogeneous structurally as well as antigenically.

3. Sequence Diversity between Subgroups A and B

Sequence analysis of genes of the subgroup B strain 18537 showed that substantial sequence diversity exists between subgroups A and B (Table II) (Johnson *et al.*, 1987b; Johnson and Collins, 1988a,b, 1989). The most highly conserved protein of the genes sequenced to date, the N protein, had only 16 amino acid differences (91% identity), while at the other extreme the most divergent protein, G, had 135 changes (53% identity). Thus, the different genes appear to have different rates of change. Comparisons of individual protein sequences are described in later sections.

Regarding nucleotide sequences, the gene-start sequences were exactly conserved between the subgroups, except for a single substitution in the *SH* gene of 18537 (3' CCCCAUUUA). The gene-end sequences were somewhat less well conserved, but retained the general organization of a conserved pentanucleotide followed by several variable nucleotides and a short run of U residues. The overlap between the *22K* and *L* genes also was present in strain 18537: between the subgroups the length of the overlap was conserved exactly and its nucleotide sequence was 92% identical (Table II).

Between the subgroups, ORFs were relatively well conserved (77–86% identity except for 67% identity for the relatively divergent G ORF), whereas noncoding gene sequences apart from the gene-start and gene-end signals were relatively poorly conserved (usually between 33% and 78% identity). Similarly, sequence analysis of six intergenic regions of strain 18537 showed that there was only 42% sequence identity with strain A2, although the lengths of the intergenic regions were exactly conserved at three junctions and were only slightly different at the other three. The general lack of conservation in the intergenic and flanking gene regions supported the interpretation that, apart from the gene-start and gene-end sequences, these regions are nonspecific spacers.

F. Structures of the RSV Proteins

1. Nucleocapsid Proteins L, N, and P

The L protein, which is presumed to be the major polymerase subunit (Banerjee, 1987), is basic and relatively hydrophobic, containing 29% hydrophobic residues (isoleucine, leucine, valine, and methionine), compared to an average content of 20% calculated from a sequence library (Dayhoff et al., 1978). Its content of isoleucine and leucine (22%) is nearly twice the average (12%). The six other L protein sequences that have been determined for non-segmented negative-strand viruses have similar characteristics and also are nearly identical in length: RSV, 2165 amino acids; SV, 2228; PIV3, 2233; MeV, 2183; NDV, 2204; VSV 2109; and rabies virus, 2142 (Schubert et al., 1984; Tordo et al., 1988; also see Chapter 5).

Preliminary pairwise comparisons between the RSV L protein and the others showed low but significant sequence relatedness spanning the center of the molecules, with the N-terminal and C-terminal regions being poorly related. Sequence relatedness was greatest between RSV and the four paramyxoviruses, especially PIV3, but significant relatedness also existed between RSV and the two rhabdoviruses (our unpublished data). Relatedness among all of the sequences was greatest in the vicinity of amino acids 700–900. In particular, this region contained two short segments of nearly exact amino acid identity between RSV and similarly spaced regions in the others. One of these highly conserved segments, amino acids 775–786 in the sequence of the RSV L protein, had 58% identity with the comparable region in the L protein of rabies virus, and 92% identity with the comparable regions in the L proteins of SV, PIV3, MeV, and NDV. The second conserved segment, amino acids 810–815 of the RSV L protein, had 67% identity with the comparable region in the rabies virus L protein, and 83% identity with comparable regions in the L proteins of SV, PIV3, MeV, NDV, and VSV (our unpublished data). These short segments of nearly exact identity are contained in a large region of sequence similarity, in which many of the changes are conservative. These observations are evidence of evolutionary relatedness among all of these L proteins, although additional comparisons will be required to estimate relative evolutionary distances.

The N protein is substantially shorter (291 amino acids) than the NP proteins of other paramyxoviruses, such as PIV3, MuV, and MeV (515,553, and 523 amino acids, respectively), and does not have detectable sequence relatedness with the others (Collins et al., 1985; Spriggs et al., 1986; also see Chapter 5). RSV N protein differs from some of the others in not being detectably phosphorylated (Lambert, 1988), and its predicted net polypeptide charge is slightly basic rather than neutral or acidic.

Treatment of nucleocapsids with trypsin cleaved the 41K N protein into fragments of 14K and 27K; the latter presumably was released or exposed, because it was degraded completely upon further incubation (Ward et al., 1984). This is reminiscent of SV NP protein, which has a large N-terminal hydrophobic domain that is trypsin resistant, and a C terminus that is trypsin sensitive and hydrophilic (Chapter 5). However, there is not sufficient information on the RSV N protein to conclude that there is analogy in functional organization.

The P protein is acidic and hydrophilic, and is the most heavily phosphorylated RSV protein (Satake et al., 1984; Lambert, 1988). It contains nearly twice (21%) the average (12%) content of acidic residues (Dayhoff et al., 1978) and also is rich in serine and theronine residues. RSV P is substantially shorter (241 amino acids) than its counterparts in PIV3, MuV, and MeV (602, 390, and 507 amino acids, respectively) (Chapter 5). RSV P appears to share a low, barely significant level of sequence relatedness with a central domain of the P proteins of PIV3 and MeV (Spriggs and Collins, 1986).

Comparison of RSV subgroups A and B showed that the P protein consists of two relatively large conserved domains (amino acids 1–58 and 86–241, with a total of 96% identity) separated by a short divergent region (amino acids 59–85, 52% identity) (our unpublished data). Similarly, the P proteins of SV and PIV3 consist of large, conserved N- and C-terminal domains that are separated by a short, divergent spacer sequence. For SV, the N-terminal domain contains most of the phosphorylation sites and can be removed from nucleocapsids by trypsin treatment. The C-terminal domain remained nucleocapsid associated following trypsinization and appeared to completely fulfill the role of P in subsequent *in vitro* transcription assays (Chinchar and Portner, 1981; also see Chapter 5). It is tempting to speculate that the P protein of RSV has a similar functional organization.

Postinfection human and animal sera contain abundant antibodies to the N and P proteins (Ward *et al.*, 1983; our unpublished observations). Also, the N protein is an important antigen for RSV-specific human and murine cytotoxic T lymphocytes and for murine helper T lymphocytes (Bangham *et al.*, 1986; Openshaw *et al.*, 1988; Gupta *et al.*, 1990). But the N protein did not induce detectable levels of RSV-neutralizing serum antibodies or significant levels of protective immunity when expressed individually in rodents by a recombinant vaccinia virus (King *et al.*, 1987; our unpublished data). Immunization of mice with SDS–PAGE-purified P protein also did not induce detectable levels of RSV-neutralizing antibodies or protective immunity (Routledge *et al.*, 1988). But the lack of response might be attributable to the use of denatured material for immunization.

2. M and 22K Proteins

The M and 22K proteins are nonglycosylated inner components of the viral envelope (Peeples and Levine, 1979; Collins *et al.*, 1984b; Huang *et al.*, 1985). They are very basic, and both have been observed to migrate in SDS–PAGE as multiple, electrophoretically-distinct species that differ in mobility apparently due to differences in intramolecular disulfide bonding (Gruber and Levine, 1983; Routledge *et al.*, 1987a). In addition, a fraction of 22K protein extracted from infected cells was in the form of disulfide-linked oligomers (Routledge *et al.*, 1987a). Both the M and 22K proteins also have been reported to exist in phosphorylated forms in infected cells and in virions (Lambert *et al.*, 1988). For both proteins, this modification involved only a subset of intracellular protein and, in the case of M, resulted in a reduction in electrophoretic mobility (Lambert *et al.*, 1988).

The M protein is smaller (256 amino acids) than its counterparts in PIV3, MuV, and MeV (353, 375, and 335 amino acids, respectively) (Chapters 2, 3, and 16). The M proteins of the *Morbillivirus* and *Paramyxovirus* genera contain a conserved hydrophobic segment in the C-terminal one-third of the molecule that is a candidate to be involved with membrane interaction (Bellini *et al.*, 1986). The RSV M protein contains a similarly located 25-amino acid segment (spanning position 200 in the amino acid sequence) that is the most hydrophobic region of the molecule and might have a similar functional role. The RSV M protein does not appear to have significant sequence relatedness with the others except for a short region of apparent relatedness with the NDV M protein (residues 150–220 of RSV M and 120–190 of NDV M) (Chambers et al., 1986). However, the region of apparent relatedness between RSV and NDV does not coincide with the hydrophobic segment in the NDV sequence, and so the significance of this limited amount of sequence similarity is unclear.

RSV is unique among the nonsegmented negative-strand viruses in having a second nonglycosylated membrane-associated protein, the 22K protein. The amino acid sequence of 22K does not contain any region of sufficient hydrophobicity to suggest membrane insertion (Collins and Wertz, 1985a; Elango et al., 1985b). In infected cells, essentially all of the 22K protein is inaccessible to extracellular trypsin, consistent with an internal location (Routledge et al., 1987b).

A small amount of 22K protein (or its fragments) could be detected on the cell surface by immunofluorescent antibody staining under conditions where the cells remained impermeable to trypan blue and where other internal antigens, namely N, P, and M, were not detected at the surface (Routledge et al., 1987b). The surface localization of 22K occurred only late in infection and might not involve intact 22K protein, and therefore this finding might be unrelated to its function.

The pattern of immunofluorescent staining of the 22K protein is permeabilized RSV-infected cells was suggestive of colocalization with the N and P proteins present in nucleocapsids (Routledge et al., 1987b). In contrast, the staining pattern of the M protein was similar to those of the G and F proteins, perhaps indicating that these proteins colocalize at cell membranes. Another possible clue to the function of 22K was that the kinetics of its phosphorylation was similar to that of the P protein, were distinct from that of the M protein, and appeared to be coordinate with virion assembly and release (Lambert et al., 1988). Perhaps the 22K protein associates with nucleocapsids relatively early in the pathway to virion assembly, whereas association with M might be a later event that immediately precedes budding.

Functions for the M proteins of paramyxoviruses probably include interaction with viral nucleocapsids to inhibit or regulate RNA synthesis, perhaps in preparation for packaging, and interaction with nucleocapsids, envelope components, and possibly the cytoskeleton during virion morphogenesis (Chapter 16). The distinction between M and 22K proteins with regard to their immunofluorescent staining pattern ad kinetics of phosphorylation suggests that they have different functional roles. Presumably, the functions fulfilled by the M protein for the other nonsegmented negative-strand viruses are divided between the M and 22K proteins for RSV.

The detection of epitopes of the 22K protein at the cell surface indicates that it might be a significant surface antigen involved in host immunity. 22K protein purified by SDS–PAGE did not induce detectable RSV-neutralizing serum antibodies or protective immunity in mice (Routledge et al., 1988), but this needs to be evaluated using native antigen. Evaluation of the antigenicity of M has been initiated (Cannon and Bangham, 1989), but is inconclusive.

3. Nonstructural 1C and 1B Proteins

Apart from sequence analysis, the 1C and 1B proteins have not been well characterized. Lambert et al. (1988) detected a M_r 14,000 phosphorylated RSV protein (in addition to P, M, and 22K) that presumably is a form of either the 1C or 1B protein. In pulse-chase radiolabeling experiments, newly-synthesized intracellular 1C protein comigrated with 1C protein made in vitro, but was subsequently reduced by 1500 in M_r (Huang et al., 1985). This processing step also was observed when the 1C protein was expressed using a vaccinia virus–1C recombinant virus (our unpublished data). Whereas the phosphorylation of the M_r 14,000 species occurred during a 15-min labeling period, the processing of 1C had a half-time of approximately 2 hr, indicating that these are separate events.

Paramyxoviruses encode one or more nonstructural proteins by several different coding strategies (Chapters 2, 3, 6, and 7). For example, SV encodes the nonstructural C protein from a separate ORF in the P/C mRNA, and also encodes truncated forms of the P and C proteins as nonstructural species by translational initiation at internal AUG triplets in the respective ORFs. In contrast, NDV does not appear to encode a unique nonstructural protein. However, translation of the NDV P ORF yields, in addition to full-length P protein, two truncated nonstructural forms of the P protein that result from translational initiation at internal AUGs (Collins et al., 1982; McGinnes et al., 1988). RSV is unique in having two separate, individually transcribed genes for nonstructural proteins (Collins et al., 1986). The functions of these various nonstructural proteins, including the 1C and 1B proteins, are unknown, and their predicted amino acid sequences do not provide clues. Perhaps the 1C and 1B proteins are functional counterparts to the nonstructural proteins of these other paramyxoviruses, although there is no evidence of structural relatedness.

The 1B and 1C proteins are among the most abundant RSV proteins of infected cells and therefore might be significant antigens involved in host immunity. The 1B protein was not efficiently recognized in vitro by RSV-specific murine helper T lymphocytes (Openshaw et al., 1988), but systematic evaluation of these proteins has not been reported.

4. Fusion Protein

A hallmark of the paramyxoviruses is a fusion protein that is separate from the attachment protein, is active at physiologic pH, and is responsible for viral penetration and syncytium formation (Chapter 13).

The RSV F protein was identified by immunoprecipitation with monoclonal antibodies that, in cell culture, inhibit syncytium formation (Walsh and Hruska, 1983). In addition, the synthesis of the F protein in tissue culture, using a cDNA of the F mRNA expressed from a recombinant simian virus 40 vector, resulted in extensive syncytium formation (our unpublished data). This confirmed the identification of the biological activity of the F protein and showed that expression of F alone was sufficient for fusion function.

The RSV F protein has general structural similarities with the F proteins of other paramyxoviruses, which form a more closely related group (Collins et al., 1984a; Elango et al., 1985a; Spriggs et al., 1986; Barrett et al., 1987). These

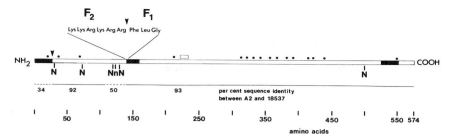

FIGURE 8. Schematic diagram of the F_0 protein of strain A2. The three hydrophobic domains are shaded: the N-terminal signal, the internal domain that becomes the N terminus of the F_1 subunit, and the membrane anchor which is close to the C terminus. Other symbols: N, potential acceptor sites for N-linked carbohydrate; n, an additional site present in the Long, SS2, and 18537 strains; dots, Cys residues; box, a major element in the binding site of an RSV-neutralizing monoclonal antibody (Trudel *et al.*, 1987). The percent amino acid sequence identity between strains A2 and 18537 (representing the two antigenic subgroups) is summarized at the bottom: each percent value refers to the polypeptide region above it, with the different regions being demarcated by continuous and discontinuous segments in the horizontal line.

similarities include overall length, the location of hydrophobic domains and cleavage site (see below), and the approximate locations of carbohydrate side chains and cysteine residues. The amino acid sequence of RSV F does not have a high degree of sequence identity (less than 20%) with the others. But unambiguous sequence relatedness was demonstrated (Spriggs *et al.*, 1986) with a scoring matrix that is based on observed patterns of amino acid conservation and substitution compiled from surveys of sequence divergence among related proteins (Dayhoff *et al.*, 1978). These observations leave little doubt that RSV F is related to the others and was derived by divergent evolution from a common ancestral protein.

The F protein (Fig. 8) contains an N-terminal hydrophobic signal peptide that directs its cotranslational translocation into the rough endoplasmic reticulum, and which is cleaved in the vicinity of amino acids 21–25. The most strongly hydrophobic domain in the F protein, located near the C terminus (amino acids 525–550), serves as a stop-transfer signal and membrane anchor, leaving a 24-amino acid cytoplasmic domain. Consistent with this, an F protein mutant that lacked the membrane anchor, expressed from a baculovirus vector containing a modified F cDNA, was processed into an antigenically authentic glycosylated form that was efficiently secreted rather than anchored in membrane (Wathan *et al.*, 1989).

As is characteristic for paramyxoviruses, RSV F protein is synthesized as a precursor, F_0, which is cleaved by a cellular protease to generate two disulfide-linked subunits, NH_2-F_2-ss-F_1-COOH (Gruber and Levine, 1985b; Fernie *et al.*, 1985; Huang *et al.*, 1985; also see Chapter 13). Amino acid sequencing showed that the proteolytic cleavage occurs following amino acid 136 (Elango *et al.*, 1985a). In pulse-chase radiolabeling experiments, F_1 and F_2 were detected 20–30 min posttranslation, and processing of the pulse-labeled pool of F_0 was completed by 1½–3 hr (Gruber and Levine, 1985a; Fernie *et al.*, 1985). Studies to map the intracellular compartment where cleavage occurs provided conflicting results [suggesting that cleavage occurs before the trans Golgi compartment in one study (Gruber and Levine, 1985b) and following the medial compartment in another study (Fernie et al., 1985)],and this issue is unresolved.

Studies with other paramyxoviruses showed that cleavage of F_O is obligatory for fusion activity and is a factor in tissue tropism (see Chapter 13). One determinant of cleavability is the number of arginine and lysine residues at the cleavage site; RSV has six such residues, compared to five, four, four, three, and one for SV5, MeV, MuV, PIV3, and SV, respectively (Glickman et al., 1988; Paterson et al., 1989; also see Chapters 2, 3, and 13). This suggests that properties of the virus other than the structure of the cleavage site of the F protein are responsible for the characteristic restriction of RSV infection to the respiratory tract.

However, when F protein was expressed in lepidopteran cells from an F cDNA in a recombinant baculovirus, most of the molecules were not cleaved (Wathen et al., 1989). Only a small fraction of molecules appeared to be cleaved authentically, and another small fraction appeared to be cleaved at an alternate site located in F_1 approximately 50 residues from its N terminus. A likely possibility is that the lepidopteran cells lack the protease required for correct, efficient processing. However, the F protein expressed from that particular construct lacked the C-terminal 23 amino acids, which comprise almost all of the cytoplasmic domain. Therefore the formal possibility exists that this resulted in subtle conformational changes that inhibited cleavage. But the lepidopteran-derived F protein appeared to be antigenically authentic, based on reactivity with a panel of monoclonal antibodies (M. Wathen and B. Murphy, unpublished data), suggesting that it was conformationally authentic. In the case of the SV F protein, cleavage of the F_O precursor results in a major conformational change (Hsu et al., 1981). Based on this precedent, it is somewhat surprising that, in the case of the RSV F protein, the lepidopteran-derived uncleaved form was antigenically similar to the cleaved form from mammalian cells, since a major conformational change might be expected to affect antigenic sites. Perhaps the conformational change that is thought to occur upon cleavage is localized to the F_2–F_1 junction and does not alter the major antigenic sites.

For paramyxoviruses in general, the N terminus of F_1, created by the activating cleavage, is hydrophobic and is thought to initiate fusion by directly interacting with the host cell membrane. Consistent with this idea, short synthetic peptide analogs of the N-terminal residues of the F_1 proteins of SV and MeV inhibit fusion, although the mechanism of the effect is unknown (Richardson and Choppin, 1983; Hull et al., 1987). However, for RSV, a series of synthetic peptide analogs of the F_1 N terminus had no activity in inhibiting RSV plaque formation (Lobl et al., 1988). This suggests that, for RSV, this might not be a suitable approach for chemotherapy or investigation of F protein function.

The sequence of the F_1 N terminus is highly conserved among the members of the *Morbillivirus* and *Paramyxovirus* genera (Chapters 2 and 3), but the corresponding sequence in RSV is distinct (Fig. 9). The first three residues of the F_1 subunit of these viruses, including RSV, are conserved: Phe-X-Gly. However, some NDV strains have Leu as the terminal amino acid, indicating that even this triplet is not obligatory (Toyoda et al., 1987). The F_1 N terminus of RSV strain A2 does contain a second triplet, Gly-Val-Ala (dotted underlining in

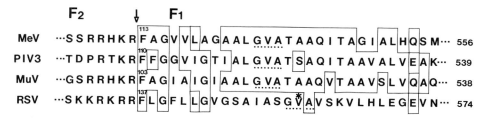

FIGURE 9. Comparison of the amino acid sequences of the cleavage site and N terminus of the F_1 proteins of RSV strain A2 and several representative paramyxoviruses. The amino acid that becomes the N terminus of the F_1 subunit is numbered above the line according to its complete unmodified F_0 sequence, whose length is shown at the right. The cleavage sites are indicated with an arrow. Sequence identities in F_1 are boxed or underlined. The asterisk indicates an amino acid (V for strain A2) that is variable among different RSV strains (I for 18537 and SS2, T for Long).

Fig. 9), that is present closely upstream in the F_1 proteins of the other paramyxoviruses. However, this triplet was not conserved exactly in other RSV strains, being Gly-Ile-Ala in strains 18537 and SS2 and Gly-Thr-Ala in strain Long (Baybutt and Pringle, 1987; Johnson and Collins, 1988a; López et al., 1988). The lack of sequence identity between RSV and the other paramyxoviruses in this region of F_1 indicates that paramyxovirus fusion activity does not require a specific fusion sequence. Instead, there probably is a more general requirement for hydrophobicity and an extended unordered or helical structure at the N terminus of F_1.

The first five residues of RSV F_1 are identical to residues 9–13 of the fusogenic gp41 proteins of human and simian immunodeficiency viruses (Gallaher, 1987; Gonzalez-Scarano et al., 1987). This is not necessarily evidence of a common ancestry, because (1) the remainder of the molecules are not significantly related and (2) these five residues are not well conserved even between RSV and the other paramyxoviruses. These residues might be important (but not obligatory) contributors to general structural features for fusion activity, and could be the result of divergent or convergent evolution.

The sequence of the strain A2 F_0 protein contains five potential acceptor sites for N-linked carbohydrate, and estimates of the carbohydrate content suggest that all are utilized (Collins et al., 1984a). The SS2 and Long strains of subgroup A and the 18537 strain of subgroup B contain a sixth potential acceptor site at position 120 (Baybutt and Pringle, 1987; Johnson and Collins, 1988a; López et al., 1988); it is not known whether this additional site is utilized. The small F_2 subunit contains four (or five) of the five (or six) acceptor sites, and the presence of these host-specific side chains might shield or alter antigenic sites in the virus-encoded polypeptide (Skehel et al., 1984; Elder et al., 1986; Sjoblom et al., 1987), and might account for the apparent poorer immunogenicity in the native protein of F_2 compared with F_1 (Walsh et al., 1986). The carbohydrate of the F protein is sulfated, and the oligosaccharide side chain of the F_1 subunit is processed into complex form, whereas no information is available on the processing of the side chains of the F_2 subunit (Cash et al., 1979; Gruber and Levine, 1985a). The carbohydrate side chains are not required for the cleavage of F_0 or for the incorporation of F protein into virions (Huang, 1983; Lambert, 1988), indicating that in their absence the polypeptide chain maintains sufficient conformational integrity for processing and transport to the cell surface.

Another feature of the paramyxovirus F protein is a cysteine-rich region in the F_1 subunit (Spriggs et al., 1986; Barrett et al., 1987). In the viruses of the Morbillivirus and Paramyxovirus genera, a 100-amino acid domain, which begins approximately 220 residues from the N terminus of F_1 and ends approximately 75 residues before the membrane anchor, contains eight cysteine residues that are exactly conserved in number and spacing (Chapters 2 and 3). The RSV F_1 protein contains an analogous domain, but it is somewhat distinct: it is 127 amino acids in length, begins 176 amino acids from the N terminus of F_1 and ends 85 amino acids before the membrane anchor (Fig. 8). It contains 11 cysteine residues, and these differ in number and sequence position from those of the other paramyxoviruses. Thus, the pattern of disulfide bonding in this region of F_1 might be exactly conserved among the morbilliviruses and paramyxoviruses, but probably is somewhat distinct for RSV.

RSV F can be recovered quantitatively as a noncovalently-linked homodimer ($2X F_2 + F_1$) when extracted from infected cells and analyzed by SDS–PAGE without reduction and heat denaturation (Walsh et al., 1986). The dimer was not observed upon treatment with N-glycanase or synthesis in the presence of tunicamycin, suggesting that its stability is reduced in the absence of the carbohydrate side chains (Lambert, 1988). SV F protein has been isolated as a tetramer consisting of two analogous dimers, and this tetramer is thought to be the membrane spike (Sechoy et al., 1987). Given the general structural similarities between the SV and RSV F proteins, the higher-order structure of RSV F might be similar. On the other hand, a small fraction of RSV F protein extracted from infected cells was found as complexes containing the F dimer linked by a disulfide bond(s) to a G protein monomer (Arumugham et al., 1989). Also, others have reported that small amounts of G protein copurify with F protein during immunoprecipitation with F-specific monoclonal antibodies (Routledge et al., 1986). This would suggest an alternative possibility, that the RSV F and G proteins are not segregated into different spikes. However, as noted above, coexpression of the G protein was not required for the assembly of F into a fusogenic unit. It will be important to examine the higher-order structures of the RSV proteins in greater detail.

The F proteins of strains A2 and 18537, representing the two antigenic subgroups, are highly conserved overall (Table II), but have two regions of divergence (Fig. 8) (Johnson and Collins, 1988b). One region, the signal peptide, had only 34% identity, indicating that its function is dependent upon hydrophobicity rather than on a specific amino acid sequence. The second region, residues 111–129, had only 50% identity as well as the additional potential site for N-linked carbohydrate noted above. This divergent region is relatively hydrophilic and has a predicted secondary structure of a coil and turn. This suggests that it might be exposed in the folded molecule, and might contribute to a subgroup-specific B-cell or T-cell epitope (Anderson et al., 1985; Mufson et al., 1985; Pemberton et al., 1987).

The F protein has been shown to be a major, independently protective antigen that efficiently induced RSV serum neutralizing antibodies, RSV-specific cytotoxic and helper T lymphocytes, and a high level of resistance to RSV challenge in rodents and monkeys (Olmsted et al., 1986; Bangham et al., 1986;

Walsh et al., 1987b; Wertz et al., 1987; Pemberton et al., 1987; Wathen et al., 1989; Gupta et al., 1990). Studies in cotton rats suggested that F protein is the most important RSV protective antigen, and the protective immunity induced by F protein was equally effective against viruses representing both subgroups, consistent with their extensive structural similarities (Olmsted et al., 1986; Johnson et al., 1987b). A major element of the binding site of an RSV-neutralizing monoclonal antibody was mapped by binding studies to positions 221–232 (Fig. 8) (Trudel et al., 1987).

5. The Attachment G Protein

In contrast to the F protein, the G protein lacks detectable sequence relatedness with its counterparts in the other paramyxoviruses, the HN and H proteins (Spriggs et al., 1986). Also, the G protein is much shorter (Fig. 14, below), lacks hemagglutination and neuraminidase activities (Richman et al., 1971), and has other major structural differences that are described below (Wertz et al., 1985; Satake et al., 1985; Spriggs and Collins, 1986; Johnson et al., 1987b; see also Chapter 14).

The G protein was identified as the viral attachment protein by the observation that G-specific polyclonal antibodies blocked the adsorption of radiolabeled RSV to tissue culture cells, whereas antibodies to the F protein inhibited subsequent syncytium formation, but not the initial step of virus adsorption (Levine et al., 1987). Furthermore, a monoclonal antibody that inhibited PVM hemagglutination activity bound to the PVM G protein, which has general structural similarities to RSV G (Ling and Pringle, 1989b).

The major features of the amino acid sequence of the RSV G protein, and the pattern of conservation between subgroups of these structural features and of the amino acid sequence, are shown in Fig. 10. The amino acid sequence of the G protein contains a single major hydrophobic domain that is located near the N terminus (residues 38–66). This appears to contain all of the structural information required for membrane insertion and anchoring, based on three observations: (1) A C-terminally truncated form of the G protein that contained only the N-terminal 71 amino acids plus a 12-amino acid heterologous reporter peptide was efficiently inserted, anchored, and transported to the cell surface when expressed from mutagenized cDNA using a recombinant expression vector (Olmsted et al., 1989); (2) the hydrophic domain of G could be exchanged, by cDNA mutagenesis and recombinant vector expression, with the analogous N-terminal signal-anchor of PIV3 HN protein, and both chimeric proteins were processed authentically (our unpublished data); and (3) a soluble form of the G protein is secreted from infected tissue culture cells, and amino acid sequencing showed that it differs from the membrane-anchored form by the absence of the proposed anchor domain (Hendricks et al., 1987, 1988). The soluble form appears to be generated by posttranslational release of the ectodomain by a signal peptidase-like activity, with the cleavage occuring at two alternate sites, between amino acids 65 and 66 or 74 and 75 (Hendricks et al., 1988). These studies support the idea that the G protein is oriented in the membrane such that the N terminus is cytoplasmic and the C-terminal three-fourths is extracellular. This is the one structural feature common to the G, HN, and H proteins.

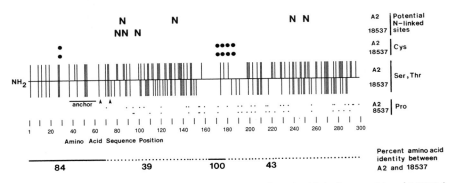

FIGURE 10. Structural features of the G glycoproteins of strain A2 (subgroup A) and 18537 (subgroup B). Amino acid sequence positions are shown for: potential acceptor sites for O-linked carbohydrates (Ser and Thr residues, vertical lines); potential acceptor sites for N-linked sugars (N); Cys residues (filled circles); Pro residues (dots); two naturally occuring cleavage sites that give rise to soluble forms of G (arrows); and the membrane anchor (underlining). Amino acid identity between the strain A2 and 18537 sequences (representing the two antigenic subgroups) is summarized at the bottom, with each percent value referring to the region above it and the different regions being demarcated by continuous and discontinuous segments in the horizontal line.

The G protein is modified intracellularly by the addition of N-linked and O-linked carbohydrate. This gives rise to a mature form that has a much larger M_r (84,000–90,000 for strain A2) than that of its polypeptide moiety (molecular weight 32,587 for strain A2, calculated from the amino acid sequence deduced from nucleotide sequencing and confirmed by SDS–PAGE of the unglycosylated form synthesized in vitro) (Collins et al., 1984b; Wertz et al., 1985; Satake et al., 1985).

The G protein of strain A2 was estimated to contain N-linked carbohydrate of M_r 8000 to 12,000 (the differences reflect the uncertainty of estimating the size of a relatively high-molecular-weight, diffusely-migrating species) (Gruber and Levine, 1985a; Fernie et al., 1985; Wertz et al., 1985; Satake et al., 1985). The sequence of A2 G contains four potential acceptor sites for N-linked carbohydrate, and the M_r estimates suggest that most or all of the sites are utilized. Surprisingly, the number and sequence positions of the potential acceptor sites were not conserved at all between subgroups (Fig. 10). The four potential acceptor sites of strain A2 were conserved in a second subgroup A strain, the Long strain (Johnson et al., 1987b). But the Long strain contained four additional potential sites, and SDS–PAGE comparisons (Routledge et al., 1986) suggest that the Long G protein contains at least one additional side chain. Thus, there appears to be both intersubgroup and intrasubgroup heterogeneity in the positions and numbers of N-linked side chains.

The presence of O-linked carbohydrate on G was first demonstrated by the finding that its M_r was greatly reduced when it was synthesized in the presence of monensin, an ionophore that inhibits transport between the medial and trans (late) compartments of the Golgi complex and thereby inhibits O-glycosylation in the latter compartment (Griffiths et al., 1983; Gruber and Levine, 1985a; Fernie et al., 1985). The presence of O-linked sugars was confirmed by digestion with O-glycanase and by lectin-binding studies (Paradiso et al., 1987; Lambert, 1988). O-linked side chains have been described for a number of cellular and viral glycoproteins, and are attached to the polypeptide chain through O-glycosidic linkage between N-acetyl galactosamine and serine or theronine residues. They do not appear to occur in any other paramyxovirus or rhabdovirus.

The total molecular weight of the O-linked carbohydrate of G has not yet been determined. The available estimates have been made by SDS–PAGE, and suggest a remarkably high content of M_r 30,000–57,000 (Gruber and Levine, 1985a; Fernie et al., 1985; Wertz et al., 1985; Satake et al., 1985). But O-linked sugars can have a disproportionate effect on electrophoretic mobility, leading to overestimates of carbohydrate content (Cummings et al., 1983, and references therein).

The O-linked side chains can vary considerably in length and can be as short as monosaccharides (Schachter and Roseman, 1980; Lundstrom et al., 1987). Therefore, the G protein might contain a few long chains or many short chains. Consistent with this latter possibility, the ectodomains of the G proteins of both subgroups contain more than 70 potential acceptor sites. Indeed, the abundance of these potential acceptor sites gives the G protein a content of hydroxy amino acids that is more than twice (30%) the average content (13%) (Dayhoff et al., 1978).

The exact sequence positions of the potential acceptor sites are poorly conserved (40% identity) between subgroups, but for both subgroups most of the potential acceptor sites are arranged in two large clusters, one proximal to the membrane and the other proximal to the C terminus (Fig. 10). Analysis of the glycosylation of C-terminally truncated G proteins, expressed from mutagenized cDNAs by recombinant simian virus 40 vectors, suggested that the O-linked sugars are divided equally between the two clusters of potential acceptor sites (Olmsted et al., 1989).

The ectodomain is very rich in proline residues (with a content of 13% compared with the average of 5%), whereas the membrane and cytoplasmic domains are devoid of proline residues (Fig. 10). Proline residues are strong terminators of ordered secondary structure and contributors to β-turns (Chou and Fasman, 1974), and might contribute to an extended nonglobular structure that perhaps is kinked by many short turns. Regions that contain serine and threonine residues in the context of high contents of proline, alanine, or glycine residues have been shown in some proteins to contain the sites of attachment of O-linked sugars (Hill et al., 1977; Fiat et al., 1980; Eckhardt et al., 1987; López et al., 1987). This probably is because these regions have extended structures that render them accessible to glycotransferases, which commonly add O-linked sugars to completed rather than nascent molecules. The structure of the extracellular domain of G might be similar to that of mucins, which have high contents of serine, threonine, and proline residues, which appear to be relatively devoid of secondary structure, and which contain numerous sugar side chains (Hill et al., 1977; Eckhardt et al., 1987).

Metabolic radiolabeling and immunoblotting studies detected a glycosylated M_r 45,000 form of the G protein that appears to be a major intermediate in posttranslational processing (Gruber and Levine, 1985a,b; Fernie et al., 1985; Hendricks et al., 1987,1988; Wertz et al., 1989). Upon treatment with endoglycosidase H, the M_r 45,000 form was converted into a species that was similar in electrophoretic mobility to the unglycosylated polypeptide chain, indicating that most or all of the carbohydrate of the M_r 45,000 species was N-linked (Gruber and Levine, 1985a,b; Wertz et al., 1989). The finding that the N-linked side chains of this species were fully sensitive to endoglycosidase H, in contrast to those of the mature form, which were fully resistant, supported the idea that the M_r 45,000 species is a processing intermediate that accumulates prior to the trans Golgi compartment (Gruber and Levine, 1985b; Fernie et al., 1985; Wertz et al., 1989). This was also supported by the finding that the accumulation of this intermediate was increased in monensin-treated cells (Gruber and Levine, 1985a; Fernie et al., 1985). Pulse-chase experiments provided evidence that the M_r 45,000 species is converted into the mature M_r 84,000–90,000 form through a series of intermediates that, when analyzed by SDS–PAGE, form a heterogeneous ladder leading up to the mature form (Fernie et al., 1985; Wertz et al., 1989). This ladder appears to consist of G protein molecules that differ in the degree of O-glycosylation. Consistent with this idea, a similar ladder of partially O-glycosylated species was generated by treatment of the mature M_r 84,000–90,000 species with O-glycanase (Lambert, 1988). Furthermore, the conversion of the M_r 45,000 species to the M_r 84,000–90,000 species was completely blocked in a mutant cell line under conditions where O-glycosylation was inhibited, and was restored under conditions that were permissive for O-glycosylation (Wertz et al., 1989).

The intracellular maturation of G is relatively rapid despite the apparently complex, multistep nature of the processing pathway. In pulse-chase experiments, mature-sized G protein was detected following a 10-min labeling period, and the half-time for the formation of mature G protein following a 10-min pulse was approximately 15 min (Fernie et al., 1985), although other workers reported a somewhat longer transit time of approximately 30 min (Gruber and Levine, 1985b).

For proteins in general, O-glycosylation has been reported to occur at several intracellular sites, and this can vary depending upon the cell type. Cotranslational O-glycosylation has been reported, but usually the sugars are added and elongated in one or more of the Golgi compartments (Strous, 1979; Roth, 1984; Elhammer and Kornfeld, 1984; Dunphy and Rothman, 1985; Jukinen et al., 1985; Roth et al., 1986; Patzelt and Weber, 1986; Lundstrom et al., 1987; Spielman et al., 1987). The results described above suggest that the M_r 45,000 species contains little or no O-linked sugars, and that most of the O-linked sugars of the G protein are added in the trans Golgi compartment. However, there is one preliminary report that the M_r 45,000 contains some O-linked sugar (Paradiso et al., 1987), and it is possible that this species might contain some side chains that are added in or before the cis or medial compartments and are elongated and augmented in the trans compartment.

As an alternate, more speculative hypothesis for the processing of the G protein, it was suggested that the M_r 45,000 species might be fully glycosylated form of G (containing only a small total amount of O-linked sugar), and that the M_r 84,000–90,000 form might consist of two M_r 45,000 molecules that either are covalently linked or that dimerize artifactually during the SDS–PAGE, perhaps as described previously for glycophorin (Marton and Gevin, 1973; Tuech and Morrison, 1974). But there is no published evidence to support this hypothesis, and it seems inconsistent with the evidence described above.

The G protein analyzed under nonreducing conditions included a M_r 175,000 form that appeared to consist of two disulfide-linked M_r 84,000–90,000 molecules, evidence that the maturation of G includes homo-oligomerization (Lambert, 1988). However, others found that a fraction of G protein extracted from infected cells was disulfide-linked to F protein (Arumugham et al., 1989), and thus the nature of the oligomerization of G remains to be clarified.

Oligomerization, O-glycosylation, and the incorporation of G into virions can occur in the absence of N-linked sugars, and even fully unglycosylated G was incorporated into virions (Gruber and Levine, 1985b; Fernie et al., 1985; Satake et al., 1985; Lambert, 1988; Wertz et al., 1989). Quantitative analysis of the cell-surface expression of the G protein in the absence of N-linked sugars, O-linked sugars, or both indicated that the absence of either type of sugar had a modest effect (>50% reduction) on G protein maturation, whereas the absence of both types resulted in an approximately ten-fold reduction in surface expression (Wertz et al., 1989). Thus, the sugars of the G protein do not appear to be obligatory for its cell-surface expression, although the reductions in the efficiency of maturation in the absence of one or both types of side chains indicate that they do make a contribution, perhaps by increasing the stability of the higher-order protein structure.

The extensive amino acid sequence divergence between the A2 and 18537 G proteins noted in Table II was localized to two large segments (amino acids 70–163, 39% identity, and amino acids 177–298, 43% identity) that comprise most of the extracellular domain (Fig. 10). Although many of the amino acid changes were nonconservative, the overall hydropathic profile was very similar for the two strains (Johnson et al., 1987b).

In contrast, the membrane anchor and cytoplasmic domains were 84% identical, a value that is similar to those of most of the other RSV proteins. However, these conserved domains do not appear to contain specific sequences required for G protein synthesis and processing: they could be replaced, by cDNA mutagenesis and expression by recombinant vectors, with the analogous domains of the PIV3 HN protein without affecting glycosylation and cell-surface expression (our unpublished data).

The ectodomain contained a single short segment of exact sequence identity between the two subgroups (amino acids 164–176). Also, the number (four) and sequence positions of the cysteine residues in the ectodomain were exactly conserved. Somewhat unusually, these are clustered within a single 14-amino acid segment (amino acids 173–186) that overlaps the conserved sequence. This is the most hydrophobic segment of the ectodomain, is relatively lacking in potential acceptor sites for carbohydrate, and is flanked by regions that are highly hydrophilic and contain numerous potential carbohydrate acceptor sites (Fig. 10). Secondary structure predictions suggest that this region contains a tight turn between two beta sheets, a structure that might be stabilized by intramolecular disulfide bonding. Alternatively, the cysteine residues might be involved in intermolecular disulfide bonding to stabilize an oligomeric structure.

Full-length G protein was shown to be an independently-protective antigen that induced RSV-neutralizing serum antibodies and resistance to RSV challenge in rodents, although it appeared to be less effective than the F protein

(Elango *et al.*, 1986; Stott *et al.*, 1986; Olmsted *et al.*, 1986; Johnson *et al.*, 1987a; Walsh *et al.*, 1987b). The immunity induced by the G protein was 13-fold less effective against challenge with RSV from the heterologous subgroup compared with the homologous subgroup (Johnson *et al.*, 1987a; Stott *et al.*, 1987). Thus, the structural dimorphism determined by the sequence analysis and the antibody-binding studies is associated with reduced efficiency in cross-protection.

We suggest that the divergence in the G protein reflects two factors: (1) selective immune pressure and (2) a relative tolerance for amino acid substitution. Analysis of the sequences of the RSV mRNAs and proteins supports the idea that the G protein can tolerate a high level of mutation compared to the other RSV proteins: 58% of the single-nucleotide differences between subgroups in the ORF for the G protein ectodomain resulted in amino acid changes, compared to 10–22% for the 1B, 1C, N, and F proteins and the G transmembrane and cytoplasmic domains (Johnson and Collins, 1989). The F protein also would be expected to be subject to immune selection, but the finding that its sequence is relatively highly conserved suggests that strict structural or functional constraints exist that reduce the rate of substitution.

The cysteine-rich sequence presumably is important for G protein structure or function, since it is the only apparent conserved sequence in the ectodomain. But the inefficiency in cross-subgroup protection would suggest that this conserved region is not part of a major neutralization epitope. We speculate that this cysteine-rich, hydrophobic region is relatively inaccessible to antibodies because it is part of a receptor-binding pocket that is shielded by the adjacent hydrophilic domains. These hydrophilic flanking structures probably would be antigenic, but structural heterogeneity could reduce the efficiency of cross-strain neutralization.

The extensive glycosylation of the ectodomain might reduce or alter its antigenicity by shielding the virus-specific protein with host-specific sugar (Skehel *et al.*, 1984; Elder *et al.*, 1986; Sjoblom *et al.*, 1987). In addition, heterogeneity in O-glycosylation could occur due to partial utilization of sites (Fiat *et al.*, 1980; Dahr *et al.*, 1982; Niemann *et al.*, 1984) and could result in an antigenically diverse population of molecules even within a genetically homogeneous virus population. This might be one factor in the poor neutralizing activity that is characteristic of G-specific monoclonal antibodies, including the common occurrence of an unusually large nonneutralizable fraction in homologous virus preparations (Fernie *et al.*, 1982; Walsh and Hruska, 1983; Örvell *et al.*, 1987; Tsutsumi *et al.*, 1987; Anderson *et al.*, 1988; Garcia-Barreno *et al.*, 1989). The G protein also has been shown to be a poor antigen for T cells in humans and mice (Bangham *et al.*, 1986; Pemberton *et al.*, 1987; Gupta *et al.*, 1990), and it is possible that this might be due to interference in processing or recognition due to the carbohydrate side chains.

In summary, the structural dimorphism between subgroups and the carbohydrate side chains of the G protein are two potential mechanisms for reducing its efficiency as a protective antigen. The significance and magnitude of these effects remain to be fully evaluated, although, as mentioned above, the subgroup dimorphism was associated with a 13-fold reduction in cross-subgroup protective efficacy in cotton rats (Johnson *et al.*, 1987a; Stott *et al.*, 1987). Despite these factors, the G protein does induce substantial amounts of homologous and heterologous immunity (Elango et al., 1986; Olmsted et al., 1986; Stott et al., 1986, 1987; Walsh et al., 1987a; Johnson et al., 1987a). Perhaps cooperative binding of poorly-neutralizing or nonneutralizing antibodies is an important contributor to protective immunity in vivo and RSV neutralization by polyclonal antibodies in vitro. Any factor that reduces the efficiency of the G protein as a protective antigen could contribute to the general observation that naturally-acquired resistance to RSV replication in the respiratory tract typically is incomplete, allowing the virus to infect repeatedly.

Epitope mapping studies, which tested synthetic peptides representing all of the G ectodomain in binding assays with convalescent and monoclonal antibodies, showed that the most reactive peptides were those that represented amino acids 134–158 and amino acids 174–208 (Norrby et al., 1987). These regions correspond to the hydrophilic regions that flank the cysteine-rich conserved segment. Analysis of deletion mutants of the G protein showed that the C-terminal 68 amino acids were not required for the induction of protective immunity or RSV-neutralizing serum antibodies (Olmsted et al., 1989). This is consistent with the finding that peptides from that region were not antigenic in binding assays (Norrby et al., 1987). A further deletion of the G protein that extended into the cysteine-rich segment resulted in a truncated form that was glycosylated and transported to the cell surface, but which was completely inactive in inducing RSV-neutralizing serum antibodies and protective immunity (Olmsted et al., 1989).

6. The SH Protein

The short SH protein contains a central core of hydrophobic amino acids flanked by short N-terminal and C-terminal domains that each contain a potential acceptor site for N-linked carbohydrate (Fig. 11) (Collins and Wertz, 1985b).

In immunoprecipitation experiments, antibodies raised against a synthetic peptide representing the C-terminal 12 amino acids of SH reacted with four electrophoretically distinct species of estimated M_r 4800, 7500, 13,000–15,000, and 21,000–30,000 (Fig. 12) (Olmsted and Collins, 1989). Antibodies against a peptide of the N-terminal eight amino acids of SH reacted with all except the M_r 4800 species (our unpublished data).

The M_r 7500 species appears to be the full-length, unglycosylated, unprocessed form of SH. The M_r 4800 species also is unglycosylated, and it appeared to lack N-terminal amino acids of the complete SH protein based on its lack of reactivity with the N-terminal-specific antibodies described above. The initiating methionine codon of the SH ORF is in a sequence context (82-CCAAATGG) which is unfavored for efficient translational initiation due to the presence of a C residue in the −3 position (Kozak, 1987). Ribosomes that bypass the first initiation site could initiate at the second methionine codon in the sequence (Met-23), which is in the same ORF. Initiation there would generate an N-terminally-truncated form of the SH protein that would be the appropriate length (42 amino acids) to be the M_r 4800 species. This model was confirmed by studies in which a cDNA of the SH gene was modified by site-directed mutagenesis and expressed in tissue culture cells using an SV40 ex-

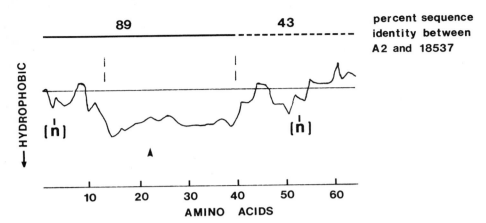

FIGURE 11. Diagram of the SH protein of RSV strain A2, showing the hydropathic profile, potential sites for the addition of N-linked carbohydrate (n), and the position (arrow) of the second methionine in the sequence that is the translational start site for the shorter M_r 4800 species. Amino acid identity between the A2 and 18537 strains (representing the two antigenic subgroups) is summarized at the top, with each percent value referring to the polypeptide region below it and the regions being demarcated by continuous and discontinuous segments in the horizontal line.

pression vector. Modification of Met-1 to Ser-1 resulted in expression of only the M_r 4800 species, and modification of Met-23 to Ser-23 eliminated the synthesis of the M_r 4800 species without affecting the synthesis of the other three species (our unpublished data).

The other two major forms of the SH protein, the M_r 13,000–15,000 and the M_r 21,000–30,000 species, are glycoproteins (Olmsted and Collins, 1989). The mutagenesis experiments described above showed that these glycosylated forms are derived from the M_r 7500 species rather than from the M_r 4800 species. [There also is evidence that the M_r 4800 species is processed into minor glycosylated forms of M_r 10,500–12,500 and M_r 18,000–30,000 that presumably are N-terminally truncated counterparts of the major glycosylated forms described above (our unpublished data). These are not discussed further here, but add to the array of structurally-distinct SH species.]

Cotranslational addition of N-linked carbohydrate to the M_r 7500 species generates the M_r 15,000 species, which is reduced in M_r to 13,000 over a 20–30 min period (Fig. 12), probably due to trimming of the carbohydrate moiety. The N-linked chain remained sensitive to endoglycosidase H (Olmsted and Collins, 1989).

The M_r 21,000–30,000 glycosylated species appeared 20–30 min after translation (Fig. 12) (Olmsted and Collins, 1989) and was generated by the addition of polylactosamine to the N-linked side chain of SH, with the substrate probably being the M_r 13,000 species (Sheshberadaran, M. A. Williams, P. L. Collins, R. A. Olmsted, and R. A. Lamb, unpublished results). Polylactosaminylation has been described for several cellular glycoproteins, such as blood group antigens, and for the small NB protein of influenza B virus, but its significance for RSV is unknown (Kornfeld and Kornfeld, 1985; Williams and Lamb, 1988).

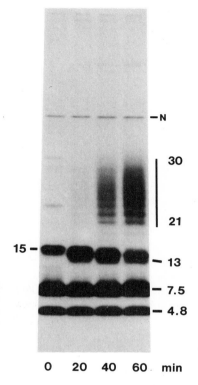

FIGURE 12. The SH protein of RSV strain A2. Time course of appearance of the different forms of the SH protein. RSV-infected HEp-2 cells were labeled with [^{35}S]methionine for 15 min followed by nonradioactive chase periods of 0, 20, 40, and 60 min, as indicated. Cell lysates were prepared and analyzed by immunoprecipitation with antibodies specific to the C terminus of SH followed by SDS–PAGE on a 17% gel. The SH species are identified by their estimated M_r in kilodaltons, and the position of a small amount of contaminating N protein is indicated N.

Although the SH protein can be glycosylated, most of the intracellular SH protein accumulated as the unglycosylated M_r 4800 and 7500 species (Fig. 12). Usually, the utilization of an acceptor site for N-linked glycosylation is an all-or-nothing event (Kornfeld and Kornfeld, 1985). The structure of SH surrounding the acceptor site might interfere with glycosylation, or perhaps the synthesis of SH is rapidly completed because of its small size and the nascent polypeptide is only briefly available for glycosylation. Addition of polylactosamine to the N-linked carbohydrate of the SH protein also appeared to be inefficient (Fig. 12). Consequently, all four major forms of the SH protein accumulated intracellularly. The same constellation of SH species accumulated when the SH cDNA was expressed using recombinant SV40 or vaccinia virus vectors, showing that the processing of SH is independent of the other viral proteins (Olmsted and Collins, 1989).

All four forms of the SH protein copurified with the membrane fraction from lysates of infected cells. They appeared to be integral membrane proteins because the stability of their association, as assayed by treatment with detergent and exposure to high pH, was similar to that of the F protein and was more stable than the membrane associations of the M and 22K proteins (Olmsted and Collins, 1989).

Immunofluorescent staining of intact infected cells showed that the SH protein is expressed abundantly at the cell surface. Immunoprecipitation of antibody–antigen complexes formed by incubating SH-specific antibodies with intact infected cells showed that the M_r 7500, 13,000–15,000, and 21,000–30,000 species, but not the M_r 4800 species, were present at the cell surface, and were in approximately the same relative abundances as found intracellularly (Olmsted and Collins, 1989). In pulse-chase experiments, the M_r 7500 protein appeared at the cell surface within 5–10 min after translation, a rapid rate that approximates the maximum flow rate of the exocytotic pathway (Wieland et al., 1987). In comparison, the glycosylated M_r 13,000–15,000 and 21,000–30,000 species were transported more slowly and appeared at the surface 30–40 min after translation. None of the SH proteins appeared to be secreted (Olmsted and Collins, 1989).

Trypsin treatment of RSV-infected, intact cells showed that the C terminus of the M_r 7500 species was trypsin sensitive and therefore extracellular, whereas the N terminus was not detectably digested and presumably was cytoplasmic (our unpublished data). This orientation predicts that the carboxy-terminal acceptor site for carbohydrate (Asn-52) is utilized, whereas the second, N-proximal site (Asn-3) is not. However, the position of the carbohydrate has not yet been confirmed directly.

The M_r 7500 and 21,000–30,000 species were present in virion preparations, suggesting that these two forms of the SH protein are structural components of the virion (Huang et al., 1985; our unpublished data). However, we cannot currently exclude the possibility that the SH protein in virion preparations represents contamination by cellular membranes, which tend to copurify with RSV virions and which would contain SH protein. Final determination of the status of SH as a structural protein must await analysis by more direct methods such as immunoelectron microscopy.

The SH proteins of strains A2 and 18537, representing the two subgroups, shared 76% identity and also shared the two potential glycosylation sites and the second methionine codon at position 23 (Table II, our unpublished data). Furthermore, the complex glycosylation pattern of SH was conserved between the subgroups (our unpublished data), suggesting that it has functional importance.

The amino acid differences between the subgroup strains were not uniform: the cytoplasmic and transmembrane regions (residues 1–41) were more highly conserved (89% identity) than was the extracellular domain (residues 42–64, 43% identity) (Fig. 11). This suggests that the transmembrane and cytoplasmic domains have functional importance, whereas the extracellular domain might exist solely to maintain SH firmly in the membrane. One possibility is that SH has a functional role in interacting with viral components at the inner face of the plasma membrane. The relatively greater divergence observed for the extracellular C-terminal domain might be a consequence of selective immune pressure, as is hypothesized for the ectodomain of the G protein.

SV5 encodes a 44-amino acid SH protein that, like RSV SH, is inserted in the membrane with the C terminus oriented extracellularly (Hiebert et al., 1988; also see Chapter 6). A similar 57-amino acid protein probably also exists for MuV, based on the identification and sequencing of a putative SH mRNA (Elango et al., 1989). The SH genes of all three viruses have the same gene map position, preceding the attachment protein gene, and this supports the idea that these are analogous proteins. However, sequence alignments did not provide unambiguous evidence of relatedness. Influenza virus types A and B also encode small integral membrane proteins, namely the 97-amino acid M2 protein and the 100-amino acid NB protein, respectively (Williams and Lamb, 1986; Zebedee and Lamb, 1988; also see Chapter 6). The existence of a small surface protein for each of these enveloped viruses is suggestive of functional similarity and could be compatible with divergent evolution from a common ancestral protein or convergent evolution such that unrelated proteins have assumed similar functions.

Postinfection human sera were shown to contain antibodies specific to synthetic peptides of the extracellular domain of the SH protein (Nicholas et al., 1988). An antibody binding site was mapped to amino acids 46–56 of the strain A2 protein. This segment also contains the predicted site (Asn-52) for the sugar side chain, which might interfere with recognition of the glycosylated species. Also, amino acids 45–62 appeared to contain an epitope for helper T cells (Nicholas et al., 1988), although other workers reported that SH protein was not significantly recognized by RSV-primed murine T cells (Pemberton et al., 1987; Openshaw et al., 1988). In a preliminary study, immunization of cotton rats with a recombinant vaccinia virus that expressed SH did not confer significant protective immunity to RSV infection, despite the fact that SH is a surface protein (our unpublished data). It might be that SH is not a major antigen for B or T cells, and the detection of the correspondingly weak responses might be variable and influenced by the experimental design.

III. RSV REPLICATION

A. Attachment, Penetration, and Growth Cycle

The G protein mediates attachment of the virion to host cells (Levine et al., 1987). The cellular receptor has not been identified, but appears to be abundant based on the high capacity of cultured cells for binding purified G protein (Walsh et al., 1984). Because the RSV G protein is structurally dissimilar to the HN or H proteins of the other paramyxoviruses, significant differences might exist in the structure of the attachment spike and cellular receptor. However, the apparent lack of hemagglutination activity for RSV probably is not a fundamental characteristic of the unique structure of G, because the PVM G protein, which appears to have structural similarities to RSV G, does have in vitro hemagglutinin activity (Ling and Pringle, 1989b).

The F protein mediates viral penetration by fusion of the viral envelope with the plasma membrane of the host cell. RSV penetration was monitored by video microscopy, and the rate of membrane flow at 30°C was 110–250 μm/sec, which would correspond to 2–4 sec for a spherical virion (Bachi, 1988). In contrast, the half-time of penetration as measured by the acquisition of antibody insensitivity by preadsorbed RSV was more than 15 min (Levine and Hamilton, 1969). Taken together, these observations indicate that the initiation of fusion by preadsorbed virions is the rate-limiting step in penetration. By video microscopy, fusion appeared to be the reverse of budding, suggesting that "membrane budding and fusion depend on common parameters controlling the flow of membrane" (Bachi, 1988). Viral penetration results in the delivery of the viral nucleocapsid to the cytoplasm, leaving the envelope antigens in the cell plasma membrane (Routledge *et al.*, 1987b).

The kinetics of RSV growth varies considerably with RSV strain, cell type, multiplicity of infection, and other factors. In HEp-2 cells infected with two to five infectious units per cell of RSV strain A2, the synthesis of viral proteins and RNAs can be detected by 2–6 hr after infection, and progeny virus by 10–12 hr. The production of RNA reaches a maximum by 16–20 hr, and protein synthesis and virus release are maximal by 20–24 hr. There is no evidence of temporal gene regulation. Cell fusion can be detected by 18–24 hr, extensive syncytia form by 30–48 hr, and destruction of the monolayer quickly follows (Levine and Hamilton, 1969; Wunner *et al.*, 1975; Lambert *et al.*, 1980,1988; Huang, 1983).

The synthesis of viral macromolecules is abundant; for example, analysis of cell lysates by SDS–PAGE and Coomassie blue staining showed that the viral proteins are among the most abundant infected-cell species (Garcia-Barreno *et al.*, 1988). The synthesis of cellular polypeptides is not specifically inhibited, and modest reductions in cellular DNA and RNA synthesis occur relatively late in infection (Levine *et al.*, 1977). Viral proteins (Kisch *et al.*, 1962; Berthiaume *et al.*, 1974; Stott *et al.*, 1984) and RNAs (Huang and Wertz, 1982) accumulate in the cytoplasm, and replication appears to be independent of nuclear function (Wunner *et al.*, 1975; Lambert *et al.*, 1980).

RSV N protein has been shown to accumulate in a cytoplasmic pool followed by association with intermediate filaments of the cytoskeleton (Garcia-Barreno *et al.*, 1988). By analogy to other paramyxoviruses and VSV, this might be indicative of an association of nucleocapsids with the cytoskeleton. This suggests that RSV, like the others, utilizes the cytoskeleton in its replicative cycle, perhaps, for example, as a structural support or transport mechanism in virion morphogenesis (Hamaguchi *et al.*, 1985; Bohn *et al.*, 1986).

B. vRNA Transcription

1. *in Vitro* Transcription

Virions of negative-strand viruses contain a viral polymerase that directs the copying of input vRNA into complementary mRNAs, an activity that can be demonstrated *in vitro* (Ball and Wertz, 1981; Iverson and Rose, 1981; Emerson, 1982, 1985; Kingsbury, 1985, 1989; Banerjee, 1987). In infected cells, transcription by the input nucleocapsid (primary transcription) and subsequent transcription by progeny nucleocapsids (secondary transcription) yield the same types and relative amounts of mRNA, and there is no evidence of temporal regulation of individual genes.

Because the proteins of RSV nucleocapsids appear to be analogous to those of other paramyxoviruses and rhabdoviruses, it seems likely that the transcriptional machinery of RSV will have general similarities. However, a preliminary report of polymerase activity in RSV virions *in vitro* has not been confirmed (Mbuy and Rochovansky, 1984). RSV transcription has been achieved in *in vitro* reactions that contain extracts from lysolecithin-treated RSV-infected cells, but the protein components of the transcriptase and its presence in the virion remain to be demonstrated (Herman, 1989).

2. Transcriptional Map

The transcriptional map was determined by comparing the rates of inactivation of individual genes by UV irradiation (Dickens et al., 1984). For the nine smaller genes, the sensitivity of each was directly proportional to its distance from the 3' vRNA end. This implied that transcription initiates at a single promoter at the 3' end, with the individual genes transcribed sequentially in their 3' to 5' order (Ball and Wertz, 1981; Emerson, 1982). The synthesis of L mRNA could not be accurately monitored due to its low abundance and large size, but the identification of a polytranscript of the 22K and L genes implied that it was transcribed as the last gene in the sequence (Collins and Wertz, 1983).

The transcriptase might initiate sequential transcription in the leader region at the exact 3' end of vRNA or, alternatively, at the 1C gene (Ball and Wertz, 1981; Iverson and Rose, 1981; Emerson, 1982; Talib and Hearst, 1983; Banerjee, 1987; Vidal and Kolakofsky, 1989). The structure and expression of this region of vRNA remain to be defined.

3. Transcriptive Signals

During sequential transcription, the polymerase initiates gene transcription at the gene-start signals (Collins et al., 1986). The gene-end signals are thought to stall the polymerase and direct the synthesis of polyadenylate by reiterative copying of the short oligo(U) tracts (Schubert et al., 1980; Iverson and Rose, 1981; Ball and Wertz, 1981). The polymerase apparently crosses each intergenic region without synthesis and resumes transcription at the next downstream start signal, but also occasionally transcribes across intergenic regions to generate polytranscripts.

It seems clear that the gene-start and gene-end signals function independently of each other, because their relative positions are reversed in the 22K–L gene junction without any apparent effect on function (Collins et al., 1987). It also is likely that neither the intergenic nor flanking gene sequences other than the start and end signals contain important conserved structural signals, because those sequences are not well conserved between the RSV antigenic subgroups (Johnson et al., 1987a; Johnson and Collins, 1988a,b, 1989; and our unpublished data). Also, inspection of several RSV genes showed that sequences closely similar to the gene-start or gene-end signals do not occur within genes, and instead are specific to the termini. All of these findings support the interpretation that the gene-start and gene-end signals are discrete, self-contained transcriptive signals.

In the rhabdovirus VSV, and in several paramyxoviruses, such as SV, PIV3, and MeV, the intergenic regions are highly conserved within each virus and between viruses. For example, most of the intergenic regions of SV, PIV3, and MeV are identical and are among the most highly-conserved nucleotide sequences of these viruses (Chapters 2 and 3). This high level of conservation implies that these sequences, together with the adjoining gene-start and gene-end sequences, have important functional roles. On the other hand, the intergenic regions of RSV, SV5, NDV, and MuV, as well as the rhabdovirus rabies virus (Collins et al., 1986; Tordo et al., 1986; Chapter 2), are generally not well conserved within or between viruses, with RSV exhibiting the greatest heterogeneity among the paramyxoviruses. It is possible that the difference in intergenic conservation between these two groups of viruses represents a difference in transcriptional mechanism, with the intergenic sequences serving as transcriptive signals in the SV group, but not in the RSV group.

4. Nonequimolar Synthesis of mRNAs

The relative molar amounts of intracellular RSV mRNAs have not been quantified. But it was apparent, from RNA blot hybridization experiments that used vRNA as an internal standard for comparing relative amounts, that the mRNAs of genes that are proximal to the 3′ end of vRNA accumulate intracellularly in greater molar amounts than do those of downstream genes (Collins and Wertz, 1983; Collins et al., 1984b). This suggests that, as is the case for nonsegmented negative-strand viruses in general, the frequency of RSV gene transcription decreases with increasing distance from the promoter, perhaps due to polymerase falloff (Fig. 13) (Iverson and Rose, 1981; Ball and Wertz, 1981; Emerson, 1985; Kingsbury, 1985, 1989). This polar effect on mRNA synthesis probably is a major factor determining the relative molar amounts of each mRNA and protein. Studies with VSV showed that polymerase falloff occurs primarily at the intergenic regions rather than within genes (Iverson and Rose, 1981). It is possible that the heterogeneous intergenic regions of RSV might influence attenuation: for example, attenuation might be more severe at longer intergenic sequences. If so, then the intergenic regions of RSV might have functional significance, and their structures might have evolved to provide optimal proportions of the mRNAs.

5. Transcription of the Overlapping 22K and L Genes

The model of sequential transcription does not provide for the transcription of overlapping genes into overlapping mRNAs (Ball and Wertz, 1981; Iverson and Rose, 1981; Banerjee, 1987). The mechanism of transcriptional initiation of the L gene remains to be defined. One possibility is that the polymerase can somehow exit from the 22K gene and gain access to the L gene without dissociating from the template. But there are no obvious structural features in this region of RNA that suggest such a mechanism. Alternatively, vRNA might contain a second, separate promoter for the L gene. However, the molar amount of the combined transcripts of the L gene was considerably lower than that of the 22K gene, which would be more consistent with sequential transcription from the 3′ vRNA end than with independent entry (Collins et al., 1987). Therefore, if an independent promoter exists for L, it must be relatively inefficient.

It also is possible that transcription of the L gene occurs by sequential transcription from the 3′ vRNA end. The synthesis of polytranscripts is evidence that the RSV polymerase occasionally fails to recognize gene-end signals and continues transcription of the downstream sequences. Perhaps, in an anal-

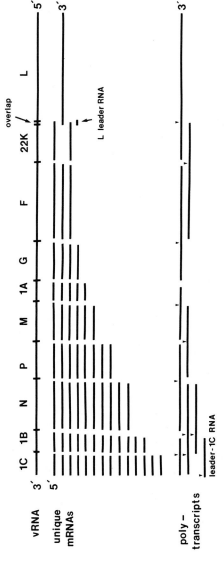

FIGURE 13. Diagram of the RNA products of RSV transcription. The top line is RSV vRNA (3' to 5'), and below are the positive-sense transcripts (5' to 3'): unique mRNAs, polytranscripts, and L leader RNA. The unique mRNAs are shown in idealized relative amounts to illustrate that the abundance of each mRNA is related to the position of its gene in the 3' to 5' map. L leader RNA, the putative leader–1C readthrough transcript, and the nine identified "dicistronic" polytranscripts are shown without representation of their relative amounts. The concept for the diagram is from published work (Kingsbury, 1990). The sizes of genes and transcripts are not to scale.

ogous way, the polymerase might occasionally fail to recognize a gene-start signal and would instead continue to move down the template without synthesis. If so, then a fraction of the polymerases that encounter the 22K gene-start sequence could fail to initiate transcription and instead would move down the 22K gene without synthesis. These polymerases would be available to initiate transcription at the next gene-start signal, that of the L gene. If the L gene is accessible only to polymerases that failed to initiate at the 22K gene-start signal, this would impose a steep gradient of attenuation between the 22K and L genes.

In summary, the simplest model for RSV transcription involves polymerase entry primarily at the 3' vRNA end, followed by sequential movement of the polymerase down the template. The polymerase appears to alternate between synthetic and nonsynthetic modes. It seems likely that recognition of the gene-start and gene-end signals is an obligatory step in switching between the two modes. The gene-start and gene-end signals might be the only nucleotide sequence signals operative during transcription.

C. vRNA Replication

Negative-strand virus vRNA replication, which has been studied most extensively with VSV and SV, involves the synthesis of (1) a replicative intermediate (vcRNA) that is a complete positive-sense copy of vRNA, and (2) progeny vRNA using vcRNA as a template. The synthesis of vcRNA and vRNA is tightly coupled to their encapsidation and is inhibited when NP is limiting. The rapid, concurrent encapsidation of product RNA is thought to alter the product RNA, template, or polymerase in such a way that gene-end signals and other attenuation signals in vcRNA and vRNA are ignored and full-length strands are made (Emerson, 1985; Kingsbury, 1985; Banerjee, 1987; Vidal and Kolakofsky, 1989; also see Chapters 7 and 8).

Consistent with this model, the synthesis of vRNA-sized RNAs in RSV-infected cells is inhibited by cycloheximide, whereas vRNA transcription is unaffected (Huang and Wertz, 1982). RSV vcRNA was identified in intracellular RSV RNA by RNA blot hybridization (Huang and Wertz, 1982). Analysis of lysates of RSV-infected cells showed that vRNA and vcRNA were present exclusively in encapsidated form (Huang, 1983). Hybridization studies using RNase A indicated that approximately 10–15% of the RNA of infected-cell nucleocapsids is vcRNA (Huang, 1983).

The in vitro system for RSV RNA synthesis mentioned above was shown to direct the encapsidation of product RNA into nucleocapsid-like structures, indicating that it performed at least some steps of vRNA replication in vitro (Herman, 1989).

Defective-interfering particles have been described for RSV, but have not been characterized molecularly (Treuhaft and Beem, 1982). One RSV cDNA, copied from mRNA synthesized by virus from a high passage level, appeared to be a copy of a fused mRNA containing the start of the N mRNA fused to the end of the G mRNA (Elango and Venkatesan, 1983; Collins et al., 1985). Such an mRNA might have been synthesized by transcription of a vRNA molecule that had an internal deletion that fused the two genes. An analogous fused transcript was described for a deletion mutant of VSV (Herman and Lazzarini, 1981a). However, it also is possible that this aberrant cDNA is due to an error in mRNA synthesis or cDNA cloning.

Thus, the study of RSV vRNA replication is at an early stage, but at least some general features appear to conform to the SV model.

D. Virion Morphogenesis

Progeny virions mature by budding through areas of the plasma membrane that contain localized accumulations of viral spikes (Armstrong et al., 1962; Norrby et al., 1970; Kalica et al., 1973; Bachi and Howe, 1973; Berthiaume et al., 1974; Bachi, 1988). Bachi (1988) noted, by video microscopy, that the "maturation of virions occurs within circumscribed regions of the cell surface overlaying regions of cytoplasm undergoing hectic motion." This motion might involve cellular components, such as elongating cytoskeleton filaments, or viral structures, such as aggregating spikes, inner membrane proteins, or nucleocapsids. Budding appeared to take place at specific sites, where several virions were observed to bud in succession (Bachi, 1988). Inhibition of protein synthesis at any time during infection stopped the release of virus (Levine and Hamilton, 1969), indicating that there is an ongoing requirement for newly synthesized viral or cellular protein despite the large intracellular accumulation of viral species.

Much of the progeny virus is in the form of empty filaments, which probably is one factor in the poor yield of infectious virus (Bachi and Howe, 1973). Furthermore, 90% or more of budded infectious virus fails to detach from the plasma membrane (Levine and Hamilton, 1969; Wunner et al., 1975; Bachi, 1988). It is possible that the apparent absence of a receptor-destroying activity in RSV virions is one factor in their inefficient release. Nascent virions also have been observed to fuse with their parental cell rather than detach (Bachi, 1988). The slow rate for the initiation of fusion noted above would make this a relatively infrequent event, allowing most of the progeny virions the opportunity to detach.

IV. EVOLUTIONARY RELATIONSHIPS

A. RSV Antigenic Subgroups

The two RSV antigenic subgroups exhibit extensive sequence dimorphism in the intergenic and noncoding gene regions and in certain polypeptide domains, such as the ectodomains of the G and SH proteins (Table II). Divergence in these regions presumably occurs because functional or structural contraints on nucleotide or amino acid substitution were less strict than elsewhere.

This showed that the antigenic diversity is associated with extensive sequence diversity rather than differences at just a few immunodominant sites. In contrast, there is only a small amount of intrasubgroup diversity. Thus, the two subgroups are distinct and might represent an early stage in divergent evolution, a process that might eventually yield two distinct RSV types analogous to the different parainfluenza types. Alternatively, the two subgroups might represent the maximum extent of divergence possible for human strains and might be at a stable endpoint. However, the unusual capacity of the G protein to tolerate amino acid substitutions probably is an ongoing characteristic, because the high ratio of amino acid change per nucleotide change noted above was observed both between and within the subgroups (Johnson and Collins, 1989).

Antigenic analysis of RSV isolated from different geographic locations over a period of 30 years revealed a low, random pattern of divergence within two relatively stable subgroups (Anderson et al., 1985; Mufson et al., 1985; Morgan et al., 1987; Hendry et al., 1988; Beeler and Coelingh, 1989). This indicates that the divergence is not rapid and suggests that the two subgroups are relatively ancient.

The rate of nucleotide substitution for RSV was estimated for the G gene, the one with the greatest extent of intrasubgroup and intersubgroup sequence diversity, by ribonuclease protection studies (Storch et al., 1989). In a preliminary study of several subgroup A strains, no nucleotide changes were detected for RSV clones that had been isolated serially during acute infection of adults or following multiple laboratory passages in tissue culture (Storch et al., 1989). The sensitivity of the assay for detecting nucleotide differences remains to be calibrated, but these results suggest that the evolution of the virus population is not rapid even during infection of adults, who almost certainly would have had previous RSV infections and would have existing immunity that might exert selective pressure on the virus.

B. Relationships with Other Paramyxoviruses

As described above, unambiguous sequence relationships between the RSV proteins and their counterparts in other paramyxoviruses was demonstrated only for the F and L proteins, with the latter also having regions of sequence relatedness with rhabdovirus L proteins. The sequence relatedness for the RSV F protein supports the classification of RSV (and the other pneumoviruses) in the paramyxovirus family, although the many other differences indicate that the pneumoviruses are relatively distinct evolutionarily from the other members of the family. The conservation of F and L protein sequences suggests that membrane fusion and aspects of polymerase function are the most highly conserved viral activities.

Comparison of paramyxovirus genetic maps suggests that they fall into four groups (Fig. 14): (1) genus *Pneumovirus*, represented by RSV; (2) genus *Morbillivirus*, represented by MeV; (3) a subgroup of genus *Paramyxovirus* that is represented here by PIV3 and also includes SV and probably human PIV1; and (4) a second subgroup of genus *Paramyxovirus* that is represented here by MuV and includes SV5 and probably NDV (Chapters 1–3). The length of vRNA is very similar for all four groups (legend to Fig. 14), consistent with a common evolutionary origin or a limitation on vRNA size.

The viral groups represented by MeV and PIV3 are almost identical with regard to gene map, arrangement and sizes of ORFs, and sequences of intergenic regions. Also, the amino acid sequences of the NP, M, F, and L proteins share unambiguous sequence relatedness. This supports the interpretation that these viruses are derived by divergent evolution, without major rearrangement of genome organization, from a common ancestor. One major difference between the two groups is that the *Morbillivirus* attachment H protein lacks the neuraminidase activity that is associated with the *Paramyxovirus* attachment HN protein.

The group represented by MuV shares many general features of gene map and protein structure with the PIV3 and MeV groups. Also, there is significant relatedness involving the amino acid sequences of the NP, M, F, and (for the MuV and PIV3 groups) HN proteins. But MuV has several significant differences in genome organization compared to MeV and PIV3. Specifically, MuV (1) has a smaller P protein, (2) lacks a C protein, (3) has an additional gene that appears to encode an SH-like protein, and (4) has intergenic regions that are structurally heterogeneous, in contrast to the highly-conserved intergenic triplets of PIV3 and MeV. Thus, whereas the genome organizations of PIV3 and MeV are closely similar, that of MuV has a number of distinct features.

These observations are consistant with all three viral groups (PIV3, MeV, and MuV) being related evolutionarily, with the MuV and PIV3 groups being somewhat more closely related with regard to amino acid sequence, but with the PIV3 and MeV groups being somewhat more closely related with regard to genome organization. It might have been expected that genome organization would have been relatively more resistant to evolutionary divergence compared to amino acid sequences, but that does not appear to have been the case.

Compared with the PIV3, MeV, and MuV groups, RSV encodes several proteins (N, P, M, SH, F, and L) that have obvious counterparts in the others. This presumption is supported by the sequence relationships of the RSV F and L proteins with the corresponding proteins of the other genera. Also, the map order of some of the RSV genes, $3'$ ... *NP-P-M* ... *F* ... *L*, is the same as the order of the corresponding genes in PIV3, MeV, and MuV. These similarities strongly support the interpretation that RSV is related evolutionarily to the other three groups.

But RSV also has a number of differences in genome organization relative to the others. Specifically, RSV (1) has three additional genes encoding separate mRNAs for the 1C, 1B, and 22K proteins, which do not have obvious counterparts in the other paramyxoviruses, (2) lacks an obvious counterpart to either the C or V proteins, (3) encodes N, P, and M proteins that are smaller than their counterparts in the other viruses, (4) has intergenic regions that are much more heterogeneous than those of MuV and do not resemble the highly conserved intergenic regions of PIV3 and MeV, (5) has two overlapping transcriptional elements, the *22K* and *L* genes, (6) has an attachment protein, G, which is the functional counterpart to the HN or H proteins of the others, but which is completely distinct structurally and does not exhibit evidence of evolutionary relatedness, (7) has glycoprotein genes that are arranged in a map order, $3'$ ... *G-F* ... , that is the reverse of the order of the corresponding genes, $3'$ ... *F-HN* ... , in the other viruses, and (8) appears to have 3' and 5' vRNA extragenic terminal regions that are, respectively, substantially shorter and substantially longer than the corresponding vRNA regions in the other genera.

The 3'-proximal fourth of the gene map is one region of extensive dif-

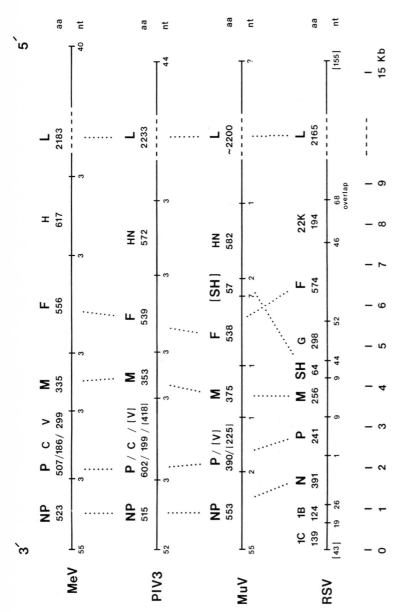

FIGURE 14. Comparison of the 3' and 5' gene maps of RSV and three representative members of the *Paramyxovirus* and *Morbillivirus* genera. The maps are drawn to scale with regard to nucleotide lengths except for the *L* genes. Nucleotide lengths of the 3' leader, intergenic, and 5' trailer regions are indicated below each map. The RSV leader and trailer region lengths are in parentheses because they are estimates, and the MuV trailer region has not yet been mapped. The numbers of amino acids in the deduced sequences of the unmodified translation products are indicated above each map. Predicted proteins that have not yet been directly identified are in parentheses. The amino acid length of MuV L is an estimate. Proteins that appear to be analogous between RSV and the others are indicated in larger letters, and these relationships are indicated in the vertical dimension with dotted lines. The vRNA nucleotide lengths are: MeV, 15,892; PIV3, 15,461; MuV, not determined; and RSV, 15,222 [see Note added in proof].

ference between RSV and the other three groups. RSV is unique compared to the others because the major nucleocapsid protein gene is not first in the 3' to 5' map, but instead is preceded by the small *1C* and *1B* genes. If these various paramyxoviruses indeed are derived from a single ancestor, then the *1C-1B-N-P* region of RSV and the *NP-P-C* region of the others might be related evolutionarily. If so, perhaps these proteins might contain analogous enzymatic or structural domains that have come to be encoded by different strategies and have been distributed among different polypeptide chains.

The other map region that has extensive diversity is the *SH-G-F-22K* region of RSV versus the *F-SH-HN* region of MuV (and the *F-HN* regions of the PIV3 and MeV groups, which lack the SH gene). One possibility is that RSV evolved from an ancestral virus that had a gene arrangement in this region that was like that of MuV. We speculate that, in this postulated predecessor, (1) the present RSV *G* and *22K* genes originally were fused as a single, larger ancestral gene *(G + 22K)* encoding an attachment protein that would be more similar in size and map position to the HN and H proteins, and (2) the RSV *F* gene originally preceded the SH gene in the map. This hypothetical map order (3'-*F-SH-G + 22K*) would be the same as that of MuV (3'-*F-SH-HN*). We speculate that a rearrangement took place involving the insertion of the *F* gene into the middle of the *G + 22K* gene, ultimately creating four genes from three and establishing the present arrangement of RSV genes. This model is attractive because it accounts for the present gene arrangement of RSV by a single-step rearrangement, but the mechanism for the rearrangement is not known. One mechanism for genome rearrangement of nonsegmented negative-strand viruses is well known (Chapter 10): the polymerase of negative-strand viruses in general can detach from the template during synthesis of vRNA or vcRNA without releasing the nascent strand, and can reinitiate at a different site and resume synthesis. This can result in either sequence deletion (such as in defective-interfering vRNA) or sequence duplication. However, this mechanism probably would not account for the rearrangement proposed here.

It's interesting to note that this same vRNA region also appears to be structurally diverse in the rhabdoviruses. For the rhabdoviruses, the junction between the attachment G protein gene and the *L* gene consists of: (1) for VSV strain Indiana, an intergenic dinucleotide; (2) for VSV strain New Jersey, an intergenic region that contains 19 nucleotides in contrast to the conserved dinucleotides at the other junctions; (3) for rabies virus, a 423-nucleotide sequence that contains a possible remnant protein gene; and (4) for infectious hematopoietic necrosis virus, an additional gene, *NV*, that encodes a nonstructural M_r 12,000 protein (Rose, 1980; Kurath *et al.*, 1985; Tordo *et al.*, 1986; Luk *et al.*, 1987).

V. CONCLUSIONS

Sequence analysis of RSV vRNA is nearly complete, and most or all of the major RNAs and proteins have been identified. Most of the RSV proteins and most of the features of the gene map are generally similar to those of other paramyxoviruses. The unambiguous sequence relationships of the RSV F protein to the corresponding proteins of the other members of the family is probably the most compelling single piece of evidence for classification of RSV as a paramyxovirus. The sequence relationships of the RSV L protein to its counterparts in other paramyxoviruses and rhabdoviruses speak of a common ancestry for all of these nonsegmented negative-strand viruses.

The presence of additional RSV genes and proteins, and the differences in some aspects of the gene map, indicate that RSV has undergone considerable divergence, including vRNA rearrangement, from the other paramyxoviruses. Additional sequence comparisons, especially of the L protein, may make it possible to construct a phylogenetic tree of the negative-strand viruses. One feature of RSV that is difficult to reconcile with the idea of a common ancestry with the other paramyxoviruses is the novel structure of the G protein: the G protein seems to be almost completely dissimilar to the HN and H proteins, and it is difficult to envision a common ancestral protein.

The molecular studies provided clues and reagents that will be useful for future studies of RSV biology and for designing vaccines. These include identification of the amino acid sequence dimorphism and distinctive glycosylation of the G protein, the identification of F as the major conserved protective antigen, the availability of cDNA clones for evaluating the contributions of individual proteins to host immunity, and the availability of complete nucleotide and amino acid sequences for designing synthetic oligonucleotide and peptide reagents.

With regard to continuing molecular studies, the nucleotide and amino acid sequences provide only the most rudimentary clues about the higher-order structures of the viral molecules and the dynamics of viral gene expression, vRNA replication, and virion assembly. The next level of investigation will include (1) determining three-dimensional polypeptide structures, especially for the virion spikes; (2) characterizing protein–protein and protein–nucleic acid interactions, especially in the assembly of nucleocapsids, virion spikes, and virions; (3) analysis of vRNA transcription and replication *in vitro* and *in vivo* with the use of specific cloned probes; and (4) investigating functional roles for vRNA sequences and viral proteins, by biochemical analysis of subviral components, by analysis of viral variants and mutagenized, expressed cDNAs, by the use of antibodies and antisense oligonucleotides to inhibit individual species *in vitro* and *in vivo*, and by the use of vector-expressed proteins to complement conditional-defective RSV mutants or subviral components such as purified nucleocapsids.

A major obstacle to molecular studies is that, for negative-strand viruses in general, methods for regenerating infectious virus from cDNA have yet to be developed. The difficulty rests in the fact that the minimum unit for infectivity is the nucleocapsid rather than naked nucleic acid. The next logical step will be to explore methods for incorporating synthetic RSV vRNA into nucleocapsids *in vitro* or *in vivo*. Suitable synthetic molecules could contain the 3' and 5' sequences of vRNA, could contain a marker gene such as chloramphenicol acetyl transferase under the control of gene-start and gene-end signals, and could be transfected into RSV-infected cells. This is one of several approaches to the difficult task of developing the capability of producing live RSV from cloned viral nucleic acid.

Molecular studies with RSV (and the other pneumoviruses) are of particular interest because RSV is somewhat distinct from the other nonsegmented negative-strand viruses. The identification of features that it shares with the others and features that are different is useful in gaining a broader understanding of this virus group.

Note Added in Proof: The 3'-terminal leader region of RSV vRNA has now been sequenced in its entirety and is 44 nucleotides in length rather than the estimate of 43 nucleotides shown in the figures (M. A. Mink and P. L. Collins, unpubished data). Although the figures have not been modified to contain this final, confirmed number, it is reflected in the value of 15,222 given in the text and figure legends for the nucleotide length of RSV vRNA of strain A2. I will not attempt to list all of the new information that has emerged since this review was submitted in August, 1989 and revised in November, 1989. But it is noteworthy that the goal of obtaining infectious virus from cDNA described in Section V has now been achieved for MeV (I. Ballart, D. Eschle, R. Cattaneo, A. Schmid, M. Metzler, J. Chan, S. Pifko-Hirst, S. A. Udem, and M. A. Billeter, 1990, *EMBO J.*, **9**:379–384; see also Chapter 12, this volume). Also, Krystal, Palese, and co-workers have introduced vRNA segments synthesized *in vitro* from cDNA into infectious influenza A virus (M. Emani, W. Luytjes, M. Krystal, and P. Palese, 1990, *Proc. Natl. Acad. Sci. USA*, **87**:3802–3807, and references therein).

ACKNOWLEDGMENTS. I thank Drs. Brian Murphy, Robert Chanock, David Stec, Michael Mink, and Geoffrey Cole for comments on the manuscript, Drs. David Stec, Michael Mink, Brian Murphy, Robert Olmsted, and Philip Johnson for unpublished data from our laboratory, Dr. Narayanasamy Elango for communicating unpublished data on MuV, and Dr. Christian Marck for providing the "DNA Strider" program (Marck, 1988). This chapter is dedicated, 8 years after the fact, with warm regards and thanks to my Ph.D. thesis advisors, Drs. Andrew Ball, Lawrence Hightower, and Philip Marcus.

VI. REFERENCES

Akerlind, B., Norrby, E., Orvell, C., and Mufson, M. A., 1988, Respiratory syncytial virus: Heterogeneity of subgroup B strains, *J. Gen. Virol.* **69**:2145–2154.

Anderson, L. J., Hierholzer, J. C., Tsou, C., Hendry, R. M., Fernie, B. F., Stone, Y., and McIntosh, K., 1985, Antigenic characterization of respiratory syncytial virus strains with monoclonal antibodies, *J. Infect. Dis.* **151**:626–633.

Anderson, L. J., Bingham, P., and Hierholzer, J. C., 1988, Neutralization of respiratory syncytial virus by individual and mixtures of F and G monoclonal antibodies, *J. Virol.* **62**:4232–4238.

Armstrong, J. A., Pereira, H. G., and Valentine, R. C., 1962, Morphology and development of respiratory syncytial virus in cell cultures, *Nature* **196**:1179–1181.

Arumugham, R. G., Hildreth, S. W., and Paradiso, P. R., 1989, Interprotein disulfide bonding between F and G glycoproteins of human respiratory syncytial virus, *Arch. Virol.* **105**:65–79.

Bachi, T., 1988, Direct observation of the budding and fusion of an enveloped virus by video microscopy, *J. Cell Biol.* **107**:1689–1695.

Bachi, T., and Howe, T., 1973, Morphogenesis and ultrastructure of respiratory syncytial virus, *J. Virol.* **12**:1173–1180.

Ball, L. A., and Wertz, G. W., 1981, VSV RNA synthesis: How can you be positive?, *Cell* **26**:143–144.

Banerjee, A. K., 1987, The transcription complex of vesicular stomatitis virus, *Cell* **48**:363–364.

Bangham, C. R. M., Openshaw, P. J. M., Ball, L. A., King, A. M. Q., Wertz, G. W., and Askonas, B. A., 1986, Human and murine cytotoxic T cells specific to respiratory synctial virus recognize the viral nucleoprotein (N), but not the major glycoprotein (G), expressed by vaccinia virus recombinants, *J. Immunol.* **137**:3973–3977.

Barrett, T., Clarke, D. K., Evans, S. A., and Rima, B. K., 1987, The nucleotide sequence of the gene encoding the F protein of canine distemper virus: A comparison of the deduced amino acid sequence with other paramyxoviruses, *Virus Res.* **8**:373–386.

Baybutt, H. N., and Pringle, C. R., 1987, Molecular cloning and sequencing of the F and 22K membrane protein genes of the RSS-2 strain of respiratory syncytial virus, *J. Gen. Virol.* **68**:2789–2796.

Beeler, J. A., and Coelingh, K. V. W., 1989, Neutralization epitopes of the F glycoprotein of respiratory syncytial virus: Effect of mutation upon fusion function, *J. Virol.* **63**:2941–2950.

Bellini, W. J., Englund, G., Richardson, C. D., Rozenblatt, S., and Lazzarini, R. A., 1986, Matrix genes of measles virus and canine distemper virus: Cloning, nucleotide sequences, and deduced amino acid sequences, *J. Virol.* **58**:408–416.

Berthiaume, L., Joncas, J., and Pavilanis, V., 1974, Comparative structure, morphogenesis and biological characteristics of the respiratory syncytial (RS) virus and the pneumonia virus of mice, *Arch. Ges. Virusforsch.* **45**:39–51.

Bohn, W., Rutter, G., Hohenberg, H., Mannweiler, K., and Nobis, P., 1986, Involvement of actin filaments in budding of measles virus: Studies on cytoskeletons of infected cells, *Virology* **149**:91–106.

Cannon, M. J., and Bangham, C. R. M., 1989, Recognition of respiratory syncytial virus fusion protein by mouse cytotoxic T cell clones and a human cytotoxic T cell line, *J. Gen. Virol.* **70**:79–87.

Cash, P., Wunner, W. H., and Pringle, C. R., 1977, A comparison of the polypeptides of human and bovine respiratory syncytial viruses and murine pneumonia virus, *Virology* **82**:369–379.

Cash, P., Pringle, C. R., and Preston, C. M., 1979, The polypeptides of human respiratory syncytial virus: Products of cell-free protein synthesis and post-translational modifications, *Virology* **92**:375–384.

Castaneda, S. J., and Wong, T. C., 1989, Measles virus synthesizes both leaderless and leader-containing polyadenylated RNAs *in vivo*, *J. Virol.* **63**:2977–2986.

Cattaneo, R., Kaelin, K., Baczko, K., and Billeter, M. A., 1989, Measles virus editing provides an additional cysteine-rich protein, *Cell* **56**:759–764.

Cavanagh, D., and Barrett, T., 1988, Pneumovirus-like characteristics of the mRNA and proteins of turkey rhinotracheitis virus, *Virus Res.* **11**:241–256.

Chambers, P., Millar, N. S., Platt, S. G., and Emmerson, P. T., 1986, Nucleotide sequence of the gene encoding the matrix protein of Newcastle disease virus, *Nucleic Acids Res.* **14**:9051–9061.

Chanock, R. M., Murphy, B. R., Collins, P. L., Coelingh, K. V. W., Olmsted, R. A., Snyder, M. H., Spriggs, M. K., Prince, G. A., Moss, B., Flores, J., Gorziglia, M., and Kapikian, A. Z., 1988, Live virus vaccines for respiratory and enteric tract diseases, *Vaccine* **6**:129–133.

Chinchar, V. G., and Portner, A., 1981, Functions of Sendai virus nucleocapsid polypeptides: Enzymatic activities in nucleocapsids following cleavage of polypeptide P by *Staphylococcus aureus* protease V8, *Virology* **109**:59–71.

Chou, P. Y., and Fasman, G. D., 1974, Conformational parameters for amino acids in helical, beta-sheet and random coil regions calculated from proteins, *Biochemistry* **13**:211–244.

Coates, H. V., Alling, D. W., and Chanock, R. M., 1966, An antigenic analysis of respiratory syncytial virus isolates by a plaque reduction neutralization test, *Am. J. Epidemiol.* **89**:299–313.

Coelingh, K. L. V., Winter, C. C., and Murphy, B. R., 1988, Nucleotide and deduced amino acid sequence of hemagglutinin-neuraminidase genes of human type 3 parainfluenza viruses isolated from 1957 to 1983, *Virology* **162**:137–143.

Collins, M. S., and Gough, R. E., 1988, Characterization of a virus associated with turkey rhinotracheitis, *J. Gen. Virol.* **69**:909–916.

Collins, P. L., and Wertz, G. W., 1983, cDNA cloning and transcriptional mapping of nine polyadenylated RNAs encoded by the genome of human respiratory syncytial virus, *Proc. Natl. Acad. Sci. USA* **80:**3208–3212.

Collins, P. L., and Wertz, G. W., 1985a, The envelope-associated 22K protein of human respiratory syncytial virus: Nucleotide sequence of the mRNA and a related polytranscript, *J. Virol.* **54:**65–71.

Collins, P. L., and Wertz, G. W., 1985b, The 1A protein gene of human respiratory syncytial virus: Nucleotide sequence of the mRNA and a related polycistronic transcript, *J. Virol.* **141:**283–291.

Collins, P. L., and Wertz, G. W., 1985c, Nucleotide sequences of the 1B and 1C nonstructural protein mRNAs of human respiratory syncytial virus, *Virology* **143:**442–451.

Collins, P. L., Wertz, G. W., Ball, L. A., and Hightower, L. E., 1982, Coding assignments of the five smaller mRNAs of Newcastle disease virus, *J. Virol.* **43:**1024–1031.

Collins, P. L., Huang, Y. T., and Wertz, G. W., 1984a, Nucleotide sequence of the gene encoding the fusion (F) glycoprotein of human respiratory syncytial virus, *Proc. Natl. Acad. Sci. USA* **81:**7683–7687.

Collins, P. L., Huang, Y. T., and Wertz, G. W., 1984b, Identification of a tenth mRNA of respiratory syncytial virus and assignment of polypeptides to the 10 viral genes, *J. Virol.* **49:**572–578.

Collins, P. L., Anderson, K., Langer, S. J., and Wertz, G. W., 1985, Correct sequence for the major nucleocapsid protein mRNA of respiratory syncytial virus, *Virology* **146:**69–77.

Collins, P. L., Dickens, L. E., Buckler-White, A., Olmsted, R. A., Spriggs, M. K., Camargo, E., and Coelingh, K. V. W., 1986, Nucleotide sequences for the gene junctions of human respiratory syncytial virus reveal distinctive features of intergenic structure and gene order, *Proc. Natl. Acad. Sci. USA* **83:**4594–4598.

Collins, P. L., Olmsted, R. A., Spriggs, M. K., Johnson, P. R., and Buckler-White, A. J., 1987, Gene overlap and attenuation of transcription of the viral polymerase L gene of human respiratory syncytial virus, *Proc. Natl. Acad. Sci. USA* **84:**5134–5138.

Cummings, R. D., Kornfield, S., Schneider, W. J., Hobgood, K. K., Tolleshaug, H., Brown, M. S., and Goldstein, J. L., 1983, Biosynthesis of N- and O-linked oligosaccharides of the low density lipoprotein receptor, *J. Biol. Chem.* **258:**15261–15273.

Dahr, W., Beyreuther, K., Kordowicz, M., and Kruger, J., 1982, N-terminal amino acid sequence of sialoglycoprotein D (glycophorin C) from human erythrocyte membranes, *Eur. J. Biochem.* **125:**57–62.

Dayhoff, M. O., Hunt, L. T., and Hurst-Calderone, S., 1978, Composition of proteins, in "Atlas of Protein Sequence and Structure," Vol. 5, Suppl. 3 (M. O. Dayhoff, ed.), National Biomedical Research Foundation, Silver Spring, Maryland.

Dickens, L. E., Collins, P. L., and Wertz, G. W., 1984, Transcriptional mapping of human respiratory syncytial virus, *J. Virol.* **52:**364–369.

Dunphy, W. G., and Rothman, J. E., 1985, Compartmental organization of the Golgi stack, *Cell* **42:**13–21.

Eckhardt, A. E., Timpte, C. S., Abernethy, J. L., Toumadje, A., Johnson, W. C., and Hill, R. L., 1987, Structural properties of porcine submaxillary gland apomucin, *J. Biol. Chem.* **262:**11339–11344.

Elango, N., and Venkatesan, S., 1983, Amino acid sequence of human respiratory syncytial virus nucleocapsid protein, *Nucleic Acids Res.* **11:**5941–5951.

Elango, N., Satake, M., Coligan, J. E., Norrby, E., Camargo, E., and Venkatesan, S., 1985a, Respiratory syncytial virus fusion glycoprotein: Nucleotide sequence of mRNA, identification of cleavage activation site and amino acid sequence of N-terminus of F_1 subunit, *Nucleic Acids Res.* **13:**1559–1573.

Elango, N., Satake, M., and Venkatesan, S., 1985b, mRNA sequence of three respiratory syncytial virus genes encoding two nonstructural proteins and a 22K structural protein. *J. Virol.* **55:**101–110.

Elango, N., Prince, G. A., Murphy, B. R., Venkatesan, S., Chanock, R. M., and Moss, B. R., 1986, Resistance to human respiratory syncytial virus (RSV) infection induced by immunization of cotton rats with a recombinant vaccinia virus expressing the RSV G glycoprotein, *Proc. Natl. Acad. Sci. USA* **83:**1906–1910.

Elango, N., Kovamees, J., Varsanyl, T. M., and Norrby, E., 1989, mRNA sequence and deduced amino acid sequence of the mumps virus small hydrophobic protein gene, *J. Virol.* **63:**1413–1415.

Elder, J. H., McGee, J. S., and Alexander, S., 1986, Carbohydrate side chains of Rauscher leukemia virus envelope glycoproteins are not required to elicit a neutralizing antibody response, *J. Virol.* **57**:340–342.

Elhammer, A., and Kornfeld, S., 1984, Two enzymes involved in the synthesis of O-linked oligosaccharides are localized on membranes of different densities in mouse lymphoma BW5147 cells, *J. Cell Biol.* **98**:327–331.

Emerson, S. U., 1982, Reconstitution studies detect a single polymerase entry site on the vesicular stomatitis virus genome, *Cell* **31**:635–642.

Emerson, S. U., 1985, Rhavdoviruses, *in:* "Virology," (B. N. Fields, ed.), pp. 1119–1132, Raven Press, New York.

Fernie, B. F., and Gerin, J. L., 1980, The stabilization and purification of respiratory syncytial virus using MgSO$_4$, *Virology* **106**:141–144.

Fernie, B. F., Cote, P. J., Jr., and Gerin, J. L., 1982, Classification of hybridomas to respiratory syncytial virus glycoproteins, *Proc. Soc. Exp. Biol. Med.* **171**:266–271.

Fernie, B. F., Dapolito, G., Cote, P. J., Jr., and Gerin, J. L., 1985, Kinetics of synthesis of respiratory syncytial virus glycoproteins, *J. Gen. Virol.* **66**:1983–1990.

Fiat, A.-M., Jolles, J., Aubert, J.-P., Loucheux-Lefebvre, M.-H., and Jolles, P., 1980, Localisation and importance of the sugar part of human casein, *Eur. J. Biochem.* **111**:333–339.

Gallaher, W. R., 1987, Detection of a fusion peptide sequence in the transmembrane protein of human immunodeficiency virus, *Cell* **50**:327–328.

Garcia-Barreno, B., Jorcano, J. L., Aukenbauer, T., Lopez-Galindez, C., and Melero, J. A., 1988, Participation of cytoskeletal intermediate filaments in the infectious cycle of human respiratory syncytial virus (RSV), *Virus Res.* **9**:307–322.

Garcia-Barreno, B., Palomo, C., Penas, C., Delgado, T., Perez-Brena, P., and Melero, J., 1989, Marked differences in the antigenic structure of human respiratory syncytial virus F and G glycoproteins, *J. Virol.* **63**:925–932.

Giminez, H. B., Cash, P., and Melvin, W. T., 1984, Monoclonal antibodies to human respiratory syncytial virus and their use in comparison of different virus isolates, *J. Gen. Virol.* **65**:963–971.

Gimenez, H. B., Hardman, N., Keir, H. M., and Cash, P., 1986, Antigenic variation between human respiratory syncytial virus isolates, *J. Gen. Virol.* **67**:863–870.

Glickman, R. L., Syddall, R. J., Iorio, R. M., Sheehan, J. P., and Bratt, M. A., 1988, Quantitative basic residue requirements in the cleavage-activation site of the fusion glycoprotein as a determinant of virulence for Newcastle disease virus, *J. Virol.* **62**:354–356.

Gonzalez-Scarano, F., Waxham, M. N., Ross, A. M., and Hoxie, J. A., 1987, Sequence similarities between human immunodeficiency virus gp41 and paramyxovirus fusion proteins, *AIDS Res. Hum. Retroviruses* **3**:245–252.

Griffiths, G., Quinn, P., and Warren, G., 1983, Dissection of the Golgi complex. I. Monensin inhibits the transport of viral membrane proteins from medial to trans Golgi cisternae in baby hamster cells infected with Semliki Forest virus, *J. Cell Biol.* **96**:835–850.

Gruber, C., and Levine, S., 1983, Respiratory syncytial virus polypeptides. III. The envelope-associated proteins, *J. Gen. Virol.* **64**:825–832.

Gruber, C., and Levine, S., 1985a, Respiratory syncytial virus polypeptides. IV. The oligosaccharides of the glycoproteins, *J. Gen. Virol.* **66**:417–432.

Gruber, C., and Levine, S., 1985b, Respiratory syncytial virus polypeptides. V. The kinetics of glycoprotein synthesis, *J. Gen. Virol.* **66**:1241–1247.

Gupta, R., Yewdell, J. W., Olmsted, R. A., Collins, P. L., and Bennink, J. R., 1990, Primary pulmonary murine cytotoxic T lymphocyte specificity in respiratory syncytial virus pneumonia, *Microb. Pathogen.* (in press).

Hamaguchi, M., Nishikawa, K., Toyoda, T., Yoshida, T., Hanaichi, T., and Nagai, Y., 1985, Transcriptive complex of Newcastle disease virus II. Structural and functional assembly associated with the cytoskeletal framework, *Virology* **147**:295–308.

Hendricks, D. A., Baradaran, K., McIntosh, K., and Patterson, J. L., 1987, Appearance of a soluble form of the G protein of respiratory syncytial virus in fluids of infected cells, *J. Gen. Virol.* **68**:1705–1714.

Hendricks, D. A., McIntosh, K., and Patterson, J. L., 1988, Further characterization of the soluble form of the G glycoprotein of respiratory syncytial virus, *J. Virol.* **62**:2228–2233.

Hendry, R. M., Talis, A. L., Godfrey, E., Anderson, L. J., Fernie, B. F., and McIntosh, K., 1986, Concurrent circulation of antigenically distinct strains of respiratory syncytial virus during community outbreaks, *J. Infect. Dis.* **153:**291–297.

Hendry, R. M., Burns, J. C., Walsh, E. E., Graham, B. S., Wright, P. F., Hemming, V. G., Rodriguez, W. J., Kim, H. W., Prince, G. A., McIntosh, K., Chanock, R. M., and Murphy, B. R., 1988, Strain-specific serum antibody responses in infants undergoing primary infection with respiratory syncytial virus, *J. Infect. Dis.* **157:**640–646.

Herman, R. C., 1989, Synthesis of respiratory syncytial virus RNA in cell-free extracts, *J. Gen. Virol.* **70:**755–761.

Herman, R. C., and Lazzarini, R. A., 1981a, Aberrant glycoprotein mRNA synthesized by the internal deletion mutant of vesicular stomatitis virus, *J. Virol.* **40:**78–86.

Herman, R. C., and Lazzarini, R. A., 1981b, Vesicular stomatitis virus RNA polymerase can read through the boundary between the leader and N genes *in vitro*, *J. Virol.* **38:**792–796.

Hiebert, S. W., Richardson, C. D., and Lamb, R. A., 1988, Cell surface expression and orientation in membranes of the 44-amino acid SH protein of simian virus 5, *J. Virol.* **62:**2347–2357.

Hill, H. D., Jr., Schwyzer, M., Steinman, H. M., and Hill, R. L., 1977, Ovine submaxillary mucin, *J. Biol. Chem.* **252:**3799–3804.

Hsu, M.-C., Scheid, A., and Choppin, P. W., 1981, Activation of the Sendai virus fusion protein (F) involves a conformational change with exposure of a new hydrophobic region, *J. Biol. Chem.* **256:**3557–3563.

Huang, Y. T., 1983, The genome and gene products of human respiratory syncytial virus, Ph.D. Thesis, University of North Carolina, Chapel Hill, North Carolina.

Huang, Y. T., and Wertz, G. W., 1982, The genome of respiratory syncytial virus is a negative-stranded RNA that codes for at least seven mRNA species, *J. Virol.* **43:**150–157.

Huang, Y. T., and Wertz, G. W., 1983, Respiratory syncytial virus mRNA coding assignments, *J. Virol.* **46:**667–672.

Huang, Y. T., Collins, P. L., and Wertz, G. W., 1985, Characterization of the 10 proteins of human respiratory syncytial virus: Identification of a fourth envelope-associated protein, *Virus Res.* **2:**157–173.

Hull, J. D., Krah, D. L., and Choppin, P. W., 1987, Resistance of a measles virus mutant to fusion inhibitory oligopeptides is not associated with mutations in the fusion peptide, *Virology* **159:**368–372.

Iverson, L. E., and Rose, J. K., 1981, Localized attenuation and discontinuous synthesis during vesicular stomatitis virus transcription, *Cell* **23:**477–484.

Johnson, P. R., and Collins, P. L., 1988a, The fusion glycoprotein of human respiratory syncytial viruses of subgroups A and B; Sequence conservation provides a structural basis for antigenic relatedness, *J. Gen. Virol.* **69:**2623–2628.

Johnson, P. R., and Collins, P. L., 1988b, The A and B subgroups of human respiratory syncytial virus: Comparison of intergenic and gene-overlap sequences, *J. Gen. Virol.* **69:**2901–2906.

Johnson, P. R., and Collins, P. L., 1989, The 1B (NS2), 1C (NS1) and N proteins of human respiratory syncytial virus (RSV) of antigenic subgroups A and B: Sequence conservation and divergence within RSV genomic RNA, *J. Gen. Virol.* **70:**1539–1547.

Johnson, P. R., Olmsted, R. A., Prince, G. A., Murphy, B. R., Alling, D. W., Walsh, E. E., and Collins, P. L., 1987a, Antigenic relatedness between glycoproteins of human respiratory syncytial virus subgroups A and B: Evaluation of the contributions of F and G glycoproteins to immunity, *J. Virol.* **61:**3163–3166.

Johnson, P. R., Spriggs, M. K., Olmsted, R. A., and Collins, P. L., 1987b, The G glycoprotein of human respiratory syncytial viruses of subgroups A and B: Extensive sequence divergence between antigenically related proteins, *Proc. Natl. Acad. Sci. USA* **84:**5625–5629.

Jukinen, M., Andersson, L. C., and Gahmberg, C. G., 1985, Biosynthesis of the major human red cell sialoglycoprotein, glycophoren A, *J. Biol. Chem.* **260:**11314–11321.

Kalica, A. R., Wright, P. F., Hetrick, F. M., and Chanock, R. M., 1973, Electron microscopic studies of respiratory syncytial temperature-sensitive mutants, *Arch. Ges. Virusforsch.* **41:**248–258.

Kennedy, H. E., Jones, B. V., Tucker, E. M., Ford, N. J., Clarke, S. W., Furze, J., Thomas, I. H., and Stott, E. J., 1988, Production and characterization of bovine monoclonal antibodies to respiratory syncytial virus, *J. Gen. Virol.* **69:**3023–3032.

King, A. M. Q., Stott, E. J., Langer, S. J., Young, K. K.-Y., Ball, L. A., and Wertz, G. W., 1987, Recombinant vaccinia viruses carrying the N gene of human respiratory syncytial virus: Studies of gene expression in cell culture and immune response in mice, *J. Virol.* **61:**2885–2890.

Kingsbury, D. W., 1985, Orthomyxo and paramyxoviruses and their replication, in "Virology" (B. N. Fields, ed.), pp. 1157–1178, Raven Press, New York.

Kingsbury, D. W., 1990, Paramyxoviridae and their replication in "Virology" (B. N. Fields, ed.), pp. 945–962, Raven Press, New York.

Kingsbury, D. W., Bratt, M. A., Choppin, P. W., Hanson, R. P., Hosaka, Y., ter Meulen, V., Norrby, E., Plowright, W., Rott, R., and Wunner, W. H., 1978, Paramyxoviridae, Intervirology 10:137–152.

Kisch, A. L., Johnson, K. M., and Chanock, R. M., 1962, Immunofluorescence with respiratory syncytial virus, Virology 16:177–189.

Kornfeld, R., and Kornfeld, S., 1985, Assembly of asparagine-linked oligosaccharides, Annu. Rev. Biochem. 54:631–664.

Kozak, M., 1987, An analysis of 5'-noncoding sequences from 699 vertebrate messenger RNAs, Nucleic Acids Res. 15:8125–8148.

Kurath, G., Ahern, K. G., Pearson, G. D., and Leong, J. C., 1985, Molecular cloning of the six mRNA species of infectious hematopoietic necrosis virus, a fish rhabdovirus, and gene order determination by R loop mapping, J. Virol. 53:469–476.

Lambden, P. R., 1985, Nucleotide sequence of the respiratory syncytial virus phosphoprotein gene, J. Gen. Virol. 66:1607–1612.

Lambert, D. M., 1988, Role of oligosaccharides in the structure and function of respiratory syncytial virus glycoproteins, Virology 164:458–466.

Lambert, D. M., and Pons, M. W., 1983, Respiratory syncytial virus glycoproteins, Virology 130:204–214.

Lambert, D. M., Pons, M. W., Mbuy, G. N., and Dorsch-Hasler, K., 1980, Nucleic acids of respiratory syncytial virus, J. Virol. 36:837–846.

Lambert, D. M., Hambor, J., Diebold, M., and Galinski, B., 1988, Kinetics of synthesis and phosphorylation of respiratory syncytial virus polypeptides, J. Gen. Virol. 69:313–323.

Lerch, R. A., Stott, E. J., and Wertz, G. W., 1989, Characterization of bovine respiratory syncytial virus proteins and mRNAs and generation of cDNA clones to the viral mRNAs, J. Virol. 63:833–840.

Levine, S., and Hamilton, R., 1969, Kinetics of the respiratory syncytial virus growth cycle in HeLa cells, Arch. Ges. Virusforsch. 28:122–132.

Levine, S., Peeples, M., and Hamilton, R., 1977, Effect of respiratory syncytial virus infection of HeLa-cell macromolecular synthesis, J. Gen. Virol. 37:53–63.

Levine, S., Klaiber-Franco, R., and Paradiso, P. R., 1987, Demonstration that glycoprotein G is the attachment protein of respiratory syncytial virus, J. Gen. Virol. 68:2521–2524.

Ling, R., and Pringle, C. R., 1988, Turkey rhinotracheitis virus: in vivo and in vivo polypeptide synthesis, J. Gen. Virol. 69:917–923.

Ling, R., and Pringle, C. R., 1989a, Polypeptides of pneumonia virus of mice: I. Immunological cross-reactions and post-translational modifications, J. Gen. Virol. 70:1427–1440.

Ling, R., and Pringle, C. R., 1989b, Polypeptides of pneumonia virus of mice: II. Characterization of the glycoproteins, J. Gen. Virol. 70:1441–1452.

Lobl, T. J., Renis, H. E., Epand, R. M., Maggiora, L. L., and Wathen, M. W., 1988, Peptides as potential virus inhibitors: Synthesis and bioassay of five respiratory syncytial virus peptide analogs with antimeasles activity, Int. J. Peptide Protein Res. 32:326–330.

López, J. A., Chung, D. W., Fujikawa, K., Hagen, F. S., Papayannopoulou, T., and Roth, G. J., 1987, Cloning of the alpha chain of human platelet glycoprotein Ib: A transmembrane protein with homology to leucine-rich alpha-2 glucoprotein, Proc. Natl. Acad. Sci. USA 84:5615–5619.

López, J. A., Villanueva, N., Melero, J. A., and Portela, A., 1988, Nucleotide sequence of the fusion and phosphoprotein genes of human respiratory syncytial (RS) virus Long strain: Evidence of subtype genetic heterogeneity, Virus Res. 10:249–262.

Luk, D., Masters, P. S., Gill, D. S., and Banerjee, A. K., 1987, Intergenic sequences of the vesicular stomatitis virus genome (New Jersey serotype): Evidence for two transcription initiation sites within the L gene, Virology 160:88–94.

Lundstrom, M., Olofsson, S., Jeanssom, S., Lycke, E., Datema, R., and Mansson, J.-E., 1987, Host cell-induced differences in O-glycosylation of Herpes simplex virus gC-1, Virology 161:395–402.

Marck, C., 1988, "DNA Strider": A "C" program for the fast analysis of DNA and protein sequences on the Apple McIntosh family of computers, Nucleic Acids Res. 16:1829–1836.

Marton, L. S. G., and Gevin, J. E., 1973, Subunit structure of the major human erythrocyte glycoprotein: Depolymerization by heating ghosts with sodium dodecyl sulfate, *Biochem. Biophys. Res. Commun.* **52:**1457–1462.

Mbuy, G. N., and Rochovansky, O. M., 1984, RNA-dependent RNA polymerase associated with respiratory syncytial virus, in "Nonsegmented Negative Strand Viruses: Paramyxoviruses and Rhabdoviruses" (D. H. L. Bishop and R. W. Compans, eds.), Academic Press, Orlando, Florida.

McGinnes, L., McQuain, C., and Morrison, T., 1988, The P protein and the nonstructural 38K and 29K proteins of Newcastle disease virus are derived from the same open reading frame, *Virology* **164:**256–264.

McIntosh, K. M., and Chanock, R. M., 1985, Respiratory syncytial virus, in "Virology" (B. N. Fields, ed.), pp. 1285–1304, Raven Press, New York.

McIntosh, K., and Fishout, J. M., 1980, Immunopathologic mechanisms in lower respiratory tract disease of infants due to respiratory syncytial virus, *Prog. Med. Virol.* **26:**94–118.

Morgan, L. A., Routledge, G., Willcocks, M. M., Samson, A. C. R., Scott, R., and Toms, G. L., 1987, Strain variation of respiratory syncytial virus, *J. Gen. Virol.* **68:**2781–2788.

Mufson, M. A., Orvell, C., Rafnar, B., and Norrby, E., 1985, Two distinct subtypes of human respiratory syncytial virus, *J. Gen. Virol.* **66:**2111–2124.

Murphy, B. R., Prince, G. A., Collins, P. L., Coelingh, K. V. W., Olmsted, R. A., Spriggs, M. K., Parrot, R. H., Hyun-Wha, K., Brandt, C. D., and Chanock, R. M., 1988, Current approaches to the development of vaccines effective against parainfluenza and respiratory syncytial viruses, *Virus Res.* **11:**1–15.

Nicholas, J. A., Mitchell, M. A., Levely, M. E., Rubino, K. L., Kinner, J. H., Harn, N. K., and Smith, C. W., 1988, Mapping an antibody-binding site and a T-cell-stimulating site on the 1A protein of respiratory syncytial virus, *J. Virol.* **62:**4465–4473.

Niemann, H., Geyer, R., Klenk, H.-D., Linder, D., Stirm, S., and Wirth, M., 1984, The carbohydrates of mouse hepatitis virus (MHV) A549: Structure of the O-glycosidically linked oligosaccharides of glycoprotein E1, *EMBO J.* **3:**665670.

Norrby, E., Marusyk, H., and Orvell, C., 1970, Morphogenesis of respiratory syncytial virus in a green monkey kidney cell line (Vero), *J. Virol.* **6:**237–242.

Norrby, E., Mufson, M. A., and Sheshberadaran, H., 1986, Structural differences between subtype A and B strains of respiratory syncytial virus, *J. Gen. Virol.* **67:**2721–2729.

Norrby, E., Mufson, M. A., Alexander, H., Hougten, R. A., and Lerner R. A., 1987, Site-directed serology with synthetic peptides representing the large glycoprotein G of respiratory syncytial virus, *Proc. Natl. Acad. Sci. USA* **84:**6572–6576.

Olmsted, R. A., and Collins, P. L., 1989, The 1A protein of respiratory syncytial virus is an integral membrane protein present as multiple, structurally distinct species, *J. Virol.* **63:**2019–2029.

Olmsted, R. A., Elango, N., Prince, G. A., Murphy, B. R., Johnson, P. R., Moss, B., Chanock, R. M., and Collins, P., 1986, Expression of the F1 glycoprotein of respiratory syncytial virus by a recombinant vaccinia virus: Comparison of the individual contributions of the F and G glycoproteins to host immunity, *Proc. Natl. Acad. Sci. USA* **83:**7462–7466.

Olmsted, R. A., Murphy, B. R., Lawrence, L. A., Elango, N., Moss, B., and Collins, P. L., 1989, Processing, surface expression, and immunogenicity of carboxy-terminally truncated mutants of G protein of human respiratory syncytial virus, *J. Virol.* **63:**411–420.

Openshaw, P. J. M., Pemberton, R. M., Ball, L. A., Wertz, G. W., and Askonas, B. A., 1988, Helper T cell recognition of respiratory syncytial virus in mice, *J. Gen. Virol.* **69:**305–312.

Örvell, C., Norrby, E., and Mufson, M. A., 1987, Preparation and characterization of monoclonal antibodies directed against five structural components of human respiratory syncytial virus subgroup B, *J. Gen. Virol.* **68:**3215–3135.

Paradiso, P., Hu, B., and Hildreth, S., 1987, Structure of the respiratory syncytial virus glycoprotein G, in "VII International Congress of Virology, Abstracts," R25.21, p. 189.

Paterson, R. G., Shaughnessy, M. A., and Lamb, R. A., 1989, Analysis of the relationship between cleavability of a paramyxovirus fusion protein and length of the connecting peptide, *J. Virol.* **63:**1293–1301.

Patzelt, C., and Weber, B., 1986, Early O-glycosidic glycosylation of proglucagon in pancreatic islets: An unusual type of prohormonal modification, *EMBO J.* **5:**2103–2108.

Peeples, M., and Levine, S., 1979, Respiratory syncytial virus polypeptides: Their location in the virion, *Virology* **95:**137–145.

Pemberton, R. M., Cannon, M. J., Openshaw, P. J. M., Ball, L. A., Wertz, G. W., and Askonas, B. A., 1987, Cytotoxic T cell specificity for respiratory syncytial virus proteins: Fusion protein is an important target antigen, *J. Gen. Virol.* **68**:2177–2182.

Richardson, C. D., and Choppin, P. W., 1983, Oligopeptides that specifically inhibit membrane fusion by paramyxoviruses: Studies on the site of action, *Virology* **131**:518–532.

Richman, A. V., Pedeira, F. A., and Tauraso, N. M., 1971, Attempts to demonstrate hemagglutination and hemadsorption by respiratory syncytial virus, *App. Microbiol.* **21**:1099–1100.

Rose, J. K., 1980, Complete intergenic and flanking gene sequences from the genome of vesicular stomatitis virus, *Cell* **19**:415–421.

Roth, J., 1984, Cytochemical localization of terminal *N*-acetyl-*D*-galactosamine residues in cellular compartments of intestinal goblet cells: Implications for the topology of O-glycosylation, *J. Cell. Biol.* **98**:399–406.

Roth, J., Taatjes, D. J., Weinstein, J., Paulson, J. C., Greenwell, P., and Watkins, W. M., 1986, Differential subcompartmentation of terminal glycosylation in the Golgi apparatus of intestinal adsorptive and goblet cells, *J. Biol. Chem.* **261**:14307–14312.

Routledge, E. G., Willcocks, M. M., Samson, A. C. R., Scott, R., and Toms, G. L., 1986, Respiratory syncytial virus glycoprotein expression in human and simian cell lines, *J. Gen. Virol.* **67**:2059–2064.

Routledge, E. G., Willcocks, M. M., Morgan, L., Samson, A. C. R., Scott, R., and Toms, G. L., 1987a, Heterogeneity of the respiratory syncytial virus 22K protein revealed by Western blotting with monoclonal antibodies, *J. Gen. Virol.* **68**:1209–1215.

Routledge, E. G., Willcocks, M. M., Morgan, L., Samson, A. C. R., Scott, R., and Toms, G. L., 1987b, Expression of the respiratory syncytial virus 22K protein on the surface of infected HeLa cells, *J. Gen. Virol.* **68**:1217–1222.

Routledge, E. G., Willcocks, M. M., Samson, A. C. R., Morgan, L., Scott, R., Anderson, J. J., and Toms, G. L., 1988, The purification of four respiratory syncytial virus proteins and their evaluation as protective agents against experimental infection in BALB/c mice, *J. Gen. Virol.* **69**:293–303.

Sanchez, A., Kiley, M. P., Holloway, B. P., McCormick, J. B., and Auperin, D. D., 1987, Sequence analysis of the Ebola virus genome indicates that it is similar in organization to paramyxoviruses and rhabdoviruses, in "VII International Congress of Virology, Abstracts," R25.32, p. 191.

Satake, M., and Venkatesan, S., 1984, Nucleotide sequence of the gene encoding respiratory syncytial virus matrix protein, *J. Virol.* **50**:92–99.

Satake, M., Elango, N., and Venkatesan, S., 1984, Sequence analysis of the respiratory syncytial virus phosphoprotein gene, *J. Virol.* **52**:991–994.

Satake, M., Coligan, J. E., Elango, N., Norrby, E., and Venkatesan, S., 1985, Respiratory syncytial virus envelope glycoprotein (G) has a novel structure, *Nucleic Acids Res.* **13**:7795–7812.

Schachter, H., and Roseman, S., 1980, Mammalian glycotransferases: Their role in the synthesis and function of complex carbohydrates and glycolipids, in "The Biochemistry of Glycoproteins and Proteoglycans" (W. J. Lenhart, ed.), pp. 85–160, Plenum Press, New York.

Schubert, M., Keene, J. D., Herman, R. C., and Lazzarini, R. A., 1980, Site on the vesicular stomatitis virus genome specifying polyadenylation and the end of the *L* gene mRNA, *J. Virol.* **34**:550–559.

Schubert, M., Harmison, G. G., and Meier, E., 1984, Primary structure of the vesicular stomatitis virus polymerase (L) gene: Evidence for a high frequency of mutations, *J. Virol.* **51**:505–514.

Sechoy, O., Philippot, J. R., and Bienvenue, A., 1987, F Protein–F protein interaction within the Sendai virus identified by native bonding or chemical cross-linking, *J. Biol. Chem.* **262**:11519–11523.

Sjoblom, I., Lundstrom, M., Sjogren-Jansson, E., Glorioso, J. C., Jeansson, S., and Olofsson, S., 1987, Demonstration and mapping of highly carbohydrate-dependent epitopes in the Herpes simplex virus type 1-specified glycoprotein C, *J. Gen. Virol.* **68**:545–554.

Skehel, J. J., Stevens, D. J., Daniels, R. S., Douglas, A. R., Knossow, M., Wilson, I. A., and Wiley, D. C., 1984, A carbohydrate side chain on hemagglutinins of Hong Kong influenza viruses inhibits recognition by a monoclonal antibody, *Proc. Natl. Acad. Sci. USA* **81**:1779–1783.

Spielman, J., Rockley, N. L., and Carraway, K. L., 1987, Temporal aspects of O-glycosylation and cell surface expression of ascites sialoglycoprotein-1, the major cell surface sialomucin of 13762 mammary ascites tumor cells, *J. Biol. Chem.* **262**:269–275.

Spraker, T. R., Collins, J. K., Adrian, W. J., and Olterman, J. H., 1986, Isolation and serological evidence of a respiratory syncytial virus in bighorn sheep from Colorado, *J. Wildl. Dis.* **22:**416–418.

Spriggs, M. K., and Collins, P. L., 1986, Sequence analysis of the P and C protein genes of human parainfluenza virus type 3: Patterns of amino acid sequence homology among paramyxovirus proteins, *J. Gen. Virol.* **67:**2705–2719.

Spriggs, M. K., Olmsted, R. A., Venkatesan, S., Coligan, J. E., and Collins, P. L., 1986, Fusion glycoprotein of human parainfluenza virus type 3: Nucleotide sequence of the gene, direct identification of the cleavage activation site, and comparison with other paramyxoviruses, *Virology* **152:**241–251.

Storch, G. A., and Park, C. S., 1987, Monoclonal antibodies demonstrate heterogeneity in the G glycoprotein of prototype strains and clinical isolates of respiratory syncytial virus, *J. Med. Virol.* **22:**345–356.

Storch, G. A., Park, C. S., and Dohner, D. E., 1989, RNA fingerprinting of respiratory syncytial virus using ribonuclease protection: Application to molecular epidemiology, *J. Clin. Invest.* **83:**1894–1902.

Stott, E. J., and Taylor, G., 1985, Respiratory syncytial virus: Brief review, *Arch. Virol.* **84:**1–52.

Stott, E. J., Bew, M. H., Taylor, G., Jebbett, J., and Collins, A. P., 1984, The characterization and uses of monoclonal antibodies to respiratory syncytial virus, *Dev. Biol. Std.* **57:**237–244.

Stott, E. J., Ball, L. A., Young, K. K., Furze, J., and Wertz, G. W., 1986, Human respiratory syncytial virus glycoprotein G expressed from a recombinant vaccinia virus vector protects mice against live-virus challenge, *J. Virol.* **60:**607–613.

Stott, E. J., Taylor, G., Ball, L. A., Anderson, K., Young, K.-Y., King, A. M. Q., and Wertz, G. W., 1987, Immune and histopathological responses in animals vaccinated with recombinant vaccinia viruses that express individual genes of human respiratory syncytial virus, *J. Virol.* **61:**3855–3861.

Strous, G. J. A. M., 1979, Initial glycosylation of proteins with acetylgalactosaminylserine linkages, *Biochemistry* **76:**2694–2698.

Talib, S., and Hearst, J. E., 1983, Initiation of RNA synthesis *in vitro* by vesicular stomatitis virus: Single internal initiation in the presence of aurintricarboxylic acid and vanadyl ribonucleoside complexes, *Nucleic Acids Res.* **11:**7031–7042.

Taylor, G., Stott, E. J., Bew, M., Fernie, B. F., Cote, P. J., Collins, A. P., Hughes, M., and Jebbett, J., 1984, Monoclonal antibodies protect against respiratory syncytial virus infection in mice, *Immunology* **52:**137–142.

Thomas, S. M., Lamb, R. A., and Paterson, R. G., 1988, Two mRNAs that differ by two nontemplated nucleotides encode the amino coterminal proteins P and V of the paramyxovirus SV5, *Cell* **54:**891–902.

Tordo, N., Poch, O., Ermine, A., Keith, G., and Rougeon, F., 1986, Walking along the rabies genome: Is the large G–L intergenic region a remnant gene?, *Proc. Natl. Acad. Sci. USA* **83:**3914–3918.

Tordo, N., Poch, O., Ermine, A., Keith, G., and Rougeon, F., 1988, Completion of the rabies virus genome sequence determination: Highly conserved domains among the L (polymerase) proteins of unsegmented negative-strand RNA viruses, *Virology* **165:**565–576.

Toyoda, T., Sakaguchi, T., Imai, K., Inocencio, N. M., Goton, B., Hamaguchi, M., and Nagai, Y., 1987, Structural comparison of the cleavage-activation site of the fusion glycoprotein between virulent and avirulent strains of Newcastle disease virus, *Virology* **158:**242–247.

Treuhaft, M. W., and Beem, M. O., 1982, Defective interfering particles of respiratory syncytial virus, *Infect. Immun.* **37:**439–444.

Trudel, M., Nadon, F., Séguin, C., Dionne, G., and Lacroix, M., 1987, Identification of a synthetic peptide as part of a major neutralization epitope of respiratory syncytial virus, *J. Gen. Virol.* **68:**2273–2280.

Tsutsumi, H., Flanagan, T. D., and Ogra, P. L., 1987, Monoclonal antibodies to the large glycoproteins of respiratory syncytial virus: Possible evidence for several functional antigenic sites, *J. Gen. Virol.* **68:**2161–2167.

Tsutsumi, H., Onuma, M., Suga, K., Honjo, T., Chiba, Y., Chiba, S., and Ogra, P. L., 1988, Occurrence of respiratory syncytial virus subgroup A and B strains in Japan, *J. Clin. Microbiol.* **26:**1171–1174.

Tuech, J. K., and Morrison, M., 1974, Human erythrocyte membrane sialo glycoproteins: A study of interconversion, *Biophys. Biochem. Res. Commun.* **59:**352–360.

Ueba, O., 1978, Respiratory syncytial virus, I. Concentration and purification of the infectious virus, *Acta Med. Okayama* **32**:265–272.

Venkatesan, S., Elango, N., and Chanock, R. M., 1983, Construction and characterization of cDNA clones for four respiratory syncytial viral genes, *Proc. Natl. Acad. Sci. USA* **80**:1280–1284.

Vidal, S., and Kolakofsky, D., 1989, Modified model for the switch from Sendai virus transcription to replication, *J. Virol.* **63**:1951–1958.

Walsh, E. E., and Hruska, J., 1983, Monoclonal antibodies to respiratory syncytial virus proteins: Identification of the fusion protein, *J. Virol.* **47**:171–177.

Walsh, E. E., Schlesinger, J. J., and Brandriss, M. W., 1984, Purification and characterization of GP90, one of the envelope glycoproteins of respiratory syncytial virus, *J. Gen. Virol.* **65**:761–767.

Walsh, E. E., Cote, P. J., Fernie, B. F., Schlesinger, J. J., and Brandriss, M. W., 1986, Analysis of the respiratory syncytial virus fusion protein using monoclonal and polyclonal antibodies, *J. Gen. Virol.* **67**:505–513.

Walsh, E. E., Brandriss, M. W., and Schlesinger, J. J., 1987a, Immunological differences between the envelope glycoproteins of two strains of human respiratory syncytial virus, *J. Gen. Virol.* **68**:2169–2176.

Walsh, E. E., Hall, C. B., Briselli, M., Brandriss, M. W., and Schlesinger J. J., 1987b, Immunization with glycoprotein subunits of respiratory syncytial virus to protect cotton rats against viral infection, *J. Infect. Dis.* **155**:1198–1204.

Ward, K. A., Lambden, P. R., Ogilvie, M. M., and Watt, P. J., 1983, Antibodies to respiratory syncytial virus polypeptides and their significance in human infection, *J. Gen. Virol.* **64**:1867–1876.

Ward, K. A., Everson, J. S., Lambden, P. R., and Watt, P. J., 1984, Antigenic and structural variation in the major nucleocapsid protein of respiratory syncytial virus, *J. Gen. Virol.* **65**:1749–1757.

Wathen, M. W., Brideau, R. J., and Thomsen, D. R., 1989, Immunization of cotton rats with the human respiratory syncytial virus F glycoprotein produced using a baculovirus vector, *J. Infect. Dis.* **159**:255–264.

Wertz, G. W., Collins, P. L., Huang, Y., Gruber, C., Levine, S., and Ball, L. A., 1985, Nucleotide sequence of the G protein gene of human respiratory syncytial virus reveals an unusual type of viral membrane protein, *Proc. Natl. Acad. Sci. USA* **82**:4075–4079.

Wertz, G. W., Stott, E. J., Young, K. K.-Y., Anderson, K., and Ball, L. A., 1987, Expression of the fusion protein of human respiratory syncytial virus from recombinant vaccinia virus vectors and protection of vaccinated mice, *J. Virol.* **61**:293–301.

Wertz, G. W., Krieger, M., and Ball, L. A., 1989, Structure and cell surface maturation of the attachment glycoprotein of human respiratory syncytial virus in a cell line deficient in O glycosylation, *J. Virol.* **63**:4767–4776.

Wieland, F. T., Gleason, M. L., Serafini, T. A., and Rothman, J. E., 1987, The rate of bulk flow from the endoplasmic reticulum to the cell surface, *Cell* **50**:289–300.

Wilde, A., and Morrison, T., 1984, Structural and functional characterization of Newcastle disease virus polycistronic RNA, *J. Virol.* **51**:71–76.

Williams, M. A., and Lamb, R. A., 1986, Determination of the orientation of an integral membrane protein and sites of glycosylation by oligonucleotide-directed mutagenesis: Influenza B virus NB glycoprotein lacks a cleavable signal sequence and has an extracellular NH2-terminal region, *Mol. Cell. Biol.* **6**:4317–4328.

Williams, M. A., and Lamb, R. A., 1988, Polylactosaminoglycan modification of a small integral membrane glycoprotein, influenza B virus NB, *Mol. Cell. Biol.* **8**:1186–1196.

Wunner, W. H., and Pringle, C. R., 1976, Respiratory syncytial virus proteins, *Virology* **73**:228–243.

Wunner, W. H., Faulkner, C. P., and Pringle, C. R., 1975, Respiratory syncytial virus: Some biological and biochemical properties, *in* "Negative Strand Viruses" (B. Mahy and R. Barry, eds.), pp. 193–201, Academic Press, New York.

Zebedee, S. L., and Lamb, R. A., 1988, Influenza A virus M2 protein: Monoclonal antibody restriction of virus growth and detection of M2 in virions, *J. Virol.* **62**:2762–2772.

CHAPTER 5

Evolutionary Relationships of Paramyxovirus Nucleocapsid-Associated Proteins

Exeen M. Morgan

I. INTRODUCTION

The nucleocapsid cores of the *Paramyxoviridae* are single, left-handed helical structures, approximately 1 μm in length, composed of protein complexed with a single-stranded RNA genome of negative polarity (Finch and Gibbs, 1970; Choppin and Compans, 1975; Mahy and Barry, 1975; Heggeness *et al.*, 1980). The three nucleocapsid-associated proteins (NP, P, and L) contribute to virion structure and play an important role in viral RNA synthesis and replication. Much remains to be learned about these three proteins and their relationships with each other and with the genomic RNA, despite the substantial amount of data obtained to date. The increasing availability of sequence information for the nucleocapsid proteins should yield new insights into the structure and function of these proteins and may even provide new clues to the genetic relationships among the *Paramyxoviridae*. This chapter represents one step toward this goal by providing analyses of the available amino acid sequences of the three nucleocapsid-associated proteins (see also Chapters 2–4 and 9, and the Appendix).

A. Paramyxovirus Nucleocapsid Structure

The major protein component of nucleocapsids is the nucleoprotein (NP protein), which ranges in size from M_r 43,000 to 57,000 (Table I). Two other proteins are associated with nucleocapsids: the P proteins (polymerase-associ-

EXEEN M. MORGAN • Biotechnology Services Division, Microbiological Associates, Inc., Rockville, Maryland 20850.

TABLE I. Molecular Weights
of Nucleocapsid-Associated Proteins

Virus	NP protein	P protein	L protein
PIV3	57,823	68,860	255,812
	57,819	67,683	
	57,809	67,541	
SV	54,543	62,011	252,846
MeV	58,111	53,900	247,611
MuV	61,792	41,587	
CDV	58,000[b]	54,936	
NDV	53,161	42,183	248,822
		42,126	
SV5		44,000	
RS	42,600	27,150	

[a]The predicted molecular weights of the proteins were derived from the nucleic acid sequences of the genes of the following viruses: PIV3 (Galinski et al., 1986a,b; Jambou et al., 1986; Luk et al., 1986; Sanchez et al., 1986; Spriggs and Collins, 1986; Galinski et al., 1988); SV (Sendai virus) (Giorgi et al., 1983; Shioda et al., 1983; Morgan et al., 1984; Shioda et al., 1986); MeV (measles virus) (Bellini et al., 1985; Rozenblatt et al., 1985; Blumberg et al., 1988); MuV (mumps virus) (Elango, 1989; Takeuchi et al., 1988); CDV (Barrett et al., 1985; Yusoff et al., 1987; McGinnes et al., 1988); SV5 (Thomas et al., 1988); and RS virus (Satake et al., 1984; Collins et al., 1985).
[b]The sequence reported for the CDV NP gene is incomplete and lacks approximately 28 nucleotides, relative to measles virus, at the 5' end of the sequence (Rozenblatt et al., 1985).

ated or phosphoproteins), ranging in size from M_r 27,000 to 68,000; and the L (large) proteins, ranging from M_r 248,000 to 255,000 (Table I). It is estimated, based upon electron microscopic observations of nucleocapsids, that there are 11–13 "structural units" per helix turn with a pitch of 5 nm, resulting in a total of 2200–2800 subunits for a 1-μm nucleocapsid with 200 turns (Finch and Gibbs, 1970; Choppin and Compans, 1975; Hosaka and Hosoi, 1983). From these data Lamb et al. (1976) estimated that there are 2600 NP molecules, 300 P molecules, and 40 L molecules per nucleocapsid.

The exact arrangement of the three proteins within nucleocapsids is not known. The nucleocapsid subunits are not arranged perpendicular to the long axis of the particle; instead, they have "a polar appearance as that of interlocking arrow-heads pointing to one end of the nucleocapsid" (Finch and Gibbs, 1970). Models of the interaction between individual NP subunits and the genomic RNA have been proposed based on the structural similarities between paramyxovirus nucleocapsids and the rod-shaped plant viruses, such as tobacco mosaic virus (Kingsbury, 1977). One difference is that paramyxovirus nucleocapsids are flexible, permitting them to be coiled within virus particles (Finch and Gibbs, 1970), whereas most of the plant viruses are rigid.

Even less is known about the location and distribution of the P and L proteins within nucleocapsids, but recent immunoelectron microscopy studies

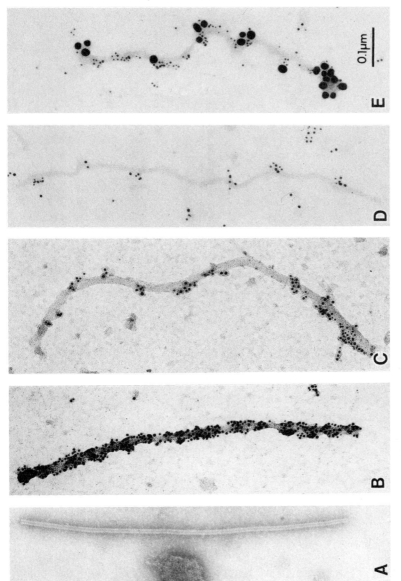

FIGURE 1. Localization of nucleocapsid-associated proteins of Sendai virus by immunogold labeling. (A) An untreated nucleocapsid. Nucleocapsids were incubated with (B) monoclonal antibodies to the NP protein (C) monoclonal antibodies to the P protein, (D) polyclonal antisera to the L protein, (E) antibodies to both P and L proteins. The preparations were then incubated with antisera conjugated with gold particles. In panel E, the large particles recognize anti-P antibodies and the smaller particles anti-L antibodies. [Photographs provided by K. G. Murti, A. Portner, and E. M. Morgan.]

(Murti *et al.*, 1985; Portner and Murti, 1986; Portner *et al.*, 1988) have provided some clues. When Sendai virus nucleocapsids were incubated with antibodies to the NP protein, the antibodies decorated the entire molecule, as might be expected. Antibodies to the P and L proteins appeared to be distributed in clusters along the length of the nucleocapsid, but when nucleocapsids were incubated with antibodies to both P and L, L was detected only in conjunction with P (Fig. 1). Both P and L proteins are required for viral RNA synthesis (see Section I.B) and these results suggest that their structural proximity may also be required to produce functional nucleocapsids (Portner *et al.*, 1988).

B. Functions of Nucleocapsid-Associated Proteins

The nucleocapsids of paramyxoviruses have at least two functions: they protect the genomic RNA, and they are the site of viral RNA synthesis. Virion-associated RNA polymerase activity in paramyxoviruses was first demonstrated by Huang *et al.* (1971), using Newcastle disease virus (NDV). Although it is known that all of the enzymatic activities required for mRNA synthesis and modification and for the replication of genomic RNA are associated with paramyxovirus nucleocapsids (see Chapter 9), the specific roles of the nucleocapsid-associated proteins, especially P and L, in RNA synthesis remain undefined.

The RNA-negative phenotypes of several temperature-sensitive mutants of NDV have been mapped to the genes coding for both the P and L proteins (Peeples and Bratt, 1982). Chinchar and Portner (1981a) found that proteolytic digestion of Sendai virus nucleocapsids removed part of the P protein, with little effect on *in vitro* transcriptase activity; similar treatment of NDV produced greater degradation of the P protein and severely inhibited RNA synthesis (Chinchar and Portner, 1981b). In addition, monoclonal antibodies directed against the NP and P proteins have been shown to inhibit *in vitro* transcriptase activity of Sendai virus nucleocapsids, but the mechanism of this inhibition is not known (Deshpande and Portner, 1984, 1985).

A more direct demonstration of the requirement for P and L proteins in RNA synthesis was provided by Hamaguchi *et al.* (1983), who showed that removal of P and L from NDV nucleocapsids resulted in the loss of *in vitro* transcriptase activity, and that the addition of purified P and L proteins to the stripped nucleocapsids restored transcriptase activity. Recently, Gotoh *et al.* (1989) reported enhanced Sendai virus yields from cells transfected with stripped nucleocapsids when the cells were also infected with recombinant vaccinia viruses carrying the *P* and *L* genes. These investigators also conclude that the products of both genes are required for viral RNA synthesis.

II. SEQUENCE ANALYSES OF NUCLEOCAPSID PROTEINS

This section summarizes analyses of the amino acid sequences of paramyxovirus nucleocapsid-associated proteins and examines their deduced evolutionary relationships. There are, however, some limitations to this approach that merit mention:

1. The sequences of the three genes that code for the NP, P, and L proteins are not available for each paramyxovirus. Additional sequence data may not have much impact on the currently proposed alignments, but such information would permit more complete sequence comparisons.

2. Only the deduced primary amino acid sequences of the three proteins can be compared. More precise relationships will become evident when the secondary and tertiary structures of the proteins have been determined.

3. The deduced amino acid sequences were analyzed using GENALIGN (IntelliGenetics, Inc., Mountain View, CA), a computer program developed by Dr. Hugo Martinez (Sobel and Martinez, 1985), and the reduced (ten-letter) alphabet of Jimez-Montano and Zamora-Cortina (1981). This reduced alphabet is designed to study evolutionary changes and uses only the most conservative amino acid replacements. Therefore, the data that are shown represent a single method of sequence alignment; as noted in the following sections, somewhat different results have been reported by investigators using other programs.

A. NP Proteins

Sequences of the *NP* genes and the deduced amino acid sequences of the corresponding NP protein have been reported for seven paramyxoviruses: human parainfluenza virus type 3 (PIV3) (Galinski *et al.*, 1986a; Jambou *et al.*, 1986; Sanchez *et al.*, 1986); Sendai virus (Shioda *et al.*, 1983; Blumberg *et al.*, 1984; Morgan *et al.*, 1984); measles virus (Rozenblatt *et al.*, 1985); canine distemper virus (CDV) (Rozenblatt *et al.*, 1985); NDV (Ishida *et al.*, 1986); mumps virus (Elango, 1989); and respiratory syncytial (RS) virus (Collins *et al.*, 1985). Comparison of the amino acid sequences of all seven proteins reveals little sequence identity, with a few exceptions (Table II). Greater than 50% sequence identity was observed between individual virus pairs (PIV3 and Sendai virus; CDV and measles virus), a finding in accord with previous reports (Rozenblatt *et al.*, 1985; Galinski *et al.*, 1986; Jambou *et al.*, 1986). The lowest degree of observed sequence identity was for RS virus; previous studies reported no NP sequence similarities between this virus and a limited number of other paramyxoviruses (Galinski *et al.*, 1986a; Jambou *et al.*, 1986).

TABLE II. Conserved Amino Acids of Paramyxovirus NP Proteins[a]

	PIV3	SV	MeV	CDV	NDV	MuV	RS
PIV 3 (515)	—	71	31	28	30	29	27
SV (517)	58	—	31	29	30	28	24
MeV (523)	20	19	—	71	33	34	21
CDV (523)	18	19	63	—	35	37	20
NDV (489)	18	19	21	22	—	47	20
MuV (553)	17	16	21	22	33	—	19
RS (391)	11	8	8	7	9	6	—

[a]The alignments (shown in Fig. 2) were scored pairwise. The total number of conserved amino acids in each pair, which includes both identical amino acids and conservative replacements (figures above the diagonal), and the total number of identical amino acids (figures below the diagonal) are expressed as a percentage of the aligned sequences. The number of amino acids in each protein is given in parentheses.

Although the overall observed sequence identity of the NP proteins is low, comparisons of several NP proteins have revealed small regions of sequence identity (Rozenblatt et al., 1985; Galinski et al., 1986a; Elango, 1989), which were still evident when all the NP sequences were aligned (Fig. 2). There are three distinct regions toward the center of the molecules where clusters of identical or conserved amino acids occur in all of the NP proteins, except that of RS virus. In RS virus, four of five residues were conserved in region 2. Pairwise analysis of the sequences in the regions delineated by arrows in Fig. 2 reveals that, with the exception of RS virus, the degree of amino acid conservation in all the NP sequences increases to approximately 50% (Table III). In contrast, the sequences toward the amino and carboxyl termini show little sequence identity or conservation.

The NP protein can be thought of as a nucleic acid-binding protein because its major function is to interact with and bind genomic RNA, other NP molecules, and, presumably, P and L proteins. Although no distinct domain has been observed in the linear sequence of any paramyxovirus NP protein that could function as a potential nucleic acid-binding site, the amino-terminal half of the molecule of some paramyxoviruses is more positively charged, suggesting that this portion may be in contact with the genomic RNA (Morgan et al., 1984; Sanchez et al., 1986).

There is also evidence that the carboxyl-terminal half of the molecule is exposed to the environment. Most of the monoclonal antibodies directed against Sendai virus NP (Deshpande and Portner, 1984; Gill et al., 1988) and measles virus NP (Buckland et al., 1989) bind to sites in the carboxyl-terminal half of the molecule. However, one measles virus monoclonal antibody-binding site is located in the amino-terminal region between amino acids 122 and 150 (Buckland et al., 1989).

Heggeness et al. (1981) showed that the carboxyl-terminal region (M_r 12,000) of the Sendai virus NP could be removed by tryptic digestion without altering the structure of the nucleocapsid. Regions in the carboxyl-terminal half of several paramyxovirus NP proteins that correspond approximately to regions 2 and 3 (Fig. 2) occur within hydrophobic domains and are predicted to represent internal regions of the protein (Galinski et al., 1986a). In addition, one monoclonal antibody to Sendai virus NP was mapped to a region between amino acids 290 and 295, which is located between regions 2 and 3 and is predicted to contain a trypsin cleavage site (Morgan et al., 1984; Gill et al., 1988). The existence of a hydrophobic domain at region 3 is also in accord with the findings of Heggeness et al. (1981) that trypsin digestion cleaved that part of the NP protein, but the resulting fragment remained bound to the nucleocapsid. It is possible that these regions of sequence identity are folded such that they are buried within the molecule and the intervening regions can loop out, exposing sites that are accessible to proteolytic cleavage and antibody binding.

B. L Proteins

Only four L gene sequences have been determined: those of PIV3 (Galinski et al., 1988), Sendai virus (Neubert and Koch, 1985; Morgan and Rakestraw,

```
PIV    mlslfdtFnarrqenitksagGaiipgqkntvsifalgptitdddekmtlallflshsldnekqhaqra   69
SV     magllstfdtFssrrsesinksggGavipgqrstvsvfvlgpsvtddadklfiattflahsldtdkqhsqrg  72
MeV    matllrslalFkrnkdkppitsgsGgairgikhiiivpipgdssittrsrlldrlvrlignpdvsgpkltga  72
CDV     lFkrtrdqpplasgsGgasrgikhviivlipgdssivtrsrlldrlvrlvgdpkingpkltgi  72
MuV    mssvlkaferFtieqqlqdrgeegsippetlksavkvfvintpnpttryqmlnfclriicsqnrrashrvga  72
NDV    mssvfdeyeqllaaqtrpngahggGekgstlkvevpvftlnsddpedrwnfavfclriavsedankplrqga  72
            *   * *                                              *

PIV    gflvslLsmayanpelylttngsnadvkyviynirkdlkrqkyggfvvktreniyekttewifgsdldydqe  141
SV     gflvslLamaysspelylttngvnadvkyviyniekdpkrtktdgfivktrdmeyerttewlfg..pmvnks  142
MeV    ligilsLfvgspgqliqritddpdvsirllevvqsdqsqsgltfasrgtnmedead..qyfshddpissdqs  142
CDV    lisilsLfvespgqliqriiddpdvsiklvevipsinsacgltfasrgaswilrad..effkivdegskaqg  142
MuV    lialfsLpsagmqnhirladrspeaqierceidgfepgtyrlipnara..nltaneiaayalladdlpptin  142
NDV    lisllcshsqvmrnhvalagkqneatlavlei....dgftngvpqfnnrsgvseeraqrfmmiagslpracs  140
          *         *         *               +     +*  *  +  *

                                       ↓
                                       1
PIV    tmlqngrnnstiedlvhtfgypsclgalii QiWivlvKAiT sisglrkgfftRleafrQdgtVqaglvlsgd  213
SV     plfqgqrdaadpdtllqtygypaclgaiiv QvWivlvKAiT ssaglrkgffnRleafrQdgtVkgalvftge  214
MeV    rfgwfenkeisdievqdpegfnmilgtila QiWvllaKAvT apdtaadselrRtikytQqrrVvgefrlerk  214
CDV    qlgwlenkdivdievdnaeqfnillasila QiWillaKAvT apdtaadsemrRwikytQqrrVvgefrmnki  214
MuV    ngtpyvhadvelqpcdeieqfldrcysvli QaWvmvcKcmT aydqpagsadrRfakyqQqgrlearymlqpe  214
NDV    ngtpfvtagveddapeditdtlerilsiqa QvWvtvaKAmT ayetadesetrRinkymQqgrVqkkyilhpv  212
         *   +  +* *  + *  +  *+       +    *   +  *

                                              2
RS                                             lrwGv
PIV    tvdqigsinrsqqslvtlmVetlitmntsrwdlttiekniqivgnYIrdaGLasFfn TiryGieT rmaALsL   285
SV     tvegigsvmrsqqslvslmVetlvtmntarsdlttlekniqlvgnYIrdaGLasFmn TikyGveT kmaALtL   286
MeV    wldvvrniiaedlslrrfmValildikrtpgnkpriaemicdidtYIveaGLasFil TikfGieT mypALgL   286
CDV    wldivrnriaedlslrrfmValildikrspgnkpriaemicdidnYIveaGLasFil TikfGieT mypALgL   286
MuV    aqrliqtairkslvvrqyltfelqlarrqgllsnryyamvgdigkYIensGLtaFfl TlfyalgT kwspLsL   286
NDV    crsaiqltirqslavriflVselkrgrntaggtstyynlvgdvdsYIrntGLtaFfl TlkyGinT ktsALaL   284
          *           *            *  *  ++    ++   +    ** ***  +      + +

                                            3
                                            ↓
PIV    stlrpdinrlkaLMeLYlskGprAPficiLrdpihgeFapGnYPa iWSYAMGVa vvqnraMqqyvtgRsyld   357
SV     snlrpdinklrsLidtYlskGprAPficiLkdpvhgeFapGnYPa lWSYAMGVa vvqnkaMqqyvtgRtyld   358
MeV    hefagelstlesLMnLYqqmGkpAPymvnLensiqnkFsaGsYPl lWSYAMGVg velensMgglnfgRsyfd   358
CDV    hefsgelttiesLMmLYqqmGetAPymvlLensvqnkFsaGsYPl lWSYAMGVg velensMgglgfgRsyfd   358
MuV    aaftgeltklrsLMmLYrdiGeqArylalLeapqimdFapGgYPl ifSYAMGVg svldvqMrnytyaRpfln   358
NDV    sslsgdiqkmkqLMrLYrmkGdnAPymtlLgdsdqmsFapaeYaq lySfAMGma vldkgtgkyqfaRdfms   356
         **  *  +*  *  *  + *   *  + +    +    **+++++++      *+ **

PIV    idmfqlGqavardaeaqmsstledeLGvTheakeslkrhirninssetsfhkptggsaiemaideepqfeh   429
SV     memfllGqavakdaeskissaledeLGvTdtakerlrhhlanlsggdgayhkptgggaievaldnadidlet  430
MeV    payfrlGqemvrrsagkvsstlaseLGiT..............aedarlvseiamhttedkisravgprqa   415
CDV    payfrlGqemvrrsagkvssalaaeLGiT.............keeaqlvseiaskttedrtiratgbpkqs   415
MuV    gyyfqiGvetarrqqgtvdnrvaddLGlTpeqrnevtqlvdrlargrgagipgggpvnpfvppvqqqqpaavy  430
NDV    tsfwrlGve.yaqaqgs...sinedma......aeL.....kl.......tpaa...................  388
          * *+        *   ***

PIV    radqeqdgepqssiiqyawaegnrsddrteqatesdniktteqqnirdrlnkrlndkkkqgsqpstnptnrtn  501
SV     eahadqdargtggesgertarqvsgghfvtlhgaerleeetndedvsdierriamrlaerrqgilqpmemka   502
MeV    qvsflqgdqsenelprlggkedrrvkqsrgearesyretgpsrasdaraahlptgtpldidtasessqdpqd  487
CDV    qitflhsersevanqqpptinkrsenqggdkypihfsderllgytpdvnssersgsrydtqivqddgndddr  487
MuV    adipaleesdddgdedggagfqngvqvpavrqggqtdfraqplqdpiqaqlfmplypqvsnipnnriirsia  502
NDV    ..rrgl.........aaaaqrvseetssidmptqqagvltglsdggsqapqgalnrsqgqpdtgdgetqfl   448

PIV    qdeiddlfnafgsn   515
SV     aitvsittkmtmpqq  517
MeV    srrsaepll....sckpwqes..rknkaqtrt.plqctmteif  523
CDV    ksmeaiakmrmltkml....sqpgtse..dns.pvyndkelln  523
MuV    sggwktliyydttrmvilnkmqqantetlsqtipikthsckwatgmskslt  553
NDV    dlmravansmreapnsaqgtpqpgpppptpgpsqdndtdwgy  489
```

FIGURE 2. Sequence alignment of paramyxovirus NP proteins. The sources of the published sequences are: PIV3 (Galinski *et al.*, 1986a); SV (Morgan *et al.*, 1984); MeV (Rozenblatt *et al.*, 1985); CDV (Rozenblatt *et al.*, 1985); MuV (Elango, 1989); NDV (Ishida *et al.*, 1986); and RS virus (Collins *et al.*, 1985). (+) Identical amino acids in that position in all the sequences; (*) conserved amino acids in that position in all the sequences based upon the reduced alphabet of Jimez-Montano and Zamora-Cortina (1981); (.) gap inserted to obtain the best alignment. The numbers at the right indicate amino acid positions in the respective sequences; for RS virus, only the portion of the sequence between amino acids 258 and 262 is shown. Boxes denote regions that exhibit a high percentage of identity. The arrows delineate the portion of the sequence used to derive the data shown in Table III.

TABLE III. Conserved Amino Acids within a Limited Region
of the NP Protein[a]

	PIV3	SV	MeV	CDV	NDV	MuV	RS
PIV3	—	92	46	47	50	46	36
SV		—	48	48	48	46	36
MeV			—	94	54	53	27
CDV				—	56	56	24
MuV					—	62	29
NDV						—	27
RS							—

[a]The region of the aligned NP sequences delineated by arrows (Fig. 2), encompassing 170 amino acids for all the sequences except RS virus (136 amino acids), was scored pairwise. The number of conserved amino acids in each pair is expressed as a percentage of this aligned portion of the sequence.

1986; Shioda et al., 1986), measles virus (Blumberg et al., 1988), and NDV (Yusoff et al., 1987). These extremely large proteins are almost identical in size, differing by 50 or fewer amino acids (Table IV). The L proteins exhibit approximately the same degree of sequence identity as the NP proteins of the same viruses. Pairwise matches again showed that the greatest sequence identity was between Sendai virus and PIV3, and that NDV was the least identical (Table IV).

The greatest degree of sequence identity occurred in the amino-terminal half of the aligned sequences (Fig. 3). This was noted previously by Galinski et al. (1988) in comparison of the sequence of PIV3, Sendai virus, and NDV. Ten clusters of six or more sequential amino acids (many of which are identical) are present in all four sequences, although additional L protein sequences are needed to determine whether these sites occur in all L proteins. However, if the overall conservation of the L proteins is like that of the NP proteins, and if NDV is the most divergent (excluding RS virus), it seems likely that these regions will be retained in most paramyxovirus L protein sequences.

TABLE IV. Conserved Amino Acids
in Paramyxovirus L Proteins[a]

	PIV3	SV	MeV	NDV
PIV3 (2233)	—	72	50	34
SV (2228)	61	—	50	33
MeV (2183)	36	37	—	32
NDV (2204)	23	21	20	—

[a]The alignments (shown in Fig. 3) were scored pairwise and the number of conserved amino acids in each pair, which includes both identical amino acids and conservative replacements (figures above the diagonal), and the number of identical amino acids (figures below the diagonal) are expressed as a percentage of the aligned sequences. The number of amino acids in each protein is given in parentheses.

Similarities between the sequence of the analogous L protein of the rhabdovirus vesicular stomatitis virus (VSV) (Schubert *et al.*, 1984) and the paramyxovirus L protein sequences have been noted (Morgan and Rakestraw, 1986; Blumberg *et al.*, 1988; Galinski *et al.*, 1988; Yusoff *et al.*, 1987). Only the sequences in regions 7 and 8 (Fig. 3) share any relationship with VSV; the other regions detected by the alignments of Galinski *et al.* (1988) and Yusoff *et al.* (1987) were not evident in the present work.

It is too early to do more than speculate about the L protein's structural and functional relationships with the NP and P proteins. The clustering of conserved regions in the amino-terminal one-third of the molecule suggests that these regions may be involved in binding to other molecules and could represent potential enzymatic sites. The carboxyl terminus of the molecule is apparently exposed when L is bound to nucleocapsids. The antibody-binding studies illustrated in Fig. 1 used polyclonal antisera raised against a synthetic peptide representing the last 30 amino acids at the carboxyl terminus of the molecule. This antibody was able to bind to L protein on native nucleocapsids and did not inhibit *in vitro* transcriptase (Portner *et al.*, 1988), suggesting that this region is exposed to the environment and is not required for enzymatic activity.

C. P Proteins

The P proteins are the most divergent of the nucleocapsid-associated proteins and, as a group, they are probably the least similar of the paramyxovirus proteins. Eight *P* gene sequences have been published: those of PIV3 (Galinski *et al.*, 1986b; Luk *et al.*, 1986; Spriggs and Collins, 1986), Sendai virus (Giorgi *et al.*, 1983), measles virus (Bellini *et al.*, 1985), CDV (Barrett *et al.*, 1985), mumps virus (Takeuchi *et al.*, 1988), simian virus 5 (SV5; Thomas *et al.*, 1988), NDV (Sato *et al.*, 1987; McGinnes *et al.*, 1988), and RS virus (Satake *et al.*, 1984). The predicted P proteins vary markedly in size, from 241 amino acids (RS virus) to 602 amino acids (PIV3), and they can be categorized into two size-based groups (Table V). One group, which ranges from 507 to 602 amino acids, includes PIV3, Sendai virus, measles virus, and CDV; the other, from 241 to 392 amino acids, includes SV5, NDV, mumps virus, and RS virus.

The division of the P proteins into two size groups also reflects differences in *P* gene organization among the paramyxoviruses. Those viruses with the larger P proteins (Sendai virus, PIV3, measles virus, and CDV) all encode at least one additional protein using alternate reading frames of the P mRNA (Giorgi *et al.*, 1983; Barrett *et al.*, 1985; Bellini *et al.*, 1985; Galinski *et al.*, 1986b; Luk *et al.*, 1986; Spriggs and Collins, 1986; see also Chapter 6). Like the P proteins, these smaller proteins also lack significant sequence identity (not shown). The smaller *P* genes of mumps virus (Takeuchi *et al.*, 1988) and RS virus (Satake *et al.*, 1984) are monocistronic, whereas NDV can code for two additional, small polypeptides, but these are in the same reading frame as the *P* gene (McGinnes *et al.*, 1988). The *P* gene of SV5 is unique among the *P* genes that have been sequenced in that it encodes a second protein using a separate mRNA (Thomas *et al.*, 1988).

```
PIV   MDtesnngtvsdILYPEcHLnpPIVkgKIaqlhtimslpqpYdmdDdsilvitrqkiklnkldkrqrsirrl   72
SV    MDgqessqnpsdILYPEcHLnsPIVrgKIaqlhvlldvnqpYrlkDdsiinitkhhkirngglsprqikirsl   72
MeV   MD.slsvnq...ILYPEvHLdsPIVtnKIvaileyarvphaYsleDptlcqnikhrlkng....fsnqmiin    64
NDV   Massgperaehq1ilPEsHLssPlVkhKl.lyywkltglplpdecDfdhlilsrq.wkkilesaspdtermi   70
      +      +* ++ ++  +*+ +*                     +        *    *

PIV   kliltekvndlgkytfirypemskemfklhipginskvtelllkadrtysqmtdglrDlwinvlsklasknd  144
SV    gkalqrtikdldrytfepyptysqellrldipeicdkirsvfavsdrltrelssgfqDlwlnifkqlgnieg  144
MeV   nvevgnvikksklrsypahshipypncnqdlfniedkestrkirellkkgnslyskvsDkvfqclrdtnsrlg  136
NDV   klgravhqtlnhnsritgvlhprcleelasievpdstnkfrkiekkiqihatrygelftrlcthiekkllgs  142

PIV   gsnydlneeinniskvhttyksdkWynPFktWFtiKydMRrlqKarnevtfnmgkdynlledqknfLlihpe  216
SV    regydplqdigtipeitdkysrnrWyrPFltWFsiKydMRwmqKtrpggpldtsnshnllecksytLvtygd  216
MeV   lgselredikekvinlgvymhssqWfePFlfWFtvKteMRsviKsqthtchrrrhtpvfftgssveLlisrd  208
NDV   swsnnvprseefnsirtdpafwfhskwstakfawlhikqiqrhlivaartrsaanklvmlthkvgqvfvtpe  214
            * *         *          *         +  *    *       * *

                           1
PIV   LVlIldKqnyngYliTpELVLMYCDVvEGRwnisacaklDpklqsmyqkgnnlWeviDklFpimGektfdvi  288
SV    LVmIlnKltltgYYlTfELVLMYCDVvEGRwnmsaaghlDkksigitskgeelWelvDslFsslGeeiynvi  288
MeV   LVaIisKesqhvYylTfELVLMYCDViEGRlmtetamtiDarytellgrvrymWkliDgfFpalGnptyqiv  280
NDV   LVivthtnenkftclTqELVLMYaDmmEGRdmvniisttavhlrslsekiddilqliDalakdlGnqvydvv  286
      ++ *        **  ++ +++++ +*++++           *    *  *  **   *+  *  **

PIV   slLEPLaLsliQthDpvkqLRGA.FlnHvlsEm.elifesresikeflsvdyidkildIFnkstIdeiaEIF  358
SV    alLEPLsLaliQlnDpvipLRGA.FmrHvltEl.qtvltsrdvytdaeadtivesllaIFhgtsIdekaEIF  358
MeV   amLEPLsLaylQlrDitveLRGA.FlnHcftEi.hdvldqngfsdegtyheliealdyIFitddIhltgEIF  350
NDV   slmEgfaygavQllep.sgrfaghFfafnlqElkdil.igllpndiaesv..thaiatvFsgleqnqaaEml  354
      **+     *+ *      *    +  *   *    +           **+           *

                                                  2
PIV   SFFRtFGHPpLEAsiAAekVRkyMyigKqlkfdTinkcHAiFCtIIINGYReRHGGqWPPvtlPdHahefii  430
SV    SFFRtFGHPsLEAvtAAdkVRahMyaqKaiklkTlyecHAvFCtIIINGYReRHGGqWPPcdfPdHvclelr  430
MeV   SFFRsFGHPrLEAvtAAenVRkyMnqpKvivyeTlmkgHAiFCgIIINGYRdRHGGsWPPltlPlHaadtir  422
NDV   cllRlwGHPlLEsriAAkaVRsqMcapKmvdfdmilqvlsfFkgtIINGYRkknaGvWPrvkvdtiygkiig  426
      + *+++ ++   ++    ++    +     *       + *  +++++  +* +* **  *+

              3
PIV   NAygSnsaisyEnaVDyyqSFiGiKFnkFiepqLDeDLTiYmKDKALspkksnWDtVYPasnLl.....Yrt  497
SV    NAqgSntaisyEcaVDnytSFiGfKFrkFiepqLDeDLTiYmKDKALsprkeaWDsVYPdsnL.....yYka  497
MeV   NAqaSgeglthEqcVDnwkSFaGvKFgcFmplsLDsDLTmYlKDKALaalqreWDsVYPkefL.....rYdp  489
NDV   qlhadsaeishdimlreykSlsaleFepcieydpvtnLsmflKDKAiahpndnW.lasfrrnLlsedqkkhv  497
      * *     ** *  **+ +       *      +++*****++++      +     +

                                              4           5
PIV   nasnesRRLveVFIaDskFdPhqildYVeSGdwLdDpEFNiSYSLkEKEIKqeGRLFAKMTYKMRAtQVlsE  569
SV    peseetRRLieVFInDenFnPeeiinYVeSGdwLkDeEFNiSYSLkEKEIKqtGRLFAKMTYKMRAcQVlaE  561
MeV   pkgtgsRRLvdVFInDssFdPydvimYVvSGayLhDpEFNlSYSLkEKEIKetGRLFAKMTYKMRAvQVlaE  561
NDV   keatstnRLlieFlesndFdPykemeYlttleyLrDddvavSYSLkEKEVKvnGRiFAKlTkKlRncQVmaE  569
                   * +++  *+   + +      * +* *  *+ + *  +++++ +++*+ +++++*+ +**  +++ +

PIV   tLlannIGkfFqeNGMvKgEieLlKrLtTisiSGVPr........ynevynnSkS.htddlktynkisnlnl  632
SV    tLlakgIGelFreNGMvKgEidLlKrLtTlsvSGVPr..tds......vynnSksSsekrnegmenknsggyw  633
MeV   nLisngIGkyFkdNGMaKdEqdLtKaLhTlavSGVPkdlkeshrggpvlktySrSpvhtstrnvraakg.fi  633
NDV   giladqIapfFqgNGvi.........qdsislt..............kstlamsqlsfnsnkkritdckerv  618
       **  ++ *  + +++           *  **

                 6
PIV   ssnqkskkfefkstdiyndgYETvScFlTTDLKKYCLNWRyEstaLFgetcNqIfGlnklFnWlHprLEgst  704
SV    dekkrsr.hefkatdsstdgYETlScFiTTDLKKYCLNWRfEstaLFgqrcNeIfGfktfFnWmHpvLErct  704
MeV   gfpqvirqdqdtdhpenmeaYETvSaFiTTDLKKYCLNWRyEtisLFaqrlNeIyGlpsfFqWlHkrLEtsv  704
NDV   ssn...rnhdpks.....knrrrvatFiTTDLqKYCLNWRyqtikLFahaiNqlmGlphfFeWiHlrLmdtt  682
              *                  +*++++ +++++++  *   +++   *  * + + +**  + +

                                   7                        8
PIV   iYVgDPyCPpsdkeHisLedhpdsgfyvhnPrGGIEGfCQKLWTliisIsaihLAAvriGVRvtamVQGDNQa  776
SV    iYVgDPyCPvadrmHrqLqdhadsgifihnPrGGIEGyCQKLWTliisIsaihLAAvrvGVRvsamVQGDNQa  776
MeV   lYVsDPhCPpdldaHipLykvpndqifikyPmGGIEGyCQKLWTistIpylyLAAyesGVRiaslVQGDNQv  776
NDV   mfVgDPfnPpsdptdcdLsrvpnddiyivsarGGIElCQKLWTmisIaaiqLAAarshcRvadmVQGDNQv  754
      ***+ ++  +      **       ***  +++++ +++++++ *+ +++     +* +++++++

      8
PIV   IAVTtRVPnnydYrvKKeivykdvvr.fFdsLRevmdDlGHeLKlNETIiSSkmFiYSKrIYYDGrilpQaL  847
SV    IAVTsRVPvaqtYkqKKnhvyeeitk.yFgaLRhvmfDvGHeLKlNETIiSSkmFvYSKrIYYDGkilpQcL  847
MeV   IAVTkRVPstwpYnlKK.reaarvtrdyFviLRqrlhDiGHhLKaNETIvSShfFvYSKgIYYDGllvsQsL  847
NDV   IAVTreVrsddspemvltql.hqasdnfFkeLihvnhliGHnLKdrETIrSdtfFiYSKrIfkDGailsQvL  825
      ++++   +          *+   *    ++  +++ ++   +*++++ +* ++ ** + +
```

FIGURE 3. Sequence alignment of paramyxovirus L proteins. The sources of the published sequences are: PIV3 (Galinski *et al.*, 1988); SV (Shioda *et al.*, 1986); MeV (Blumberg *et al.*, 1988); and NDV (Yusoff *et al.*, 1987). (+) Identical amino acids in that position in all the sequences; (*) conserved amino acids in that position in all the sequences, based upon the reduced alphabet of Jimez-Montano and Zamora-Cortina (1981); (.) gap inserted to obtain the best alignment. The numbers at the right indicate amino acid positions in each sequence. Boxes denote regions that exhibit a high percentage of identity.

```
PIV  KalsrCVFWSETviDEtRsAsSNlaTsfAKaIEnGYspvLgYacsifKniQQlyIaLGmnINpTitqnikdq  919
SV   KaltkCVFWSETlvDEnRsAcSNisTsiAKaIEnGYspiLgYcialyKtcQQvcIsLGmtINpTisptvrdq  919
MeV  KsiarCVFWSETivDEtRaAcSNiaTtmAKsIErGYdryLaYslnflKviQQilIsLGftINsTmtrdvvip  919
NDV  KnssklVmvSgdlsentvmscaNiastvAricEnGlpkdfcYylnyimscvQtyfdsefsyNnnshpdlnqs  897
     +   * *  +      +          +* **  +*   +       *        +         *
                                     9
PIV  yfrnpnwmqyasLiPAsvGGfNYmamSRcFVRNIGDPsvaalADiKRfIkAnLldrsvLyriMnQePGeSSF  991
SV   yfkgknwlrcavLiPAnvGGfNYmstSRcFVRNIGDPavaalADlKRfIrAdLldkqvLyrvMnQePGdSSF  991
MeV  lltnndllirmaLlPApiGGmNYlnmSRlFVRNIGDPvtssiADlKRmIlAsLmpeetLhqvMtQqPGdSSF  991
NDV  wiedisfvhsyvLtPAqlGGlsnlqySRlytRNIGDPgttafAeiKRleavgLlspnimttniltrpPGngdw  969
         *   +  ++ *++  + ++  ++ +  ++++++      +***+   +*   *   **  ++   *
                10
PIV  LdnSlTGiRnaIAGMLDTTKsLiRvgi.nrGGLtysllr.kisNYDlvQyetlsrtLrlivsdkiryedmCS  1133
SV   LgnSlTGvReaIAGMLDTTKsLvRasv.rkGGLsygilr.rlvNYDliQyetltrtLrkpvkdnieyeymCS  1133
MeV  LdhSvTGaResIAGMLDTTKgLiRasm.rkGGLtsrvi.trlsNYDyeQfragmviLtgrkrnvlidkesCS  1133
NDV  meaSsvGrRkqIqGlvDTTntvikialtrrplgikrlmrivnyssmhamlfrddvfssnrsnhplvssnmCS  1113
     *  +  +  + +  +*++++    *** + ++    **                 *        ++
PIV  VdLAiaLRqkMWihLsgGRmIsGLEtPDpLEllsGviItgsEhCkiCyssdgtnpYtWmylPgnikigsaet  1205
SV   VeLAvgLRqkMWihLtyGRpIhGLEtPDpLEllrGifIegsEvCklCrsegadpiYtWfylPdnidldtltn  1205
MeV  VqLAraLRshMWarLarGRpIyGLEvPDvLEsmrGhlIrrhEtCviC..ecgsvnYgWffvPsgcqlddidk  1203
NDV  ltLAdyaRnrsWspLtgGRkIlGvsnPDtiElveGeils..vsggctrcdsgdeqftWfhlPsnieltddts  1183
     * ++   +   +   ++  +*   ++ *+ + *             *   * +  *+    *
PIV  gisslRvPYfGSvTDERseaqLgyiknlSkpakaAiRIAmiYtWAfGnDeiSWmEAsqiAqtRANftLdsLk  1277
SV   gcpaiRiPYfGSaTDERseaqLgyvrnlSkpakaAiRIAmvYtWAyGtDeiSWmEAaliAqtRANlsLenLk  1277
MeV  etsslRvPYiGStTDERtdmkLafvrapSrslrsAvRIAtvYsWAyGdDdsSWnEAwllArqRANvsLeeLr  1275
NDV  knppmRvPYlGSkTqERraasLakiahmSphvkaAlRassvliWAyGdnevnWtaAltiAksRcNinLeyLr  1255
     *+*++ ++ + ++       ++   *+* ** + +   +++*  +   +   +  *+  + + +*   +*
PIV  ilTPvaTSTNLsHRLkDtaTQmKfSstsLiRvsRfiTmSNDNmsikeanetkDTNliYQQiMLtGLsvfEyl  1349
SV   llTPvsTSTNLsHRLkDtaTQmKfSsatLvRasRfiTiSNDNmalkeageskDTNlvYQQiMLtGLslfEfn  1349
MeV  viTPisTSTNLaHRLrDrsTQvKySgtsLvRvaRytTiSNDNlsfvisdkkvDTNfiYQQgML1GLgvlEtl  1347
NDV  llsPlpTagNLqHRLdDgiTQmtftpasLyRchltfTypmilkgyslkkeskrgmwfinrvMLiGLsliEsi  1327
     ***+*  +  ++ +++ +  +++ *+  *+    +       +             ++ ++ +*  +
PIV  sRleettghnpivmHLHiedeCCikesfndehInpestLeLirypesNefIYDkdPLkDvDlsklmvikdhs  1421
SV   mRykkgslgkplilHLHlnngCCimespqeanIpprstLdLeitqenNklIYDpdPLkDvDlelfskvrdvv  1421
MeV  fRlekdtgssntvlHLHvetdCCvipmidhprIpssrkLeLraelctNplIYDnaPLiDrDatrlytqshrr  1419
NDV  fpmtttrtydeitlHLHskfsCCire......apvavpfeLlgvapelrtvtsnkfmyDpspvsegdfarld  1393
                   *+++   ++*        +  *+ +  *  +        *
PIV  ytidmnyWddtdiihaisicTAitiaDtmsqldrDnlkEiivianDDDiNSlITEFltldilvFlktfGgll  1493
SV   htvdmtyWsddeviratsicTAmtiaDtmsqldrDnlkEmialvnDDDvNSlITEFmvidvplFcstfGgil  1493
MeV  hlvefvtWstpqlyhilaksTAlsmiDlvtkfekDhmnEisaligDDDiNSfITEFllieprlFtiylGqca  1491
NDV  laifksyelnlesyptielmnilsissgkligqsvvsydedtsiknDaiivydntrnwiseaqnsdvvrlfe  1465
         *            ***           *      + *     *
PIV  vnqfAytlyslkieGrdliwdyimrtLrdtShsilKVLsNALSHPKvfKrFWdcGvlnPi..ygpntasqdq  1563
SV   vnqfAyslyglnirGreeiwghvvriLkdtShavlKVLsNALSHPKifKrFWnaGovePv..ygpnlsnqdk  1563
MeV  ainwAfdvhyhrpsGkyqmgellssfLsrmSkgvfKVLvNALSHPKiyKkFWhcGiiePihgpsldaqnlht  1563
NDV  yaalevlldcsyqlyylrvrgldnivLymgdlyknmpgillsniaatishpvihsrlhavglvnhngshqla  1537
         *        *            +                              * *
PIV  iklalsiceYsldLfmrewlngvsleiyiCdsDmeVandRkqafisrHLsfvccLaeiasfgPnllnltylE  1635
SV   illalsvceYsvdLfmhdwqggvpleifiCdnDpdVadmRrssflarHLaylcsLaeisrdgPrlesmnslE  1635
MeV  tvcnmvytcYmtyLdlllneeleeftfllCesDedVvpdRfdniqakHLcvladLycqpgtcPpiqglrpvE  1635
NDV  dtdfiemsakllvsctrrvisglysgnkydllfpsVlddnlnekmlqlisrlccLytvlfattreipkirgl  1609
         *
PIV  rldlLkqylelnikedPtlkyvqisglliksfpstvTYvRktaIKylRirgisppeviddwdpiedenmlDn  1707
SV   rlesLksyleltflddPvlrysqltglvikvfpstlTYiRkssIKvlRtrgigvpevledwdpeadnallDg  1707
MeV  kcavLtdhikaeamlsPagsswninpiiivdhysclTYlRrgsIKqiR.......lrvd......pgfifDa  1694
NDV  saeekcsvlteyllsdavkpllspdqvssimspniiTfpanlyymsrkslnlireredkdsilallfpqepl  1681
```

FIGURE 3. (*continued*)

```
PIV  ivktindncnkdnkgnkinnfwglalknyqvlkir..sitsdsdnndrldaStgGltlpqggnylshqlRlf  1777
SV   iaaeiq..qniplghqtrapfwglrvsksqvlrlr..gykei..trgeigrSgvGltlpfdgrylshqlRlf  1773
MeV  l.aevn..vsqpkigsnnisnmsik.afrpphddvakllkdintskhnlpiSg.Gnlanye...ihafrRi.  1757
NDV  lefpsvqdigarvkdpftrqpaaflqeldlsaparydaftlsqihpeltspnpeedylvrylfrgigtasss  1753
          *

PIV  GINStsClKAlElsqilmkevnkdkDrLfLGEGaGaMLacYdatLgpavnyYNSGlnitdviGQRELkifPs  1849
SV   GINStsClKAlEltyllsplvdkdkDrLyLGEGaGaMLscYdatLgpcinyYNSGvyscdvnGQRELniyPa  1845
MeV  GINSsaCyKAvEistlirrclepgeDgLfLGEGsGsMLitYkeiLklskcfYNSGvsansrsGQRELapyPs  1829
NDV  wykashllsvpEvrcarhgnslylaegsgaimsllelhvphetiyyntlfsnemnpppqrhfgptptqflnsv  1825
          *    +*        *        *        *

PIV  EVsLVgkklgnvtqilnrVKVLFNGnPnsTWiGnmeCeslIwselndkSiGlvHcDmEgaigKseetvLhEh  1921
SV   EVaLVgkklnnvtslgqrVKVLFNGnPgsTWiGndeCeallIwnelqnsSiGlvHcDmEggdhKddqvvLhEh  1917
MeV  EVgLVehrm....gvgniVKVLFNGrPevTWvGsvdCfnfIvsniptsSvGfiHsDiEtlpdKdtiekLeEl  1897
NDV  vyrnlqaevtckdgfvqefrpLwrenteesdltsdkavgyItsavpyrSvsllHcDiEippgsnqslldqla  1897
          *   *     * *+*     * *      +    *   +*  *+ +*+

PIV  ysvirityLiGdddvvLisKiiPtitpnwsrilylyklywkdVsiislktsnpaStelyliSKdayctimep  1993
SV   ysviriayLvGdrdvvLidKiaPrlgtdwtrqlslylrywdeVnlivlktsnpaStemyplSrhpksdiied  1989
MeV  aailsmalLlGkigsiLviKlmPfsgdfvqgfisyvgshyreVnlv....yprySnfisteSylvmtdlkan  1965
NDV  inlsliamhsvreggvviiKvlyamgyyfhllmnlfapcstkgyilsngyacrgdmecylvfvmgylggptf  1969
          *  *       *** +*     *          **

PIV  sevvlsklkrlslleennllkwiilskkknnewlhheikeGerdygvmrpyhmalqifgfqinlnhlakefl  2065
SV   sktvlasllplskedsikiekwiliekakahewvtrelreGsssssgmlrpyhqalqtfgfepnlyklsrdfl  2061
MeV  rlmnpekikqqiiessvrtspglighilsikqlsciqaivGdavsrgdinptlkkltpieqvlincglaing  2037
NDV  vhevvrmaktlvqrhgtllsksdeitltrlftsqrqrvtdilssplprlikylrknidtalieaggqpvrpf  2041
          *

PIV  stsdltninniiqsfqrtikdvlfewinithddKrhklgggrynifplknkgklrllsrrlvlswislslstr  2137
SV   stmniadthncmiafnrvlkdtifewaritesdKrlkltgkydlypvrdsgklktisrrlvlswislsmstr  2133
MeV  pklckelihhdvasgqdgllnsililyrelarfKdnqrsqqgmfhaypvlvssrqrelisritrkfwghill  2109
NDV  caeslvstladitqitqiiashidtvirsviymeaegdladtvflftpynlstdgkkrtslkqctrqilevt  2113

PIV  lltgrfpdeKFehraqtGyvsLadtdleslkllsKntiknyrecigsisywflktkevkIlMkliGgaKllgi  2209
SV   lvtgsfpdqKFearlqlGivsLssreirnlrvitKtlldrfediihsityrfltkeikIlMkilGavKm.fg  2204
MeV  ysgnrklinKFiqnlksGyiLdlhqnifvknlsKs..ek..................qIiM..tGglK...r  2157
NDV  ilglrvedlnkigdvislvlkgmismedliplrtylkhstcpkylkavlgitklkemftdtsvlyltraqqk  2185
          *         *         *  *                                       *

PIV  prqykepeeqlled..ynqhdefdid                                              2233
SV   arq..neyttviddgslgdiepydss                                             2228
MeV  ewvfkvtvketkewyklvgysalikd                                             2183
NDV  fymktignavkgyysncds                                                    2204
```

FIGURE 3. (continued)

Comparison of the sequence alignments of all eight P proteins yielded no more relationships than were found in comparison to computer-generated random sequences. When the two size groups were aligned separately, some matches were obtained. However, unlike the alignments of the NP and L proteins, there were not significant clusters of amino acids, although scattered individual matches (either identical or conservative amino acid replacements) were seen more frequently in the carboxyl-terminal region of the protein.

Pairwise comparison of the aligned sequences reveals some interesting relationships (Table V). For example, PIV3 and Sendai virus exhibit greater than 50% sequence identity of the NP and L proteins (Tables II and IV), but only 19% identity of the P protein sequences. In contrast, measles virus and CDV, which have greater than 60% identity between their NP sequences, also have 43% identity of their P protein sequences. The greatest percentage of conservation of P sequences was between mumps virus and SV5 (71%). It will be interesting to learn the degree of sequence identity between the NP and L proteins of these two viruses.

The precise functions of the P genes and their various protein products are not yet known. The lack of significant sequence identity among the various P

TABLE V. Conserved Amino Acids
within Paramyxovirus P Proteins[a]

Group 1				
	PIV3	SV	MeV	CDV
PIV3 (602)	—	27	19	21
SV (568)	19	—	15	14
MeV (507)	9	7	—	52
CDV (507)	8	6	43	—
Group 2				
	NDV	SV5	MuV	RS
NDV (395)	—	26	31	31
SV5 (392)	15	—	71	18
MuV (391)	18	33	—	17
RS (241)	12	6	5	—

[a]The published P protein sequences used to compile this table are from: PIV3 (Spriggs and Collins, 1986); SV (Giorgi et al., 1983); MeV (Bellini et al., 1985); CDV (Barrett et al., 1985); NDV (McGinnes et al., 1988); MuV (Takeuchi et al., 1988); SV5 (Thomas et al., 1988); and RS virus (Satake et al., 1984). The alignments were scored pairwise and the number of conserved amino acids in each pair, which includes both identical amino acids and conservative replacements (figures above the diagonal), and the number of identical amino acids (figures below the diagonal), are expressed as a percentage of the aligned sequences. The P proteins are arranged into two groups based on size, as discussed in the text. The number of amino acids in each protein is given in parentheses.

TABLE VI. Average Percentages of Amino Acid Groups in P Proteins[a]

Virus	Acidic (Asp + Glu)	Basic (Arg + Lys + His)	Hydroxyl (Ser + Thr)
PIV3	16.2	15.8	23.2
SV	16.2	15.4	18.2
MeV	14.4	12.6	15.2
CDV	15.8	13.6	16.0
NDV	10.6	11.7	18.7
MuV	10.5	10.7	16.1
SV5	11.0	11.5	18.1
RS	21.1	13.3	17.4
Average protein[b]	11.6	13.5	13.1
VSV[c]	21.5	11.4	18.1

[a]The published P protein sequences used in Table V were used to derive this table.
[b]These figures represent the percent composition of amino acid groups in a pool of 314 protein sequences (Dayhoff et al., 1978).
[c]From Gill and Banerjee (1985).

proteins has led to the suggestion that their functions "apparently do not involve active sites and/or structural features of conserved sequences" (Spriggs and Collins, 1986). In this respect, the P proteins are like the phosphoproteins of the rhabdoviruses (Gill and Banerjee, 1985). Most of the P proteins are acidic. In addition, all these proteins have a greater than average serine plus threonine content (Table VI). These amino acids provide numerous potential sites for posttranslational phosphorylation, which can increase the overall negative charge of the proteins. In Sendai virus most of those phosphorylation sites have been mapped to the amino-terminal half of the P protein (Hsu and Kingsbury, 1982).

Data suggest that the carboxyl-terminal portion of the P protein, which has greater sequence conservation and is often more hydrophobic than the amino terminus, is required for binding of P protein to nucleocapsids. Proteolytic digestion of Sendai virus nucleocapsids produced a fragment of the P protein (M_r 40,000) that remained attached to nucleocapsids, and these treated nucleocapsids retained most of their RNA synthesizing capability (Chinchar and Portner, 1981a). Deshpande and Portner (1985) mapped monoclonal antibody-binding sites to this protease-resistant region; more recently, these binding sites have been mapped within the last 224 amino acids of the carboxyl terminus (K. W. Ryan and A. Portner, unpublished observations). This same region of the P protein is required for binding to nucleocapsids *in vitro* (Ryan and Kingsbury, 1988). As in the case of the NP protein, it seems likely that the carboxyl terminus of the P protein is folded in such a way that it can interact with nucleocapsids and with the external environment.

III. CONCLUSIONS

Many of the paramyxovirus genomes should be completely sequenced within the next few years. What will this wealth of data reveal? There is already sufficient evidence for conservation of regions of sequence identity among the nucleocapsid-associated proteins to confirm that paramyxoviruses and morbilliviruses are related, that pneumoviruses is less related, and that all these viruses, along with the *Rhabdoviridae*, have evolved from an ancestral negative-stranded prototype. The regions of conserved sequences should provide targets for studies of *in vitro* deletions and site-specific mutations to determine if alterations in these areas affect enzymatic activity or protein–protein or protein–RNA interactions. The next major advance will probably occur when the three-dimensional structures of these proteins are determined.

IV. REFERENCES

Barrett, T., Shrimpton, S. B., and Russell, S. E. H., 1985, Nucleotide sequence of the entire protein coding region of canine distemper virus polymerase-associated (P) protein mRNA, *Virus Res.* **3**:367–372.

Bellini, W. J., Englund, G., Rozenblatt, S., Arnheiter, H., and Richardson, C. D., 1985, Measles virus P gene codes for two proteins, *J. Virol.* **53**:908–919.

Blumberg, B. M., Giorgi, C., Rose, K., and Kolakofsky, D., 1984, Preparation and analysis of the nucleocapsid proteins of vesicular stomatitis virus and Sendai virus, and analysis of the Sendai virus leader–*NP* gene region, *J. Gen. Virol.* **65**:769–779.

Blumberg, B. M., Crowley, J. C., Silverman, J. I., Menonna, J., Cook, S. D., and Dowling, P. C., 1988, Measles virus *L* gene sequence evidences elements of ancestral RNA polymerase, *Virology* **164**:487–497.

Buckland, R., Giraudon, P., and Wild, F., 1989, Expression of measles virus nucleoprotein in *Escherichia coli:* Use of deletion mutants to locate the antigenic sites, *J. Gen. Virol.* **70**:435–441.

Chinchar, V. G. and Portner, A., 1981a, Functions of Sendai virus nucleocapsid polypeptides: Enzymatic activities in nucleocapsids following cleavage of polypeptide P by *Staphylococcus aureus* protease V8, *Virology* **109**:59–71.

Chinchar, V. G. and Portner, A., 1981b, Inhibition of RNA synthesis following proteolytic cleavage of Newcastle disease virus P protein, *Virology* **115**:192–202.

Choppin, P. W., and Compans, R. W., 1975, Reproduction of paramyxoviruses, *in* "Comprehensive Virology," Vol. 4 (H. Fraenkel-Conrat and R. R. Wagner, eds.), pp. 95–178, Plenum Press, New York.

Collins, P. L., Anderson, K., Langer, S. J., and Wertz, G. W., 1985, Correct sequence for the major nucleocapsid protein mRNA of respiratory syncytial virus, *Virology* **146**:69–77.

Dayhoff, M. O., Hunt, L. T., and Hurst-Calderone, S., 1978, Composition of proteins, *in* "Atlas of Protein Sequence and Structure," Vol. 5, Suppl. 3 (M. O. Dayhoff, ed.), pp. 363–373, National Biomedical Research Foundation, Silver Spring, Maryland.

Deshpande, K. L., and Portner, A., 1984, Structural and functional analysis of Sendai virus nucleocapsid protein NP with monoclonal antibodies, *Virology* **139**:32–42.

Deshpande, K. L., and Portner, A., 1985, Monoclonal antibodies to the P protein of Sendai virus define its structure and role in transcription, *Virology* **140**:125–134.

Elango, N., 1989, The mumps virus nucleocapsid mRNA sequence and homology among the *Paramyxoviridae* proteins, *Virus Res.* **12**:77–86.

Finch, J. T., and Gibbs, A. J., 1970, Observations on the structure of the nucleocapsids of some paramyxoviruses, *J. Gen. Virol.* **6**:141–150.

Galinski, M. S., Mink, M. A., Lambert, D. M., Wechsler, S. L., and Pons, M. W., 1986a, Molecular cloning and sequence analysis of the human parainfluenza 3 virus RNA encoding the nucleocapsid protein, *Virology* **149**:139–151.

Galinski, M. S., Mink, M. A., Lambert, D. M., Wechsler, S. L., and Pons, M. W., 1986b, Molecular cloning and sequence analysis of the human parainfluenza 3 virus mRNA encoding the P and C proteins, *Virology* **155**:46–60.

Galinski, M. S., Mink, M. A., and Pons, M. W., 1988, Molecular cloning and sequence analysis of the human parainfluenza 3 virus gene encoding the L protein, *Virology* **165**:499–510.

Gill, D. S., and Banerjee, A. K., 1985, Vesicular stomatitis virus NS proteins: Structural similarity without extensive sequence homology, *J. Virol.* **55**:60–66.

Gill, D. S., Takai, S., Portner, A., and Kingsbury, D. W., 1988, Mapping of antigenic domains of Sendai virus nucleocapsid protein expressed in *Escherichia coli, J. Virol.* **62**:4805–4808.

Giorgi, C., Blumberg, B. M., and Kolakofsky, D., 1983, Sendai virus contains overlapping genes expressed from a single mRNA, *Cell* **35**:829–863.

Gotoh, H., Shioda, T., Sakai, Y., Mizumoto, K., and Shibuta, H., 1989, Rescue of Sendai virus from viral ribonucleoprotein-transfected cells by infection with recombinant vaccinia viruses carrying Sendai virus L and *P/C* genes, *Virology* **171**:434–443.

Hamaguchi, M., Yoshida, T., Nishikawa, K., Naruse, H., and Nagai, Y., 1983, Transcriptive complex of Newcastle disease virus. I. Both L and P proteins are required to constitute an active complex, *Virology* **128**:105–117.

Heggeness, M. H., Scheid, A., and Choppin, P. W., 1980, Conformation of the helical nucleocapsids of paramyxoviruses and vesicular stomatitis virus: Reversible coiling and uncoiling induced by changes in salt concentration, *Proc. Natl. Acad. Sci. USA* **77**:2631–2635.

Heggeness, M. H., Scheid, A., and Choppin, P. W., 1981, The relationship of conformational changes in the Sendai virus nucleocapsid to proteolytic cleavage of the NP polypeptide, *Virology* **114**:555–562.

Hosaka, Y., and Hosoi, J., 1983, Study of negatively stained images of Sendai virus nucleocapsids using minimum-dose system, *J. Ultrastruct. Res.* **84**:140–150.

Hsu, C.-H., and Kingsbury, D. W., 1982, Topography of phosphate residues in Sendai virus proteins, *Virology* **120**:225–234.

Huang, A. S., Baltimore, D., and Bratt, M. A., 1971, Ribonucleic acid polymerase in virions of Newcastle disease virus: Comparison with the vesicular stomatitis virus polymerase, *J. Virol.* **7**:389–394.

Ishida, N., Taira, H., Omata, T., Mizumoto, K., Hattori, S., Iwasaki, K., and Kawakita, M., 1986, Sequence of 2,617 nucleotides from the 3' end of Newcastle disease virus genome RNA and the predicted amino acid sequence of viral NP protein, *Nucleic Acids Res.* **14**:6551–6564.

Jambou, R. C., Elango, N., Venkatesan, S., and Collins, P. L., 1986, Complete sequence of the major nucleocapsid protein gene of human parainfluenza type 3 virus: Comparison with other negative strand viruses, *J. Gen. Virol.* **67**:2543–2584.

Jimez-Montano, M., and Zamora-Cortina, L., 1981, Evolutionary model for the generation of amino acid sequences and its application to the study of mammal alpha-hemaglobin, *in* "Proceedings VII International Biophysics Congress" (Mexico City).

Kingsbury, D. W., 1977, Paramyxoviruses, *in* "The Molecular Biology of Animal Viruses," Vol. 1 (D. P. Nayak, ed.), pp. 349–382, Marcel Dekker, New York.

Lamb, R. A., Mahy, B. W. J., and Choppin, P. W., 1976, The synthesis of Sendai virus polypeptides in infected cells, *Virology* **69**:116–131.

Luk, D., Sanchez, A., and Banerjee, A. K., 1986, Messenger RNA encoding the phosphoprotein (P) gene of human parainfluenza virus 3 is bicistronic, *Virology* **153**:318–325.

McGinnes, L., McQuain, C., and Morrison, T., 1988, The P protein and the nonstructural 38K and 29K proteins of Newcastle disease virus are derived from the same open reading frame, *Virology* **164**:256–264.

Mahy, B. W. J., and Barry, R. N., 1975, Preface, *in* "Negative Strand Viruses," Vol. 1 (B. W. J. Mahy and R. N. Barry, eds.), p. xi, Academic Press, London.

Morgan, E. M., and Rakestraw, K. M., 1986, Sequence of the Sendai virus L gene: Open reading frames upstream of the main coding region suggest that the gene may be polycistronic, *Virology* **154**:31–40.

Morgan, E. M., Re, G. G., and Kingsbury, D. W., 1984, Complete sequence of the Sendai virus NP gene from a cloned insert, *Virology* **135**:279–287.

Murti, K., Portner, A., Troughton, K., and Deshpande, K., 1985, Localization of proteins on viral nucleocapsids using immunoelectron microscopy, *J. Electron Microsc. Tech.* **2**:139–146.

Neubert, W. J., and Koch, E. M., 1985, Molecular clones representing Sendai virus L gene: Sequence of 1,179 nucleotides from the 3' end of the L gene, *Zentrabl, Bakteriol. Microbiol. Hyg. [B]* **260**:498–499.

Peeples, M. E., and Bratt, M. A., 1982, UV irradiation analysis of complementation between, and replication of, RNA-negative temperature-sensitive mutants of Newcastle disease virus, *J. Virol.* **41**:965–973.

Portner, A., and Murti, K. G., 1986, Localization of P, NP and M proteins on Sendai virus nucleocapsids using immunogold labeling, *Virology* **150**:469–478.

Portner, A., Murti, K. G., Morgan, E. M., and Kingsbury, D. W., 1988, Antibodies against Sendai virus L protein: Distribution of the protein in nucleocapsids revealed by immunoelectron microscopy, *Virology* **163**:236–239.

Rozenblatt, S., Eizenberg, O., Ben-Levy, R., Lavie, V., and Bellini, W. J., 1985, Sequence homology within the morbilliviruses, *J. Virol.* **53**:684–690.

Ryan, K. W., and Kingsbury, D. W., 1988, Carboxyl-terminal region of Sendai virus P protein is required for binding to viral nucleocapsids, *Virology* **167**:106–112.

Sanchez, A., Banerjee, A. K., Furuichi, Y., and Richardson, M. A., 1986, Conserved structures among the nucleocapsid proteins of the *Paramyxoviridae*: Complete nucleotide sequence of human parainfluenza virus type 3 NP mRNA, *Virology* **152**:171–180.

Satake, M., Elango, N., and Venkatesan, S., 1984, Sequence analysis of the respiratory syncytial virus phosphoprotein gene, *J. Virol.* **52**:991–994.

Sato, H., Oh-hira, M., Ishida, N., Imamura, Y., Hattori, S., and Kawakita, M., 1987, Molecular cloning and nucleotide sequence of P, M and F genes of Newcastle disease virus avirulent strain D26, *Virus Res.* **7**:241–255.

Schubert, M., Harmison, G. G., and Meier, E., 1984, Primary structure of the vesicular stomatitis virus polymerase *(L)* gene: Evidence for a high frequency of mutations, *J. Virol.* **51**:505–514.

Shioda, T., Hidaka, Y., Kanda, T., Shibuta, H., Nomoto, A., and Iwasaki, K., 1983, Sequence of 3,687 nucleotides from the 3' end of Sendai virus genome RNA and the predicted amino acid sequences of viral NP, P and C proteins, *Nucleic Acids Res.* **11:**7317–7330.

Shioda, T., Iwasaki, K., and Shibuta, H., 1986, Determination of the complete nucleotide sequence of the Sendai virus genome RNA and the predicted amino acid sequence of the F, HN and L proteins, *Nucleic Acids Res.* **14:**1545–1563.

Sobel, E., and Martinez, M. H., 1985, A multiple sequence alignment program, *Nucleic Acids Res.* **14:**363–374.

Spriggs, M. K., and Collins, P. L., 1986, Sequence analysis of the P and C protein genes of human parainfluenza virus type 3: Patterns of amino acid sequence homology among paramyxovirus proteins, *J. Gen. Virol.* **67:**2705–2719.

Takeuchi, K., Hishiyama, M., Yamada, A., and Sugiura, A., 1988, Molecular cloning and sequence analysis of the mumps virus gene encoding the P protein: Mumps virus *P* gene is mono-cistronic, *J. Gen. Virol.* **69:**2043–2049.

Thomas, S. M., Lamb, R. A., and Paterson, R. G., 1988, Two mRNAs that differ by two nontem-plated nucleotides encode the amino coterminal proteins P and V of the paramyxovirus SV5, *Cell* **54:**891–902.

Yusoff, K., Millar, N. S., Chambers, P., and Emmerson, P. T., 1987, Nucleotide sequence analysis of the *L* gene of Newcastle disease virus: Homologies with Sendai and vesicular stomatitis viruses, *Nucleic Acids Res.* **15:**3961–3976.

CHAPTER 6

The Nonstructural Proteins of Paramyxoviruses

Robert A. Lamb and Reay G. Paterson

I. INTRODUCTION

This chapter describes properties of the paramyxovirus genes and their mRNAs which encode polypeptides known as nonstructural proteins. The designation nonstructural is used here in a loose sense to describe a virus-encoded polypeptide that is synthesized in virus-infected cells but is greatly underrepresented in or absent from purified virions. In virus-infected CV1 cells the nonstructural polypeptides B, C′, and C synthesized by Sendai virus and the nonstructural cysteine-rich polypeptide V synthesized by SV5 can be identified (Fig. 1). In addition to nonstructural polypeptide V, a small hydrophobic (SH) integral membrane protein is synthesized in SV5-infected cells, but it cannot be resolved on the gel shown in Fig. 1. All these polypeptides will be discussed in detail in the ensuing sections.

As the gene structure and mRNA structure of these nonstructural polypeptides of paramyxoviruses is discussed in other chapters of this volume, and as the proteins probably do not have common properties, this review is by necessity a little eclectic. Much of this chapter is written from a historical perspective, as the identification of each of the four known types of nonstructural protein encoded by paramyxoviruses was made in the authors' laboratories.

II. PARAMYXOVIRUS C PROTEINS

A. Identification in Infected Cells

The first unambiguous finding of the synthesis of nonstructural polypeptides in paramyxovirus-infected cells was made with the identification of the

ROBERT A. LAMB and REAY G. PATERSON • Department of Biochemistry, Molecular Biology, and Cell Biology, Northwestern University, Evanston, Illinois 60208–3500.

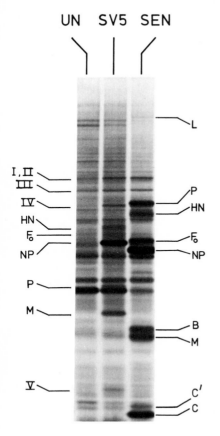

FIGURE 1. Polypeptides synthesized in Sendai virus- and SV5-infected cells. CV1 cells were infected with 10 pfu/cell of Sendai virus or SV5. At 14 hr after infection, cells were labeled for 30 min with 30 μCi/ml [35S]-methionine and the polypeptides were subjected to electrophoresis on a 13% polyacrylamide gel as described (Lamb et al., 1976). UN, Uninfected control cells; SV5, SV5-infected cell lysate; SEN, Sendai virus-infected cell lysate. L, Large transcriptase-associated polypeptide; P, polymerase/phosphoprotein/transcriptase polypeptide; HN, hemagglutinin-neuraminidase; F_O, precusor uncleaved fusion polypeptide; NP, nucleocapsid polypeptide; B, Sendai virus nonstructural polypeptide B (phosphorylated form of the matrix, M, protein); M, matrix or membrane protein; C' and C, nonstructural polypeptides derived from the Sendai virus P gene; V, SV5 cysteine-rich polypeptide derived from the SV5 P gene; I, II, III, IV, cellular polypeptides, originally termed glucose regulatory polypeptides (GRP), whose synthesis is induced by paramyxovirus infection (Peluso et al., 1979). Polypeptide IV is known to be identical to GRP78/BiP, a resident component of the endoplasmic reticulum, and polypeptides I and II are known to be the glycosylated and unglycosylated forms, respectively, of the resident endoplasmic reticulum polypeptide known as GRP96.

C polypeptide ($M_r \approx 22,000$) in Sendai virus-infected cells (Lamb et al., 1976). Polypeptide C was shown to be a discrete virus-encoded polypeptide by demonstrating that it contained unique peptide fragments in comparison with the other viral polypeptides on partial proteolysis and had a distinct tryptic peptide map (Lamb and Choppin, 1978). Further evidence that C is encoded by the Sendai virus genome, and is not a cellular polypeptide whose synthesis is induced by viral infection, was provided by finding differences in the electrophoretic mobility on polyacrylamide gels of the C polypeptide from different strains of Sendai virus (Etkind et al., 1980). Sendai virus C polypeptide was found to be synthesized at the same time after infection as the other viral polypeptides and to accumulate in a stable manner in infected cells. However, on disruption of the cells, polypeptide C is rapidly degraded unless protease inhibitors are present (Lamb and Choppin, 1977a). The C polypeptide is not phosphorylated under conditions where phosphorylation of P, NP, and M can readily be detected (Lamb and Choppin, 1977b). When Sendai virus poly(A)-containing RNAs were translated in vitro in a wheat germ cell-free system, in addition to C, a second polypeptide of slightly slower electrophoretic mobility than C was readily detected and was designated C' (Etkind et al., 1980). The two polypeptides, C and C', have essentially the same tryptic peptide maps and thus it was clear that C and C' were derived from the same gene (Etkind et al., 1980). However, it was not determined whether C and C' differed by amino acid sequence or by a posttranslational modification. On retrospective analysis, it was realized that C' was also synthesized in infected cells, but in amounts significantly less than those observed *in vitro* (Etkind et al., 1980).

B. The P and C Proteins Are Encoded in Overlapping Reading Frames

The size of the mRNA encoding C was originally examined by fractionating mRNAs isolated from Sendai virus-infected cells on sucrose density gradients and translating the mRNA from individual fractions in an *in vitro* system. It was found that the C mRNA was much larger than expected, as it cofractionated with that for P ($M_r \approx 79,000$) and NP ($M_r \approx 60,000$) (Dethlefsen and Kolakofsky, 1983). When cDNA clones of the Sendai virus mRNAs became available, it became possible to explore further the nature of the C mRNA. A P-specific cDNA clone was used in mRNA hybrid-selection experiments to purify specific mRNAs which on subsequent translation *in vitro* yielded both P and C (+C') (Dowling et al., 1983). The nucleotide sequence of the P + C cDNA indicated that there are two overlapping open reading frames: a large one of 568 amino acids that encodes P (calculated M_r 62,011) and a second open reading frame of 204 amino acids that has its AUG initiation codon separated from the first AUG initiation codon by 7 nucleotides (Giorgi et al., 1983; Shioda et al., 1983) (see Fig. 2). The calculated size for the protein derived from the second open reading frame (M_r 24,015) is compatible with the observed size of C (or C'). However, the calculated size of P (M_r 62,011) is much smaller than the observed size of P (M_r 79,000) and they probably differ because of the extensive phosphorylation of P in infected cells, which decreases its ability to bind SDS and thus alters its mobility on polyacrylamide gels. To show more directly that the second open reading fame encodes C (+C'), antisera were generated to a synthetic peptide predicted to be encoded in the C open reading frame and it was found that these antisera would immunoprecipitate the C (+C') polypeptide (Portner et al., 1986; Curran et al., 1986). Thus, the accumu-

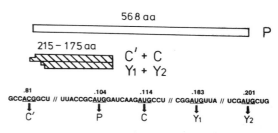

FIGURE 2. Representation of the open reading frames of the Sendai virus *P* gene that are used to synthesize the P, C', C, Y1, and Y2 Polypeptides, The P open reading frame is shown as an open box and the open reading frames that encode C', C, Y1, and Y2 are shown as cross-hatched boxes. The nucleotide sequences surrounding the initiation codons for the polypeptides are shown. Data derived from Shioda et al. (1983), Giorgi et al. (1983), Curran and Kolakofsky (1988a), and Gupta and Patwardhan (1988).

lated data indicated that the Sendai virus P and C (+C') polypeptides are
translated from overlapping reading frames on a functionally bicistronic
mRNA. This is an analogous situation to the coding strategy elucidated for
influenza B virus, where it was found that the glycoproteins NB and NA are
translated in overlapping reading frames from a bicistronic mRNA with the
NB and NA AUG initiation codons separated by 4 nucleotides (M. W. Shaw *et
al.*, 1982, 1983; Williams and Lamb, 1989).

C. Multiple Initiation Codons on One mRNA

Although the accumulated data showed that the two open reading frames
encode P- and C-related proteins, the difference between C and C' was only
elucidated recently. A precursor–product relationship could not be demonstrated between C and C' either in pulse-chase-type analysis or in experiments
in which the amino acid analogue canavanine was incorporated into the C and
C' proteins in the expectation of making the proteins nonfunctional and un-
cleavable by cellular proteases (Lamb *et al.*, 1976, Curran *et al.*, 1986). In
addition, although the ratio of C' to C synthesized *in vitro* varied between
experiments (Curran *et al.*, 1986), the kinetics of accumulation of C and C'
suggested that independent initiation events were occurring rather than the
polypeptides having a precursor–product relationship. One possibility that was
considered was that the synthesis of C' initiated at the second AUG codon on
the mRNA at nucleotides 114–117 and that the smaller C protein is the result
of translation initiating at the third AUG codon (nucleotides 183–185) on the
mRNA in the same reading frame as C'. Indirect data to support the concept
that this third internal initiation codon is used for the synthesis of either C or
C' was provided (Gupta and Kingsbury, 1985; Curran *et al.*, 1986). However,
more recently, using a combination of site-specific mutagenesis of P + C
cDNAs, transcription of synthetic mRNA from the altered cDNA templates,
and *in vitro* translation of the RNAs, data have been obtained which indicate
that initiation of C' occurs at an ACG codon at nucleotides 81–83, that the
initiation codon for P is at nucleotides 104–107, and the initiation codon for C
is at nucleotides 114–117 (Curran and Kolakofsky, 1988a; Gupta and Patwardhan, 1988). Formal proof that C' initiates its synthesis using an ACG
codon was shown by N-terminal amino acid sequencing of the polypeptide
(cited in Curran and Kolakofsky, 1988a). Initiation of protein synthesis using
non-AUG codons is uncommon, but not unprecedented, and has, for example,
been found to occur with adeno-associated-virus (AAV) capsid protein (ACG)
(Becerra *et al.*, 1985), for c-*myc* (CUG) (Hann *et al.*, 1988), and basic fibroblast
growth factor (CUG) (Prats *et al.*, 1989). Recent studies have identified two
smaller polypeptides (Y and Y') that are carboxy-coterminal with C both in
infected cells and when *in vitro* translation systems are programmed with
Sendai virus poly(A)-containing mRNAs, and these probably arise by internal
initiation within the C open reading frame (Curran and Kolakofsky, 1988a,
Gupta and Patwardhan, 1988) (Fig. 2). It has also been recently reported that a
polypeptide X which represents approximately the last 95 residues of P can be
detected at about 0.5 times the abundance level of P in Sendai virus-infected
cells (Curran and Kolakofsky, 1987, 1988b). Protein X does not seem to be

derived from P by proteolytic cleavage and data have been obtained which have been interpreted to suggest that it is synthesized from the P mRNA by a scanning-independent but 5′ cap-dependent initiation process (see below) (Curran and Kolakofsky, 1988b). However, the interpretation of these experiments has recently been challenged (Kozak, 1989).

D. Initiation Codon Consensus Sequences and the Scanning Hypothesis

It has been determined from surveys of the nucleotide sequences of eukaryotic cellular and viral mRNAs that protein synthesis usually initiates at the AUG codon nearest the 5′ end of the mRNA [reviewed in Kozak (1987)]. However, approximately 10% of mRNAs do not fit this pattern, as protein synthesis begins in these mRNAs at initiation codons that are downstream of the first AUG codon (Kozak, 1987). The consensus nucleotide sequence containing the AUG codon for the initiation of protein synthesis is 5′-CC^A_GCC-AUGG-3′ (Kozak, 1986a), although less than 5% of natural eukaryotic mRNAs have the ideal sequence. Nonetheless, these sequences are thought to be an important determinant for translation initiation (Kozak, 1981a,b). The cumulative data derived from experiments involving site-specific mutagenesis of cloned cDNAs and expression of the mutants in eukaryotic cells indicate that a large variability in protein production occurs, depending on the nucleotide sequence context of the AUG initiation codon (Kozak, 1983, 1984, 1986b; Lomedico and McAndrew, 1982; Johansen et al., 1984). Both the A or G residue at position −3 (the A of the initiation codon is designated as position +1) and the G at position +4 have been found to be the nucleotides of most importance in determining initiation, but their contributions are not simply additive. For example, when an A is at −3, then the presence of a G versus a U in the +4 position made less than a twofold difference, but the positive effect of a G at +4 is more significant when an A residue is not at position −3 (Kozak, 1986b).

A scanning model has been introduced to describe the mechanism by which eukaryotic mRNAs initiate translation (Kozak, 1981b). Ribosomes and initiation factors are proposed to bind at or near the 5′ end of the mRNA in a process facilitated by the 5′ cap structure and then migrate in a linear manner on the mRNA, "scanning" until reaching the first AUG initiation codon. Two modifications were later made to the model because several eukaryotic viral mRNAs have been reported to be polycistronic (Kozak, 1984, 1986a). It is now thought likely that ribosomes can terminate translation of an upstream open reading frame, resume "scanning," and then initiate translation at the AUG codon of the downstream reading frame (Liu et al., 1984; Kozak, 1986b). In addition, some viral mRNAs are functionally bicistronic, with protein synthesis occurring at two different AUG codons that are not separated by a termination codon, either using the same open reading frame or two different overlapping reading frames [reviewed in Kozak (1986a)]. In these mRNAs, the 5′-proximal initiation codon is found to be in a suboptimal nucleotide sequence context compared to the consensus sequence. It is presumed that initiation occurs at the second AUG in addition to the first AUG because some ribosome preinitiation complexes bypass the 5′-proximal AUG ("leaky scanning") and initiate protein synthesis at the second AUG codon (reviewed in Kozak, 1986a, 1986c, 1989).

The Sendai virus P/C mRNA initiation codons for C' at nucleotides 81–83, for P at nucleotides 104–106, and for C at nucleotides 114–116 are found in increasingly favorable contexts for ribosomal initiation by a ribosome "scanning" from the 5' end of the mRNA. The first ACG codon functions presumably because it lies in an excellent context (GCCACGG), but works poorly because of the ACG in place of an AUG; the next AUG codon is in a poor context (CGGAUGG) and functions poorly; and the third AUG condon is in a stronger, but still not perfect context (AAGAUGC). Thus, it is possible that initiation of the Y proteins could occur by continued scanning of ribosomes past the third AUG codon until reaching the AUGs at positions 183–186 and 201–203, which are believed to be responsible for the initiation of the Y proteins (Curran and Kolakofsky, 1988a).

E. Subcellular Localization and Possible Function of Sendai Virus C Proteins

Although it has long been speculated that the Sendai virus C proteins would play a role in the replication of the virion RNA, no data have been forthcoming to suggest that this is the case. The C proteins are highly basic (Giorgi et al., 1983; Shioda et al., 1983), but do not colocalize with the P protein on nucleocapsids and are found evenly distributed in the cytoplasm, as determined by immunofluorescence experiments using antisera to the C protein C terminus (Portner et al., 1986). In addition, the anti-C-protein peptide sera did not inhibit viral RNA synthesis when added to an extract of infected cells. However, this result should not be interpreted to mean that the C proteins have no role in RNA replication, as this negative result could be due to a variety of factors, such as a low affinity of antibody for the antigen or shielding of the antibody-combining site by other proteins.

When a recombinant cDNA clone was engineered to express only the C protein using an SV40-based vector, it was observed that in cells expressing the C mRNA there was a high level of interferon induction and an extensive cytopathic effect (Taira et al., 1987). Although no direct evidence could be obtained to show that it was the synthesis of the C proteins rather than a feature of the mRNA that caused the interferon induction, the introduction of a translational stop codon in the C-protein coding region abolished the effect (Taira et al., 1987). Thus, these data indicate that the C proteins are largely responsible for the induction of interferon production during Sendai virus infection.

In speculating on the function of the C proteins in infected cells, it has to be borne in mind that if all paramyxoviruses use the same basic enzymatic functions for the replication of their RNA, then the C proteins do not provide a common activity, because SV5, NDV, and mumps virus do not have C proteins expressed from the 5' end of the P gene. It is not known whether another viral protein compensates for the function of C in these cases, or whether the C proteins have a "luxury function" involved in a specialized role in the infection of an animal.

F. When Is a Nonstructural Protein a Structural Protein?

Based on the amounts of C and C' expressed in Sendai virus-infected cells, there is no doubt that C and C' are greatly underrepresented in virions compared to the structural polypeptides P, HN, F_0, NP, and M. Thus, either the C proteins fail to interact with the other virion components, or if they do, there must be a mechanism by which the C proteins are excluded from the assembly process. When the polypeptides of purified preparations of [^{35}S]-methionine-labeled virions were analyzed on gels and autoradiographs deliberately overexposed, the C polypeptide could be detected in virions (Fig. 3). The amount of C has not been determined on a molar basis and it is not known if the C protein

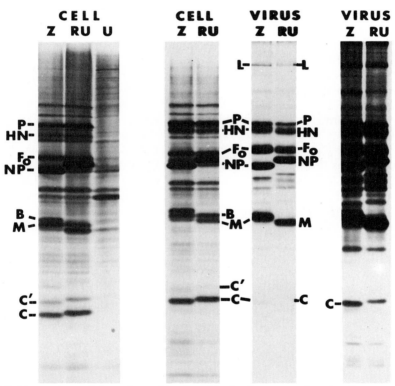

FIGURE 3. Comparison of the polypeptides found in Sendai virus-infected cells and in purified virions. To examine the polypeptides synthesized in Sendai virus-infected cells, primary chick embyro fibroblasts were infected and labeled with either [^{3}H]-leucine or [^{35}S]-methionine at 14 hr after infection for 1 hr. To examine the polypeptides found in virions, Sendai virus-infected chick embyro fibroblasts were labeled with 25 μCi/ml [^{35}S]-methionine from 14 to 24 hr after infection in Dulbecco's modified Eagle's medium containing one-tenth the normal concentration of methionine. Virions were purified on two sucrose density gradients as described (Lamb et al., 1976). Left three lanes: Polypeptides labeled with [^{3}H]-leucine. Middle four lanes: polypeptides labeled with [^{35}S]-methionine. Right two lanes: overexposure of autoradiograph shown in middle panel for polypeptides in purified virions to emphasize the presence of the C polypeptide. Cell: polypeptides synthesized in infected cells; virus: polypeptides of purified virions; Z and RU: Z and Rockefeller University strains of Sendai virus, respectively; U: uninfected cells.

observed is due to contamination of the virion preparation with cellular vesicles that contain C protein or whether the protein is a genuine component of the virion particles. This could be elucidated by examining virions using immune electron microscopy with an appropriate C antibody, but unless a function of the virion could be inhibited by the antibody, it would still not be known if the association of C with virions was necessary or adventitious.

G. Identification of C Proteins of Parainfluenza Virus 3, Measles Virus, and CDV

A nonstructural polypeptide designated C of $M_r \approx 22,000$ has been identified in cells infected with human parainfluenza virus type 3 (PI-3) (Sanchez and Banerjee, 1985a; Spriggs and Collins, 1986). In hybrid-arrest translation experiments cDNA fragments derived from the P gene inhibited the translation in vitro of both the P and C polypeptides (Sanchez and Banerjee, 1985b), making it likely that the genomic organization for P and C of PI-3 would be analogous to that of Sendai virus. The nucleotide sequence of the PI-3 P gene was obtained (Luk et al., 1986; Galinsky et al., 1986; Spriggs and Collins, 1986) and it was found that the two proteins are encoded by overlapping reading frames. The PI-3 P-protein open reading frame contains 602 amino acids and the C-protein open reading frame consists of 199 amino acids. The two AUG initiation codons for P and C are separated by 7 nucleotides and neither is in a strong context. With PI-3, no evidence has been obtained for an equivalent of the Sendai virus C' protein. A comparison of the C protein sequences from PI-3 and Sendai virus has shown quite extensive regions of amino acid identity and further regions of amino acid similarity (Luk et al., 1986; Galinsky et al., 1986; Spriggs and Collins, 1986).

The measles virus P gene also encodes two proteins, P (507 amino acids) and C (186 amino acids), using overlapping reading frames (Bellini et al., 1985). The two AUG initiation codons for P and C are separated by 19 nucleotides, with neither being in a strong context (Fig. 4). The unequivocal assignment of the presumptive C protein identified in measles virus-infected cells to the C open reading frame was made using antisera raised to peptides predicted from the nucleotide sequence (Bellini et al., 1985). The measles virus C-specific peptide sera were also used to investigate the subcellular localization of C, and it was found that the C protein colocalizes with the nucleocapsids in the cytoplasm and the nucleus (Bellini et al., 1985). It was proposed that the mea-

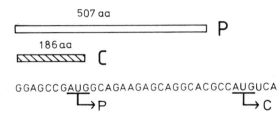

FIGURE 4. Representation of the open reading frames of the measles virus P gene. The P polypeptide open reading frame is shown as an open box and the C polypeptide is shown as a cross-hatched box. The nucleotide sequences surrounding the initiation codons are shown. Data derived from Bellini et al. (1985).

sles virus C protein might play a role in nucleocapsid assembly or in RNA synthesis. In recent experiments in which the subcellular localization of the measles virus C protein expressed by an adenovirus-based eukaryotic vector was investigated, it was found that the C protein exhibited a diffuse cytoplasmic immunofluorescent staining pattern in both measles virus-infected and vector-infected cells (Alkhatib *et al.*, 1988). The staining pattern observed was more like that reported for the Sendai virus C protein (Portner *et al.*, 1986) and the reason for the differences between the data for the subcellular localization of measles virus C protein will need further investigation.

The CDV *P* gene also contains two overlapping reading frames for P (507 amino acids) and C (174 amino acids), with 19 nucleotides separating the two AUG initiation codons (Barrett *et al.*, 1985). Although the existence of a CDV nonstructural polypeptide had been previously reported (Hall *et al.*, 1980), the identification of the CDV C polypepide was definitively established when it was found that it was cross-reactive with the measles virus C peptide sera (Richardson *et al.*, 1985). A comparison of the CDV C protein amino acid sequence with the measles virus C protein sequence shows extensive regions of amino acid identity and further regions of amino acid similarity (Barrett *et al.*, 1985).

III. PARAMYXOVIRUS CYSTEINE-RICH PROTEINS

A. Identification of the Polypeptide and Its Gene in SV5

A study of the polypeptides synthesized in paramyxovirus SV5-infected cells detected a nonstructural protein, designated V ($M_r \approx 24,000$) (Peluso *et al.*, 1977). More recently, the SV5 "*P*" gene was shown to encode both the P protein ($M_r \approx 44,000$) and protein V by the arrest of translation *in vitro* of both P and V using a cDNA clone derived from SV5-specific mRNAs (Paterson *et al.*, 1984). Separation of the specific mRNAs on methylmercury–agarose gels and translation *in vitro* of the mRNAs in the gel fractions indicated that the mRNA for P and V cofractionated (Paterson *et al.*, 1984). The P and V proteins were found to have tryptic peptides in common, although no precursor–product relationship could be demonstrated (Paterson *et al.*, 1984; R. G. Paterson, unpublished data).

To investigate the coding strategy used to express the P and V proteins from the SV5 "*P*" gene, the nucleotide sequence of the "*P*" gene was determined from cDNA clones (Thomas *et al.*, 1988). The gene was shown to be 1298 nucleotides in length and to contain an untranslated region of 60 nucleotides preceding the first AUG initiation codon. Unexpectedly, it was found that there were two overlapping open reading frames (Fig. 5) capable of encoding proteins of 222 amino acids (reading frame 0) and 250 amino acids (+1 reading frame). Either of these reading frames is large enough to encode protein V ($M_r \approx 24,000$) (Paterson *et al.*, 1984), but neither is large enough to encode protein P ($M_r \approx 44,000$), assuming that the electrophoretic mobility of the P protein is not aberrant. Sequence analysis of two other cDNA clones confirmed the presence of two protein-coding regions. Exhaustive nucleotide se-

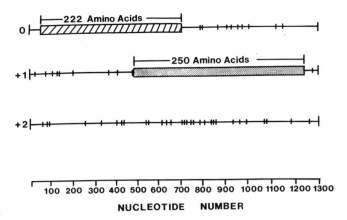

FIGURE 5. Representation of the three reading frames of the SV5 *P* gene determined from nucleotide sequence analysis of cDNA clone P203-1 and SV5 virion RNA. The overlapping open reading frames in the 0 and +1 reading frames are indicated by the hatched and stippled boxes, respectively. The 5′ and 3′ ends of the clone are defined by the large vertical bars, and the termination codons in all three reading frames are indicated by the small vertical bars. [From Thomas *et al.* (1988).]

quence analysis made it unlikely that a simple sequencing error had been made. Although the same error in more than one clone could have arisen due to reverse transcriptase errors during synthesis of the cDNAs, biological data to be discussed below eliminated this possibility.

B. Assignment of Coding Regions

Examination of the predicted amino acid sequences of the two reading frames indicates a region from residues 190 to 218 in the 0 reading frame that contains seven cysteine residues, whereas the +1 reading frame only contains one cysteine at residue 357. Conversely, the +1 reading frame contains six methionine residues, while the 0 reading frame has only one methionine in addition to the initiation methionine. Given the difference in the methionine and cysteine contents of the 0 and +1 reading frames, it was expected that an examination of the relative ease with which the P and V proteins could be selectively radiolabeled would give an indication of which reading frames encoded them. When SV5-infected cells were labeled with either [^{35}S]-methionine or [^{35}S]-cysteine, polypeptide V was detected with either radioisotopic precursor, whereas the P protein was poorly labeled with [^{35}S]-cysteine, although readily labeled with [^{35}S]-methionine (Fig. 6). These observations suggested that protein V possesses the cysteine-rich region encoded by the 0 reading frame, while the P protein apparently does not.

To further understand the coding strategy for P and V, use was made of P- and V-specific monoclonal antibodies which recognize three nonoverlapping

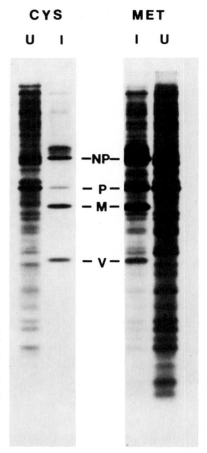

FIGURE 6. Comparison of the efficiency of labeling of SV5 cysteine-rich polypeptide V with radioactive cysteine or methionine. SV5-infected CV1 cells were labeled with either [^{35}S]-cysteine or [^{35}S]-methionine as described (Thomas *et al.*, 1988) and the polypeptides were analyzed by polyacrylamide gel electrophoresis. Cys: [^{35}S]-cysteine-labeled polypeptides; Met: [^{35}S]-methionine-labeled polypeptides; I: SV5-infected cells; U: uninfected cells.

antigenic sites (Randall *et al.*, 1987). Antibodies from all three groups were found to be capable of immunoprecipitating P from SV5-infected cell lysates, whereas protein V was only recognized by one group of antibodies (Thomas *et al.*, 1988). Recognition of both P and V polypeptides by a single group of monoclonal antibodies indicated that they have amino acid sequences in common.

Protein V was shown to be encoded by the 0 reading frame (see Fig. 5) by two lines of evidence.

1. The epitopes that the monoclonal antibodies recognized were mapped to the two open reading frames. Full-length and truncated synthetic RNA transcripts were made from the P + V cDNA using a T7 promoter and T7 RNA polymerase. The RNA transcripts were translated *in vitro* and the translation products were immunoprecipitated using the three groups of monoclonal antibodies. The group I antibody was found to recognize protein products encoded by the 5' end of the gene (the 0 reading frame), the largest of which was protein V. The group II antibodies were found to recognize a region from the N terminus of the +1 open reading frame, and the group III antibodies recognized a C-

terminal region of the +1 reading frame (Thomas *et al.*, 1988). Thus, these data indicate that protein V is the product of the 0 reading frame and that because both P and V are immunoprecipitated by the group I antibodies, and V cannot be immunoprecipitated by the group II and group III antibodies, P and V are N-coterminal. In addition, antigenic region mapping of the group II and group III antibodies indicated that protein P is derived from amino acid residues encoded by a large part of the +1 reading frame.

2. When the stop codon that terminates translation in the 0 reading frame was eliminated by site-specific mutagenesis, a larger form of protein V could be synthesized from a synthetic RNA transcribed from the mutated template DNA (Thomas *et al.*, 1988), indicating that the deleted stop codon is normally used to terminate translation of protein V.

C. Strategy by Which P and V Are Encoded

The possibility that the mechanism used to translate P was ribosomal frameshifting which would occur at a position before the protein V stop codon was considered (Thomas *et al.*, 1988). In other known cases of frameshifting, it has always been found to occur using *in vitro* translation systems (e.g., Jacks *et al.*, 1988). However, in the case of the P and V proteins it was found that although the P protein could be translated efficiently *in vitro* using poly(A)-containing mRNAs isolated from SV5-infected cells, it was not possible to detect the synthesis of the P protein when *in vitro* runoff transcripts of P+V cDNA were used to program the cell-free translation system. Therefore, no positive data to implicate ribosomal frameshifting as the mechanism for the generation of the SV5 P protein could be provided.

The next most plausible mechanism by which P and V could be encoded was that a second mRNA species existed that differed from the mRNA that had been cloned so that it could be translated to yield the P protein. Nuclease S_1 protection analysis using poly(A)-containing mRNA from SV5-infected cells and DNA fragments from the P + V cDNA provided data which strongly indicated that a second mRNA population exists in SV5-infected cells and a second species of P-specific cDNA clone was isolated (Thomas *et al.*, 1988). The nucleotide sequences of 22 P- and V-specific cDNA clones were obtained over the region of overlap between the 0 and +1 reading frames. Of the 22 clones sequenced, 12 had the same sequence as the clones originally sequenced and 10 differed from these clones by having two additional bases between nucleotides 548 and 551. The nucleotide sequences over the relevant regions of a P mRNA clone and of a V mRNA clone are shown in Figure 7. The sequence of the SV5 genomic template RNA (vRNA) is shown in the message sense for comparison (Fig. 7). It can be seen that whereas the V cDNA has four G residues between nucleotides 548 and 551 and the SV5 vRNA has four C residues (shown here as G residues in the mRNA sense), the P cDNA has six G residues. The two extra G residues cause a switch from the 0 reading frame to the +1 reading frame, and the predicted amino acid sequences are shown in Fig. 7B. Thus, the SV5 P mRNA contains two G residues that are not present in the genomic vRNA, and this mRNA species has the capacity to encode a

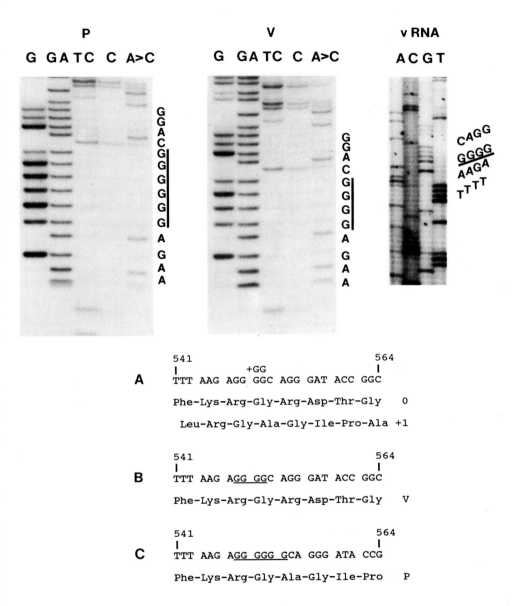

FIGURE 7. Sequence of a P mRNA cDNA clone in comparison to the sequence of a V mRNA cDNA clone and the sequence of the SV5 genomic RNA. Top: The nucleotide sequences of a P cDNA clone and a V cDNA clone in the region of nucleotides 541–564 are shown to illustrate the six G or four G residues in the P cDNA and V cDNA, respectively. The sequence of the SV5 genomic template RNA (vRNA) is shown in the message sense, as determined by dideoxy primer extension sequencing using reverse transcriptase. Bottom: The predicted amino acid sequences of the P and V proteins in the region of the six G or four G residues. [From Thomas et al. (1988).]

polypeptide of 392 amino acids (predicted $M_r \approx 44,000$), which is compatible with the observed size of P ($M_r \approx 44,000$) (Thomas et al., 1988).

D. Mechanism for the Addition of Extra Nucleotides to mRNAs

It has not been formally shown whether the addition of the extra G residues is a cotranscriptional or posttranscriptional process. However, polyadenylation of the mRNAs of negative-strand RNA viruses is thought to occur by a "slippage" or "stuttering" mechanism involving the reiterative copying of a stretch of U residues located at the end of each gene and this process is presumed to be mediated by the virus-encoded RNA polymerase. The extra G residues are added to the P transcript at a position in the template vRNA where there is a run of four C residues and it is therefore conceivable that the two G residues are added to the P mRNA by a mechanism identical or similar to that envisaged for polyadenylation. It is of interest to note that immediately upstream of the four C residues on the SV5 genomic RNA the sequence 3'-AAAAUUCU-5' is found. This sequence resembles the putative polyadenylation signal found at the end of all SV5 genes and is identical to the sequence at the end of the SV5 *HN* gene (Hiebert et al., 1985a). Thus, "stuttering" or "slippage" is an attractive model for the mechanism by which the extra G residues are added to the P mRNA and we postulated that this would be testable experimentally by analyzing the products of the in vitro RNA transcriptase reaction, and it is now clear that with Sendai virus such a process can occur with transcriptase reactions using purified virions (see below). Recently the term "RNA editing" has been used to describe a phenomenon observed in mitochondrial transcripts from trypanosomes which is characterized by the presence in the mature mRNA of uridine residues that are not encoded in the gene (Benne et al., 1986; Feagin et al., 1987, 1988; J. M. Shaw et al., 1988), and a similar phenomenon has been found in the mammalian apolipoprotein-B mRNA, where a U residue replaces a templated C (Powell et al., 1987; Chen et al., 1987). Therefore, as "RNA editing" implies a mechanism that is posttranscriptional, it is strictly inappropriate to describe the addition of extra nucleotides to a paramyxovirus mRNA by the term "editing" unless the elucidation of the mechanism indicates that this is the case.

Whatever the mechanism involved in the addition of the two extra G residues in the SV5 mRNA transcript, it occurs at a specific site and is limited as to the number of nucleotides added. Although the sample size may not be large enough to draw statistically valid conclusions, of 22 clones sequenced over the region of the extra nucleotides, only clones with either four or six G residues were found (Thomas et al., 1988). Whether other mRNA species with one, three, four, etc. extra nucleotides exist has not been determined for SV5. This strategy for the expression of the P and V proteins of SV5 provided the first evidence in paramyxoviruses of two functional mRNAs being produced from a single gene (Thomas et al., 1988). However, as detailed below, we predicted that all paramyxoviruses would encode an unrecognized cysteine-rich protein derived from an mRNA different from that of the P protein and with

```
                                                        *   *
SV5      562-726    G G F H R R E Y S I G W V G D E V K V T E W C N P S C S P I T A A A R R
MUMPS    663-833    G G - H R R E Y S L S Y V Q G E V R V F E W C N P I C S P I T A A A R F
NDV      605-799    P G - H R R E H T I S W T M G G V T T I S W C N P S C S P I R A E P R Q
SENDAI   1048-1254  K G - H R R E H I I Y E R D G Y I V D E S W C N P V C S R I R I I P R R
PI-3     1132-1332  E G - Y R R E Q S I Y K E G N Y S I A E S W C N S I Y T K T R F I S R Q
MEASLES  749-958    K G - H R A Q I S L I W N G D R V F I D R W C N P M C S K V T L G T I R

                        *   *     *       *     *
SV5                 F E C T C H Q C P V T C S E C E R D T *
MUMPS               H S C K C G N C P A K C D Q C E R D Y G P P *
NDV                 Y S C T C G S C P A T C R L C A G D D L Y D G G D I T E G K *
SENDAI              E L C V C K T C P K V C K L C R D D I Q C M R P D P F C R E T F R S *
PI-3                T S C M C S K C T K Q C R Y C I K D R L P S R I S H R G F N G Q *
MEASLES             A R C T C G E C P R V C E Q C R T D T G V D T R I W Y H N L P E I P E *
```

FIGURE 8. Amino acid sequence homology between the cysteine-rich region of protein V of SV5 and amino acid sequences predicted from open reading frames of other paramyxovirus *P* genes. The published nucleotide sequences of the *P* genes of several paramyxoviruses and the morbiliviruses measles virus and CDV were translated in all three reading frames. In each case, in a reading frame overlapping that for the P protein, a cysteine-rich region was identified and is listed in the single-letter amino acid code. Only the region of significant conservation of sequence is shown with its corresponding nucleotide number: the N-terminal region of the open reading frame is omitted. The small star at the end of each amino acid sequence represents a translation termination codon. The boxes identify positions where three or more amino acids have been conserved in all six viruses. A dash indicates that a gap was placed in the alignment and the large stars above the sequences identify the seven conserved cysteine residues. Sources for the *P* gene nucleotide sequences are as follows: SV5, Thomas *et al.* (1988); mumps virus, Takeuchi *et al.* (1988); NDV, Sato *et al.* (1987); Sendai virus, Shioda *et al.* (1983) and Giorgi *et al.* (1983); parainfluenza virus 3 (PI-3), Galinski *et al.* (1986) and Luk *et al.* (1986); measles virus, Bellini *et al.* (1985). The homology with CDV using the *P* gene sequence of Barrett *et al.* (1985) is not shown here, but is almost identical to that shown for measles virus. The numbering of the mumps virus nucleotides starts with the first nucleotide shown in the sequence by Takeuchi *et al.* (1988) and not with the 5′ end of the mRNA. [From Thomas *et al.* (1988).]

one of the two mRNAs containing extra nucleotides not found in the genomic RNA (Thomas *et al.*, 1988; Paterson *et al.*, 1989).

E. Conservation of the Cysteine-Rich Region of Protein V in Paramyxoviruses

Examination of the predicted amino acid sequences of the SV5 P and V proteins indicated that the cysteine-rich C-terminal portion of protein V has a great similarity to the conserved cysteine-rich regions (sometimes known as zinc fingers) found in many proteins involved in the regulation of transcription, including the adenovirus E1A protein [for review see Moran and Matthews (1987)], the yeast transcription factor GAL4 (Johnston and Dover, 1987), and proteins belonging to the steroid hormone receptor superfamily [for review see Evans (1988)]. The binding of metal ions by the cysteine-rich regions of these proteins has an important role in either binding of nucleic acid by the protein, mediating protein–protein interactions, or stabilizing oligomeric forms of a protein, as in the Tat protein of human immunodeficiency virus (Frankel *et al.*, 1988). As the cysteine-rich regions in these proteins play such an important role in mediating their regulatory activities, we thought it likely that the cysteine-rich region of SV5 protein V would be conserved in the genome of other paramyxoviruses and that it would probably be located within the *P* genes. The cysteine-rich region from protein V was therefore compared with the protein sequences predicted from all three reading frames of the *P* genes of mumps virus, NDV, Sendai virus, PIV-3, measles virus, and CDV (Takeuchi *et al.*, 1988; Sato *et al.*, 1987; McGinnes *et al.*, 1988; Shioda *et al.*, 1983; Giorgio *et al.*, 1983; Galinski *et al.*, 1986; Luk *et al.*, 1986; Bellini *et al.*, 1985; Barrett *et al.*, 1985). As shown in Fig. 8, a highly conserved cysteine-rich region was identified in an open reading frame, different from that of the P protein in all the paramyxovirus *P* gene sequences examined (Thomas *et al.*, 1988; Paterson *et al.*, 1989). The location of the open reading frames containing the cysteine-rich region in relation to the *P* gene nucleotide sequences is shown in Fig. 9. A striking feature of this cysteine-rich region is that it is more conserved in sequence between the different paramyxoviruses than is the amino acid sequence of the P protein encoded by the same nucleotides but translated in another reading frame.

F. Prediction of Cysteine-Rich Polypeptides and mRNAs with Extra Nucleotides in All Paramyxoviruses

If all paramyxoviruses and morbilliviruses encode cysteine-rich proteins; then it seemed reasonable to expect that the phenomenon observed with SV5 of the addition of nontemplated nucleotides would be general to all of these viruses (Paterson *et al.*, 1989). For the purposes of this discussion, the assumption has been made that the published sequences of the *P* gene (derived mostly from cDNA copies of mRNAs, but in some cases confirmed on vRNA) faithfully reflect those in each virus genome [with mumps virus it is clear that the

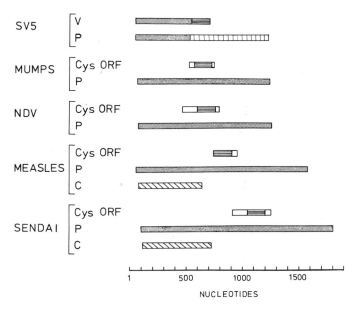

FIGURE 9. Representation of the position of the open reading frames containing the cysteine-rich regions in paramyxovirus P genes. For mumps virus, NDV, measles virus, and Sendai virus, the P protein open reading frames described in the papers cited in the legend to Fig. 8 and presumed to be encoded by the vRNA are shown as stippled boxes. For measles virus and Sendai virus, the C protein overlapping reading frames are shown as diagonally hatched boxes. To simplify this diagram, the multiple starts of translation initiation of the C protein of Sendai virus (Curran et al., 1988a) are omitted. The open reading frames containing the cysteine-rich regions are shown as open boxes, with the region of high similarity among all paramyxoviruses embodying the cysteine-rich domain represented by horizontal bars. In the case of SV5, where the structure of the mRNA encoding P is known, the stippled box represents the amino acids shared with protein V and the vertically cross-hatched box is the region specific to P. [From Paterson et al. (1989).] Since the time that this diagram was originally drawn, it has become apparent that the mumps virus genome, like the SV5 genome, encodes protein V containing the "cys ORF" and that extra nucleotides have to be added to create the mRNA that encodes P. Thus, mumps virus should have been illustrated akin to SV5 and the mumps virus open reading frames are analogous to that shown for SV5 in Fig. 5.

original sequence as proposed for the SBL-1, Miyahara, and Enders strains (Takeuchi et al., 1988; Elango et al., 1989) is at least not correct for the RW strain of mumps virus (Paterson and Lamb, 1990; see below)]. Then, in all cases, the P protein is encoded by the genome, except in the cases of SV5 and mumps virus, where the genomes encode protein V. To gain access to the open reading frame containing the cysteine-rich domain, a single nucleotide would need to be added to the mRNA transcript in the case of NDV, measles virus, CDV, Sendai virus, and PIV-3. The nucleotide sequence of the P genes in the small region between the 5' end of the open reading frame and the beginning of the conserved cysteine-rich domain were examined for a feature, such as sequences bearing a resemblance to paramyxovirus polyadenylation signals followed by sequences containing consecutive C residues (vRNA sense), that might make it possible to predict the precise site of the adition of extra nucleotides.

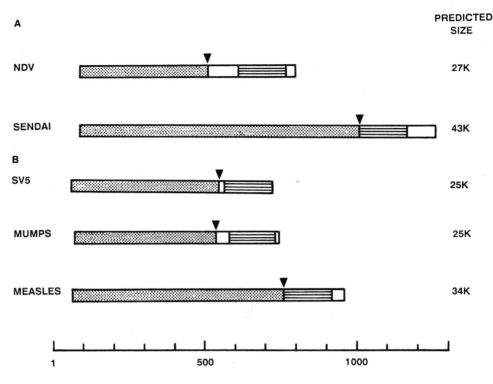

FIGURE 10. Representation of the structure of the presumptive (A) and identified (B) paramyx-ovirus cysteine-rich polypeptides derived from the *P* genes. For NDV and Sendai virus the stippled box indicates the region shared between the V protein and the P protein, and the horizontal bars and open boxes represent the region after the site (indicated by the arrow) of the predicted addition of nontemplated G residues to the mRNA. For SV5 and mumps virus, protein V is known to be encoded by the vRNA and the arrow denotes the position at which the two nongenomic G residues are added to switch reading frames to encode the C-terminal half of the P protein (Thomas *et al.*, 1988; Paterson *et al.*, 1989; Paterson and Lamb, 1990). For measles virus, the stippled box indicates the region shared between the V protein and the P protein, and the horizontal bars and open boxes represent the region after the site (indicated by the arrow) that nontemplated G residues are added to create the V mRNA [data from Cattaneo *et al.* (1989)]. Although not shown, CDV would be predicted to have an almost identical arrangement to the measles virus hybrid P–cysteine-rich protein and PIV-3 would encode a protein of similar structure to the presdumptive protein of Sendai virus [adapted from Paterson *et al.* (1989)]. Since the time this diagram was originally drawn the Sendai virus V polypeptide mRNA has been identified and it can be translated *in vitro* to yield a V polypeptide (Vidal *et al.*, 1990).

In the mumps virus (strains Miyahara, Enders, and SBL-1) *P* gene sequence (Takeuchi *et al.*, 1988; Elango *et al.*, 1989a), approximately 50 nucleotides upstream of the locus encoding the cysteine-rich region there is the vRNA sequence 3'-AAAUUCUCCCCCCCC-5' (nucleotides 524–536) (Elango *et al.*, 1989a), which in the RW strain has been found to be 3'-AAAUUCUCCCC-

CC-5′ (Paterson and Lamb, 1990). This sequence in the mumps genome bears a striking resemblance to the sequence in the SV5 *P* gene (3′-AAAUUCUC-CC-5′) where the addition of the nontemplated G residues occurs in the P mRNA. In the NDV *P* gene sequence (Sato *et al.*, 1987; McGinnes *et al.*, 1988), the vRNA sequence 3′-AGUUCUUCCC-5′ at nucleotides 501–509 is the closest match to a polyadenylation signal followed by three C residues in the appropriate region of the gene. In measles virus, speculation about the position at which extra nucleotides might be added is severely limited, due to the finding that the conserved region of the cysteine-rich open reading frame is flanked by stop codons. However, immediately at the beginning of the reading frame (nucleotides 740–749) (Bellini *et al.*, 1985), the vRNA sequence 3′-AAUUUUUCCC-5′ can be found, making this region the most likely candidate for the addition of nongenomic nucleotides. As will be described below, measles virus mRNAs derived from the *P* gene and containing extra G residues at this precise point have been identified (Cattaneo *et al.*, 1989). Similarly, in Sendai virus, the vRNA sequence 3′-UGUUUUUUCCC-5′ (nucleotides 1042–1053) (Giorgi *et al.*, 1983) can be identified, making it the possible site for addition of extra nucleotides, and it has recently been found that Sendai virus mRNAs exist that contain an extra G residue at this precise point (Vidal *et al.*, 1990; see below). The consequence of the addition of nucleotides in the mRNA transcripts at these positions is to be able to translate hybrid P-cysteine-rich polypeptides of the structure shown in Fig. 10. The calculated molecular weights of these polypeptides are as follows: mumps, 25,000; NDV, 27,000; measles virus, 34,000; and Sendai virus, 43,000. However, given that the P proteins are phosphoproteins, the hybrid P-cysteine-rich proteins when identified experimentally may be observed to have larger molecular weights.

G. Identification of the Nonstructural Protein V and Its mRNAs in Other Paramyxoviruses

The nature of the measles virus mRNAs expressed in the brains of patients with either subacute sclerosing panencephalitis (SSPE) or measles inclusion body encephalitis was examined by cloning the mRNAs from human brain autopsy material and obtaining their nucleotide sequences (Cattaneo *et al.*, 1989). In a survey of *P*-gene-specific mRNAs it was found that transcripts existed which contained a differing number of G residues at *P* gene nucleotide 751, the predicted position of addition of extra G residues (Paterson *et al.*, 1989). Seven *P*-gene-specific cDNA clones were isolated which contained three G residues complementary to the three C residues in the measles virus virion RNA. In addition, ten cDNA clones were isolated which contained one extra G residue, and two clones were isolated with three extra G residues. Translation of synthetic RNA derived from cDNAs with the three G residues yielded the P polypeptide. However, translation of RNA that contains one extra G residue at nucleotide 751 and which switches the reading frame yielded, as expected, a polypeptide of $M_r \approx 46,000$ that could be preferentially labeled with cysteine relative to methionine labeling (Cattaneo *et al.*, 1989). As this polypeptide is thought to be the measles virus counterpart to SV5 virus polypeptide V, it has been designated V (Cattaneo *et al.*, 1989). Unfortunately, although measles virus mRNAs isolated from infected cells can be translated to yield polypeptide V, it has not been possible to identify the polypeptide in infected cells (Cattaneo et al., 1989).

In a study of Sendai virus mRNAs derived from the P/C gene to search for species with extra nucleotides, it was found that two major species existed, one an exact copy of the viral genome and the other containing a single G insertion within the run of three Gs at the site predicted above (P/C gene nucleotide 1053) (Vidal et al., 1990). The mRNAs or portions of them containing the insertion site were directly cloned, and the bacterial colonies obtained were screened by hybridization using oligonucleotides which could distinguish between the presence of three or four G nucleotides at this position. It was found that ~62% of the cDNAs contained no added G residues, ~31% contained one extra G residue, and ~7% contained more than 1 G residue (2–8 G residues). No evidence could be found for variation in the sequence of the Sendai virus genome at the insertion site in a study of more than 300 cDNAs. The frequency of insertion was relatively fixed, as it did not vary significantly between two different Sendai virus strains, and it could not be altered by coinfection with defective-interfering particles, a viral species that has long been postulated to modulate transcription/replication processes. The addition of extra G nucleotides occurred at a similar frequency when mRNAs were synthesized by the virion RNA-dependent RNA transcriptase in vitro (Vidal et al., 1990). This latter result strongly suggests that the insertion of G residues at this specific site is catalyzed by the viral transcriptase. Synthetic RNAs derived from the cDNAs containing one extra G residue could be translated in vitro to yield polypeptide V ($M_r \approx 60,000$), which comigrates with Sendai virus nucleocapsid polypeptide. Synthetic RNAs derived from the cDNAs containing two extra G residues yielded a polypeptide designated W ($M_r \approx 55,000$) (Vidal et al., 1990). No evidence has been obtained for expression of V and W in Sendai virus-infected cells, but when the appropriate antisera are generated, they are expected to be identified.

In mumps virus-infected cells, two nonstructural polypeptides ($M_r \approx$ 23,000–28,000 and 17,000–19,000) were observed to be synthesized (Rima et al., 1980; Herrler and Compans, 1982). Tryptic peptide mapping of these polypeptides indicated that they were related to the P protein (Herrler and Compans, 1982). Hybrid-selection of mRNA using a P-gene-specific cDNA and in vitro translation experiments have resulted in the synthesis of P and other products estimated to be of $M_r \approx 30,000, 20,000,$ and 14,500.

The nucleotide sequences of cDNAs derived from mRNA and vRNA for the Miyahara, Enders, and SBL-1 strains of mumps virus have been published, and in all cases a single open reading frame sufficient to encode the P polypeptide was identified (Takeuchi et al., 1988., Elango et al., 1989a). However, in the SBL-1 strain, six C residues were found in the vRNA starting at nucleotide 531, whereas in the Miyahara and Enders strains, eight C residues were reported, and in the SBL-1 strain a compensating change occurred upstream to maintain the P protein reading frame. This is precisely the region described above where it was predicted that extra G residues would be added in the mRNA (Paterson et al., 1989). Moreover, the mRNA cDNA clone of the SBL-1 strain contained ten G residues at this location (Elango et al., 1989a). As discussed above, the sequence of this region of the RW strain of mumps virus has been obtained (Paterson and Lamb, 1990) and it is clear in this strain of mumps virus that there are six C residues, which means that the mumps virus RW strain genome, like SV5, encodes protein V. The simplest explanation for the conclusion that mumps virus strains Enders, Miyahara, and SBL-1 encode P, which now seems unlikely to be correct, is that at the time the work was done there was no reason to suspect that the genome could encode V. The high G/C content of this region makes it prone to sequencing artifacts unless special care is taken.

Work in our laboratory has indicated that the pMP1 cDNA sequenced by Elango et al. (1989a) arrests the translation in vitro of the $M_r \approx 30,000$ and $M_r \approx 20,000$ nonstructural proteins in addition to P. The $M_r \approx 30,000$ polypeptide synthesized in infected cells can be more heavily labeled with cysteine than with methionine and has been designated V, whereas the $M_r \approx 20,000$ nonstructural polypeptide cannot be labeled with cysteine (Paterson and Lamb, 1990). The $M_r \approx 20,000$ polypeptide is the only polypeptide translated from synthetic RNA transcripts derived from the pMP1 cDNA and this is compatible with the size of the predicted open reading frame (Paterson and Lamb, 1990). After the ten G residues beginning with nucleotide 531 in clone pMP1, translation is predicted to continue for 15 amino acid residues in a reading frame not used for either P or V.

To understand the mRNA species that are produced from the mumps virus P/V gene, cDNAs were made to the region between P/V gene nucleotides 500 and 600 using oligonucleotide primers, and the nucleotide sequences of 54 cDNAs were obtained. In addition, the sequences of 70 cDNA clones of this region of the vRNA were obtained. The vRNA sequences all contained six C residues beginning at nucleotide 531. The mRNA sequences indicated that 34 (63%) of the cDNAs had six G residues beginning at nucleotide 531, none had seven G residues, ten (19%) had eight G residues, seven (13%) had nine G residues, one (2%) had ten G residues, and two (4%) had 11 G residues (Paterson and Lamb, 1990). Thus, mRNAs with six G residues would encode the cysteine-rich protein V and mRNAs with nine G residues would encode the same protein, but containing an extra glycine residue. Messenger RNAs containing eight G residues (i.e., containing two more nucleotides than a direct copy of the genome) encode the P protein and, as discussed above, mRNAs containing ten G residues encode the $M_r \approx 20,000$ nonstructural protein. In vitro transcription of these cDNAs into synthetic RNA using bacteriophage T7 DNA-dependent RNA polymerase and translation of the RNAs using reticulocyte lysates confirmed the coding assignment. Thus, these data suggest that mumps virus is very much like SV5 in that both viruses have the cysteine-rich protein V encoded by their genomes, with the mRNAs for the P proteins containing the nontemplated nucleotides, and this is in contrast to the reverse situation found with measles virus and Sendai virus.

With NDV, two nonstructural polypeptides of $M_r \approx 33,000–36,000$ and 28,000–33,000 (Collins et al., 1982) have been identified and they have been shown to have tryptic peptide maps in common with P (Collins et al., 1982). The nucleotide sequence of the NDV P gene (Sato et al., 1987; McGinnes et al., 1988) predicts only one open reading frame which would encode P. Translation of synthetic RNA derived from the P cDNA clone yielded polypeptides of $M_r \approx$ 38,000 and 29,000 in addition to P, and polypeptides of identical mobility on gels could also be identified in infected cells (McGinnes et al., 1988). When truncated P cDNA clones containing 5'-terminal deletions were transcribed into RNA and the RNA was translated, it was found that the polypeptides of $M_r \approx$ 38,000 and 29,000 were still synthesized, and thus it was thought likely that these polypeptides are derived as a result of internal initiation on the mRNAs both in vitro and in vivo (McGinnes et al., 1988). The relationship of these $M_r \approx$ 38,000 and 29,000 species to those described by Collins et al. (1982) is not known. A cysteine-rich protein V of NDV has not yet been described and the complete characterization of P-gene-related nonstructural proteins is worthy of continued investigation.

H. Function of the Paramyxovirus Cysteine-Rich Protein V?

As the paramyxovirus P gene is believed to be an essential component of the RNA transcriptase complex (Hamaguchi et al., 1983) and as the cysteine-rich region of protein V encoded by the overlapping reading frame is more conserved between paramyxoviruses than is the sequence of P, it seems reasonable to predict that protein V has an essential role in replication of the virus.

The function of V is unknown, but many other proteins that contain the cysteine-rich motif identified in protein V are found to interact with DNA or RNA [reviewed in Klug and Rhodes (1987) and Frankel and Pabo (1988)] and the role of protein V could be to mediate any one or more of the uncharacterized steps in transcription, polyadenylation, and replication of the vRNA. The SV5 protein V can be detected as a component of purified SV5 virions by immune blotting analysis, using a P- and V-specific monoclonal antibody (R. G. Paterson and R. A. Lamb, unpublished observations). However, the amount of V present in virions is at low levels compared to the abundance of the other virion structural polypeptides and the argument discussed above for the C proteins concerning the relevance of finding low amounts of "nonstructural" polypeptides in virions is applicable here.

IV. PARAMYXOVIRUS SMALL HYDROPHOBIC (SH) PROTEINS

A. Identification of the Polypeptide and Its Gene in SV5

Nucleotide sequencing of the SV5 50S genome RNA and of polycistronic HN mRNAs indicated that there are 297 nucleotides between the F and HN genes and these were designated the SH gene (Hiebert et al., 1985a, 1985b) (Fig. 11). An SH mRNA of 292 nucleotides (plus polyadenylate residues) that is transcribed from the SH gene was identified in infected cells by RNA blot hybridization and nuclease S1 mapping (Hiebert et al., 1985b).

The SH mRNA contains a single open reading frame of 44 amino acids (Hiebert et al., 1985b). The predicted amino acid sequence (Fig. 12) indicates that the SH protein could be divided into two domains, an N-terminal hydro-

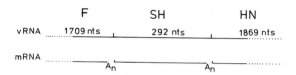

FIGURE 11. Diagram showing the region of the SV5 50S genome RNA encoding SH, and the mRNAs transcribed from this region. Nts, nucleotides; A_n, polyadenylate residues. [From Hiebert et al. (1985b).]

philic region of 16 residues and a C-terminal hydrophobic region of 23 amino acids with a hydrophobic index of greater than 2, a value normally found for regions of proteins which interact with membranes (Kyte and Doolittle, 1982) (Fig. 13). The amino acid sequence of SH predicts that the protein would contain six phenylalanine residues and four isoleucine residues, and when SV5-infected cells were labeled with [3H]-phenylalanine and [3H]-isoleucine a previously unrecognized polypeptide of $M_r \approx 5000$ was readily detected. This polypeptide species was shown to be encoded by the SH mRNA by hybrid-selection and hybrid-arrest translation experiments and translation of size-selected SH mRNA (Hiebert et al., 1985b).

To facilitate the localization of the SH protein in SV5-infected cells, an antiserum to an SH-specific synthetic oligopeptide (SH residues 2–18) was prepared in rabbits. This antiserum is capable of immunoprecipitating the SH protein from lysates of SV5-infected cells (Hiebert et al., 1988) (Fig. 14). The SH protein was found to have properties of an integral membrane protein, as it is resistant to extraction from membranes by alkali (Hiebert et al., 1988), a strong protein denaturant which has been shown to extract peripheral membrane proteins, while integral membrane proteins remain associated with the lipid bilayer (Steck and Yu, 1973; Gilmore and Blobel, 1985; Paterson and Lamb, 1987). Cell fractionation studies indicated that SH is inserted into the endoplasmic reticulum and transported along the exocytic pathway to the Golgi apparatus (Hiebert et al., 1988). Immunofluorescent staining of SV5-infected cells permeabilized with the cholesterol-like detergent saponin and using affinity-purified antibody to SH showed a diffuse punctate staining in the focal plane above the nucleus, a result compatible with SH being expressed at the plasma membrane with its N terminus in the cytoplasm (Fig. 15). The SH antibody could not stain the surface of unpermeabilized infected cells, which is an expected negative result if the SH N terminus is cytoplasmic (Hiebert et al., 1988). Biochemical enrichment of plasma membrane fractions showed a corresponding enrichment of SH, which supports the immunofluorescent data for the subcellular localization of SH. Protease treatment of infected cell surfaces did not affect the electrophoretic mobility of SH, whereas protease treatment of intracellular microsomes yielded an SH-protected fragment which could not be immunoprecipitated with the SH N-terminal-specific antiserum. These data are compatible with SH being oriented in membranes such that its N terminus is cytoplasmic (Hiebert et al., 1988), and a diagram indicating the orientation of SH is shown in Fig. 16.

The role of SH in the SV5 replicative cycle is not known. The evidence

```
.1           .20          .40          .60          .80
AGGACCGAACCTAGTAGTATTGAAAGAACCGTCTCGGTCAATCTAGGTAATCGAGCTGATACCGTCTCGGAAAGCTCAAATC    ATG CTG
                                                                               Met-Leu-

CCT GAT CCG GAA GAT CCG GAA AGC AAG AAA GCT ACA AGG AGA GCA GGA AAC CTA ATT ATC TGC TTC
Pro-Asp-Pro-Glu-Asp-Pro-Glu-Ser-Lys-Lys-Ala-Thr-Arg-Arg-Ala-Gly-Asn-Leu-Ile-Ile-Cys-Phe-
        .100         .120         .140

CTA TTC ATC TTC TTT CTG TTT GTA ACC TTC ATT GTT CCA ACT CTA AGA CAC TTG CTG TCC TAA CAC
Leu-Phe-Ile-Phe-Phe-Leu-Phe-Val-Thr-Phe-Ile-Val-Pro-Thr-Leu-Arg-His-Leu-Leu-Ser ***
    .160         .180         .200

CTGCTATAGGCTATCCACTGCATCATCTCTCCTGCCATACTTCCTACTCACATCATATCTATTTAAAGAAAAAA
    .220         .240         .260         .280
```

FIGURE 12. Nucleotide sequence of the SH mRNA and the predicted amino acid sequence of the SH protein. The extensively hydrophobic region in the SH protein that is thought to be the membrane-spanning domain is underlined. [From Hiebert et al. (1985b).]

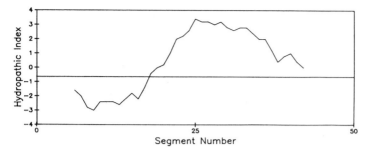

FIGURE 13. Hydropathy plot of the predicted SH protein. The relative hydrophobicity and hydrophilicity of the protein along its amino acid sequence were calculated as described (Kyte and Doolittle, 1982) with a segment length of seven amino acids. The consecutive scores are plotted from the NH$_2$ to the COOH terminus of the protein. The midpoint line corresponds to the grand average of the hydropathy of the amino acid compositions found in most sequenced proteins (Kyte and Doolittle, 1982). [From Hiebert *et al.* (1985b).]

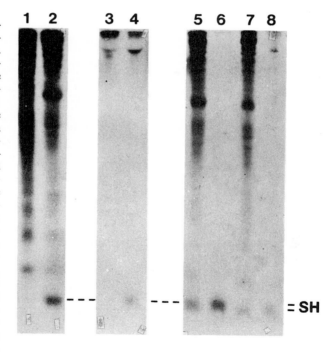

FIGURE 14. Immunoprecipitation of SH with N-terminal-domain-specific peptide antisera. SV5-infected CV1 cells were labeled with [³H]-phenylalanine plus [³H]-isoleucine, or [³⁵S]-methionine, and lysed in 1% SDS. Appropriate samples were immunoprecipitated with an antiserum raised to an oligopeptide containing residues 2–18 of SH using the amino acid sequence predicted from the nucleotide sequence of the SH gene (see Fig. 12). Samples were subjected to SDS–PAGE using 9% highly cross-linked polyacrylamide gels containing 8 M urea. Lanes 1 and 2: [³H]-labeled lysates analyzed directly; lane 1, uninfected cells; lane 2, SV5 strain W3A-infected cells. Lanes 3 and 4: [³⁵S]-methionine-labeled lysates immunoprecipitated with the peptide antisera; lane 3, uninfected cells; lane 4, SV5 strain W3A-infected cells. Lanes 5–8: [³H]-labeled samples; lanes 5 and 7 are direct lysates and lanes 6 and 8 are immunoprecipitated samples using the peptide antisera. Lanes 5 and 6: SV5 strain W3A; lanes 7 and 8: SV5 strain 2WR. Two different strains of SV5 were used in the experiment, as the difference in mobility of the polypeptide on gels is a further indication that the polypeptide is virus encoded and not a cellular polypeptide that is nonspecifically immunoprecipitated. [From Hiebert *et al.* (1988).]

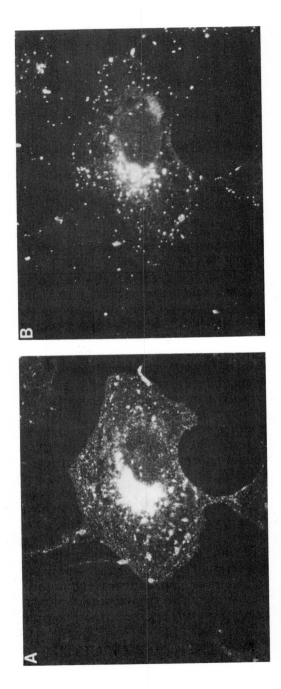

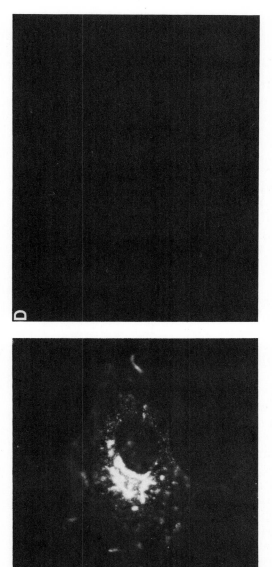

FIGURE 15. Indirect immunofluorescence of saponin-permeabilized SV5-infected cells. Mock-infected or SV5-infected cells were fixed with 0.5% paraformaldehyde and permeabilized with 0.1% saponin. Saponin was maintained in all wash steps and during antibody binding. Cells were stained with SH-specific IgG plus fluoresceine-conjugated goat anti-rabbit IgG and HN-specific mouse ascites fluid and rhodamine-conjugated goat anti-mouse IgG. A and C: SH staining at two different focal levels to show fluorescence of proteins expressed at the cell surface, by focusing (A) above the nucleus, and (C) in the plane of the nucleus. B: HN staining on the same field of cells as in panels A and C and focused as in panel A. D: Mock-infected cells photographed using the filters for rhodamine staining. The exposure for panel D was manually adjusted to be the same as that for panel B. [From Hiebert *et al.* (1988).]

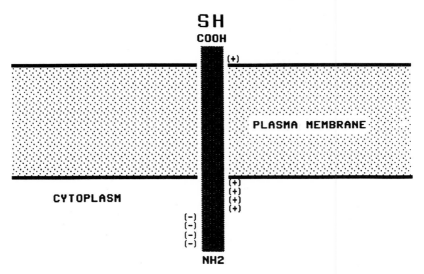

FIGURE 16. Diagram indicating the orientation of SH in membranes. The relative locations of the acidic (−) and basic (+) amino acid residues in the polypeptide are indicated. [From Hiebert *et al.* (1988).]

suggests that SH is preferentially excluded from virions. However, the assays are not sensitive enough to exclude the possibility that there is a small amount of SH in virions. The absence of a significant ectodomain in SH suggests that the functional domain may be the highly charged N-terminal region exposed at the cytoplasmic face of the plasma membrane. It is possible that SH has a role in organizing the assembly of viral proteins to form a patch at the plasma membrane. Several other enveloped viruses also express small integral membrane proteins of unproven function in infected cells: e.g., M_2 of influenza A virus (Lamb *et al.*, 1985), NB of influenza B virus (Williams and Lamb, 1986), 1A of respiratory syncytial virus (Olmsted and Collins, 1989), gene 5 ORF2 of mouse hepatitis virus (Leibowitz *et al.*, 1988), and the protein encoded by *vpu* of human immunodeficiency virus (HIV) (Strebel *et al.*, 1988). The structural similarities make it possible that these types of proteins may have a common function. This concept is given credence by finding that frameshift mutants in *vpu* cause a reduction in progeny virion yield, and it has been suggested that the *vpu* protein may be required for efficient virus assembly or maturation of progeny virions (Strebel *et al.*, 1988). Further circumstantial evidence for a role of the small hydrophobic proteins in assembly or budding was provided by finding that a monoclonal antibody to the N-terminal ectodomain of influenza A virus M_2 protein is capable of restricting the growth of the virus in a plaque assay (Zebedee and Lamb, 1988, 1989).

However, in speculating on the role of the SH protein in the replicative cycle of SV5, it has to be borne in mind that the only other paramyxovirus in which a related gene has been found is mumps virus (see below). The nucleotide sequences of the Sendai virus, measles virus, and NDV genomes indicate that an SH gene is not present in these viruses. However, it has not been ruled out that an SH-type protein is not encoded by an overlapping reading frame of another gene. In this context it is interesting to note that M_2 of influenza A virus and NB of influenza B virus are polypeptides with very similar structural features, yet they are encoded in different influenza virus RNA segments and by different mechanisms: M_2 is encoded by a spliced mRNA and NB is encoded in an overlapping open reading frame of a functionally bicistronic mRNA (Lamb et al., 1981; Shaw et al., 1983).

B. The SH Gene of Mumps Virus

Nucleotide sequencing of the mumps virus genome RNA and its mRNAs has indicated that mumps virus contains an *SH* gene located between *F* and *HN* on the genome (Elango et al., 1989b; Elliott et al., 1989; Neal Waxham, personal communication). An SH mRNA of 310 nucleotides (excluding polyadenylate residues) has been identified in mumps virus-infected cells (Elango et al., 1989b). This SH mRNA contains an open reading frame of 57 amino acids. The predicted amino acid sequence indicates that the presumptive SH polypeptide would contain an N-terminal hydrophobic domain of 25 amino acids which could interact in a stable manner with a lipid membrane (Elango et al., 1989b). Unfortunately, it has not yet been possible to identify the SH polypeptide synthesized either *in vitro* or in infected cells.

V. SENDAI VIRUS NONSTRUCTURAL POLYPEPTIDE B: INTRACELLULARLY PHOSPHORYLATED MATRIX PROTEIN

When the polypeptides synthesized in Sendai virus-infected cells were examined, a nonstructural polypeptide designated B ($M_r \approx 36,000$) was identified in addition to the C and C' polypeptides (Lamb et al., 1976). Polypeptide B has a slightly reduced electrophoretic mobility in comparison to the matrix protein (M) on polyacrylamide gels. In cell fractionation studies, polypeptide B was apparently unstable even in the presence of protease inhibitors (Lamb and Choppin, 1977b). However, pulse-label-chase protocols of Sendai virus-infected cells indicated that M was rapidly converted to polypeptide B. In a 30-min chase period, over 90% of M becomes converted to polypeptide B (Lamb and Choppin, 1977b). Tryptic peptide mapping studies confirmed that M and B contain the same polypeptide backbone. It was shown by labeling Sendai virus-infected cells with [^{32}P]-orthophosphate that polypeptide B is a phosphorylated form of the matrix protein (Lamb and Choppin, 1977b): the addition of phosphate residues causes the quantum shift of electrophoretic mobility observed. Purified Sendai virions contain very little phosphorylated M protein, but M protein in virions can be converted to its phosphorylated form (B) by the virion-associated protein kinase (Lamb, 1975; Lamb and Choppin, 1977b). Thus, the data suggest that M is phosphorylated and dephosphorylated before it is found in mature virions. Phosphorylation and dephosphorylation, which bring about conformational changes in polypeptides, represent a major mechanism of signal transduction in cells, whether it be a step in the conversion of a normal cell to a malignant cell or in the process of memory and learning, such as in sensitization of sensory neuron/motor neuron control with the sea snail *Aplysia punctata* (Alberts *et al.*, 1989). It is possible that the phosphorylated form of the Sendai virus matrix protein might interact with the plasma membrane of the infected cell in a manner different from that of the unphosphorylated form and that phosphorylation/dephosphorylation could thus influence events at the plasma membrane during virus assembly, such as the association of the viral glycoproteins with the M protein, redistribution of cellular membrane proteins, and the alteration of microtubules and microfilaments in the budding process.

VI. PROSPECTS

One of the next major goals in the area of research concerning paramyxovirus nonstructural proteins should be the elucidation of their role in the paramyxovirus replicative cycle both in tissue culture cells and in animal infections. From their amino acid sequence it is apparent that the nonstructural proteins fall into several groups that are likely to function at different stages in the paramyxovirus life cycle. While protein V is predicted to exist in all paramyxoviruses, other nonstructural proteins such as the C and C' polypeptides of Sendai and measles viruses appear to be encoded by only a subset of paramyxoviruses. How can one rationalize such observations? The fact that protein V appears to be common to all paramyxoviruses would implicate it as having an indispensable function in paramyxovirus replication. The predicted C-terminal cysteine-rich region of protein V that has the hallmarks of a "zinc-finger" motif suggests that it might be involved in the transcription and/or replication of paramyxovirus RNA in infected cells. It will be necessary to determine whether protein V has the ability to interact with RNA and to see if it is capable of binding heavy metals. Perhaps in the case of the C proteins, their role is only required in the specific pathology of a measles or Sendai virus infection in the whole animal and therefore they are dispensable in other paramyxoviruses. It will be interesting to investigate whether paramyxoviruses other than mumps virus and SV5 can dispense with an SH protein. Alternatively, an *SH* gene/protein in the other viruses may remain to be detected, especially as small integral membrane proteins are encoded by many other enveloped RNA viruses.

ACKNOWLEDGMENTS. We are grateful to Carolyn Jenkins for carefully typing the review. Research in the authors' laboratories was supported by research grants AI-20201 and AI-23173 from the National Institutes of Health.

VII. REFERENCES

Alberts, B., Bray, D., Lewis, J., Raff, M., Roberts, K. and Watson, J. D., 1989, "Molecular Biology of the Cell," 2nd ed., Garland Publishing, New York.
Alkhatib, G., Massie, B., and Briedis, D. J., 1988, Expression of bicistronic measles virus P/C mRNA by using hybrid adenoviruses: Levels of C protein synthesized *in vivo* are unaffected by the presence or absence of the upstream P initiator codon, *J. Virol.* **62**:4059–4069.

Barrett, T., Shrimpton, S. B., and Russell, S. E. H., 1985, Nucleotide sequence of the entire protein coding region of canine distemper virus polymerase-associated (P) protein mRNA, *Virus Res.* **3:**367–372.

Becerra, S. P., Rose, J. A., Hardy, M., Baroudy, B. M., and Anderson, C. W., 1985, Direct mapping of adeno-associated virus capsid proteins B and C: A possible ACG initiation codon, *Proc. Natl. Acad. Sci. USA* **82:**7919–7923.

Bellini, W. J., Englund, G., Rozenblatt, S., Arnheiter, H. and Richardson, C. D., 1985, Measles virus P gene codes for two proteins, *J. Virol.* **53:**908–919.

Benne, R., Van Den Burg, J., Brackenhoff, J. P. J., Sloof, P., Van Boom, J. H., and Tromp, M. C., 1986, Major transcript of the frameshifted *cox*II gene from trypanosome mitochondria contains four nucleotides that are not encoded in the DNA, *Cell* **46:**819–826.

Cattaneo, R., Kaelin, K., Baczko, K., and Billeter, M. A., 1989, Measles virus editing provides an additional cysteine-rich protein, *Cell* **56:**759–764.

Chen, S.-H., Habib, G., Yang, C.-Y., Gu, Z.-W., Lee, B. R., Weng, S. A., Silberman, S. R., Cai, S.-J., Deslypere, J. P., Rosseneu, M., Gotto, Jr., A. M., Li, W.-H., and Chan, L., 1987, Apolipoprotein B-48 is the product of a messenger RNA with an organ-specific in-frame stop codon, *Science* **238:**363–366.

Collins, P. L., Wertz, G. W., and Hightower, L. E., 1982, Coding assignments of the five smaller mRNAs of Newcastle disease virus, *J. Virol.* **43:**1024–1031.

Curran, J. A., and Kolakofsky, D., 1987, Identification of an additional Sendai virus non-structural protein encoded by the P/C mRNA, *J. Gen. Virol.* **68:**2515–2519.

Curran, J., and Kolakofsky, D., 1988a, Ribosomal initiation from an ACG codon in the Sendai virus P/C mRNA, *EMBO J.* **7:**245–251.

Curran, J., and Kolakofsky, D., 1988b, Scanning independent ribosomal initiation of the Sendai virus X protein, *EMBO J.* **7:**2869–2874.

Curran, J. A., Richardson, C., and Kolakofsky, D., 1986, Ribosomal initiation at alternate AUGs on the Sendai virus P/C mRNA, *J. Virol.* **57:**684–687.

Dethlefsen, L. A., and Kolakofsky, D., 1983, In vitro synthesis of the nonstructural C protein of Sendai virus, *J. Virol.* **46:**321–324.

Dowling, P. C., Giorgi, C., Roux, L., Dethlefsen, L. A., Galantowicz, M. E., Blumberg, B. M., and Kolakofsky, D., 1983, Molecular cloning of the 3′ proximal third of Sendai virus genome, *Proc. Natl. Acad. Sci. USA* **80:**5213–5216.

Elango, N., Kovamees, J., and Norrby, E., 1989a, Sequence analysis of the mumps virus mRNA encoding the P protein, *Virology* **169:**62–67.

Elango, N., Kovamees, J., Varsanyi, T. M., and Norrby, E., 1989b, mRNA sequence and deduced amino acid sequence of the mumps virus small hydrophobic protein gene, *J. Virol.* **63:**1413–1415.

Elliott, G. D., Afzal, M. A., Martin, S. J., and Rima, B. K., 1989, Nucleotide sequence of the matrix, fusion and putative SH protein genes of mumps virus and their deduced amino acid sequences, *Virus Res.* **12:**61–76.

Etkind, P. R., Cross, R. K., Lamb, R. A., Merz, D. C., and Choppin, P. W., 1980, In vitro synthesis of structural and nonstructural proteins of Sendai and SV5 viruses, *Virology* **100:**22–33.

Evans, R. A., 1988, The steroid and thyroid hormone receptor superfamily, *Science* **240:**889–895.

Feagin, J. E., Jasmer, D. P., and Stuart, K., 1987, Developmentally regulated addition of nucleotides within the apocytochrome b transcripts in *Trypanosoma brucei*, *Cell* **49:**337–345.

Feagin, J. E., Abraham, J. M., and Stuart, K., 1988, Extensive editing of the cytochrome c oxidase III transcript in *Trypanosoma brucei*, *Cell* **53:**413–422.

Frankel, A. D., and Pabo, C. O., 1988, Fingering too many proteins, *Cell* **53:**675.

Frankel, A. D., Bredt, D. S., and Pabo, C. O., 1988, Tat protein from human immunodeficiency virus forms a metal-linked dimer, *Science* **240:**70–73.

Galinski, M. S., Mink, M. A., Lambert, D. M., Wechsler, S. L., and Pons, M. W., 1986, Molecular cloning and sequence analysis of the human parainfluenza 3 virus mRNA encoding the P and C proteins, *Virology* **155:**46–60.

Gilmore, R., and Blobel, G., 1985, Translocation of secretory proteins across the microsomal membrane occurs through an environment accessible to aqueous perturbants, *Cell* **42:**939–950.

Giorgi, C., Blumberg, B. M., and Kolakofsky, D., 1983, Sendai virus contains overlapping genes expressed from a single mRNA, *Cell* **35:**829–836.

Gupta, K. C., and Kingsbury, D. W., 1985, Translational modulation *in vitro* of a eukaryotic viral mRNA encoding overlapping genes: Ribosome scanning and potential roles of conformational changes in the P/C mRNA of Sendai virus, *Biochem. Biophys. Res. Commun.* **131:**91–97.

Gupta, K. C., and Patwardhan, S., 1988, ACG, the initiator codon for Sendai virus C protein, *J. Biol. Chem.* **263:**8553–8556.

Hall, W. W., Lamb, R. A., and Choppin, P. W., 1980, The polypeptides of canine distemper virus: Synthesis in infected cells and relatedness to the polypeptides of other mobilliviruses, *Virology* **100:**433–449.

Hamaguchi, M., Yoshida, T., Nishikawa, K., Naruse, H., and Nagai, Y., 1983, Transcriptive complex of Newcastle disease virus. I. Both L and P proteins are required to constitute an active complex, *Virology* **128:**105–117.

Hann, S. R., King, M. W., Bentley, D. L., Anderson, C. W., and Eisenman, R. L., 1988, A non-AUG translation initiation in c-*myc* exon 1 generates an N-terminally distinct protein whose synthesis is disrupted in Burkitt's lymphomas, *Cell* **52:**185–195.

Herrler, G., and Compans, R. W., 1982, Synthesis of mumps virus polypeptides in infected vero cells, *Virology* **119:**430–438.

Hiebert, S. W., Paterson, R. G., and Lamb, R. A., 1985a, Hemagglutinin-neuraminidase protein of the paramyxovirus simian virus 5: Nucleotide sequence of the mRNA predicts an N-terminal membrane anchor, *J. Virol.* **54:**1–6.

Hiebert, S. W., Paterson, R. G., and Lamb, R. A., 1985b, Identification and predicted sequence of a previously unrecognized small hydrophobic protein, SH, of the paramyxovirus simian virus 5, *J. Virol.* **55:**744–751.

Hiebert, S. W., Richardson, C. D., and Lamb, R. A., 1988, Cell surface expression and orientation in membranes of the 44-amino-acid SH protein of simian virus 5, *J. Virol.* **62:**2347–2357.

Jacks, T., Madhani, H. D., Masiarz, F. R., and Varmus, H. E., 1988, Signals for ribosomal frameshifting in the Rous sarcoma virus *gag–pol* region, *Cell* **55:**447–458.

Johansen, H., Schumperli, D., and Rosenberg, M., 1984, Affecting gene expression by altering the length and sequence of the 5' leader, *Proc. Natl. Acad. Sci. USA* **81:**7698–7702.

Johnston, M., and Dover, J., 1987, Mutations that inactivate a yeast transcriptional regulatory protein cluster in an evolutionary conserved DNA binding domain, *Proc. Natl. Acad. Sci. USA* **84:**2401–2405.

Klug, A., and Rhodes, D., 1987, 'Zinc fingers': A novel protein motif for nucleic acid recognition, *Trends Biochem. Sci.* **12:**464–469.

Kozak, M., 1981a, Possible role of flanking nucleotides in recognition of the AUG initiator codon by eucaryotic ribosomes, *Nucleic Acids Res.* **9:**5233–5252.

Kozak, M., 1981b, Mechanism of mRNA recognition by eukaryotic ribosomes during initiation of protein synthesis, *Curr. Top. Microbiol. Immunol.* **93:**81–123.

Kozak, M., 1983, Translation of insulin-related polypeptides from messenger RNAs with tandemly reiterated copies of the ribosome binding site, *Cell* **34:**971–978.

Kozak, M., 1984, Point mutations close to the AUG initiator codon affect the efficiency of translation of rat preproinsulin *in vivo*, *Nature* **308:**241–246.

Kozak, M., 1986a, Regulation of protein synthesis in virus-infected animal cells, *Adv. Virus Res.* **31:**229–292.

Kozak, M., 1986b, Point mutations define a sequence flanking the AUG initiator codon that modulates translation by eukaryotic ribosomes, *Cell* **44:**283–292.

Kozak, M., 1986c, Bifunctional messenger RNAs in eukaryotes, *Cell* **47:**481–483.

Kozak, M., 1987, An analysis of 5'-noncoding sequences from 699 vertebrate messenger RNAs, *Nucleic Acids Res.* **15:**8125–8148.

Kozak, M., 1989, The scanning model for translation: An update, *J. Cell. Biol.* **108:**229–241.

Kyte, J., and Doolittle, R. F., 1982, A simple method for displaying the hydropathic character of a protein, *J. Mol. Biol.* **157:**105–132.

Lamb, R. A., 1975, The phosphorylation of Sendai virus proteins by a virus particle-associated protein kinase, *J. Gen. Virol.* **26:**249–263.

Lamb, R. A., and Choppin, P. W., 1977a, The synthesis of Sendai virus polypeptides in infected cells. II. Intracellular distribution of polypeptides, *Virology* **81:**371–381.

Lamb, R. A., and Choppin, P. W., 1977b, The synthesis of Sendai virus polypeptides in infected cells. III. Phosphorylation of polypeptides, *Virology* **81:**382–397.

Lamb, R. A., and Choppin, P. W., 1978, Determination by peptide mapping of the unique polypeptides in Sendai virions and infected cells, *Virology* **84**:469–478.

Lamb, R. A., Mahy, B. W. J., and Choppin, P. W., 1976, The synthesis of Sendai virus polypeptides in infected cells, *Virology* **69**:116–131.

Lamb, R. A., Lai, C.-J., and Choppin, P. W., 1981, Sequences of mRNAs derived from genome RNA segment 7 of influenza virus: Colinear and interrupted mRNAs code for overlapping proteins, *Proc. Natl. Acad. Sci. USA* **78**:4170–4174.

Lamb, R. A., Zebedee, S. L., and Richardson, C. D., 1985, Influenza virus M_2 protein is an integral membrane protein expressed on the infected-cell surface, *Cell* **40**:627–633.

Leibowitz, J. L., Perlman, S., Weinstock, G., DeVries, J. R., Budzilowicz, C., Weissemann, J. M., and Weiss, S. R., 1988, Detection of a murine coronavirus nonstructural protein encoded in a downstream open reading frame, *Virology* **164**:156–164.

Liu, C.-C., Simonsen, C. C., and Levinson, A. D., 1984, Initiation of translation at internal AUG codons in mammalian cells, *Nature* **309**:82–85.

Lomedico, P. T., and McAndrew, S. J., 1982, Eukaryotic ribosomes can recognize preproinsulin initiation codons irrespective of their position relative to the 5' end of mRNA, *Nature* **299**:221–226.

Luk, D., Sanchez, A., and Banerjee, A. K., 1986, Messenger RNA encoding the phosphoprotein (P) gene of human parainfluenza virus 3 is bicistronic, *Virology* **153**:318–325.

McGinnes, L., McQuain, C., and Morrison, T., 1988, The P protein and the nonstructural 38K and 29K proteins of Newcastle disease virus are derived from the same open reading frame, *Virology* **164**:256–264.

Moran, E., and Mathews, M. B., 1987, Multiple functional domains in the adenovirus E1A gene, *Cell* **48**:177–178.

Olmsted, R. A., and Collins, P. L., 1989, The 1A protein of respiratory syncytial virus is an integral membrane protein present as multiple, structurally distinct species, *J. Virol.* **63**:2019–2029.

Paterson, R. G., and Lamb, R. A., 1987, Ability of the hydrophobic fusion-related external domain of a paramyxovirus F protein to act as a membrane anchor, *Cell* **48**:441–452.

Paterson, R. G., and Lamb, R. A., 1990, RNA editing by G nucleotide insertion in mumps "P" gene mRNA transcripts, *J. Virol.* September issue.

Paterson, R. G., Harris, T. J. R., and Lamb, R. A., 1984, Analysis and gene assignment of mRNAs of a paramyxovirus, simian virus 5, *Virology* **138**:310–323.

Paterson, R. G., Thomas, S. M., and Lamb, R. A., 1989, Specific non-templated nucleotide addition to a simian virus 5 mRNA: Prediction of a common mechanism by which unrecognized hybrid P-cysteine-rich proteins are encoded by paramyxovirus "P" genes, In "Genetics and Pathogenicity of Negative Strand Viruses" (D. Kolakofsky and B. W. J. Mahy, eds.), pp. 232–245, Elsevier, London.

Peluso, R. W., Lamb, R. A., and Choppin, P. W., 1977, Polypeptide synthesis in simian virus 5-infected cells, *J. Virol.* **23**:177–187.

Peluso, R. W., Lamb, R. A., and Choppin, P. W., 1979, Infection with paramyxoviruses stimulates synthesis of cellular polypeptides which are also stimulated in cells transformed by Rous sarcoma virus or deprived of glucose, *Proc. Natl. Acad. Sci. USA* **75**:6120–6124.

Portner, A., Gupta, K. C., Seyer, J. M., Beachey, E. H., and Kingsbury, D. W., 1986, Localization and characterization of Sendai virus nonstructural C and C' proteins by antibodies against synthetic peptides, *Virus Res.* **6**:109–121.

Powell, L. M., Wallis, S. C., Pease, R. J., Edwards, Y. H., Knott, T. J., and Scott, J., 1987, A novel form of tissue-specific RNA processing produces apolipoprotein-B48 in intestine, *Cell* **50**:831–840.

Prats, H., Kaghad, M., Prats, A. C., Klagsbrun, M., Lelias, J. M., Liauzun, P., Chalow, P., Tauber, J. P., Amalric, F., Smith, J.. A., and Caput, D., 1989, High molecular mass forms of basic fibroblast growth factor are initiated by alternative CUG codons, *Proc. Natl. Acad. Sci. USA* **86**:1836–1840.

Randall, R. E., Young, D. F., Goswami, K. K. A., and Russell, W. C., 1987, Isolation and characterization of monoclonal antibodies to simian virus 5 and their use in revealing antigenic differences between human, canine and simian isolates, *J. Gen. Virol.* **68**:2769–2780.

Richardson, C. D., Berkovich, A., Rosenblatt, S., and Bellini, W. J., 1985, Use of antibodies directed against synthetic peptides in identifying cDNA clones, establishing reading frames, and deducing the gene order of measles virus, *J. Virol.* **53**:186–193.

Rima, B. K., Robert, M. W., McAdam, W. D., and Martin, S. J., 1980, Polypeptide synthesis in mumps virus infected cells, *J. Gen. Virol.* **46**:501–505.

Sanchez, A., and Banerjee, A. K., 1985a, Studies on human parainfluenza virus 3: Characterization of the structural proteins and *in vitro* synthesized proteins coded by mRNAs isolated from infected cells, *Virology* **143**:45–54.

Sanchez, A., and Banerjee, A. K., 1985b, Cloning and gene assignment of mRNAs of human parainfluenza virus 3, *Virology* **147**:177–186.

Sato, H., Oh-hira, M., Ishida, N., Imamura, Y., Hattori, S., and Kawakita, M., 1987, Molecular cloning and nucleotide sequence of *P, M* and *F* genes of Newcastle disease virus avirulent strain D26, *Virus Res.* **7**:241–255.

Shaw, J. M., Feagin, J. E., Stuart, K., and Simpson, L., 1988, Editing of kinetoplastid mitochondrial mRNAs by uridine addition and deletion generates conserved amino acid sequences and AUG initiation codons, *Cell* **53**:401–411.

Shaw, M. W., Lamb, R. A., Erikson, B. W., Briedis, D. J., and Choppin, P. W., 1982, Complete nucleotide sequence of the neuraminidase gene of influenza B virus, *Proc. Natl. Acad. Sci. USA* **79**:6817–6821.

Shaw, M. W., Choppin, P. W., and Lamb, R. A., 1983, A previously unrecognized influenza B virus glycoprotein from a bicistronic mRNA that also encodes the viral neuraminidase, *Proc. Natl. Acad. Sci. USA* **80**:4879–4883.

Shioda, T., Hidaka, Y., Kanda, T., Shibuta, H., Nomoto, A., and Iwasaki, K., 1983, Sequence of 3,687 nucleotides from the 3′ end of Sendai virus genome RNA and the predicted amino acid sequences of viral NP, P and C proteins, *Nucleic Acids Res.* **11**:7317–7330.

Spriggs, M. K., and Collins, P. L., 1986, Sequence analysis of the P and C protein genes of human parainfluenza virus type 3: Patterns of amino acid sequence homology among paramyxovirus proteins, *J. Gen. Virol.* **67**:2705–2719.

Steck, T. L., and Yu, J., 1973, Selective solubilization of proteins from red blood cell membranes by protein perturbants, *J. Supramol. Struct.* **1**:220–248.

Strebel, K., Klimkait, T., and Martin, M. A., 1988, A novel gene of HIV-1, *vpu*, and its 16-kilodalton product, *Science* **241**:1221–1223.

Taira, H., Kanda, T., Omata, T., Shibuta, H., Kawakita, M., and Iwasaki, K., 1987, Interferon induction by transfection of Sendai virus C gene cDNA, *J. Virol.* **61**:625–628.

Takeuchi, K., Hishiyama, M., Yamada, A., and Sugiura, A., 1988, Molecular cloning and sequence analysis of the mumps virus gene encoding the P protein: Mumps virus P gene is monocistronic, *J. Gen. Virol.* **69**:2043–2049.

Thomas, S. M., Lamb, R. A., and Paterson, R. G., 1988, Two mRNAs that differ by two non-templated nucleotides encode the amino co-terminal proteins P and V of the paramyxovirus SV5, *Cell* **54**:891–902.

Vidal, S., Curran, J., and Kolakofsky, D., 1990, Editing of the Sendai virus P/C mRNA by G insertion occurs during mRNA synthesis via a virus-coded activity, *J. Virol.* **64**:239–246.

Williams, M. A., and Lamb, R. A., 1986, Determination of the orientation of an integral membrane protein and sites of glycosylation by oligonucleotide-directed mutagenesis: Influenza B virus NB glycoprotein lacks a cleavable signal sequence and has an extracellular NH_2-terminal region, *Mol. Cell. Biol.* **6**:4317–4328.

Williams, M. A., and Lamb, R. A., 1989, Effect of mutations and deletions in a bicistronic mRNA on the synthesis of influenza B virus NB and NA glycoproteins, *J. Virol.* **63**:28–35.

Zebedee, S. L., and Lamb, R. A., 1988, Influenza A virus M_2 protein: Monoclonal antibody restriction of virus growth and detection of M_2 in virions, *J. Virol.* **62**:2762–2772.

Zebedee, S. L., and Lamb, R. A., 1989, Growth restriction of influenza A virus by M_2 protein antibody is genetically linked to the M_1 protein, *Proc. Natl. Acad. Sci. USA* **86**:1061–1065.

CHAPTER 7

Paramyxovirus RNA Synthesis and *P* Gene Expression

DANIEL KOLAKOFSKY, SYLVIA VIDAL, AND JOSEPH CURRAN

I. PARAMYXOVIRUS RNA SYNTHESIS IN RELATION TO THAT OF OTHER (−) RNA VIRUSES

Five RNA virus families have now been classified as negative stranded: paramyxoviruses and rhabdoviruses, which contain nonsegmented genomes, and arenaviruses, bunyaviruses, and influenza viruses, which contain two, three, and seven to eight segments, respectively. The name was coined to distinguish this group from that of plus-strand RNA viruses, such as poliovirus, when it became clear that the free genomic RNAs of the former were not infectious, their mRNAs were the complements of the genomes, and these virions contained an RNA-dependent RNA polymerase. Although some genome segments of arenaviruses and some bunyaviruses code for mRNAs of both polarities and are referred to as ambisense (Auperin *et al.*, 1984), the term negative stranded applies here as well, to the extent that all these viruses follow a basic replication strategy. For all these viruses, the genomic RNA is tightly bound with at least ten times its weight of the major structural protein (NP or N) in the form of a helical nucleocapsid (NC) and cannot be translated (Leung *et al.*, 1977), even in the case of ambisense genomes which contain protein-coding sequences near their 5′ ends. Genome replication therefore begins with mRNA synthesis from the infecting genomes, by the associated viral polymerase. The proteins resulting from these primary transcripts are required for genome amplification, which takes place via a full-length complement of the (−) genome, the antigenome, which is also found only as an NC structure.

A central event in the replication of all (−) RNA viruses is the balance between genome transcription to yield mRNAs and genome replication, i.e.,

---

DANIEL KOLAKOFSKY, SYLVIA VIDAL, AND JOSEPH CURRAN • Department of Microbiology, University of Geneva Medical School, C.M.U., 1211 Geneva 4, Switzerland.

antigenome synthesis, as the genome template is used for both events. In the case of segmented viruses such as influenza virus and bunyavirus for which information is available, this control takes place, at least in part, at the level of chain initiation, as the initiation of genome transcription and replication are clearly different events. Viral mRNAs start on capped primers derived from host-cell mRNAs, whereas genome replication begins with a triphosphate (Krug, 1981; Lamb and Choppin, 1983; Bishop *et al.*, 1983; Patterson and Kolakofsky, 1984). For the nonsegmented viruses, on the other hand, these events cannot be distinguished at the level of initiation, as both are considered to start with ATP at the precise 3′ end of the template. Of course, what distinguishes these events is that transcription produces a short leader RNA molecule and several mRNAs (see below), and that transcription terminates before the end of the template [see Banerjee (1987), Emerson and Schubert (1987), and Kolakofsky and Roux (1987), for recent reviews]. These multiple transcripts are presumably the result of the polymerase stopping and restarting at each of the gene junctions. The mRNAs are capped and polyadenylated during synthesis, whereas the leader RNA is unmodified at either end, and the junctions contain conserved sequences which specify these events in part.

In summary, replication is thought to be the same for all (−) RNA viruses, in that it starts with a triphosphate at the 3′ end of the template and produces an uninterrupted, exact copy. Transcription, on the other hand, is quite different for the segmented and nonsegmented viruses. For the segmented viruses, there is no leader region which separates the beginning of the mRNA from the 3′ end of the template, mRNAs begin at this end on a capped primer, and the polymerase never reinitiates after terminating to form the 3′ end of the mRNA. These viral polymerases do not contain capping activity per se, i.e., guanylyl and methyl transferases, but a cap-dependent endonuclease which generates primers and a primer-dependent transcriptase (Bouloy *et al.*, 1978; Patterson *et al.*, 1984). For the nonsegmented viruses, mRNA synthesis is primer independent, and is thought to take place only upon reinitiation of the polymerase at each junction.

In this review, we will also cite experiments done with the rhabdovirus vesicular stomatitis virus (VSV), mostly when similar work has not been done with paramyxoviruses. We appreciate that these are two different virus families, and the differences between them have been brought into sharper focus of late. However, VSV remains of strong interest to those who work with paramyxoviruses, as the differences as well as similarities are informative.

II. THE SWITCH FROM SENDAI VIRUS TRANSCRIPTION TO REPLICATION

Upon entering the cytoplasm, the replicative cycle begins with the synthesis of viral mRNAs by the infecting NCs. Purified virion NCs, which are complexes of the P and L proteins in addition to the tightly associated genomic RNA : NP template, also carry out mRNA synthesis *in vitro*, which is dependent on the presence of both the L and P proteins (Hamaguchi *et al.*, 1983). The P and L proteins have been found by immunogold staining to colocalize in clusters at sites which are randomly distributed on the NCs (Portner et al., 1988). For VSV, some of the analogous complexes (NS–L proteins) which are bound to internal sites on the viral genome are also active, giving rise to short transcripts from several of the mRNA start sites in vitro (Testa et al., 1980). These mRNA leaders, however, appear to be dead-end transcripts, in that they cannot be chased into mRNAs (Chanda and Banerjee, 1981; Lazzarini et al., 1982). Initiation at the internal mRNA start sites can also be seen in vitro using only ATP and CTP, and by combining this restriction of substrates with reconstitution studies, Emerson (1982) has shown that if these internally bound complexes are removed and added back to the core template, they only initiate at the 3' end of the template. Emerson has suggested that the internally bound polymerases were trapped during maturation by M protein interactions with progeny NCs, and that this might be a way of ensuring that multiple copies of the polymerase reside on each virion NC. Analogous experiments have not been done with paramyxoviruses, but we assume that here, too, the polymerase enters its template at the 3' end and RNA synthesis begins with the leader region.

The virion polymerase reaction is thought to represent primary transcription in vivo, which is experimentally defined by carrying out the infection in the absence of protein synthesis. No antigenome RNA synthesis occurs under these conditions, presumably because polymerases which initiate at the 3' end of the genome template terminate and reinitiate at the leader/NP gene (l/NP) junction, and at each subsequent junction, to produce the leader and the monocistronic mRNAs sequentially. At each junction, there is a certain probability that the polymerase will not reinitiate, leading to a gradient of mRNA abundance in which the more distal cistrons are progressively underrepresented. The junctions contain conserved sequences which are thought to specify termination (via polyadenylation) and reinitiation, or mRNA stop and start signals (Giorgi et al, 1983; Gupta and Kingsbury, 1984). These junctional signals are probably not the only determinants acting to stop polymerase readthrough. In some persistent measles virus infections, only P–M bicistronic mRNAs are made, even though the consensus sequence at the junction is unaltered (Cattaneo et al., 1986). For antigenome synthesis to occur, the polymerase must disregard all the termination signals, and this requires de novo viral protein synthesis. There is also a requirement for on-going protein synthesis throughout the infection, as genome and antigenome synthesis are thought to occur coordinately with their encapsidation, and about 2000 copies of NP are required for each assembly.

Antigenomes (as well as genomes) are found only in NCs, and NP is the only viral protein known to be required for genome replication that is not also required for transcription. NP abundance therefore appears to be at least part of the mechanism which switches the polymerase from transcription to genome replication. This has given rise to a minimal model for this switch (Blumberg et al., 1981; Kolakofsky and Blumberg, 1982; Arnheiter et al., 1985), diagrammed in Fig. 1, whose general features were first suggested by Kingsbury (1974). We know that at least for VSV, leader RNA first accumulates in vivo free of assembled NP, but later in infection this RNA is found assembled as mini-NCs (Blumberg and Kolakofsky, 1981). The leader RNA must therefore contain

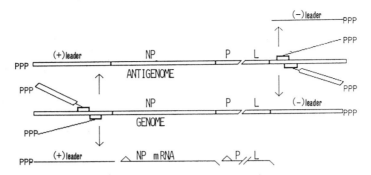

FIGURE 1. Representation of the proposed model for Sendai virus transcription and genome replication. Genome and antigenome RNAs assembled with NP protein as NCs are shown as horizontal open boxes. Unencapsidated RNAs such as (+)- and (−)-leader RNAs and mRNAs are shown as thin horizontal lines. Vertical lines within the boxes designate gene junctions. The genes are not drawn to scale, and the lengths of the (+)- and (−)-leader regions have been exaggerated for clarity. The triangle at the left and the slanted line at the right end of each mRNA refer to the 5' cap group and the poly(A) tail, respectively. Small boxes above or below the templates refer to polymerases which carry nascent (+)- or (−)-leader chains which have either already begun NC assembly (open boxes) or remain unencapsidated (thin lines). For further details, see text.

the site for the initiation of NC assembly, and this site maps to the first 14 bases at the 5' end of the leader chain in assembly studies *in vitro* (Blumberg *et al.*, 1983). The presence of this site, which is unique to the leader regions, then explains why the antigenome (as well as the genome) is encapsidated, whereas the viral mRNAs are not. The minimal model proposes that the switch is effected entirely by the initiation of NC assembly on the nascent leader chain, the polymerase and the template being the same for both processes. NC assembly on the nascent chain somehow prevents the polymerase from responding to the leader stop signal, and all subsequent stop signals, as assembly of the nascent antigenome RNA presumably follows the advancing polymerase. The ratios of transcription to replication would then be determined by the levels of unassembled NP intracellularly. The model suggests a self-regulatory system where insufficient NP promotes mRNA synthesis to provide more viral proteins, and excess NP promotes genome replication, which lowers the pool of unassembled NP.

The model is also supported by the presence of a VSV (−) leader RNA *in vivo* (Leppert *et al.*, 1979), which can also be made *in vitro*, templated from the 3' end of the antigenome template (and copy-back defective-interfering templates). Like (+) leader and mRNA synthesis from genome templates, (−)-leader synthesis *in vivo* does not require on-going protein synthesis (Blumberg *et al.*, 1981). No other subgenomic RNAs are made from antigenomes, presumably because there are no reinitiation sites following the (−)-leader stop signal. The existence of the (−) leader is important for the model. It suggests that the polymerase which initiates on antigenomes, *which are templates only for genome replication*, is also not controlled at the level of chain initiation. The requirement for on-going protein synthesis for genome synthesis therefore also acts at the level of polymerase readthrough of a stop site on the antigenome. We have recently found that (−)-leader RNAs also exist for Sendai virus, and that their synthesis is similarly independent of on-going protein synthesis, at least *in vitro* (our unpublished data).

The model was derived in large part with VSV, but what information is available for Sendai virus is entirely consistent with it. However, as its name implies, the minimal model is likely to be a first approximation, and we already know that things are not that simple. Unassembled NP may not act in isolation during NC assembly, as VSV N is suggested to act as a complex with NS (the VSV equivalent of the Sendai P protein) (Howard *et al.*, 1986; Peluso and Moyer, 1988), and Baker and Moyer (1988) have shown that something in addition to purified NP is required to initiate (but not elongate) Sendai genome replication *in vitro*. Moreover, mutants of VSV (polR) have been described in which changes in the N protein (as part of the template) dramatically increase the frequency with which the polymerase reads through the first junction in the absence of concurrent assembly (Perrault *et al.*, 1983). Accommodation of such mutants within the minimal model is difficult.

We have recently found, unexpectedly, that the Z strain of Sendai virus behaves very similarly to the polR mutants of VSV, even though this strain hade been considered to be a wild-type (wt) strain. Another Sendai virus strain (H), on the other hand, behaved similarly to wt VSV in this respect. These experiments (Vidal and Kolakofsky, 1989) were done by examining the levels of leader readthrough RNAs, i.e., RNAs containing the leader region joined to the beginning of the NP mRNA, relative to the levels of the NP mRNA (which begins at nucleotide 56), in intracellular RNAs which had not been encapsidated (they had been centrifuged through CsCl density gradients). The CsCl pellet RNAs were examined by RNase mapping so that the readthrough RNAs and the NP mRNAs could be quantitated simultaneously with the same riboprobe. A staggered set of riboprobes was also used, both to help identify the protected RNA fragments and to determine how far the readthrough RNAs had extended. Consistent with the minimal model, leader readthrough RNAs were very scarce in H strain-infected cell CsCl pellet RNA, but were found to be relatively abundant in Z-strain infections. Figures 2 and 3 show some of these results using the Z strain. Leader RNAs *per se* are hard to detect in these intracellular samples, possibly because they turn over more rapidly or are encapsidated and therefore no longer pellet through CsCl. However, the levels of the more abundant leader readthrough RNAs relative to NP mRNA can be accurately determined by using increasing amounts of the viral RNAs to protect the riboprobe. As the sequences of the protected fragments are known, the results are expressed on a molar basis. Summing all our results, there is at least 20 times more readthrough RNA in Z- than in H-infected cells. In the former case, readthrough RNAs can represent as much as 40% of the total NP mRNA. However, the relative levels of readthrough RNAs fall sharply as a function of their chain length, and very few of them have extended more than 300 nucleotides from the first junction.

The Z-strain polymerase therefore reads through the first junction at a relatively high frequency, even in the absence of concurrent assembly of the nascent chain. However, having read through this junction, these polymerases abort shortly afterwards. In contrast, polymerases which have reinitiated at

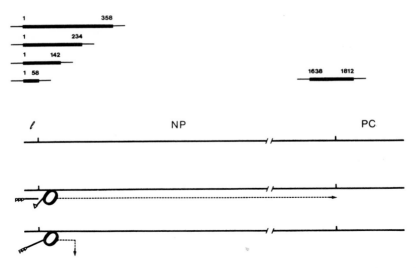

FIGURE 2. Diagram of riboprobes used to measure RNA synthesis and the modified model. The middle line represents the 3' or left end of the minus-strand genome with the leader, NP, and PC regions separated by vertical bars representing the junctions. Map positions are numbered from this end, which is spatially equivalent to the 5' end of the antigenome. Above are shown the Z-strain riboprobes used. The extents of the viral sequences are indicated by heavy lines and map positions; the thin lines at the ends indicate transcribed vector sequences. Below are shown the two situations of plus-strand RNA synthesis found in Z-strain-infected cells in the absence of concurrent assembly. In each case the polymerase is shown just after having crossed the first junction, but with or without having reinitiated, and the consequences for continued elongation are shown by the dashed lines. [From Vidal and Kolakofsky (1989).]

this junction continue on to the next junction at more than 90% frequency. When the frequency of polymerase readthrough of the next junction (*NP/PC*) was measured in these same experiments, the Z and H polymerases could no longer be distinguished, as both polymerases read through the second junction at the same low frequency (0.5%), similar to the H polymerase at the first junction.

   These leader readthrough RNAs are unlikely to be true replication intermediates for several reasons. (1) They are far too abundant to be transient intermediates (40% of NP mRNA at 4 hr after infection). (2) They accumulate at the same rate relative to NP mRNA even when protein synthesis is blocked, and therefore clearly do not require new viral proteins for their synthesis. (3) They are also very stable; they still represent as much as 20% of NP mRNA at 18 hr. (4) We have seen only (+)- but not (−)-leader readthrough RNAs in Z-infected cells (unpublished results), even though (−) genomes are the predominant products of replication. All of these findings distinguish these readthrough RNAs from *bona fide* replication intermediates described for VSV (Hill *et al.*, 1979; Simonsen *et al.*, 1979).

   The unassembled leader readthrough RNAs therefore appear to be dead-end transcripts, of no particular value to the infection. They are also of little harm, as the Z strain grows almost as quickly as H, again similar to polR and

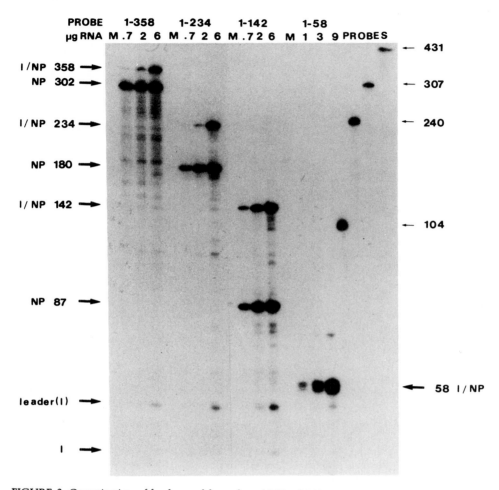

FIGURE 3. Quantitation of leader readthrough and NP mRNAs in strain Z-infected cells with a nested set of riboprobes. (A) Three different amounts of CsCl pellet RNA (taken 18 hr after infection), each a threefold increment as indicated, were used to protect riboprobes representing positions 1–58, 1–142, 1–234, and 1–358 (Fig. 2). Including their transcribed vector sequences, these riboprobes were 104, 240, 307, and 431 nucleotides long, respectively, and are shown on the right (probes). The RNAs remaining after RNase digestion were separated by electrophoresis. The autoradiogram from one such experiment is shown; the lengths (in nucleotides) of the various probes used and RNAs detected are indicated on the sides. The probes used in each experiment are indicated above, as is the amount of protecting RNA for each lane. M refers to 6 or 9 μg of RNA from mock-infected cells. The first eight lanes, containing the larger probes, were exposed for shorter times than those containing the smaller probes. (B) Bands representing the leader read-through RNAs (l/NP) and the NP mRNAs (NP) were excised from a parallel gel which contained blank lanes in between and their Cerenkov radiation was determined. The counts present in each band were divided by the number of uridines and plotted as a function of the amount of protecting RNA for the three smallest riboprobes. Only the NP mRNA bands from riboprobe 1–142 are shown. (C) Results obtained with the lowest amount of protecting RNA from three separate experiments, one of which is shown in panels A and B, plotted according to the 3′ end of the plus-stranded fragment protected in the various hybrids. To normalize the values for the NP mRNA, the average of the values from the three riboprobes in each experiment (115, 34, and 189 cpm/U

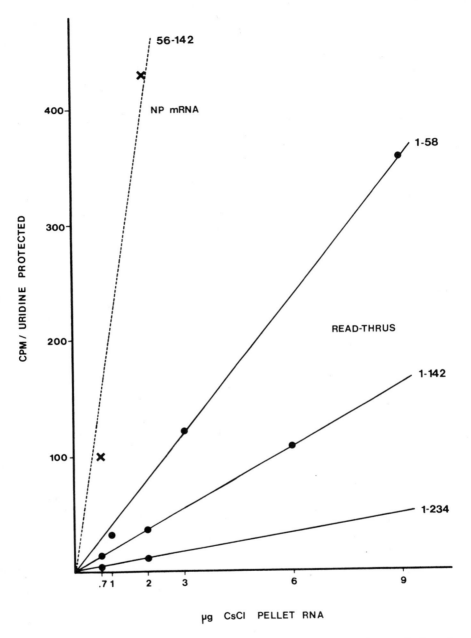

FIGURE 3. (*continued*)   protected) have been set to 100, and the results from each riboprobe are plotted relative to this value. The readthrough RNAs, on the other hand, are plotted relative to the NP mRNA detected by the same probe in the same experiment, except for that detected by ribo-probe 1–58, which is plotted relative to the average value for the NP mRNA. Large and small crosses and circles indicate the mean and individual values for NP mRNA and the leader readthrough RNAs, respectively. [From Vidal and Kolakofsky (1989).]

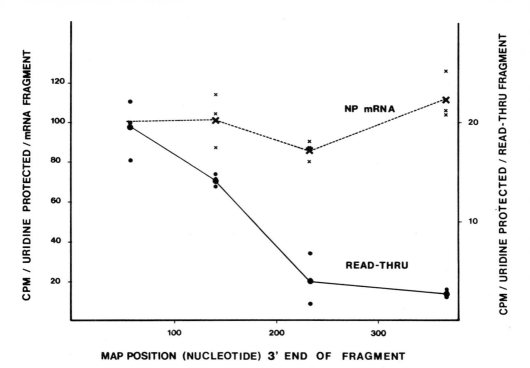

FIGURE 3. (continued)

wt VSV. By analogy with VSV, strain H would be wt, and strain Z would be mutant. However, both Sendai virus strains are, as far as we know, natural isolates. We also note that although VSV polR was selected by a heat resistance protocol, polR is not heat resistant, and the manner in which this phenotype was selected is unclear (J. Perrault, personal communication). The readthrough phenotype is not particularly detrimental to the abilities of these viruses to replicate, and might be selected for under either natural or unnatural circumstances whenever slightly attenuated viral growth is advantageous.

Regardless of their role in viral replication, which remains unclear, the leader readthrough RNAs provide insight into the mechanism of genome transcription and replication. They speak directly to the question of whether the polymerases which transcribe or replicate are in fact identical, as the minimal model proposes, or are different. One conclusion of the above work is that the Z polymerase which makes mRNAs can be distinguished from that which makes antigenomes, even in the absence of concurrent NC assembly. The former continues down the genome template efficiently without concurrent assembly once it has crossed the first junction, whereas the latter requires NC assembly. These two polymerases are therefore clearly different, but can only be distinguished once they have crossed the first junction. Before this junction, they both carry the same leader chain (with a 5′ triphosphate). After this junction, one carries the nascent leader readthrough RNA, whereas the other carries the nascent NP mRNA (having reinitiated at position 56). The question now arises as to the stage at which they became different. One possibility is that they (or their templates) were in fact different from the start, but this difference is only manifest after crossing the first junction. The other possibility, which we favor, is that in the act of reinitiating at the *NP* gene, the polymerase is converted from a form which requires concurrent NC assembly to continue down the template efficiently, to a form which does not.

The role in the model of NP protein assembling with progeny RNA must also be modified, but less significantly. The minimal model suggests that assembling NP acts only at the junctions. The data suggest that when the polymerase has not been converted to an assembly-independent mode, concurrent assembly is required for the polymerases to proceed between junctions as well. The leader stop site would then be only the first of multiple sites where the polymerase will stop without concurrent assembly.

The notion that there are many sites where the polymerase will stop when it has not been converted to an assembly-independent mode may perhaps explain those other curious dead-end transcripts, the VSV mRNA leaders. These transcripts result primarily from polymerases which have been packaged on virion NCs at or near an mRNA start site. They start the mRNA chains at the correct positions with a triphosphate, but terminate prematurely at discrete sites soon afterward, and the majority of these chains remain uncapped. The internally bound polymerases then appear to be in the assembly-dependent mode for elongation, like those that start at the 3' end of the template, even though we assume they reached their internal positions via 3' end loading, and possibly also by reinitiation. One explanation for why they are in the assembly-dependent mode is that during virion maturation, or simply with time extracellularly, they reverted from their assembly-independent mode by the undoing of whatever modification that had taken place. Both examples of premature termination by the Sendai and VSV polymerases then could result from the failure to convert to an assembly-independent mode or to revert from this mode.

We know very little about what changes in the polymerase complex might take place concomitant with reinitiation which could render the complex independent of concurrent assembly for efficient elongation. Work with VSV virions has shown that polymerases which initiate at leader or mRNA start sites can be distinguished by a variety of experimental conditions (Banerjee, 1987), including ATP concentration (Perrault and McClear, 1984; Beckes *et al.*, 1987). Cleavage of the beta/gamma bond of ATP is known to be required for initiation (and termination?), and this might be related to protein phosphorylation. The VSV NS protein, like the paramyxovirus P protein, is the major viral phosphoprotein, and NS proteins with different amounts of phosphate have different specific activities for RNA synthesis *in vitro* (Kingsford and Emerson, 1980). Phosphorylation (or dephosphorylation) of P could therefore also be involved in modifying paramyxovirus polymerase activity, and such modifications could take place when the polymerase has paused at the junction during termination/reinitiation. The protein kinase activity of Sendai virions, which heavily phosphorylates P and NP, is very tightly associated with the viral NCs (Vidal *et al.*, 1988). It is also concievable that polymerase components can be exchanged during pausing at the junctions.

A slightly more elaborate model now takes shape. The old focus of how concurrent assembly of the nascent chain prevents premature termination when the polymerase is in the assembly-dependent mode remains, but there is a new focus. What is the modification that presumably takes place during reinitiation that converts the polymerase to an assembly-independent mode for elongation, and how does it operate? Moreover, for paramyxoviruses, a complete description of genome replication will probably have to include the V (and C) proteins, also coded for by the P gene, and which do not appear to be present in rhabdoviruses. The unusual ways in which the genetic information for these proteins is expressed are detailed below.

III. TRANSCRIPTIONAL CHOICE WITHIN THE SENDAI VIRUS *P/C* GENE

The viral polymerase has other choices to make besides whether to obey or read through junctional signals, at least within the *P* gene. This paramyxovirus gene, the second in the transcriptional map, stands apart from the others in that it always codes, or appears to code, for more than one primary translation product (see Chapter 6). (The pneumovirus RSV *P* gene is probably an exception here.) For Sendai virus (Giorgi *et al.*, 1983; Shioda *et al.*, 1983), human and bovine PIV3 (Galinski *et al.*, 1986; Luk *et al.*, 1986, H. Shibuta, personal communication), and the morbilliviruses measles (Bellini *et al.*, 1985) and CDV (Barrett *et al.*, 1985), the *P* gene codes for a 600-amino acid P protein and contains an overlapping gene (*C*, coding for a 200-amino acid protein) within the N terminus of the P coding region. Here, the P and C proteins are made from the same mRNA by ribosomal choice of their respective initiation codons (see below). For SV5 (Thomas *et al.*, 1988), NDV (Sato *et al.*, 1987), and mumps virus (Takeuchi *et al.*, 1988), the P protein is only about 400 amino acids long and the overlapping *C* gene at the N terminus is missing, yet this gene still codes for other primary translation products. An unusual manner in which this happens has recently become clear for SV5 (Thomas *et al.*, 1988; also see Chapter 6). The SV5 *P* gene is interrupted in the middle by a stop codon, and the mRNA, which is an exact complement of the genome, is translated into the shorter V protein. A second mRNA with a precise two-G insertion upstream of the stop codon is also made, and translated into the longer P protein (the insertion opens the reading frame to near the end of the altered mRNA).

The C-terminal region of V, which is absent in P, is unusually rich in Cys, and remarkably conserved in all other paramyxovirus *P* genes. This is in sharp contrast to the *P* (and *C*) coding regions, which appear to be the least conserved among all the paramyxovirus coding regions. Thomas *et al.*, (1988) predicted that V proteins, i.e., proteins containing the same N terminus as P but with the Cys-rich C-terminal domain, would be expressed in all paramyxoviruses (except RSV). By the time the work of Thomas *et al.* appeared in print, Cattaneo *et al.* (1989) had independently found a second *P* gene mRNA in measles virus, coding for a V protein. Because this second measles *P* gene mRNA also contained a G insertion within a run of G's, and runs of G's with similar flanking sequences are found near the required frameshift position in the other paramyxoviruses, one could now predict where these other insertions would be.

The circumstantial evidence that a V protein would also be expressed in Sendai virus was compelling, and we had by this time seen more than our share of primary translation products from the Sendai virus *P* gene (see below). However, neither we nor others had ever noticed a V protein, either in infected cells or by *in vitro* translation of the total mRNA population. We had also done a good deal of work with genomic *P* gene clones, but these would yield only mRNAs without insertions by *in vitro* transcription via the SP6 polymerase. However, the predicted Sendai V protein would be M_r 46,000 and if it migrated anomalously on gels similar to the behavior of P, it could comigrate with NP or F_0. We therefore looked for a V mRNA, assuming that it would differ from the P mRNA as predicted, i.e., a one-G insertion within a run of three G's, at position 1053.

cDNA was made from infected cell RNA with a *P*-gene-specific primer, digested with enzymes which would cut on either side of the expected insertion and only once within the *P* gene, and then cloned. Colonies were screened with three oligonucleotides. The first oligonucleotide, referred to as the neutral oligo, is complementary to a region just downstream of the predicted insertion site where no mRNA modifications are expected. The neutral oligo identifies those clones which carry the P gene fragment, without discriminating whether the insertions have or have not taken place. The other two oligos are complementary to the insertion site, and can distinguish whether there are three (genomic oligo, P mRNA) or four (+G oligo, V mRNA) G's at this position. In preliminary experiments, we found that one-fourth to one-third of the *P* gene mRNAs were V mRNAs by this test, and representative clones from this group did in fact translate *in vitro* into V (and C), but not P, proteins, as expected, when the requisite subcloning was carried out. (The V protein does in fact comigrate with NP on SDS gels.) Numerous examples of "misediting" were also found, including the insertion of two, four, five, and eight G's at the same site. In terms of "misediting," Sendai virus and measles virus are thus similar, and perhaps distinct from SV5, where this event appears to be less frequent. Both Sendai and measles virus insertions involve a single G, whereas two G's are inserted for SV5, and the mechanism for adding only a single G may be inherently more leaky.

The insertions are thought to be due to a very precise stuttering, or reiterative copying, of the viral polymerase during transcription of the short run of C's on the template. Polyadenylation of mRNAs is thought to take place by a similar, but much less precise, mechanism at the short run of U's at each mRNA stop site. However, we would like to know whether the insertions are due entirely to the viral polymerase, or whether cellular factors are also involved, or whether the insertions could possibly even be due to some form of editing which is nonviral. This question can be addressed with purified Sendai virions, which make significant amounts of mRNA *in vitro*. In experiments where such mRNA was examined by directed cloning as above, about one-fifth of the *P*-gene mRNAs were also found to have a precise one-G insertion.

The insertions are then likely to be the result of a virus-coded activity, but there is a *caveat* to this conclusion. Cellular actin is visible in stained protein gels of our purified virions, and host tRNAs, as well as RNase NU and RNase P, can be detected enzymatically. It is therefore not impossible that a cellular editing activity has been similarly included in virions. If this were so, however, we would expect that this activity would also edit the P/C mRNA expressed from a recombinant vaccinia virus carrying a genomic copy of this gene (VV–P/C). Cytoplasmic RNA was therefore prepared from BHK cells infected with VV–P/C when P and C protein expression is maximal, by sedimentation through a cushion of 5.7 M CsCl to remove DNA. We controlled the absence of VV DNA from our sample by showing that no neutral oligo⁺ colonies could be generated by cloning without prior cDNA synthesis. The insertion region of the recombinant mRNA was then cloned and examined. Of the 198 neutral oligo⁺ clones examined, all were also genomic oligo⁺. We found only three ambiguous clones which were also +G oligo⁺, but which on sequencing had three G's at nucleotide 1053. The P/C mRNA expressed from VV DNA is therefore not modified in the same cell line in which the mRNA expressed from Sendai virus is modified by a single G insertion 31% of the time.

If the insertions result from the polymerase stuttering during transcription, then the insertion activity would not operate on preformed mRNAs. This could be tested by examining the VV-derived P/C mRNA from cells that were coinfected with Sendai virus if the natural and the VV mRNA could be distinguished. Serendipitously, we found a single base change within the XbaI to EcoRI region between VV–P/C which was constructed from a clone obtained in 1981 (a G at nucleotide 1035) and our current Sendai H virus stocks (a C at nucleotide 1035). We assume that this base difference does not affect the frequency of insertion at nucleotide 1053, since this virus strain has had the same growth characteristics in different cell lines over these years. Oligonucleotides were then made which could distinguish a G or a C at nucleotide 1035 and were used to first separate cDNA clones of mRNA from coinfected cells according to the viral genome from which they had been transcribed. These were then examined for G insertion at nucleotide 1053. We found that none of the VV mRNAs were altered during the same infection in which 36% of the natural Sendai virus mRNA had insertions. As the coinfection was started at the same time and there were five times less VV mRNA than the natural mRNA as judged by the number of clones of each type, it seems unlikely that the lack of insertions in the VV mRNA was due to limiting insertion activity. It therefore appears that the insertion activity acts only during RNA synthesis by the Sendai virus polymerase. This experiment also rules out the possibility that the activity is host coded, but only induced by Sendai virus infection.

Taken together, these results indicate that the activity responsible for the insertions is virus coded, and cannot function in *trans*. However, the somewhat lower insertion frequency *in vitro* could mean that cellular proteins help in this process. Alternatively, it might reflect other differences between the *in vitro* and *in vivo* conditions for mRNA synthesis. We also note that the term "editing" is therefore not an exact description of this process, as the changes appear to be made during and not after mRNA synthesis.

The Sendai virus V protein is composed of the first 316 residues of P, followed by 68 amino acids containing the Cys-rich domain. The Cys motif here is reminiscent of "zinc fingers" or other metal-binding sites such as those found in the HIV tat protein (Evans, 1988; Frankel *et al.*, 1988), and these are associated with nucleic acid-binding proteins in other systems. One therefore

expects that this protein will be involved in viral RNA synthesis, and antibodies specific to the SV5 P and V proteins also precipitate L and NP as a
complex (Thomas *et al.*, 1988). However, as the Sendai V protein comigrates
with NP on SDS gels and we have no antibodies which are specific to V (nor to
the N-terminal half of P), we have been unable to locate V unambiguously.

However, the Sendai V protein is unlikely to be associated with NCs in the
same manner as P, from what is known about the region of P responsible for its
binding to NCs. Using P proteins with various deletions and cell extracts in
which the NCs are associated with their endogenous P and L proteins, Ryan
and Kingsbury (1988) mapped the NC attachment site *in vitro* to two contiguous restriction fragments, one specifying amino acids 345–538, the other
specifying residue 539 to the C terminus at residue 568. Their data do not
distinguish between a single site bridging residues 538/539 or two or more
separate sites. Using NCs purified by CsCl density gradients, we have found
that neither residues 1–538 nor residues 473–568 (the X protein, see below) by
themselves bind to NCs. These experiments presumably measure binding to
NP, the only protein remaining on these NCs. This suggests that residues 345–
472 in addition to residues 539–568 are required for NC attachment, and that
there are two or more sites on P which constitute the attachment region. The V
protein, however, contains only residues 1–316 of P, and is thus missing this
region entirely.

The N-terminal half of the P protein sequence is shared with V, and within
this region are found the sites at which P is highly phosphorylated (Vidal *et al.*,
1988). We assume that V will also be highly phosphorylated, but this remains
to be determined. The possible functions of the N-terminal half of P are difficult to even guess at this time. In a remarkable experiment, Chinchar and
Portner (1981) removed about half of the P sequences by proteolysis from the
protein bound to NCs, yet found that this had little effect on their ability to
make RNA *in vitro.* Later work showed that what remained attached to NCs
was indeed the C-terminal half of the P chain (Deshpande and Portner, 1985;
Vidal *et al.*, 1988). The phosphorylated N-terminal half therefore does not
appear to participate directly in RNA synthesis, at least as measured in their *in
vitro* system. For VSV, when the highly phosphorylated N-terminal half of NS
is removed, RNA synthesis is lost, but can be restored when this domain or a
structurally similar one is added back in *trans* (Chattopadhyay and Banerjee,
1987, 1988). This may represent a real difference between these two systems,
but it is also possible that V was present and resistant to proteolysis in the
experiment of Chinchar and Portner, but hiding under NP (which is protease
resistant), and substituted functionally for the lost acidic domain of P. Many
experiments will have to be reexamined now that V is known to exist.

IV. RIBOSOMAL CHOICE DURING SENDAI VIRUS
P GENE EXPRESSION

A plethora of ribosomal initiation sites are used during translation of the
Sendai *P*-gene mRNAs. Site-directed mutagenesis has shown that the initiation codon closest to the 5' end of the capped mRNA is in fact a non-AUG

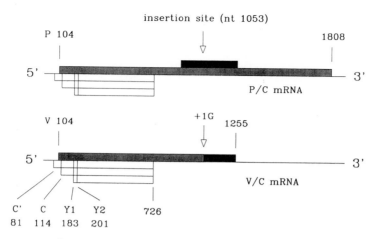

FIGURE 4. Protein-coding regions of the Sendai virus P/C and V/C mRNAs. The mRNAs are shown as a horizontal lines with their 5' and 3' ends indicated. The *P, C,* and *V* open reading frames are shown as stippled, open, and black boxes, respectively. The position (in nucleotides) of each start codon and the protein which results is indicated; numbers refer to the map position of the first base of the start codon relative to the 5' end of the mRNA. The V protein is presumably made by ribosomes which start at AUG/104, as all other ways of acessing this ORF are blocked by stop codons. Note that both the unaltered (P) and inserted (V) mRNAs would translate the C proteins identically.

codon (ACG/81) which starts the C' protein (Curran and Kolakofsky, 1988a; Gupta and Patwardham, 1988) (see Fig. 4), followed by AUGs at positions 104, 114, 183, and 201, which start the P (and V), C, Y1, and Y2 proteins, respectively (Curran and Kolakofsky, 1988a; Patwardham and Gupta, 1988). Proteins C', C, Y1, and Y2 therefore have staggered N termini and are carboxy coterminal, and are all translated from the C open reading frame (ORF). When each of these AUGs is changed to a noninitiator codon, only its cognate protein is ablated in translations *in vitro*. A protein representing the last 95 amino acids of the P ORF, the X protein, also appears to be expressed independently during infection, presumably from ATG/1523; however, this has not been confirmed by mutational analysis (Curran and Kolakofsky, 1988b). Since each of these initiation codons functions independently, they provide internal controls for each other in experiments examining the manner in which the initiation codons are selected. In particular, we would like to know whether all these initiation sites are selected by a particularly leaky ribosomal scanning mechanism, or whether other mechanisms operate on this mRNA, which is so unusual in this respect.

Two experimental approaches were adopted for *in vitro* studies using mRNAs from the genomic *P* gene clone (Curran and Kolakofsky, 1989). The first approach was to either create a new ATG upstream of ACG/81 such that a P protein with an N-terminal extension of 25 amino acids would result (CAG/29–ATG), or to change ACG/81 to ATG. In this approach it is assumed that at least the first two of the natural initiation sites (ACG/81 and ATG/104) are designed to be inherently leaky for scanning ribosomes. Introduction of a nonleaky site upstream should then limit scanning ribosomes to initiation at this 5' nonleaky site. Proteins which continue to be made from the downstream sites under these conditions would presumably initiate by a scanning-independent mechanism.

Upon translation of CAG/29–ATG mRNA, C' and P synthesis could no longer be detected, C synthesis was reduced to 30–50% of control wt mRNA, and Y synthesis was reduced by less than 20%. Upon translation of ACG/81–ATG, C' was increased (sevenfold) as expected, P was again undetectable, C was reduced to 16% of control levels, but again Y continued to be made at near normal levels (75%). In both these mutant mRNAs, if the total absence of P synthesis indicates that scanning ribosomes cannot bypass the new 5'-proximal initiation codon, then the remaining synthesis of C and Y is unlikely to be the result of leaky scanning.

In a second *in vitro* approach, oligonucleotides complementary to the 5' untranslated region were used to specifically cleave the mRNA with RNase H. Translation of these mRNAs which contained a physical break in the chain between the cap and the initiation codons yielded results very similar to the above. C' and P synthesis were completely ablated, significant C synthesis continued, and Y synthesis was mostly unaffected. As it is difficult to see how a scanning ribosome can cross a physical gap in the chain, some C synthesis and most Y synthesis must be scanning independent, at least *in vitro*.

To examine whether scanning-independent initiation also took place *in vivo*, the ACG/81–ATG mutant and wt constructs were transferred to vaccinia virus and protein expression was monitored in recombinant infected cells by immunoblotting (Curran and Kolakofsky, 1989). We found that under these conditions C as well as P synthesis was completely suppressed by the new ATG/81, whereas Y synthesis actually increased severalfold relative to the wt construct. Y initiation, *in vivo* as *in vitro*, therefore apparently takes place with ribosomes that do not scan from the 5' end of the mRNA, but somehow bind internally. Why C initiation is partly scanning independent *in vitro* but not *in vivo* is unclear. However, one possibility is that ribosomes can bind internally near the Y initiation sites and scan upstream to a limited extent *in vitro*, but not *in vivo*.

V. THE UNKNOWN FUNCTIONS OF THE VARIOUS C PROTEINS

We know little about the possible functions of the C proteins, or why in Sendai virus they should be present as an N-terminally nested set. These proteins are clearly underrepresented in virions relative to what is present intracellularly (although some C protein can always be detected in virions by sensitive immunoblotting). They are therefore unlikely to be important in primary transcription. In measles virus-infected cells, where only a single species of the C protein has been detected and in relatively low amounts, C colocalizes with P on NCs (Bellini *et al.*, 1985). In Sendai virus-infected cells, where the C proteins are much more abundant (relative to the other viral proteins), most of the C proteins do not colocalize with P (Portner *et al.*, 1986;

J. Curran and R. Compans, unpublished). However, C proteins made *in vitro* bind efficiently to CsCl-banded NCs under conditions where the X protein, or the P protein missing the 30 C-terminal amino acids, do not (our unpublished results). NCs therefore appear to contain specific binding sites for C, but these would be insufficient to bind all the C that is made during Sendai virus infection. In contrast to the P proteins, which are highly acidic, the C proteins are highly basic, but we have seen no evidence of specific association between them. Although it is tempting to speculate that the C protein(s) play a role in genome replication, there is little evidence that this is so. On the other hand, one report has suggested that the Sendai virus C protein is involved in interferon induction (Taira *et al.*, 1987).

SV5, mumps virus, and NDV do not appear to contain C proteins, in that there is no C ORF overlapping the N-terminal end of P, as found in the other paramyxoviruses. This might represent a true difference between these two groups. However, it is also possible that there is such a domain elsewhere in their genomes, which could be expressed, for example, as the C terminus of a protein made from an altered mRNA. Considering the diversity with which genetic information is expressed from the paramyxovirus *P* gene, which now includes base insertions in the mRNA, non-AUG initiation codons, and scanning-independent ribosomal initiation, it would be rash to exclude such possibilities out of hand.

VI. REFERENCES

Arnheiter, H., Davis, N. L., Wertz, G., Schubert, M., and Lazzarini, R. A., 1985, Role of nucleocapsid protein in regulating vesicular stomatitis virus RNA synthesis, *Cell* **41**:259–267.

Auperin, D. D., Romanowski, V., Galinski, M., and Bishop, D. H. L., 1984, Sequencing studies of Pichinde arenavirus S RNA indicate a novel coding strategy, an ambisense viral S RNA, *J. Virol.* **52**:897–904.

Baker, S. C., and Moyer, S. A., 1988, Encapsidation of Sendai virus genome RNAs by purified NP protein during *in vitro* replication, *J. Virol.* **62**:834–838.

Banerjee, A. K., 1987, Transcription and replication of rhabdoviruses, *Microbiol. Rev.* **51**:66–87.

Barrett, T., Shrimpton, S. B., and Russell, S. E. H., 1985, Nucleotide sequence of the entire protein coding region of canine distemper virus polymerase-associated (P) protein mRNA, *Virus Res.* **3**:367–372.

Beckes, J. D., Haller, A. A., and Perrault, J., 1987, Differential effect of ATP concentration on synthesis of vesicular stomatitis virus leader RNAs and mRNAs, *J. Virol.* **61**:3470–3478.

Bellini, W. J., Englund, G., Rozenblatt, S., Arnheiter, H., and Richardson, C. D., 1985, Measles virus P gene codes for two proteins, *J. Virol.* **53**:908–919.

Bishop, D. H. L., Gay, M. E., and Matsuoko, Y., 1983, Nonviral heterogeneous sequences are present at the 5' ends of one species of snowshoe hare bunyavirus S complementary RNA, *Nucleic Acids Res.* **11**:6409–6418.

Blumberg, B. M., and Kolakofsky, D., 1981, Intracellular vesicular stomatitis virus leader RNAs are found in nucleocapsid structures, *J. Virol.* **40**:568–576.

Blumberg, B. M., Leppert, M., and Kolakofsky, D., 1981, Interaction of VSV leader RNA and nucleocapsid protein may control VSV genome replication, *Cell* **23**:837–845.

Blumberg, B. M., Giorgi, C., and Kolakofsky, D., 1983, N protein of vesicular stomatitis virus selectively encapsidates leader RNA *in vitro*, *Cell* **32**:559–567.

Bouloy, M., Plotch, S. J., and Krug, R. M., 1978, Globin mRNAs are primers for the transcription of influenza viral RNA *in vitro*, *Proc. Natl. Acad. Sci. USA* **75**:4886–4890.

Cattaneo, R., Schmid, A., Rebmann, G., Baczko, K., ter Meulen, V., Bellini, W. J., Rozenblatt, S., and Billeter, M. A., 1986, Accumulated measles virus mutations in a case of subacute sclerosing panencephalitis: Interrupted matrix protein reading frame and transcription alteration, *Virology* **154**:97–107.

Cattaneo, R., Kaelin, K., Baczko, K., and Billeter, M. A., 1989, Measles virus editing provides an additional cysteine-rich protein, *Cell* **56**:759–764.

Chanda, P. K., and Banerjee, A. K., 1981, Identification of promoter-proximal oligonucleotides and a unique dinucleotide, pppGpC, from *in vitro* transcription products of vesicular stomatitis virus, *J. Virol.* **39**:93–103.

Chattopadhyay, D., and Banerjee, A. K., 1987, Two separate domains within vesicular stomatitis virus phosphoprotein support transcription when added in *trans*, *Proc. Natl. Acad. Sci. USA* **84**:8932–8936.

Chattopadhyay, D., and Banerjee, A. K., 1988, NH$_2$-terminal acidic region of the phosphoprotein of vesicular stomatitis virus can be functionally replaced by tubulin, *Proc. Natl. Acad. Sci. USA* **85**:7977–7981.

Chinchar, V. G., and Portner, A., 1981, Functions of the Sendai virus nucleocapsid polypeptides: Enzymatic activities in nucleocapsids following cleavage of polypeptide P by *Staphylococcus aureus* protease V8, *Virology* **109**:59–71.

Curran, J. A., and Kolakofsky, D., 1988a, Ribosomal initiation from an ACG codon in the Sendai virus P/C mRNA, *EMBO J.* **7**:245–251.

Curran, J. A., and Kolakofsky, D., 1988b, Scanning independent ribosomal initiation of the Sendai virus X protein, *EMBO J.* **7**:2869–2874.

Curran, J. A., and Kolakofsky, D., 1989, Scanning independent ribosomal initiation of the Sendai virus Y proteins *in vitro* and *in vivo*, *EMBO J.* **8**:521–526.

Deshpande, K. L., and Portner, A., 1985, Monoclonal antibodies to the P protein of Sendai virus define its structure and role in transcription, *Virology* **140**:125–134.

Emerson, S. U., 1982, Reconstitution studies detect a single polymerase entry site on the vesicular stomatitis virus genome, *Cell* **31**:635–642.

Emerson, S. U., and Schubert, M., 1987, Molecular basis of rhabdovirus replication, in "The Molecular Basis of Viral Replication" (P. Bercoff, ed., pp. 255–276, Plenum Press, New York.

Evans, R. A., 1988, The steroid and thyroid hormone receptor super-family, *Science* **240**:889–895.

Frankel, A. D., Bredt, D. S., and Pabo, C. O., 1988, Tat protein from human immunodeficiency virus forms a metal-linked dimer, *Science* **240**:70–73.

Galinski, M. S., Mink, M. A., Lambert, D. M., Wechsler, S. L. and Pons, M. W., 1986, Molecular cloning and sequence analysis of the human parainfluenza 3 virus mRNA encoding the P and C proteins, *Virology* **155**:46–60.

Giorgi, C., Blumberg, B. M., and Kolakofsky, D., 1983, Sendai virus contains overlapping genes expressed from a single mRNA, *Cell* **35**:829–836.

Gupta, K. C., and Kingsbury, D. W., 1984, Complete sequences of the intergenic and mRNA start signals in the Sendai virus genome: Homologies with the genome of vesicular stomatitis virus, *Nucleic Acids Res.* **12**:3829–3841.

Gupta, K. C., and Patwardhan, S., 1988, ACG, the initiator codon for a Sendai virus protein, *J. Biol. Chem.* **263**:8553–8556.

Hamaguchi, M., Yoshida, T., Nishikawa, K., Naruse, H., and Nagai, Y., 1983, Transcriptive complex of Newcastle disease virus. I. Both L and P proteins are required to constitute an active complex, *Virology* **128**:105–117.

Hill, V. M., Simonsen, C. C., and Summers, D. F., 1979, Characterization of vesicular stomatitis virus replicating complexes isolated in renografin gradients, *Virology* **99**:75–83.

Howard, M., Davis, N. L., Patton, J., and Wertz, G. W., 1986, Roles of vesicular stomatitis virus N and NS proteins in viral RNA replication, in "The Biology of Negative Strand Viruses" (B. Mahy and D. Kolakofsky, eds.), pp. 134–140, Elsevier Science Publishing, New York.

Kingsbury, D. W., 1974, The molecular biology of paramyxoviruses, *Med. Microbiol. Immunol.* **160**:73–83.

Kingsford, L., and Emerson, S. U., 1980, Transcriptional activities of different phosphorylated species of NS protein purified from vesicular stomatitis virions and cytoplasm of infected cells, *J. Virol.* **33**:1097–1105.

Kolakofsky, D., and Blumberg, B. M. 1982, A model for the control of non-segmented negative strand viruses genome replication, in "Virus Persistence Symposium 33. Society for General Microbiology" (B. W. J. Mahy, A. C. Minson, and G. K. Darby, eds.), pp. 203–213, Cambridge University Press, Cambridge.

Kolakofsky, D., and Roux, L., 1987, The molecular biology of paramyxoviruses, in "The Molecular Basis of Viral Replication" (P. Bercoff, ed.), pp. 277–297, Plenum Press, New York.

Krug, R. M., 1981, Priming of influenza viral RNA transcription by capped heterologous RNAs, Curr. Top. Microbiol. Immunol. 93:125–149.

Lamb, R. A., and Choppin, P. W., 1983, The gene structure and replication of influenza virus, Annu. Rev. Biochem. 52:467–506.

Lazzarini, R. A., Chien, I., Yang, F., and Keene, J. D., 1982, The metabolic fate of independently initiated VSV mRNA transcripts, J. Gen. Virol. 58:429–441.

Leppert, M., Rittenhouse, L., Perrault, J., Summers, D. F., and Kolakofsky, D., 1979, Plus and minus strand leader RNAs in negative strand virus-infected cells, Cell 18:735–747.

Leung, W. C., Ghosh, H. P., and Rawls, W. E., 1977, Strandedness of Pichinde virus RNA, J. Virol. 22:235–237.

Luk, D., Sanchez, A., and Banerjee, A. K., 1986, Messenger RNA encoding the phosphoprotein (P) gene of human parainfluenza virus 3 is bicistronic, Virology 153:318–325.

Patterson, J. L., and Kolakofsky, D., 1984, Characterization of La Crosse virus small-genome transcripts, J. Virol. 49:680–685.

Patterson, J. L., Holloway, B., and Kolakofsky, D., 1984, La Crosse virions contain a primer-stimulated RNA polymerase and a methylated cap-dependent endonuclease, J. Virol. 52:215–222.

Patwardhan, S., and Gupta, K. C., 1988, Translation initiation potential of the 5' proximal AUGs of the polycistronic P/C mRNA of Sendai virus: A multipurpose vector for site specific mutagenesis, J. Biol. Chem. 263:4907–4913.

Peluso, R. W., and Moyer, S. A., 1988, Viral proteins required for the in vitro replication of vesicular stomatitis virus defective interfering particle genome RNA, Virology 162:369–376.

Perrault, J., and McClear, P. W., 1984, ATP dependence of vesicular stomatitis virus transcription initiation and modulation by mutation in the nucleocapsid protein, J. Virol. 51:635–642.

Perrault, J., Clinton, G. M., and McClure, M. A., 1983, RNP template of vesicular stomatitis virus regulates transcription and replication functions, Cell 35:175–185.

Portner, A., Gupta, K. C., Seyer, J. M., Beachy, E. H., and Kingsbury, D. W., 1986, Localization and characterization of Sendai virus nonstructural C and C' proteins by antibodies against synthetic peptides, Virus Res. 6:109–121.

Portner, A., Murti, K. G., Morgan, E. M., and Kingsbury, D. W., 1988, Antibodies against Sendai virus L protein: Distribution of the protein in nucleocapsids revealed by immunoelectron microscopy, Virology 163:236–239.

Ryan, K. W., and Kingsbury, D. W., 1988, Carboxy-terminal region of Sendai virus P proteins is required for binding to viral nucleocapsids, Virology 167:106–112.

Sato, H., Oh-hira, M., Ishida, N., Imamura, Y., Hattori, S., and Kawakita, M., 1987, Molecular cloning and nucleotide sequence of P, M and F genes of Newcastle disease virus avirulent strain D26, Virus Res. 7:241–255.

Shioda, T., Hidaka, Y., Kanda, T., Shibuta, H., Nomoto, A., and Iwasaki, K., 1983, Sequence of 3,687 nucleotides from the 3' end of Sendai virus genome RNA and the predicted amino acid sequences of viral NP, P and C protein, Nucleic Acids Res. 11:7317–7330.

Simonsen, C. C., Hill, V. M., and Summers, D. F., 1979, Further characterization of the replicative complex of vesicular stomatitis virus, J. Virol. 31:494–505.

Taira, H., Kanda, T., Omata, T., Shibuta, H., Kawakita, M., and Iwasaki, K., 1987, Interferon induction by transfection of Sendai virus C gene cDNA, J. Virol. 61:625–628.

Takeuchi, K., Hishiyama, M., Yamada, A., and Sugiura, A., 1988, Molecular cloning and sequence analysis of the mumps virus gene encoding the P protein: Mumps virus P gene is monocistronic, J. Gen. Virol. 69:2043–2049.

Testa, D., Chanda, P. K., and Banerjee, A. K., 1980, Unique mode of transcription in vitro by vesicular stomatitis virus, Cell 21:67–275.

Thomas, S. M., Lamb, R. A., and Paterson, R. G., 1988, Two mRNAs that differ by two nontemplated nucleotides encode the amino coterminal proteins P and V of the paramyxovirus SV5, Cell 54:891–902.

Vidal, S., and Kolakofsky, D., 1989, Modified model for the switch from Sendai virus transcription to replication. J. Virol. 63:1951–1958.

Vidal, S., Curran, J., Orvell, C., and Kolakofsky, D., 1988, Mapping of monoclonal antibodies to the Sendai virus P protein and the location of its phosphates, J. Virol. 62:2200–2203.

CHAPTER 8

Function of Paramyxovirus 3' and 5' End Sequences
In Theory and Practice

BENJAMIN M. BLUMBERG, JOHN CHAN, AND STEPHEN A. UDEM

I. THE KINGSBURY–KOLAKOFSKY MODEL FOR TRANSCRIPTION AND REPLICATION

A. Features of the Kingsbury Model

A clear and detailed model for paramyxovirus transcription and replication was first presented by Kingsbury (1974, 1977). This early model envisioned a promoter sequence for polymerase entry located at the genomic and antigenomic 3' ends, and a polarity gradient of transcription of the viral mRNAs by the viral polymerase operating on the helical nucleocapsid, the physical properties of which are determined by the nucleocapsid protein. A key feature of this model was a suggestion that the relative amounts of polymerase products (mRNAs or full-length genomes) would be modulated by the availability of a virus-encoded protein, possibly the structural nucleocapsid protein itself, in the switchover from transcription to replication (see also Chapters 7 and 9).

B. Development of the Model in Vesicular Stomatitis Virus (VSV)

Until very recently, this model was tested mainly in the rhabdovirus vesicular stomatitis virus (VSV), where the polarity gradient was also observed and

BENJAMIN M. BLUMBERG • Departments of Neurology and Microbiology and Immunology, University of Rochester Medical Center, Rochester, New York 14642. JOHN CHAN AND STEPHEN A. UDEM • University of Medicine and Dentistry New Jersey—New Jersey Medical School, Newark, New Jersey 07103.

where the discovery of leader RNAs, the initial product of viral transcription, proved to be the royal road to the study of mechanism [masterfully reviewed by Banerjee (1987)]. In VSV, the first 14 nucleotides at the 5' end of the nascent leader RNA were shown to contain a nucleation site for encapsidation by the nucleocapsid (N) protein (Blumberg *et al.*, 1983). After engaging the genome 3'-end promoter, the viral polymerase synthesizes a short RNA transcript: it becomes encapsidated and replication of a full-length genome ensues if sufficient N protein is available; if not, RNA synthesis is interrupted with release of a free leader RNA, and transcription of viral mRNA commences. The stoichiometric requirement for N protein in nucleocapsid assembly thus implies homeostatic regulation of transcription and replication by its concentration in the cytoplasm of the infected cell (Blumberg *et al.*, 1981; Kolakofsky and Blumberg, 1982).

The cutoff of the nascent transcript to give a leader RNA suggested the operation of a termination signal (Leppert *et al.*, 1979). Its existence was proven indirectly in an analytical review of VSV infections involving defective-interfering (DI) particles (Blumberg and Kolakofsky, 1983). The identity of the termination sequence in the genome is unknown, although it seems reasonable that it would be located near the leader cutoff point, just before the start site for the N mRNA encoded by the genome 3'-end sequence. Thus, three operational elements in transcription and replication have been defined in the VSV model: promoter and leader RNA terminator sites in the genomic RNA, and an encapsidation/nucleation site in the nascent RNA strand. By segregating the encapsidation/nucleation site into the cutoff leader transcript, leader RNAs function to prevent encapsidation of viral mRNAs or wasteful synthesis of unencapsidated genomic RNAs.

II. TESTING OF THE MODEL IN PARAMYXOVIRUSES SHOWS DIFFERENCES FROM VSV

A. Leader RNAs Are Not an Essential Operational Element

Testing of the model in paramyxovirus systems shows several differences from VSV. Although (+)-leader RNAs are present in Sendai virus- (Leppert *et al.*, 1979) and Newcastle disease virus (NDV)-infected cells (Kurilla *et al.*, 1985), (−) leaders encoded by sequences of the antigenome 3' end have not been documented, in contrast to the situation in VSV-infected cells, where leader RNAs derived from both genome and antigenome are abundant (Leppert and Kolakofsky, 1980). Instead, there are leader–NP and higher-order readthrough mRNAs in paramyxovirus-infected cells. Early experiments (Leppert *et al.*, 1979) showed that oligo-dT-bound RNAs from Sendai virus-infected cells contained sequences that annealed with a probe made from 3'-end-labeled 50S genome RNA, although the nonpolyadenylated fraction contained four times the concentration of these sequences. Later, primer extension experiments with Sendai virus *in vitro* transcripts gave a product thought to reflect leader–NP readthrough RNAs, although this material constituted less than 1% of the NP mRNA made *in vivo* (Blumberg *et al.*, 1984). Recent, more elegant RNase protection experiments revealed a range of leader–NP read-through transcripts that is virus strain dependent (Vidal and Kolakofsky, 1989). Leader–NP transcripts have also been observed in mumps virus (Elango et al., 1988) and human parainfluenza virus 3 (HPV3; K. Dimock, personal communication).

From the perspective of the VSV model, these results imply that leader–NP readthrough RNAs represent sloppy replicational starts under conditions of limiting NP protein. However, the situation in measles virus (MV)-infected cells is more extreme: (+)-leader RNAs have not been detected with either DNA or RNA probes (Billeter et al., 1984; Crowley et al., 1988). Moreover, (−) leaders also were not detected among the RNAs extracted from MV-infected cells, even under vigorous annealing conditions, using high-specific-activity riboprobes (Crowley et al., 1988). Two groups have further found that leader–NP readthroughs comprise 5–7.5% of all MV intracellular transcripts, and estimated that (+)-leader RNAs do not constitute more than 0.25–0.43% of NP mRNAs (Castaneda and Wong, 1989; Chan et al., 1989).

Vidal and Kolakofsky (1989) suggest that rapid turnover accounts for a decrease of Sendai virus (+)-leader RNA found in vivo relative to the amount made in vitro; nonetheless, only low amounts of Sendai (+) leaders are made relative to those of VSV. Rapid degradation of MV leader RNAs has not been ruled out, although (+)-leader RNAs were not detectable by direct ^{3}H-UTP pulse labeling in RNAs extracted from MV-infected cells (Chan et al., 1989). We therefore interpret these results to imply that the leader termination signal is weak or absent in paramyxoviruses, and that leader RNAs are not an essential operational element in the transcription and replication of all paramyxoviruses.

B. Experimental Evidence and Sequence Comparison of Paramyxovirus 3' and 5' Ends Suggest a Possible Auxiliary Internal Promoter Site

Also in contrast to the VSV model, two types of experiments suggest the existence of an extended promoter region in paramyxoviruses. R-loop spreads of measles virus 50S genomic RNA show rare panhandle structures (arrow in Fig. 1) of about 160 nucleotides (S. A. Udem, unpublished data). A similar observation was previously made in Sendai virus (Kolakofsky, 1976), but these structures were not stable to RNase treatment in 0.3 M NaCl (Leppert et al., 1977). Since the antigenome does not function as a template for synthesis of viral mRNAs, the 3'- and 5'-terminal partial complementarities that give rise to the unstable panhandle structures most likely reflect similar promoter sequences at the genome and antigenome 3' termini. In line with these data, primer extension experiments using an oligonucleotide (BB3) representing a sequence 79 nucleotides from the MV (−) genome 5' end unexpectedly gave bands representing the 3' end as well, by virtue of a 14/16 base match with a sequence 85 nucleotides from the antigenomic 5' end (Crowley et al., 1988) (see Fig. 2).

Comparison of all known paramyxovirus 3' and 5' end sequences (Fig. 2)

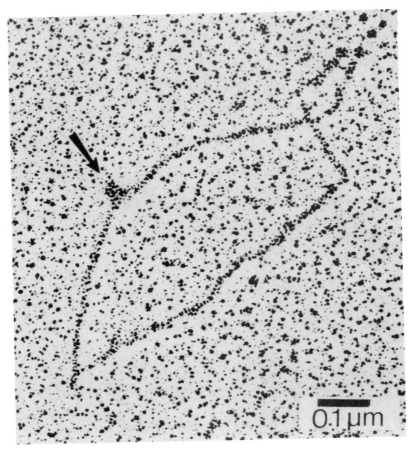

FIGURE 1. Circular form assumed by MV genome RNA suggests regions of terminal complementary sequences. Measles virus 50S genome RNA purified by buoyant density centrifugation in cesium trifluoroacetate gradients was spread, shadowed, and examined by electron microscopy (Udem and Cook, 1984). Occasional unit-length, single-stranded, circular forms demonstrating a terminal duplexed region (panhandle) of about 1% of the total contour length (15,889 nucleotides) are found.

shows that the first 11 nucleotides at the 3' ends of all (−) and (+) template sequences are highly conserved. The canonical NP mRNA start sequences [R1; designation of Shioda *et al.* (1983)] are also highly conserved, with strikingly precise spacing. The canonical mRNA termination (R2) sequences on the antigenome template are also conserved, although their spacing varies. Since replication begins at the precise 3' end of the genome, the replicase needs to engage the template nucleocapsid upstream (i.e., 5') of the actual polymerization site; thus, beyond their role in transcription, the R1 and R2 regions may function in concert with the 3' end during replication. We propose that the BB3-equivalent sequences present in Sendai virus, NDV, HPV3, bovine parainfluenza virus 3 (BPV3), and MV may provide an additional upstream site for polymerase engagement at the start of transcription and/or replication. The

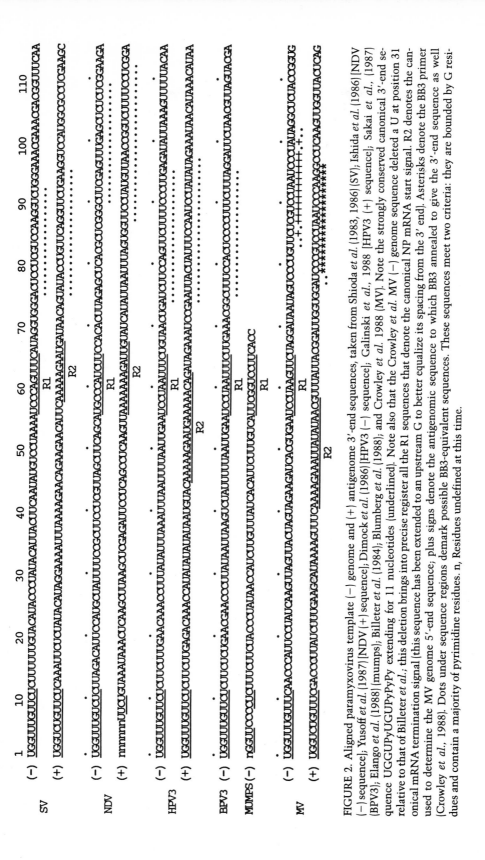

FIGURE 2. Aligned paramyxovirus template (−) genome and (+) antigenome 3′-end sequences, taken from Shioda et al. (1983, 1986)[SV]; Ishida et al. (1986)[NDV (−)sequence]; Yusoff et al. (1987)[NDV (+) sequence]; Dimock et al. (1986)[HPV3 (−) sequence]; Galinski et al., 1988 [HPV3 (+) sequence]; Sakai et al., (1987) [BPV3]; Elango et al. (1988)[mumps]; Billeter et al. (1984); Blumberg et al. (1988); and Crowley et al. 1988 [MV]. Note the strongly conserved canonical 3′-end sequence UGGUPyUGUPPyPyPy extending for 11 nucleotides (underlined). Note also that the Crowley et al. MV (−) genome sequence deleted a U at position 31 relative to that of Billeter et al.; this deletion brings into precise register all the R1 sequences that denote the canonical NP mRNA start signal. R2 denotes the canonical mRNA termination signal (this sequence has been extended to an upstream G to better equalize its spacing from the 3′ end). Asterisks denote the BB3 primer used to determine the MV genome 5′-end sequence; plus signs denote the antigenomic sequence to which BB3 annealed to give the 3′-end sequence as well (Crowley et al., 1988). Dots under sequence regions demark possible BB3-equivalent sequences. These sequences meet two criteria: they are bounded by G residues and contain a majority of pyrimidine residues. n, Residues undefined at this time.

BB3-equivalent sequences are bounded by multiple sterically bulky G residues and contain a majority of smaller pyrimidine residues internally; thus, they could both guide and facilitate polymerase attachment. Speculatively, the relatively high content of G residues within the SV and NDV BB3-equivalent sequences may correlate with the finding of leader RNAs only in these viruses, and the fact that the positions of the BB3-equivalent sequences vary (as do the R2 sequences) may reflect differences in the active site of the L proteins.

C. Host-Cell-Encoded Proteins May Mediate Encapsidation

Host-range restriction experiments early suggested specific host-cell involvement in the replication of VSV (Pringle, 1978). Paramyxoviruses exhibit narrow host tropism, and paramyxovirus replication is intimately tied to host factors. Host-cell proteins have been shown to stimulate transcription and replication of NDV *in vitro* (Hamaguchi *et al.*, 1985), and the NDV transcriptive complex is functional only in association with the cytoskeleton (Yoshida *et al.*, 1986). MV transcription and replication are also stimulated by a cellular cytoplasmic extract *in vitro* (Fig. 3) (Chan *et al.*, 1989). Recently, tubulin and

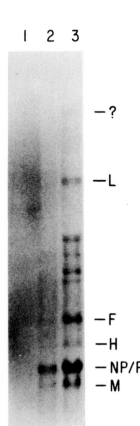

FIGURE 3. Stimulation of MV *in vitro* transcription by host-cell factors. A standard *in vitro* transcription assay system (Seifried *et al.*, 1978; Ray and Fujinami, 1987) containing nonionic-detergent-treated, highly-purified, infectious measles virions (Udem, 1984) was supplemented with small amounts of an unfractionated extract prepared from uninfected HeLa cells. The radiolabeled (³H-UTP) transcription products generated were analyzed by HCHO–agarose gel electrophoresis. Lane 1 documents the absence of any detectable MV transcription products generated by purified virus alone. With the addition of 2 μg of uninfected cell extract (lane 2), synthesis of transcripts encoded by the 3′ end of the MV genome (N, P, M) clearly proceeds, while with 10 μg of extract (lane 3), all the known MV polyadenylated monocistronic and polycistronic transcripts are produced, as well as a putative full-length antigenome-sized transcript denoted by "?".

microtubule-associated proteins (MAPs) were shown to be involved in the synthesis of VSV and Sendai virus RNAs (Hill *et al.*, 1986; Moyer *et al.*, 1986). Sendai virus M protein also associates with actin (Guiffre *et al.*, 1982). Indeed, actin, tubulin, and MAPs interact (Arakawa and Frieden, 1984; Littauer *et al.*, 1986) and are likely to play a role in cytoplasmic transport of nascent virus. Finally, active nuclear transport of viral components is suggested by the formation of inclusion bodies in MV encephalitides such as SSPE (ter Meulen *et al.*, 1983), and the transient intranuclear localization of VSV leader RNA (Kurilla *et al.*, 1982).

We employed northwestern blots to examine the binding of MV RNAs to components of the infected cell. In this technique, a crude protein mixture (e.g., MV-infected cell cytoplasmic extract) is electrophoresed on an SDS–polyacrylamide gel and the separated proteins are then blotted onto a nitrocellulose membrane, as for a Western blot. The blot is then probed with a radiolabeled nucleic acid containing the sequences of interest. For this purpose, we subcloned the inserts of several MV 3'- and 5'-end clones into pGEM vectors (Promega Corp. Madison, Wisconsin) in order to make riboprobes. Unexpectedly, northwestern blot experiments described in Fig. 4 suggest that host-cell proteins (particularly a M_r 90,000 protein), but not MV proteins, strongly and

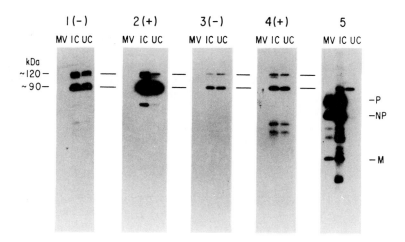

FIGURE 4. MV nascent strand RNA sequences bind preferentially to an M_r 90,000 protein endoded by the host cell. Replica nitrocellulose membrane blots, each containing the SDS–PAGE resolved proteins in 1 µg of purified measles virions (MV), 50 µg of uninfected cell protein (UC), or 50 µg of MV-infected cell protein (IC), were probed with (+)- or (−)-sense RNA transcripts generated from pGEM subclones of MV clones #124 and #160 [containing, respectively, the precise MV genome 3' end and leader region plus 239 nt of the *NP* gene, and a nonoverlapping region of similar length in the middle of the *NP* gene (Crowley *et al.*, 1987, 1988)]. At least 90% of the SP6- and T-7 generated *in vitro* transcripts were intact as assessed by denaturing gel electrophoresis. Transcripts were of similar specific activities (cpm/µg) and all binding reactions were performed using identical molar inputs of the transcript probes. Panel 1: probe #124 (−) sense; panel 2: probe #124 (+) sense; panel 3: probe #160 (−) sense; panel 4: probe #160 (+) sense. Panel 5 shows the result of reprobing the blot in panel 3 with a hyperimmune anti-MV serum recognizing all the MV structural proteins except H (i.e., a Western blot analysis). As expected, MV proteins are found only in the MV and IC lanes.

specifically bind RNAs containing MV nascent leader sequences; both (+) and (−) MV leader-containing transcripts are bound (our unpublished data).

Nonspecific binding of VSV mRNAs, possibly to RNA-binding proteins normally present in cytoplasm, has been described (Grubman and Shafritz, 1981; Rosen *et al.*, 1982). At least eight unidentified host proteins were shown to interact with VSV mRNAs by photochemical cross-linking (Adam *et al.*, 1986). This finding may be due to gratuitous binding of poly(A) tails by cytoplasmic VSV N protein (Blumberg and Kolakofsky, 1983); however, the MV transcripts used as probes in our experiments (Fig. 4) lack poly(A) tails and do not assemble into nucleocapsid structures with MV NP protein under *in vitro* conditions previously used for assembly of VSV leader RNAs (Blumberg *et al.*, 1983, 1984; B. M. Blumberg, unpublished data). Interestingly, the results of the northwestern blot experiments suggest the presence of a protein-binding site within the first 200–300 nucleotides of the nascent MV RNA strand, although the triplet phase repeat of A residues postulated to represent the encapsidation/nucleation signal in VSV leader RNAs (Blumberg *et al.*, 1983) has no evident equivalent in the nascent RNA transcripts of paramyxoviruses (see Fig. 2).

III. A MODIFIED MODEL FOR PARAMYXOVIRUS TRANSCRIPTION AND REPLICATION

A. Elements of the Modified Model

To account for the many differences from VSV, we postulate a radically modified model for paramyxovirus transcription and replication, schematically illustrated in Fig. 5. This new model has three distinctive features.

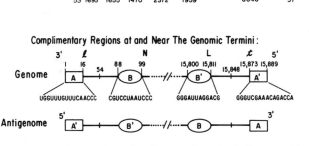

FIGURE 5. Map of the MV genome showing putative promoter and/or regulatory regions at and near the genome and antigenome termini. Top: a schematic map of the MV genome, beginning at the 3′ end with 53 nucleotides of leader sequence (1) and ending at the 5′ terminus with 37 nucleotides of trailer sequence (t). Gene boundaries are denoted by vertical bars; below each gene is the number of coding nucleotides (Billeter *et al.*, 1984; Blumberg *et al.*, 1988; Cattaneo *et al.*, 1987b; Crowley *et al.*, 1988). Bottom: an expanded schematic view of the complementary sequences at the MV genome and antigenome ends that may function as promoter sites in transcription and replication. (A) The conserved 3′ end sequence. (B) The BB3-equivalent auxiliary promoter sequence. A third promoter region between (A) and (B), consisting of the conserved R1 or R2 sequences, has been omitted for clarity (see Fig. 2).

1. In transcription the polymerase starts at an auxiliary internal promoter, whereas in replication it starts at the 3' end promoter. This extends our previous suggestion of separate transcriptional and replicative promoters that was based on the apparent absence of MV leader RNAs (Crowley et al., 1988).

2. Host-cell proteins serve to nucleate assembly by NP protein. We postulate that both paramyxovirus nascent RNA transcripts and NP proteins specifically interact with one or several endogenous host-cell-encoded RNA-binding proteins.

3. Nucleocapsid assembly proceeds from an internal nucleation site. This point is more speculative, being based on the absence of an evident triplet phase repeat at the precise 5' end of paramyxovirus nascent RNAs, the results shown in Fig. 4 suggesting a protein-binding site within the first 200–300 nucleotides of the nascent MV RNA strands, and postulate 2 above.

B. Reassortment of Functional Elements in the Modified Model

Some observations on what has really changed from the VSV model may help put the modified model in perspective. We think it likely that both VSV and paramyxovirus polymerases require two separate binding sites for stabilization at the initiation step of either transcription or replication. For example, the 3' end and the R1 or R2 sequence may function together as the promoter for replication, while the R1 sequence and the BB3 equivalent sequence may function as the promoter for transcription. By splitting the promoter into two sites, the number of operational sequence elements has thus increased by one over the VSV model—there are now three polymerase-binding sites and a protein-binding site for encapsidation/nucleation. In the absence of experimental testing, we cannot yet define the function of these sites. Also, we have only compared the first 115 nucleotides of the viral genomes; other polymerase stabilizer sites may lie beyond this region. However, there would appear to have been a reassortment of functional elements in that the leader RNA terminator site in VSV has become part of a second promoter in some paramyxoviruses.

Vidal and Kolakofsky (1989) have proposed a different modified model based on reinitiation of the Sendai polymerase at the leader–NP junction. Interestingly, both modified models suggest that the VSV polR mutants of Perrault et al. (1983) and Perrault and McClear (1984), which exhibit high levels of leader–N readthrough mRNAs due to alteration of the N protein, may be highly relevant to the polymerase mechanism in paramyxoviruses. In these VSV mutants, differences in the ATP requirement for transcription and replication map to the mutant polR N protein. This may imply that, in addition to the concentration of free N protein in the cytoplasm (Kolakofsky and Blumberg, 1982), the state of phosphorylation of the N protein (which is functionally part of the polymerase) partly determines its mode of activity (Beckes et al., 1987; Helfman and Perrault, 1988).

In the context of our modified model, the increased readthrough of polR mutants implies that changes in the nucleocapsid protein simultaneously act to weaken the leader RNA terminator signal and increase its activity as a promoter. This dual effect suggests that the leader RNA terminator site in VSV may correspond, in part, to the BB3-equivalent sequence in paramyxoviruses.

Also, in our modified model, encapsidation/nucleation in paramyxoviruses depends strongly on host-cell factors. Nonetheless, the central role of NP protein in the switchover from transcription to replication is maintained. This is supported by the loss of MV leader–NP readthrough mRNA and antigenome synthesis in the presence of cycloheximide (Castaneda and Wong, 1989), which most likely reflects shutoff of replicative starts due to the unfulfilled stoichiometric requirement for NP protein in nucleocapsid assembly. Now, however, the cytoplasmic concentration of free NP affects the polymerase's choice of initiation site, rather than whether or not it reads through the terminator. This new element of choice at the initiation step may imply distinct transcriptional and replicational forms of the polymerase in paramyxoviruses. It may also help explain why paramyxovirus DI particles containing only 5' "copyback" termini do not rapidly outgrow "fusion" DIs as is found in VSV, since the enormous amplification of copyback genomes due to choice at the leader RNA termination step is no longer a factor (Re *et al.*, 1983; cf. Blumberg and Kolakofsky, 1983).

IV. CONCLUDING REMARKS

Although a radical departure from the VSV model, our modified model for paramyxovirus transcription/replication seems worthy of consideration for two reasons. First, it provides a natural context for the intimate involvement of tissue-specific host factors in paramyxovirus transcription and replication through their function in encapsidation/nucleation. For example, MV gene expression is strongly restricted in nervous tissue relative to cultured cells (Cattaneo *et al.*, 1987a,b). Second, it makes strong predictions that are readily testable. In particular, our modified model predicts backward as well as forward assembly from a postulated internal encapsidation/nucleation site(s). We recall that tobacco mosaic virus (TMV) provided a clue for identifying the VSV encapsidation/nucleation site (Blumberg *et al.*, 1983). In the context of our modified model, it is noteworthy that the TMV encapsidation/nucleation site lies some 900 nucleotides from the 3' end of the genome, and that TMV nucleocapsid assembly proceeds bidirectionally from this site [reviewed by Butler (1984)]. Furthermore, the specificity of TMV nucleocapsid protein for this site is not absolute (Atreya and Siegel, 1989).

We bear firmly in mind that our present modified model is little more than an outline. However, the data suggest that no single mechanism can describe transcription/replication in all paramyxoviruses. Indeed, there may be a continuum of transcriptional mechanisms, with rhabdoviruses (VSV) lying at one extreme and morbilliviruses (MV) at the other, with lyssaviruses and other paramyxoviruses in between, as previously suggested by Tordo *et al.* (1986, 1988) on finding a pseudogene in rabies virus. Certainly, the data show that the VSV model is incomplete, that an adequate description of transcription and replication in paramyxoviruses requires a modified model, and that further work will be required to fill in the outlines of either system.

V. REFERENCES

Adam, S. A., Choi, Y. D., and Dreyfuss, G., 1986, Interaction of mRNA with proteins in vesicular stomatitis virus-infected cells, *J. Virol.* **57:**614–622.

Arakawa, T., and Frieden, C., 1984, Interaction of microtubule-associated proteins with actin filaments, *J. Biol. Chem.* **259:**11730–11734.

Atreya, C. D., and Siegel, A., 1989, Localization of multiple TMV encapsidation initiation sites on *rbcL* gene transcripts, *Virology* **168:**388–392.

Banerjee, A. K., 1987, Transcription and replication of rhabdoviruses, *Microbiol. Rev.***51:**66–87.

Beckes, J. D., Haller, A. A., and Perrault, J., 1987, Differential effect of ATP concentration on synthesis of vesicular stomatitis virus leader RNAs and mRNAs, *J. Virol.* **61:**3470–3478.

Billeter, M. A., Baczko, K., Schmid, A., and ter Meulen, V., 1984, Cloning of DNA corresponding to four different measles virus genomic regions, *Virology* **132:**147–159.

Blumberg, B. M., and Kolakofsky, D., 1983, An analytical review of defective infections of vesicular stomatitis virus, *J. Gen. Virol.* **64:**1839–1847.

Blumberg, B. M., Leppert, M., and Kolakofsky, D., 1981, Interaction of VSV leader RNA and nucleocapsid protein may control VSV genome replication, *Cell* **23:**837–845.

Blumberg, B. M., Giorgi, C., and Kolakofsky, D., 1983, N protein of vesicular stomatitis virus selectively encapsidates leader RNA *in vitro*, *Cell* **32:**559–567.

Blumberg, B. M., Giorgi, C., Kolakofsky, D., Rose, K., and Kocher, H., 1984, Preparation and analysis of the nucleocapsid proteins of vesicular stomatitis virus and Sendai virus, and analysis of the Sendai virus leader–*NP* gene region, *J. Gen. Virol.* **65:**769–779.

Blumberg, B. M., Crowley, J. C., Silverman, J. I., Menonna, J., Cook, S., and Dowling, P. C., 1988, Measles virus L protein evidences elements of ancestral RNA polymerase, *Virology* **164:**487–497.

Butler, P. J. G., 1984, The current picture of the structure and assembly of tobacco mosaic virus, *J. Gen. Virol.* **65:**253–279.

Castaneda, S. J., and Wong, T. C., 1989, Measles virus synthesizes both leaderless and leader-containing polyadenylated RNAs *in vivo*, *J. Virol.* **63:**2977–2986.

Cattaneo, R., Rebmann, G., Baczko, K., ter Meulen, V., and Billeter, M. A., 1987a, Altered ratios of measles virus transcripts in diseased human brains, *Virology* **160:**523–526.

Cattaneo, R., Rebmann, G., Schmid, A., Baczko, K., ter Meulen, V., and Billeter, M. A., 1987b, Altered transcription of a defective measles virus genome derived from a diseased human brain, *EMBO J.* **6:**681–688.

Chan, J., Pifko-Hirst, S. and Udem, S. A., 1989, Where is the measles virus (+) strand leader RNA transcript?, *in* "Genetics and Pathogenicity of Negative Strand Viruses" (B. W. J. Mahy and D. Kolakofsky, eds.), pp. 222–231, Elsevier Biomedical Press, Amsterdam.

Crowley, J., Dowling, P., Menonna, J., Schanzer, B., Young, E., Cook, S., and Blumberg, B., 1987, Molecular cloning of 99% of measles virus genome, positive identification of 5' end clones, and mapping of the *L* gene region, *Intervirology* **28:**65–77.

Crowley, J. C., Dowling, P. C., Menonna, J., Silverman, J. I., Schuback, D., Cook, S. D., and Blumberg, B. M., 1988, Sequence variability and function of measles virus 3' and 5' end and intercistronic regions, *Virology* **164:**498–506.

Dimock, K., Rud, E. W., and Yong Kang, C., 1986, 3'-Terminal sequence of human parainfluenza virus 3 genomic RNA, *Nucleic Acids Res.* **11:**4694.

Elango, N., Varsanyi, T. M., Kovamees, J., and Norrby, E., 1988, Molecular cloning and characterization of six genes, determination of gene order and intergenic sequences and leader sequence of mumps virus, *J. Gen. Virol.* **69:**2893–2900.

Galinski, M. S., Mink, M. A., and Pons, M. W., 1988, Molecular cloning and sequence analysis of the human parainfluenza 3 virus gene encoding the L protein, *Virology* **165:**499–510.

Grubman, M. S., and Shafritz, D. A., 1977, Identification and characterization of messenger ribonucleoprotein complexes from vesicular stomatitis virus-infected HeLa cells, *Virology* **81:**1–16.

Guiffre, R. A., Tovel, D. R., Kay, C. M., and Tyrrell, D. L. J., 1982, Evidence for an interaction between the membrane protein of a paramyxovirus and actin, *J. Virol.* **42:**963–968.

Hamaguchi, M., Nishikawa, K., Toyoda, T., Yoshida, T., Hanaichi, T., and Nagai, Y., 1985, Transcriptive complex of Newcastle disease virus. II. Structural and functional assembly associated with the cytoskeletal framework, *Virology* **147:**295–308.

Helfman, W. B., and Perrault, J., 1988, Altered ATP utilization by the polR mutants of vesicular stomatitis virus maps to the N-RNA template, *Virology* **167**:311–313.

Hill, V. M., Harmon, S. A., and Summers, D. F., 1986, Stimulation of vesicular stomatitis virus *in vitro* RNA synthesis by microtubule-associated proteins, *Proc. Natl. Acad. Sci. USA* **83**:5410–5413.

Ishida, N., Taira, H., Omata, T., Mizumoto, K., Hattori, S., Iwasaki, K., and Kawakita, M., 1986, Sequence of 2,617 nucleotides from the 3' end of Newcastle disease virus genome RNA and the predicted amino acid sequence of viral NP protein, *Nucleic Acids Res.* **15**:6551–6564.

Kingsbury, D. W., 1974, The molecular biology of paramyxoviruses, *Med. Microbiol. Immunol.* **160**:73–83.

Kingsbury, D. W., 1977, Paramyxoviruses, in "The Molecular Biology of Animal Viruses" (D. P. Nayak, ed.), pp. 349–382, Marcel Dekker, New York.

Kolakofsky, D., 1976, Isolation and characterization of Sendai virus DI-RNAs, *Cell* **8**:547–555.

Kolakofsky, D., and Blumberg, B. M., 1982, A model for the control of non-segmented negative strand virus genome replication, in "Virus Persistence" (B. W. J. Mahy, A. C. Minson, and G. K. Darby, eds.), pp. 203–213, Cambridge University Press, Cambridge.

Kurilla, M. G., Piwnica-Worms, H., and Keene, J. D., 1982, Rapid and transient localization of the leader RNA of vesicular stomatitis virus in the nuclei of infected cells, *Proc. Nat. Acad. Sci. USA* **79**:5240–5244.

Kurilla, M. G., Stone, H. O., and Keene, J. D., 1985, RNA sequence and transcriptional properties of the 3' end of the Newcastle disease virus genome, *Virology* **145**:203–212.

Leppert, M., and Kolakofsky, D., 1980, Effect of defective interfering particles on plus- and minus-strand leader RNAs in vesicular stomatitis virus infected cells, *J. Virol.* **35**:704–709.

Leppert, M., Kort, L., and Kolakofsky, D., 1977, Further characterization of Sendai virus DI-RNAs: A model for their generation, *Cell* **12**:539–552.

Leppert, M., Rittenhouse, L., Perrault, J., Summers, D. F., and Kolakofsky, D., 1979, Plus and minus strand leader RNAs in negative strand virus-infected cells, *Cell* **18**:735–747.

Littauer, U. Z., Giveon, D., Thierauf, M., Ginzburg, I., and Ponstingl, H., 1986, Common and distinct tubulin binding sites for microtubule-associated proteins, *Proc. Natl. Acad. Sci. USA* **83**:7162–7166.

Moyer, S. A., Baker, S. C., and Lessard, J. L., 1986, Tubulin: A factor necessary for the synthesis of both Sendai virus and vesicular stomatitis virus RNAs, *Proc. Natl. Acad. Sci. USA* **83**:5410–5413.

Perrault, J., and McClear, P. W., 1984, ATP dependence of vesicular stomatitis virus transcription initiation and modulation by mutation in the nucleocapsid protein, *J. Virol.* **51**:635–642.

Perrault, J., Clinton, G. M., and McClure, M. A., 1983, RNP template of vesicular stomatitis virus regulates transcription and replication functions, *Cell* **35**:175–185.

Pringle, C. R., 1978, The tdCE and hrCE phenotypes: Host range mutants of vesicular stomatitis virus in which polymerase function is affected, *Cell* **15**:597–606.

Ray, J., and Fujinami, R. S., 1987, Characterization of *in vitro* transcription and transcriptional products of measles virus, *J. Virol.* **61**:3381–3387.

Re, G. G., Gupta, K. C., and Kingsbury, D. W., 1983, Genomic and copy-back 3' termini in Sendai virus defective interfering RNA species, *J. Virol.* **45**:659–664.

Rosen, C. A., Ennis, H. L., and Cohen, P. S., 1982, Translational control of vesicular stomatitis virus protein synthesis: Isolation of an MRNA sequestering particle, *J. Virol.* **44**:932–938.

Sakai, Y., Suzu, S., Shioda, T., and Shibuta, H., 1987, Nucleotide sequence of the bobine parainfluenza 3 virus genome: Its 3' end and the genes of NP, P, C and M proteins, *Nucleic Acids Res.* **15**:2927–2944.

Seifried, A. S., Albrecht, P., and Milstein, J. B., 1978, Characterization of an RNA-dependent RNA polymerase activity associated with measles virus, *J. Virol.* **25**:781–787.

Shioda, T., Hidaka, Y., Kanda, T., Shibuta, H., Nomoto, A., and Iwasaki, K., 1983, Sequence of 3,687 nucleotides from the 3' end of Sendai virus genome RNA and the predicted amino acid sequence of viral NP, P and C proteins, *Nucleic Acids Res.* **11**:7317–7338.

Shioda, T., Iwasaki, K., and Shibuta, H., 1986, Determination of the complete nucleotide sequence of Sendai virus genome RNA and the predicted amino acid sequences of the F, HN and L proteins, *Nucleic Acids Res.* **14**:1545–1563.

Ter Meulen, V., Stephenson, J. R., and Kreth, H. W., 1983, Subacute sclerosing panencephalitis.

Virus–host interactions: Receptors, persistence, and neurological diseases, *in* "Comprehensive Virology," Vol. 18 (H. Fraenkel-Conrat and R. R. Wagner, eds.), pp. 105–159, Plenum Press, New York.

Tordo, N., Poch, O., Ermine, A., Keith, G., and Rougeon, F., 1986, Walking along the rabies genome: Is the large *G–L* intergenic region a remnant gene?, *Proc. Natl. Acad. Sci. USA* **83:**3914–3918.

Tordo, N., Poch, O., Ermine, A., Keith, G., and Rougeon, F., 1988, Completion of the rabies virus genome sequence determination: highly conserved domains along the L (polymerase) proteins of unsegmented negative strand RNA viruses, *Virology* **165:**565–576.

Udem, S. A., 1984, Measles virus: Conditions for the propagation and purification of infectious virus in high yield, *J. Virol. Meth.* **8:**123–136.

Udem, S. A., and Cook, K. A., 1984, Isolation and characterization of Measles virus intracellular nucleocapsid RNA, *J. Virol.* **49:**57–65.

Vidal, S., and Kolakofsky, D., 1989, Modified model for the switch from Sendai virus transcription to replication, *J. Virol.* **63:**1951–1958.

Yoshida, T., Nakayama, Y., Nagura, H., Toyoda, T., Nishikawa, K., Hamaguchi, M., and Nagai, Y., 1986, Inhibition of the assembly of Newcastle disease virus by monensin, *Virus Res.* **4:**179–195.

Yusoff, K., Millar, N. S., Chambers, P., and Emmerson, P. T., 1987, Nucleotide sequence analysis of the *L* gene of Newcastle disease virus: Homologies with Sendai and vesicular stomatitis viruses, *Nucleic Acids Res.* **15:**3961–3976.

CHAPTER 9

The Role of Viral and Host Cell Proteins in Paramyxovirus Transcription and Replication

SUE A. MOYER AND SANDRA M. HORIKAMI

I. INTRODUCTION

A. General Scheme of Paramyxovirus RNA Synthesis

The viruses of the *Paramyxoviridae* family contain a nonsegmented RNA genome of the negative (−)-strand sense [for earlier reviews see Kingsbury (1977), Choppin and Compans (1975), Kolakofsky and Roux (1987)]. In the case of Sendai virus, the prototype of the family, the genome RNA and its full-length complement, the antigenome RNA, are found both in the virion and in the infected cell as an RNase-resistant nucleocapsid due to the tight association of the major nucleocapsid protein, NP (M_r 57,000) with the RNA. Two other viral proteins, the P (M_r 79,000) and L (M_r 240,000) proteins, are associated less tightly with the nucleocapsid and function as subunits of the RNA-dependent RNA polymerase. Three additional viral proteins, the hemagglutinin-neuraminidase (HN, M_r 72,000, fusion (F_0, M_r 65,000), and matrix (M, M_r 34,000) proteins, are associated with the lipid envelope of the virion.

During the reproduction of paramyxoviruses in the cytoplasm of the infected cell, the nucleocapsid (RNA–NP) serves as the template for transcription by the viral RNA polymerase which synthesizes individual mRNAs coding for each viral protein. Transcription initiates at the exact 3' end of the nucleocapsid RNA with the synthesis of a short leader RNA, which is not thought to be further modified (Leppert *et al.*, 1979), followed by the sequential synthesis of the six mRNAs. Each of the mRNAs is capped and methylated at its 5' end (Colonno and Stone, 1975, 1976b; Chinchar and Portner, 1981;

SUE A. MOYER AND SANDRA M. HORIKAMI • Department of Immunology and Medical Microbiology, University of Florida College of Medicine, Gainesville, Florida 32610.

Yoshikawa *et al.*, 1986) and polyadenylated at the 3' end (Pridgen and Kings-bury, 1972; Marx *et al.*, 1975; Hall and ter Meulen, 1977). The nucleocapsid RNA must also serve as the template for RNA replication, a process which generates nucleocapsids containing the full-length complementary plus (+)-strand RNA, which in turn serves as the template for the synthesis of the progeny nucleocapsids containing (−)-strand genome RNA. The unusual fea-ture of RNA replication in these viruses is that concomitant synthesis of viral proteins is required (Robinson, 1971b), since RNA replication is obligatorily linked to its simultaneous encapsidation with the nucleocapsid protein. In our research we have been interested in defining both the mechanisms employed in the related but distinct processes of viral transcription and RNA replication and the functions that the viral and host cell proteins play in each of these processes.

B. Structure of the Genome RNAs of Sendai and Measles Viruses

The paramyxovirus genome is about 15,400 nucleotides in length and the genes for both Sendai and measles viruses have been mapped in the order 3'-leader-*NP-P+C-M-F$_0$-HN-L*-5' (Glazer *et al.*, 1977; Leppert *et al.*, 1979; Shioda *et al.*, 1983, 1986; Richardson *et al.*, 1985; Dowling *et al.*, 1986; Rima *et al.*, 1986). The complete sequences have been determined for both Sendai virus (Shioda *et al.*, 1983, 1986; Gupta and Kingsbury, 1984; Morgan and Rakestraw, 1986) and measles virus (Richardson *et al.*, 1986; Alkhatib and Briedis, 1986; Bellini *et al.*, 1985, 1986; Gerald *et al.*, 1986; Rozenblatt *et al.*, 1985; Blumberg *et al.*, 1988). Analysis of the data suggests that several consensus sequences are present in the genome RNAs that may serve important regulatory functions in RNA synthesis. For example, in Sendai virus each gene junction has a consen-sus sequence 3'-UNAUUCUUUUU GAA UCCCANUUUC-5', where the first 11 nucleotides represent the 3' end of the previous gene and are thought to specify the putative polyadenylation signal; the next three nucleotides are a nontranscribed intergenic spacer sequence conserved in all but the *HN−L* junc-tion; and the last ten nucleotides are thought to encode the start signal for the next viral gene and probably contain the capping and methylation signals (Gupta and Kingsbury, 1982, 1984). Sequences of approximately 55 nucleotides at the 3' end of the (+) and (−) genome RNAs are considered to be nontrans-lated leader sequences. In the case of measles virus, similar, but less well conserved junction sequences occur: polyadenylation 3'-AAUAUUUUUU-5'; intergenic region, 3'-GAA-5' or 3'-GCA-5'; and start site 3'-UCC(U/C)NN (G/U)A(U/C)C(U/A)-5' (Cattaneo *et al.*, 1987; Bellini *et al.*, 1985). Similar sequence motifs are found at the gene junctions of other paramyxoviruses, although in some instances they diverge more extensively than those of Sendai and measles viruses.

A second family of viruses, the rhabdoviruses, also contains a nonseg-mented, single-stranded RNA genome of the negative-strand sense. Both the paramyxoviruses and rhabdoviruses have a similar genome structure and share many similarities in their replicative and reproductive strategies [for a review, see Banerjee (1987)]. Vesicular stomatitis virus (VSV), the prototype of the rhabdovirus family, has been studied extensively and has served as the model for RNA synthesis in all negative-strand viruses. We shall compare similarities as well as differences in RNA synthesis between these two families where appropriate.

C. Viral Proteins Required for RNA Synthesis

1. Overall Nucleocapsid Structure

The paramyxovirus particle and isolated intracellular nucleocapsids contain the components required for viral mRNA synthesis, which are the RNA–NP template and the RNA polymerase, which consists of the viral L and P proteins (Buetti and Choppin, 1977; Colonno and Stone, 1976a). It has been estimated that associated with each nucleocapsid there are approximately 2600 copies of NP protein and 300 and 30 molecules of the P and L proteins, respectively (Lamb et al., 1976). Portner and co-workers (Portner and Murti, 1986; Portner et al., 1988) have described the direct localization of the viral nucleocapsid proteins using immunoelectron microscopy of Sendai virus nucleocapsids isolated from either infected cells or purified virus. Utilizing monoclonal antibodies to the Sendai virus NP protein, they determined that the NP proteins were evenly distributed along the nucleocapsids from either source, while the P and L proteins were found in random clusters on the intracellular nucleocapsids. With virus-derived nucleocapsids, which are metabolically inactive, there was a more uniform distribution of the L and P proteins. Furthermore, simultaneous labeling of the same nucleocapsid with both anti-L and anti-P antibodies confirmed that both proteins were present in the same clusters, directly demonstrating their cooperative interaction in RNA synthesis.

2. NP Protein

The Sendai virus NP protein completely encapsidates the genome RNA, rendering it ribonuclease resistant. The NP protein is tightly bound to the RNA such that it is not removed in the presence of high salt, but bands in CsCl as a ribonucleoprotein particle. The NP protein is the most abundant viral protein in the infected cell, since it is the first and most abundant mRNA transcribed. The NP gene has been sequenced for a number of paramyxoviruses, including Sendai and measles viruses (Shioda et al., 1983; Morgan et al., 1984; Rozenblatt et al., 1985), but there are no identifiable RNA-binding domains in the deduced sequences of these proteins despite their strong RNA-binding capacity. Both biochemical and monoclonal antibody epitope mapping experiments indicate that the 220 C-terminal amino acids of the NP protein reside on the surface of the nucleocapsid (Heggeness et al., 1981; Gill et al., 1988). These data suggest that the RNA-binding domain resides in the N-terminal portion of the protein.

Blumberg et al. (1984) developed methodology to isolate the NP and N proteins from Sendai virus and VSV, respectively, and found that both are blocked at their N termini. In the case of VSV, they also showed that the purified N protein will encapsidate isolated VSV leader RNA in the 5' to 3' direction *in vitro* (Blumberg et al., 1983) and that leader RNA is found encapsidated by N protein in the infected cell (Blumberg and Kolakofksy, 1981). These studies led to a model for RNA replication where the N protein modulates viral transcription by its ability to bind to nascent leader RNA. Following binding, continued RNA synthesis and encapsidation block the processing signals at all the gene junctions, yielding a full-length encapsidated (+)-strand RNA (Blumberg et al., 1981). The model predicts that the availability of the nucleocapsid protein regulates the switch from transcription to replication, which explains the observation that replication does not occur in the infected cell in the presence of the protein synthesis inhibitor cycloheximide (Robinson, 1971b).

3. Viral RNA Polymerase. L and P Proteins

The virion-associated RNA polymerase activity was identified in a number of paramyxoviruses by transcription of purified viruses, including Newcastle disease virus (NDV) (Huang et al., 1971), Sendai virus (Robinson, 1971a; Stone et al., 1971), mumps virus (Bernard and Northrop, 1974), and measles virus (Seifried et al., 1978). In the only reported reconstitution experiments, Hamaguchi et al. (1983), using NDV, purified each subunit and showed that both the P and L proteins are required for RNA synthesis from the purified viral RNA–NP template.

The L protein is very large (M_r ~240,000) and although no individual activities have been identified, it is presumed to have multiple functions by analogy with the VSV L protein, which has been shown to catalyze polyadenylation (Hunt et al., 1984), methylation (Hercyk et al., 1988), and polymerization (De and Banerjee, 1985; Ongradi et al., 1985). The L protein probably also participates in mRNA capping and in the coordinate RNA replication and encapsidation reactions. Several paramyxovirus L-protein gene sequences have been determined, including those of Sendai virus (Shioda et al., 1986; Morgan and Rakestraw, 1986), NDV (Yusoff et al., 1987), measles virus (Blumberg et al., 1988), and parainfluenza 3 virus (Galinski et al., 1988). In addition, several rhabdovirus *L* genes have been sequenced: those of VSV Indiana serotype (Shubert et al., 1984), VSV New Jersey serotype (Feldhaus and Lesnaw, 1988), and rabies virus (Tordo et al., 1988). Comparisons of the sequences have revealed several highly conserved regions within the *L* gene among both groups of viruses, which suggests that both families may have a common ancestral *L* (polymerase) gene, with the conserved domains being those which catalyze identical activities in RNA synthesis. These conserved sequences are therefore of great interest and deserve further investigation.

The second polymerase subunit, P (M_r 79,000), is also essential for RNA synthesis; however, its exact role is unknown. In Sendai virus-infected cells, the P protein is not found in the soluble protein pool, but appears to enter directly into nucleocapsids (Kingsbury et al., 1978; Kristensson and Orvell, 1983). The P protein is highly phosphorylated, although the function of this modification is not known. Hsu and Kingsbury (1982) reported that 80% of the phosphates are located in the N terminus of the P protein, which is removed when intact nucleocapsids are digested with V8 protease. Interestingly, the C-terminal portion of V8-treated P protein (about 400 amino acids) remains attached to the nucleocapsid and with L protein is capable of synthesizing properly modified mRNAs, albeit at a slightly reduced level. These data suggest that the phosphorylated N terminus of the P protein is not required for RNA synthesis (Chinchar and Portner, 1981).

Monoclonal antibodies against the Sendai virus P protein have been mapped to epitopes found only on the C terminus and inhibit RNA synthesis *in vitro*, again suggesting that this is the domain of P protein which is essential for polymerase activity (Deshpande and Portner, 1985; Vidal *et al.*, 1988). In more recent experiments, Ryan and Kingsbury (1988), using truncated synthetic P proteins, showed directly that only the C-terminal 224 amino acids of P protein are essential for the binding of the protein to nucleocapsids. The unrelated C proteins also bind to nucleocapsids *in vitro*. A similar C-terminal nucleocapsid-binding domain of the analogous VSV protein, NS, has been reported (Gill *et al.*, 1986; Emerson and Schubert, 1987).

The Sendai virus nonstructural proteins C and C' (M_r ~22,000) were first found in infected cells by Lamb and co-workers (Lamb *et al.*, 1976; Lamb and Choppin, 1977; Etkind *et al.*, 1980). Subsequent studies (see Chapters 6 and 7) have shown that the *P* gene of paramyxoviruses is very complex and encodes a variety of nonstructural proteins, including C and C' (Curran and Kolakofsky, 1988b; Bellini *et al.*, 1985) and Y_1 and Y_2 (~11 kD) (Curran and Kolakofsky, 1988a; Patwardhan and Gupta, 1988; Dillon and Gupta, 1989), which are all encoded from an alternate reading frame at the 5' end of the P mRNA, and the X protein (M_r 10,000) (Curran and Kolakofsky, 1987, 1988b, which is encoded from the 3' end of the P open reading frame. The Y_1, Y_2, and X proteins have been identified in virus-infected cells (Curran and Kolakofsky, 1987, 1988a, 1988b; Patwardhan and Gupta, 1988; Dillon and Gupta, 1989), with, however, unknown function(s).

Immunofluorescence studies have shown that the C protein is present throughout the cytoplasm in Sendai virus-infected cells as well as associated with nucleocapsids (Portner *et al.*, 1986; Omata-Yamada *et al.*, 1988). Similarly, in measles virus-infected cells, the C protein is concentrated in inclusions containing viral nucleocapsids (Bellini *et al.*, 1985). These data suggest that C protein may have a role in paramyxovirus RNA synthesis, yet attempts to inhibit Sendai virus RNA synthesis *in vitro* with an anti-peptide antibody against the C protein failed (Portner *et al.*, 1986). Taira *et al.* (1987) have presented evidence that interferon is induced by the Sendai C protein; however, the significance of this finding remains to be established.

The other viral structural proteins do not seem to be involved in RNA synthesis, with the possible exception of the M protein. The VSV M protein has been shown to have a negative regulatory role in viral mRNA synthesis (Clinton *et al.*, 1978; Martinet *et al.*, 1979; Carroll and Wagner, 1979) and there is indirect evidence that the Sendai virus M protein may have a similar role during virus infection (Marx *et al.*, 1974), although this has not yet been tested directly.

II. SENDAI VIRUS RNA SYNTHESIS

A. Transcription and the Effect of Host Cell Proteins

In order to study Sendai virus transcription and replication in detail, a cell-free system was needed that would support RNA synthesis. Portner (1982) had earlier described methodology that yielded extracts from Sendai virus-infected LLMCK2 cells that gave viral RNA synthesis; however, the system was not very efficient. We have described new methodology for the *in vitro* transcription and replication of the genome RNAs of Sendai virus and its defective-interfering particle, DI-H (Carlsen *et al.*, 1985). In developing this system, we adapted the methodology we developed for an *in vitro* RNA synthesis system for VSV (Peluso and Moyer, 1983). The key feature is the gentle, fast preparation of cytoplasmic extracts, where infected cells are permeabilized with lysolecithin and lysed by pipetting, and the nuclei are removed by low-speed centrifugation.

Figure 1 compares Sendai virus RNAs synthesized *in vivo* with those synthesized *in vitro* in these cytoplasmic extracts. We employ infection with

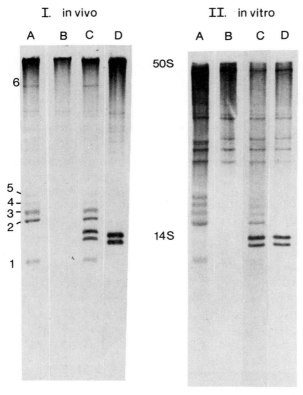

FIGURE 1. Agarose gel analysis of viral RNA synthesized *in vitro* in extracts of cells infected with Sendai virus in the presence or absence of DI-H. Panel I: BHK cells infected with Sendai virus or Sendai virus plus DI-H were labeled with [³H]uridine (50 μCi/ml) from 2 to 12 hr after infection. Cytoplasmic cell extracts were then prepared and one-half of each sample was treated with micrococcal nuclease (10 μg/ml). The RNA was isolated and analyzed by acid/8 M urea/1.5% agarose gel electrophoresis. *In vivo* RNA products: Sendai virus infection untreated (A) or treated with nuclease (B); Sendai virus plus DI-H coinfection untreated (C) or treated with nuclease (D). Panel II: Cytoplasmic extracts were prepared at 12 hr after infection from BHK cells infected with Sendai virus or Sendai virus plus DI-H. The extracts were incubated in the presence of [³H]UTP (250 μCi/ml) for 2 hr; one-half of each sample was treated with micrococcal nuclease, and the RNA was isolated as described above. *In vitro* RNA products in extracts of: Sendai virus untreated (A) or treated with nuclease (B) or Sendai virus plus DI-H untreated (C) or treated with nuclease (D) are shown. Numbers 1 through 6 indicate the positions of the Sendai virus mRNA species, and 50S and 14S indicate the positions of the genome RNAs of wild-type Sendai virus and the DI-H particle, respectively. [Reprinted from Carlsen *et al.* (1985), with permission.]

wild-type Sendai virus to study both mRNA synthesis and replication and coinfection with Sendai virus and DI-H to study replication of a smaller genomic RNA. Extracts prepared from Sendai virus-infected cells at 12 hr after infection, which is the time of maximal RNA replication *in vivo*, faithfully catalyzed the synthesis of the viral mRNAs, as well as some replication and encapsidation of the wild-type genome RNA in the absence of *de novo* protein synthesis (Fig. 1, lanes IA and IIA). The six viral mRNAs have been identified by their ribonuclease sensitivity (lanes IB and IIB) and labeled 1–6 by increasing size. Based on the sizes of the proteins and RNAs, these mRNAs probably code for the M (M_r 34,000), NP (M_r 57,000), F_0 (M_r 65,000), HN (M_r 74,000), P + C (M_r 79,000 + M_r 22,000), and L (M_r 240,000) proteins, respectively. See Section II.B for further discussion of RNA replication.

It has been known for many years that *in vitro* transcription of purified paramyxoviruses is quite poor compared to that of VSV (Huang *et al.*, 1971; Stone *et al.*, 1971). Various components such as polyanions (Stone and Kingsbury, 1973) or uninfected cell extracts (Portner, 1982) have been found to stimulate Sendai virus RNA synthesis. The latter suggested to us that a host cell factor(s) might play a role in this process. We have, in fact, shown that tubulin acts as such a positive transcription factor for *in vitro* RNA synthesis by two different negative-strand viruses, Sendai virus and VSV (Moyer *et al.*, 1986). A monoclonal antibody directed against β-tubulin completely inhibited both mRNA synthesis and RNA replication catalyzed *in vitro* by extracts of cells infected with either virus, and also mRNA synthesis by detergent-disrupted purified VSV. For detergent-disrupted purified Sendai virus, the normally low synthesis of both leader RNA and the NP mRNA was markedly stimulated by the addition of purified soluble tubulin, although there was no detectable synthesis of the other mRNAs.

While we observed stimulation of VSV by the addition of tubulin, Hill *et al.* (1986) reported that the microtubule-associated proteins (MAPs) greatly stimulated VSV RNA synthesis. In more recent experiments we have studied the effects of several components of the cellular microtubule system on the transcription of purified Sendai virus. We show here that compared with no additions (Fig. 2, A1 and B2), both leader RNA and NP mRNA synthesis were stimulated in purified detergent-disrupted Sendai virus by the addition of soluble tubulin (Fig. 2, A2 and B3, respectively) as we reported earlier (Moyer *et al.*, 1986), while RNA synthesis was affected only minimally by MAPs (Fig. 2, A3 and B4). In these experiments, the product leader RNAs were identified by hybridization to a leader-specific probe, and are about 40 and 33–35 nucleotides in length. Leppert *et al.* (1979) reported earlier the synthesis of Sendai leader RNAs of 55 and 31 nucleotides. The reason for the discrepancy in the sizes is unclear. There seem to be a number of pause sites during transcription of the *NP* gene in the presence of tubulin that give rise to significant levels of early termination products (Fig. 2, B3).

Analysis of the molar ratios of the viral RNAs synthesized in several experiments under different conditions shows several striking features of these reactions (Table I). First, under all conditions Sendai virus transcription exhibited strong polarity, where the amount of leader RNA greatly exceeded that of the next product, the NP mRNA. This finding probably accounts, in part, for

A.   B.

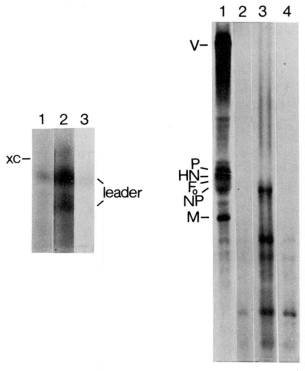

FIGURE 2. RNA synthesis by purified wild-type Sendai virus *in vitro*. Detergent-disrupted Sendai virus (10 μg) was incubated *in vitro* in a reaction mixture for 3 hr at 30°C as described by Carlsen *et al.* (1985) with the addition of 10% glycerol and the components indicated below. Panel A: leader RNA was synthesized in the presence of [α-^{32}P]UTP (300 μCi/ml), and the RNA was isolated and analyzed by electrophoresis on a 12% polyacrylamide/8 M urea gel at 350 V for 3 hr. Leader RNA synthesized by: virus alone (1), or in the presence of 500 μg/ml soluble bovine brain tubulin (2), or 400 μg/ml of bovine brain microtubule-associated proteins (MAPs) (3). The position of xylene cyanol (xc) marker is indicated on the gel. Panel B: Sendai virus RNA was synthesized with [^{3}H]UTP (250 μCi/ml) for 3 hr at 30°C, isolated, and analyzed by electrophoresis on 1.5% agarose/acid/8 M urea gels. mRNA synthesized by virus alone (2), or in the presence of 500 μg/ml of soluble tubulin (3) or 200 μg/ml MAPs (4). Lane 1: ^{3}H-labeled RNA from Sendai virus-infected cells, with the letters identifying the mRNAs.

the previous difficulty in establishing an effective *in vitro* transcription system from virus. The addition of MAPs had little effect on the synthesis of either leader or NP RNAs, in contrast to its stimulation of VSV RNA synthesis (Hill *et al.*, 1986). The data suggest that with the addition of lower concentrations of tubulin both leader and NP mRNAs were equally stimulated, but with higher levels there was a preferential stimulation of the synthesis of the NP mRNA with little further enhancement of leader RNA. The mechanism of tubulin stimulation is unknown. One possibility may be that tubulin duplicates the function of the acidic N-terminal domain of the P protein, as was described for VSV. In the latter case, tubulin was able to substitute in *trans* or as a chimeric protein with the C-terminal domain of the VSV NS protein (Chattopadhyay and Banerjee, 1987, 1988).

We show here that poly-L-glutamate, which had been earlier reported to stimulate Sendai virus *in vitro* RNA synthesis (Stone and Kingsbury, 1973), did so to a lesser extent than tubulin (Table I). The carboxy termini of the tubulin subunits are glutamate rich (Krauhs *et al.*, 1981; Ponstingl *et al.*, 1981), and

TABLE I. Molar Ratios of Leader and NP RNAs
Synthesized *in Vitro* by Sendai Virus

Addition[a]	Molar ratio[b]	
	Leader RNA	NP mRNA
None	158	1.0
100 µg/ml MAPs	276	1.2
200 µg/ml MAPs	247	2.0
400 µg/ml MAPs	344	ND[c]
250 µg/ml Tb	862	5.6
500 µg/ml Tb	948	12.1
250 µg/ml poly-L-glutamate	695	4.0
500 µg/ml poly-L-glutamate	755	5.4
1000 µg/ml poly-L-glutamate	743	3.0
250 µg/ml MT	973	6.0
500 µg/ml MT	814	12.8

[a]MAP, Microtubule-associated proteins; Tb, soluble tubulin; MT, taxol-assembled microtubules.
[b]The molar amount of each RNA was calculated from the radioactivity incorporated into the larger leader RNA and NP mRNA and normalized to the value for the NP mRNA in the absence of additions (1.0). The data represent the average of three experiments.
[c]Not done.

this domain is obviously important for the stimulatory effect; however, tubulin itself had a greater stimulatory effect on RNA synthesis, suggesting that the entire molecule is needed. As we described earlier (Moyer *et al.*, 1986), additional host cell factors seemed to be required for complete transcription of the Sendai virus nucleocapsid, since the soluble protein fraction from uninfected cells was required for the synthesis of the more distal mRNAs, which were not detectable during tubulin stimulation. The addition of cellular proteins also had the concomitant, unexplained interesting effect of stimulating synthesis of a slightly longer leader RNA (Moyer *et al.*, 1986).

In further experiments, we have found that microtubules irreversibly assembled with taxol from soluble tubulin also stimulated Sendai virus transcription equally as well as soluble tubulin (Table I). This is a rather surprising result, but it suggests that *in vivo* an interaction occurs between the microtubules of the cytoskeleton and the Sendai virus nucleocapsid. What is the nature of this interaction? We previously reported that a β-tubulin monoclonal antibody specifically immunoprecipitated tubulin and the VSV L protein from VSV-infected cell extracts (Moyer *et al.*, 1986), suggesting that the tubulin αβ dimer may actually be a subunit of the VSV RNA polymerase. In similar experiments with Sendai virus, however, no L protein could be immunoprecipitated with the anti-tubulin antibodies, indicating that tubulin is not a subunit of the polymerase. In this case, tubulin appears to function as a cofactor necessary for RNA synthesis, perhaps acting to facilitate binding to and/or stability of the polymerase on the nucleocapsid.

It is particularly interesting that various microtubule components of the cytoskeletal system have become part of the enzymatic machinery for RNA synthesis by both Sendai virus and VSV, although by apparently different mechanisms. Indeed, the cytoskeleton had previously been reported to be the site of the physical association of VSV, Sendai virus, and NDV transcriptive complexes in infected cells (Cervera et al., 1981; Chatterjee et al., 1984; Kingsbury et al., 1978; Hamaguchi et al., 1985). In the latter case, upon infection the uncoated NDV nucleocapsids initially asociated with the cytoskeleton and all of the progeny genome-length RNA as well as a large proportion of the NDV mRNAs were also associated with the cytoskeleton. These data, as well as our own, suggest that both the infecting and progeny paramyxovirus nucleocapsids attach to the cytoskeleton matrix of their host, where components of the framework also act as subunits or cofactors of the viral RNA polymerases. Hence, these viruses utilize the cytoskeleton for both a physical anchoring or positioning of the nucleocapsid as well as for enzymatic activity. This phenomenon is an example of how exquisitely these viruses have evolved to favor their own propagation by utilizing basic, fundamental structural components of the living cell for their reproduction.

B. RNA Replication and the Effect of Host Cell Proteins

We would like to now discuss studies of Sendai virus RNA replication. The infected cell extracts that we described above were also capable of supporting the synthesis and encapsidation of the 50S Sendai virus genome RNA in vitro (Fig. 1, IIB). The replication products appeared as a ladder of partially completed, micrococcal nuclease-resistant RNAs up to the size of genome RNA. The majority of the RNA replication products from Sendai virus nucleocapsids were, in fact, smaller than genome length. Similar results were also observed earlier in a different Sendai virus in vitro replication system (Portner, 1982). It appears that these incomplete, yet distinct RNAs may represent specific pause sites in the replication process in vitro, and they were also observed to a lesser extent in vivo (Fig. 1, IB). In contrast, only full-length, nuclease-resistant RNAs were synthesized in vitro as well as in vivo, in Sendai virus and DI-H coinfected cells (Fig. 1, IID and ID, respectively). This difference in the wild-type and the DI-H RNA replication reactions is perhaps due to the small size of the DI-H RNA or to the nature of the sequences contained in the DI-H genome RNA.

An important feature of this in vitro RNA replication system is that the essential components in the cytoplasmic extract can be separated, then recombined and still retain activity. The cytoplasmic extract can be fractionated by centrifugation into two components: the nucleocapsid with the associated RNA polymerase, and the soluble protein fraction. Figure 3 shows that DI-H nucleocapsids alone did not replicate and the soluble protein fraction was required for RNA replication and encapsidation in vitro (Fig. 3a and 3b). RNA replication did not require de novo protein synthesis, but instead relied on the soluble proteins already present in the cytoplasm of the infected cells. Robinson (1971b) first showed that significant Sendai virus genome replication util-

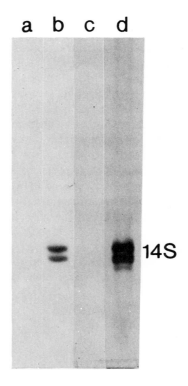

FIGURE 3. Reconstitution of DI-H RNA replication and encapsidation *in vitro*. At 12 hr after infection, cytoplasmic extracts were prepared from BHK cells coinfected with Sendai virus plus DI-H. Extracts were separated by centrifugation into the nucleocapsid pellet and soluble protein fraction. Alternatively, a mixture of purified standard Sendai virions and predominately DI-H virions (65 μg per reaction) was disrupted with Triton X-100. The samples were incubated as described below in the presence of [³H]UTP for 2 hr, treated with micrococcal nuclease, and the remaining RNA was isolated and analyzed by acid/urea/agarose gel electrophoresis. Nucleocapsid RNA synthesized *in vitro* from the intracellular nucleocapsids alone (a), or in the presence of the soluble protein fraction from 10⁷ Sendai virus-infected cells (b). RNA synthesized *in vitro* from purified virus alone (c), or in the presence of the soluble protein fraction from 10⁷ Sendai virus-infected cells (d). The positions of 14S DI-H RNAs are indicated. [Redrawn from Carlsen *et al.* (1985), with permission.]

izing a pool of virus proteins could occur in infected cells when *de novo* viral protein synthesis was inhibited by cycloheximide. Similarly, the soluble protein fraction of infected cells was also necessary and sufficient to support RNA replication and encapsidation from purified, detergent-disrupted virus (Fig. 3c and 3d).

The reconstitution assay was then used to identify the specific viral and possibly host proteins required for Sendai virus RNA replication. For the purposes of discussion, RNA replication can be divided into two steps: (1) the first step is the initiation of encapsidation, where NP protein is thought to bind to specific sequences in nascent leader RNA, and (2) an elongation step involving the continued RNA synthesis and simultaneous encapsidation, presumably mediated by cooperative protein–protein interactions. The RNA replication system that we have developed allows separate analysis of the protein requirements for both the initiation and elongation steps of encapsidation during replication. Intracellular DI-H nucleocapsids (containing the associated RNA polymerase) represent a population of molecules in all stages of replication with various lengths of preinitiated replicative RNA. Such templates allow analysis of the elongation step of encapsidation. On the other hand, detergent-disrupted DI-H virions do not contain any nascent product RNAs and can serve as templates to measure both initiation and elongation.

According to the proposed model of negative-strand RNA virus replication, the nucleocapsid protein would be the major protein required for the encapsidation of replicative RNA (Blumberg *et al.*, 1983, also see Chapter 8). We purified the Sendai virus NP protein from intracellular nucleocapsids as described by Blumberg *et al.* (1984) and tested its ability to support the replication of the RNA of the intracellular DI-H nucleocapsids. NP protein alone was, in fact, able to support the synthesis and encapsidation of DI-H genome RNA with the level of replication directly dependent on the concentration of NP protein in the reaction (Fig. 4). These data show that the NP protein is the only additional protein required for the elongation step of encapsidation. The purified NP protein was shown to be a soluble monomer under the conditions used for the *in vitro* RNA replication reactions.

In contrast, for VSV, the corresponding major nucleocapsid protein (N) cannot be isolated as a soluble monomer protein, but is found in infected cells as a soluble component only when it is complexed to the VSV NS protein (Peluso and Moyer, 1988). In the case of VSV, it is the N : NS complex which serves as the substrate for VSV RNA replication, whereas N protein alone does not.

We have used several monoclonal antibodies to the Sendai virus NP and P proteins [generously provided by Deshpande and Portner (1984, 1985)] to test

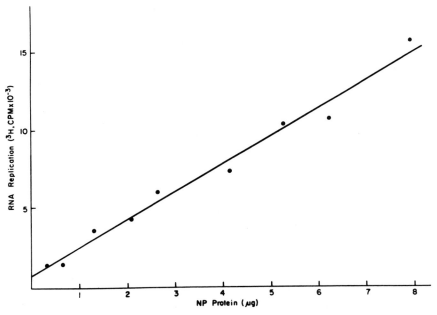

FIGURE 4. Effect of NP protein concentration on *in vitro* DI-H RNA replication. The micrococcal nuclease-resistant RNA products from *in vitro* reactions containing intracellular DI-H nucleocapsids and increasing concentrations of purified NP protein were analyzed by acid/urea/agarose gel electrophoresis as described in the legend to Fig. 3. The 14S DI-H replication products were identified by fluorography, excised together from the gel, dissolved in 1 ml of 30% H_2O_2 for 17 hr at 60°C, and quantitated by liquid scintillation counting. The data are the cumulative results from several different experiments with two different preparations of purified NP protein. The line represents the best fit to the experimental points. [Reprinted from Baker and Moyer (1988), with permission.]

for possible interactions of these two viral proteins in the soluble protein fraction of Sendai virus-infected cells. No stable association (i.e., complex) of the NP and P proteins was detected (our unpublished data) under conditions which readily allowed detection of the VSV N : NS complex (Peluso and Moyer, 1988). Each of these antibodies did, however, inhibit DI-H RNA replication (our unpublished data), as well as transcription (Deshpande and Portner, 1984, 1985).

We have found that the initiation step of encapsidation is a complex process. Purified detergent-disrupted DI-H virions alone (Fig. 5A, lane 1) synthesized only a small amount of leader RNA. The addition of purified tubulin stimulated (threefold) synthesis of leader RNAs of the same sizes that are synthesized with wild-type Sendai virus in the presence of tubulin (Fig. 5A, lanes 2 and 3). The leader RNA synthesized in the presence of tubulin was, however, not encapsidated when NP protein was also added to the reaction (data not shown).

RNA replication was also assayed in the in vitro reactions with purified DI-H. Little or no RNA replication was observed with DI-H along (Fig. 5B, lane 1), or with the addition of tubulin or NP (data not shown), or when the soluble

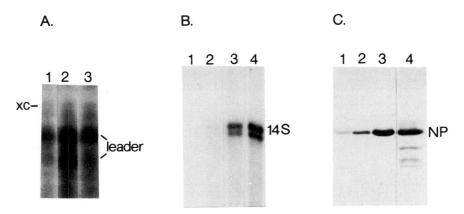

FIGURE 5. In vitro replication and encapsidation of purified DI-H. The soluble protein fraction of uninfected BHK cells (USP) was isolated as described earlier (Carlsen et al., 1985). Purified virus was incubated with 0.1% Triton X-100 for 20 min at 4°C and added to a 100-µl reaction in the presence of [α-^{32}P]UTP (350 µCi/ml) as described by Carlsen et al. (1985), except for the addition of 10% glycerol, tRNA (100 µg/ml), and 10 µM UTP. The RNA products were isolated and analyzed in panel A on a 12% polyacrylamide–8 M urea gel. Leader RNA synthesized from purified DI-H (22 µg) either alone (lane 1), or in the presence of 50 µg of tubulin (lane 2), or purified wild-type Sendai virus (21 µg) in the presence of 50 µg of tubulin (lane 3). In panel B, nuclease-resistant RNA products synthesized from purified DI-H (10 µg) were analyzed on an acid/1.5% agarose/urea gel with no additions (lane 1), or the addition of USP from 5 × 10^6 cells (lane 2), USP and purified NP protein (2.7 µg) (lane 3), and USP, NP, and tubulin (50 µg) (lane 4). Panel C: Purified DI-H (10 µg) was incubated in vitro as described above except with unlabeled UTP (1 mM) and the addition of ^{35}S-labeled purified NP protein (5514 cpm/µg). The product was treated with micrococcal nuclease and banded on CsCl. The nucleocapsid fraction was pelleted and the associated protein was analyzed by 11% polyacrylamide–SDS gel electrophoresis. Nucleocapsid products synthesized with the addition of ^{35}S-NP (2.7 µg) alone (lane 1), USP from 5 × 10^6 cells and ^{35}S-NP (lane 2), or USP, ^{35}S-NP, and tubulin (25 µg) (lane 3), marker ^{35}S-labeled Sendai virus nucleocapsids isolated from infected cells (lane 4).

protein fraction from uninfected cells (USP) (Fig. 5B, lane 2) was added. Interestingly, when USP was added together with NP there was a 27-fold stimulation of DI-H replication and an even greater increase (42-fold) with the addition of tubulin with USP and NP (Fig. 5B, lanes 3 and 4 respectively). Total leader RNA synthesis was not changed, nor was leader RNA encapsidated under these latter conditions (data not shown). There appears to be a component other than tubulin in USP which is required along with NP for replication, since the addition of just tubulin and NP yields no replicative products (data not shown).

As an alternative method of measuring progeny nucleocapsid formation, we used encapsidation with radiolabeled, purified NP protein to confirm these results. The addition of NP alone to DI-H virions gave no RNA replication (see above, data not shown), but did result in a small amount of exchange of the labeled NP protein with NP associated with the nucleocapsid template (Fig. 5C, lane 1). The addition of NP and USP stimulated replication and encapsidation and the further addition of tubulin gave an even greater increase (Fig. 5C, lanes 2 and 3, respectively), confirming the experiments with RNA labeling (Fig. 5B). From these data we conclude that the initiation of RNA replication requires the viral NP protein and at least two host cell proteins, one of which is tubulin. The identity of the second cell protein is currently under investigation.

The Sendai virus nonstructural proteins C and C', whose exact functions are unknown, may have some role(s) in viral RNA synthesis, since they are associated with nucleocapsids in the infected cell and bind to nucleocapsids *in vitro* (Ryan and Kingsbury, 1988). To test a possible role of the C proteins in RNA synthesis, we prepared a monoclonal antibody against the N-terminal peptide of the C protein (generously donated by Dr. W. Bellini). This antibody immunoprecipitated C and C' from Sendai virus-infected cells; however, it had no effect on Sendai virus transcription or replication when added to the *in vitro* reactions described here (data not shown). Similar results were reported by Portner *et al.* (1986) utilizing a rabbit anti-C-terminal peptide antibody.

C. Effect of Heterologous Viral Proteins

The replication of the RNA of negative-strand viruses requires a source of the viral nucleocapsid protein to allow the coordinate synthesis and encapsidation of RNA into nuclease-resistant nucleocapsids. Since the nucleocapsid proteins of viruses from different families perform similar functions in RNA replication, we wanted to determine if heterologous viral proteins from either the same or different families of viruses could substitute for the homologous protein *in vitro*. Intracellular DI-H nucleocapsids incubated with the homologous soluble protein fraction from Sendai virus-infected cells gave significant RNA replication and, interestingly, the measles virus soluble protein fraction, i.e., N protein (see Section III), also supported a significant level of replication (44%) (Table II). The soluble protein fractions from cells infected with either of two VSV serotypes, however, were unable to support DI-H replication. Although the sequence similarity of the Sendai and measles virus NP proteins is low (~20%), it is statistically significant, unlike the similarity to VSV N pro-

TABLE II. The Effect of Virus Proteins
from Different Families on Sendai Virus
DI-H RNA Replication

Conditions[a]	DI-H RNA replication	
	cpm	Percent
DI-H intracellular nucleocapsids		
+ Sendai virus soluble protein	2048	100.0
+ measles virus soluble protein	901	44.0
+ VSV (Indiana) soluble protein	52	2.5
+ VSV (New Jersey) soluble protein	241	11.8

[a]DI-H nucleocapsids isolated from Sendai virus and DI-H coinfected cells (1×10^7 cells) were incubated in the presence of [^{3}H]UTP (200 μCi/ml) with the soluble protein fraction from the indicated wild-type homologous or heterologous virus-infected cells (1×10^7 cells) under the reaction conditions of Carlsen et al. (1985). Nuclease-resistant DI-H RNA replication products were quantitated following acid/agarose/urea gel electrophoresis by cutting out the 14S products and determining the radioactivity, as described in Fig. 4. The percentages are related to the homologous combination (100%).

tein (~10%) (Jambou et al., 1986). Our data imply that there may be a conserved domain(s) which is specifically utilized for the elongation reaction in the nucleocapsid proteins of viruses of the same family, even though Sendai and measles virus NP proteins are not closely related overall.

III. MEASLES VIRUS RNA SYNTHESIS

A. Transcription and RNA Replication

To confirm and generalize our observations on the mechanism of Sendai virus transcription and replication, we wanted to study RNA synthesis in another paramyxovirus. We have chosen measles virus, since this is one of the most medically relevant RNA viruses, being involved not only in acute, but also slow virus infections of humans. We have developed an efficient cell-free transcription/replication system to study the components, both viral and host, required for measles virus RNA synthesis. The basic methodology involving lysolecithin permeabilization of infected cells which had previously been successful for the establishment of in vitro RNA synthesis systems for both VSV and Sendai virus also worked very well for measles virus. Recently, Herman (1989) has also employed this general procedure for successful respiratory syncytial virus RNA synthesis in vitro. Perhaps the key feature yielding an efficient measles transcription system was the selection of a monolayer human cell line, A549, which gave a high level of viral RNA synthesis in vivo. This appears to be achieved, in part, by the fact that progeny measles virus formation is very limited, resulting in the maximal intracellular accumulation of the viral nucleocapsids and RNA synthesis by 17 hr after infection.

Cytoplasmic extracts of infected cells prepared at 17 hr after infection synthesized measles virus mRNAs of the same sizes as observed *in vivo* (Fig. 6, lanes B, C, and A, respectively) and supported at least some replication of the genome RNA (Fig. 6). As we observed for VSV and Sendai virus, measles virus RNA replication *in vitro* did not require *de novo* protein synthesis, but utilized the preformed proteins present in the infected cell at the time of extract preparation. We have found that the ability to replicate genome RNA, however, appeared to be substantially reduced in the reconstitution reactions, although transcription could be fully restored (data not shown). Nonetheless, a small amount of virion-length RNA was synthesized and encapsidated, as determined by banding of progeny nucleocapsids on CsCl gradients and gel electrophoresis of the RNA product (data not shown).

The measles virus nonstructural C protein has been found by immunofluorescence to be associated with viral nucleocapsids in infected cells (Bellini *et al.*, 1985), suggesting its possible role in RNA synthesis. The addition of a rabbit anti-measles C peptide antibody (generously provided by Dr. W. Bellini) to these *in vitro* reactions, however, had no effect on measles virus transcription or replication (data not shown). Thus, the role of the C protein in both measles and Sendai virus infections remains unknown.

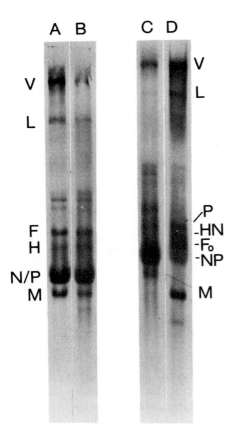

FIGURE 6. Measles virus RNA synthesis *in vitro*. Cytoplasmic extracts of measles virus-infected A549 cells were prepared by lysolecithin permeabilization at 17 hr after infection in a reaction mixture containing 0.1 M HEPES (pH 8.1), 0.05 M NH_4Cl, 7 mM KCl, 4.5 mM MgOAc, 1 mM each of spermidine, DTT, CTP, and GTP, 10 μM UTP, 2 mM ATP, 2 μg/ml actinomycin D, and 3.3 mg/ml creatine phosphate and 40 U/ml creatine phosphokinase as an ATP regenerating system. The extracts were incubated at 30°C for 3 hr in the presence of [³H]UTP (250 μCi/ml). The RNAs were purified and analyzed by acid/8 M urea/1.5% agarose gel electrophoresis and fluorography. Measles virus RNAs synthesized in two different experiments (lanes B and C) are compared with either measles virus ³H-labeled RNAs from infected A549 cells (A) or with Sendai virus ³H-labeled RNAs from infected BHK cells (D).

B. Role of Tubulin

Having established a measles virus transcription system, we were particularly interested to test if components of the cytoskeleton might be required. Tubulin, in fact, also appears to be an essential component for measles virus RNA synthesis, as it is for Sendai virus. An anti-β-tubulin monoclonal antibody inhibited measles virus transcription and repliction *in vitro*, whereas antimyeloma or antiactin antibodies did not (Fig. 7). The antitubulin antibody also immunoprecipitated both tubulin and the measles virus L protein from the soluble protein fraction of measles virus-infected cells (data not shown). This antibody also coprecipitated tubulin and the VSV L protein (Moyer *et al.*, 1986), as discussed earlier, although we were unable to detect any interaction of tubulin with Sendai virus L protein in similar experiments.

The inhibition data suggest that tubulin is important for measles virus RNA synthesis. In testing this hypothesis, we found that the addition of the soluble protein fraction from infected cells stimulated transcription from intracellular nucleocapsids of all the measles virus RNAs (Fig. 8, A–E). At least one component involved in this stimulation was tubulin, since soluble tubulin alone stimulated the synthesis of N and P mRNAs, but not the other mRNAs (Fig. 8, F–I). While the level of tubulin stimulation of N and P mRNAs was lower (about threefold) relative to that observed with purified Sendai virus (tenfold; Table I), it is likely that in the former case significant amounts of host

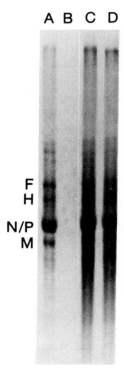

FIGURE 7. The effect of monoclonal antibodies on measles virus *in vitro* RNA synthesis. Samples of an extract of measles virus-infected A549 cells were incubated as described in Fig. 6 at 4°C for 30 min and then at 30°C for 2 hr in the presence of 980 units/ml of RNasin (Promega Biotech, Madison, WI) with no additions (A), 2 μl anti-β-tubulin monoclonal antibody (B), 2 μl antimyeloma control ascites fluid (C), and 2 μl antiactin monoclonal antibody (D). The RNA products were purified and analyzed as described in Fig. 6. The measles virus mRNAs are identified on the left.

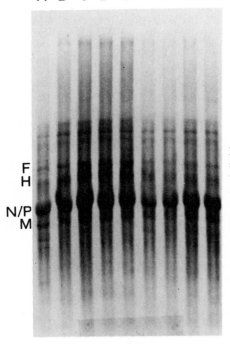

FIGURE 8. Effects of cell protein and tubulin on measles virus *in vitro* RNA synthesis. Extracts of measles virus-infected cells were separated into nucleocapsid and soluble protein fractions as described by Carlsen *et al* (1985). The measles virus intracellular nucleocapsids from 3×10^6 cells (3.75 mg/ml) were added to transcription reactions described in Fig. 6 containing no additions (A) or the soluble protein fraction from infected cells at protein concentrations of 0.625 (B), 1.25 (C), 2.50 (D), and 3.75 (E) mg/ml, respectively, or soluble tubulin at 42 (F), 107 (G), 212 (H), and 318 (I) μg/ml for 3 hr at 30°C. The RNA products were purified and analyzed as described in Fig. 6. The measles virus mRNAs are identified on the left.

cell proteins (i.e., tubulin) remained attached to the measles nucleocapsid fraction perhaps via the association of tubulin with the L subunit of the nucleocapsid-associated RNA polymerase. It appears that other viral or host factors might also be required for complete transcription, and future experiments employing this system should help elucidate this complex process. Ray and Fujinami (1987) have reported the *in vitro* synthesis of measles mRNAs employing intracellular nucleocapsids isolated from infected HeLa cells. The reaction conditions in their system and ours are generally similar, but some differences in results are perhaps attributable to the different methodologies employed in each case. As in our reconstitution experiments, Ray and Fujinami (1987) did not observe *in vitro* RNA replication from isolated nucleocapsids by RNA labeling experiments. However, in contrast to our results, they reported no stimulation of measles RNA synthesis by an extract of uninfected cells. The reason for this discrepancy is unclear; however, they measured total acid-precipitable counts rather than the synthesis of individual mRNAs as presented here.

C. Role of Actin

Previous experiments have sugested that another cytoskeleton component, actin, is essential for the maturation of progeny paramyxoviruses by budding. First, actin was identified as a packaged component of both measles and Sendai viruses (Wang et al., 1976; Tyrrell and Norrby, 1978) and biochemical data indicated that actin bound specifically to the M protein of Sendai and Newcastle disease viruses (Giuffre et al., 1982). Stallcup et al. (1983) showed, moreover, that treatment of measles virus-infected cells with cytochalasin B, which disrupts actin filaments, prevented virion formation and resulted in the accumulation of viral nucleocapsids within the infected cell. From these data, it is postulated that actin filaments might be essential for the transport of the nucleocapsid from the cytoplasm to the cell surface in preparation for budding.

Further support for this hypothesis comes from electron microscopic studies by Bohn et al. (1986), who showed that in cytoskeleton preparations of measles virus-infected cells, the growing end of the actin filament protruded into budding virus particles and was in close association with viral nucleocapsids. These authors suggest that vectorial growth of the actin filaments is involved both in the transport of the nucleocapsid to the surface of the cell and for budding itself.

Given the above results, the obvious question is how the measles nucleocapsid becomes and remains attached to the actin filament. We present data for one possibility in the following experiment. Figure 9 (panel B) shows reconstitution experiments to measure the association of radiolabeled proteins with template and progeny measles virus nucleocapsids under various reaction conditions. When RNA replication was induced by the addition of the radiolabeled soluble protein fraction of infected cells and nucleotide precursors, there was incorporation of some N protein into progeny nucleocapsids (panel B, lane 2), although it was not much above the background level of exchange seen in the absence of RNA synthesis (i.e., with no added nucleotides) (panel B, lane 1). This correlates with a similar low level of RNA replication as assayed by RNA labeling as described above. The interesting feature, however, of this experiment is that in the absence of RNA synthesis the actin in the infected cell extract bound to the measles nucleocapsids, although little actin binding was seen when RNA synthesis was occurring. Actin in an uninfected cell extract also bound to measles nucleocapsids when RNA synthesis could not occur, but not when transcription was allowed (Fig. 9B, lanes 3 and 4, respectively). Similar results were seen for another paramyxovirus, Sendai virus (Fig. 9, panel A). The actin binding to silent, nonsynthesizing nucleocapsids was extremely stable, since the CsCl centrifugation step employed here for purification normally removes all the viral proteins except the nucleocapsid protein. Based on these results, we propose that it is the inactive paramyxovirus nucleocapsids, that is, those which have completed RNA synthesis, which bind to the actin filaments for transport and budding. The actin remains tightly associated with the progeny virions, i.e., it is packaged, until it is released when new RNA synthesis begins upon subsequent infection.

How is viral RNA synthesis terminated in the cell? In the case of VSV, the M protein appears to function, in part, to inhibit transcription both in vivo and in vitro [for a review, see Banerjee (1987)]. VSV, however, does not appear to utilize actin for maturation; since actin was not packaged in the virion, cytochalasin B had no effect upon virus yield (Gentry and Busserau, 1980), and

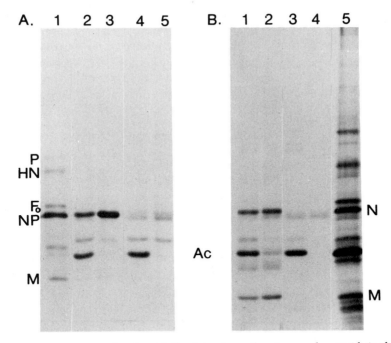

FIGURE 9. The association of actin with Sendai and measles virus nucleocapsids in the absence of RNA synthesis. Unlabeled nucleocapsids were isolated from extracts of virus-infected cells and incubated with a [³H]leucine and [³⁵S]methionine double-labeled soluble protein fraction from virus-infected or uninfected cells under various reaction conditions. The product nucleocapsids were nuclease treated and purified by banding on CsCl gradients, and the nucleocapsid-associated proteins were analyzed by electrophoresis on an 11% polyacrylamide–SDS gel. Panel A: The markers are ³H-labeled purified Sendai virus proteins (lane 1). Ac indicates actin (M_r 43,000). Unlabeled intracellular Sendai virus nucleocapsids were incubated in Sendai virus reaction mix either with a ³H- and ³⁵S-labeled soluble protein fraction from Sendai virus-infected cells in the absence (lane 2) or presence (lane 3) of all ribonucleoside triphosphates or with a ³H- and ³⁵S-labeled soluble protein fraction from uninfected cells in the absence (lane 4) or presence (lane 5) of all ribonucleoside triphosphates. Panel B: The markers are ³H-labeled purified measles virus proteins (lane 5). Unlabeled intracellular measles virus nucleocapsids were incubated in measles virus reaction mix either with a ³H- and ³⁵S-labeled soluble protein fraction from measles virus-infected cells in the absence (lane 1) or presence (lane 2) of all ribonucleoside triphosphates or with a ³H- and ³⁵S-labeled soluble protein fraction from uninfected cells in the absence (lane 3) or presence (lane 4) of all ribonucleoside triphosphates. Samples incubated in the absence of added substrate contained 20 mM EDTA to inhibit incorporation of endogenous nucleotides.

finally, we found no actin binding to VSV nucleocapsids under any conditions *in vitro* (data not shown). Indirect evidence (Marx *et al.*, 1974) suggests that the paramyxovirus M protein, which is known to interact with actin *in vitro* (Giuffre *et al.*, 1982), may also inhibit RNA synthesis *in vivo* in preparation for budding. We have shown *in vitro*, however, that M protein need not necessarily be present, since when Sendai or measles virus RNA synthesis was artificially prevented by removing the nucleotide precursors, actin in an uninfected cell extract bound tightly to silent nucleocapsids (Fig. 9). The nature of the association of actin and paramyxovirus nucleocapsids is under further investigation.

IV. CONCLUSIONS AND PROSPECTS

The establishment of Sendai and measles virus *in vitro* transcription/replication systems described here has allowed a preliminary understanding of the components and mechanisms which govern these processes. A great deal of additional work will be required, however, to understand the molecular basis of negative-strand virus RNA synthesis. Particularly intriguing will be an elucidation of the role of the host cell components in these processes. There are, in addition, recent new observations that can be explored with the availability of these *in vitro* RNA synthesis systems. First, there is growing evidence that the paramyxovirus polymerase is not always accurate, as shown by the synthesis of readthrough polycistronic transcripts (Dowling *et al.*, 1986; Wilde and Morrison, 1984) and leader–NP transcripts, some of which are encapsidated and polyadenylated (Jambou *et al.*, 1986; Elango *et al.*, 1988; Vidal and Kolakofsky, 1989; Castaneda and Wong, 1989). Whether these products have functional significance or are the result of polymerase errors remains to be determined. Second, measles virus appears not to synthesize either (+)- or (−)-strand leader RNA *in vivo*, suggesting that, unlike other paramyxoviruses, in this case different initiation sites are used for transcription and replication (Crowley *et al.*, 1988; Castaneda and Wong, 1989). Finally, several paramyxoviruses appear to use RNA editing to alter the nature of the trancripts from the *P* gene (Thomas *et al.*, 1988; Cattaneo *et al.*, 1989). Exciting new developments are sure to follow.

ACKNOWLEDGMENTS. The research from our laboratory reported here was supported by a grant from the National Institute of Allergy and Infectious Diseases (AI-14594). We would like to acknowledge Dr. Robley Williams, Jr. (Vanderbilt University) as our collaborator in the unpublished tubulin experiments. We thank Pat Austin for the excellent preparation of the manuscript.

V. REFERENCES

Alkhatib, G., and Briedis, D. J., 1986, The predicted primary structure of the measles virus hemagglutinin, *Virology* **150**:479–490.

Baker, S. C., and Moyer, S. A., 1988, Encapsidation of Sendai virus genome RNAs by purified NP protein during *in vitro* replication, *J. Virol.* **62**:834–838.

Banerjee, A. K., 1987, Transcription and replication of rhabdoviruses, *Microbiol. Rev.* **51**:66–87.

Bellini, W. J., Englund, G., Rozenblatt, G., Arnheiter, H., and Richardson, C. D., 1985, Measles virus *P* gene codes for two proteins, *J. Virol.* **53**:908–919.

Bellini, W. J., Englund, G., Richardson, C. D., Rozenblatt, S., and Lazzarini, R. A., 1986, Matrix genes of measles virus and canine distemper virus: Cloning, nucleotide sequences and deduced amino acid sequences, *J. Virol.* **58**:408–416.

Bernard, J. P., and Northrop, R. L., 1974, RNA polymerase activity in mumps virus, *J. Virol.* **14**:183–186.

Blumberg, B. M., and Kolakofsky D., 1981, Intracellular vesicular stomatis virus leader RNAs are found in nucleocapsid structures, *J. Virol.* **40**:568–576.

Blumberg, B. M., Leppert, M., and Kolakofsky, D., 1981, Interaction of VSV leader RNA and nucleocapsid protein may control VSV genome replication, *Cell* **23**:837–845.

Blumberg, B. M., Giorgi, C., and Kolakofsky, D., 1983, N protein of vesicular stomatitis virus selectively encapsidates leader RNA *in vitro*, *Cell* **32**:559–657.

Blumberg, B. M., Giorgi, C., Rose, K., and Kolakofsky, D., 1984. Preparation and analysis of the nucleocapsid proteins of vesicular stomatitis virus and Sendai virus, and analysis of the Sendai virus leader *NP* gene region, *J. Gen. Virol.* **65**:769–779.

Blumberg, B. M., Crowley, J. C., Silverman, J. I., Menonna, J., Cook, S. D., and Dowling, P. C., 1988, Measles virus L protein evidences elements of ancestral RNA polymerase, *Virology* **164**:487–497.

Bohn, W., Rutter, G., Hohenberg, H., Mannweiler, K., and Nobis, P., 1986, Involvement of actin filaments in budding of measles virus: Studies on cytoskeletons of infected cells, Virology **149**:91–106.

Buetti, E., and Choppin, P. W., 1977, The transcriptase complex of paramyxovirus SV5, *Virology* **82**:493–508.

Carlsen, S. R., Peluso, R. W., and Moyer, S. A., 1985, *In vitro* replication of Sendai virus wild type and defective interferring particle genome RNAs, *J. Virol.* **54**:493–500.

Carroll, A. R., and Wagner, R. R., 1979, Role of the membrane (M) protein in endogenous inhibition of *in vitro* transcription of vesicular stomatitis virus, *J. Virol.* **29**:134–142.

Castaneda, S. J., and Wong, T. C., 1989, Measles virus synthesizes both leaderless and leader-containing polyadenylated RNAs *in vivo*, *J. Virol.* **63**:2977–2986.

Cattaneo, R., Rebmann, G., Schmid, A., Baczko, K., ter Meulen, V., and Billeter, M. A., 1987, Altered transcription of a defective measles virus genome derived from a diseased human brain, *EMBO J.* **6**:681–688.

Cattaneo, R., Kaelin, K., Baczko, K., and Billeter, M. A., 1989, Measles virus editing provides an additional cysteine-rich protein, *Cell* **56**:759–764.

Cervera, M., Dreyfuss, G., and Penman, S., 1981, Messenger RNA is translated when associated with the cytoskeletal framework in normal and VSV-infected HeLa cells, *Cell* **23**:113–120.

Chatterjee, P. K., Cervera, M. M., and Penman, S., 1984, Formation of vesicular stomatitis virus nucleocapsid from cytoskeletal framework-bound N protein: Possible model for structural assembly, *Mol. Cell. Biol.* **4**:2231–2234.

Chattopadhyay, D., and Banerjee, A. K., 1987, Two separate domains within vesicular stomatitis virus phosphoprotein support transcription when added in *trans*, *Proc. Natl. Acad. Sci. USA* **84**:8932–8936.

Chattopadhyay, D., and Banerjee, A. K., 1988, NH_2-Terminal acidic region of the phosphoprotein of vesicular stomatitis virus can be functionally replaced by tubulin, *Proc. Natl. Acad. Sci. USA* **85**:7977–7981.

Chinchar, V. G., and Portner, A., 1981, Functions of Sendai virus nucleocapsid polypeptides: Enzymatic activities in nucleocapsids following cleavage of polypeptide P by *Staphylococcus aureus* protease V8, *Virology* **109**:59–71.

Choppin, P. W., and Compans, R. W., 1975, Reproduction of paramyxoviruses, *in* "Comprehensive Virology," Vol. 4 (H. Fraenkel-Conrat and R. R. Wagner, eds.), pp. 95–178, Plenum Press, New York.

Clinton, G. M., Little, S. P., Hagen, F. S., and Huang, A. S., 1978, The matrix (M) protein of vesicular stomatitis virus regulates transcription, *Cell* **15**:1455–1462.

Colonno, R. J., and Stone, H. O., 1975, Methylation of messenger RNA of Newcastle disease virus *in vitro* by a virion associated enzyme, *Proc. Natl. Acad. Sci. USA* **72**:2611–2615.

Colonno, R. J., and Stone, H. O., 1976a, Isolation of a transcriptive complex from Newcastle disease virus, *J. Virol.* **19**:1080–1089.

Colonno, R. J., and Stone, H. O., 1976b, Newcastle disease virus messenger RNA lacks 2'-O-methylated nucleotides, *Nature* **261**:611–614.

Crowley, J. C., Dowling, P. C., Menonna, J., Silverman, J. I., Schuback, D., Cook, S. D., and Blumberg, B. M., 1988, Sequence variability and function of measles virus 3' and 5' ends and intercistronic regions, *Virology* **164**:498–506.

Curran, J., and Kolakofsky, D., 1987, Identification of an additional Sendai virus non-structural protein encoded by the P/C mRNA, *J. Gen. Virol.* **68**:2515–2519.

Curran, J., and Kolakofsky, D., 1988a, Ribosomal initiation from an ACG codon in the Sendai virus P/C mRNA, *EMBO J.* **7**:245–251.

Curran, J., and Kolakofsky, D., 1988b, Scanning independent ribosomal initiation of the Sendai virus X protein, *EMBO J.* **7**:2869–2874.

De, B. P., and Banerjee, A. K., 1985, Requirements and functions of vesicular stomatitis virus L and NS proteins in the transcription process *in vitro*, *Biochem. Biophys. Res. Commun.* **126**:40–49.

Deshpande, K. L., and Portner, A., 1984, Structural and functional analysis of Sendai virus nucleocapsid NP protein with monoclonal antibodies, *Virology* **139**:32–42.

Deshpande, K. L., and Portner, A., 1985, Monoclonal antibodies to the P protein of Sendai virus define its structure and role in transcription, *Virology* **140**:125–134.

Dillon, P. J., and Gupta, K. C., 1989, Expression of five proteins from the Sendai virus P/C mRNA in infected cells, *J. Virol.* **63**:974–977.

Dowling, P. C., Blumberg, B. M., Menonna, J., Adamus, J. E., Cook, P. J., Crowley, J. C., Kolakofsky, D., and Cook, S. D., 1986, Transcriptional map of the measles virus genome, *J. Gen. Virol.* **67**:1987–1992.

Elango, N., Varsanyi, T., Kovamees, J., and Norrby, E., 1988, Molecular cloning and characterization of six genes, determination of gene order and intergenic sequences and leader sequence of mumps virus, *J. Gen. Virol.* **69**:2893–2900.

Emerson, S. U., and Schubert, M., 1987, Location of the binding domains for the RNA polymerase L and the ribonucleocapsid template within different halves of the NS phosphoprotein of vesicular stomatitis virus, *Proc. Natl. Acad. Sci. USA* **84**:5655–5659.

Etkind, P. R., Cross, R. K., Lamb, R. A., Merc, D. C., and Choppin, P. W., 1980, *In vitro* synthesis of structural and nonstructural proteins of Sendai and SVS viruses, *Virology* **100**:22–33.

Feldhaus, A. L., and Lesnaw, J. A., 1988, Nucleotide sequence of the *L* gene of vesicular stomatitis virus (New Jersey): Identification of conserved domains in the New Jersey and Indiana L proteins, *Virology* **163**:359–368.

Galinski, M. S., Mink, M. A., and Pons, M. W., 1988, Molecular cloning and sequence analysis of the human parainfluenza 3 virus gene encoding the L protein, *Virology* **165**:499–510.

Gentry, N., and Busserau, F., 1980, Is cytoskeleton involved in vesicular stomatitis virus reproduction?, *J. Virol.* **34**:777–781.

Gerald, C., Buckland, R., Barker, R., Freeman, G., and Wild, T. F., 1986, Measles virus haemagglutinin gene: Cloning, complete nucleotide sequence analysis and expression in COS cells, *J. Gen. Virol.* **67**:2695–2703.

Gill, D. S., Chattopadhyay, D., and Banerjee, A. K., 1986, Identification of a domain within the phosphoprotein of vesicular stomatitis virus that is essential for transcription *in vitro*, *Proc. Natl. Acad. Sci. USA* **83**:8873–8877.

Gill, D. S., Takai, S., Portner, A., and Kingsbury, D. W., 1988, Mapping of antigenic domains of Sendai virus nucleocapsid protein expressed in *Escherichia coli*, *J. Virol.* **62**:4805–4808.

Giuffre, R. M., Tovell, D. R., Kay, C. M., and Tyrrell, D. L. J., 1982, Evidence for an interaction between the membrane protein of a paramyxovirus and actin, *J. Virol.* **42**:963–968.

Glazier, K., Raghow, R., and Kingsbury, D. W., 1977, Regulation of Sendai virus transcription: Evidence for a single promoter *in vivo*, *J. Virol.* **21**:863–871.

Gupta, K. C., and Kingsbury, D. W., 1982, Conserved polyadenylation signals in two negative-strand RNA virus families, *Virology* **120**:518–523.

Gupta, K. C., and Kingsbury, D. W., 1984, Complete sequences of the intergenic and mRNA start signals in the Sendai virus genome: Homologies with the genome of vesicular stomatitis virus, *Nucleic Acids Res.* **12**:3829–3841.

Hall, W. W., and ter Meulen, V., 1977, Polyadenylic acid [poly(A)] sequences associated with measles virus intracellular ribonucleic acid (RNA) species, *J. Gen. Virol.* **35**:497–510.

Hamaguchi, M., Yoshida, T., Nishikawa, K., Naruse, H., and Nagai, Y., 1983, Transcriptive complex of Newcastle disease virus. I. Both L and P proteins are required to reconstitute an active complex, *Virology* **128**:105–117.

Hamaguchi, M., Nishikawa, K., Toyoda, T., Yoshida, T., Hanaichi, T., and Nagai, Y., 1985, Transcriptive complex of Newcastle disease virus II. Structural and functional assembly associated with the cytoskeletal framework, *Virology* **147**:295–308.

Heggeness, M. H., Scheid, A., and Choppin, P. W., 1981, The relationship of conformational changes in the Sendai virus nucleocapsid to proteolytic cleavage of the NP polypeptide, *Virology* **114**:555–562.

Hercyk, N., Horikami, S. M., and Moyer, S. A., 1988, The vesicular stomatitis virus L protein possesses the mRNA methyltransferase activities, *Virology* **163**:222–225.

Herman, R. C., 1989, Synthesis of respiratory syncytial virus RNA in cell-free extracts, *J. Gen. Virol.* **70:**755–761.

Hill, V. M., Harmon, S. A., and Summers, D. F., 1986, Stimulation of vesicular stomatitus virus *in vitro* RNA synthesis by microtubule-associated proteins, *Proc. Natl. Acad. Sci. USA* **83:**5410–5413.

Hsu, C.-H., and Kingsbury, D. W., 1982, Topography of phosphate residues in Sendai virus proteins, *Virology* **120:**225–234.

Huang, A. S., Baltimore, D., and Bratt, M. A., 1971, Ribonucleic acid polymerase in virions of Newcastle disease virus: Comparison with vesicular stomatitis virus polymerase, *J. Virol.* **7:**389–394.

Hunt, D. M., Smith, E. G., and Buckley, D. W., 1984, Aberrant polyadenylation by a vesicular stomatitis virus mutant is due to an altered L protein, *J. Virol.* **52:**515–521.

Jambou, R. C., Elango, N., Vankatesan, S., and Collins, P. L., 1986, Complete sequence of the major nucleocapsid protein gene of human parainfluenza type 3 virus: Comparison with other negative strand viruses, *J. Gen. Virol.* **67:**2543–2548.

Kingsbury, D. W., 1977, Paramyxoviruses, "The Molecular Biology of Animal Viruses," Vol. 1, (D. P. Nayak, ed.), pp. 349–382, Marcel Dekker, New York.

Kingsbury, D. W., Hsu, C. H., and Murti, K. G., 1978, Intracellular metabolism of Sendai virus nucleocapsids, *Virology* **91:**86–94.

Kolakofsky, D., and Roux, L., 1987, The molecular basis of viral replication, *in* "The Molecular Biology of Paramyxoviruses (R. Perez Bercoff, ed.), pp. 277–297, Plenum Press, New York.

Krauhs, E., Little, M., Kempf, T., Hofer-Warbinek, R., Ade, W., and Ponstingl, H., 1981, Complete amino acid sequence of beta-tubulin from porcine brain, *Proc. Natl. Acad. Sci. USA* **78:**4156–4160.

Kristensson, K., and Orvell, C., 1983, Cellular localization of five structural proteins of Sendai virus studied with peroxidase-labeled Fab fragments of monoclonal antibodies, *J. Gen. Virol.* **64:**1673–1678.

Lamb, R. A., and Choppin, P. W., 1977, The synthesis of Sendai virus polypeptides in infected cells. II. Intracellular distribution of polypeptides, *Virology* **81:**371–381.

Lamb, R. A., Mahy, B. W. J., and Choppin, P. W., 1976, The synthesis of Sendai virus polypeptides in infected cells, *Virology* **69:**116–131.

Leppert, M., Rittenhouse, L., Perrault, J., Summers, D. F., and Kolakofsky, D., 1979, Plus- and minus-strand leader RNAs in negative strand virus infected cells, *Cell* **18:**735–747.

Martinet, C., Combard, A., Printz-Arne, C., and Printz, P., 1979, Envelope proteins and replication of vesicular stomatitis virus: *In vivo* effects of RNA+ temperature-sensitive mutations on viral RNA synthesis, *J. Virol.* **29:**123–133.

Marx, P. A., Pridgen, C., and Kingsbury, D. W., 1975, Location and abundance of poly(A) sequences in Sendai virus messenger RNA molecules, *J. Gen. Virol.* **27:**247–250.

Marx, P. A., Portner, A., and Kingsbury, D. W., 1974, Sendai virus transcriptase complex: Polypeptide composition and inhibition by virion envelope proteins, *J. Virol.* **13:**107–112.

Morgan, E., and Rakestraw, K., 1986, Sequence of the Sendai L gene: Open reading frames upstream of the main coding region suggest that the gene may be polycistronic, *Virology* **154:**31–40.

Morgan, E. M., Re, G. G., and Kingsbury, D. W., 1984, Complete sequence of the Sendai virus *NP* gene from a cloned insert, *Virology* **135:**279–287.

Moyer, S. A., Baker, S. C., and Lessard, J. L., 1986, Tubulin: A factor necessary for the synthesis of both Sendai virus and vesicular stomatitis virus RNAs, *Proc. Natl. Acad. Sci. USA* **83:**5405–5409.

Omata-Yamada, T., Hagiwara, K., Katoh, K., Yamada, H., and Iwasaki, K., 1988, Purification of the Sendai virus nonstructural C protein expressed in *E. coli*, and preparation of antiserum against C protein, *Arch. Virol.* **103:**61–72.

Ongradi, J., Cunningham, C., and Szilagyi, J. F., 1985, The role of polypeptides L and NS in the transcription process of vesicular stomatitis virus New Jersey using the temperature sensitive mutant *ts* El, *J. Gen. Virol.* **66:**1011–1023.

Patwardhan, S., and Gupta, D. C., 1988, Translation initiation potential of the 5' proximal AUGs of the polycistronic P/C mRNA of Sendai virus, *J. Biol. Chem.* **263:**4907–4913.

Peluso, R. W., and Moyer, S. A., 1983, Initiation and replication of vesicular stomatitis virus genome RNA in a cell-free system, *Proc. Natl. Acad. Sci. USA* **80:**3198–3202.

Peluso, R. W., and Moyer, S. A., 1988, Viral proteins required for the *in vitro* replication of vesicular stomatitis virus defective interfering particle genome RNA, *Virology* **162**:369–376.

Ponstingl, H., Krauhs, E., Little, M., and Kempf, T., 1981, Complete amino acid sequence of alpha-tubulin from porcine brain, *Proc. Natl. Acad. Sci. USA* **78**:2757–2761.

Portner, A., 1982, Synthesis of message and genome RNAs *in vitro* by Sendai virus-infected cell nucleocapsids, *J. Gen. Virol.* **60**:67–75.

Portner, A., and Murti, K. G., 1986, Localization of P, NP and M proteins on Sendai virus nucleocapsid using immunogold labeling, *Virology* **150**:469–478.

Portner, A., Gupta, K. C., Seyer, J. M., Beachy, E. H., and Kingsbury, D. W., 1986, Localization and characterization of Sendai virus nonstructural C and C' proteins by antibodies against synthetic peptides, *Virus Res.* **6**:109–121.

Portner, A., Murti, K. G., Morgan, E. M., and Kingsbury, D. W., 1988, Antibodies against Sendai virus L protein: Distribution of the protein in nucleocapsids revealed by immunoelectron microscopy, Virology 163:236–239.

Pridgen, C., and Kingsbury, D. W., 1972, Adenylate-rich sequences in Sendai virus transcripts from infected cells, *J. Virol.* **10**:314–317.

Ray, J., and Fujinami, R. S., 1987, Characterization of *in vitro* transcription and transcriptional products of measles virus, *J. Virol.* **61**:3381–3387.

Richardson, C. D., Berkovich, A., Rozenblatt, S., and Bellini, W. J., 1985, Use of antibodies directed against synthetic peptides for identifying cDNA clones, establishing reading frames and deducing the gene order of measles virus, *J. Virol.* **54**:186–193.

Richardson, C., Hull, D., Greer, P., Hasel, K., Berkovich, A., Englund, G., Bellini, W., Rima, B., and Lazzarini, R., 1986, The nucleotide sequence of the mRNA encoding the fusion protein of measles virus (Edmonston strain): A comparison of fusion proteins from several different paramyxoviruses, *Virology* **155**:508–523.

Rima, B. K., Baczko, K., Clarke, D. K., Curran, M. D., Marten, S. J., Billeter, M. A., and ter Meulen, V., 1986, Characterization of clones for the sixth (*L*) gene and a transcriptional map for morbilliviruses, *J. Gen. Virol.* **67**:1971–1978.

Robinson, W. S., 1971a, Ribonucleic acid polymerase in Sendai virions and nucleocapsids, *J. Virol.* **8**:81–86.

Robinson, W. S., 1971b, Sendai virus RNA synthesis and nucleocapsid formation in the presence of cyclohexamide, *Virology* **44**:494–502.

Rozenblatt, S., Eizenberg, O., Ben-Levy, R., Lavie, V., and Bellini, W. J., 1985, Sequence homology with the morbilliviruses, *J. Virol.* **53**:684–690.

Ryan, K. W., and Kingsbury, D. W., 1988, Carboxyl-terminal region of Sendai virus P protein is required for binding to viral nucleocapsids, *Virology* **167**:106–112.

Seifried, A. S., Albrecht, P., and Melstien, J. B., 1978, Characterization of an RNA-dependent RNA polymerase activity associated with measles virus, *J. Virol.* **25**:781–787.

Shioda, T., Hidaka, Y., Kanda, T., Shibuta, H., Nomoto, A. and Iwasaki, K., 1983, Sequence of 3687 nucleotides from the 3' end of Sendai virus genome RNA and the predicted amino acid sequences of viral NP, P and C proteins, *Nucleic Acid Res.* **11**:7317–7330.

Shioda, T., Iwasaki, K., and Shibuta, H., 1986, Determination of the complete nucleotide sequence of the Sendai virus genome RNA and the predicted amino acid sequences of the F, HN and L proteins, *Nucleic Acid Res.* **14**:1545–1563.

Shubert, M., Harmison, G. G., and Meier, E., 1984, Primary structure of the vesicular stomatitis virus polymerases (*L*) gene: Evidence for a high frequency of mutations, *J. Virol.* **51**:505–514.

Stallcup, K. C., Raine, C. S., and Fields, B. N., 1983, Cytochalasin B inhibits the maturation of measles virus, *Virology* **124**:59–74.

Stone, H. O., and Kingsbury, D. W., 1973, Stimulation of Sendai virion transcriptase by polyanions, *J. Virol.* **11**:243–249.

Stone, H. O., Portner, A., and Kingsbury, D. W., 1971, Ribonucleic acid transcriptases in Sendai virions and infected cells, *J. Virol.* **8**:174–180.

Taira, H., Kanda, T., Omata, T., Shibuta, H., Kawakita, M., and Iwasaki, K., 1987, Interferon induction by transfection of Sendai virus *C* gene cDNA, *J. Virol.* **61**:625–628.

Thomas, S. M., Lamb, R. A., and Paterson, R. G., 1988, Two mRNAs that differ by two nontemplated nucleotides encode the amino coterminal proteins P and V of the paramyxovirus SV5, *Cell* **54**:891–902.

Tordo, N., Poch, O., Ermine, A., Keith, G., and Rougeon, F., 1988, Completion of the rabies virus genome sequence determination: Highly conserved domains among the (L) polymerase proteins of unsegmented negative-strand RNA viruses, *Virology* **165**:565–576.

Tyrrell, D. L. J., and Norrby, E., 1978, Structural polypeptides of measles virus, *J. Gen. Virol.* **39**:219–229.

Vidal, S., and Kolakofsky, D., 1989, Modified model for the switch from Sendai virus transcription to replication, *J. Virol.* **63**:1951–1958.

Vidal, S., Curran, J., Orvell, C., and Kolakofsky, D., 1988, Mapping of monoclonal antibodies to the Sendai virus P protein and the location of its phosphates, *J. Virol.* **62**:2200–2203.

Wang, E., Wolf, B. A., Lamb, R. A., Choppin, P. W., and Goldberg, R. A., 1976, The presence of actin in enveloped viruses, *in* "Cell Motility" (R. Goldman, T. Pollard, and J. Rosenbaum, ed.), pp. 589–599, Cold Spring Harbor Laboratory, Cold Spring Harbor, New York.

Wilde, A., and Morrison, T., 1984, Structural and functional characterization of Newcastle disease virus polycistronic RNA species, *J. Virol.* **51**:71–76.

Yoshikawa, Y., Mizumoto, K., and Yamanouchi, K., 1986, Characterization of messenger RNAs of measles virus, *J. Gen. Virol.* **67**:2807–2812.

Yusoff, K., Millar, N. S. Chambers, P., and Emmerson, P. T., 1987, Nucleotide sequence analysis of the L gene of Newcastle disease virus: Homologies with Sendai and vesicular stomatitis virus, *Nucleic Acids Res.* **15**:3961–3976.

CHAPTER 10

Deletion Mutants of Paramyxoviruses

Gian Guido Re

I. INTRODUCTION AND HISTORICAL PERSPECTIVE

Defective-interfering (DI) particles represent a class of virus deletion mutants that require coinfecting, nondefective, homologous (standard) helper virus to survive. In turn, DI particles usually interfere with the reproduction of the helper virus. These mutants arise spontaneously and are amplified to detectable levels by repeated passage of virus stocks at high multiplicities of infection. They were first described in influenza virus infections by von Magnus (1952), who called them "incomplete viruses," and this terminology was generally applied to other examples of spontaneously generated noninfectious viruses until Huang and Baltimore (1970) coined the designation "DI particles."

When influenza virus stocks containing DI particles were serially passaged undiluted in embryonated eggs, yields of infectious virus declined progressively and then rose to former levels in a cyclic pattern. The amount of defective virus (as measured by hemagglutinin titer) followed this pattern out of phase, declining and increasing after the decline and increase of infectious virus. This cyclic phenomenon, the von Magnus effect, is a characteristic feature of the interaction between DI and standard virus genomes. Bellet and Cooper's (1959) observation of such cycling led to the identification of vesicular stomatitis virus (VSV) DI particles. Subsequently, DI particles were described in an increasing number of virus systems, indicating that they occur generally among both RNA and DNA viruses.

Huang and Baltimore (1970) set the following criteria for their identification: (1) they contain normal viral structural proteins; (2) they contain only part of the viral genome; (3) they can replicate only in the presence of homologous virus; and (4) they interfere specifically with the intracellular replication

---

GIAN GUIDO RE • Pediatric Leukemia Research Program, Division of Pediatrics, University of Texas M. D. Anderson Cancer Center, Houston, Texas 77030.

of nondefective homologous virus. DI particles often mediate the establishment and maintenance of persistent viral infections, since even partial interference with the replication of the helper virus may reduce its cytopathogenicity. Thus, Huang and Baltimore (1970) postulated that DI particles might play a part in some chronic human diseases of unknown etiology, such as subacute sclerosing panencephalitis (SSPE) and other chronic neurological diseases, including multiple sclerosis (MS). These ideas stimulated research on DI particles in the following years.

In this chapter, I will not attempt a comprehensive review of the literature, but will focus on a few examples that illustrate some of the most compelling questions about DI paramyxoviruses: mechanisms underlying their generation, survival, and interference. In addition, their role in persistent infections and their possible involvement in human diseases will be discussed.

II. DEFECTIVE-INTERFERING PARAMYXOVIRUSES

Usually, the first indication of the presence of DI particles in a virus stock is presentation of one or more of the following related biological phenomena. During repeated passage, a virus preparation exhibits reduced growth, displaying a decreased ratio of infectivity to total virus particles, and cytopathology (CPE) is reduced. Virus stocks containing DI particles usually mediate the establishment of persistent infections and reduce the growth of fully infectious virus in mixed infections. A number of paramyxoviruses have displayed these properties.

A. Sendai Virus

Among paramyxoviruses, DI particles of Sendai virus have been the most fully characterized. Sendai virus DI particles were originally referred to as modified virus particles or incomplete virions (Tadokoro, 1958; Sokol et al., 1964; Kingsbury et al., 1970). Typically, they arose spontaneously from undiluted passage of the virus in ovo and displayed a series of properties including decreased hemolytic activity, decreased infectivity to hemagglutinin ratio, smaller size, as reflected by a lower sedimentation constant, and increased fragility relative to standard virions (Todokoro, 1958; Sokol et al., 1964).

Incomplete virions generated elsewhere (Kingsbury et al., 1970) were shown to contain 19S and 25S subgenomic RNAs which had the same (negative) polarity as the 50S virus genome, indicating their deleted nature, and these truncated virus genomes were encapsidated into short rods that otherwise looked structurally identical to full-length standard virus nucleocapsids in the electron microscope. As is the case for standard virus nucleocapsids, the RNA in DI nucleocapsids was efficiently protected from ribonuclease digestion, including even the terminal nucleotides of the RNA within them (Lynch and Kolakofsky, 1978). From the biological standpoint, the incomplete virions were not infectious and required coinfecting infectious virus for replication; in mixed infections, the DI virus interfered with the replication of the standard virus (Kingsbury et al., 1970; Portner and Kingsbury, 1971). The active interfering component of the DI particles was the RNA, since UV irradiation destroyed their ability to interfere with the replication of standard virus (Kingsbury and Portner, 1970). Sendai virus DI particles were later shown to mediate persistent infections in BHK cells (Roux and Holland, 1979). A more extensive discussion of Sendai virus DI RNA species is given below (Section III).

B. Parainfluenza Virus 3

DI RNA species have recently been detected in human parainfluenza virus type 3 (HPIV3) after serial undiluted passage in LLC-MK2 cells (Murphy et al., 1987). During these passages, infectious HPIV3 growth displayed the cyclic von Magnus effect, consistent with interference by DI particles. Of the three subgenomic RNA species detected, two were shown by hybridization with virus-specific probes to contain sequences of the parental virus genome.

C. Mumps Virus

Two laboratories have reported the existence of DI particles in mumps virus serially passaged in tissue cultures. The chicken embryo-adapted Enders strain of mumps virus, a prototype vaccine strain, readily established persistent infection in Vero cells (McCarthy et al., 1981). Subgenomic 10–18S and 20S RNA species, recovered as intracellular encapsidated RNA, were associated with this persistent infection. By pancreatic and T1 ribonuclease fingerprint analyses, these RNA species shared numerous oligonucleotides with the 50S nondefective virus genome. Furthermore, the virus containing these RNAs interfered with other strains of mumps virus.

The Leningrad-3 vaccine strain of mumps virus also established persistent infections in human cells (HEp-2 and L-41) which manifested viral antigen over a period of 3 years (Andzhaparidze et al., 1982a,b, 1983). Although L-41 cells produced interferon, the persistent infection appeared to be mediated by DI particles. Virus particles from the chronically infected cultures sedimented slightly faster than standard mumps virus, indicating that most of them were smaller than normal. This virus contained predominantly low-molecular-weight RNA (approximately 10S), while standard mumps virus infections yielded 50S genomic RNA. By hybridization with mumps virus cDNA, the low-molecular-weight RNA contained virus-specific sequences (Andzhaparidze et al., 1982a). UV irradiation of the virus from persistently infected cultures greatly reduced its interfering activity. Persistently infected HEp-2 cultures were resistent to superinfection with standard mumps virus, but were susceptible to heterologous viruses. However, persistently infected L-41 cultures were resistant also to heterologous viruses because they produced interferon.

D. Simian Virus 5

Observations of Azadova and Zhdanov (1982) are consistent with the existence of DI particles in their simian virus 5 (SV5; canine parainfluenza virus) preparations. They found that after six undiluted passages in chick embryos of a virulent stock of SV5, the infectivity-to-hemagglutinin ratio decreased 10,000-fold. This virus caused limited CPE in L cells compared to the original virus, mediated persistent infections, and reduced CPE in mixed infections with standard virus. Analysis of intracellular nucleocapsid RNA revealed that while the standard virus directed synthesis primarily of 50S viral RNA, the attenuated virus mainly produced a low-molecular-weight, virus-specific RNA (approximately 20S) and almost completely suppressed the synthesis of 50S RNA.

Evidence for the existence of DI particles in Vero cells persistently infected with a street isolate of canine parainfluenza virus was provided by Baumgartner et al. (1987). This virus was isolated from an encephalitis experimentally induced with a neurotropic strain of canine parainfluenza virus recovered from a paralytic dog, suggesting a connection between these disease processes and DI particles.

E. Newcastle Disease Virus

Maeda et al. (1978) found that propagation of Newcastle disease virus (NDV) in embryonated eggs gave rise to DI particles. By sucrose gradient centrifugation, the egg-grown virus was separable into two distinct types of particles: normal, infectious B particles and small, noninfectious T particles. The latter contained only 13S RNA, as opposed to B particles, which contained predominantly 57S RNA. The T particles interfered with the replication of standard NDV in primary cultures of chick embryo cells. More recently, NDV has been shown to establish persistent infections in the murine osteosarcoma cell line OGS (Degré and Glasgow, 1983). The properties of the persistently infected cells were consistent with the activity of DI particles.

F. Measles Virus

A substantial body of literature describes the detection of measles virus DI particles. The involvement of measles virus in human diseases and the role of DI particles has been extensively reviewed by Morgan and Rapp (1977). Rima et al. (1977) reported establishment of persistent infections in Vero cells by undiluted passage of a Vero cell-adapted Edmonston strain of measles virus. Consistent with the activity of DI particles, the persistently infected cells displayed virus antigen by immunofluorescence microscopy, underwent crisis (i.e., typical measles virus CPE) at various periods, and were resistent to standard measles virus superinfection, but susceptible to heterologous virus. Furthermore, the virus released by the persistently infected cultures had the ability to readily establish persistent infections and to reduce standard virus yield in mixed infections. Temperature-sensitive virus mutants were absent from these cultures. Interestingly, Rima and Martin (1979) also found that measles virus DI particles selectively inhibited synthesis of certain viral protein species relative to N, the nucleocapsid structural protein. They argued that this altered gene expression could be implicated in the interference with 50S virus genome replication and in turn be responsible for the shift from a lytic to a persistent infection. In this context they also contended that DI particles can be responsible for the altered gene expression often observed in SSPE (Eron et al., 1978; Hall et al., 1979).

DI particle involvement in rare human diseases has been frequently suspected. Viola et al. (1978) reviewed a series of persistent measles virus infections in humans, including SSPE, MS, systemic lupus erythematosus, and Paget's disease, a proliferative disease of osteoclasts, and entertained DI particles as one of the possible determinants of the persistent infection. More recently, Cernescu and Sorodoc (1980) presented experimental and clinicoepidemiological evidence pointing to the involvement of measles virus DI particles in SSPE.

G. Canine Distemper Virus

Tobler and Imagawa (1984) were able to establish persistent infections in Vero cells with canine distemper virus (CDV). Using the Onderstpoort strain of CDV (CDV/Ond), a laboratory strain, establishment of persistent infection was achieved after seven undiluted passages of the virus. The persistently infected cultures were resistent to challenge with standard CDV. Furthermore, the virus released from these cultures was able to interfere with the replication of CDV/Ond, but not with heterologous viruses, such as VSV. The interfering activity was sedimentable and was destroyed by UV irradiation. Interestingly, a CDV strain (CDV/ODE) isolated from a dog with old dog encephalitis, a chronic neurological disease, did not require undiluted passage, but readily established persistent infections in Vero cells. However, the properties of these persistently infected cells were not fully consistent with the presence of DI particles: the cells were resistant to challenge with standard CDV and susceptible to VSV, but the interfering activity was not released into the medium and remained cell associated. In the electron microscope, large amounts of nucleocapsids were visible in the cytoplasm. It appeared in this case that a defect in virus assembly and maturation, presumably, as the authors speculated, a mutation in the M gene, was associated with the intefering component.

H. Respiratory Syncytial Virus

Entities that may represent DI particles have been detected after passage of the Randal strain of respiratory syncytial virus (RSV) in HEp-2 cells. The interfering factor was particulate, was inactivated by UV irradiation, and interfered with the replication of standard RSV, but not with heterologous VSV. The generation of interfering particles was multiplicity dependent; the particles appeared after four undiluted passages of cloned RSV or after eight passages at low multiplicity of infection (Treuhaft and Beem, 1982).

III. GENOMES OF DEFECTIVE-INTERFERING (DI) PARAMYXOVIRUSES

A. DI Genome Structure

The majority of information on the structure of paramyxovirus DI RNA has been derived from the Sendai virus system. Definition of the structure of the genomes of Sendai virus DI particles started when Kingsbury et al. (1970) demonstrated that the 19S and 25S RNA species they found in virions had the negative polarity of the 50S virus genome. Later, Kolakofsky (1976) showed that some of the Sendai virus DI particles he had generated contained RNAs that ranged in size from 1200 to 4500 nucleotides and were able to form panhandled circular structures, as visualized by electron microscopy, suggesting that the ends of these RNAs were complementary for an estimated extent of approximately 150 bases. The termini of these RNAs were related exclusively to the 5' end of the virus genome; the 3' end of each RNA was complementary to the genomic 5' terminus (Fig. 1). This type of 3' end was termed a copyback sequence based on a proposed mechanism for its generation (Leppert et al., 1977), which is discussed in Section D, below. The double-stranded panhandles (also termed stems) were resistant to ribonuclease digestion. By gel electrophoresis, they had sizes between 110 and 150 base pairs, in good agreement with the electron microscopic estimates (Leppert et al., 1977). Re et al. (1983b), however, upon examination of DI viruses derived from the strains generated by Kolakofsky, found that those RNAs had stems approximately 50 bases longer (210 for Rb and 155 base pairs for Ha) as shown in Table I, probably reflecting differences in the extents of terminal annealing that had occurred in the two laboratories. Another independently generated copy-back DI RNA, 11a, had a stem of 190 base pairs (Re et al., 1983b). The smallest Sendai virus copy-back RNA characterized so far, Wa, is 450 bases long; the extent of its terminal complementarity is not known (Re and Kingsbury, 1988).

By T1 ribonuclease oligonucleotide fingerprinting, Amesse et al. (1982) made the fundamental observations that DI RNAs invariably retain the 5' end of the virus genome and that in addition to copy-back RNA species a number of Sendai virus DI RNAs conserved both 3' and 5' ends of the virus genome and were extensively deleted internally. These RNA species were originally termed fusion DI RNAs to indicate juxtaposition of 3' and 5' genome fragments, as indicated in the general scheme of Fig. 1. By direct RNA sequencing, Re et al. (1983a) formally proved that a number of DI RNA species (1, 1c, 7a, 7b, 7c, and Ra) had the same 3' end as the 50S virus genome.

Re et al. (1983a,b) and Re and Kingsbury (1988) also determined the 3'-end sequence of several copyback RNA species, Rb, Ha, 11a, and Wa (Fig. 2). Because of the terminal complementarity in these RNAs, the 5'-end sequence of the DI genome (identical to that of the virus genome) could be deduced, leading

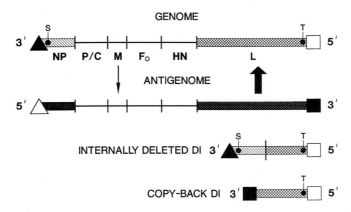

FIGURE 1. Structures of Sendai virus internally deleted and copy-back RNAs and relationships between DI and virus genome termini. (▲, □) The 3' and 5' ends of the virus genome, respectively; (△, ■) their complementary forms, the 5' and 3' ends of the antigenome, respectively. Internally deleted DI RNAs retain 3' and 5' ends of the virus genome and contain *NP* and *L* gene fragments juxtaposed at the deletion site (vertical bar); they retain the *NP* gene transcription start signal (S) and the *L* gene termination and polyadenylation signal (T). Copy-back RNAs consist of 5'-terminal fragments of the virus genome and have the 3' end of the antigenome. The arrows represent the reactions of viral RNA replication using genomes or antigenomes as templates. The larger arrow denotes that the antigenome template is used more frequently by the RNA polymerase.

to identification of the polyadenylation signal of the *L* gene, the last gene of the virus genome, and to definition of the 57-base untranslated sequence at the 5' terminus of the virus genome (Re *et al.*, 1983b). Thus, the 5' end of any DI genome includes a 5'-terminal portion of the *L* gene comprising its polyadenylation signal, followed by a trinucleotide, GAA, possibly involved in transcription termination, and by a 54-base 5'-terminal sequence comprising the putative negative-strand leader RNA sequence (Re *et al.*, 1983b). Extensive sequence information at the 3' end of the major internally deleted RNA species

TABLE I. Sendai Virus Copy-Back
RNA Structure[a]

RNA species	Stem size (base pairs)	Genome size (bases)
Rb	210	2600
Ha	155	1400
11a	190	1200
Wa	ND[b]	450

[a]Data from Amesse *et al.* (1982), Re *et al.* (1983b), and Re and Kingsbury (1988).
[b]Not determined.

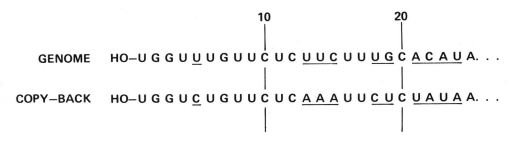

FIGURE 2. Comparison between Sendai virus genomic and copy-back 3'-end sequences. Of the first 25 nucleotides in these sequences, 15 are identical. Divergent nucleotides are underlined. [From Re *et al.* (1983a).]

contained in one of the DI strains of Sendai virus, strain 7, was obtained by dideoxynucleotide RNA sequencing (Re *et al.*, 1984, 1985). These internally deleted RNAs contained at their 3' ends the 51-base positive-strand leader RNA template sequence followed by a putative intervening tetranucleotide $(A)_4$ and by the *NP* gene transcription initiation signal, with adjacent variable amounts of the *NP* gene (Re *et al.*, 1984). The general structure of internally deleted and copy-back Sendai virus DI RNAs is illustrated in Fig. 1. The Sendai virus DI RNAs characterized to date range in size between 30 and 3% of the virus genome, which is 15,000 bases long. Table II lists them according to type; RNAs 1a, 1c, 7a, 7b, 7c, 7d, and Ra are internally deleted; RNAs Rb, Ha, 11a, and Wa are of the copy-back type.

Since the copy-back RNAs contain sequences from the 5' end of the virus genome only and since the complete sequence of the entire Sendai virus genome is known, the genomic structure of the copy-back RNAs is automatically defined if size and amount of terminal complementarity are known (Table I).

The structures of the four internally deleted RNA species of Sendai virus strain 7 have been resolved in detail (Re *et al.*, 1985). The deletion point was

TABLE II. Sendai Virus DI RNA Species

Internally deleted	Copy-back	Size (nucleotides)
1a		4100
7a		3500
Ra		3200
1c, 7d	Rb	2600
7b		2000
7c		1600
	Ha	1400
	11a	1200
	Wa	450

[a]DI RNAs are designated by a two-digit code system: a number or capital letter identifies the virus strain and a lower case letter identifies an individual RNA among multiple species. Data from Amesse *et al.* (1982), Re *et al.* (1983a,b), and Re and Kingsbury (1988).

identified and considerable sequence information was obtained. Figure 3 shows their structures, their sizes, and the position of the deletion site relative to the 3' end of the genome. All of them were deleted within the *NP* gene and, with the exception of 7c, had *L* gene sequences downstream of the deletion site. Interestingly, DI RNA 7c had sequences of the matrix protein gene *M* downstream of the deletion point (not shown in Fig. 3), indicating that a deletion event had juxtaposed *NP* with *M* gene sequences. Since DI RNA must retain the 5' end of the virus genome to survive, this implied that 7c had sustained an additional deletion between the *M* and *L* genes. All strain 7 DI RNAs retained the transcription initiation signal provided by the *NP* gene and the polyadenylation signal of the *L* gene, as depicted in the general scheme of Fig. 1. Therefore, it was predicted that these RNAs were able to transcribe DI-specific transcripts; properties and sizes of the translation product were also predicted (Fig. 3). In cells infected by strain 7, transcripts were found that indeed qualified as transcription products of the DI RNAs, and proteins that qualified as some of the predicted translation products were detected (Hsu *et al.*, 1985). DI RNA 7a is of special interest, because it contained the complete *NP* gene sequence except the last two U residues of the polyadenylation signal, as shown in Fig. 7 (Re *et al.*, 1985) and was therefore potentially capable of transcribing *NP* gene mRNAs. This RNA, however, was found to transcribe only full-length DI transcripts (Hsu *et al.*, 1985), providing some understanding of the sequences required for transcription termination: the two missing U's of the polyadenylation signal and possibly the intragenic purine trinucleotide

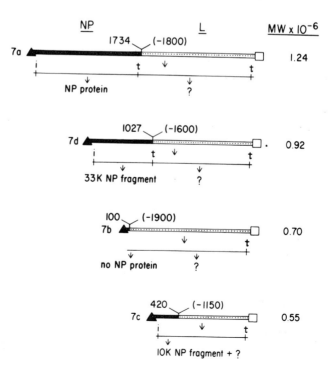

FIGURE 3. Structures of internally deleted RNAs of Sendai virus strain 7. (▲, □) The 3' and 5' ends of the virus genome, respectively. Filled bars and dotted bars represent *NP* and *L* gene sequences, respectively. Numbers above each genome specify the position of the deletion site relative to the 3' end; the size of the 5'-terminal portion of each genome, estimated from its molecular weight, is given in parentheses. Thin lines below each genome indicate transcription products (DI mRNAs) on which the positions of the translation initiation (i) and termination (t) codons are given. Below the DI mRNAs possible translation products are indicated. [From Re *et al.* (1985).]

GAA and the downstream transcription initiation signal are implicated in transcript termination and polyadenylation.

Less structural information is available for the genomes of other paramyxoviruses. DI RNA species of mumps virus display a degree of ribonuclease resistance which is almost twice that of nondefective viral RNA and is not concentration dependent. This property can be accounted for by extensive intramolecular base pairing to form double-stranded RNA and possibly panhandlelike structures. Solely based on this argument, mumps virus DI RNAs appear to have the copy-back structure (McCarthy et al., 1981).

The structure of the DI RNAs of HPIV3 is somewhat unusual (Murphy et al., 1987). Two of them were uncommonly large (64.8 and 51.4% of the virus genome) and by hybridization with virus-specific cDNAs they derived sequences from the 5' half of the virus genome, but were devoid of NP gene-specific sequences, suggesting a possible copy-back structure. However, since the NP probe did not contain the 3' terminus of the positive-strand leader RNA template, an internally-deleted nature was not ruled out. A copy-back structure for such large RNAs containing F, HN, and L gene sequences would be unprecedented among paramyxoviruses, but would resemble the structure of an unusual DI RNA of VSV, DI LT2 (Keene et al., 1981). The smallest RNA species found in these HPIV3 stocks represented 4.9% of the virus genome, but failed to hybridize to any virus-specific probe. Thus, this RNA is probably cellular in origin (Murphy et al., 1987).

B. DI RNA Survival

DI particles survive because their genomes have features that are recognized by the replicative machinery provided by helper virus. The termini of the virus genome are the vital sites for the replication of nonsegmented negative-strand RNA viruses: the 3' end represents the entry site for the viral RNA polymerase, while the 5' end represents the template for the 3' end of the antigenome, the complementary copy of the virus genome (i.e., the positive-strand RNA intermediate in viral RNA replication). The antigenomic 3' end is in turn recognized by the RNA polymerase as the entry site for the synthesis of the negative-stranded viral RNA progeny. DI RNAs invariably retain the 5' end of the virus genome (Leppert et al., 1977; Amesse et al., 1982). Direct RNA sequencing of the 3' ends of both internally deleted and copyback RNAs has elucidated the relationships between virus genome and DI RNA 3' termini (Re et al., 1983b). Internally deleted RNAs have a 3' end identical to that of the virus genome, whereas copy-back RNAs have a 3' terminus which is the complementary copy of the genomic 3' end and by definition represents the 3' end of the antigenome (Fig. 1); these replication-competent termini enable both internally deleted and copy-back RNAs to be replicated. The comparison between genomic and copy-back 3' ends is shown in Fig. 2: the replication start signal presumably resides in the first 25 3'-terminal nucleotides, where the two strands, genomic and antigenomic (copy-back), display the maximum homology; the two sequences are identical in 15 positions out of 25, accounting for their common property to be recognized by the RNA polymerase, but they also diverge in ten positions. These sequence divergences may account for the differential affinity for the RNA polymerase that these two termini display during viral RNA replication. Ability to be replicated, however, is not a sufficient condition for survival. DI RNAs must replicate more efficiently than standard virus genomes or in the long run they will be outgrown by the helper virus. Some understanding of the factors responsible for the replication efficiency of DI RNAs has been achieved through studies of DI RNA selection, as discussed below.

C. DI RNA Selection

A fundamental property of DI RNAs is their ability to reproduce at the expense of the helper virus genome. DI RNAs must therefore enjoy a selective advantage over virus genomes. Some insight into the factors responsible for this advantage has been obtained for Sendai virus (Fig. 4). DI RNAs of different genotypes interfere with each other in mixed infections (Perrault and Semler, 1979; Perrault, 1981; Rao and Huang, 1982). Taking advantage of this property and of the availability of a series of well-characterized DI RNA species both of the internally deleted and copy-back type, it has been possible to study the

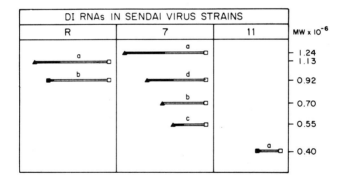

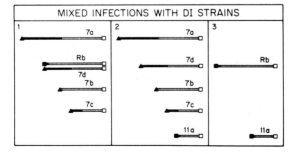

FIGURE 4. (Top) DI genomes in Sendai virus strains and (bottom) combinations used in competition experiments. Genomes have 3′ ends of (▲) internally deleted and (■) copy-back type, respectively: their estimated molecular weights are given on the right side of the top panels.

effects of the nature of the 3' terminus, ability to be transcribed, and RNA size on survival. The experimental strategy entailed passaging natural or artificial mixtures of DI RNAs with different characteristics (Fig. 4) repeatedly in eggs and analyzing the virion RNAs at each passage.

Copy-back RNAs were replicated more efficiently than internally deleted RNAs species and consistently eliminated the latter type (strain R and combinations 1 and 2 of Fig. 4), indicating that the copy-back (antigenomic) 3' end is a better promoter of replication than its genomic counterpart (Re and Kingsbury, 1986). This superior promoter activity probably reflects the effects of the few sequence divergences existing between genomic and antigenomic 3' ends (Fig. 2) and is consistent with the sequence of this terminus being responsible for preferential synthesis of genomes during virus replication (Fig. 1). Two other properties place copy-back RNAs at a selective advantage. As illustrated in Fig. 1, copy-back RNAs are transcriptionally inactive for lack of the *NP* gene transcription start signal. As discussed below, transcriptional inertness is a second selective advantage. Finally, because of terminal complementarity, negtive-strand copy-back genomes direct synthesis of positive-strand intermediates, which also have copy-back 3' ends and similarly enjoy the same replicative advantage.

In these experiments an anomaly was observed: an internally deleted RNA, Ra, was somewhat more resistant to out-competition by the copy-back RNA Rb (strain R in Fig. 4) than other internally deleted RNA species, the RNAs of strain 7 (combination 1 in Fig. 4). To identify possible structural features accounting for this behavior, DI RNA Ra was sequenced and two mutations were found at positions 47 and 51 from the 3' terminus at the 5' end of the positive leader RNA template and immediately before the transcription initiation signal of the *NP* gene. These mutations had functional consequences in that Ra was found to be unable to direct synthesis of transcripts (Re and Kingsbury, 1986). It appeared that these mutations abrogated termination of positive-strand leader RNA synthesis, causing the RNA polymerase to read through the leader RNA–*NP* gene junction and to proceed in a replicative mode, providing a glimpse into the sequence requirements for leader RNA synthesis termination. Therefore, transcriptional inertness was identified to be another factor affecting DI RNA selection: DI genomes that do not transcribe are engaged exclusively in replication.

RNA size was also found to be a determinant in DI RNA selection. During passage of strain 7 DI RNAs which have analogous structure and transcriptional capacity but different sizes (3500, 2600, 2000, and 1600 bases), as shown in Fig. 4, it was observed that only the 1600-base RNA species, 7c, survived. This indicates that small size confers a selective advantage, consistent with the hypothesis that small RNAs should be replicated faster than larger species if the rate of RNA synthesis is the same on every RNA template (Re and Kingsbury, 1988). Interestingly, in this set of experiments another apparent anomaly was encountered: DI RNAs smaller than 1600 bases were at a selective disadvantage and were eliminated by larger species. The reasons for this phenomenon were found to depend on a relative inability of small nucleocapsids to enter virions. This infirmity became evident when an efficiently replicated 1200-base RNA species, 11a, in competition with the 2600-base RNA, Rb (combination 3 of Fig. 4), was found to be more abundant relative to Rb inside the infected cells than in the released virus. For the same reasons, in a separate competition experiment (Re and Kingsbury, 1988), 11a eliminated the smallest DI RNA of Sendai virus, the 450-base Wa (not shown in Fig. 4).

Table III ranks the selective factors identified by DI RNA competition according to their importance and indicates their site of action in virus replication. It appears that DI RNA selection is controlled at two levels of virus reproduction: the level of intracellular RNA synthesis and the level of virus maturation at the cell surface. Intracellularly, copy-back 3' termini, transcriptional inertness, and small size represent selective advantages. At the cell surface, on the contrary, small size is detrimental to survival. Thus, the effect of size reduction is dual and opposite: reduction of size favors survival by increasing the rate of RNA replication, but at the same time it hinders survival by limiting nucleocapsid envelopment. These two opposite effects at some point balance each other out and the evidence discussed above indicates that a size ranging between 1600 and 1200 bases represents the inversion point below which DI particle production undergoes a relative reduction.

An important feature of the competition experiments discussed above was that long-term passage of a mixture of various Sendai virus DI genomes invariably resulted in the selection of a single RNA species (Re and Kingsbury, 1986; Re and Kingsbury, 1988). The conclusions of Kolakofsky (1979) with Sendai virus and McCarthy et al. (1981) with mumps virus contrast with this observation. They contended that serial passage caused shifting relative abundances of the existing DI RNA population, without elimination of any species. However, in retrospect, the analysis of intracellular nucleocapsid RNA from serial passage of Sendai virus (Kolakofsky, 1979) displayed a trend toward the elimination of very small DI RNA species and the enrichment of medium-size RNAs, but this trend might only have become clearer after additional passages. The data of McCarthy et al. (1981) were derived exclusively from changes in the virion RNA profile of DI mumps virus in sucrose velocity gradients upon passage; from early to late passages a shift from high- to low-molecular-weight RNAs was recorded. Since the virion RNA was conspicuously enriched in 20S RNA species at later passages, in retrospect one could argue that the selection of a single RNA species might have taken place.

TABLE III. Factors Affecting Sendai Virus DI RNA Selection and Their Site of Action on Virus Reproduction[a]

	Intracellular RNA replication	Cell surface virus maturation
3' end	Copy-back > genomic	—
Transcription	Inert > active	—
Size	Small > large	Greater > smaller than 1600 bases

[a]Factors are ranked in descending order of importance, based on the findings of Re and Kingsbury (1986, 1988). The characteristic on the left of the ">" symbol is more favorable than the characteristic on the right.

D. DI RNA Generation

Models for the generation of copy-back and internally deleted DI genomes have been proposed by Leppert *et al.* (1977) and Amesse *et al.* (1982), respectively. Both models have in common a polymerase-jumping event. For internal deletions, the polymerase is thought to terminate RNA synthesis prematurely, abandon the RNA template, carrying along the nascent RNA strand, and resume RNA synthesis at another site on the same or equivalent template (Fig. 5). In copy-back RNA generation, instead, the free polymerase that carries the nascent RNA strand would choose as template for resumption of RNA synthesis the nascent RNA strand itself and copy it back in the opposite direction (Leppert *et al.*, 1977).

The hypothesis has been put forward that Sendai virus internal deletions are generated by a copy-choice mechanism. As discussed above, the deletion site in four internally deleted RNA species (stain 7 RNAs) has been identified and the sequence at the deletion boundaries has been determined (Re *et al.*, 1985). The sequence downstream of the deletion site, the presumptive reinitiation site in the *L* gene, showed similarity with the 3' end sequence of both genome and antigenome, the replication promoters, implicating them as possible reinitiation sites (Fig. 6). On the contrary, at the *NP* gene side of the deletion, the presumptive termination site, no consensus sequence was found. Since the *NP* gene sequence had been determined (Morgan *et al.*, 1984) and the position of each deletion is known, it was possible to deduce the sequences of the *NP* gene immediately downstream of the deletion site that had been removed by each deletion event (Re *et al.*, 1985). As shown in Fig. 7, similarity was found with a transcription regulatory sequence that is found at the beginning of each virus gene, where each transcript is initiated (Gupta *et al.*, 1984). This signal is thought to play a part not only in transcript initiation, but also in

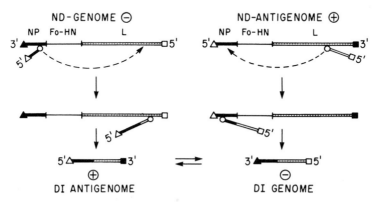

FIGURE 5. Model for copy-choice generation of internal deletions. The open circle represents the viral RNA polymerase with attached nascent RNA strand, complementary to the template. After premature termination of RNA replication in the *NP* gene, the polymerase with the nascent strand jumps to the *L* gene, where it resumes RNA synthesis. In principle, the deletion event can occur during replication of either the genome or antigenome of nondefective (ND) parental virus. [From Amesse *et al.* (1982).]

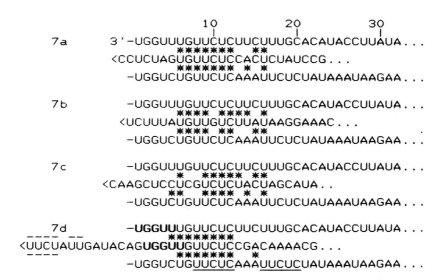

```
                    10        20        30
                    |         |         |
   7a       3'-UGGUUUGUUCUCUUCUUUGCACAUACCUUAUA...
                    *******  **
            <CCUCUAGUGUUCUCCACUCUAUCCG...
                    ******* * *
              -UGGUCUGUUCUCAAAUUCUCUAUAAAUAAGAA...

   7b         -UGGUUUGUUCUCUUCUUUGCACAUACCUUAUA...
                    ****  **** *
            <UCUUUAUGUUGUCUUAUAAGGAAAC...
                    ****  **   **
              -UGGUCUGUUCUCAAAUUCUCUAUAAAUAAGAA...

   7c         -UGGUUUGUUCUCUUCUUUGCACAUACCUUAUA...
                    *    *****  **
            <CAAGCUCCUCGUCUCUACUAGCAUA..
                    **   **** * *
              -UGGUCUGUUCUCAAAUUCUCUAUAAAUAAGAA...

   7d         -UGGUUUGUUCUCUUCUUUGCACAUACCUUAUA...
            ----  --    ********
   <UUCUAUUGAUACAGUGGUUGUUCUCCGACAAAACG...
            ----        *******   *
              -UGGUCUGUUCUCAAAUUCUCUAUAAAUAAGAA...
```

FIGURE 6. Consensus sequences downstream of internal deletions in Sendai virus strain 7. In 7a, 7b, and 7d, these sequences are in the *L* gene, but they are in the *M* gene for 7c (see text). The deletion site is indicated by the bracket on the left side of each DI 7 RNA sequence. Above and below each DI RNA sequence are given the genomic and antigenomic 3' ends, where replication of each strand initiates. Stars indicate identities between DI and promoter sequences. [From Re *et al.* (1985).]

termination of the upstream transcript (Re *et al.*, 1985). Associated with this consensus sequence in each case there is a purine triplet which also is hypothesized to play a role in transcription termination (Fig. 7). Presumably, internal deletions arise from premature termination at these sites of growing positive-stranded RNA intermediates in virus genome replication. This also implies

```
NP gene:     3'...UNAUUCUUU/UUGAAUCCCNNUUUCNNN
                                ****

7a:    ...AACUAGGCAUCAUUCUUU/UUGAAUCCCACUUUCAAGUAGGUGA
                                ****   *

7b:    ...CGUCCAAGGUCUGGGAAACG/AAACGACGGUUUCAAGUGCUACC
                               ****   *

7c:    ...UACAGUUUA/UACACUAGAUGUUGUAUCUCUUUCUGGGAUUCUC
                          ****   *

7d:    ...AAUCUUCGGAGUAUCUGUGGAU/GGACAGUUUUCCGGGGUCUCG
                               ****   *

Consensus:                     (UUUCNNG)
```

FIGURE 7. Presumptive RNA synthesis termination sites in the *NP* gene. The site of deletion is denoted by the slash. Sequences on the left of each slash were determined experimentally; sequences on the right were deduced from the known sequence of the *NP* gene (Shioda *et al.*, 1983; Morgan *et al.*, 1984). As indicated by the identities (stars), the deduced sequences contained regions similar to the conserved Sendai virus transcription start signal given at the top (specifically, the sequence downstream of the underlined triplet, GAA). The consensus sequence is given in parentheses at the bottom. [From Re *et al.* (1985).]

that the RNA polymerase possesses a distinct site that screens template sequences downstream of its catalytic site for chain elongation. Thus, a copy-choice mechanism driven by cryptic signals resembling control elements of transcription and replication may indeed be responsible for generation of internally deleted DI molecules.

IV. INTERFERENCE BY DEFECTIVE-INTERFERING PARTICLES

The evidence examined above supports the notion that interference by DI particles is achieved primarily at the level of RNA replication (Re and Kingsbury, 1986, 1988). The selective factors identified by competition between DI genomes, the sequence at the 3' end of the RNA, its transcriptional capacity, and its size all relate intimately to the process of RNA synthesis and are therefore presumed to determine the efficiency of interference. DI genomes would compete with standard genomes for the polymerase, the nucleocapsid, and envelope proteins expressed by helper virus, and effectively consume them, preventing the standard virus genome from replicating and maturing into virus particles. The von Magnus effect derives its basis from this special interaction between DI and standard virus genomes. Similar to the predators in a predator–prey ecosystem, DI RNAs, being absolutely dependent upon the nondefective helper virus, can never eliminate it. By interfering with the replication of the helper, DI particles deprive themselves of the supplied replicative machinery and limit their own replication. At low levels of replication, the standard virus gains a selective advantage because low multiplicity of infection reduces the probability of double infection with DI and standard virus and tends to eliminate DI particles. Consequently, the standard virus will grow preferentially and an increase in the multiplicity of infection will follow. This will give the residual DI particles the chance to recover and dominate the standard virus in turn, repeating the cycle (Cave et al., 1985).

Other mechanisms might account for reduction of virus production by DI particles. Altered expression of the virus genome has been often observed in cells infected with DI particles (Rima and Martin, 1979; Roux and Waldvogel, 1982, 1983; Tuffereau and Roux, 1988). Tuffereau and Roux (1988) observed that along with a reduced yield of standard Sendai virus, an increased instability of the M protein occurred. This instability was thought to derive as the indirect effect of unsuccessful budding of DI nucleocapsids. Since the M protein is presumed to act in the interaction between nucleocapsid and cell membrane during formation of the prebud complex, unsuccessful budding was thought to cause inactivation and degradation of the M protein. Budding abortion was independently supported by the observation that due to size constraints, small DI nucleocapsids are indeed restricted in their ability to enter virions at the cell surface (Re and Kingsbury, 1988). This increased turnover would deplete the intracellular supply of M protein and affect standard nucleocapsid budding, thus diminishing the overall production of virus. This secondary effect of DI particles on budding would add to the primary effect on virus genome replication.

V. PERSISTENT INFECTIONS

Many DI particles have been identified by their ability to mediate persistent infections. In cell cultures, persistent infections are achieved in the absence of interferon and the action of an immune system, indicating that DI particles *per se* are able to cause the condition and that interference with the replication of standard virus is the determining factor. By reducing the replication and spread of standard virus, DI particles mitigate the CPE that standard virus alone would cause and allow the infected cells to survive.

In the whole animal, the mechanism is complicated by interferon, which DI particles are known to induce (Johnston, 1981), and the immune response. With the arenavirus lymphocytic choriomeningitis virus (LCMV), DI particles are known to reduce viral gene expression and depress the immune response, allowing infected cells to escape immune surveillance (Welsh *et al.*, 1977; Welsh and Oldstone, 1977). There are indications that the same could happen with paramyxoviruses. Rima and Martin (1979) reported altered gene expression in cells superinfected with measles virus DI particles. Roux and Waldvogel (1983) found that the membrane glycoprotein HN undergoes a faster turnover in cells infected with Sendai virus DI particles, in connection with an instability of the M protein. It appears that the two proteins M and HN interact during virus assembly and maturation. In Sendai virus, the mature form of HN is a disulfide-linked dimer or tetramer and a drastic immunoreactivity change occurs in the conversion of the immature (monomer) into the mature forms of HN (Mottet *et al.*, 1986; Vidal *et al.*, 1989); antibodies that recognize one form ignore the other. Similarly, T. Morrison (personal communication) found that in Newcastle disease virus infections the HN glycoprotein is presented at the cell surface both as a monomer and as a disulfide-linked multimer (likely a trimer). In this case also, the monomeric form of HN is antigenically distinct from the multimeric form and is not recognized by the same antibodies. Intriguingly, the HN protein is the major virus component involved in eliciting host immunity. How the immune system sometimes fails to eliminate a virus is not fully understood. It is conceivable, however, that alteration of the rate and/or the level at which a viral antigen (HN in this case) is presented at the surface of an infected cell may have an impact on the outcome of the immune response, as reported for LCMV (Welsh *et al.*, 1977). DI particles are known to increase the turnover of HN molecules on cell surfaces (Roux *et al.*, 1985). Loss of this immunologically relevant surface antigen could allow infected cells to escape surveillance by cytotoxic T cells and/or allow the virus to evade the action of neutralizing antibodies.

VI. MEDICAL RELEVANCE OF DEFECTIVE-INTERFERING VIRUSES

As mentioned above, it was speculated in the 1970s that DI viruses might modulate viral disease expression and be involved in rare human illnesses such as SSPE and MS. By virtue of their peculiar biological properties, DI viruses might not only modulate the duration and severity of a viral disease, but might even radically change its outcome. Today, animal models have been developed supporting this view.

Atkinson *et al.* (1986) have shown, for instance, that virulent Semliki forest virus (SFV) a togavirus, coinoculated with DI particles causes persistent infections in mouse brains. The persistently infected animals did not show the clinical signs of the disease that SFV alone would cause. However, in sub-clinical infections, neurochemical abnormalities in the central nervous system were observed that persisted after infectious virus could no longer be detected (Barrett *et al.*, 1986). Welsh *et al.* (1977) were able to prevent the cerebellar disease caused by LCMV by coinoculation of DI particles in rats. The protection of the animals was accompanied by reduced expression of viral antigen in the brain. In an experimental neuropathy induced with VSV, Cave *et al.* (1984) were able to detect genomic and DI RNAs in individual mouse brains by RNA blot hybridization and for the first time were able to show the von Magnus effect *in vivo* (Cave *et al.*, 1985). In the brains of mice coinfected with standard and DI virus particles, virus growth underwent a cyclic pattern with a periodicity of about 5 days. It is also of interest that DI particles have been derived from a neurotropic canine parainfluenza virus recovered from a dog with chronic neurological disease (Baumgartner *et al.*, 1987).

Chambers and Webster (1987) reported compelling evidence for the involvement of DI particles in the control of pathogenicity in a natural virus infection. They found DI RNA species in an avirulent strain of avian influenza virus, A/Chicken/Pennsylvania/83, which caused a mild epidemic in poultry in the spring of 1983. Following this epidemic, a second influenza outbreak occurred in the fall, causing disastrously high mortality in chicken flocks. The virulent virus isolated from this epidemic was closely related to A/Chicken/Pennsylvania/83, but did not contain DI RNAs. Presumably, the RNA polymerase of the avirulent strain is more error-prone than the RNA polymerase of the virulent strain.

In human pathologies, DI particles have not been directly detected or shown to play a part so far. However, there are a number of atypical human paramyxovirus infections the mechanisms of which are not fully understood and that are potential candidates for DI particle intervention.

Paramyxoviruses have been detected in MS. Titers of anti-measles virus antibodies in the cerebrospinal fluid (CSF) higher than in the serum are commonly found in MS patients (Adams and Imagawa, 1962), indicating that the virus may reside in the brain. Structures that resemble paramyxovirus nucleocapsids have been observed in MS brain sections (Narang and Field, 1973), but attempts to isolate virus from MS biopsies have met with limited success. Nevertheless, measles virus has been recovered from MS brain material (Field *et al.*, 1972). Recently, antibodies against SV5 have also been detected in the CSF of some MS patients (Goswami *et al.*, 1987).

Measles virus has been found in the brains of SSPE patients: a variant has been characterized (Cattaneo *et al.*, 1986, 1988) whose genome contains lesions accounting for the abnormal expression of the *M* gene observed earlier by numerous investigators (Eron *et al.*, 1978; Hall *et al.*, 1979; Wechsler *et al.*, 1979).

Measles virus has been recovered from the brain of a patient with immunosuppressive measles encephalitis. Two variants were isolated which had properties similar to the SSPE variant. One of these did not produce cell-free virions (infectivity remained cell associated), the other was defective in the synthesis of M protein, and each had a different neurotropic effect in mice (Ohuchi *et al.*, 1987).

In this connection, it must be pointed out that sequences of the measles virus *P*, *M*, and *H* genes could not be detected in postmortem MS or control brain samples after sensitive probing with the relevant cDNAs by dot-blot and *in situ* hybridization (Dowling *et al.*, 1986). However, since the *P*, *M*, and *H* genes are located in the interior of the measles virus genome, those tests might have missed DI genomes containing only terminal sequences of the genome. For this reason, and because SV5 antigens have been found in MS (Goswami *et al.*, 1987), the involvement of paramyxovirus DI particles in this disease has not been ruled out.

Finally, measles virus RNA has been detected also in bone tissue in patients with Paget's disease by *in situ* hybridization (Basle *et al.*, 1986).

The etiological bases for those diseases are still unclear and again DI particles have escaped detection so far in all cases. This failure, however, may reflect the inadequacy of the procedures used to recover virus from diseased tissue; these methods entail fusion of the infected cells *in vitro* or cocultivation of the infected cells with permissive cells. Both techniques imply diluted passages of the original virus, a condition adverse to the survival of DI particles.

New methods have provided opportunities for more conclusive tests. The complete nucleotide sequence of at least one member of each paramyxovirus genus is now available and therefore oligonucleotides complementary to specific regions of these virus genomes can be constructed. These oligonucleotides can be used as primers for the amplified synthesis of specific sequences by the polymerase chain reaction (PCR) technique, permitting far more sensitive detection of target sequences (such as those retained by DI RNAs) than any previous method.

Although it appears today that the causative agent of SSPE is an attenuated measles virus defective in the expression of the matrix protein M and is therefore restricted in virus budding (Cattaneo *et al.*, 1986, 1988), the participation of DI particles in this or other related diseases has not been eliminated. DI particles might act in either of two ways.

First, SSPE is associated with reduction or lack of anti-M antibody reactivity in SSPE patients (Eron *et al.*, 1978; Hall *et al.*, 1979; Wechsler *et al.*, 1979), which can be explained by the structural defects discussed above (Cattaneo *et al.*, 1986, 1988). It is provocative that reduced levels of M protein can be achieved also by the effect of DI particles (Rima and Martin, 1979; Roux and Waldvogel, 1982; Tuffereau and Roux, 1988). Thus, DI particles might not only establish the persistent infection, but might also account for the lack of M antibody reactivity in SSPE or other related diseases. Furthermore, it appears that in SSPE and in measles inclusion body encephalitis altered gene expression is achieved through a derangement of the transcription process whereby the genes at the 5' end of the genomes are transcribed less efficiently than in

standard measles virus infections (Cattaneo et al., 1987). Again, it is remarkable that transcriptional alterations can be achieved by the effects of DI particles, as observed in cells infected with Sendai virus DI particles (Hsu et al.,
1985). Association between altered M gene expression and disease processes is
a recurrent theme with paramyxoviruses; it has been observed in immunosuppressive measles encephalitis (Ohuchi et al., 1987) and in CDV old dog encephalitis (Tobler and Imagawa, 1984), In the light of the fact that the CDV/ODE
strain is able to establish persistent infections (Tobler and Imagawa, 1984), DI
particles must be taken into serious consideration when examining the factors
responsible for this altered gene expression and the etiology of the disease.

Second, DI particles may be involved in the generation of SSPE variants. It
has been shown that in persistent infections maintained by DI particles the
parental standard virus genome undergoes changes that sometimes allow it to
escape interference by DI particles (Horodyski and Holland, 1980; DePolo et
al., 1987). It is conceivable that the M gene mutations in SSPE viruses are the
result of the selective pressure that DI particles exert on the parental standard
virus genome. Since small nucleocapsids enter virions inefficiently (Re and
Kingsbury, 1988) and turn over a lot of M protein in trying to do so (Tuffereau
and Roux, 1988), a mutation that would limit the expression of the M gene
would represent a means by which measles virus could escape from and eliminate DI particles. This scenario would explain the findings of Cattaneo et al.
(1986, 1988). Others (Huang, 1988) have also recognized the possibility that by
the time a variant is detected in diseased tissue the DI particles which initiated
the chronic infection may no longer be present.

VII. CONCLUDING REMARKS

Generated by random errors in viral RNA replication, DI particles are
deletion mutants having replicative advantages over parental helper virus.
They are selected and enriched by high-multiplicity passage and ultimately
interfere with standard virus replication. Although work with animal models
suggests that DI particles harbor the potential for participating in chronic
human diseases, they have escaped detection and their role in natural disease is
elusive. Nevertheless, since DI particles interfere with the replication of standard virus, mediate establishment of persistent infections, induce interferon,
alter gene expression, and deceive the host immune system, it is hard to escape
the conclusion that DI particles are important agents in viral pathogenesis.

Studies on paramyxovirus DI RNAs have elucidated many structural and
biological features that can lead to predictions relevant to the detection of DI
particles in diseased tissue. The competition experiments discussed above predict that in an evolving mixed DI population the surviving RNA will be ultimately a copy-back RNA with a size in the range of 1600 bases. If the same
selective mechanism operates in natural persistent infections, a DI-specific
probe can be engineered for the detection of DI RNA by in situ or dot-blot
hybridization. A cDNA representing the 1600 5'-terminal bases of the virus
genome would be ideal, because copy-back RNAs contain L gene sequences
only. Since this probe also detects nondefective virus genomes, another probe

representing an internal gene has to be used in a parallel assay to allow a differential diagnosis. This probe will detect virus genomes only because copy-back RNAs are devoid of such sequences. Since DI RNAs are the major product of a mixed infection, the 5'-terminal probe will give a hybridization signal stronger than the internal probe if DI RNAs are present or will give an identical signal if only standard genomes are present and DI RNAs are absent. As discussed earlier, the application of PCR technology to the detection of these sequences will make the task even easier.

Studies on the molecular biology of the genomes of DI particles have significantly contributed to the understanding of the structure and functions of the standard virus genome. With Sendai virus, the junction structures between the positive-strand leader RNA template and the *NP* gene, and between the *L* gene and the negative-strand leader RNA, were resolved by sequencing DI genomes before the complete sequence of the virus genome was obtained; evidence for a superior promoter activity of the copy-back terminus was substantiated by competition experiments; the notion that the transcription initiation signal is also involved in RNA synthesis termination and that the RNA polymerase may screen template sequences downstream of its catalytic site has stemmed from studies on DI RNA generation; some understanding of the sequence requirement for mRNA synthesis termination and polyadenylation as well as for leader RNA synthesis termination and suppression of transcription of the *NP* gene has been derived from competition studies among DI genomes. Finally, DI RNAs are tools of special interest to the molecular biologist who wants to study the *cis*-acting functions of the virus genome involved in replication, encapsidation, and transcription. DI RNAs representing simplified and yet replicable versions of the virus genome are the best-suited entities to manipulate for introducing sequence changes by site-directed mutagenesis and to study the effects of these mutations on replication, encapsidation, and transcription. It is very challenging, using recombinant DNA technology, to construct an artificial DI particle that would make those studies possible.

VIII. REFERENCES

Adams, J. M., and Imagawa, D. T., 1962, Measles antibodies in multiple sclerosis, *Proc. Soc. Exp. Biol. Med.* **3**:562–566.

Amesse, L. S., Pridgen, C. L., and Kingsbury, D. W., 1982, Sendai virus DI RNA species with conserved virus genome termini and extensive internal deletions, *Virology* **118**:17–27.

Andzhaparidze, O. G., Boriskin, Yu. S., Bomogolova, N. N., and Drynov, I. D., 1982a, Mumps virus-persistently infected cell cultures release defective interfering virus particles, *J. Gen. Virol.* **63**:499–503.

Andzhaparidze, O. G., Boriskin Yu. S., and Bomogolova, N. N., 1982b, Defective interfering mumps virus produced by chronically infected cell cultures, *Vopr. Virusol.* **27**:405–408.

Andzhaparidze, O. G., Bomogolova, N. N., Boriskin, Yu. S., and Drynov, I. D., 1983, Chronic non-cytopathic infection of human continuous cell lines with mumps virus, *Acta Virol.* (Prague) **27**(4):318–328.

Atkinson, T., Barrett, A. D. T., Mackenzie, A., and Dimmock, N. J., 1986, Persistence of virulent Semliki forest virus in mouse brain following co-inoculation with defective interfering particles, *J. Gen. Virol.* **67**:1189–1194.

Azadova, N. B., and Zhdanov, V. M., 1982, RNA synthesis in the L cell-SV5 system, *Acta Virol.* **26**:427–431.

Barrett, A. D. T., Cross, A. J., Crow, T. J., Johnson, J. A., Guest, A. R., and Dimmock, N. J., 1986, Subclinical infections in mice resulting from the modulation of a lethal dose of Semliki forest virus with defective interfering viruses: Neurochemical abnormalities in the central nervous system, *J. Gen. Virol.* **67**:1727–1732.

Basle, M. S., Fournier, J. G., Rozenblatt, S., Rebel, A., and Bouteille, M., 1986, Measles virus RNA detected in Paget's disease bone tissue by *in situ* hybridization, *J. Gen. Virol.* **67**:907–913.

Baumgartner, W., Krakowka, S., and Blakeslee, J., 1987, Evolution of *n vitro* persistence of two strains of canine parainfluenza virus, *Arch. Virol.* **93**:147–154.

Bellet, A. J. D., and Cooper, P. D, 1959, Some properties of the transmissible interfering component of vesicular stomatitis virus preparations, *J. Gen. Microbiol.* **21**:498–509.

Cattaneo, R., Schmid, A., Rebman, G., Baczko, K., ter Meulen, V., Bellini, W., Rozenblatt, S., and Billeter, M., 1986, Accumulated measles virus mutations in a case of subacute sclerosing panencephalitis: Interrupted matrix protein reding frame and transcription alteration, *Virology* **154**:97–107.

Cattaneo, R., Rebman, G., ter Meulen, V., and Billeter, M., 1987, Altered ratios of measles virus transcripts in diseased human brains, *Virology* **160**:523–526.

Cattaneo, R., Schmid, A., Billeter, M., Sheppard, R., and Udem, S. A., 1988, Multiple viral mutations rather than host factors cause defective measles virus gene expression in a subacute sclerosing panencephalitis cell line, *J. Virol.* **62**:1388–1397.

Cave, D. R., Hagen, F. S., Palma, E. L., and Huang, A. S., 1984, Detection of vesicular stomatitis virus RNA and its defective-interfering particles in individual mouse brains, *J. Virol.* **50**:86–91.

Cave, D. R., Hendrickson, F. M., and Huang, A. S., 1985, Defective interfering virus particles modulate virulence, *J. Virol.* **55**:366–373.

Cernescu, C., and Sorodoc Y., 1980, Subacute sclerosing panencephalitis and defective interfering measles virus particles, *Virologie* **31**:3–8.

Chambers, T. M., and Webster, R. G., 1987, Defective interfering virus associated with A/Chicken/Pennsylvania/83 influenza virus, *J. Virol.* **61**:1517–1523.

Degré, M., and Glasgow, L. A., 1983, Establishment and preliminary characterization of a cell line persistently infected with Newcastle disease virus, *Acta Pathol. Microbiol. Scand. B* **91**:389–394.

DePolo, N. J., Giachetti, C., and Holland, J. J., 1987, Continuing coevolution of virus and defective interfering particles and of viral genome sequences during undiluted passages: Virus mutants exhibiting nearly complete resistance to formerly dominant defective interfering particles, *J. Virol.* **61**:454–464.

Dowling, P. C., Blumberg, B. M., Kolakofsky, D., Cook, P., Jotkowitz, A., Prineas, J. W., and Cook, S. D., 1986, Measles virus sequences in human brain, *Virus Res.* **5**:97–107.

Eron, L., Sprague, J. A., Albrecht, F., Dunlop, R. C., Hicks, J. T., and Aulakh, G. S., 1978, Subacute sclerosing panencephalitis: An abortive infection by a measles-like virus, *in* "Negative Strand Viruses and the Host Cell" (B. W. J. Mahy and R. D. Barry, eds.), pp. 157–167, Academic Press, London.

Field, E. J., Cowshall, S., Narang, H. K., and Bell, T. M., 1972, Viruses in multiple sclerosis? *Lancet* **2**:280–281.

Goswami, K. K. A., Randal, R. E., Lange, L. S., and Russell, W. C., 1987, Antibodies against the paramyxovirus SV5 in the cerebrospinal fluids of some multiple sclerosis patients, *Nature* **327**:244–247.

Gupta, K. C., Re, G. G., and Kingsbury, D. W., 1984, Non-coding regulatory sequences of the Sendai virus genome, *in* "Nonsegmented Negative Strand Viruses," pp. 11–15, Academic Press, New York.

Hall, W. W., Lamb, R. A., and Choppin, F. W., 1979, Measles and subacute sclerosing panencephalitis virus proteins: Lack of antibodies to the M protein in patients with subacute sclerosing panencephalitis, *Proc. Natl. Acad. Sci. USA* **76**:2047–2051.

Horodyski, F. M., and Holland, J. J., 1980, Viruses isolated from cells persistently infected with vesicular stomatitis virus show altered interaction with defective-interfering particles, *J. Virol.* **36**:627–631.

Hsu, C.-H., Re, G. G., Gupta, K. C., Portner, A., and Kingsbury, D. W., 1985, Expression of Sendai virus defective-interfering genomes with internal deletions, *Virology* **146**:38–49.

Huang, A. S., 1988, Modulation of viral disease processes by defective interfering particles, in "RNA Genetics," Vol. III (E. Domingo, J. J. Holland, and P. Ahlquist, eds.), pp. 195–208, CRC Press, Boca Raton, Florida.

Huang, A. S., and Baltimore, D., 1970, Defvective viral particles and viral disease processes, *Nature* **226**:325–327.

Johnston, M. D., 1981, The characteristics required for a Sendai virus preparation to induce a high level of interferon in human lymphoblastoid cells, *J. Gen. Virol.* **56**:175–184.

Keene, J. D., Chien, I. M., and Lazzarini, R. A., 1981, Vesicular stomatitis virus defective interfering particle containing a muted internal leader RNA gene, *Proc. Natl. Acad. Sci. USA* **78**:2090–2094.

Kingsbury, D. W., and Portner, A., 1970, On the genesis of incomplete Sendai virions, *Virology* **42**:872–879.

Kingsbury, D. W., Portner, A., and Darlington, R. W., 1970, Properties of incomplete Sendai virions and subgenomic viral RNAs, *Virology* **42**:857–871.

Kolakofsky, D., 1976, Isolation and characterization of Sendai virus DI-RNAs, *Cell* **8**:547–555.

Kolakofsky, D., 1979, Studies on the generation and amplification of Sendai virus defective-interfering genomes, *Virology* **93**:589–593.

Leppert, M., Kort, L., and Kolakofsky, D., 1977, Further characterization of Sendai virus DI-RNA: A model for their generation, *Cell* **12**:539–552.

Lynch, S., and Kolakofsky, D., 1978, Ends of the RNA within Sendai virus defective interfering nucleocapsids are not free, *J. Virol.* **28**:584–589.

Maeda, A., Suzuki, Y., and Matsumoto, M., 1978, Isolation and characterization of defective interfering particles of Newcastle disease virus, *Microbiol. Immunol.* **22**:775–784.

McCarthy, M., Wolinsky, J. S., and Lazzarini, R. A., 1981, A persistent infection of Vero cells by egg-adapted mumps virus, *Virology* **114**(2):343–356.

Morgan, E. M., and Rapp, F., 1977, Measles virus and its associated diseases, *Bacteriol. Rev.* **41**:636–666.

Morgan, E. M., Re, G. G., and Kingsbury, D. W., 1984, Complete sequence of the Sendai virus *NP* gene from a cloned insert, *Virology* **135**:279–287.

Mottet, G., Portner, A., and Roux, L., 1986, Drastic immunoreactivity changes between the immature and mature form of the Sendai virus HN and F_0 glycoproteins, *J. Virol.* **59**:132–141.

Murphy, D. G., Dimock, K., and Kang, C. Y., 1987, Defective interfering particles of human parainfluenza virus 3, *Virology* **158**:439–443.

Narang, H. K., and Field, E. J., 1973, Paramyxovirus-like tubules in multiple sclerosis biopsy material, *Acta Neuropathol.* **25**:281–290.

Ohuchi, M., Ohuchi, R., Mifune, K., Ishihara, T., and Ogawa, T., 1987, Characterization of the measles virus isolated from the brain of a patient with immunosuppressive measles encephalitis, *J. Infect. Dis.* **156**:(3):436–441.

Perrault, J., 1981, Origin and replication of defective interfering particles, *Curr. Top. Microbiol. Immunol.* **93**:151–207.

Perrault, J., and Semler, B. L., 1979, Internal genome deletions in two distinct classes of defective interfering particles of vesicular stomatitis virus, *Proc. Natl. Acad. Sci. USA* **76**:6191–6195.

Portner, A., and Kingsbury, D. W., 1971, Homologous interference by incomplete Sendai virus particles: Changes in virus specific ribonucleic acid synthesis, *J. Virol.* **8**:388–394.

Rao, D. D., and Huang, A. S., 1982, Interference among defective interfering particles of vesicular stomatitis virus, *J. Virol.* **41**:210–221.

Re, G. G., and Kingsbury, D. W., 1986, Nucleotide sequences that affect replicative and transcriptional efficiencies of Sendai virus deletion mutants, *J. Virol.* **58**:578–582.

Re, G. G., and Kingsbury, D. W., 1988, Paradoxical effect of Sendai virus DI RNA size on survival: Inefficient envelopment of small nucleocapsids, *Virology* **165**:331–337.

Re, G. G., Gupta, K. C., and Kingsbury, D. W., 1983a, Genomic and copy-back 3' termini in Sendai virus defective interfering RNA species, *J. Virol.* **45**:659–664.

Re, G. G., Gupta, K. C., and Kingsbury, D. W., 1983b, Sequence of the 5' end of the Sendai virus genome and its variable representation in complementary form at the 3' ends of copy-back defective interfering RNA species: Identification of the L gene terminus, *Virology* **130**:390–396.

Re, G. G., Morgan, E., Gupta, K. C., and Kingsbury, D. W., 1984, Sendai virus DI RNA species containing 3'-terminal genome fragments, in "Nonsegmented Negative Strand Viruses," pp. 483–488, Academic Press, New York.

Re, G. G., Morgan, E. M., and Kingsbury, D. W., 1985, Nucleotide sequences responsible for generation of internally deleted Sendai virus defective interfering genomes, *Virology* **146**:27–37.

Rima, B. K., and Martin, S. J., 1979, Effect of undiluted passage on the polypeptides of measles virus, *J. Gen. Virol.* **44**:135–144.

Rima, B. K., Davidson, W. B., and Martin, S. J., 1977, The role of defective interfering particles in persistent infection of Vero cells by measles virus, *J. Gen. Virol.* **35**:89–97.

Roux, L., and Holland, J. J., 1979, Role of defective interfering particles of Sendai virus in persistent infections, *Virology* **93**:91–103.

Roux, L., and Waldvogel, F. A., 1982, Instability of the viral M protein in BHK-21 cells persistently infected with Sendai virus, *Cell* **28**:293–302.

Roux, L., and Waldvogel, F., 1983, Defective interfering particles of Sendai virus modulate HN expression at the surface of infected BHK cells, *Virology* **130**:91–104.

Roux, L., Beffy, P., and Portner, A., 1985, Three variations in the cell surface expression of the hemagglutinin-neuraminidase glycoprotein of Sendai virus, *J. Gen. Virol.* **66**:987–1000.

Shioda, T., Hidaka, Y., Kanda, T. Shibuta, H., Nomoto, A., and Iwasaki, K., 1983, Sequence of 3,687 nucleotides from the 3' end of the Sendai virus genome RNA and the predicted amino acid sequence of viral NP, P and C proteins, *Nucleic Acids Res.* **11**:7317–7330.

Sokol, F., Neurath, A. R., and Vilcek, J., 1964, Formation of incomplete Sendai virus in embryonated eggs, *Acta Virol.* **8**:59–67.

Tadokoro, J., 1958, Modified virus particles in undiluted passages of HVJ. II. The characteristics of modified particles, *Biken's J.* **1**:118–128.

Tobler, L. H., and Imagawa, D. T., 1984, Mechanism of persistence with canine distemper virus: Difference between a laboratory strain and an isolate from a dog with chronic neurological disease, *Intervirology* **21**:77–86.

Treuhaft, M. W., and Beem, M. O., 1982, Defective interfering particles of respiratory syncytial virus, *Infect. Immun.* **37**(2):439–444.

Tuffereau, C., and Roux, L., 1988, Direct adverse effects of Sendai virus DI particles on virus budding and on M protein fate and stability, *Virology* **162**:417–426.

Vidal, S., Mottet, G., Kolakofsky, D., and Roux, L., 1989, Addition of high-mannose sugars must precede disulfide bond formation for proper folding of Sendai virus glycoproteins, *J. Virol.* **63**:892–900.

Viola, M. V., Scott, C., and Duffy, F. D., 1978, Persistent measles virus infections *in vitro* and in man, *Arthritis Rheumatism* **21**(5):46–51.

Von Magnus, P., 1952, propagation of the PR8 strain of influenza A virus in chick embryos. IV. Studies on the factors involved in the formation of incomplete virus upon serial passage of undiluted virus, *Acta Pathol. Microbiol. Scand.* **30**:311–335.

Wechsler, S. L., Weiner, H. L., and Fields, B. N., 1979, Immune response in subacute sclerosing panencephalitis: Reduced antibody response to the matrix protein of measles virus, *J. Immunol.* **123**:884–889.

Welsh, R. M., and Oldstone, M. B. A., 1977, DI particles suppress expression of cell surface ag's, *J. Exp. Med.* **145**:1449–1468.

Welsh, R. M., Lampert, F. W., and Olstone, M. B. A., 1977, Prevention of virus-induced cerebellar disease by defective-interfering lymphocytic choriomeningitis virus, *J. Infect. Dis.* **136**:391–399.

CHAPTER 11

Paramyxovirus Persistence
Consequences for Host and Virus

R. E. RANDALL AND W. C. RUSSELL

I. INTRODUCTION: DEFINITION OF PERSISTENCE *IN VIVO*

Paramyxoviruses (a term used here to cover all members of the *Paramyxoviridae*) cause a wide variety of diseases in humans and other animals, ranging from uncomplicated respiratory illness to diseases such as mumps and measles, in which fatal sequelae can sometimes ensue. It has been recognized for a number of years that certain paramyxoviruses can establish persistent infections *in vitro* and it is now becoming clear that such persistence can also be seen *in vivo*. For the purpose of this chapter we will define virus persistence *in vivo* as the persistence of the virus in an infected host for periods longer than would normally be expected from an acute infection, regardless of the molecular mechanisms required for the establishment of that persistence. Consequently, virus persistence may simply be the establishment of a low-grade infection, possibly mediated via the production of defective-interfering particles or mutants (Rima and Martin, 1976; Holland *et al.*, 1978; Holland *et al.*, 1980; Young and Preble, 1980), but nevertheless requiring the continuous production of some infectious (helper) virus for the maintenance of the infection. In such situations the low fidelity rate of RNA replication may, in part, be responsible for the generation of mutants and thus be a factor contributing to the establishment of persistent infections (Holland *et al.*, 1982; Steinhauser and Holland, 1987; Cattaneo *et al.*, 1988). More recently it has also been suggested that hypermutation in RNA viruses may be promoted by a host cell's double-stranded RNA unwinding/modifying activity (Bass and Weintraub, 1988; Lamb and Dreyfuss, 1989; also see Chapter 12). Paramyxovirus persistence may also be achieved by the maintenance of virus genetic material in the absence of infectious virus replication and may possibly involve only the

R. E. RANDALL AND W. C. RUSSELL • Department of Biochemistry and Microbiology, University of St. Andrews, St. Andrews, Fife KY16 9AL, Scotland, United Kingdom.

persistence of partial or defective virus genomes (see Chapter 10). In such infections the virus genome would still have to be transcribed and replicated, although the levels of replication and the patterns of transcription may be much reduced and altered compared to those seen in an acute infection with the same virus. Consequently, in all persistent paramyxovirus infections it is to be expected that the virus must retain the capacity to encode for functional proteins involved in virus transcription and replication. However, mutations or deletions occurring in other virus-specified proteins may be tolerated. Persistent infections involving only incomplete genomes either would necessitate the establishment of the infection in long-lived cells (e.g., in nervous tissues), or the defective genomes would have to be transported from cell to cell in a manner which did not require them to be packaged in complete virus particles. Involvement of the host's immune surveillance in all these situations is obviously of crucial importance to the maintenance of persistence (see Chapter 18).

Persistent infections are thus different from latent infections, established by viruses such as herpesviruses and retroviruses, in which replication of the latent virus genome either does not occur or, if it does, is linked to the replication of the host cell's DNA and may not be mediated by virus-encoded enzymes. Furthermore, in latent infections, virus transcription is restricted in a specific manner and the maintenance of the latent state may be controlled by virus-encoded gene products. However, regardless of the mechanisms involved in the establishment of persistent paramyxovirus infections, there are clearly a number of consequences both for the virus and the host. It is the purpose of this chapter to review and *speculate* on some of the potential and actual consequences of establishing persistent paramyxovirus infections *in vivo*. It will become apparent, however, that this is a very complex area of research involving a great number of variables and there are many more questions than answers.

II. CONSEQUENCES FOR THE VIRUS

A. Possible Effects of Persistent Infections on the Epidemiology of Paramyxoviruses

There are obvious selective advantages for viruses in being able to establish persistent or latent infections in which they avoid elimination from a host by the immune response and yet retain the property of producing sufficient amounts of infectious virus (either continuously or at regular intervals thereafter) to infect susceptible members of a population. Consequently, many viruses have evolved specific modes of replication to initiate such infections, e.g., herpesviruses, retroviruses, and hepadnaviruses. However, while it has been known for some time that certain paramyxoviruses can cause persistent infections *in vivo*, it is still not clear whether there are any significant selective advantages for any paramyxoviruses in establishing such infections and, if there are, whether the genome structure and mode of replication which permit the establishment of such infections have been specifically selected during evolution.

The main consequences for any virus in establishing persistent infections in a large number of individuals within a population will be the effect that such infections have on the epidemiology of infection. Persistent virus infections may have two basic and opposing effects. First, persistent infections may increase the time an individual remains infectious, and second, persistent infections may affect and prolong the immune response of an individual, thus altering the susceptibility of that individual to future infections with the same or closely related viruses. There is a certain amount of evidence that both of these possibilities may affect the epidemiology of a number of paramyxovirus infections (see also Chapter 19).

B. Evidence for Prolonged Production of Infectious Virus by Persistently Infected Individuals

Parainfluenza virus type 3 (PIV-3) and respiratory syncytial virus (RSV) are ubiquitous viruses, and antibodies to these viruses are found throughout the human population. After infection with these viruses illness is usually short, ranging from 2 to 8 days. Infections with RSV and PIV-3 normally occur very early in life, with most children becoming seropositive within their first 2 years. Both PIV-3 and RSV must therefore be endemic within a community. Part of the explanation of why these viruses can continuously circulate within a population is that protective immunity appears to be short lived and older children and adults can be repeatedly infected (Knight, 1973; Wright, 1984; Belshe et al., 1984), thus ensuring that there is always a high percentage of susceptible individuals. However, it also appears that there may be individuals who can become persistently infected with these viruses who may secrete infectious virus into the environment for long periods and thus contribute to the endemic nature of these viruses. Immunocompromised patients, including those with cellular immunodeficiencies, appear to be particularly susceptible to prolonged infections with parainfluenza viruses and RS (Karp et al., 1974; Jarvis et al., 1979; Delage et al., 1979; Fishaut et al., 1980). Fishaut et al. (1980) reported periods of virus shedding in immunocompromised children for PIV-3 of from 20 to 235 days, for PIV-2 of greater than 79 days (Karp et al., 1974), and for RSV of from 40 to 112 days. Once infected, these children never appeared to rid themselves of their respective viruses. Similarly, in a study of patients with chronic bronchitis and emphysema, Gross et al. (1973) described an outbreak of PIV-3 in which a prolonged carrier state was observed in several patients. The duration of persistent virus shedding was as long as 3–5 months. It has also been suggested that patients with chronic bronchitis are susceptible to persistent infections with RS virus (Mikhalchenkova et al., 1987; Iakovleva et al., 1987). There is also some direct evidence that under certain specific environmental conditions healthy adults may be persistently infected with PIV-3 and shed sufficient levels of infectious virus to infect susceptible individuals (Parkinson et al., 1980). In a study of an isolated community at the American Antarctic Research Station,

Muchmore and co-workers (1981) showed that healthy asymptomatic individuals could shed virus over a 6- to 8-month period. Furthermore, they concluded that two episodes of respiratory illness occurring within that 8-month period could best be explained as being initiated by persistently infected asymptomatic carriers of the virus. However, even if there are certain individuals within normal communities in temperate and tropical climates who persistently shed infectious virus, it is not clear whether this significantly affects the epidemiology of RSV or PIV-3. Nevertheless, it has been suggested that a persistent state of infection in adults may explain the ability of parents to transmit infection to infants (Glezen et al., 1976). If these viruses do establish such infections in a significant percentage of a population, it would ensure that the virus would remain endemic within a community and may explain some of the epidemiological observations associated with PIV-3 and RSV infections.

C. Possible Effect of Persistent Infections on Herd Immunity

It is known that other paramyxoviruses, such as measles virus, can also establish persistent infections in vivo (ter Meulen and Cartner, 1982). Although in the majority of these situations it is unlikely that persistently infected individuals can transmit infectious virus to susceptible people (Panum, 1847), it is still possible that the ability of certain paramyxoviruses to establish persistent infections can influence the epidemiology of infection with these viruses. Unlike the short-lived protective immunity induced by respiratory paramyxoviruses such as PIV-3 and RSV, immunity to the diseases of measles and mumps is usually life-long (Krugman et al., 1965; Panum, 1847). A number of explanations have been put forward for this life-long immunity to these viruses. One factor may be that both measles virus and mumps virus have a relatively long incubation period before onset of clinical symptoms and also have a viremic phase in their infectious course. Consequently, although immune individuals may not be able to prevent repeated infections with these viruses, they should be able to mount a secondary immune response in time to prevent the development of clinical disease (Krugman et al., 1965). Furthermore, if these viruses are endemic, individuals may be repeatedly infected, thus keeping their levels of immunity high (booster effect). In fact, subclinical reinfection has been reported in children with either vaccine-induced or natural immunity (Krugman et al., 1965; Watson, 1965; Chang, 1971; Linneman et al., 1972, 1973). Whether individuals with subclinical reinfections are infective is not known. However, if they are infective, such a situation may contribute to the epidemic nature of these virus infections. On the other hand, in the classic description of a measles outbreak in the Faroe Islands, Panum (1847) showed that natural immunity may be life-long even in the apparent absence of subclinical reinfections. Measles disappeared from the Faroe Islands in 1781 following an epidemic, but when it was reintroduced in 1846 there were 6000 cases in a population of 7782 inhabitants. However, Panum observed that all who had experienced measles 65 years previously were immune to the disease. This is in contrast to the complete lack of immunity to measles in "virgin" populations in which the disease attacks persons of all ages from early infancy to old age (Christensen *et al.*, 1952; Peart and Nagler, 1954).

In order to explain life-long immunity to measles in the absence of repeated subclinical infections, it has been suggested that measles virus establishes persistent nonproductive infections *in vivo*. Continuous production of small amounts of virus antigen would restimulate the immune response to measles virus (Burns and Allison, 1975; ter Meulen and Cartner, 1982). Circumstantial evidence that this may be the case is presented later. There are other possible explanations for life-long immunity induced by infection with systemic viruses. It has been suggested that primary infection with these viruses produces a broad immune response with the production of many specific memory clones of lymphocytes. These clones would then be activated at periods throughout the life span of the individual, perhaps as a consequence of broad nonspecific mitogenic effects, possibly caused by infection and immunization with unrelated microorganisms or antigens (Norrby, 1987). Alternatively, life-long immunity may be induced by a combination of both these mechanisms. For example, it is thought that free virus can be captured and internalized by B cells for antigen presentation (Scherle and Gerhard, 1988). Furthermore, *in vitro* studies have shown that while measles virus can infect resting lymphocytes, greater amounts of infectious virus are produced when these cells are induced by mitogens to proliferate and differentiate (Joseph *et al.*, 1975; Lucas *et al.*, 1978; Huddlestone *et al.*, 1980). Consequently, it may be that some B cells destined to become long-term measles virus-specific memory cells may, during the acute phase of the disease, have specifically captured the virus. Because the cells are quiescent, it may be that the virus is more likely to establish a persistent infection in them. However, if these cells are then activated in some way (possibly by nonspecific mitogens), there may be subsequent production of measles virus antigens and restimulation of the immune response. Nevertheless, it is clear from these discussions that there is insufficient data to allow any firm conclusions about the mechanisms involved in the induction of such life-long immunity.

III. CONSEQUENCES FOR THE HOST

It is possible that many paramyxoviruses can cause persistent infections in humans and other animals which remain silent because they do not induce any harmful pathology. As we have just pointed out, silent infections may in fact be of potential benefit to the host if they result in the maintenance of protective immunity to the virus. However, there may also be a number of serious consequences. These will obviously be dependent to a large extent upon the type of cell infected, the influence that such infections have upon the social behavior of the cell, and on the nature of the immune response to these events. In fact, a number of important chronic human diseases have either been shown to be caused by persistent paramyxovirus infections [e.g., subacute sclerosing panencephalitis (ter Meulen *et al.*, 1983)] or linked with such infections [e.g., Paget's bone disease (Rebel *et al.*, 1977; Singer, 1980) and autoimmune chronic active hepatitis (Robertson *et al.*, 1987)], and it is instructive to consider the possible reasons for the unfavorable outcome of the infection in such cases.

A. Importance of Immune Response in the Establishment and Maintenance of Persistent Infections

To establish a persistent virus infection *in vivo* a virus must avoid elimination from the body by the immune response. It has been suggested that in order to fulfill this objective there has to be a reduction of virus-encoded antigens expressed on the plasma membrane of persistently infected cells, thereby decreasing the chances of the cell being killed by a specific antiviral immune response (Oldstone and Fujinami, 1982). Such a reduction in viral antigen expression on the cell surface may be due to the selection of persistently infected cells in which the virus can only make reduced amounts of viral proteins, e.g., because of the presence of defective genomes.

Alternatively, viral antibody-induced antigenic modulation has been proposed as a mechanism by which persistently infected cells can avoid elimination by the immune response (Joseph and Oldstone, 1975; Oldstone and Tishon, 1978; Oldstone and Fujinami, 1982). Here, it is suggested that the level of virus antigens on the surface of a persistently infected cell is kept below a critical point required for cell lysis with antibody and complement or immune lymphocytes, because the virus glycoproteins are continuously removed by their interaction with antibody. In other words, there is insufficient antibody binding to cause cell lysis, but antibody which does bind strips the virus glycoproteins from the cell surface. Regardless of the mechanism employed, persistently infected cells would, in these situations, continue to make other viral polypeptides, the synthesis of which could affect the physiological functions of the persistently infected cells, leading in turn to specific diseases (Oldstone and Fujinami, 1982). In addition, it has also been shown *in vitro* that antibodies binding with virus proteins on the surface of infected cells may reduce the actual level of other intracellular virus polypeptides, thus enhancing the chances of establishing a persistent virus infection (Fujinami and Oldstone, 1979; Fujinami and Oldstone, 1980).

Some experimental evidence that antibody-induced antigenic modulation may be important in the establishment of persistent measles virus infections *in vivo* has been presented. Wear and Rapp (1971) demonstrated that while the presence of maternal neutralizing antibody to measles virus in newborn hamsters could prevent the development of acute encephalitis following intracerebral inoculation of virus, these animals still developed a persistent measles virus infection in their central nervous system. Similarly, inoculation of either polyvalent antibodies or monoclonal antibodies to measles virus hemagglutinin led to the establishment of a chronic infection in mice, whereas the inoculation of antibody to measles virus nucleocapsid protein did not modify the course of the acute infection (Rammohan *et al.*, 1981). Furthermore, Albrecht *et al.* (1977) could only establish a persistent measles virus infection in monkeys which had preexisting antibodies to measles virus and not in nonimmune monkeys. It has been suggested that if such a mechanism could also operate with cross-reacting antibodies against other viruses, it might lead to the situation where a virus could establish a persistent infection in a host without there being any obvious specific antibody response to that virus (Russell and Goswami, 1984).

However, it now seems to be well established that cell-mediated immunity appears to be critical in controlling many paramyxovirus infections, as in measles and mumps (Burnet, 1968; Ruckdeschel et al., 1975). Given that T cells may potentially recognize target antigens on any virus protein, including internal virus structural proteins (Randall and Young, 1988b) in conjunction with the HLA antigens (Townsend et al., 1984; Delovitch et al., 1988; Bevan, 1987; Germain, 1986), it seems unlikely that antibody-induced antigenic modulation can be the complete explanation of how persistently infected cells avoid elimination by the immune response. In this respect it is of interest that Maehlen et al. (1989) showed that $CD8^+$ cytotoxic T cells were required to eliminate persistent measles virus infections from the brains of Lewis rats. In addition, interactions of measles virus with cellular constituents of the immune system, disabling specific immune responsiveness, may be of critical importance in the establishment of persistent infection with the virus (McChesney and Oldstone, 1987; Oldstone, 1989).

Because SSPE develops in patients who have overcome an initial acute measles virus infection and who appear to mount normal immune responses to other infectious agents (Blaese and Hofstrand, 1975; Sell and Ahmed, 1975), an impaired or inappropriate T-cell response specific for measles virus may have developed in these patients (Dhib-Jalbut et al., 1989a). This could arise by a combination of different mechanisms. There may, for example, be an inappropriate measles virus-specific T suppressor cell activity and the induction of tolerance to measles virus in these patients (Oldstone, 1989). Alternatively, persistently infected cells may have been selected by the immune response simply because in those cells the majority of the virus genomes present are defective and have been mutated to such an extent that the sequences coding for the major target antigens recognized by that individual's cytotoxic T lymphocytes (CTLs) have been deleted or altered. This may be compounded in the CNS, where low levels of MHC antigens are normally expressed (Lampson, 1987).

As different individuals of a heterozygous population may recognize different T-cell target antigens, depending on their histocompatability status (McMichael et al., 1986; Gotch et al., 1987), it may be that SSPE can only develop in patients with specific HLA repertoires. In other words, SSPE may only develop in individuals whose CTLs are mainly targeted to antigens on those virus proteins not required for the maintenance of cell-associated persistent infections. In fact, in SSPE it has been demonstrated that there may be defects in the synthesis of the matrix, hemagglutinin, or fusion proteins; envelope proteins that are not required for the transcription or replication of the virus genome (Cattaneo et al., 1988; Wechsler and Fields, 1978; Hall et al., 1979; Carter et al., 1983; Norrby et al., 1985; Baczko et al., 1986). Thus, in SSPE patients the acute measles infection may have been originally controlled by CTLs which recognized target antigens on virus proteins (H, F, or M) that were subsequently mutated or deleted in persistently infected cells. A prediction of this hypothesis is that SSPE (or any other chronic disease in which large amounts of paramyxovirus antigens are continuously being synthesized) would not develop in individuals whose cytotoxic T lymphocytes recognize epitopes on virus proteins critical for the maintenance of the persistent infection (e.g., NP, P, or L). Most studies on cell-mediated immune responses to measles virus in SSPE have relied on measuring the relative levels of stimulation of lymphocytes with virus or purified antigens rather than directly measuring the CTL response against cells expressing specific viral antigens (e.g., Ilonen et al., 1980; Dhib-Jalbut et al., 1988). However, the hypothesis outlined above is open to experimental analysis once target cells which synthesize individual measles virus proteins can be constructed, such as through the use of vaccinia virus vectors.

B. Spread of Paramyxoviruses in an Infected Host and Their Interaction with Hemopoietic Cells

The cycle of acute infection by some paramyxoviruses includes a viremic phase (e.g., mumps and measles), whereas other paramyxoviruses appear only to replicate locally in the respiratory tract (human parainfluenza viruses and RS). Obviously, viruses which have a viremic phase will be spread throughout the body and thus have the opportunity to infect many different cell types. This is reflected in the diverse complications which may ensue following acute measles and mumps virus infections, encephalitis, myocarditis, orchitis, and pancreatitis. Furthermore, host cells may differ in their ability to support virus replication. Thus, some types of cell may support normal virus replication, while others may not support virus replication at all or may so modulate the replication of the virus as to make the establishment of a persistent virus infection more likely.

For a paramyxovirus to establish persistent infections at sites other than at the primary site of infection it is obvious that the virus must somehow be transported to these sites. While this might be achieved by the liberation of free virus into blood or lymphatic fluid from the original site of virus replication, some paramyxoviruses may also be transported throughout the body via infected lymphoid cells. After infection with measles virus, for example, it appears that there is minimal replication of the virus at the primary site of infection, but from there the virus enters the lymphatics and infects regional lymphoid tissue. Here there is slow multiplication of virus, and the infection spreads, probably by infected leukocytes, to other lymphoreticular and related tissues, including tonsils, lymph nodes, spleen, gastrointestinal tissues, and lungs (Sergiev et al., 1960; Ono et al., 1970). Even at early times of infection, low titers of virus can be found in the blood and primarily in the leukocyte fraction. Mononuclear cells and activated lymphocytes all appear to be susceptible to infection with the virus (Gresser and Chany, 1963; Sullivan et al., 1975; Joseph et al., 1975; Osunkoya et al., 1974; Salonen et al., 1988). During virus multiplication in lymphoreticular tissues larger amounts of virus are released, resulting in a higher-grade (secondary) viremia and the onset of clinical symptoms. A similar pattern of spread also probably occurs with other paramyxoviruses, such as mumps virus, which cause generalized infections.

In SSPE patients, measles virus has been shown to persist in lymphoid cells. Recovery of the virus from such patients required the cocultivation of mixed cultures of lymph node cells with permissive tissue culture cells (Horta-Barbosa et al., 1971). More recently it has been claimed by Fournier et al. (1985) that measles virus RNA can be detected by in situ hybridization in a high percentage (70–90%) of peripheral blood mononuclear cells from SSPE patients. In the same article, the authors reported the detection of a small number of measles virus-infected cells (0.1–5%) in four seropositive control adults and in three age-matched children (10–15%). A later study (Fournier et al., 1986) on tissue sections of an appendix removed 15 days before onset of SSPE showed that many cells of the lymphoid tissue contained measles virus RNA. Control studies on three of six seropositive controls also indicated the presence of measles virus RNA, although the number of positive cells was considerably reduced. If these studies can be confirmed, it would appear that measles virus can establish persistent infection in healthy individuals and would add support to the hypothesis that persistent measles virus infections may play a role in the induction of life-long immunity to this virus.

Interestingly, Robbins et al. (1981), in a well-documented case, isolated another paramyxovirus, simian virus 5 (SV5), from the peripheral blood leukocytes of a patient suffering from SSPE. As SV5 does not appear to play a causal role in SSPE, they concluded that other paramyxoviruses may persist in human lymphoid tissues. Mitchell et al. (1978) also reported the isolation of SV5 from bone marrow aspirates of multiple sclerosis patients, and claims have been made that SV5 can be detected directly in bone marrow cells of both MS and control patients by immunofluorescence (Goswami et al., 1984a). It is interesting that specific antibodies to SV5 detected by neutralization, by immune precipitation, and by an ELISA technique can be found in a significant proportion of human sera, although this paramyxovirus has not been directly implicated in any acute disease (Goswami et al., 1984b). These and other observations suggest that SV5 may cause widespread inapparent infections in humans (Hsiung, 1972).

It is known that SV5 naturally causes persistent infections in monkeys, with the virus regularly recoverable from the kidneys long after the animals are infected (Atoynatan and Hsiung, 1969; Hsiung, 1972; Tribe, 1966). Since SV5 also naturally causes a respiratory infection in dogs (Appel and Percy, 1970; Hsiung, 1972) and can be used experimentally to infect mice and hamsters (Chang and Hsiung, 1965; Randall et al., 1988), it may be that this paramyxovirus readily crosses species barriers, causing disease in one host, but being relatively benign in others (Randall et al., 1987). Pringle and Eglin (1986) also suggested that pneumovirus of mice (PVM) or a closely related virus may cause widespread infections in humans, but whether other paramyxoviruses can readily cross species barriers requires further investigation.

Paramyxoviruses that do not appear to have a viremic phase may also be transported to different tissues by infected lymphoid cells. Thus, it has been demonstrated that RSV can infect human lymphocytes in vitro (Roberts, 1982; Domurat et al., 1985) and although viremia is unusual, RSV infections have been associated with nonrespiratory conditions, including bone, heart, neurological, and urinary tract diseases. Furthermore, RSV antigens have been detected in mononuclear leukocytes obtained from children in the acute phase of the respiratory disease (Domurat et al., 1985). Consequently, it appears clear that many paramyxoviruses can infect leukocytes and hence be transported to many different tissues. What is unknown, however, is how often paramyxoviruses establish long-lived persistent infections in lymphoid cells and, if they do, whether the majority of such infections are silent or whether under certain circumstances persistent paramyxovirus infections can lead to the induction of particular disease states.

IV. PERSISTENT PARAMYXOVIRUS INFECTIONS AND THEIR POSSIBLE ROLE IN CHRONIC HUMAN DISEASES

A. Paget's Bone Disease

In the case of measles virus, it is known from in vitro studies that infection of human lymphocytes can impair killer-cell activity and immunoglobulin synthesis (Casali et al., 1984; McChesney et al., 1986) without significant effects on cell morphology and with only minimal expression of measles-virus gene products. The question therefore arises of whether persistent paramyxovirus infections of lymphoid cells in vivo can lead to particular disease states. A possible example of such a chronic disease involving hemopoietic cells induced by persistent paramyxovirus infections may be Paget's bone disease (Hamdy, 1981). In Pagetic bone, the osteoclasts, which are highly specialized bone-resorbing cells, are altered in such a way that they become overactive and reabsorb bone in an uncontrolled manner. Osteoclasts are actually hematopoietic cells, possibly of similar lineage to macrophages (Mariano and Spector, 1979; Burger et al., 1982; Scheven et al., 1986); they appear to be long lived, and are multinuclear, being formed by fusion of blood-borne precursor cells (Scheven et al., 1986; Marks, 1983). In Pagetic bone, they increase in numbers and are larger than normal, having up to ten nuclei per cell (Hamdy, 1981; Rebel et al., 1980a). Suspicion that a persistent paramyxovirus infection of Pagetic osteoclasts may be related to this disease arose from observations that the characteristic nuclear and cytoplasmic inclusion bodies present in the abnormal oesteoclasts had appearances similar to the nucleocapsids of measles virus seen in SSPE (Mills and Singer, 1976; Rebel et al., 1977; Harvey et al., 1982). On this basis, bone sections were analyzed using measles antisera by immunofluorescence and peroxidase techniques with positive results (Rebel et al., 1980a,b; Basle et al., 1985). In later experiments, using the technique of in situ hybridization, Basle et al. (1986) demonstrated measles virus nucleic acid in both osteoclasts and osteoblasts (bone-forming cells) of Pagetic bone, but not in control bone sections. On the other hand, RSV antigens have been detected both directly in Pagetic osteoclasts and in cells derived from Pagetic bone cultivated in vitro (Mills et al., 1981; Singer and Mills, 1983). Mills et al. (1984) later reported the simultaneous detection of both RSV and measles virus in osteoclasts of the same patients. There have also been reports of the detection of SV5 and PIV-3 in Pagetic bone osteoclasts by immunofluorescence with specific monoclonal antibodies (Basle *et al.*, 1985). Seroepidemiological evidence has also been presented for (O'Driscoll and Anderson, 1985) and against (Hamill *et al.*, 1986) the involvement of another paramyxovirus, canine distemper virus, in the etiology of Paget's bone disease.

A number of important questions arise if the observations that multiple paramyxoviruses can persist in osteoclasts of Pagetic bone can be confirmed. Namely, (1) does the presence of persistent virus alter the social behavior of the cells so that they no longer function normally, and, if so, can the presence of different persistent paramyxovirus infections directly lead to the disease? Or (2) is the presence of these paramyxoviruses within osteoclasts simply a reflection that these cells are particularly long-lived cells that are capable of surviving persistent infections with multiple paramyxoviruses? And (3) do Pagetic osteoclasts become infected with free virus during a viremia or do they acquire infection through fusion with bone marrow-derived precursor cells, some of which may be persistently infected with specific paramyxoviruses? These questions will be very difficult to answer without a good animal model system for Paget's bone disease. However, in this respect it is of interest to note that it has been reported that canine distemper virus (CDV) has been detected in the osteoclasts of gnotobiotic dogs infected with CDV (Axthelm and Krakowka, 1986), although it was not stated whether such dogs showed any sign of bone deformation.

B. Autoimmune Chronic Active Hepatitis

Besides causing persistent infections in lymphoid tissues, it is clear that some paramyxoviruses may cause persistent infections in other cell types and that in certain circumstances the persistent infection and the immune response to the infection may lead to chronic disease. One such disease that has been linked to a persistent measles virus infection of the liver is autoimmune chronic active hepatitis (AICAH). Although jaundice is seldom a feature of acute measles virus infection, the virus is hepatotropic and subclinical hepatitis may be detected in as many as 80% of adults admitted to hospital with measles (Gavish *et al.*, 1983). AICAH antibody titers to measles virus are higher than those in natural measles infection and similar to levels found only in SSPE (Triger *et al.*, 1972; Robertson *et al.*, 1987). As in SSPE, some of the antibody to measles virus in AICAH patients is of the IgM class, suggesting the continuous production of virus antigen in this condition (Christie and Haukenes, 1983). Epidemiological studies on AICAH have also suggested that since the introduction of measles vaccines the age of onset of the disease has increased and that AICAH in childhood or early adolescence (the age groups vaccinated against measles) is now rare (Robertson *et al.*, 1990). Furthermore, radiolabeled oligonucleotide probes have been used to detect persistent measles virus genomes in the leukocytes of a high proportion (~70%) of patients with AICAH (Robertson *et al.*, 1987). Interestingly, in the same report it was stated that measles virus genetic information was detected in one of three patients with systemic lupus erythematosus and two of four patients with cryptogenic cirrhosis. Similarly, using the polymerase chain reaction, Kalland et al. (1989) detected measles virus RNA in seven out of eight AICAM patients and four out of six controls. However, in their studies they only detected low levels of RNA (less than one RNA copy per lymphocyte) with no quantitative differences between AICAM patients and controls. Nevertheless, these results again suggest that measles virus persistence may be common, but such infections are obviously very difficult to detect in normal individuals.

C. Multiple Sclerosis

Nervous tissue consists of a highly complex network of cells, many of which do not divide, and which may also function throughout the lifetime of the animal. In addition, most of these cells have very specialized features with intricate transport mechanisms and varying degrees of morphological and functional polarity. These characteristics may be favorable to the induction of localized virus persistence and raise the possibility that virus components can be transported readily to various parts of the nervous system and thereby impair function (Kristensson and Norrby, 1986). Most of the interest in virus persistence has been focused on the central nervous system (CNS) following the observations that a persistent measles virus infection is the cause of SSPE. The first symptoms of this fatal human disease appear on average about 6 years after the onset of a primary measles infection. The precise nature of the measles persistence and the factors inducing the disease are not immediately apparent, but the various scenarios which can be inferred based on the properties of the virus-infected brain cells are discussed in detail in Chapter 12. Most of the speculation on the role of paramyxoviruses in CNS diseases has centered on the possibility that they may be involved in the pathogenesis of multiple sclerosis (Norrby, 1978; ter Meulen and Stephenson, 1981; Russell, 1983; Goodman and McFarlin, 1987). This disease of the CNS is particularly prevalent in temperate, more affluent societies. The first symptoms of MS usually appear on average at about 30 years of age and result in multiple loss of CNS functions. The disease can be very variable in its effects and characteristically takes a relapsing and remitting course. The lesions appear as plaques within the CNS and can be quite extensive, although they are normally confined to areas of white matter (McFarlin and McFarland, 1982; Waksman and Reynolds, 1984). Closer examination of the lesions reveals significant breakdown of myelin, with relative sparing of the axons. There also appears to be extensive immunological involvement in the diseased areas (Whitaker and Snyder, 1984). Many immune cells (T and B cells, macrophages) can be identified in the lesions (Traugott et al., 1983; Hauser et al., 1986) and the immunoglobulin content of the cerebrospinal fluid is significantly increased, with indications of intrathecal synthesis of IgG. An intriguing characteristic of MS and of some other neurological diseases is that this IgG appears to be of an oligoclonal nature [for a review see Whitaker and Snyder (1984)]. This can be demonstrated by electrophoresis techniques, which show that the IgG can form discrete oligoclonal banding patterns of a cathodic character.

In the cases of SSPE and virus encephalitides, such as those caused by mumps and herpes viruses, most of the CSF oligoclonal bands appear to be directed against the relevant virus antigens (Vandvik et al., 1976, 1985; Nordal et al., 1978). In view of these findings, many attempts have been made to relate the oligoclonal IgGs found in MS to antigens of particular viruses and indeed to brain antigens on the supposition that MS may be an autoimmune disease possibly induced by a virus infection. The possibility that viruses may be a factor in the disease induction had been indicated both by epidemiological studies and by the isolation of viruses from a variety of MS tissues. Paramyxoviruses [measles virus, parainfluenza virus type 1 (69/4), SV5] have all been isolated from tissues of MS patients and have therefore been implicated as possible causative agents of MS [for reviews see ter Meulen and Stephenson (1981) and Russell (1983)].

However, in spite of earlier optimism in this area, it is probable that virus isolation from MS tissues reflects the ability of these viruses to establish persistent infections in human tissues. Analysis of the virus antibody patterns in CSF immunoglobulins has also indicated that in MS, paramyxovirus antibodies (measles virus, parainfluenza viruses 1,2,3, SV5) can be detected, although there are differences of opinion with respect to their significance. The role of measles virus has been especially studied since it has been demonstrated that MS patients have significantly higher levels of humoral antibody against measles virus than do matched controls (Adams and Imagawa, 1962; Albrecht et al., 1983). On the other hand, Goswami et al. (1987) showed that some MS CSFs appeared to contain a significant proportion of oligoclonal antibodies against the paramyxovirus SV5, suggesting that this virus may be involved in the pathogenesis of the disease. However, in subsequent experiments with another series of CSFs it was not possible to show that these antibodies comprise a significant portion of the oligoclonal antibody (Russell et al., 1989). Furthermore, Vandvik and Norrby (1989) also failed to confirm the results of Goswami et al., demonstrating that although some of the oligoclonal antibodies reacted with SV5 antigens [or antigens that may have been common to SV5 and PIV 2 or mumps (Randall and Young, 1988a)], the bulk of the oligoclonal antibodies had unknown reactivities. However, in an earlier study on the specificity of intrathecally synthesized antibodies in MS patients, Salmi et al. (1983) demonstrated the presence of antibodies to 17 viruses in at least some of the patients examined. The highest number of patients showed intrathecal synthesis of antibodies to measles virus and other paramyxoviruses as well as to rubella virus and the lowest frequency against herpes simplex virus, adenovirus, and cytomegalovirus. They also demonstrated that intrathecal antibody to particular viruses may not necessarily be in the same proportion as that found in the extrathecal circulation. Similarly, Goswami et al. (1987) showed that while antibodies to adenovirus could be readily detected in the peripheral system of many MS patients, there was relatively little adenovirus antibody in their CSF.

There have been a number of explanations put forward for the differences in the proportion of antibody within the CSF compared to the peripheral system. One explanation suggests that in MS patients there may be a nonspecific clonal activation of B cells within the CNS. Salmi et al. (1983) suggested that CSF antibody represents immunity which entered the CNS early in the life of these patients and that committed B-cell clones can reside within the CNS for long periods of time. This hypothesis is supported by the observation that the oligoclonal antibody banding pattern in MS patients does not change significantly during the course of MS, although the intensity of the oligoclonal bands may vary (Olsson and Link, 1973). Alternatively, specific lymphocytes may be immunologically sequestered into the CNS by the presence of the appropriate antigen, and the high proportion of antibodies to paramyxoviruses may therefore simply reflect the ability of these viruses to invade the CNS. For example, it has been suggested that silent invasion of the brain is a fairly common event in uncomplicated measles infection (Hanninen et al., 1980). Whether persistent silent infections are established in a significant proportion of individuals after acute infections of the CNS with paramyxoviruses remains speculative. However, in this respect it is of interest to note that measles virus-specific sequences have been detected in brain autopsies from MS patients by in situ hybridization (Haase et al., 1981; Cosby et al., 1989).

In another approach to the significance of measles virus in MS, the T-cell responses of MS patients to different measles virus antigens have been analyzed. Significantly, there appears to be an impairment in the generation of measles virus-specific cytotoxic T lymphocytes (CTLs) in a number of patients with MS (Jacobson et al., 1985). It has been suggested that this impaired response may be due to a reduction in the precursor frequency of virus-specific CTLs in MS patients compared to healthy individuals. Dhib-Jalbut et al. (1989b) suggested that this defect may be caused by abnormalities in the induction or maintenance of the measles virus CTL response or by the sequestration of measles virus CTLs within the CNS. They concluded that if sequestration of measles virus CTLs did occur, it was probable that this was due to the presence of measles virus antigens within the CNS rather than cellular cross-reactivity between a measles virus antigen and a brain antigen such as myelin (Liebert et al., 1987). The former explanation is compatible with a persistent measles virus infection occurring in the CNS in these patients. Nevertheless, even if measles virus CTLs are primarily sequestered into the CSF by the presence of measles virus antigens, this would not necessarily preclude the subsequent cross-reaction of some of these CTLs with brain antigens.

Since cell-mediated immune responses appear to be of significance in the disease progression in MS (Waksman and Reynolds, 1984; Hafler and Weiner, 1987), further studies on the nature and specificity of these responses will be crucial to any understanding of the disease and may indeed provide a pointer to the role of viruses such as paramyxoviruses and HTLV-1 (Reddy et al., 1989) in the disease etiology. Given that persistent paramyxovirus infections may induce a restricted T-cell response, mainly directed at viral proteins required for the maintenance of the persistent infection (NP, P, and L), it may be worth further considering the hypothesis that the disease is an autoimmune one induced as the result of cross-reactivity between viral antigens and a component of the brain. The consequential immunological response to a virus may then result in disease exacerbation via secondary inflammatory responses against the cross-reacting brain epitope(s). If this were the case, then there should be some indication of "mimicry" between brain and virus antigens. Serological studies have indicated that there may be cross-reacting antigens between brain antigens and measles virus and SV5 (Rastogi et al., 1979; Friedman *et al.*, 1987; Goswami *et al.*, 1985). Molecular mimicry has also been demonstrated between a number of different viruses and various tissues using monoclonal antibodies, but none of these relationships has been directly implicated in disease (Srinivasappa *et al.*, 1986). Nevertheless, it is pertinent to note that activation of autoantibodies in paramyxovirus infections has been reported (Fagraeus *et al.*, 1983; Toh *et al.*, 1979). Whether any such cross-reacting T-cell epitopes exist between virus antigens and components of the brain remains speculative. However, it is of interest to note that Ziola and Smith (1988) presented data showing that there may be T-cell target antigens common to many paramyxoviruses, including those as far apart as mumps virus and RSV. Presumably, such epitopes would be present on functionally conserved proteins such as those involved in virus transcription and replication, proteins which are likely to be required for the maintenance of persistent infections. It would thus be of particular significance if any such epitopes also cross-reacted with components of the brain.

V. ANIMAL MODELS AND THE CONSEQUENCES OF PARAMYXOVIRUS PERSISTENCE

Studies on animal model systems for paramyxovirus persistence within the CNS have demonstrated the tremendous complexities involved in establishing such infections, with the outcome depending on a multitude of host and viral factors. Canine distemper (CD), for example, commonly presents as a respiratory infection in dogs, but can recur as a chronic neurological syndrome. It is claimed that chronic encephalomyelitis with inflammatory demyelinative changes develops in about one-third of dogs experimentally infected with CD virus and that these animals harbor CD virus persistently both in the CNS and in the anterior uvea (Summers *et al.*, 1983). In a further study, inflammatory demyelinating lesions were observed in affected dogs, and in some, but not all, of these lesions, virus antigen could be detected. As virus antibodies could also be detected in CSF and sera of these animals, it was concluded that the inflammatory response in distemper could result in the clearance of the virus from some lesions. Nevertheless, despite the presence of an apparently effective intrathecal antiviral immune response, fresh noninflammatory lesions arose apparently as a result of virus replication and spread in the white matter, and these lesions appeared to coexist with inflammatory lesions in which viral clearance had taken place (Bollo *et al.*, 1986). Canine distemper virus has also been studied in a hamster model system and it is clear that different strains of the virus can produce very different disease syndromes (Cosby *et al.*, 1981).

Measles virus has also been used in animals to produce a number of interesting experimental CNS diseases. Carrigan (1987) has produced chronic relapsing encephalomyelitis in hamsters which bears some similarity to multiple sclerosis. ter Meulen and his colleagues (e.g., Liebert and ter Meulen, 1987) have also developed a rat model system which illustrated that different neurological diseases can be induced in two different rat strains by a single neurovirulent measles virus isolate, reflecting the importance of the host's immunogenetic background. This group has also shown restricted measles virus gene expression in both acute and subacute encephalitis of Lewis rats, with transcription of the virus glycoprotein genes being particularly badly affected (Schneider-Schaulies et al., 1989). Also of special note was the observation (Liebert et al., 1987) that in one strain of rat the persistent measles virus infection resulted in the induction of cell-mediated autoimmune responses against myelin basic protein. These studies demonstrated that in the appropriate genetic environment virus infection can trigger an autoimmune disease, which may therefore be of relevance to MS.

Other paramyxovirus/animal model systems which have been studied include Sendai virus in mice, where the induction of persistence apparently by defective-interfering virus (Ruthkay-Nedecka et al., 1987) has been demonstrated and where evidence of virus infection of neurones can be shown (Chan, 1985). Nevertheless, while these and other animal model systems clearly demonstrate the potential that paramyxoviruses have for causing chronic disease, apart from the clear involvement of measles virus in SSPE, the role of paramyxovirus infections in other chronic human diseases remains uncertain.

VI. CONCLUSIONS

A property of paramyxoviruses is that they readily establish persistent infections in vitro. Such a general biological property suggests that their mode of replication and genome structure may specifically facilitate the establishment of such infections. However, it is not clear whether the ability to cause persistent infections was positively selected during the evolution of ancestral paramyxoviruses. If it was, the fact that persistent paramyxovirus infections may lead to chronic disease may largely be an unfortunate consequence of the mode of replication of these viruses.

While our knowledge of paramyxovirus molecular virology has increased dramatically over the last 10–15 years, the application of this knowledge to the more complex in vivo systems is only just beginning. Nevertheless, to understand fully how persistent paramyxovirus infections are established in vivo and to unravel the biological consequences of such infections, it will be necessary to define not only the molecular events of virus–host cell interactions, but also the humoral and cell-mediated immune responses to such infections.

ACKNOWLEDGMENTS. We are indebted to Dr. Bernard Souberbielle for critical comment and to the three Margarets (Bell, Wilson, and Smith) for typing.

VII. REFERENCES

Adams, J. M., and Imagawa, D. T., 1962, Measles antibodies in multiple sclerosis, Proc. Soc. Exp. Biol. Med. 111:562–566.

Albrecht, P., Burnstein, T., Klutch, M. J., Hicks, J. J., and Ennis, F. A., 1977. Subacute sclerosing panencephalitis: Experimental infection in primates, Science 195:64–66.

Albrecht, P., Toutellotte, W. W., Hicks, J. T., Hato, S., Boone, E. J., and Potvin, J. R., 1983, Intrablood–brain barrier measles virus antibody synthesis in multiple sclerosis patients, Neurology 33:45–50.

Appel, M. J. G., and Percy, D. H., 1970, SV-5 like parainfluenza virus in dogs, *J. Am. Vet. Med. Assoc.* **156:**1778–1781.

Atoynatan, T., and Hsiung, G. D., 1969, Epidemiological studies of latent virus infections in captive monkeys and baboons, *Am. J. Epidemiol.* **8:**472–479.

Axthelm, M. K., and Krakowka, S., 1986, Immunocytochemical methods for demonstrating canine distemper virus antigen in aldehyde-fixed paraffin-embedded tissue, *J. Virol. Meth.* **13:**215–229.

Baczko, K., Liebert, U. G., Billeter, M. A., Cattaneo, R., Budka, H., and ter Meulen, V., 1986, Expression of defective measles virus genes in brain tissues of patients with subacute sclerosing panencephalitis, *J. Virol.* **59:**472–478.

Basle, M. F., Russell, W. C., Goswami, K. K. A., Rebel, A., Giraudon, P., Wild, F., and Filmon, R., 1985, Paramyxovirus antigens in osteoclasts from Paget's bone tissue detected by monoclonal antibodies, *J. Gen. Virol.* **66:**2103–2110.

Basle, M. F., Fournier, J. G., Rozenblatt, S., Rebel, A., and Bouteille, M., 1986, Measles virus RNA detected in Paget's disease bone tissue by *in situ* hybridization, *J. Gen. Virol.* **67:**907–913.

Bass, B. L., and Weintraub, H., 1988, An unwinding activity that covalently modifies its double-stranded RNA substrate, *Cell* **55:**1089–1098.

Belshe, R. B., Bernstein, J. M., and Dansby, K. N., 1984, Respiratory syncytial virus, in "Textbook of Human Virology" (R. B. Belshe, ed.), pp. 361–383, PSG Publishing Company, Littleton, MA.

Bevan, M. J., 1987, Class discrimination in the world of immunology, *Nature* **325:**192–194.

Blaese, R. M., and Hofstrand, H., 1975, Immunocompetence of patients with SSPE, *Arch. Neurol.* **32:**494–495.

Bollo, E., Zurbriggen, A., Vandevelde, M., and Faukhauser, R., 1986, Canine distemper virus clearance in chronic inflammatory demyelination, *Acta Neuropathol.* (Berlin) **72:**69–73.

Burger, E. H., Van der Meer, J. W. M., Van de Genel, J. S., Gribnau, J. C., Thesinjh, C. W., and Van Furth, R., 1982, *In vitro* formation of osteoclasts from long term cultures of bone marrow mononuclear phagocytes, *J. Exp. Med.* **156:**1604–1614.

Burnet, F. M., 1968, Measles as an index of immunological function, *Lancet* **ii:**610–613.

Burns, W. H., and Allison, A. C., 1975, Virus infections and the immune response they elicit, in "The Antigens," Vol. III (M. Sela, ed.), pp. 479–574, Academic Press, New York.

Carrigan, D. R., 1987, Chronic relapsing encephalomyelitis associated with experimental measles virus infection, *J. Med. Virol.* **21:**223–230.

Carter, M. J., Willcocks, M. M., and ter Meulen, V., 1983, Defective translation of measles virus matrix protein in subacute sclerosing panencephalitis, *Nature* **305:**153–155.

Casali, P., Rice, G. P. A., and Oldstone, M. B. A., 1984, Viruses disrupt functions of human lymphocytes. Effects of measles virus and influenza virus on lymphocyte-mediated killing and antibody production, *J. Exp. Med.* **159:**1322–1337.

Cattaneo, R., Schmid, A., Eachle, D., Baczko, K., ter Meulen, V., and Billeter, M. A., 1988, Biased hypermutation and other genetic changes in defective measles viruses in human brain infections, *Cell* **55:**255–265.

Chan, S. P. K., 1985, Induction of chronic measles encephalitis in C57BL/6 mice, *J. Gen. Virol.* **66:**2071–2076.

Chang, P. W., and Hsiung, G. D., 1965, Experimental infection of parainfluenza virus type 5 in mice, hamsters and monkeys, *J. Immunol.* **95:**591–601.

Chang, T. W., 1971, Recurrent viral infection (reinfection), *N. Eng. J. Med.* **284:**765–773.

Christensen, P. E., Schmidt, H., Jensen, O., Bang, H. O., Anderson, V., and Jordal, B., 1952, An epidemic of measles in Southern Greenland, 1951: Measles in virgin soil, *Acta Med. Scand.* **144:**313–322.

Christie, K. E., and Haukenes, G., 1983, Measles virus-specific IgM antibodies in sera from patients with chronic active hepatitis, *J. Med. Virol.* **12:**267–272.

Cosby, S. L., Lyons, C., Fitzgerald, S. P., Martin, S. J., Pressdee, S., and Allan, I. V., 1981, The isolation of large and small plaque canine distemper viruses which differ in their neurovirulence for hamsters, *J. Gen. Virol.* **52:**34.

Cosby, S. L., McQuaid, S., Taylor, M. J., Bailey, M., Rima, B. K., Martin, S. J., and Allen, I. B., 1989, Examination of eight cases of multiple sclerosis and fifty six neurological and non-neurological controls for genomic sequences to measles virus, canine distemper virus, simian virus 5 and rubella virus, *J. Gen. Virol.* **70:**2027–2036.

Delage, G., Brochu, P., Pelletier, M., Jasmin, G., and Lapointe, N., 1979, Giant-cell pneumonia caused by parainfluenza viruses, *J. Pediatr.* **94:**426–429.

Delovitch, T. L., Semple, J. W., and Phillips, M. L., 1988, Influence of antigen processing on immune responsiveness, *Immunol. Today* **9:**216–217.

Dhib-Jalbut, S., McFarland, H. F., Mingioli, E. S., Sever, J. L., and McFarlin, D. E., 1988, Humoral and cellular immune responses to matrix protein of measles virus in subacute sclerosing panencephalitis, *J. Virol.* **62:**2483–2489.

Dhib-Jalbut, S., Jacobson, S., McFarlin, D. E., and McFarland, M. F., 1989a, Impaired human leukocyte antigen-restricted measles virus-specific cytotoxic T-cell response in subacute sclerosing panencephalitis, *Ann. Neurol.* **25:**272–280.

Dhib-Jalbut, S., McFarlin, D. E., and McFarland, H. F., 1989b, Measles virus-polypeptide specificity of the cytotoxic T-lymphocyte response in multiple sclerosis, *J. Neuroimmunol.* **21:**205–212.

Domurat, F., Roberts, N. J., Jr., Walsh, E. E., and Dagan, R., 1985, Respiratory syncytial virus infection of human mononuclear leukocytes *in vitro* and *in vivo, J. Infect. Dis.* **152:**895–902.

Fagraeus, A., Orvell, C., Norberg, R., and Norrby, E., 1983, Monoclonal antibodies to epitopes of actin and vimentin obtained by paramyxovirus immunisation, *Exp. Cell Res.* **145:**425–432.

Fishaut, M., Tubergen, D., and McIntosh, K., 1980, Cellular response to respiratory viruses with particular reference to children with disorders of cell-mediated immunity, *J. Pediatr.* **96:**179–186.

Fournier, J.-G., Tardieu, M., Lebon, P., Robain, O., Ponsot, G., Rozenblatt, S., and Bouteille, M., 1985, Detection of measles virus RNA in lymphocytes from peripheral blood and brain perivascular infiltrates of patients with subacute sclerosing panencephalitis, *N. Engl. J. Med.* **313:**910–915.

Fournier, J.-G., Lebon, P., Bouteille, M., Goutieres, F., and Rozenblatt, S., 1986, Subacute sclerosing panencephalitis: Detection of measles virus RNA in appendix lymphoid tissue before clinical signs, *Br. Med. J.* **293:**523–524.

Friedman, J., Buskirk, D., Marino, I. J., Jr., and Zabriskie, J. B., 1987, The detection of brain antigens within the circulating immune complexes of patients with multiple sclerosis, *J. Neuroimmunol.* **14:**1–17.

Fujinami, R. S., and Oldstone, M. B. A., 1979, Antiviral antibody reacting on the plasma membrane alters measles virus expression inside the cell, *Nature* **279:**529–530.

Fujinami, R. S., and Oldstone, M. B. A., 1980, Alterations in expression of measles virus polypeptides by antibodies: Molecular events in antibody-induced antigenic modulation, *J. Immunol.* **125:**78–85.

Gavish, D., Kleinman, Y., Morag, A., and Chajek-Shaul, T., 1983, Hepatitis and jaundice associated with measles in young adults, *Arch. Internal Med.* **143:**674–677.

Germain, R. N., 1986, The ins and outs of antigen processing and presentation, *Nature* **322:**687–689.

Glezen, W. P., Loda, F. A., and Denny, F. W., 1976, The parainfluenza viruses, *In* "Viral Infections of Humans: Epidemiology and Control" (A. E. Evans, ed.), pp. 337–349, Plenum Press, New York.

Goodman, A., and McFarlin, D. E., 1987, Multiple sclerosis, *Curr. Neurol.* **7:**91–128.

Goswami, K. K. A., Cameron, K. R., Russell, W. C., Lange, L. S., and Mitchell, D. N., 1984a, Evidence for the persistence of paramyxoviruses in human bone marrows, *J. Gen. Virol.* **65:**1881–1888.

Goswami, K. K. A., Lange, L. S., Mitchell, D. N., Cameron, K. R., and Russell, W. C., 1984b, Does simian virus 5 infect humans?. *J. Gen. Virol.* **65:**1295–1303.

Goswami, K. K. A., Morris, R. J., Rastogi, S. C., Lange, L. S., and Russell, W. C., 1985, A neutralizing monoclonal antibody against a paramyxovirus reacts with a brain antigen, *J. Neuroimmunol.* **9:**99–108.

Goswami, K. K. A., Randall, R. E., Lange, L. S., and Russell, W. C., 1987, Antibodies against the paramyxovirus SV5 in the cerebrospinal fluids of some multiple sclerosis patients, *Nature* **327:**244—247.

Gotch, F., Rothbard, J., Howland, K., Townsend, A., and McMichael, A., 1987, Cytotoxic T lymphocytes recognise a fragment of influenza virus matrix protein in association with HLA-A2, *Nature* **326:**881–882.

Gresser, I., and Chany, C., 1963, Isolation of measles virus from the washed leukocytic fraction of blood, *Proc. Soc. Exp. Biol. Med.* **113:**695–697.

Gross, P. A., Green, R. H., and McCrea Curren, M. G., 1973, Persistent infection with parainfluenza type 3 virus in man, *Am. Rev. Respir. Dis.* **108**:894–898.

Haase, A. T., Ventura, P., Gibbs, C. J., Jr., and Tourtellotte, W. W., 1981, Measles virus nucleotide sequences: Detection by hybridisation *in situ*, *Science* **212**:672–674.

Hafler, D. A., and Weiner, H. L., 1987, T cells in multiple sclerosis and inflammatory central nervous system disease, *Immunol. Rev.* **100**:307–333.

Hall, W. W., Lamb, R. A., and Choppin, P. W., 1979, Measles and subacute sclerosing panencephalitis virus protein: Lack of antibodies to the M protein in patients with subacute sclerosing panencephalitis, *Proc. Nat. Acad. Sci. USA* **76**:2047–2051.

Hamdy, R. C., 1981, "Paget's Disease of Bone," Praeger, New York.

Hamill, R. J., Baughn, R. E., Mallette, L. E., Musher, D. M., and Wilson, D. B., 1986, Serological evidence against role for canine distemper virus in pathogenesis of Paget's disease of bone, *Lancet* **2**:1399.

Hanninen, P., Arstila, P., Lang, H., Salmi, A., and Panelius, H., 1980, Involvement of the central nervous sytem in acute, uncomplicated measles virus infection, *J. Clin. Microbiol.* **11**:610–613.

Harvey, L., Gray, T., Beneton, M. N., Douglas, D. C., Kanis, J. A., and Russell, R. G. G., 1982, Ultrastructural features of the osteoclast from Paget's disease of bone in relation to a viral aetiology, *J. Clin. Pathol.* **35**:771–779.

Hauser, S. L., Bhan, A. K., Gilles, E., Kemp, M., Kerr, C., and Weiner, H. L., 1986, Immunohistochemical analysis of the cellular infiltrate in multiple sclerosis lesions, *Ann. Neurol.* **19**:578–587.

Holland, J. J., Semler, B. L., Jones, C., Perrault, J., Reid, L., and Roux, L., 1978, Role of DI, virus mutation and host response in persistent infections by envelope RNA viruses, *in* "Persistent Viruses" (J. Stevens, G. Todaro, and C. F. Fox, eds.), pp. 57–73, Academic Press, New York.

Holland, J. J., Kennedy, I. T., Semler, B. L., Jones, C. L., Roux, L., and Grabau, E. A., 1980, Defective interfering RNA viruses and the host-cell response, *in* "Comprehensive Virology," Vol. 16 (H. Fraenkel-Conrat and R. R. Wagner, eds.), pp. 137–192, Plenum Press, New York.

Holland, J., Spindler, K., Horodyski, F., Grabau, E., Nichol, S., and Vandepol, S., 1982, Rapid evolution of RNA genomes, *Science* **215**:1577–1585.

Horta-Barbosa, L., Hamilton, R., Witting, B., Fuccillo, D. A., Sever, J. L., and Vernon, M. L., 1971, Subacute sclerosing panencephalitis: Isolation of suppressed measles virus from lymph node biopsies, *Science* **173**:840–841.

Hsiung, G. D., 1972, Parinfluenza-5 virus. Infection of man and animals, *Prog. Med. Virol.* **14**:241–274.

Huddlestone, J. R., Lampert, P. W., and Oldstone, M. B. A., 1980, Virus lymphocyte interactions: Infection of Tx and T$_H$ subsets by measles virus, *Clin. Immunol. Immunopathol.* **15**:502–509.

Iakovleva, N. V., Pokhodzei, I. V., Tout-Korshinskaia, M. I., and Sukhovskaia, O. A., 1987, Characteristics of viral infections in patients with chronic obstructive bronchitis, *Terapevticheskii Arkhiv* **59**(7):47–50.

Ilonen, J., Reunanen, M., Herva, G., Ziola, B., and Salmi, A., 1980, Stimulation of lymphocytes from subacute sclerosing panencephalitis patients by defined measles virus antigens, *Cell. Immunol.* **51**:201–214.

Jacobson, S., Flerlage, M. I., and McFarland, H. F., 1985, Impaired measles virus specific cytotoxic T cell responses in multiple sclerosis, *J. Exp. Med.* **162**:839–850.

Jarvis, W. R., Middleton, P. J., and Gefand, E. W., 1979, Parainfluenza pneumonia in severe combined immunodeficiency diseases, *J. Pediatr.* **94**:423–425.

Joseph, B. S., and Oldstone, M. B. A., 1975, Immunologic injury in measles virus infection. II. Suppression of immune injury through antigenic modulation, *J. Exp. Med.* **142**:864–876.

Joseph, B. S., Lampert, P. W., and Oldstone, M. B.A., 1975, Replication and persistence of measles virus in defined subpopulations of human leukocytes, *J. Virol.* **16**:1638–1649.

Kalland, K.-H., Endresen, C., Haukenes, G., and Schrumpt, E., 1989, Measles-specific nucleotide sequences and autoimmune chronic active hepatitis, *Lancet* **i**:1390–1391.

Karp, P., Willis, J., and Wilfert, C., 1974, Parinfluenza virus II and the immunocompromised host, *Am. J. Dis. Child.* **127**:592–593.

Knight, V., 1973, Respiratory syncytial viruses, *In* "Viral and Mycoplasmal Infections of the Respiratory Tract" (V. Knight, ed.), pp. 131–140, Lea and Febiger, Philadelphia, Pennsylvania.

Kristensson, K., and Norrby, E., 1986, Persistence of RNA viruses in the central nervous system, *Annu. Rev. Microbiol.* **40**:159–184.

Krugman, S., Gibs, J. P., Friedman, H., and Stone, S., 1965, Studies on immunity to measles, *J. Pediatr.* **66**:471–488.

Lamb, R. A., and Dreyfuss, G., 1989, Unwinding with a vengeance, *Nature* **337**:19–20.

Lampson, L. A., 1987, Molecular basis of the immune response to neural antigens, *Trends Neurosci.* **10**:211–216.

Liebert, U. G., and ter Meulen, V., 1987, Virological aspects of measles virus-induced encephalomyelitis in Lewis and BN rats, *J. Gen. Virol.* **68**:1715–1722.

Liebert, U. G., Livington, C., and ter Meulen, V., 1987, Induction of autoimmune reactions to myelin basic protein in measles virus encephalitis in Lewis rats, *J. Neuroimmunol.* **17**:103–118.

Linnenman, C. C., Rotte, T. C., Schiff, G. M., and Youtsey, J. L., 1972, A seroepidemiologic study of a measles epidemic in a highly immunized population, *Am. J. Epidemiol.* **95**:238–246.

Linnenman, C. C., Hegg, M. E., Rotte, T. C., Plair, J. P., and Schiff, G. M., 1973, Measles IgM response during reinfection of previously vaccinated children, *J. Pediatr.* **82**:798–801.

Lucas, C. J., Ubels-Postma, J. C., Rezee, A., and Galama, J. M. D., 1978, Activation of measles virus from silent infected human lymphocytes, *J. Exp. Med.* **148**:940–952.

Maehlen, J., Olsson, T., Love, A., Klareskog, L., Norrby, E., and Kristensson, K., 1989, Persistence of measles virus in rat brain neurons is promoted by depletion of CD8+T cells, *J. Neuroimmunol.* **21**:149–155.

Mariano, M., and Spector, W. G., 1979, The formation and properties of macrophage polykaryons, *J. Pathol.* **113**:1–19.

Marks, S. C., Jr., 1983, The origin of the osteoclast, *J. Oral Pathol.* **12**:226–256.

McChesney, M. B., and Oldstone, M. B. A., 1987, Viruses perturb lymphocyte functions: Selected principles characterising virus-induced immunosuppression, *Annu. Rev. Immunol.* **5**:279–304.

McChesney, M. B., Fujinami, R. S., Lampert, P. W., and Oldstone, M. B. A., 1986, Viruses disrupt functions of human lymphocytes II. Measles virus suppresses antibody production by acting on β lymphocytes, *J. Exp. Med.* **163**:1331–1336.

McFarlin, D. E., and McFarland, H., 1982, Multiple sclerosis, *N. Engl. J. Med.* **307**:1183–1188.

McMichael, A. J., Gotch, F. M., and Rothbard, J., 1986, HLA B37 determines an influenza A virus nucleoprotein epitope recognised by cytotoxic T lymphocytes, *J. Exp. Med.* **164**:1397–1406.

Mikhalchenkova, N. N., Kniazeva, L. D., and Slepushkin, A. N., 1987, Respiratory syncytial virus infection in chronic bronchitis patients, *Terapevticheskii Arkhiv* **59**(7):50–52.

Mills, B. G., and Singer, F. R., 1976, Nuclear inclusions in Paget's disease of bone, *Science* **194**:201–202.

Mills, B. G., Singer, F. G., Weiner, L. P., and Holst, P. A., 1981, Immunohistological demonstration of respiratory syncytial virus antigens in Paget's disease of bone, *Proc. Nat. Acad. Sci. USA* **78**:1209–1213.

Mills, B. E., Singer, F. R., Werner, L. P., Suffen, S. C., Stabile, E., and Holst, P., 1984, Evidence for both respiratory syncytial virus and measles virus antigens in the osteoclasts of patients with Paget's disease of bone, *Clin. Orthopaed. Rel. Res.* **183**:303–311.

Mitchell, D. N., Porterfield, J. S., Mitcheletti, R., Lange, L. S., Goswami, K. K. A., Taylor, P., Jacobs, J. P., Hockley, D. J., and Salsbury, A. J., 1978, Isolation of an infectious agent from bone marrow of patients with multiple sclerosis, *Lancet* **ii**:387–391.

Muchmore, H. G., Parkinson, A. J., Humphries, J. E., Scott, E. N., McIntosh, D. A., Scott, L. V., Cooney, M. K., and Miles, J. A. R., 1981, Persistent parainfluenza virus shedding during isolation at the South Pole, *Nature* **239**:187–189.

Nordal, H. J., Vandvik, D., and Norrby, E., 1978, Demonstration of electrophoretically restricted virus-specific antibody in serum and cerebrospinal fluid by imprint electroimmunofixation, *Scand. J. Immunol.* **7**:381–396.

Norrby, E., 1978, Viral antibodies in multiple sclerosis, *Prog. Med. Virol.* **24**:1–39.

Norrby, E., 1987, Towards new viral vaccines for man, *Adv. Virus Res.* **32**:1–34.

Norrby, E., Kristensson, K., Brzosko, W. J., and Kapsenberg, J. G., 1985, Measles virus matrix protein detected by immunoflorescence with monoclonal antibodies in the brain of patients with subacute sclerosing panencephalitis, *J. Virol.* **56**:337—340.

O'Driscoll, J. B., and Anderson, D. C., 1985, Past pets and Paget's disease, *Lancet* ii:907–913.

Oldstone, M. B. A., 1989, Viral persistence, *Cell* 56:517–520.

Oldstone, M. B.A., and Fujinami, R. S., 1982, Virus persistence and avoidance of immune surveillance: How measles can be induced to persist in cells, escape immune assault and injure tissues, *in* "Virus Persistence" (B. W. J. Mahy, A. C. Minson, and G. K. Darby, ed.), Cambridge University Press, Cambridge.

Oldstone, M. B. A., and Tishon, A., 1978, Immunologic injury in measles virus infection. IV. Antigenic modulation and abrogation of lymphocyte lysis of virus-infected cells, *Clin. Immunol. Immunopathol.* 9:55–62.

Olsson, J.-E., and Link, H., 1973, Immunoglobulin abnormalities in multiple sclerosis. Relation to clinical parameters: Exacerbations and remissions, *Arch. Neurol.* 28:392–399.

Ono, K., Iwa, N., Kato, S., and Konobe, R., 1970, Demonstration of viral antigens in giant cells formed in monkeys experimentally infected with measles virus, *Biken J.* 13:329–337.

Osunkoya, B. O., Cooke, A. R., Ayeni, O., and Adejumo, T. A., 1974, Studies on leukocyte cultures in measles. II. Detection of measles virus antigens in human leukocytes by immunofluorescence, *Arch. Ges. Virusforsch.* 44:313–322.

Panum, P. L., 1847, Iagttageiser anstillede under maeslingeapidemieu paa faeröernei aaret 1846, *Bibliothek Laeger* 1:270–344.

Parkinson, A. J., Muchmore, H. G., McConnell, T. A., Scott, L. V., and Miles, J. A. R., 1980, Serological evidence for parainfluenzavirus infection during isolation at South Pole Station, Antarctica, *Am. J. Epidemiol.* 112:334–340.

Peart, A. F. W., and Nagler, F. P., 1954, Measles in the Canadian Artic, 1952, *Can. J. Public Health* 45:146–156.

Pringle, C. R., and Eglin, R. P., 1986, Murine pneumonia virus: Seroepidemiological evidence of widespread human infections, *J. Gen. Virol.* 67:975–982.

Rammohan, K. W., McFarland, H. F., and McFarlin, D. E., 1981, Induction of subacute murine measles encephalitis by monoclonal antibody to virus haemagglutinin, *Nature* 290:588–589.

Randall, R. E., and Young, D. F., 1988a, Comparison between parainfluenza virus type 2 and simian virus 5: Monoclonal antibodies reveal major antigenic differences, *J. Gen. Virol.* 69:2051–2060.

Randall, R. E., and Young, D. F., 1988b, Humoral and cytotoxic T cell responses to internal and external structural proteins of simian virus 5 induced by immunization with solid matrix–antibody–antigen complexes, *J. Gen. Virol.* 69:2505–2516.

Randall, R. E., Young, D. F., Goswami, K. K. A., and Russell, W. C., 1987, Isolation and characterisation of monoclonal antibodies to simian virus 5 and their use in revealing antigenic differences between human, canine and simian isolates, *J. Gen. Virol.* 68:2769–2780.

Randall, R. E., Young, D. F., and Southern, J. A., 1988, Immunization with solid matrix–antibody–antigen complexes containing surface or internal virus structural proteins protects mice from infection with the paramyxovirus, simian virus 5, *J. Gen. Virol.* 69:2517–2526.

Rastogi, S. C., Clausen, J., Offner, H., Konat, G., and Fog, T., 1979, Partial purification of MS specific brain antigens, *Acta Neurol. Scand.* 59:281–296.

Rebel, A., Malkani, K., Basle, M., and Bregon, C., 1977, Is Paget's disease of bone a viral infection?, *Calc. Tiss. Res.* 22:283–286.

Rebel, A., Basle, M., Pouplard, A., Malkani, K., Filmon, R., and Lepatezour, A., 1980a, Bone tissue in Paget's disease of bone, *Ultrastruct. Immunocytol. Arthritis Rheumatism* 23:1104–1114.

Rebel, A., Basle, M. F., Pouplard, A., Kouyoumdjian, S., Filmon, R., and Lepatezour, A., 1980b, Viral antigens in osteoclasts from Paget's disease of bone, *Lancet* i:344–346.

Reddy, E. P., Sandberg-Wollheim, M., Mettus, R. V., Ray, P. E., de Freitas, E., and Koprowski, H., 1989, Amplification and molecular cloning of HTLV-1 sequences from DNA of multiple sclerosis patients, *Science* 243:529–533.

Rima, B. K., and Martin, S. J., 1976, Persistent infection of tissue culture cells by RNA viruses, *Med. Microbiol. Immunol.* 162:89–118.

Robbins, S. J., Wrzas, H., Kline, A. L., Tenser, R. B., and Rapp, F., 1981, Rescue of a cytopathic paramyxovirus from peripheral blood leukocytes in subacute sclerosing panencephalitis, *J. Infect. Dis.* 143:396–403.

Roberts, N. J., Jr., 1982, Different effects of influenza virus, respiratory syncytial virus and Sendai virus on human lymphocytes and macrophages, *Infec. Immun.* 35:1142–1146.

Robertson, D. A. F., Zhang, S. L., Guy, E. C., and Wright, R., 1987, Persistent measles virus genome in autoimmune chronic active hepatitis, *Lancet* **2**:9–11.

Robertson, D. A. F., Zhang, S.-L., Fawkner, K., Alveyn, C., Hu, X.-Y., and wright, R., 1990, Measles vaccination may have protected young people from autoimmune chronic active hepatitis, Submitted for publication.

Ruckdeschel, J. C., Graziano, K. D., and Mardiney, M. R., 1975, Additional evidence that the cell-associated immune system is the primary host defence against measles, *Cell. Immunol.* **17**:11–18.

Russell, W. C., 1983, Paramyxovirus and morbillivirus infections and their relationship to neurological disease, *Prog. Brain Res.* **59**:113–132.

Russell, W. C., and Goswami, K. K. A., 1984, Antigenic relationships in the *Paramyxoviridae*— Implications for persistent infections in the central nervous system, *in* "Viruses and Demyelinating Diseases" (C. Mims, M. L. Cuzner, and R. E. Kelly, eds.), pp. 89–99, Academic Press, London.

Russell, W. C., Randall, R. E., and Goswami, K. K. A., 1989, Multiple sclerosis and paramyxoviruses, *Nature* **340**:104.

Ruttkay-Nedecka, S., Rajcani, J., Eleckova, E., and Ruttkay-Nedecka, G., 1987, Infection of the central nervous system of mice by standard Sendai virus, defective interfering Sendai virus and the mixture of both; Comparison of virus multiplication and pathogenicity, *Acta Virol.* **31**:78–82.

Salmi, A., Reunanan, M., Ilonen, J., and Panelius, M., 1983, Intrathecal antibody synthesis to virus antigens in multiple sclerosis, *Clin. Exp. Immunol.* **52**:241–249.

Salonen, R., Ilonen, J., and Salmi, A., 1988, Measles virus infection of unstimulated blood mononuclear cells *in vitro:* Antigen expression and virus production preferentially in monocytes, *Clin. Exp. Immunol.* **71**:224–228.

Scherle, P. A., and Gerhard, W., 1988, Differential ability of B cells specific for external vs internal influenza virus proteins respond to help from influenza virus-specific T-cell clones *in vivo*, *Proc. Nat. Acad. Sci. USA* **85**:4446–4450.

Scheven, B. A. A., Visser, J W. M., and Nijweide, P. J., 1986, *In vitro* osteoclast generation from different bone marrow fractions, including a highly enriched haematopoeitic stem cell population, *Nature* **321**:79–81.

Schneider-Schaulies, S., Liebert, U. G., Baczko, K., Cattaneo, R., Billeter, M., and ter Meulen, V., 1989, Restriction of measles virus gene expression in acute and subacute encephalitis of Lewis rats, *Virology* **171**:525–534.

Sell, K. W., and Ahmed, A., 1975, Humoral and cellular immune responses in patients with SSPE, *Arch. Neurol.* **32**:496.

Sergiev, P. G., Ryazantseva, N. E., and Shroit, I. G., 1960, The dynamics of pathological processes in experimental measles in monkeys, *Acta Virol.* **4**:265–273.

Singer, F. R., 1980, Paget's disease of bone: A slow virus infection?, *Calc. Tiss. Int.* **31**:185–187.

Singer, F. R., and Mills, G. B., 1983, Evidence for viral etiology of Paget's disease of bone, *Clin. Orthopaed. Rel. Res.* **178**:245–251.

Srinivasappa, J., Saegusa, J., Probhakar, B. S., Gentry, M. K., Buchmeier, M. J., Wiktor, T. J., Koprowski, H., Oldstone, M. B. A., and Notkins, A. N., 1986, Molecular mimicry: Frequency of reactivity of monoclonal antiviral antibodies with normal tissues, *J. Virol.* **57**:397–401.

Steinhauser, D. A., and Holland, J. J., 1987, Rapid evolution of RNA viruses, *Annu. Rev. Microbiol.* **41**:409–433.

Sullivan, J. L., Barry, D. W., Lucas, S. J., and Albrecht, P., 1975, Measles infection of human mononuclear cells. I. Acute infections of peripheral blood lymphocytes and monocytes, *J. Exp. Med.* **142**:773–784.

Summers, B. A., Greisen, H. A., and Appel, M. J G., 1983, Does virus persist in the avea in multiple sclerosis, as in canine distemper encephalomyelitis, *Lancet* **ii**:372–375.

ter Meulen, V., and Cartner, M. J., 1982, Morbillivirus persistent infections in animals and man, *in* "Virus Persistence" (W. J. Mahy, A. C. Minson, and G. K. Darby, eds.), pp. 99–132, Cambridge University Press, Cambridge.

ter Meulen, V., and Stephenson, J. R., 1981, The possible role of virus infections in multiple sclerosis and other related demyelinating diseases, *in* "Multiple Sclerosis: Pathology, Diagnoses and Management" (J. F. Haupike, C. W. M. Adams, and W. W. Rourtellotte, eds.), pp. 379–399, Williams and Wilkins, Baltimore, Maryland.

ter Meulen, V., Stephenson, J. R., and Kreth, H. W., 1983, Subacute sclerosing panencephalitis, in "Comprehensive Virology," Vol. 18 (H. Fraenekl-Conrat and R. R. Wagner, eds.), pp. 105–159, Plenum Press, New York.

Toh, B. H., Yildiz, A., Sotelo, J., Osung, O., and Holboraw, E. J., 1979, Viral infections and IgM autoantibodies to cytoplasmic intermediate filaments, Clin. Exp. Immunol. 37:76–82.

Townsend, A. R. M., McMichael, A. J., Carter, N. P., Huddleston, J. A., and Brownlee, G. G., 1984, Cytotoxic T cell recognition of influenza nucleoprotein and hemagglutinin expressed in transfected mouse L cells, Cell 39:13–25.

Traugott, U., Reinherz, E. L., and Raine, C. F., 1983, Multiple sclerosis: Distribution of T cell subsets with active chronic lesions, Science 219:308–310.

Tribe, G. W., 1966, An investigation of the incidence, epidemiology and control of simian virus 5, Br. J. Exp. Pathol. 47:472–479.

Triger, D. R., Kurtz, J. B., MacCallum, F. O., and Wright, R., 1972, Raised antibody titres to measles and rubella viruses in chronic active hepatitis, Lancet i:665–667.

Vandvik, B., and Norrby, E., 1989, Paramyxovirus SV5 and multiple sclerosis, Nature 338:769–992.

Vandvick, B., Norrby, E., Nordal, H., and Degré, M., 1976, Oligoclonal measles virus-specific IgG antibodies isolated by virus immunoadsorption of cerebrospinal fluids, brain extracts, and sera from patients with subacute sclerosing panencephalitis and multiple sclerosis, Scand. J. Immunol. 5:979–992.

Vandvik, B., Sholdenberg, B., Forsgen, M., Steirnstedt, G., Jeanson, S., and Norrby, E., 1985, Long term persistence of intrathecal virus-specific antibody response after herpes simplex virus encephalitis, J. Neurol. 231:307–312.

Waksman, B. H., and Reynolds, W. E., 1984, Multiple sclerosis as a disease of immune regulation, Proc. Soc. Exp. Biol. Med. 175:282–294.

Watson, G. I., 1965, Serological studies on second attacks of measles and rubella, Lancet 1:80–81.

Wear, D. J., and Rapp, F., 1971, Latent measles virus infection of the hamster central nervous system, J. Immunol. 107:1593–1598.

Wechsler, S. L., and Fields, B. N., 1978, Differences between intracellular polypeptides of measles and subacute sclerosing panencephalitis virus, Nature 272:458–460.

Whitaker, J. N., and Snyder, D. S., 1984, Studies of autoimmunity in multiple sclerosis, CRC Crit. Rev. Clin. Neurobiol. 1:45–82.

Wright, P. F., 1984, Parainfluenza viruses, in "Textbook of Human Virology" (R. B. Belshe, ed.), pp. 299–309, PSG Publishing Company, Littleton, MA.

Younger, J. S., and Preble, O. T., 1980, Viral persistence: Evolution of viral populations, in "Comprehensive Virology," Vol. 16 (H. Fraenkel-Conrat and R. R. Wagner, eds.) pp. 73–135, Plenum Press, New York.

Ziola, B., and Smith, R. H., 1988, T cell cross reactivity among viruses of the Paramyxoviridae, Viral Immunol. 1:111–119.

CHAPTER 12

Molecular Biology of Defective Measles Viruses Persisting in the Human Central Nervous System

MARTIN A. BILLETER AND ROBERTO CATTANEO

I. INTRODUCTION

A. General Features of Subacute Sclerosing Panencephalitis (SSPE) and Measles Inclusion Body Encephalitis (MIBE)

SSPE, a disease of the human central nervous system (CNS) clearly triggered by a persisting virus, has been intensively studied despite its rare occurrence, as it may provide insight into the generation of similar, more common human syndromes of suspected viral etiology, such as multiple sclerosis. SSPE develops in children many years after an acute measles virus (MV) infection consequent to unnoticeable latent persistence of the virus. Starting with intellectual deterioration and uncontrolled movements, accompanied by typical electroencephalographic patterns, the disease progresses through steady cerebral degeneration and leads to coma and death within months or years. Several pathological features are typical of the disease [for reviews see Wolinsky and Johnson (1980), Wechsler and Meissner (1982), ter Meulen *et al.* (1983), and Kristensson and Norrby (1986)], including (1) a strong inflammatory response accompanied by massive leukocyte infiltrations and some demyelination; (2) high antibody titers against most MV antigens in the cerebrospinal fluid (CSF) as well as in the serum; and (3) viral ribonucleocapsid structures in the form of inclusion

MARTIN A. BILLETER AND ROBERTO CATTANEO • Institute for Molecular Biology I, University of Zürich, CH 8093 Zürich, Switzerland bodies in different cell types both in the gray and the white matter, in the absence of infectious virus or budding viral particles. The last feature has been proposed to be causally related to a defect of the persisting MV, specifically to the absence or a defect of the matrix (M) protein, which takes a key role in the formation of virions [reviewed by Banerjee (1987)]. The experimental support for this suggestion was first the failure to find antibodies directed against MV M protein in both the CSF and serum of SSPE patients (Wechsler and Fields, 1978; Hall et al., 1979) and then the inability to detect M protein in SSPE brain material (Hall and Choppin, 1981). Although, indeed, multiple defects of the MV M gene have been described in the meantime, it is nevertheless clear now that SSPE cannot be explained simply by the elimination of the MV M protein.

MIBE can develop in immunosuppressed individuals, such as young leukemia patients treated with cytostatic drugs (Roos et al., 1981; Ohuchi et al., 1987). The clinical and virological manifestations closely resemble those typical of SSPE, but the silent phase of MV persistence is generally shorter, and due to the antiproliferative treatment the patients do not mount high antibody titers against MV antigens.

B. Aspects of Disease Development

The development of the fatal diseases might be subdivided into three different aspects: (1) establishment of MV persistence, (2) transition to extensive replication of defective genomes, and (3) pathogenic consequences. It should be stressed that the current understanding of the disease is far from being complete; therefore, the distinction of these aspects is somewhat arbitrary.

1. Establishment of MV Persistence

The establishment of MV persistence might not be a rare event, but actually might be the rule after any acute MV infection. This is mainly suggested by the life-long immune protection typically ensuing from either a normal infection or from immunization with live MV vaccine, which could be easily explained if in some restricted area of the host a subliminal MV replication were permanently established. In principle, the suspected replication could take place anywhere in the body, and the transport of MV genomes to the brain, e.g., through lymphocytes as transporting agent, could take place later. However, the CNS seems particularly suited for subliminal MV replication for reasons outlined in Section I.C.3. In addition, some suggestive evidence for the presence of MV sequences in autopsy brain material has been obtained on the basis of in situ hybridization. Haase et al. (1984) detected MV sequences in 13 of 25 cases of multiple sclerosis (MS) and in 4 of 7 control brains, whereas Cosby et al. (1989) found MV sequences in 2 of 8 MS patients and in 1 of 56 controls. However, Dowling et al. (1986) failed to detect MV RNA in both MS and control brains. In our opinion, the experimental evidence for the presence of MV sequences is not compelling, and in particular, we do not regard the somewhat higher incidence of MV RNA detection in MS patients as very significant. Unfortunately, a potentially more conclusive demonstration of MV sequences by PCR amplification followed by sequence determination has not yielded meaningful results. The great difficulty underlying all these experiments consists in the very local occurrence of potentially ongoing viral replication. The one clear conclusion to be drawn so far is that the total amount of MV RNA in either MS or control brains must be orders of magnitude lower than in SSPE cases.

2. Transition from the Latent State to Extensive Replication and Spread of Defective MV Genomes

In all SSPE and MIBE cases examined by us, representing the final stage of the disease, MV-specific RNA was easily detected (Baczko et al., 1986, 1988; Cattaneo et al., 1987b). It seems highly unlikely that during the long latency, viral replication and transcription occur at a level nearly as high. In fact, a study in which biopsy material was analyzed at various stages of the clinically manifest disease indicated that at relatively early stages of CNS deterioration the level of MV expression is quite low (Haase et al., 1985). Thus, it appears that the progression of the disease is accompanied by involvement of a steadily increasing number of cells in viral replication. Analysis of the implicated MV genomes has shown that they had undergone extensive mutational changes, as discussed below. It seems likely that these genetic alterations of the virus are of prime importance for the late, accelerated spread of persisting MV genomes; a precise knowledge of the viral genotypes present at the end of the disease and of the phenotypic consequences of mutations introduced during persistence appears as a crucial prerequisite for understanding the mechanisms involved in the conversion from a latent into a proliferative, progressive viral persistence.

3. Pathogenic Consequences of Progressing Viral Persistence

MV replication does not directly interfere with housekeeping functions of the host cells, and cytopathic effects such as syncytia formation, typical for an advanced stage of lytic MV infections, are not observed in SSPE brains. It is reasonable to assume that MV replication interferes with specialized cellular functions, such as synthesis of transmitters, receptors, or myelin components, as indicated by the observed limited demyelination. Alternatively, or in addition, it seems likely that autoimmune reactions are stimulated, as suggested by studies with a model system involving rats intracerebrally inoculated with a neuroadapted MV strain (Liebert and ter Meulen, 1988; Dörries et al., 1989). These pathogenic reactions thought to mimic features implicated in SSPE are summarized in more detail elsewhere (Billeter et al., 1989).

C. Experimental Systems

1. Detection of MV Transcripts in Brain Autopsy Material and SSPE-Derived Cell Lines

Investigation of brain autopsy material from SSPE cases allows the characterization of MV genomes and gene products directly implicated in the terminal stage of the disease. Since such material is limited and does not allow analysis of ongoing viral replication, persistently infected cell lines, derived from SSPE brains by cocultivation with fibroblast cell lines normally supporting lytic MV replication, were also used [for reviews see Wechsler and Meissner (1982) and ter Meulen and Carter (1984)]. Although the possibility cannot be excluded that the cell-associated viruses recovered in such cell lines may be biased by selection during the cocultivation procedure and thus do not truly represent the genomes responsible for the disease, we found that very similar defects and a similar overall mutation load were present in the persisting nonproductive viruses analyzed.

RNA blot hybridization permits identification of the mono- and polycistronic viral mRNA species as well as the genomic and antigenomic RNAs. Single-stranded RNA probes are used preferably and inclusion of synthetic RNAs shorter than the monocistronic transcripts allows precise quantitation of the species analyzed (Cattaneo et al., 1987b). The first gene (N) was shown to be transcribed at a fairly high rate (about 7- to 20-fold lower than in lytically infected cell cultures), whereas the distal genes were transcribed with a very reduced efficiency.

The capability of the majority of MV transcripts to encode proteins can be tested by in vitro translation followed by precipitation with monospecific sera or monoclonal antibodies and resolution by electrophoresis. Lack of M and H protein synthesis was shown for several SSPE cases (Baczko et al., 1986). However, for some proteins no specific antibodies suitable for immunoprecipitation exist, most notably for the nonglycosylated form of the fusion (F) protein and for the V protein, consisting of the N-terminal portion of the phosphoprotein (P) followed by a cysteine-rich domain, encoded by edited P transcripts (Cattaneo et al., 1989a). The F protein and most other MV proteins can be detected directly with high sensitivity using immunofluorescence techniques (Hall and Choppin, 1981; Norrby et al., 1985; Baczko et al., 1986; Liebert et al., 1986); however, quantitation is not possible. Exact quantitation is also difficult to achieve by immunoprecipitation of proteins translated in vitro, and only antibody (Western) blotting appears suitable for this purpose (Swoveland and Johnson, 1989).

The most striking feature of SSPE resulting from these investigations is the extensive difference between individual cases. A wide repertoire of mutations bears upon the formation of single viral gene products both at the transcriptional and the translational level, abolishing or reducing below detection sensitivity viral envelope components, or altering their structure such that recognition by monospecific antibodies is hampered. Conversely, N and P proteins are clearly identified in all cases, although different electrophoretic mobilities mainly of P proteins indicate mutational changes.

2. Characterization of MV Genes with Full-Length cDNA Clones

Subsequent to the first cloning of MV-specific sequences using, respectively, polyadenylated RNA from infected cells (Gorecki and Rozenblatt, 1980) or genomic RNA (Billeter et al., 1984), a large variety of measles virus-specific clones has been isolated. DNA sequences representing full-length MV gene transcripts suitable for expression in cell cultures have been recovered either by assembling cloned DNA containing partial sequences or by using different procedures designed for full-length cloning (Alkhatib and Briedis, 1986; Ayata et al., 1989). An experimental protocol for this purpose using specially designed primers (Schmid et al., 1987) has yielded cDNAs of the five major MV transcripts of an SSPE cell line (Cattaneo et al., 1988a) and SSPE and MIBE autopsy material (Cattaneo et al., 1988b). This technique requires only minute amounts of RNA and directly yields very high enriched libraries suitable for the isolation of all desired specific full-length cDNAs simultaneously. Direct in vitro transcription of these clones followed by in vitro translation (Cattaneo et al., 1988b) allows one to verify quickly whether the protein reading frames in the isolated clones are intact. Sequence comparisons of cDNA clones of the MV Edmonston strain isolated in different laboratories showed variations below 0.2% (A. Schmid, unpublished results), indicating that most of the sequence variations (1–3%) identified between clones from different SSPE and MIBE cases are not due to cloning artifacts.

3. Animal Model Systems

A relatively high proportion of Lewis rats infected intracerebrally with a neuroadapted MV strain develop a subacute measles encephalitis (SAME) which resembles SSPE in some of its manifestations (Liebert and ter Meulen, 1987). This system appears very useful to study aspects of the establishment of persistent MV infections in the CNS as well as pathogenic consequences of progressive persistence, but it is not suited to investigation of the transition from latency to progressive viral persistence. Nevertheless, it should be mentioned here that as soon as MV-specific transcripts are detectable in the rat brain, they show quantitative ratios very reminiscent of those observed in SSPE and MIBE autopsy material as mentioned above: transcripts of distal genes are much reduced in proportion to transcripts of the first gene (Schneider-Schaulies et al., 1989). The same phenomenon is observed in some cell lines of CNS origin, but generally not in cell lines from other sources (S. Schneider-Schaulies, unpublished results). Since such altered ratios of MV mRNAs result in decreased expression of viral surface antigens in these infected cells, the viral infection would tend to be hidden from immune attack. Thus, the brain appears particularly predisposed to the establishment of persistent infections, providing a possible explanation for the hypothesized frequent MV latency in human CNS.

It is not clear yet which host factors are responsible for the down-regulation of the distal MV genes in the CNS. Modulation by the immune response is excluded in the initial phase of infection. Modulation of MV infection by interferon has been shown in another animal model system using hamsters (Carrigan and Kabacoff, 1987). The finding that primary cell lines from brain show altered ratios of MV transcripts suggests that some cell types either contain a factor interfering with efficient transcription of the MV genome or lack a factor required for efficient viral transcription in addition to the virus-encoded proteins.

II. ALTERATIONS DOCUMENTED IN PERSISTENT MV GENOMES

A. Basis of Sequence Comparisons

The great difficulty encountered in the definition of mutational changes fixed during the period of persistence is the lack of a reliable MV sequence for comparison, since the sequence of the originally infecting (lytic) virus is not available for any of the SSPE or MIBE cases analyzed. It is not suitable to base comparisons on one sequence of a lytic MV, since differences between lytic MV genomes obviously exist; it would be particularly biased to use the sequence of the extensively passaged, attenuated vaccine Edmonston strain [for an overview see Calain and Roux (1988)], the only one that has been completely sequenced. The most reliable, but clearly not ideal comparison is thus a "consensus" sequence, represented by the nucleotides present most often at each position among all known MV sequences, considering lytic as well as persistent strains. It was possible to construct such a sequence for all five major genes; it should be noted that different numbers of sequences from independent strains are available for the individual genes, including either 6 (*N, P, F,* and *H* genes) or 13 (*M* gene). In the following discussion, nucleotides deviating from the consensus are generally referred to as "mutations," although this term is correct only for an unknown fraction of deviations. For the *L* gene, only partial sequences are generally available and can be compared only with the complete Edmonston strain sequence. The primary data on which all sequence comparisons in the following sections are based are published to a large extent for the *M* genes in Cattaneo *et al.* (1988b) and are made available for the other genes in Cattaneo *et al.* (1989b).

To illustrate our approach, 200 nucleotides, including the start of the 13 known MV *M* genes, are represented in Fig. 1. Inspection of the differences in the individual strains from the consensus sequence reveals that they occur more or less at random; furthermore, in general only a minority of differences are simultaneously present in more than one of the sequences. These common differences are probably due predominantly to strain variances of the original infecting MV and might be mostly neutral in terms of disease development; however, it cannot be excluded that some of them play a pathogenic role. A computer-generated comparison of the *M* genes of lytic and persistent strains, constructed by use of a program to detect ancestral relationships, is presented in Fig. 2. It must be stressed, however, that this scheme should only facilitate the strain overview and does not define the genealogy of the strains. It is striking that the few lytic viruses examined are grouped very closely together, but possibly this reflects more a bias in the strains analyzed rather than a general close relatedness of all lytic viruses. In contrast, the persistent virus disease strains are all far removed, loosely falling into different subgroups. Moreover, contrary to the lytic strains, the persistent strains are located very distantly from branching points, related to the fact that about half of the deviations from the consensus sequence are present only in single strains; most of these "unique" differences are probably due to mutations accumulated during the long period of persistence.

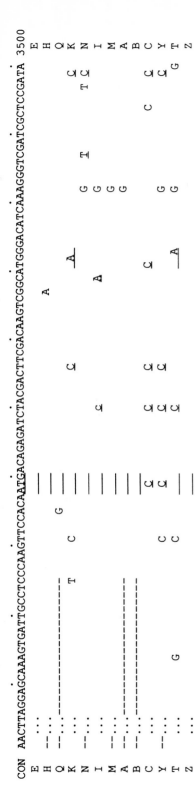

FIGURE 1. Sequences of the first 200 nucleotides of measles virus (MV) *M* genes, including the *P/M* gene boundary. Of the 13 MV strains shown, the first 3 are lytic viruses. The other 10 are defective persisting viruses related to SSPE and MIBE. E: Edmonston (Bellini *et al.*, 1986); H: street virus Human 2 (Curran and Rima, 1988); Q: Toyoshima CAM-RB (Cattaneo *et al.*, 1988b); K: SSPE case K (Cattaneo *et al.*, 1986); N: SSPE line Biken (Ayata *et al.*, 1989); I: SSPE line IP-3-Ca (Cattaneo *et al.*, 1988a); M: SSPE line MF (Cattaneo *et al.*, 1988b); A, B: SSPE cases A and B (Cattaneo *et al.*, 1988b); C: MIBE case C (Cattaneo *et al.*, 1988b); Y: SSPE line Yamagata-1I (Wong *et al.*, 1989); T and Z: SSPE lines Niigata-1 and ZH (Enami *et al.*, 1989). CON: consensus sequence, defined by the nucleotides occurring most frequently in the compiled 13 sequences. The T residues of the cDNAs represent U residues of MV transcripts (plusstrand polarity). Underlined nucleotides indicate replacement sites, specifying amino acids that differ from those specified by the consensus sequence. Underlined codons are translation initiation or termination sites. The second AUG codon is underlined in cases C and Y, where the first AUG has been mutated to ACG. The nontranscribed nucleotides at the *P/M* gene boundary are marked with a group of three dots. Sequences of the primers used for cloning are indicated by dashes. For most strains, several independent cDNA clones have been analyzed; lower case letters indicate nucleotides found only in one clone.

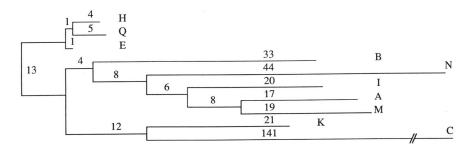

FIGURE 2. Relationships between MV strains visualized as a network. For strain designation the same letters are used as in the legend of Fig. 1. A computer program designed to minimize mutational changes between strains was used. The length of the horizontal portions of the lines connecting the calculated branch points either with other branch points or with the strain endings is proportional to the number of nucleotide differences indicated. [Reproduced with permission from Cattaneo *et al.* (1989b).]

B. Nucleotide Substitutions Due to Polymerase Errors

1. Ratio of Individual Nucleotide Substitutions

Most nucleotide substitutions are presumably due to errors of the polymerase; quantitatively, substitutions prevail by far over deletions or insertions. Among the substitution mutations as defined above (i.e., differences from the consensus sequence) transitions prevail by a factor of almost ten over transversions (Table I), as might be expected for errors due to a relatively error-prone viral replicase [for reviews, see Domingo and Holland (1988) and Steinhauer and Holland (1987)]. No major bias is apparent for individual nucleotide transitions or transversions, although replacements of U residues by C are significantly more frequent than the other transitions, and C to A as well as G to U

TABLE I. Nucleotide Substitution
Frequencies in Measles Virus[a]

Transitions		Transversions			
U → C	166	U → A	8	U → G	9
A → G	103	A → U	6	A → C	14
C → U	124	C → A	31	C → G	9
G → A	117	G → U	24	G → C	8

[a]Number of nucleotides deviating from the consensus sequence (see legend to Fig. 1 for definition of consensus sequence). The compilation of sequences in Cattaneo *et al.* (1989b) was used, omitting those genes in which a hypermutation event was obviously suggested. Identical deviations occurring in more than one sequence are considered only once, to minimize the effect of strain differences present before the onset of latency. Of the 619 substitutions, 510 are transitions and 109 are transversions. Thus, for each nucleotide the transition is on average 9.4 times more frequent than one of the two transversions.

seem to predominate among the transversions. (Note, however, that suspected hypermutation events referred to below in Section II.C have been excluded from the table. The statistically relevant prevalence of U-to-C transitions might in fact be explained on the basis of minor hypermutation events.) The ratios of the nucleotide substitutions recorded probably reflect the ratios of polymerase errors quite faithfully, although obviously a large quantitative difference exists between total misincorporated nucleotides and mutations fixed in the progeny. Despite the fact that a small minority of all misincorporated nucleotides is acceptable for a viable progeny and that only a fraction of selectively neutral misincorporations is expected to be fixed in the progeny, it seems unlikely that this creates a large bias in terms of the nucleotide substitution pattern. Nevertheless, there might be some particular error frequencies of the enzyme which are not apparent from the table. Errors presumably arise at equal frequency during plus- and minus-strand synthesis, but the deviations are recorded for the plus polarity. Thus, nucleotide substitutions involving complementary nucleotides (e.g., the pair of transitions C to U and G to A or the pair of transversions C to A and G to U), expected to be equal statistically, could theoretically arise exclusively by one substitution type of the pair.

2. Distribution of Mutations in Genomes and Individual Genes

Figure 3 presents the distribution of mutations, defined above as deviations of individual MV genomes from the consensus sequence, in sequence blocks of 100 consecutive nucleotides, covering altogether the first five genes, N, P, M, F, and H of the MV genome (E: lytic Edmonston strain; I: SSPE line IP-3-Ca; A and B: SSPE cases A and B; C: MIBE case C). Differences resulting in a different encoded amino acid (replacement site mutations) are represented by the black portions of the columns, whereas silent mutations are represented by the lightly stippled areas; presumed nontranslated regions outside the major coding regions are represented by heavily stippled areas. Apart from the striking accumulation of mutations in the whole M gene of the MIBE case (C) and in a segment of the second half of the H gene of the SSPE line IP-3-Ca (I), which are most likely due to hypermutation events discussed below and which should not be considered here, this representation shows a nonrandom distribution of replacement sites, in contrast to a more uniform distribution of silent sites. Note that the number of mutations accumulating in the long region around the M/F gene border, which is presumably not translated, is particularly high, indicating that the selective pressure to maintain this region is particularly low.

The lytic virus Edmonston (E) also appears to be affected by mutations, of course, although to a lesser degree than the SSPE- and MIBE-derived genomes. Possibly, those mutations which accumulated during persistence might be reflected more faithfully by basing the comparison on a consensus sequence in which only the lytic MV strains are considered; such a representation would lower the number of alterations in the lytic strains, whereas the alterations of the persistent strains would appear increased. However, since our database of lytic viruses is very small and might reflect only a fraction of the circulating viruses, the representation as shown is preferable since it is less biased. In addition, it should be borne in mind that the Edmonston strain is not a wild-

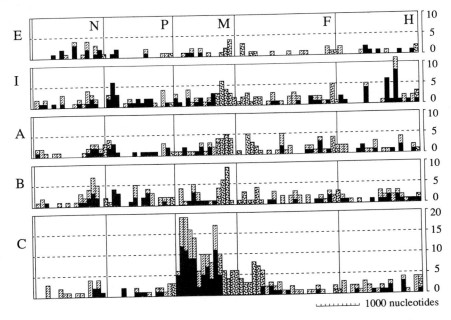

FIGURE 3. Deviations of persistent MV genomes from the consensus sequence. The top panel represents a lytic virus (E: Edmonston); the lower panels depict SSPE-related persisting MV genomes (I: IP-3-Ca; A and B: SSPE cases A and B; C: MIBE case C). Blocks of 100 successive nucleotides are symbolized by the columns. The height of each column (scale indicated on the ordinate on the right) represents the sum of replacement-site mutations (black areas) and silent-site mutations (lightly stippled areas). For segments outside the major coding regions, presumed to be nontranslated, heavily stippled areas are used. It should be noted that 38–41 bases at some gene junctions could not be determined because in monocistronic clones the gene border regions are obscured by the primers used for cloning (Schmid et al., 1987) and bicistronic clones covering the intergenic regions were not obtained throughout.

type virus. Some of its mutations probably reflect the attenuated phenotype obtained by multiple passages through nonnatural hosts and cell lines.

The more or less random accumulation of silent-site mutations in the different genes (comprising about 0.9% nucleotide changes throughout all coding regions) might be expected for an RNA genome which, apart from its coding function, is not subject to extensive structural requirements, given the fact that it is functional as an encapsidated entity. It should be realized that generally in coding regions slightly less than one-third of the nucleotide changes will lead to silent mutations, since almost exclusively the third bases of each codon are possible targets for silent mutations and since the number of transversions is low (Table I) in comparison to transitions, which in the last position of codons result only rarely in amino acid substitutions. The number of 3.2% for (total) nucleotide changes in nontranslated regions is consistent with the view that equal low overall selective pressure operates to satisfy particular RNA structure requirements. In contrast, replacement site mutations, which should amount to about 2% by these considerations if they were not selected against, are quantitatively different for each gene: the M gene shows the heaviest load of replacement mutations (almost 1.2%), followed by a load about equal for *P*, *C*, and *H* (average about 0.8%); *N* and *F* genes show the highest constancy, amounting to less than 0.5% replacement site mutations. (Note that in the case of the *P* gene the replacement site mutations of the *P*, *C*, and *V* reading frames were considered; thus, black areas are exaggerated in the first half of the gene in Fig. 3.) Interestingly, among the different members of the *Paramyxoviridae* a different pattern of protein variability is observed, in the order of *P* > *H* = *N* > *F* = *M* > *L* (Rima, 1989). This appears to the reflect the selective pressures operating on the gene products of lytic viruses, which must be quite different from those relevant for persistent propagation. For the latter, particular proteins are probably dispensable altogether and others might have to maintain only a part of their functional roles. Not much is known about the *L* genes of persistent MV strains; only partial *L* gene sequences of SSPE case K, comprising about 2000 nucleotides, have been determined, and they can be compared only with the Edmonston strain sequence (B. Blumberg, A. Schmid, and R. Cattaneo, unpublished observations). This comparison suggests that in persisting MV as well, the accumulation of replacement site mutations in *L* is the lowest of all MV genes.

Even within the individual coding regions, the replacement site mutations are not distributed randomly, as can be guessed from Fig. 3 and as emerges more clearly from Fig. 4, where the cumulative deviations of the gene products of the four persistent MV strains (I, A, B, and C taken together) from the protein consensus sequences are represented in consecutive regions of 50 amino acids. Increasing darkness of shadings indicates increasing degrees of protein variability. The M protein of the persistent strains show high variability over all its length. Both the P and the C proteins, encoded in different reading frames of the *P* gene, show the highest variability in the amino-terminal region. In contrast, the cysteine-rich region of the V protein, specified by edited *P*-gene transcripts in a different reading frame after the one inserted G residue (Cattaneo *et al.*, 1989a), is remarkably conserved in a region where the *P* reading frame is again quite variable. This suggests an important interaction of this V region with a constant target also in persistent infections. The N protein is quite conserved, except for its carboxyl-terminal region, which might not bind to the RNA, but instead might interact with the P protein, which is also quite variable. Interestingly, in clinical isolates of lytic MV strains, the only variable immunogenic epitopes are clustered in the carboxyl-terminal region of the N protein; no variability detectable by immunological methods was revealed in the M, F, and H proteins (Giraudon *et al.*, 1988; Buckland *et al.*, 1989). The H protein is about as variable as the P protein, although some of the variability in the second half of the molecule must probably be ascribed in part to a hypermutation event. The constancy of the F protein over its entire length except the very carboxyl-terminal region is quite remarkable.

C. Hypermutation of Genome Segments

The M gene of the MIBE case C shows an exceptionally high level of transitions from U to C, about 50%, as read in the plus-strand polarity. Since various cDNA clones derived from this brain did not show more variability

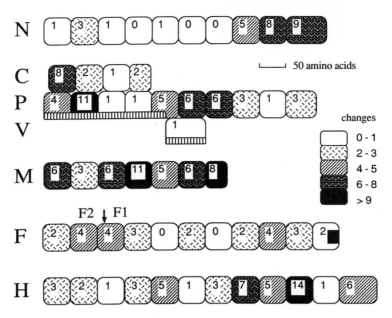

FIGURE 4. Variability of the proteins in persisting MV strains. The derived sequences of the proteins (as indicated on the left) of each of the SSPE cases A and B, the MIBE case C, and the SSPE cell line IP-3-Ca were compared with the consensus protein sequence [six sequences were used for constructing the consensus sequence of proteins N, P/C/V, F, and H, ten sequences for M; for the strains used see Cattaneo *et al.* (1989b), Table I]. For segments of 50 consecutive amino acids, represented as boxes, the sum of all differences of the four sequences from the consensus is given precisely by a number and approximately, to facilitate the overview, by different shading as specified on the right. In case of the M protein, strain K has been taken instead of the hypermutated MIBE M gene; some of the variability in the carboxyl-terminal part of the H protein is due to the supposed segmental hypermutation of the *H* gene in IP-3-Ca. The V protein, identical with the P protein for the first 231 amino acids and differing in the last 68 amino acids from P due to a frameshift caused by editing (insertion of one G residue), is indicated by a ribbon marked with vertical dashes. The small black square at the end of the F protein symbolizes premature termination. [Modified reproduction, with permission, from Cattaneo *et al.* (1989b).]

among themselves than that apparent in other cases, it is highly unlikely that the accumulation of all these mutations occurred in repeated steps of single mutations; a hypermutation in a single step appears much more likely. Initially, we were not able to explain this event other than by a biased polymerase copying the *M* gene segment of the genome (or the antigenome) in a very odd fashion, causing multiple misincorporations of the same kind (Cattaneo *et al.,* 1988b). However, the discovery by Bass and Weintraub (1988) that an enzymatic activity previously termed "double-stranded RNA unwindase" converts up to 50% of the A residues in duplex RNA in inosine (I) provides a much more plausible explanation for the hypermutation event, as shown in Fig. 5 and discussed in more detail elsewhere (Bass *et al.,* 1989). In brief, we suggest the aberrant formation of a double-stranded RNA structure involving the MV genome and an M mRNA. After unwinding/modification, many biased mutations would have arisen concomitantly with the first replication round by introduction of C residues opposite to I residues. The hypermutated MV ge-

FIGURE 5. Model for hypermutation events. *Transcription:* Normally, measles virus transcripts, which are not enwrapped by nucleocapsid protein (gray circles) covering the genomic template, are displaced from the template as they are being synthesized, and are polyadenylated (symbolized by two A residues, light type) at the intergenic boundaries, probably by a slippage mechanism. *"Collapsed" transcription:* Rarely, the newly synthesized transcript might not be unwound from the template, or collapses back on it (the three letters A in bold type represent some of the A residues of the template in the abnormally transcribed region). Alternatively, hybrids might also form between genomes or antigenomes, possibly in regions not completely encapsidated, and short (degraded?) complementary RNA. *Unwinding/modification:* RNA "unwindase" (Bass and Weintraub, 1988) converts some of the A residues in both template and transcript into inosine (I). Only I residues generated in the template are relevant in the further steps. *Replication, plus strand:* In a subsequent round of replication, a full-length plus strand (antigenome) is generated and encapsidated concomitantly; the "collapsed" hybrid strand is thereby removed from the template, and opposite to the I residues of the template usually cytosine (C) residues are incorporated. *Replication, minus strand:* The

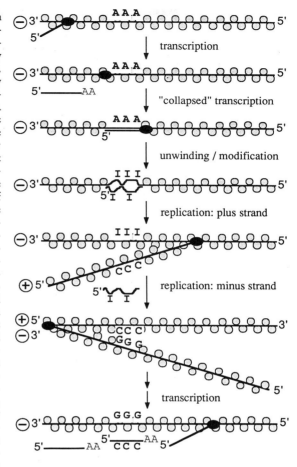

minus strand (genome) copy of the hypermutated antigenome contains G residues in most positions where A residues in the original genomic template were converted to I. *Transcription:* Further transcripts (and antigenomes) contain C in many positions where normal transcripts contained U. [Reproduced with permission from Bass *et al.* (1989)].

nome, defective in M protein function, would then have been selected in the persistently infected brain.

It is interesting to note that other cases of possible hypermutation events have been identified in MV genomes (Figure 6). The *M* genes of three SSPE-derived strains Biken (Ayata *et al.*, 1989), Yamagata-1 (Wong *et al.*, 1989), and ZH (Enami *et al.*, 1989) and of the SSPE case K (Cattaneo *et al.*, 1986) all show clusterings of U to C which are, although not as extensive as in the MIBE case, at least three times higher than the other types of transitions. The Yamagata case is particularly interesting: two substrains differing in the amount of C substitutions exist. Whereas both substrains show the transitions depicted by black columns, only one substrain shows in addition the substitutions indicated by dark stippling. This suggests two superimposed hypermutation events; one probably occurred in the patient, whereas the second most likely

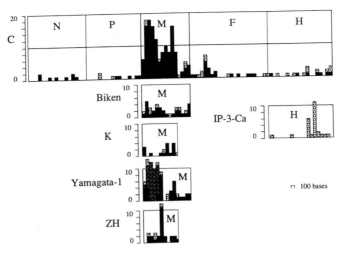

FIGURE 6. Biased hypermutation events in persistent MV genomes. Segments of 100 successive nucleotides are symbolized by the columns. Mutations converting U to C (black portions) and A to G (stippled portions) are represented; the number of mutations is indicated by the ordinate on the left. The antigenome of case C is represented on the top, the M genes of cases Biken, K, Yamagata-1, and ZH underneath the M gene of case C. In the case of Yamagata-1, black areas represent U-to-C mutations present in both Yamagata-1V and Yamagata-1I substrains, whereas heavy shading indicates U-to-C mutations present only in the Yamagata-1I substrain. [Modified reproduction, with permission, from Cattaneo et al. (1989b).]

occurred during propagation of the strain, altering the M gene more thoroughly, but in a more restricted area. Finally, in the H gene of the SSPE-derived strain IP-3-Ca a segment of 276 nucleotides shows conversion of 19 A residues to G, whereas no other type of nucleotide substitution is present. The latter mutations might be explained by formation of a partial duplex of an MV antigenome with a minus-strand segment covering a part of the H gene.

We think that such hypermutation events probably arise also during "normal" lytic infections and are not mechanistically related to a persistent replication type or to localization in the CNS. However, in normal infections, such events, at least if they are as drastic as in the cases mentioned above, cannot become apparent, because such hypermutated genomes could never survive as progeny. As far as we know, the first report of a multiply mutated RNA which might be explained by the proposed mechanism concerns a defective-interfering RNA of vesicular stomatitis virus (O'Hara et al., 1984). If our hypothesis is correct, one might expect to find traces of such hypermutations also in other RNA viruses, characterized by multiple U-to-C or A-to-G mutations clustered in segments of the viral genomes.

Mutations of this kind, on a smaller scale which does not strike the eye, might have been responsible for the unproportionally high number of the apparently "single" replacements of U by C, as mentioned above (Table I). If this suggestion is correct, some of the U-to-C transitions would not be due to polymerase errors. Their prevalence over A-to-G transitions could be explained by the much more abundant synthesis of plus strands in comparison with minus strands, giving a higher probability for plus strands to collapse with minus-strand templates than the opposite. The number of hypothesized hyper-mutation events occurring predominantly in genomes (five cases, in *M* genes) rather than in antigenomes (one case, in gene *H*) is in agreement with this explanation.

D. Nucleotide Deletions

In contrast to the very large number of nucleotide substitutions (more than 600 deviations from the consensus have been identified in the sequences considered in this review; see Table I), only two deletion events have been observed. A deletion of one nucleotide in the *F* gene of SSPE case A shifts the reading frame 26 codons prior to the normal termination codon, causing incorporation of ten "wrong" amino acids before a stop in that reading frame (Cattaneo *et al.*, 1988b). A two-nucleotide deletion in SSPE case K theoretically prolongs the *M* gene reading frame by 17 codons; however this remains without biological consequences, at least in the final version of the persisting MV genome, since the *M* reading frame is interrupted shortly after the initiation codon (Cattaneo *et al.*, 1986). Omission of nucleotides by the viral polymerase appears to be a much less frequent event than nucleotide misincorporation; although deletions in open reading frames would usually be eliminated by selection, only one has been observed in all untranslated regions, where point deletions are expected to be generally less harmful.

E. Nucleotide Insertions

In *Paramyxoviridae*, G residues are inserted at a particular position in the coding region of a large fraction of the P mRNAs, allowing production of an additional protein termed V (Thomas *et al.*, 1988; Cattaneo *et al.*, 1989a). In our view, this editing event is most likely related to a natural "slippage" or "stuttering" function of the enzyme, which probably mediates the poly-adenylation at the 3' terminus of transcripts by repeatedly copying a region of consecutive U residues at the end of each gene (Iverson and Rose, 1981; Gupta and Kingsbury, 1985). A more thorough discussion of paramyxovirus editing is found elsewhere (Thomas *et al.*, 1988; Cattaneo *et al.*, 1989a; also see Chapters 6 and 7), but it is important to note here that the editing function seems to be required both for lytic and persistent infections, since in both instances about half of the P transcripts recovered by cloning are edited.

Despite this proposed "regular" insertion function of the MV transcription complex, only one "irregular" insertion event, probably related mechanistically to the editing function of the enzyme, has been detected. In a cDNA clone derived from the MV Edmonston strain, 19 consecutive C residues were inserted adjacent to a series of five preexisting consecutive C residues in the 5' untranslated region of the *F* gene. The low incidence of insertion mutations can be rationalized by a strict confinement of slippage functions to the transcription complex, whereas the replication complex is devoid of such functions.

F. Formation of Defective-Interfering (DI) RNA

Although in several cell lines persistently infected with MV we have identified small RNA species containing sequences of both polarities derived from the 5' terminal region of the viral genome (A. Schmid and M. A. Billeter, unpublished results), DI RNAs are not a regular occurrence in such cell lines. In some brain specimens, small RNAs were detected with certain probes (K. Baczko, unpublished observations), but we interpret these findings as hybridization with degraded RNA often present in such material. Even if DI RNAs could be clearly identified in SSPE material, a fortuitious association would appear more likely rather than a causal relationship with the development of the disease. In numerous experimental infections of cell lines and also of whole animals, DI RNA has been shown to diminish the impact of the infection. In some vaccine strain material, DI RNA has been detected which might additionally attenuate the properties of the strain (Calain and Roux, 1988), but functional MV vaccine without any DI RNA also exists (L. Roux, personal communication). Since triggering of the disease appears to involve an up-scaling of latent persistence rather than a down-regulation (which had to occur invariably at the establishment of persistence), a dependence of disease development on the presence of DI RNA seems unlikely.

III. ALTERATIONS OF GENE FUNCTIONS

A. Matrix Protein

In most of the cases analyzed the M protein is either completely eliminated or functionally inactivated. Different mechanisms of functional elimination were encountered. The most spectacular case is that of the MIBE patient C, where the massive hypermutation described above abolished both the start and the termination codons and altered the theoretical protein in an additional 73 positions (Cattaneo et al., 1988b). In SSPE Case K and in the SSPE-derived cell-associated viruses Niigata and ZH, the M reading frame is interrupted 11, 11, and 21 codons after the initiation, respectively, as shown in Fig. 1 (Cattaneo et al., 1986; Enami et al., 1989). Moreover, both in case K and in a SSPE-derived cell-associated virus (MF) no monocistronic M gene transcript is formed, as discussed in Section III.E, precluding efficient translation of the M reading frame (Cattaneo et al., 1987a). In the other SSPE-derived lines IP-3-Ca and Biken, a very unstable M protein is produced (Sheppard et al., 1986; Cattaneo et al., 1988a; Ayata et al., 1989; Enami et al., 1989). In the Biken strain, the canceling of the stop codon elongates the protein by eight amino acids, and 15 other amino acid changes render the M protein difficult to recognize by normal hyperimmune sera (Ayata et al., 1989), accounting for the failure to detect the M protein in cells harboring the Biken strain (Sato et al., 1985). Given this fact, the reported absence of M protein in autopsy material from many SSPE cases might in some instances be due to lack of immunological recognition.

It should be mentioned here that in the case of vesicular stomatitis virus the M protein plays an apparently important regulatory role by acting as a transcriptional repressor late in the infection cycle, as first defined on the basis of M protein mutants and more recently investigated at the molecular level mainly by Wagner and his co-workers [reviewed by Wagner (1987) and Banerjee (1987)]. Such a function of M protein has neither been identified in *Paramyxoviridae* nor has it been rigorously excluded. It is interesting to speculate that the possible lack of M repressor function might also be related to the progressive nature of MV persistence in SSPE and MIBE cases.

No particular defect has been identified in the M proteins of SSPE cases A and B, but no functional test has been carried out. Thus, it is not clear whether the functional absence of the M protein, as has been clearly demonstrated in the majority of the cases, is an absolute requirement for the development of SSPE.

B. Fusion Protein

A much smaller number of *F* genes of persistent MV strains has been analyzed by complete sequence determination. Moreover, the analysis of F proteins as produced by *in vitro* translation of autopsy-derived mRNAs has been hampered because this nonglycosylated form of the protein is difficult to immunoprecipitate (Baczko *et al.*, 1986). It seems all the more significant that three of four disease-associated *F* genes analyzed in such detail (SSPE autopsy A and B and the SSPE-derived line IP-3-Ca) show a truncation of the protein (Cattaneo *et al.*, 1989b). This was first suggested by the slightly higher electrophoretic mobility of these proteins obtained by transcription and translation *in vitro* and was then confirmed at the sequence level: in line IP-3-Ca, a newly generated stop codon results in the omission of the last 24 amino acids; in case A, a deletion of one nucleotide results in the replacement of the last 27 amino acids by another 11; and in case B a newly generated stop codon results in the omission of the last 19 amino acids (Cattaneo *et al.*, 1989b). It is unlikely that such important alterations of the cytoplasmic domain of the MV F proteins, comprising normally a total of merely 33 amino acids, has no functional consequences. For the G protein of VSV, the functional equivalent of MV F protein, it has been shown that less important alterations of the short cytoplasmic domain severely interfere with the assembly of virions and/or already with the surface expression of G protein, probably mainly by slowing down its transport through the Golgi apparatus (Whitt *et al.*, 1989). On the other hand, it is probably not fortuitous that among all F proteins investigated not a single alteration affecting the site required for fusion activation by cleavage into F1 and F2 subunits was encountered; the series of hydrophobic residues at the N terminus of F1 thought to initiate the fusion process itself (Glickman *et al.*, 1988) is also strictly conserved. In addition, the whole protein is very well conserved, as mentioned above. Thus, the fusion function appears to be intact in all cases, as it is probably required for cell fusion "from within" to allow the spread of MV genomes through the affected brains. In the MIBE case C, the cytoplasmic domain is not truncated.

With the present state of knowledge, it is not clear whether a functional alteration of the F protein is required for disease development. In any event, in contrast to the M protein, F seems clearly indispensable for the propagation and spread of persisting MV genomes in the brain.

C. Hemagglutinin

The hemagglutinin seems to be more faithfully conserved than M protein, as revealed by the maintenance of the entire reading frame and the relatively small number of replacement site mutations mentioned above. On the other hand, the *H* gene of strain IP-3-Ca, apparently affected by some hypermutation event in one segment, produces a protein which is glycosylated but shows low hemadsorption and no hemagglutination activity (S. A. Udem, unpublished results). The fact that the *H* gene is more conserved than *M* is surprising, in view of the fact that *H* mRNA is expressed at extremely low levels in both SSPE and MIBE brains (Cattaneo et al., 1987b); in addition, this protein might be expected to be altogether superfluous in infections which apparently spread in the brain without involvement of virions. Although it cannot be excluded that the maintenance is fortuitous in the small sample of cases analyzed, our results argue in favor of a requirement of this protein also in persistent infections. The question of whether hemagglutinin might fulfill roles unrecognized so far, other than binding to cell surface receptors, cannot be answered at present.

D. The Transcription–Replication Functions: N, P(C,V), and L Proteins

Since transcription as well as replication of the viral genome have to proceed efficiently both in progressive persistent and in lytic infections, the viral proteins involved in these processes are expected to be conserved. Indeed, L protein, as far as it has been analyzed, appears as the most conserved gene product (B. Blumberg, R. Cattaneo, and A. Schmid, unpublished results), and N is also well conserved, except for its C-terminal portion. P is the most variable of these proteins, in agreement with the general high variability of P in the different paramyxoviruses (Rima, 1989). The two accessory proteins C and V encoded by the *P* gene might also be involved somehow in the complex polymerase functions; their degree of conservation, which is rather low except for the almost strict maintenance of the cysteine-rich portion of V (Thomas et al., 1988; Cattaneo et al., 1989a), follows the overall pattern observed throughout the *Paramyxoviridae* (Rima, 1989).

One peculiarity of the MV transcription machinery, as revealed by the analysis of SSPE case K and of the SSPE-derived cell line MF, should be mentioned here: as outlined above, in one of these cases are P- and M-gene monocistronic transcripts made at a level exceeding 1% of the normal amount, but bicistronic P–M transcripts accumulate at about the usual level of the single-gene transcripts. Surprisingly, no difference in comparison with the Edmonston strain sequence, including 30 nucleotides upstream and 20 nucleotides downstream of the gene boundary, is present in these strains (Cattaneo *et al.*, 1986, 1987a). Furthermore, comparison with additional MV strains which are perfectly normal in terms of *P* and *M* transcription revealed not a single deviation from a functionally competent sequence for more than 100 nucleotides both upstream and downstream of the gene boundary. Thus, the lack of monocistronic transcripts is not due to a *cis*-acting defect, but to the failure of the transcription complex to recognize the *P–M* boundary. This implies that in the case of MV, individual gene boundary regions are recognized differently. Other bicistronic transcripts, particularly the *M–F* transcript, have also been found to be increased considerably at the expense of monocistronic RNAs in SSPE-related measles viruses (Cattaneo *et al.*, 1987b). However, cases as drastic as those described above can probably never be found for other gene boundaries, since the virtually complete inactivation of genes other than *M* will most likely abolish production of progeny virus.

IV. CONCLUSIONS AND OUTLOOK

A. The Suggested Role of Viral Mutations in Disease Development

The main purpose of this review is the characterization of mutations encountered in MV genomes related to SSPE and MIBE, in view of their possible role in the triggering of the diseases. It is possible to advance the hypothesis that all the mutations described here are only a gratuitous byproduct of the diseases and do not play a decisive role in the etiology of these persistent infections. The properties of the mutations described above and the considerations outlined below do not support such a hypothesis. On the other hand, it is likely that additional factors, more difficult to identify than viral mutations, also play an important role in the generation of the disease. The facts that boys are affected at a frequency about three times higher than girls and that rural populations seem to be affected at a higher level than those in metropolitan areas show clearly that both the genetic background of the host and environmental factors are likely to modulate the outcome of the persistent infections.

Why are persisting MV genomes more heavily altered by mutations than lytic viruses, even if the latter can be estimated to have undergone at least as many successive replication cycles as a persisting virus during its latency? The most important factor is probably the relaxed selective constraint imposed on gene products of persisting viruses, which might have to fulfill only a fraction of the functional requirements indispensable for a lytic virus. In addition to virion formation, lytic virus requires abundant liberation of virions from the host, survival outside the host in an adverse environment, and the ability to infect new hosts from the outside. It is probably true that the majority of the mutations accumulating during persistence are indeed irrelevant for the development of the disease. Nevertheless, there is an important argument favoring the concept that certain mutations are of positive selective value and thus a required condition for the diseases: the clonal nature of the viral genomes encountered.

Two observations strongly suggest a clonal origin of the viral genomes replicating in the terminal stage of the disease. First, the hypermutated *M* gene detected in the MIBE case analyzed must have arisen in one single MV genome after the initial infection. It is difficult to argue that the drastic alteration did not enable this genome to prevail over all the competing "normal" genomes. Second, the number of differences between cDNA clones recovered from one case is about ten times lower than the large number of differences between cases. In the absence of any positive selection for the prevalence of one successful genomic clone, the progeny of which then accumulates only a few additional mutations, a much broader distribution of mutations would be expected in cDNA clones recovered from one case.

It is difficult at present to identify properties of mutations required for selection. Hiding the infection from immune attack would obviously constitute such an advantage and would be brought about by virtually abolishing normal viral expression at the cell surface (provided that the infected cells do not efficiently present processed antigens). This might be accomplished by the complete or functional elimination of M protein required for the local clustering and stabilization of the envelope proteins in the cytoplasmic membrane, or by mutation of the segments of the envelope proteins interacting with M. The complete or functional absence of M protein documented in a majority of SSPE and MIBE cases would fit such a scheme. However, it is obvious that a requirement as simple as this can never constitute a condition sufficient for progressive persistence, which most likely demands a number of additional functional modifications, such as a way to fuse cell membranes without recognizable exposure of viral components at the cell surface. In this perspective, disease development would require a combination of functional changes which might not be similar in all cases of disease. Each functional change must be envisaged to be mediated by mutation events which differ from case to case, as suggested by the variety of documented modes by which *M* gene inactivation is achieved. A requirement of multiple combinatorial alterations of gene functions would necessarily result in a low incidence of SSPE; the very low incidence actually observed might be influenced additionally by host and environmental factors.

B. Outlook

Among the very numerous deviations of the individual persisting MV genome sequences from a "consensus" sequence, only a few are characterized unambiguously as mutations conferring particular properties to the phenotype of the affected genome. To obtain more insight into the properties of the various mutated genes, which are at hand in the form of full-length cDNA clones, a functional characterization is absolutely required. A few of these particular properties might be discovered after individual expression of these genes either *in vitro* or *in vivo*. However, a more thorough understanding requires the elaboration of biological tests such as complementation and rescue. The reconstitution of an entire MV genome sequence from cDNA components and the ability to obtain replicating measles virus from it, as recently established in our laboratory (Ballart *et al.*, 1990), appears as a useful start in this direction.

V. REFERENCES

Alkhatib, G., and Briedis, D. J., 1986, The predicted primary structure of the measles virus hemagglutinin, *Virology* **150**:479–490.

Ayata, M., Hirano, H., and Wong, T. C., 1989, Structural defect linked to nonrandom mutations in the matrix gene of Biken strain subacute sclerosing panencephalitis virus defined by cDNA cloning and expression of chimeric genes, *J. Virol.* **63**:1162–1173.

Baczko, K., Liebert, U. G., Billeter, M. A., Cattaneo, R., Budka, H., and ter Meulen, V., 1986, Expression of defective measles virus genes in brain tissues of patients with subacute sclerosing panencephalitis. *J. Virol.* **59**:472–478.

Baczko, K., Liebert, U. G., Cattaneo, R., Billeter, M., Roos, R. P., and ter Meulen, V., 1988, Restriction of measles virus gene expression in measles inclusion body encephalitis, *J. Infect. Dis.* **158**:144–150.

Ballart, I., Eschle, D., Cattaneo, R., Schmid, A., Metzler, M., Chan, J., Pifko-Hirst, S., Udem, S. A., and Billeter, M. A., 1990, Infectious measles virus from cloned cDNA, *EMBO J.* **9**:379–384.

Banerjee, A. K., 1987, Transcription and replication of rhabdoviruses, *Microbiol. Rev.* **51**:66–87.

Bass, B. L., and Weintraub, H., 1988, An unwinding activity that covalently modifies its double-stranded RNA substrate, *Cell* **55**:1089–1098.

Bass, B. L., Weintraub, H., Cattaneo, R., and Billeter, M. A., 1989, Biased hypermutation of viral RNA genomes could be due to unwinding/modification of double-stranded RNA, *Cell* **56**:331.

Bellini, W. J., Englund, G., Richardson, C. D., Rozenblatt, S., and Lazzarini, R. A., 1986, Matrix genes of measles virus and canine distemper virus: Cloning, nucleotide sequences, and deduced amino acid sequences, *J. Virol.* **58**:408–416.

Billeter, M. A., Baczko, K., Schmid, A., and ter Meulen, V., 1984, Cloning of cDNA corresponding to four different measles virus genomic regions, *Virology* **132**:147–159.

Billeter, M. A., Cattaneo, R., Schmid, A., Eschle, D., Kaelin, K., Rebmann, G., Udem, S. A., Sheppard, R. D., Baczko, K., Liebert, U. G., Schneider-Schaulies, S., Brinckmann, U., and ter Meulen, V., 1989, Host and viral features in persistent measles virus infections of the brain, in "Genetics and Pathogenicity of Negative Strand Viruses" (B. W. J. Mahy and D. Kolakofsky, eds.), pp. 356–366, Elsevier, Amsterdam.

Buckland, R., Giraudon, P., and Wild, F., 1989, Expression of measles virus nucleoprotein in *Escherichia coli:* Use of deletion mutants to locate the antigenic sites, *J. Gen. Virol.* **70**:435–441.

Calain, P., and Roux, L., 1988, Generation of measles virus defective interfering particles and their presence in a preparation of attenuated live-virus vaccine, *J. Virol.* **62**:2859–2866.

Carrigan, D. R., and Kabacoff, C. M., 1987, Identification of a nonproductive, cell-associated form of measles virus by its resistance to inhibition by recombinant human interferon, *J. Virol.* **61**:1919–1926.

Cattaneo, R., Schmid, A., Rebmann, G., Baczko, K., ter Meulen, V., Bellini, W. J., Rozenblatt, S., and Billeter, M. A., 1986, Accumulated measles virus mutations in a case of subacute sclerosing panencephalitis: Interrupted matrix protein reading frame and transcription alteration, *Virology* **154**:97–107.

Cattaneo, R., Rebmann, G., Schmid, A., Baczko, K., ter Meulen, V., and Billeter, M. A., 1987a, Altered transcription of a defective measles virus genome derived from a diseased human brain. *EMBO J.* **6**:681–688.

Cattaneo, R., Rebmann, G., Baczko, K., ter Meulen, V., and Billeter, M. A., 1987b, Altered ratios of measles virus transcripts in diseased human brains, *Virology* **160**:523–526.

Cattaneo, R., Schmid, A., Billeter, M. A., Sheppard, R. D., and Udem, S. A., 1988a, Multiple viral mutations rather than host factors cause defective measles virus gene expression in a subacute sclerosing panencephalitis cell line, *J. Virol.* **62**:1388–1397.

Cattaneo, R., Schmid, A., Eschle, D., Baczko, K., ter Meulen, V., and Billeter, M. A., 1988b, Biased hypermutation and other genetic changes in defective measles viruses in human brain infections, *Cell* **55**:255–265.

Cattaneo, R., Kaelin, K., Baczko, K., and Billeter, M. A., 1989a, Measles virus editing provides an additional cysteine-rich protein, *Cell* **56**:759–764.

Cattaneo, R., Schmid, A., Spielhofer, P., Kaelin, K., Baczko, K., ter Meulen, V., Pardowitz, J.,

Flanagan, S., Rima, B., Udem, S. A., and Billeter, M. A., 1989b, Mutated and hypermutated genes of persistent measles viruses which caused lethal human brain diseases, *Virology,* **173:**415–425.

Cosby, S. L., McQuaid, S., Taylor, M. J., Bailey, M., Rima, B., Martin, S. J., and Allen, I. V., 1989, Examination of eight cases of multiple sclerosis and 56 neurological and non-neurological controls for genomic sequences of measles virus, canine distemper virus, simian virus 5 and rubella virus, *J. Gen. Virol.* **70:**2027–2036.

Curran, M. D., and Rima, B. K., 1988, Nucleotide sequence of the gene encoding the matrix protein of a recent measles virus isolate. *J. Gen. Virol.* **69:**2407–2411.

Domingo, E., and Holland, J. J., 1988, High error rates, population equilibrium and evolution of RNA replication systems, *in* "RNA Genetics" (E. Domingo, J. J. Holland, and P. Ahlquist, eds.), pp. 3–36, CRC Press, Boca Raton, Florida.

Dörries, R., Liebert, U. G., and ter Meulen, V., 1989, Comparative analysis of virus-specific antibodies and immunoglobulins in serum and cerebrospinal fluid of subacute measles virus induced encephalomyelitis (SAME) in rats and subacute sclerosing panencephalitis (SSPE), *J. Neuroimmunol.* **19:**339–352.

Dowling, P. C., Blumberg, B. M., Kolakofsky, D., Cook, P., Jotkowitz, A., Prineas, J. H., and Cook, S. D., 1986, Measles virus nucleic acid sequences in human brain, *Virus Res.* **5:**97–107.

Enami, M., Sato, T. A. and Sugiura, A., 1989, Matrix Protein of cell-associated subacute sclerosing panencephalitis viruses, *J. Gen. Virol.* **70:**2191–2196.

Giraudon, P., Jacquier, M. F., and Wild, T. F., 1988, Antigenic analysis of African measles virus field isolates: Identification and localization of one conserved and two variable epitope sites on the NP protein, *Virus Res.* **18:**137–152.

Glickman, R. L., Syddal, R. J., Iorio, R. M., Sheenan, J. P., and Bratt, M. S., 1988, Quantitative basic residue requirements in the cleavage-activation site of the fusion glycoprotein as a determinant of virulence for Newcastle disease virus, *J. Virol.* **62:**354–356.

Gorecki, M., and Rozenblatt, S., 1980, Cloning of DNA complementary to the measles virus mRNA encoding nucleocapsid protein, *Proc. Natl. Acad. Sci. USA* **77:**3686–3690.

Gupta, K. C., and Kingsbury, D. W., 1985, Polytranscripts of Sendai virus do not contain intervening polyadenylate sequences, *Virology* **141:**102–109.

Haase, A. T., Stowring, L., Ventura, P., Burks, J., Ebers, G., Tourtellotte, W. W., and Warren, K., 1984, Detection by hybridization of viral infection of the CNS, *Ann. N.Y. Acad. Sci.* **436:**103–108.

Haase, A. T., Gantz, D., Eble, B., Walker, D., Stowring, L., Ventura, P., Blum, H., Wietgrefe, S., Zupancic, M., Tourtellotte, W., Gibbs, C. J., Jr., Norrby, E., and Rozenblatt, S., 1985, Natural history of restricted synthesis and expression of measles virus genes in subacute sclerosing panencephalitis, *Proc. Natl. Acad. Sci. USA* **82:**3020–3024.

Hall, W. W., and Choppin, P. W., 1981, Measles virus proteins in the brain tissue of patients with subacute sclerosing panencephalitis, *N. Engl. J. Med.* **304:**1152–1155.

Hall, W. W., Lamb, R. A., and Choppin, P. W., 1979, Measles and subacute sclerosing panenecephalitis virus protein: Lack of antibodies to the M protein in patients with subacute sclerosing panenecephalitis, *Proc. Natl. Acad. Sci. USA* **76:**2047–2051.

Iverson, L. E., and Rose, J. K., 1981, Localized attenuation and discontinuous synthesis during vesicular stomatitis virus transcription, *Cell* **23:**477–484.

Kristensson, K., and Norrby, E., 1986, Persistence of RNA viruses in the central nervous system, *Annu. Rev. Microbiol.* **40:**159–184.

Liebert, U. G., and ter Meulen, V., 1987, Virological aspects of measles virus induced encephalomyelitis in Lewis and BN rats, *J. Gen. Virol.* **68:**1715–1722.

Liebert, U. G., Baczko, K., Budka, H., and ter Meulen, V., 1986, Restricted expression of measles virus proteins in brains from cases of subacute sclerosing panencephalitis, *J. Gen. Virol.* **67:**2435–2444.

Liebert, U. G., Linington, C., and ter Meulen, V., 1988, Induction of autoimmune reactions to myelin basic protein in measles virus encephalitis in Lewis rats, *J. Neuroimmunol.* **19:**103–118.

Norrby, E., Kristensson, K., Brzosko, W. J., and Kapsenberg, J. G., 1985, Measles virus matrix protein detected by immune fluorescence with monoclonal antibodies in the brain of patients with subacute sclerosing panencephalitis, *J. Virol.* **56:**337–340.

O'Hara, P. J., Nichol, S. T., Horodyski, F. M., and Holland, J. J., 1984, Vesicular stomatitis virus defective interfering particles can contain extensive genomic sequence rearrangements and base substitutions, *Cell* **36**:915–924.

Ohuchi, M., Ohuchi, R., Mifune, K., Ishihara, T., and Ogawa, T., 1987, Characterization of the measles virus isolated from the brain of a patient with immunosuppressive measles encephalitis, *J. Infect. Dis.* **156**:436–441.

Rima, B. K., 1989, Comparison of amino acid sequences of the major structural proteins of the paramyxo- and morbilliviruses, *in* "Genetics and Pathogenicity of Negative Strand Viruses" (B. W. J. Mahy and D. Kolakofsky, eds.) pp. 254–263, Elsevier, Amsterdam.

Roos, R. P., Graves, M. C., Wollmann, R. L., Chilcote, R. R., and Nixon, J., 1981, Immunologic and virologic studies of measles inclusion body encephalitis in an immunosuppressed host: The relationship to subacute sclerosing panencephalitis, *Neurology* **31**:1263–1270.

Sato, T. A., Fukuda, A., and Sugiura, A., 1985, Characterization of major structural proteins of measles virus with monoclonal antibodies, *J. Gen. Virol.* **66**:1397–1409.

Schmid, A., Cattaneo, R., and Billeter, M. A., 1987, A procedure for selective full length cDNA cloning of specific RNA species, *Nucleic Acids Res.* **15**:3987–3996.

Schneider-Schaulies, S., Liebert, U. G., Baczko, K., Cattaneo, R., Billeter, M. A., and ter Meulen, V., 1989, Restriction of measles virus gene expression in acute and subacute encephalitis of Lewis rats, *Virology* **171**:525–534.

Sheppard, R. D., Raine, C. S., Bornstein, M. B., and Udem, S. A., 1986, Rapid degradation restricts measles virus matrix protein expression in a subacute sclerosing panencephalitis cell line, *Proc. Natl. Acad. Sci. USA* **83**:7913–7917.

Steinhauer, D. A., and Holland, J. J., 1987, Rapid evolution of RNA viruses, *Annu. Rev. Microbiol.* **41**:409–433.

Swoveland, P. T., and Johnson, K. P., 1989, Host age and cell type influence measles virus protein expression in the central nervous system, *Virology* **170**:131–138.

ter Meulen, V., and Carter, M., 1984, Measles virus persistency and disease, *Prog. Med. Virol.* **30**:44–61.

ter Meulen, V., Stephenson, J. R., and Kreth, H. W., 1983, Subacute sclerosing panencephalitis, *Compr. Virol.* **18**:105–159.

Thomas, S. M., Lamb, R. A., and Paterson, R. G., 1988, Two mRNAs that differ by two nontemplated nucleotides encode the amino coterminal proteins P and V of the paramyxovirus SV5, *Cell* **54**:891–902.

Wagner, R. R., 1987, Rhabdovirus biology and infection: An overview, *in* "The Rhabdoviruses" (R. R. Wagner, ed.), pp. 9–74, Plenum Press, New York.

Wechsler, S. L., and Fields, B. N., 1978, Differences between the intracellular polypeptides of measles and subacute sclerosing panencephalitis virus, *Nature* **272**:458–460.

Wechsler, S. L., and Meissner, H. C., 1982, Measles and SSPE viruses: Similarities and differences, *Prog. Med. Virol.* **28**:65–95.

Whitt, M. A., Chong, L., and Rose, J. K., 1989, Glycoprotein cytoplasmic domain sequences required for rescue of a vesicular stomatitis virus glycoprotein mutant, *J. Virol.* **63**:3569–3578.

Wolinsky, J. S., and Johnson, R. T., 1980, Role of viruses in chronic neurological diseases, *Compr. Virol.* **16**:257–296.

Wong, T. C., Ayata, M., Hirano, A., Yoshikawa, Y., Tsuruoka, H., and Yamanouchi, K., 1989, Generalized and Localized Biased Hypermutation Affecting the Matrix Gene of a Measles Virus Strain That Causes Subacute Sclerosing Panencephalitis, *J. Virol.* **63**:5464–5468.

CHAPTER 13

Structure, Function, and Intracellular Processing of the Glycoproteins of *Paramyxoviridae*

Trudy Morrison and Allen Portner

I. INTRODUCTION

The membranes of *Paramyxoviridae* virions are studded with spike structures which can be readily visualized in electron micrographs. Biochemical characterization of these structures has shown that there are two different kinds of spikes on the surfaces of virions. One spike structure serves to attach the virus to the surfaces of the cells and is composed of a protein termed the hemagglutinin-neuraminidase (HN) in the case of the *Paramyxovirus* genus, the hemagglutinin (H) in the case of *Morbillivirus,* and the G protein in the case of *Pneumovirus.* As reflected in the nomenclature, the attachment protein of *Paramyxovirus* also contains a neuraminidase activity not demonstrated for the attachment proteins of the other genera (Scheid *et al.,* 1972). The other spike is composed of the fusion protein and mediates the fusion of the viral membrane with that of the host cell, a step essential to the penetration and uncoating of the viral genetic information (Scheid and Choppin, 1974, 1977). This protein also mediates fusion between an infected cell and an adjacent cell, termed syncytium formation, a property that facilitates cell-to-cell spread of the viral genetic information.

These two glycoproteins are inserted into the plasma membrane of in-

TRUDY MORRISON • Department of Molecular Genetics and Microbiology, University of Massachusetts Medical School, Worcester, Massachusetts 01655.   ALLEN PORTNER • Department of Virology and Molecular Biology, St. Jude Children's Research Hospital, Memphis, Tennessee 38101-0318.

fected cells and are incorporated into virions as the virus buds from the cell surface. For insertion of these proteins into the plasma membrane, the virus utilizes the resident host-cell pathways for the synthesis, transport, and post-translational modifications of surface glycoproteins. This pathway is critical not only for incorporation of the viral glycoproteins into the plasma membrane prior to budding, but also for modifications of the structure of these proteins crucial to their biological function.

Recently, the genes for the attachment protein and the fusion protein of many members of the *Paramyxoviridae* family have been cloned and sequenced. The availability of sequence information from a number of these viruses allows direct comparison of the deduced amino acid sequences and determinants of conformation across the family of viruses. Such a comparison highlights sequences and structural determinants that are likely to be central to the function and intracellular processing of the proteins. In addition, the availability of monoclonal antibodies to many of these proteins has allowed the selection of mutants defective in specific functions. The correlation of specific amino acid changes with altered function also points to important structural properties of the polypeptides critical to function. The expression of cDNA clones of these genes both in cell-free systems and in eucaryotic cells is also revealing a great deal about the synthesis and intracellular processing of these proteins. Applications of monoclonal antibodies and mutational analysis are disclosing structural determinants required for the proper processing of the proteins. This chapter is a discussion of the attachment proteins and fusion proteins of a number of different members of the *Paramyxoviridae* family, with an emphasis on current understanding of the structural basis for function and intracellular processing pathways.

II. PARAMYXOVIRUS ATTACHMENT PROTEINS

A. Structure, Function, and Evolution

1. Primary Sequence Comparisons and Evolution

Early studies divided the *Paramyxoviridae* family into three subgroups based on the biological, antigenic, and biochemical properties of the viruses (Mathews, 1982). Recent sequence analysis of eight attachment protein genes has provided new insight into their protein structure, biological activity, and evolutionary relatedness (Blumberg *et al.*, 1985b; Hiebert *et al.*, 1985; Miura *et al.*, 1985; Satake *et al.*, 1985; Wertz *et al.*, 1985; Alkhatid and Briedis, 1986; Elango *et al.*, 1986; Gerald *et al.*, 1986; Millar *et al.*, 1986; Jorgenson *et al.*, 1987; McGinnes *et al.*, 1987; Tsukiyama *et al.*, 1987; Waxham *et al.*, 1988). In general, comparisons of sequences support the division of the *Paramyxoviridae* into three genera and also suggest evolutionary relationships of the viruses within a genus, as well as between genera (see also Chapters 1–4). Table I shows the percent of identical and most conservative amino acid replacements in matched pairs. Matched pairs are derived from the eight protein sequences aligned simultaneously to maximize sequence homologies. Al-

TABLE I. Conservation of *Paramyxoviridae* Attachment Proteins[a]

	RV	MV	SV	PIV3	NDV	MuV	SV5	RSV
RV		68	32	26	19	18	16	14
MV			37	30	22	19	17	12
SV				62	35	33	30	30
PIV3					43	35	31	22
NDV						47	44	27
MuV							62	39
SV5								60
RSV[b]								

[a]RV, Rinderpest virus; MV, measles virus; SV, Sendai virus; PIV3, parainfluenza virus 3; NDV, Newcastle disease virus; MuV, mumps virus; SV5, simian virus 5; RSV, respiratory syncytial virus. Numbers in the body of the table are the percent of identical plus most conservative (Jimenz-Montano and Zamora-Cortina, 1981) amino acid replacements in the primary sequence of aligned pairs. Matched pairs are derived from the eight protein sequences aligned simultaneously as described in Fig. 2.
[b]To align RSV, the alighment program inserted numerous gaps to accommodate the small size of the RSV G protein (298 amino acids).

though the *Paramyxovirus* genus shows varying degrees of conserved sequences among its members, an interesting pattern emerges. On one end of a spectrum of similarities, the HN proteins of Sendai virus (SV) and parainfluenza virus 3 (PIV3) are highly conserved (62%), while on the other end, conservation of Simian virus (SV5) and mumps virus (MuV) sequences (62%) suggests that these two viruses are most closely related. Interestingly, here are four paramyxoviruses which replicate in mammalian hosts, and yet SV and PIV3 share little sequence similarity with SV5 and MuV. On the other hand, the avian paramyxovirus Newcastle disease virus (NDV) shares considerable sequence similarities with all of the other four virus attachment proteins (Table I). Thus, the NDV sequence bridges the ends of the spectrum of more distantly related sequences.

Surprisingly, comparisons of the RSV G-protein sequences with members of the *Paramyxovirus* suggest that RSV G is related to the SV5 HN sequence (60%) and the MuV HN sequence (39%) (Table I). Applying the same criteria for comparisons, the measles virus (MV) and rinderpest virus (RV) H protein sequences (68% similarity with each other) show some sequence conservation with the HN protein of SV (37%) and PIV3 (30%). However, little or no sequence conservation exists between the H protein of MV and RV and any other attachment proteins (Table I).

The conservation of sequence information of the attachment proteins of the *Paramyxoviridae* is complex; however, they appear to have evolved in two lines of descent from a common progenitor. One lineage is suggested by the conservation of sequences of the MuV, SV5, and RSV attachment proteins, and another by the sequence relatedness of SV, PIV3, MV, and RV attachment proteins. The NDV attachment protein shares sequence similarities with all members of the *Paramyxovirus* genus. Perhaps the NDV sequence is closer to an early progenitor and therefore shares its more primitive HN characteristics with other members of the genus. It has been suggested that the HN proteins of *Paramyxovirus* were formed by a concatenation of the neuraminidase and hemagglutinin gene segments of an influenza virus-like ancestor (Blumberg *et al.*, 1985b). Perhaps this ancestor existed in an avian host, as does NDV, and the mammalian paramyxoviruses are descended from subsequent ancestors whose evolutionary pathway is closer to NDV. Indeed, it is thought that influenza viruses which exist in mammalian species may have been derived from influenza viruses which replicate in avian hosts (Webster and Laver, 1975; Webster *et al.*, 1982).

2. Conserved Structural Determinants

Although there are sequence similarities between some members of different paramyxovirus genera, it is limited. Even within each genus, sequence homologies vary considerably, with some viruses within a genus showing little similarity. However, there is a striking conservation of structural determinants in the attachment proteins of both the *Paramyxovirus* and *Morbillivirus* genera which strongly support their evolutionary relatedness as well as providing clues to structure–function relationships. In comparing the sequence alignment of any family of proteins, low variability at a given residue position indicates structural and functional importance of the position. In particular, structurally important glycine residues, cysteine residues, and hydrophobic regions tend to be well conserved. In families of proteins, glycine residues are comparatively well conserved during evolution. Without side-chain hindrance, glycine residues can adopt unusual dihedral angles generating kinks in the main chain and wind the main chain through tight places in the molecule. In some cases, the absence of a side chain at a conserved glycine position is crucial for the α helices to pack tightly (Schulz and Schirmer, 1979). Similarly, cysteine residues important for folding are well conserved. Thus, the conservation of glycine and cysteine residues at precise locations suggests that these residues are of critical importance in maintaining the structural framework of a protein. The positional conservation of cysteine residues and hydrophobic regions in all attachment proteins of the *Paramyxoviridae* is shown diagrammatically in Fig. 1. When sequences are aligned to maximize homologies as described above, most of the cysteine residues of viruses in the *Paramyxovirus* and *Morbillivirus* genera are clustered into five regions (shown as boxed areas) and align either precisely or within four amino acids. In addition, a strong conservation is shown for three glycine residues located in the two genera at positions that align precisely with the Sendai virus sequence at residues 191, 327, and 409 in Fig. 2.

All the *Paramyxoviridae* attachment proteins have only one long, nonpolar, mainly hydrophobic stretch of amino acids long enough to serve as a transmembrane anchor region. What is unusual is that this transmembrane region is located at the amino terminus of the molecule, and that in virus particles the amino terminus is located inside the virus and the carboxy terminus on the outside (Schuy *et al.*, 1984). Conservation of this structural property along with conservation of cysteine and glycine residues is consistent with the derivation of the attachment protein genes from a common progenitor.

A comparison of the glycosylation sites within the attachment protein sequences suggests that glycosylation sequences have evolved along with the

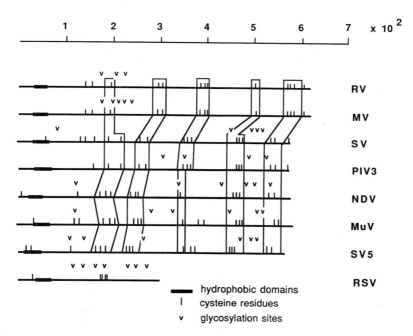

FIGURE 1. Comparisons of structural determinants of *Paramyxoviridae* attachment proteins. Each protein is diagrammed with the amino terminus on the left and the carboxy terminus on the right. Hydrophobic regions are shown as thick lines, cysteine residues as short vertical lines, and glycosylation sites as v's. Potential glycosylation sites in the presumed cytoplasmic domain are not indicated. The scale at the top is ×10² amino acids. When sequences were aligned for similar and identical amino acids using Needleman–Wunsch (1970) analysis (see Fig. 2), the cysteine residues shown in the boxed regions aligned precisely or within four amino acids (see Fig. 2).

division of the family *Paramyxoviridae* into three genera. The *Paramyxovirus* HN proteins have similar locations of glycosylation sites and these sites are distinct from the *Morbillivirus* H proteins, which have their own characteristic pattern of glycosylation (Fig. 1). The differences in the two genera may relate in some way to the presence or absence of neuraminidase. However, the variability in the number and location of glycosylation sites suggests that their conservation is not under strong selection pressure and thus precise positioning may not be an important determinant of structure or function of attachment proteins.

3. Conserved Primary Sequences

In contrast to the conservation of determinants of protein structure, the general overall lack of conservation of sequence between genera supports separation of the *Paramyxoviridae* into three genera. The conservation of the sequences in the HN proteins of members of *Paramyxovirus* provides direct evidence of their relatedness. Similarly, the H proteins of members of *Morbillivirus* are very similar. Figure 2 shows a direct comparison of the sequences of five *Paramyxovirus* HN proteins. The boxed regions enclose the sequences

```
NDV  --------M-DRA--VSQVAL--ENDEREAKNTWRLIF-RIAI-LILT----V---VTLATSV-AS--LVY    46
SV5  --------M--VA--EDA-PV-RA-TCRVLFRTTTLIFLCT-L-LALS----I--SI-LYESL-ITQKQIM    46
MuV  MEPSKLFTISDNATFAPG-PVNNA-ADKKTFRTCFRILVLS-VQ-AVTL--ILV-I-LTLGELVRMINDQGL  64
SV   MDGDRGKRDSYWSTSPSGSTTKLASGWERSSKVDTWLLILSFTQWALSIATVIICI-IISARQGYSTKEYSM  71
PI3  MEYWK-HTNHGKDAGNELETS-MATHGNKITNKITY-ILWTIIILVLLSIVFIIVLINSIKSEKAHESLLQDB 69

NDV  SMGASTPS--DLVGIPTRIS-RAEEKITSALGSNQDVVDRIYKQVALESPLALLNTETTIMNAITSLSYQIN   115
SV5  SQAGSTGS---NSGLGS-ITDLL-NNILS-VA-NQ--II--IY-NSAVALPLQLDTLESTILTAIKSLQTSDK  106
MuV  SNQLSSITDK--I-RES-AT-MI-ASAV-GV-MNQ--V--I-HGVTVSLPLQIEGNQNQLLSTLATICTSKK   123
SV   TVEALNMSSR-EV-KESLTS-LIRQEVI-A-RAVN--II---QSSVQTGIPVLLNKNSRDVIQMIDKSC-SRQ  132
PI3  NNEFMEVITEKIQMASDNI-NDLIQSGV--NTRLL--TII---QSHVQNYIPISLTQQMSDIRKFII-SEITIRN 132

NDV  GAAN-NSGWGA-PIHD-PDFII--GGI-GKELIVDDASDVTSFYPSAFQEHLNFIPAPTTGSGCTRIPSFDMS  181
SV5  LEQN-CSWSAAL-IND-NRYII--NGINQFYFSIAEGRNL-TLGP-LLN-MPSFIIPTATTPEGCTRIPSFSLT 170
MuV  QISN-CSTNIPL-VND-LRFII--NGINKF-IIEDYANHDFSIG-HPLN-MPSFIIPTATSPNGCTRIPSFSLG 187
SV   ELTQLCESTIAVHHAEGIAPLEPHSFWRCPVGEPYLSSDPKI--SLL-LGPSLILSGSTIISGCVRLPSLSIG  201
PI3  D--NQEVPPQRITHDVGIKPLNPDDFWRCTSGLPSLMKTPKI--RLN-PGPGLLAMPTTVDGCVRTPSLVIN   199

NDV  ATHYCYTHNVILSGCRDHSHSHQYLALG-VLRTTATGRIFFSTLRSISLDDTQNRKSCS-VSATPLGCDMLC   251
SV5  KTHWCYTHNVILNGCQDHVSSNQFVSMG-IIEPTSAGFPFFRTLKTLYLSDGVNRKSCS-ISTVPGGCMMYC   240
MuV  KTHWCYTHNVINANCKDHTSSNQYVSMG-IILVQTASGYPMFKTLKIQYLSDGLNRKSCS-IATVPDGCAMYC   258
SV   EAIYAMSSNLITQGCADIGKSYQVLQLGYISLNS-DMFPDLNPVVSHTYDINDNRKSCSVVATGTRGYQL-C   271
PI3  DLIYAYTSNLITRGCQDIGKSYQVLQIGIITVNS-DLVPDLNPRISHTFNINDNRKSCSLALLNTDVYQL-C   269

NDV  SKVTETEEEDYNSAVPTLNAHGRL-GFDGQYHEK-DLDVTTLFEDWVANYPGVGGGSFIDGRVWFSVYGGLK   321
SV5  FVSTQPERDDYFSAAPPEQRIIIM-YYNDTIVER-IINPPGVLDVWATLNPGTGSGVYYLGWVLFPIYGGVI   310
MuV  YVSTQLETDDYAGSSPPTQKLTLL-FYNDTVTER-TISPSGLEGNWATLVPGVGSGIYFENKLIFPAYGGVL   327
SV   SMPTVDERTDYSSDGIEDLVLDVLDLKGSTKSHRYRNSEVDLDHPFSALYPSVGNGIATEGSLIFLGYGG-L   342
PI3  STPKVDERSDYASSGIEDIVLDIVNHDGSISTTRFKNNNISFDQPYAALYPSVGPGIYYKGKIILFLGYGG-L  340

NDV  PNSPSDTVQEGKYVIYKRYNDTCPDEQDYQIRMAKSSYKPGRFGGK-RIQQAILSIKV-STSLGEDPVLTVP   391
SV5  KGTSLWNNQANKYFIPQMVAALCSQNQATQVQNAKSSYYSSWFGNR-MIQSGILACPLRQ-DLTNECLVLPF   380
MuV  PNSTLGVKLAREFFRPVNPYNPCSGPQQDLDQRALRSYPSYLSNR-RVQSAFLVCAWNQILVTNCELVVP-   397
SV   --TT-PL-QG-D-TK--CRTQGCQQVSQDTCNEALK---ITWLGGKQVVNVIIRVNDYLSERPKIRVTTIP-   402
PI3  ---EHPI-N----ENAICNTTGCPGKTQRDCNQA---SHSPWFSDRRMVNSIIVDKGLNSIPKLRVWTI-S   400

NDV  PNTVTLMGAEGRILTVGTSHFLYQRGSSYFSPALLYPMTV--S----MKTATLHSPYTFNAFTRPGSIPGQA   457
SV5  SNDQVLMGAEGRLYMYGDSVYYQRSNSWWPMTMLYKVTITFTN--GQPSAISAQNVPTQQVPRPGTGDCSA   450
MuV  SNNQTLMGAEGRVLLINNRLLYYQRSTSWWPYELLYEISFTFTNS--GQSSVNMSWIPIVSFTRPGSGKCSG   467
SV   ITQN-YLGAEGRLLKLGDRVYIYTR-SSGW-HSQL-QI-GVLDVS--HPLTIN--WTPHEALSRPGNKEQNW   465
PI3  MRQN-YWGSEGRLLLLGNKIYIYTR-STSW-HSKL-QL-GIIDITDYSDIRI--KWTWHNVLSRPGNNECPW   465

NDV  SARCPNSCVTGVYTDPYPL-IFYRNHT--LRGVF-GTMLDSEQARLNPTSAVFDSTSRSRITRVSSSSTKAA   525
SV5  TNRCPGFCLTGVYADAWLLTMPSSTSTFGSEATFTGSYLNTATQRINPTMYIANNTQIISSQQFGSSGQEAA   524
MuV  ENVCPIACVSGVYLDPWPLTPYSHQSGINRRNFYFTGALLNSSTTRVNPTLYVSALNNLKVLAPYGTQGLSAS  539
SV   YNTCPKECISGVYTDAYPLSPDAANVATVTLYANT-SRVNPTIMYSN-TTNI--INMLRIKDVQLEVAYTTI   533
PI3  GHSCPDGCIIGVYTDAYPLNPDGSIVSSVILDSQ-KSRVNPVITYS-TST--ERVNELAIRNKTLSAGYTT-   532

NDV  YTTSTICFKVVKTNKTYCLSIAEIS-NTLFGEFRIVPLLVEILKNDGVREARS
SV5  YGHTTICFRDTGSVMVYCIYIIELSS-SLLGQFQIVP----FIRQ--VTLS
MuV  YTTTICFQDTGDASVVYCVYIMELAS-NIVGEFQILP----VL--TRLTIT
SV   SSCITHFGKGYCFHIIEINQKSLNTLQPMLFKTSIP----KL--CK-AES
P13  TSCIIHYNKGYCFHIVEINHKSLDTFQPMLFKTEIP----K--SCS
```

FIGURE 2. Comparison of the primary amino acid sequences of the HN glycoprotein of the *Paramyxovirus* genus. Identical and conservative amino acid replacements in all five proteins are boxed. Positions of amino acid changes in monoclonal antibody (MAb) escape mutants, temperature-sensitive mutants, and mutants derived by other selective conditions are indicated by triangles. The inhibition of a specific biological activity by the selecting MAb or a decrease in activity in the mutant itself are indicated as follows: neuraminidase activity, △; hemagglutination, ▲; hemagglutination and neuraminidase, ▲; and hemolysis, ●. Sequences were aligned by the Needleman–Wunsch (1970) method as an alignment option of the GENALIGN program. The alignment was with a reduced alphabet to identify common regions (Jimenz-Montano and Zamora-Cortina, 1981). The reduced alphabet used ten symbols to group the most conservative amino acid replacements based on similarities of the functional groups. The ten amino acid groupings were as follows: [P], [E,D], [V, L, I, M], [A, G], [H], [F, Y, W], [S, T], [K,R], [Q,N], [C]. From the reduced alphabet alignment, the corresponding match in the original 20-amino acid alphabet is given. [GENALIGN is a copyrighted software product of IntelliGenetics, Inc.; the program was developed by Dr. Hugo Mastinex of the University of California at San Francisco.]

that are identical or have the most conservative amino acid substitutions. Although conserved primary sequences are found throughout the molecule, Morrison (1988) has noted two regions that show a clustering of conserved sequences. Using the NDV sequence in Fig. 2, one can see one region on the amino-terminal half of the sequence extending from residue 163 to 319, and a second region on the carboxy-terminal side extending from 399 to 476. Within each region are stretches of a particularly impressive concentration of conserved sequences. On the amino-terminal side there are five areas of highly conserved sequences and on the carboxy-terminal side there are two. Particularly noteworthy are the adjacent GC residues (NDV positions 171 and 172) which are conserved at similar positions in the *Morbillivirus* H proteins as well as in the *Paramyxovirus* HN proteins. Further downstream is the conserved sequence NRKSCS, which has been implicated in neuraminidase activity (Jorgensen *et al.*, 1987). The sequence extending from 300 to 319 contains four conserved glycines and one proline, all important structural determinants. The residue at position 305 is another glycine which is conserved in all paramyxovirus attachment proteins, with the exception of the RSV G protein. Much further downstream in a short span of conserved sequences is a third glycine (399), which is conserved in both the HN and H attachment proteins. A final region that shows a high density of conserved sequences lies between residues 449 and 476. Of particular interest is the location of three cysteine residues within a short span of amino acids, which increases the possibility of disulfide linkages. Overall, about 15% of the sequences are conserved in all the HN proteins. This includes ten glycines, eight cysteines, and four prolines, all important determinants of protein structure. Thus, the conservation of these short sequences suggests that they are indeed important structural elements or centers of biological activity. These short sequences are not found in the MV, RV, and RSV sequences. Perhaps this is the structural basis for the lack of a neuraminidase function in the *Morbillivirus* and *Pneumovirus* attachment proteins, since it is likely that some of these conserved sequences form the enzymatic active site.

4. Functional Domains

The attachment proteins are multifunctional molecules that are responsible for binding the virus to sialic acid-containing receptors on host cells (Chapter 15), agglutination of erythrocytes, neuraminidase activity (Scheid *et al.*, 1972; Scheid and Choppin, 1973; Tozawa *et al.*, 1973), antigenic properties, and possibly a role in cell fusion (Miura *et al.*, 1982; Örvell and Grandien, 1982; Heath *et al.*, 1983; Merz and Wolinsky, 1983; Nishikawa *et al.*, 1983; Gitman and Loyter, 1984; Iorio and Bratt, 1984; Portner, 1984; Ray and Compans, 1986; Portner *et al.*, 1987b). Although no direct evidence exists that establishes the structural correlates of these biological properties, the conserved sequences within the paramyxovirus subgroup are beginning to provide some insight. The regions of the HN molecule responsible for hemagglutination and neuraminidase activities have not yet been located on the amino acid sequence. There is some evidence that a single site in the molecule is responsible for both activities (Scheid and Choppin, 1974); however, the preponderance of evidence suggests that cell-binding and neuraminidase activities are associated with independent sites (Smith and Hightower, 1980, 1982; Portner, 1981; Sugawara *et al.*, 1982; Örvell and Grandien, 1982; Nishikawa *et al.*, 1983; Iorio and Bratt, 1984; Ray and Compans, 1986; Portner *et al.*, 1987b). Several investigators have attempted to assign the biological activities of HN to regions of its sequence. Blumberg *et al.* (1985b), using partial sequence homologies and predicted secondary structure similarities between the SV HN and the influenza neuraminidase and hemagglutinin, have predicted that the SV neuraminidase active site lies between amino acid residues 163 and 382 and the hemagglutination receptor pocket between 458 and 547. Jorgensen *et al.* (1987) have proposed that the sequence NRKSCS conserved in five paramyxoviruses has structural similarities to the known structure of the influenza virus neuraminidase, and corresponds to the neuraminidase of HN. In Sendai virus, this region is located at amino acids 254–260 (Fig. 2).

Analysis of mutants selected with monoclonal antibodies that inhibit the various functions of HN, temperature-sensitive mutants defective in hemagglutination and cell binding, and mutants selected with substrate analogues of neuraminidase have helped define regions on the primary sequence involved in the biological activities of HN. Figure 2 shows the location of mutations on the primary sequences of HN. Analysis of SV antibody escape mutants deficient in neuraminidase activity led to the proposal that the neuraminidase active site is located in a region of HN between amino acids 184 and 260 (Thompson and Portner, 1987). This corresponds to the same general region proposed by Blumberg *et al.* (1985b) and Jorgenson *et al.* (1987). NDV escape mutants selected with antibodies that inhibit neuraminidase have amino acid substitutions at residues 193, 194, and 201 (R. M. Iorio, personal communication). These mutations are in the same vicinity as the neuraminidase-deficient mutants of SV. Furthermore, both a neuraminidase-deficient mutant in the HN protein of NDV isolated by Smith and Hightower (1982) and a mumps virus neuraminidase-defective HN protein (Waxham and Aronowski, 1988) have an amino acid change in the very same position, at residue 175 (Fig. 2). All these mutations that effect neuraminidase activity are concentrated in a highly conserved region extending from residue 163 to 210, suggesting that this region forms part of the neuraminidase active site (Fig. 2). Interestingly, all of the mutations are located between the conserved sequences (boxed areas), suggesting that the conserved residues are not readily replaced without losing biological activity. The NDV and MuV mutations at position 175 are particularly interesting because they are located just two residues from a GC sequence which is conserved in both the *Paramyxovirus* and *Morbillivirus* genera. A possible scenario for the structure of the neuraminidase is that the conserved NRKSCS sequence forms the enzyme active site and the long stretch of conserved sequences interlaced with nonconserved sequences from 163 to 210 forms the walls of the neuraminidase sialic acid substrate-binding pocket. Antibodies that inhibit neuraminidase activity select mutations in nonconserved residues which may surround the rim of the pocket. Other mutants with decreased enzyme activity may have substitutions such as that at residue 175 that perturb the structure of the pocket.

Analysis of amino acid changes in escape mutants selected with antibodies that inhibit hemagglutination (SV residues 277 and 279) and a *ts* mutant defective in hemagglutination and binding to host cell receptors (SV residues 262 and 264) has led Thompson and Portner (1987) to place the Sendai virus erythrocyte-binding site close to the carboxy side of the neuraminidase site. However, their results suggest that the host-cell receptor-binding pocket of HN is formed from several regions brought together by protein folding. In studies of the PIV3 HN protein, Coelingh *et al.* (1986, 1987) found amino acid substitutions at numerous positions on the carboxy side of the proposed neuraminidase site (Fig. 2) in escape mutants selected with antibodies that inhibit both hemagglutination and neuraminidase. R. M. Iorio (personal communication) and Yusoff *et al.* (1988) obtained similar results with the NDV HN protein. Thus, these studies with PIV3 and NDV also suggest that the receptor-binding pocket of HN is formed by different regions of the primary sequence brought into proximity by protein folding.

A role for HN in fusion is suggested by studies in which various biochemical and immunological treatments of HN caused the loss of fusion (hemolysis activity), but did not inhibit hemagglutination (Miura *et al.*, 1982; Heath *et al.*, 1983; Portner, 1984; Gitman and Loyter, 1984; Portner *et al.*, 1987b). Thompson and Portner (1987), using a monoclonal antibody that inhibits only fusion, selected an SV escape mutant with a substitution in the HN protein at residue 420 (filled circle, Fig. 2). This site is well separated from those sites involved in hemagglutination and neuraminidase activities. Whether this portion of the HN molecule plays some yet undefined role in the fusion process awaits analysis of other antibody escape mutants. It also has been reported that the bovine PIV3 fusion protein expressed from a vaccinia vector will not mediate fusion unless the cells are coinfected with a vector containing the *HN* gene (Sakai and Shibuta, 1989). In addition, a single amino acid change in the HN sequence, at position 539, severely diminishes the ability of these cells to fuse.

As noted, all of the amino acid substitutions with one exception are outside conserved sequences (Fig. 2). Mutations in the proposed neuraminidase region are between a long segment of closely spaced conserved sequences. However, mutations which are carboxy-terminal to the neuraminidase site in the proposed hemagglutination region are either at residues well removed from or close to highly conserved sequences (SV sequence 277–283 and NDV sequence 449–462). Mutations that map far from conserved sequences may not be part of the hemagglutination active site, but were selected instead by antibodies that sterically interfere with virus attachment. On the other hand, antibodies that select neutralization escape mutants that map close to conserved regions may interfere directly with the receptor-binding pocket. Alternatively, the conserved sequences may represent essential determinants of stability or folding of the molecule which if perturbed may indirectly alter the active site. In addition, mutations that are well removed from conserved sequences may be close to biologically active sites in the three-dimensional structures. The NDV, SV, and PIV3 HN proteins contain several mutations selected with antibodies that inhibit hemagglutination or both hemagglutination and neuraminidase activity. These mutations are close to three conserved cysteine residues which lie in a stretch of conserved sequences (NDV sequence 449–465). The closeness of these cysteine residues suggests that they may form loop structures or stabilize interchain loops (Schulz and Schirmer, 1979). This area may be highly folded and antibodies that bind to this region may destabilize the molecule and disrupt the binding pocket.

5. Functional Domains and Conformation of the Mature Proteins

Typical of transmembrane viral proteins, the attachment proteins of paramyxoviruses have a three-domain structure. This includes a large, hydrophilic, carbohydrate-containing domain external to the membrane, a small, uncharged, and primarily hydrophobic peptide spanning the membrane, and a small, hydrophilic domain internal to the membrane. Although each domain exists in a different environment, it is not known whether they are folded separately to assume a functionally stable structure or if the folding of one domain influences the folding of another.

To understand fully the structural correlates of the binding, neuraminidase, and fusion functions, to expand our understanding of the antigenic structure, and to explore the evolutionary relatedness of the different paramyxovirus genera as well as between the orthomyxoviruses and paramyxoviruses, a description of the attachment protein at the atomic level is necessary. Thus far, the three-dimensional structure has not been established for any paramyxovirus attachment proteins. However, recent studies by Thompson et al. (1988) have led to the production of a soluble form of the SV HN protein which is suitable for crystallization and three-dimensional structural determination. Their analysis of the biological and structural properties of this nonaggregating form of HN has provided new insight into the morphology of the SV HN protein. The soluble form of HN was generated by the enzymatic removal of the hydrophobic membrane-spanning region, creating a new amino terminus at residue 132. The homogeneity of the cleaved product suggests that the HN molecule is composed of a trypsin-sensitive stalk and a trypsin-resistant globular head. Electron microscopy of the solubilized HN suggests that the morphological features of Sendai virus HN and influenza virus neuraminidase may be similar, supporting a possible evolutionary relationship. The overall dimensions of the influenza neuraminidase tetrameric head are 10 by 10 by 6 nm (Varghese et al., 1983). Two-dimensional electron micrographs of the SV HN soluble tetramers reveal a similar size of approximately 10 by 10 nm (Thompson et al., 1988). Thus, the Sendai virus HN tetramers and the influenza virus neuraminidase both have a similar box-shaped head appearance made from four identical subunits. However, the determination of the three-dimensional structure of HN is necessary to confirm structural similarity.

Thompson et al. (1988) have provided evidence that the amino-terminal 131 amino acids are not necessary for the biological and antigenic capabilities of HN dimers and tetramers, and that all the biological activities are located in a proposed globular head region. This is in agreement with mutational analysis in which no neutralization escape mutants isolated so far in four different paramyxoviruses map within the amino-terminal 131 residues (Fig. 2). Thus, the overall morphology of the Sendai virus HN protein is a long, slender stalk of at least 131 residues topped with a globular head structure. The primary purpose of the stalk is to connect the box-shaped head to the virus membrane. The biological activities reside in the globular head structure.

B. Intracellular Processing of the Attachment Protein

1. Introduction

The mature attachment protein which is incorporated into virions with full biological activity is formed through a complex series of steps as it is transported from its site of synthesis to the cell surface (see also Chapter 17). It is likely that some of the conserved structural determinants described above are not only required for maintenance of the mature structure, but are necessary for directing the correct series of steps required for transport of the protein to the cell surface and the proper folding of the protein. Indeed, it is important in considering the function of individual posttranslational modifications to consider not only their importance in the final form of the protein, but also their importance in formation of that final form. Thus, a full understanding of the structural determinants of function requires an understanding of the steps involved in acquiring proper structure.

2. Membrane Insertion

As described above, one feature common to all *Paramyxoviridae* attachment proteins is a hydrophobic region very near the amino terminus and the absence of such a sequence near the carboxy terminus (Fig. 1). This distribution of hydrophobic domains is common to type 2 glycoproteins, which are proteins inserted in membranes with their amino terminus located on the cytoplasmic face and their carboxy terminus located in the lumen of internal membranes or on the external surface of cells or virions (Blobel, 1980; Wickner and Lodish, 1985). By analogy with other such glycoproteins, the hydrophobic domain located near the amino terminus must serve a dual function, that of a signal sequence as well as a membrane anchor (Wickner and Lodish, 1985). Both of these properties of the NDV HN-protein amino-terminal sequences have been directly demonstrated. Using truncated mRNAs derived in a cell-free system by transcribing a truncated cDNA copy of the HN gene, it has been shown that the first 91 amino acids (including the 26-amino acid cytoplasmic tail and the 22-amino acid hydrophobic sequence; Fig. 1) are sufficient to insert (Wilson *et al.*, 1988) and anchor (C. Wilson, unpublished observations) the nascent chain into microsomal membranes in a cell-free system. In addition, elimination of the first 47 amino acids from the sequence by mutating the cDNA templates used for the generation of mRNA results in a truncated protein which will not bind to membranes (C. Wilson and T. Morrison, unpublished observations). This result shows that these amino-terminal sequences are also necessary for membrane insertion. While not yet shown with other paramyxovirus attachment proteins, these properties of the amino terminus of the NDV protein are presumably typical of the family.

Studies of membrane insertion of the NDV HN glycoprotein have shown that the steps in this process are typical of most eucaryotic glycoproteins and therefore are likely to be common to all paramyxovirus attachment proteins. The signal recognition particle (SRP) binds to the amino terminus of the nascent chain, inhibiting further translation until the ribosome–SRP-nascent chain complex interacts with the SRP receptor present in microsomal membranes (Wilson et al., 1987). Upon binding to the SRP receptor, translation resumes and translocation of the protein proceeds through a GTP- and ribosome-dependent step (Wilson et al., 1988), resulting in the insertion of the protein into the rough endoplasmic reticulum membrane with the amino terminus on the cytoplasmic side of the membrane and the carboxy terminus in the lumen (Wilson et al., 1987). Trypsin digestion of vesicles isolated from infected cells results in a slight reduction in the size of the nascent HN protein, consistent with the removal of the short M_r 2000 cytoplasmic tail. Surprisingly, HN protein inserted in microsomal vesicles in a cell-free system is completely resistant to protease digestion (Wilson et al., 1987). This difference from HN protein formed in infected cells may be due to peculiarities of the cell-free system. Alternatively, the cell-free product may represent early stages in folding not seen in pulse-labeled infected cells. Indeed, the product of the cell-free reaction does not contain intramolecular disulfide bonds present in the nascent product made in infected cells (Wilson et al., 1987). This different conformation may influence the accessibility of the cytoplasmic tail to protease.

The specific amino-terminal sequences which direct SRP binding, insertion into, and translocation across the membrane are not well defined in any system. A comparison of the hydrophobic domains of all the Paramyxovirus attachment protein sequences shows no significant sequence homology (Fig. 2). Hydrophobicity plots of this region also show different patterns (McGinnes et al., 1987). Thus, the detailed structural determinants at the amino terminus required for this process are not clear. No mutational analysis of these sequences has been reported.

The cytoplasmic domain of these proteins may play some role in proper membrane insertion. Deletion of the cytoplasmic tail of the NDV HN protein results in less efficient correct membrane insertion and translocation (C. Wilson et al., 1990). In a cell-free system, wild-type protein is translocated across microsomal membranes on average twice as efficiently as the mutant lacking the cytoplasmic tail. Although a small percent of the mutant product does appear to be translocated and glycosylated, the majority of mutant protein molecules are not glycosylated nor translocated across the membrane, thus remaining accessible to protease digestion. However, much of the unglycosylated trypsin-sensitive protein is inserted into membranes in an alkali-resistant and therefore quite stable fashion. These results suggest that in the absence of a cytoplasmic tail, much of the HN protein buries its amino terminus in the membrane and is not translocated.

The cytoplasmic tail may also serve to stabilize the membrane association of the wild-type HN protein. As noted above, a small fraction of NDV HN protein lacking the cytoplasmic tail is properly translocated across the membrane and glycosylated. This population of molecules is much less stably associated with membranes than the wild-type glycosylated protein, as measured by stability to alkaline extraction. While 70% of the glycosylated wild-type protein remains associated with the membrane under alkaline conditions, only 30% of the glycosylated protein missing a cytoplasmic tail remains associated with membranes (C. Wilson, unpublished observations). However, a comparison of the sequences of the cytoplasmic tails of the *Paramyxovirus* attachment proteins does not reveal structural clues to the role of this region of the protein (Fig. 2): there is very little sequence homology. In addition, there are few common properties among all the sequences, except for the presence of positively charged residues. Even the length of this region varies from a maximum of 40 residues in the case of RSV to a minimum of 17 residues in the case of SV5 (Fig. 2).

3. Intracellular Transport of the Attachment Protein

a. Kinetics of Transport

Once inserted into the membrane of the rough endoplasmic reticulum, the attachment protein is transported to the cell surface by the complex pathway used by cellular secretory and membrane proteins. The kinetics of the transport of the attachment protein has been determined in several systems. Sendai virus HN protein was shown to reach the cell surface with a half-time of approximately 40 min (Blumberg *et al.*, 1985b; Bowen and Lyles, 1982), while the NDV HN protein has a half-time of 78 min (Morrison and Ward, 1984). Although the precise kinetics of the mumps virus HN protein was not determined, data presented by Herrler and Compans (1983) also suggest a slow transit of this protein to the cell surface. In the case of the NDV HN protein, it was shown that a significant amount of the transit time is required for the passage of the protein from the rough endoplasmic reticulum (RER) to the medial Golgi membranes, as measured by the time required for the acquisition of resistance to endoglycosidase H (Morrison and Ward, 1984). The reason for the slow transit of the attachment proteins is unclear, although it has been suggested that the orientation of the protein in membranes may somehow play a role (Blumberg *et al.*, 1985b). Slow transit may also be related to the finding by Peluso *et al.* (1978) and Collins and Hightower (1982) that paramyxovirus infection results in induction of the p78 stress protein. This protein has been since identified as Bip (Munro and Pelham, 1986; Hendershot *et al.*, 1988), a protein thought to be involved in the processing of proteins in the RER (Gething *et al.*, 1986; Kozutsumi *et al.*, 1988; Bole *et al.*, 1986). Inhibition of Bip synthesis has been reported to speed the transport of a secreted protein (Dorner *et al.*, 1988). Perhaps increased synthesis of Bip in paramyxovirus infection slows the transit of the viral glycoproteins.

b. Glycosylation

During membrane translocation and subsequent transit through the membrane systems of the cell, oligosaccharides are added and modified (Kornfeld and Kornfeld, 1985; Rose and Doms, 1988). The attachment proteins of the

Paramyxovirus and *Morbillivirus* contain only N-linked carbohydrate side chains (Nakamura *et al.*, 1982; Stallcup and Fields, 1981; Morrison *et al.*, 1981), whereas the G protein of *Pneumovirus* contains both N-linked and O-linked sugars (Wertz *et al.*, 1985; Lambert, 1988). N-linked carbohydrate side chains are first added to glycoproteins as they are translocated (Kornfeld and Kornfeld, 1985; Rose and Doms, 1988; Vidal *et al.*, 1989). *Paramyxoviridae* attachment protein sequences vary from a minimum of three potential N-linked glycosylation sites in the case of rinderpest virus to a maximum of six sites present in NDV, SV5, and mumps sequences (see Fig. 1). In most cases, it is unknown if all sites are utilized. In the case of SV5, it is reported that there are three carbohydrate side chains (Prehm *et al.*, 1979), although there are six potential sites. Indeed, the difference in molecular weights between the glycosylated and unglycosylated forms of the NDV HN protein suggests that not all sites are used (Clinkscales *et al.*, 1978). However, a similar comparison of the two forms of the mumps HN protein suggests that all sites are used (Herrler and Compans, 1983). In no case has it been directly shown which carbohydrate addition sites are used and which are not. Nor are the rules for the use of carbohydrate addition sites known.

Core oligosaccharides added in the RER to nascent chains are potentially susceptible to further processing as the protein moves through the cell (Kornfeld and Kornfeld, 1985). In most paramyxovirus systems studied, a single mature attachment protein molecule contains side chains which are processed to the complex form and side chains which remain as simple high mannose forms as shown by the fact that the protein remains partially endoglycosidase H-sensitive (Schwalbe and Hightower, 1982; Herrler and Compans, 1983; Yoshima *et al.*, 1981; Waxham *et al.*, 1986; Mottet *et al.*, 1986). The factors which determine which side chains are processed and which are not remain undefined. In another system it has been suggested that side chains which remain simple are buried in the secondary structure of the molecule and are inaccessible to processing enzymes (Hsieh *et al.*, 1983). A test of this idea has not been reported for any *Paramyxoviridae* protein. In addition, the importance of the side-chain processing in the transport and final conformation of the protein has not been assessed.

There is abundant evidence that the carbohydrate side-chain composition depends upon the host cell [reviewed in Kornfeld and Kornfeld (1985)]. However, there appears to be some as yet undefined influence of the particular viral protein on the final structure of the sugar side chains in several systems, including a paramyxovirus system (Prehm *et al.*, 1979; Yoshima *et al.*, 1981).

The role of N-linked oligosaccharides in intracellular transport is likely a secondary role related to the effect of the side chains on the conformation of the protein. Indeed, the addition of oligosaccharide side chains to nascent chains as the protein emerges on the lumenal side of the RER may help direct the proper folding of the molecule. Inhibition of glycosylation by tunicamycin treatment of infected cells has had variable results on intracellular transport. The unglycosylated HN protein of NDV is transported through the cells and into virus particles (Morrison *et al.*, 1981). In contrast, in SV, MV, and MuV, virion formation is inhibited by tunicamycin (Herrler and Compans, 1983;

Nakamura *et al.*, 1982; Stallcup and Fields, 1981), although Nakamura *et al.* (1982) showed that the unglycosylated HN protein of Sendai virus was able to migrate from the rough endoplasmic reticulum to smooth membranes in the absence of glycosylation. Unglycosylated protein was not, however, detected at the cell surface. These results are reminiscent of the variable results obtained with different strains of VSV which were attributed to variable effects on the overall conformation of the protein (Gibson *et al.*, 1978; Leavitt *et al.*, 1977; Chatis and Morrison, 1982). Indeed, as discussed above, the positions of the glycosylation sites in the different *Paramyxoviridae* attachment protein sequences are not at all well conserved (Fig. 1). It is possible, therefore, that the positions of carbohydrate side chains have evolved to accommodate changes in the amino acid sequence which slightly alter overall conformation. Alternatively, the precise position of the side chains may not be critical. Resolution of the three-dimensional structure of these proteins will likely shed light on this issue. The role of individual side chains in transport or activity of any *Paramyxoviridae* attachment protein has not yet been reported.

The O-linked sugars found on the RSV G protein are presumably added in the Golgi membranes (Kornfeld and Kornfeld, 1985). The role of these sugars in the transport of this protein is unknown. Further, it is totally unclear why the RSV attachment protein is subjected to this modification while the others are not.

c. Disulfide Bond Formation

Another significant posttranslational modification in most *Paramyxoviridae* attachment proteins is the formation of disulfide-linked oligomers (Markwell and Fox, 1980; Smith and Hightower, 1981). In the NDV system, it has been shown that these oligomers form posttranslationally, but in the RER (T. Morrison, in preparation). In the measles virus system, it has been suggested that oligomer formation occurs in the Golgi membranes (Yamada *et al.*, 1988).

Not all HN protein made in NDV-infected cells is converted to the disulfide-linked form. Furthermore, formation of these disulfide bonds is not a prerequisite for the transport of the protein through the cell, since the non-disulfide-linked form of the NDV HN protein is transported to the cell surface with the same kinetics as the disulfide-linked form (Morrison *et al.*, 1990). Similar results are seen in the mumps system (Herrler and Compans, 1983). Indeed, HN protein expressed from a retrovirus vector (Morrison and McGinnes, 1989) is primarily in the nondisulfide-linked form, but it is transported normally to the cell surface (Morrison and McGinnes, 1989; Morrison *et al.*, 1990). It has been found in the NDV system that the level of expression of the HN protein determines how much of the protein is converted to a disulfide-linked oligomer. Under circumstances where the expression level is low, such as very early in infection, or during expression of a cDNA using a retrovirus vector, 70–80% of the HN protein exists in the nondisulfide-linked form. However, later in infection, when the level of expression is much higher, up to 80% of the HN protein forms disulfide-linked oligomers (Morrison *et al.*, 1990).

The nondisulfide-linked form of the HN protein detected late in infection is not a precursor to the oligomeric form (T. Morrison, et al., 1990). This was shown by determining the kinetics of appearance of antigenic sites on the two forms of the protein in infected cells and by differences in the posttranslational modifications of the two forms. Rather, the nascent protein is folded in two different ways, one resulting in disulfide-linked oligomers and the other in the nondisulfide-linked form. The level of expression determines which pathway is favored. Thus, Fig. 3 shows two different folding pathways for the attachment protein, based on results with the NDV HN protein. These findings have interesting implications for infections in which the level of gene expression is lowered, such as persistent infections, infections with DI particles, and infections with mutants defective in RNA synthesis. The nondisulfide-linked HN protein detected in other viral systems (Mottet *et al.*, 1986; Herrler and Compans, 1983) has not been characterized.

Not all *Paramyxoviridae* attachment proteins exist as disulfide-linked oligomers. The PIV3 HN protein (Jambou *et al.*, 1985; Storey *et al.*, 1984) and the HN proteins of some strains of NDV (Moore and Burke, 1974) do not form disulfide-linked oligomers. However, it is likely that these attachment proteins form oligomers which are held together by noncovalent interactions. Indeed, such nondisulfide-linked oligomers have been detected with a strain of NDV that contains no disulfide-linked oligomers (T. Morrison, in preparation).

It has been found that the strains of NDV which do not form disulfide-linked oligomers are missing a cysteine residue near the hydrophobic transmembrane domain (Shehan *et al.*, 1987). The PIV3 sequence is also missing a comparable cysteine residue (Fig. 1). Thus, Shehan *et al.* (1987) suggested that it is the presence or absence of this residue that determines the formation of the disulfide-linked oligomer. However, protease digestion of the Sendai virus HN oligomer isolated from virions cleaves the protein at a point which would remove the comparable cysteine residue and yet the oligomer remains intact (Thompson *et al.*, 1988). Thus, the presence or absence of this residue cannot totally explain the existence of the disulfide-linked oligomer. Further, the presence or absence of one cysteine residue cannot explain the formation of tetramers reported in the SV system (Markwell and Fox, 1980; Thompson *et al.*, 1988).

The formation of intramolecular disulfide bonds in the attachment protein remains largely unexplored. In other systems, these bonds form on the nascent chain in the RER (Freedman, 1984). Indeed, use of monoclonal antibody has shown that the attachment protein is subjected to several conformational changes during transport. The clearest indication of these changes comes from the finding that the nascent HN proteins of NDV (Nishikawa *et al.*, 1986), of mumps virus (Waxham *et al.*, 1986), and of Sendai virus (Mottet *et al.*, 1986) are unreactive to most monoclonal antibodies and even to polyclonal sera from several sources (T. Morrison, unpublished observations). Only later do the molecules gain reactivity. Furthermore, different antigenic sites of the NDV HN protein appear with slightly different kinetics, suggesting that more than one change occurs (T. Morrison, in preparation). By analogy with the fusion protein (see below), some of these changes may represent folding of the nascent protein accompanied by alterations in intramolecular disulfide bonding (Morrison *et al.*, 1985, 1987). The role of Bip in this process remains to be defined.

d. Fatty Acid Acylation

Another posttranslational modification of glycoproteins is fatty acid acylation, first described as a modification of the VSV G protein (Schmidt and Schlesinger, 1979b). This modification, which occurs in the RER (Berger and Schmidt, 1985) or in *cis* Golgi membranes (Schmidt and Schlesinger, 1980), is not thought to be involved in transport (Rose *et al.*, 1984). Indeed, the HN proteins of NDV (Chatis and Morrison, 1982) and Sendai virus (Veit *et al.*, 1989) do not have this modification. However, the HN of SV5 is fatty acid-acylated (Veit *et al.*, 1989). It was suggested that the presence of cysteine residue on the cytoplasmic leaflet of the membrane is the signal for fatty acid addition (Veit *et al.*, 1989). The variability of addition of fatty acid among very closely related viruses raises the question of a role for this modification. It has been suggested that the fatty acid helps stabilize the glycoprotein in membranes (Schmidt and Schlesinger, 1979a). In this context, it is interesting that SV5 has the shortest cytoplasmic tail of all reported attachment proteins (Fig. 1). Addition of a fatty acid to this molecule could stabilize the interaction of the protein with membranes.

e. Proteolytic Cleavage

Another form of posttranslational modification common to glycoproteins is proteolytic processing. Most paramyxovirus attachment proteins are not subjected to this change. They do not even contain a cleavable signal sequence. However, some strains of NDV synthesize the HN protein as a precursor HN_0, which is processed to HN by the removal of approximately 90 amino acids from the carboxy terminus of the protein (Nagai *et al.*, 1976; Schuy *et al.*, 1984). These NDV strains are missing a polypeptide chain termination signal in the *HN* gene sequence that is present in other NDV strains (Sato *et al.*, 1987a; Gotoh *et al.*, 1988; Gorman *et al.*, 1988). Processing of HN_0 requires the addition of protease to activate hemagglutinating activity. Thus, this modification is related to the activity of the protein.

4. Summary of Attachment Protein Intracellular Processing

Figure 3 summarizes current understanding as described above of the steps involved in the intracellular processing of *Paramyxoviridae* attachment proteins. Not all steps have been explored for all proteins; thus, the figure is an amalgam of data from a number of different systems. In addition, two pathways are shown based on results with the NDV HN protein. As indicated above, there are many questions to be answered concerning the steps in this pathway and the function of the numerous posttranslational modifications. However, because of the structural similarities in this family of proteins, it is likely that many of the general principles established for one protein will apply to the others.

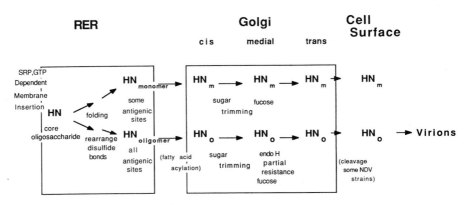

FIGURE 3. Intracellular processing of the *Paramyxoviridae* attachment proteins. A summary of current understanding of the steps involved in the processing and transport of the attachment proteins of *Paramyxoviridae*. Not all steps have been shown for all viruses. Steps shown in parentheses are known not to be common to all attachment proteins. Most of the information on which this diagram is based comes from studies of the HN protein of the paramyxovirus subgroup; thus, the designation HN is utilized. Two pathways are shown, based on work with the NDV HN protein (see text). SRP, signal recognition particle; endo H., endoglycosidase H; RER, rough endoplasmic reticulum; HN_m, HN monomer; HN_o, HN disulfide-linked oligomer.

III. PARAMYXOVIRUS FUSION PROTEINS

A. Structure, Function, and Evolution

1. Sequence Comparisons and Evolution

The other surface protein of *Paramyxoviridae* mediates fusion between the virus and host-cell membrane and between an infected cell and an adjacent cell (Bratt and Gallagher, 1969). The fusion (F) proteins are synthesized as an inactive precursor F_0, which is subsequently cleaved to produce the biologically active form of two disulfide-linked subunits F_1 and F_2 (Homma and Ohuchi, 1973; Scheid and Choppin, 1973, 1974, 1977; Hardwick and Bussell, 1978; Kohama *et al.*, 1981; Hsu *et al.*, 1983).

A comparison of the fusion protein sequence similarities (Table II) and structural determinants (Fig. 4) reveals a significant pattern of relatedness. Fusion protein sequences have been conserved to a much greater extent across the entire *Paramyxoviridae* than among the attachment proteins (Fig. 5 and Table II). This conservation suggests that either the fusion genes have diverged more recently from a common progenitor than the attachment proteins or the fusion proteins are less tolerant to variation. The sequences of nine fusion proteins have been determined (Collins *et al.*, 1984; Paterson *et al.*, 1984b; Blumberg *et al.*, 1985a; Elango *et al.*, 1985; Miura *et al.*, 1985; Chambers *et al.*, 1986; McGinnes and Morrison, 1986; Richardson *et al.*, 1986; Shioda *et al.*, 1986; Spriggs *et al.*, 1986; Barrett *et al.*, 1987; Waxham *et al.*, 1987; Hsu *et al.*, 1988; Tsukiyama *et al.*, 1988).

Table II compares the conservation of fusion protein sequences among all members of the *Paramyxoviridae* that have been sequenced. In *Paramyxovirus*, the fusion proteins are highly conserved between MuV and SV5 (60%) and SV

TABLE II. Conservation of *Paramyxoviridae* Fusion Proteins[a]

	MuV	SV5	NDV	RV	MV	SV	PIV3	RSV
MuV		60	48	40	38	33	35	23
SV5			53	42	40	35	34	23
NDV				50	46	40	39	26
RV					87	47	41	28
MV						48	43	27
SV							64	31
PIV3								43
RSV								

[a]See footnote to Table I for key to virus abbreviations. Values are given as the percent of identical and most conservative amino acid replacements in the primary sequence of aligned pairs. Matched pairs are derived from the eight protein sequences aligned simultaneously and described in Fig. 2.

and PIV3 (64%). The conservation is considerably less between these pairs of viruses, suggesting a more distant evolutionary origin. This result is reminiscent of what was seen in the attachment proteins; members of the paramyxovirus group that show a high degree of conserved sequences between their attachment proteins also have closely related fusion protein sequences. The

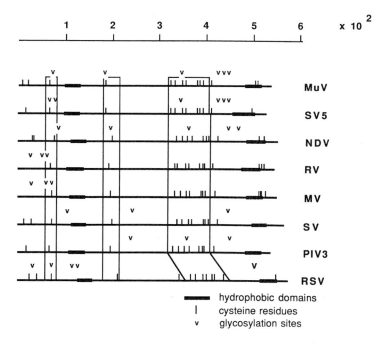

FIGURE 4. Comparisons of structural determinants of *Paramyxoviridae* fusion proteins. Each protein is diagrammed with the amino terminus on the left and the carboxy terminus on the right. Hydrophobic regions are shown as thick lines, cysteine residues as short vertical lines, and glycosylation sites as v. Potential glycosylation sites in the presumed cytoplasmic tail are not indicated. The scale at the top is ×10² amino acids. When sequences were aligned for similar and identical amino acids using Needleman–Wunsch (1970) analysis (see Fig. 5), the cysteine residues shown in the boxed regions aligned precisely or within four amino acids, except the last cysteine residue in the RSV sequence, which is within seven amino acids of the other sequences (see Fig. 5).

NDV sequence shows considerable conservation with both the *Paramyxovirus* (39–53%) and the *Morbillivirus* (46–50%) genera. The *Morbillivirus* MV and RV also show the same high degree of conservation of fusion protein sequences (87%) that was seen in their attachment proteins (68%). Perhaps the fusion proteins are not under the same selection intensity as the attachment proteins or, alternatively, the fusion proteins may not be able to tolerate changes as readily as the attachment proteins without disrupting their structure and/or function.

2. Conserved Structural Domains

Even more prominent is the conservation of residues which tend to be important determinants of structure and/or function. Figure 4 illustrates how precise this conservation is with respect to the position of cysteine residues, glycosylation sites, and the consistency in the sizes and locations of hydrophobic sequences. Moreover, two glycine and two proline residues are conserved in all sequences (Fig. 5, proline 81 and 276, and glycine 109 and 313).

3. Conserved Primary Sequences

Figure 5 shows a comparison of eight published fusion protein sequences. The sequences that are identical or have only the most conservative amino acid replacement at a particular position are boxed. Comparison of all fusion protein sequences revealed that overall about 10% of the sequences are conserved. The region around the cleavage site (indicated by an arrow at MuV residue 102) is the most prominent area of conserved sequences. The portion of the molecule to the right of the cleavage site is the F_1 subunit and to the left is F_2. Conserved sequences are located in both subunits in the vicinity of the cleavage site. The highly conserved sequence at the amino terminus of F_1 is of considerable biological importance, first, because of its hydrophobic nature, a characteristic of proteins which interact with the lipid portion of membranes, and second, because synthetic peptides representing portions of this sequence inhibited virus-induced fusion of host cells (Richardson *et al.*, 1980). Thus, the sequence spanning approximately the first 25 amino acids at the amino terminus of F_1 is presumed to be in the region primarily responsible for the fusion activity of the F molecule.

Conserved at the carboxy terminus of all F_2 subunits is a basic amino acid, either an arginine or a lysine, which forms a part of the cleavage activation site, and is recognized by proteolytic enzymes with trypsinlike specificity. The amino terminus of F_0 is conspicuous in the absence of conserved sequences, suggesting that this region may not contain framework structural elements. The next 50 residues are punctuated by conservation of predominantly hydrophobic sequences (boxed residues, Fig. 5) which tend to be located in the interior of the folded protein molecules. Particularly striking in this region is the conservation of a cysteine residue (positioned at MuV residue 64) in all but the RSV fusion sequence. However, even the RSV sequence contains a cysteine in this region just four residues to the amino side of this exact alignment of cysteines (Figs. 4 and 5). The structural importance of this conserved cysteine

```
MuV   M---------KAFSVTCLGFAV----FSSSICV------NINILQQIGYI----KQQVRQL SYYSQSSSSY      48
SV5   MG----TIIQ--FL----VVSCLLA---GAGS-LDPA----A--LMQIGVI-PT---NVRQL-MYYTEASSAF      48
NDV   MGPRPSTKNPVPM----MLTVRVA-L--VLSCICPANSIDGRPLAAAGIV-VTG---DKAVNI-YTSSQTGS      60
RV    MG--------I-L----FAAL-LA-M--TNPHLATGQ-IHWGNLSKIGVV-GTGSAS-YK--V-MTQSSHQS      50
MV    MG--------LKV--NVS-AIFMAVLLTL-QT-PTGQ-IHWGNLSKIGVV-GIGSAS-YK--V-MTRSSHQS      53
SV    M---------TAYI--QRSQCISTSLLVVL-TTLVSCQ-IPRDRLSNIGVIV-DEGKSL-K--IAGSHESRY-      55
PI3   M---------PTSI--L--LIITT--MIMA-S--FCQ-IDITKLQHVGVLVNSS-KGM-K--ISQNFETRY-      48
RSV   M---------ELLI--LKANAITT--ILTAVT--FC---FASGQNITEEFYQSTCSAVSKGYLSALRTGWYT      54

MuV   IVVKL-LPNI--QPTDNSC-EFKSV----TQY-NKTL-SNLL---LP-IAE---NINNI-----ASPSPGS      97
SV5   IVVKL-MPTI--DSPISGC-NITSI----SSY-NATV-TKLL----QP-IGE---NLETI---RNQL--IPT      97
NDV   IIVKL-LPNL--PKDKEAC-AKAPL---DAY-NRTL-TTLL----TP-LGD---SIRRI--Q-ESV-T-TS     110
RV    LVIKL-MPNI--TAI-DNC-TKTEI----MEYK-RLL-GTVL----KP-IREA---LNAITKNIKPIQS-ST    101
MV    LVIKL-MPNI--TLLN-NC-TRVEI----AEYR-RLL-RTVL----EP-IRHA---LNAMTQNIRPVQS VA    105
SV    IVLSL-VPGV-D-FEN-GCG-TAQVIQ----YK-SLL-NRLLI----P-LRDAL-DLQEALITVTNDTTQNA    110
PI3   LILSL-IPKI-E-DSN-SCG-DQQIKQ----YK-RLLD-RLII----P-LYDGLR-LQKDVI-VSNQESNEN    102
RSV   SVITIELSNIKENKCNGTDAKVKLIKQELDKYKNAVTELQLLMQSTPPTNNRARRELPRFM-NYTLNNAKKT    125

MuV   --RRHKRFAGIAIGIAALGVATAAQVTAAVSLVQAQTNARAIAAMKNSIQATNRAIFEVKEGTQQLAIAVQA    167
SV5   --RRRRRFAGVVIGLAALGVATAAQVTAAVALVKANENAAAILNLKNAIQKTNAAVADVVQATQSLGTAVQA    167
NDV   GGRRQKRFIGAIIGGVAALGVATAAQITAAAALIQAKQNAANILRLKESIAATNEAVHEVTDGLSQLAVAVGK    181
RV    TSRRHKRFAGVVLAGAALGVATAAQITAGIALHQSMMNSQAIESLKASLETTNQAIEEIRQAGQEMVLAVQG    173
MV    SSRRHKRFAGVVLAGAALGVATAAQITAGIALHQSMLNSQAIDNLRASLETTNQAIEAIRQAGQEMILAVQG    177
SV    GAPQS-RFFGAVIGTIALGVATSAQITAGIALAEAREAKRDIALIKESMTKTHKSIELLQNAVGEQILALKT    181
PI3   TDPRTKRFFGGVIGTIALGVATSAQITAAVALVEAKQARSDIEKLKEAIRDTNKAVQSVQSSIGNLIVAIKS    174
RSV   NVTLSKKRKRRFLGFL-LGVG-SA-IASGVAVSKVLHLEGEVNKIKSALLSTNKAVVSLSNGVSVLTSKVLD    194

MuV   IQDHI-NTIMNTQLNNMSCQILDNQLATYLGLYLTELTTVF--QPQLINPALSPISIQALRSLLGSMTPAVV    236
SV5   VQDHI-NSVVSPAITAANCKAQDAIIGSILNLYLTELTTIF--HNQITNPALSPITIQALRILLGSTLPTVV    236
NDV   MQQFV-NDQFNKTAQELGIIRIAQQVGVELNLYLTELTTVFG--PQITSPALNKLTIQALYNLAGGNMDYLL    250
RV    VQDYINNELVPAMGQ-LSCEIVGQKLGLKLLRYYTEILSLFG--PSLRDPVSAELSIQALSYALGGDINKIL    242
MV    VQDYINNELIPSMNQ-LSCDLIGQKLGLKLLRYYTEILSLFG--PSLRDPISAEISIQALSYALGGDINKVL    247
SV    LQDFVNDEIKPAI-SELGCETAALRLGIKLTQHYSELLTAFG-SN-FGTIGEKSLTLQALSSLYSANITEIM    250
PI3   VQDYVNKEIVPSI-ARLGCEAAGLQLGIALTQHYSELTNIFGD-NI-GSLQEKGIKLQGIASLYRTNITEIF    243
RSV   LKNYIDKQLLPIV-NKQSCSISNIETVIEFQQKNNRLLEITREFSVNAGVTTPVSTYMLTNSELLSLINDMP    265

MuV   QATLSTSISAAEILSAGLMEGQIVSVLLDEMQMIVK-INIPTI-VTQSNALVIDFYSIS-SFINNQESIIQL    305
SV5   EKSFNTQISAAEILSSGLLTGQIVGLDLTYMQMVV-IKIELPTL-TVQPATQIIDLATIS-AFINNQEVMAQL    305
NDV   TKLGVGNNQLSSLIGSGLITGNPIL-LYDSQTQLGIQVTLPSVGNLNNM-RATYLETLS-VSTTRGFASALV    319
RV    EKLGYSGSDLLAILESKGIKA-KITYVDIESYFIVLSIAYPSLSEIKGVIVHR-LESVSYNIGSQEWYT-TV    311
MV    EKLGYSGGDDLLGILESRGIKA-RITHVDTESYFIVLSIAYPTLSEIKGVIVHR-LEGVSYNIGSQEWYT-TV    315
SV    TTIKTGQSNIYDVIIYTEQIKG-TVIDVDLERYMVTLSVKIPILSEVPGVLIHK-ASSISYNIDGEEWY-VTV    319
PI3   TTSTVDKYDIYDLLFTESIK-VRVIDVDLNDYSIALQVRLPLLTRLLNTQIYRV-DSISYNIQNREWY-IPL    312
RSV   ITNDQKKLMSNNV-QIVRQQSYSIMSI-IKEEVLAYVVQLP-LYGVIDTPCWKLHTSPLCTTNTKEGSNICL    334

MuV   -P---DRILEIGNEQWSYPAKNCKLT-RHHIFCQYNEAERLSLESKLC--LAGNISACVFS-PIAGSYMRRF    369
SV5   -P---TRVMVTGSLIQAYPASQCTIT-PNTVYCRYNDAQVLSDDTMAC--LQGNLTRCTFS-PVVGSFLTRF    369
NDV   -P---KVVTQVGSVIEELDTSYCIEVDLDLYCTRIVTFPMSPGIYSC--LSGNTSACMYS-KTEGALTTPY    383
RV    -P---RYVATQGYLISNFDDTPCAFTPEGTI-CSQNAIYPMSPLLQEC--FRGSTRSCART-LVLGSIGNRF    375
MV    -P---KYVATQGYLISNFDESSCTFMPEGTV-CSQNALYPMSPLLQEC-L-RGSTKSCART-LVSGSFGNRF    379
SV    -PS---HILSRASFLGGADITDCVESRLTYI-CPRDPAQLIPDSQQKC-I-LGDTTRCPVT-KVVDSLIPKF    383
PI3   -PS---HIMTKGAFLGGADVKECIEAFSSYI-CPSDPGFVLNHEMESC-L-SGNISQCPRTV-VKSDIVPRY    376
RSV   TRTDRGWYCDNAGSVSFFPQAETCKVQSNRVFCDTMNSLTLPSEINLCNVDIFNPKYDCKIMTSKTDVSSSV    406

MuV   VALDGTIIVANCRSLTCLC-KSPSYPIYQPDHHAVTTIDLTTCQTLSLDGLDFSI----VSLSNITY-AENLT    435
SV5   VLFDGIMYANCRSMLCKC-MQPAAVILQPSSSPVTVIDMYKCVSLQLDNLRFTI-T---QLANVTY-NSTIK    435
NDV   MTIKGSMIANCKMTTCRC-VNPPGIISQNYGEAVSLIDKQSCNVLSLDGITLRL-S--GEF-DATY-QKNIS    449
RV    ILSKGNLIIGNCASILCKC-YTTGSIISQDPDKILTYIAADQCPVVEVNGVTIQVGSR--EYSDAVY-LHEID    443
MV    ILSQGNLIANCASILCKC-YTTGTIINQDPDKILTYIAADHCPVVEVNGVTIQVGSR-R-YPDAVY-LHRID    447
SV    AFVNGGVVANCIASTCTC-GTGRRPISQDRSKGVVFLTHDNCGLIGVNGVEL-YANR-RG-HDATWGVQNLT    451
PI3   AFVNGGVVANCITTTCTCNGIGNR-INQPPPDQGVKIITHKECNTIGINGM-LFNTNK-EG-TLAFYTPNDIT    444
RSV   ITSLGAIVSCYGKTKCT-ASNKNRGIIKTFSNGCDYVSNKGMDTVSM-GNTLYYVNKQEGKSLYVKGEPII-    475
```

FIGURE 5. Comparison of the primary amino acid sequences of the fusion glycoprotein of the *Paramyxoviridae*. Identical and the most conservative amino acid replacements in all eight proteins are boxed. The positions of amino acid changes in MAb escape mutants and mutants derived with other selective conditions are indicated by triangles. The small arrow above MuV residue 102 designates the aligned sites of proteolytic cleavage in all proteins, with the F₁ portion of each molecule to the right and F₂ to the left. Sequences were aligned as described in the legend to Fig. 2.

```
MuV   ISLSQTINTQPIDISTELSKVNASLQNAVKYIKESNHQLQSV---SVNS-KIGAIIVAALVL-SIL-SIHIS    501
SV5   LESSQILSIDPLDISQNLAAVNKSLSDALQHLAQSDTYLSAI---TSATTTS-VLSIIAICLGS-L-GLIL-    500
NDV   IQDSQVIITGNLDISTELGNVNNSISNALNKLEESNSKLDKVNV-K-LTSTSALITYIVLTIISLVFGIL-S    518
RV    L-GPPISL-EKLDVGTNLWNAVTKLEKA-KDLLDS-SDLILENI-K-GVSVT-NTGYILV-GVGLI-AVW-G    505
MV    L-GPPISL-ERLDVGTNLGNAIAKLEDA-KELLES-SDQILRSM-K-GLSST-SIVYILI-AVCL-GGLI-G    509
SV    V-GPAIAI-RPIDISLNLADATNFLQD-SKAELE-KARKILSEVGRWYNS-RETVITIIV-VMVV-ILVW-I    515
PI3   L-NNSVAL-DPIDISIELNKAKSDLEE-SKEWI-RRSNQKLDSIGNWHQS-STTII-IVL-IMII-ILFI-I    507
RSV   NFYDPL-VFPSDEFDASLSQVNEKI-NQSLAFI-RKSDELLHNV-NAGKS-TTNIM-I-TTIIIV-II-M-I    537

MuV   LL--FCCWAYI-----ATK-.-EIRRI---NFKTNHINTI--SSSVDDLIR--Y.
SV5   II--L--LSVVVW-----K---LLTI-V--V-AN-RNRMEN--FV--YHK---.
NDV   LV--L--ACYLMYK---QKAQQK-TL-L-WL-GN--NTLDQ-M--RATTK--M.
RV    IL--II-TCCCK-KR---RTDNKVS-TMV-L---NPGLR-PDLT--GTSKSYVRSL
MV    I--PALI-CCCR-GR-CNKKG---EQVGM---S-RPGLK-PDLT--GTSKSYVRSL
SV    I--VIIIVL-YRLRR-SMLMGNPDDRIPRDTYTLEPKIRHM-YTNGGFDAMAEKR.
PI3   --NVTIIII-A-V-K-YYRI-QKRNRV--D--QND-K-PYV-LTN--------K
RSV   --LLSLIAV-GLL--LYCKARSTPVTLSKD--Q-LSGINNIAFSN.
```

FIGURE 5. (*continued*)

position is suggested by the fact that antibodies which neutralize NDV infectivity map with one, two, and four residues of this position (Toyoda *et al.*, 1988; Neyt *et al.*, 1989). It is likely that this conserved cysteine position links the F_2 subunit with F_1.

The conservation of many cysteine residues at specific positions in all F_1 sequences is remarkably precise. The majority of the cysteines (seven) are concentrated in a small area of just 65 residues (Fig. 4). Using the mumps virus sequence, one finds that the limits of the region are from residue 324 to 389 (Fig. 5). The positions of three of the cysteines are conserved in all F sequences (residues 333, 348, and 385), while at four other positions (324, 356, 380, and 387) cysteines are conserved in all sequences except in RSV, and here a cysteine is present within two residues. In families of proteins, cysteine residues important for disulfide bridges tend to be well conserved. Therefore, the precise conservation of these cysteine residues in a background in which only 10% of the total F sequences are conserved suggests that these cysteines participate in disulfide bridges as covalent cross-links between parts of the polypeptide. The concentration of cysteines in this relatively short region suggests the possibility that the folding arrangement of this segment of the molecule is a series of cross-linked loop structures. The fact that monoclonal antibodies targeted to this cysteine-rich portion of the molecule neutralize infectivity (Portner *et al.*, 1987a; Toyoda *et al.*, 1988; Neyt *et al.*, 1989; K. Coelingh, personal communication) provides strong evidence of its importance in the structure and function of the molecule (Fig. 5).

4. Functional Domains of the Mature Protein

The fusion proteins of paramyxoviruses are simple integral membrane proteins which possess only a single membrane spanning domain. These proteins have a typical three-domain structure, consisting of a large, relatively hydrophilic domain external to the virion, a second domain of 20 or more uncharged amino acids that anchor the proteins to the lipid bilayer, and, immediately adjacent, a hydrophilic carboxyl-terminal domain which exists on the inner side of the virion bilayer or the host-cell plasma membrane. Besides electron micrographs of the Sendai virus F protein which reveal a stalklike structure anchoring a globular head region to the virion membrane (K. Murti, personal communication), little is known about their shape.

a. Proteolytic Cleavage and Activation of the Fusion Function

The fusion protein of paramyxoviruses is responsible for virus penetration into the host cell, hemolysis activity, and cell fusion (Homma and Ohuchi, 1973; Scheid and Choppin, 1974). The cleavage of F_0 to form the active molecule is an important determinant of virus host range, tissue tropism, and pathogenicity (Scheid and Choppin, 1976; Nagai et al., 1976; Nagai and Klenk, 1977).

The proteolytic cleavage of F_0 results in a new N terminus on F_1 and a conformational change in the molecule which exposes a previously hidden hydrophobic region (Hsu et al., 1981; Kohama et al., 1981). Comparisons of the N terminus of F_1 by either amino acid sequencing (Gething et al., 1978; Scheid and Choppin, 1977; Richardson et al., 1980; Server et al., 1985; Varsanyi et al., 1985) or nucleotide sequencing of cDNA clones (Fig. 5) have shown that this region is highly conserved. These results have led to the concept that the new N terminus of F_1 interacts directly with the target host-cell membrane and thereby facilitates the fusion process (Gething et al., 1978; Scheid et al., 1978; Richardson et al., 1980; Richardson and Choppin, 1983). Recent studies using hydrophobic photoaffinity-labeling of SV proteins during fusion provide direct evidence that hydrophobic penetration of the F protein into the target membrane is responsible for triggering the fusion event (Novick and Hoekstra, 1988).

b. Cysteine-Rich Region

Recent evidence from different laboratories implicates the cysteine-rich region located near the membrane anchor region in membrane fusion. Portner et al. (1987a) selected SV escape mutants with monoclonal antibodies to the fusion protein. The antibody used in this selection inhibited membrane fusion and neutralized wild-type virus infectivity. Surprisingly, the amino acid substitution that conferred antibody resistance on the mutant was located in the cysteine-rich region at position 407, far removed from the proposed fusion sequence at the amino terminus of the F_1 polypeptide (Fig. 5). In studies of NDV (Toyoda et al., 1988; Neyt et al., 1989) and PIV3 (K. Coelingh, personal communication), mutants were also isolated that were resistant to the fusion-inhibitory and neutralizing effects of the selecting antibody. Analysis of the mutants also revealed changes that were not in the fusion sequence, but in the cysteine-rich area. In the sequence presented in Fig. 5, the NDV substitutions are located at residues 343 and 379, and the PIV3 changes at 337, 396–398, and 452. Using a synthetic peptide that inhibits fusion, Hull et al. (1987) isolated MV mutants with changes only in the conserved cysteine-rich region (Fig. 5). What are the implications of these findings? First, it is apparent that the fusion sequence is not the only determinant of fusion activity and other sequences are likely to play critical roles in maintaining a biologically active structure. Second, disulfide cross-links generated from the cysteine-rich region may provide conformational stability required to maintain the fusion protein in the correct orientation. Alternatively, the fusion sequences may not act alone, but require the participation of distant regions of the molecules to form the fusion-active site.

c. Cleavage Site

In addition to the conservation of the N terminus of F_1, sequences amino-terminal to the F_0-precursor cleavage activation site (indicated by arrows in Fig. 5) contain a stretch of basic amino acid residues. The number of basic residues varies from one in the F_0 protein of SV to six in RSV. Apparently, efficient cleavage of F_0 depends on the number of basic residues at the cleavage activation site, since the SV F_0 protein contains only one basic amino acid and is cleaved in only a limited number of cell types, whereas other paramyxoviruses have a broader host range. Moreover, it has been shown that the pathogenicity of NDV in cell culture is related to the cleavability of the F_0 protein; pathogenic strains with four basic residues at or near the cleavage site are readily cleavable, while avirulent strains with only two are cleaved in only a limited number of cell types (Nagai et al., 1976; Nagai and Klenk, 1977; Toyoda et al., 1987; Glickman et al., 1988). Recent results by Paterson et al. (1989) also show that the number of basic amino acids at the cleavage site of SV5 is important for cleavage, but apparently this is not the only structural determinant involved. Their results show that cleavage of F_0 may not necessarily produce a biologically active molecule and suggest that the sequence of the cleavage site may be important in determining whether cleavage results in a conformation that is biologically active.

Much remains to be learned before the structural correlates of membrane fusion and antigenic structure of fusion proteins can be fully understood. However, the tools are at hand to learn more through further studies of mutants selected by monoclonal antibodies and the analysis of fusion proteins altered by site-directed mutagenesis. Moreover, success in establishing the structures of the hemagglutinin and neuraminidase proteins of influenza virus makes us hopeful that eventually the three-dimensional structure of both the fusion and attachment proteins of the *Paramyoxviridae* will be solved.

B. Intracellular Processing of the Fusion Protein

1. Membrane Insertion

The fusion proteins of *Paramyxoviridae* are typical of type 1 glycoproteins (Wickner and Lodish, 1985). These proteins have an amino-terminal signal sequence which directs membrane insertion and is cleaved, and a hydrophobic domain located near the carboxy terminus which serves as the stop-transfer sequence and membrane anchor. There is direct evidence of signal sequence cleavage in both the NDV and SV systems (Gorman et al., 1988; Blumberg et al., 1985a). Direct peptide sequencing of the mature protein shows that 31 amino acids are removed from the amino terminus of the NDV fusion protein and 25 amino acids are removed from SV fusion protein. A comparison of the predicted primary sequences of the amino termini of eight paramyxoviruses shows little sequence similarity, but all are typical of most signal sequences: there are charged residues near the amino terminus followed by a hydrophobic domain (Von Heijne, 1983) (Figs. 4 and 5). The carboxy-terminal hydrophobic domains are located from 20 to 42 residues from the ends of the molecules (Fig. 5). These sequences vary from 24 to 38 residues in length and share little sequence similarity. Beyond the hydrophobic domain is the region of the molecule which presumably remains on the cytoplasmic side of the membrane. There is little sequence similarity in this region as well. These regions of the proteins may not have as stringent sequence requirements for proper processing and function of the molecules as more amino-terminal regions. Alternatively, these regions may be subject only to constraints specific for individual viruses.

Membrane insertion of *Paramyxoviridae* fusion proteins is likely very typical of most type 1 eucaryotic glycoproteins. Indeed, the membrane insertion of the NDV fusion protein requires SRP (Reitter *et al.*, submitted). Further, SRP inhibits translation in the absence of membranes. The protein inserted into membranes becomes protease resistant, except for the removal of a short segment of one end of the protein (Bowen and Lyles, 1981; J. Reitter and T. Morrison, in preparation), a result consistent with the localization of the carboxy terminus on the cytoplasmic side of the membrane.

One element of the fusion protein sequence that represents a puzzle in terms of membrane insertion is the fusion peptide. This is a hydrophobic sequence which begins at residue 108–116 (depending upon the virus) from the amino terminus of all fusion proteins (Fig. 5, arrow indicates start of fusion sequence). The sequence is 26 amino acids long and resembles a typical stop-transfer sequence. Yet there is no evidence that this sequence functions as a stop-transfer/membrane anchor sequence or a membrane anchor. Work of Davis and Model (1985) suggested that the position of a hydrophobic domain plays a role in its function. Paterson and Lamb (1987) tested this idea and found that the SV5 fusion sequence would serve as a membrane anchor if placed at the end of an influenza virus hemagglutinin sequence missing its own membrane anchor. They argued that the fusion sequence domain is just hydrophobic enough to stop translocation and anchor a protein in membranes if it is at the end of a sequence. At an internal location, its hydrophobicity is not sufficient to function in this way. Further, Paterson and Lamb (1987) showed that removal of the preceding arginines which serve as the cleavage site still does not convert the sequence to a membrane anchor during translocation of the protein. Similar conclusions have been reached by Davis and Hsu (1986).

A unique and puzzling feature of the morbillivirus fusion protein gene is the existence of an extremely long 5' end of the mRNA. This region of the measles virus sequence contains no open reading frames (Richardson *et al.*, 1986); however, both the RV and CDV mRNA sequences contain an open reading frame in the same frame as the fusion protein and could potentially encode a fusion protein with a much larger F_2 than is typical of members of *Paramyxovirus* and RSV. However, there is no evidence for a larger fusion protein in these virus infections, and it is proposed that the fusion protein synthesis in RV and CDV infections is initiated at the second and third AUG, respectively, in the open reading frames of the mRNAs (Hsu *et al.*, 1988; Barrett *et al.*, 1987).

2. Intracellular Transport of the Fusion Protein

a. Kinetics of Transport

The kinetics of transport of the fusion protein to the cell surface has been examined in several systems with variable results. Blumberg *et al.* (1985b) report that the SV fusion protein rapidly appears at the cell surface with a half-time of transport of 8–10 min, much less than that of the HN protein in the same experiment. In contrast, Bowen and Lyles (1982) report the appearance of the SV fusion protein at cell surfaces, and Portner and Kingsbury (1976) report the incorporation of the protein into virions with a half-time approximately 40 min. In addition, the fusion proteins of NDV (Morrison *et al.*, 1985) and mumps virus (Herrler and Compans, 1983) are reported to appear at the cell surface with the same slow kinetics as the HN protein. The reasons for these variable results remain unclear.

b. Glycosylation

The fusion proteins are all subjected to N-linked glycosylation (Morrison and Simpson, 1980; Nakamura *et al.*, 1982; Herrler and Compans, 1982; Lambert, 1988). There is no evidence of O-linked forms of carbohydrate in any of these proteins. There is variation in the number of potential N-linked oligosaccharides in the different sequences, from a minimum of three in the MV sequence to a maximum of six in the SV and MuV sequences. The distribution of these sites is also not well conserved (Fig. 4). All but the PIV3 sequence contains at least one site in the F_2 portion of the sequence, but the position of the site is quite variable. All but the *Morbillivirus* sequences contain sites in the F_1 portion of the molecule. These sites are located primarily near the hydrophobic membrane anchor region, but their precise location is not well conserved. The lack of conservation of positions of glycosylation sites, particularly in the face of the impressive conservation in positions of cysteine residues, is surprising. Clearly, the role of sugar side chains in the structure, function, folding, and transport of the molecules is not well understood.

There is no evidence concerning which glycosylation sites are utilized and which are not. It has been reported that the F_1 portion of the SV5 protein contains three carbohydrate side chains and the F_2 portion contains one side chain (Yoshima *et al.*, 1981). This result indicates that one potential glycosylation site in both portions of the molecule is not used. Formulation of the rules for the use of carbohydrate addition sites awaits further studies.

The carbohydrate side chains on the mumps virus fusion protein are a mixture of complex and simple types (Herrler and Compans, 1983), whereas those on the NDV fusion protein are all simple (T. Morrison, unpublished observations). Furthermore, there is evidence in the mumps virus system that F_2 contains only endo-H sensitive forms of carbohydrate (Herrler and Compans, 1983). Again, as in the case of the attachment proteins, the reasons for variations in the representation of complex and simple carbohydrate side chains in the mature proteins are unclear.

c. Disulfide Bond Formation

The conservation in the positions of cysteine residues, particularly within the F_1 portion of all the sequences (Fig. 4), suggests that folding of the molecule and intramolecular disulfide bonds are crucial to the function of the molecule. However, these bonds remain undefined. It has not even been directly demonstrated which cysteines function to hold the F_1 and F_2 portions of the molecule together.

The formation of the correct intramolecular disulfide bonds is not a simple process. It was noted several years ago that the fusion protein of NDV undergoes a dramatic conformational change soon after its synthesis (McGinnes *et al.*, 1985, 1987). This change was detected by migration on polyacrylamide gels in the absence of reducing agent, by sizes of polypeptides generated by partial proteolysis in the absence of reducing agent, and by reactivity to monoclonal antibody. A characterization of this change showed that the nascent fusion protein contains intramolecular disulfide bonds which are subsequently rearranged as the protein moves from the RER (Morrison *et al.*, 1987). Further, it was found that the change is independent of any conformational change that occurs upon cleavage of the molecule (Morrison *et al.*, 1985). Thus, for this molecule, and perhaps glycoproteins in general, an important posttranslational modification is the disruption and rearrangement of intramolecular disulfide bonds.

d. Proteolytic Cleavage

As discussed above, one of the most significant alterations in the fusion protein during its maturation is the cleavage of F_0 into the F_1 and F_2 polypeptides (Scheid and Choppin, 1977). With the exception of the SV sequence and some avirulent strains of NDV, the sequence amino-terminal to the cleavage site of the F_0 polypeptide contains closely positioned pairs of basic amino acid residues (Fig. 5), a sequence motif similar to cleavage sites in precursors to peptide hormones (Undenfriend and Kilpatrick, 1983). It is well documented that precursors to hormones are processed in or just beyond the *trans* Golgi membranes (Steiner *et al.*, 1980). It is likely that the *Paramyxoviridae* fusion proteins with these double basic amino acid residues are cleaved by the same enzymes used to cleave these hormone precursors. Indeed, the NDV fusion protein is cleaved in the vicinity of the *trans* Golgi membranes (Morrison *et al.*, 1985). However, the SV fusion protein, which has a single basic amino acid at the cleavage site, is cleaved external to the cell (Lamb *et al.*, 1976) and requires the presence of a protease in the overlay for plaquing in most cell types.

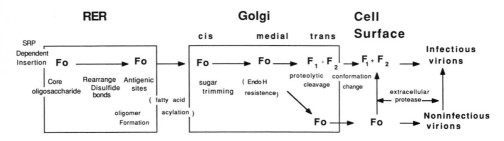

FIGURE 6. Intracellular processing of the *Paramyxoviridae* fusion proteins. A summary of current understanding of the steps involved in the processing and transport of the fusion proteins of *Paramyxoviridae*. Not all steps have been shown with all proteins. Steps shown in parentheses are known not to occur with all fusion proteins.

e. Fatty Acid Acylation

The fusion protein of NDV is fatty acid-acylated (Chatis and Morrison, 1982), whereas the fusion proteins of the closely related SV and SV5 are not (Veit *et al.*, 1989). Thus, this modification may not play a significant role in the intracellular processing of these proteins.

f. Oligomer Formation

On the basis of the structure of purified fusion protein reconstituted into lipid vesicles, Sechoy *et al.* (1987) suggest that the SV fusion protein consists of a tetrameric structure consisting of two identical dimers than can be chemically cross-linked. In this context, it has been noted by Buckland and Wild (1989) that a leucine zipper motif is present in all reported fusion protein sequences and they suggest that fusion protein oligomers may be formed due to this sequence.

3. Summary of Fusion Protein Intracellular Processing

Figure 6 summarizes current understanding of the steps involved in the synthesis, membrane insertion, transport, and posttranslational processing of *Paramyxoviridae* fusion proteins. The pathway presented is a summary of data obtained in a number of different systems. While not all aspects of the pathway have been documented for each protein, the overall similarities of the proteins of all members of the family in terms of structural determinants and activity make it likely that they all follow the same general principles. As in the case of the attachment proteins, there are still many details in this process that are unclear and await further study.

IV. CONCLUSIONS

In the past few years, there has been an explosion of information concerning the surface of glycoproteins of the *Paramyxoviridae* due in large part to the participation of many laboratories in the cloning and sequencing of the genes from many members of the family and to the development of monoclonal antibody panels specific for these proteins. This rather happy situation has allowed comparisons of a number of different members of the family, providing insights into the important structural determinants of the glycoproteins as well as the evolutionary relatedness of the viruses. When comparisons are made that allow for conservative amino acid replacements as has been done here, there is a clear relationship between all attachment proteins and between the fusion proteins. In addition, these comparisons reveal conserved determinants of structure across the family, such as specific cysteine residues, glycines, prolines, and hydrophobic domains, which are likely to be fundamental to the biological function of the protein. Sequence analysis of mutants selected with monoclonal antibodies is beginning to provide a picture of the active sites of the proteins and areas of the molecules important for the function of each protein. Future studies, using the tools developed in the past several years and insights gained from these comparative analyses, can now focus on specific questions about the structural basis for the functions of these proteins.

The importance of conserved structural determinants in the proper processing of the proteins must also be kept in mind. Indeed, a complete understanding of the mature forms of the proteins requires an understanding of the steps involved in their formation. Studies of the intracellular processing of these proteins are beginning to yield insights into this process. Furthermore, since these proteins utilize resident host-cell pathways for the transport and processing of glycoproteins, information gained from the study of these viruses will be important to general questions of protein processing.

ACKNOWLEDGMENTS. Work from the authors' laboratories cited here was supported by grants from the National Institutes of Health GM-37745 (T.M.) and AI-11949 (A.P.).

V. REFERENCES

Alkhatib, G., and Briedis, D. J., 1986, The predicted primary structure of the measles virus hemagglutinin, *Virology* **150:**479–490.

Barrett, T., Clarke, D. K., Evans, S. A., and Rima, B. K., 1987, The nucleotide sequence of the gene encoding the F protein of canine distemper virus: A comparison of the deduced amino acid sequence with other paramyxoviruses, *Virus Res.* **8:**373–386.

Berger, M., and Schmidt, M. F., 1985, Protein fatty acyltransferase is located in the rough endoplasmic reticulum, *FEBS Lett.* **187:**289–294.

Blobel, G., 1980, Intracellular protein topogenesis, *Proc. Natl. Acad. Sci. USA* **80:**7249–7253.

Blumberg, B. M., Giorgi, C., Rose, K., and Kolakofsy, D., 1985a, Sequence determination of the Sendai virus fusion protein gene, *J. Gen. Virol.* **60:**317–331.

Blumberg, B., Girogi, C., Roux, L., Raju, R., Dowling, P., Chollet, A., and Kolakofsky, D. 1985b, Sequence determination of the Sendai virus HN gene and its comparison to influenza glycoproteins, *Cell* **41:**269–278.

Bole, D. G., Hendershot, L. M., and Kearney, J. F., 1986, Posttranslational association of immunoglobulin heavy chain binding protein with nascent heavy chains in nonsecreting and secreting hydridomas, *J. Cell Biol.* **102:**1558–1566.

Bowen, H. A., and Lyles, D. S., 1981, Structure of Sendai viral proteins in plasma membranes of virus infected cells, *J. Virol.* **37:**1079–1082.

Bowen, H. A., and Lyles, D. S., 1982, Kinetics of incorporation of Sendai virus proteins into host plasma membrane and virus, *Virology* **121**:1–11.

Bratt, M. A., and Gallagher, W. R., 1969, Preliminary analysis of the requirements for fusion from within and fusion from without by Newcastle disease virus, *Proc. Natl. Acad. Sci. USA* **64**:536–540.

Buckland, R., and Wild, F., 1989, Leucine zipper motif extends, *Nature* **338**:547–547.

Chambers, P., Millar, N. S., and Emmerson, P. T., 1986, Nucleotide sequence of the gene encoding the fusion glycoprotein of Newcastle disease virus, *J. Gen. Virol.* **67**:2685–2694.

Chatis, P. A., and Morrison, T. G., 1981, Mutational changes in the vesicular stomatitis virus glycoprotein affect the requirement of carbohydrate in morphogenesis, *J. Virol.* **37**:307–316.

Chatis, P. A., and Morrison, T. G., 1982, Fatty acid modification of Newcastle disease virus glycoproteins, *J. Virol.* **43**:342–347.

Clinkscales, C. W., Bratt, M. A., and Morrison, T. G., 1978, Synthesis of Newcastle disease virus polypeptides in a wheat germ cell free system, J. Virol. **22**:97–101.

Coelingh, K. J., Winter, C. C., Murphy, B. R., Rice, J. M., Kimball, P. C., Olmsted, R. A., and Collins, P. L., 1986, Conserved epitopes in the hemagglutinin-neuraminidase proteins of human and bovine parainfluenza type 3 viruses: Nucleotide sequence analysis of variants selected with monoclonal antibodies, *J. Virol.* **60**:90–96.

Coelingh, K. W., Winter, C. C., Jorgenson, E. D., and Murphy, B. R., 1987, Antigenic and structural properties of the hemagglutinin-neuraminidase glycoprotein of human parainfluenza type 3 sequence analysis of variants selected with monoclonal antibodies which inhibit infectivity, hemagglutination and neuraminidase activities, *J. Virol.* **61**:1473–1477.

Collins, P. L., and Hightower, L. E., 1982, Newcastle disease virus stimulates the cellular accumulation of stress (heat shock) mRNAs and proteins, *J. Virol.* **44**:703–707.

Collins, P. L., Huang, Y. T., and Wertz, G. W., 1984, Nucleotide sequence of the gene encoding the fusion (F) glycoprotein of human respiratory syncytial virus, *Proc. Natl. Acad. Sci. USA* **81**:7687–7691.

Davis, N. G., and Hsu, M.-C., 1986, The fusion related hydrophobic domain of Sendai (F) protein can be moved through the cytoplasmic membrane of *E. coli*, *Proc. Natl. Acad. Sci. USA* **83**:5091–5095.

Davis, N. G., and Model, P., 1985, An artificial anchor domain: Hydrophobicity suffices to stop transfer, *Cell* **41**:607–614.

Dorner, A. J., Krane, M. G., and Kaufman, R. J., 1988, Reduction of endogenous GRP78 levels improves secretion of a heterologous protein in CHO cells, *Mol. Cell. Biol.* **8**:4063–4070.

Elango, N., Satake, M., Coligan, J. E., Norrby, E., Camargo, E., and Venkatesan, S., 1985, Respiratory syncytial virus fusion glycoprotein: Nucleotide sequence of mRNA, identification of cleavage activation site and amino acid sequence of N-terminus F_1 subunit, *Nucleic Acids Res.* **13**:1559–1574.

Elango, N., Coligan, J. E., Jambou, R. C., and Venkatesan, S., 1986, Human parainfluenza type 3 virus hemagglutinin-neuraminidase glycoprotein: Nucleotide sequence of the mRNA and limited amino acid sequence of the purified protein, *J. Virol.* **57**:481–489.

Freedman, R. B., 1984, Native disulfide bond formation in protein biosynthesis, evidence for the role of protein disulfide isomerase, *Trends Biochem. Sci.* **9**:438–441.

Gerald, C., Buckland, R., Barker, R., Freeman, G., and Wild, T. F., 1986, Measles virus haemagglutinin gene: Cloning, complete nucelotide sequence analysis and expression in COS cells, *J. Gen. Virol.* **67**:2695–2703.

Gething, M. J., White, J. M., and Waterfield, M. D., 1978, Purification of the fusion protein of Sendai virus: Analysis of the NH2-terminal sequences generated during precursor activation, *Proc. Natl. Acad. Sci. USA* **75**:2737–2740.

Gething, M. J., McCannon, K., and Sambrook, J., 1986, Expression of wild type and mutant forms of influenza hemagglutinin: The role of folding in intracellular transport, *Cell* **46**:938–952.

Gibson, R., Leavitt, R., Kornfeld, S., and Schlesinger, S., 1978, Synthesis and infectivity of vesicular stomatitus virus containing nonglycosylated G protein, *Cell* **13**:671–679.

Gitman, A. G., and Loyter, A., 1984, Construction of fusogenic vesicles bearing specific antibodies: Targeting of reconstituted Sendai virus envelopes towards neuraminidase-treated human erythrocytes, *J. Biol. Chem.* **259**:9813–9820.

Glickman, R. L., Syddall, R. J., Iorio, R. M., Sheehan, J. P., and Bratt, M. A., 1988, Quantitative basic residue requirements in the cleavage activation site of the fusion glycoprotein as a determinant of virulence of Newcastle disease virus, *J. Virol.* **62**:354–356.

Gorman, J. J., Nestorowicz, A., Mitchell, S. J., Corino, G. L., and Selleck, P. W., 1988, Characterization of the sites of proteolytic activation of Newcastle disease virus membrane glycoprotein precursors, *J. Biol. Chem.* **263**:12522–12531.

Gotoh, B., Sakaguchi, T., Nishikawa, K., Inocencio, N. M., Hamaguchi, M., Toyoda, T., and Nagai, Y., 1988, Structural features unique to each of the three antigenic sites on the hemagglutinin-neuraminidase protein of Newcastle disease virus, *Virology* **163**:174–182.

Hardwick, J. M., and Bussell, R. H., 1978, Glycoproteins of measles virus under reducing and nonreducing conditions, *J. Virol.* **25**:687–692.

Heath, T. D., Martin, F.J., and Macher, B. A., 1983, Association of ganglioside–protein conjugates into cell and Sendai virus. Requirement for the HN subunit in viral fusion, *Exp. Cell Res.* **149**:163–175.

Hendershot, L. M., Ting, J., and Lee, A. S., 1988, Identity of the immunoglobulin heavy-chain-binding protein with the 78,000-Dalton glucose regulated protein and the role of posttranslational modifications in its binding function, *Mol. Cell. Biol.* **8**:4250–4256.

Herrler, G., and Compans, R. W., 1982, Synthesis of mumps virus polypeptides in infected Vero cells, *Virology* **119**:431–438.

Herrler, G., and Compans, R. W., 1983, Posttranslational modification and intracellular transport of mumps virus glycoproteins, *J. Virol.* **47**:345–362.

Hidaka, Y., Kanda, T., Iwasaki, K., Nomoto, A., Shioda, T., and Shibuta, H., 1984, Nucleotide sequence of a Sendai virus genome region covering the entire *M* gene and the 3′ proximal 1013 nucleotides of the *F* gene, *Nucleic Acids Res.* **12**:7965–7973.

Hiebert, S. W., Paterson, R. G., and Lamb, R. A., 1985, The hemagglutinin-neuraminidase protein of the paramyxovirus SV5: Nucleotide sequence of the mRNA predicts an N-terminal membrane anchor, *J. Virol.* **54**:1–6.

Homma, M., and Ohuchi, M., 1973, Trypsin action on the growth of Sendai virus in tissue culture cells. III. Structural differences of Sendai viruses grown in eggs and in tissue culture cells, *J. Virol.* **12**:1457–1465.

Hsieh, P., Rosner, M. R., and Robbins, P. W., 1983, Selective cleavage by endo-β-acetylglucosaminidase H at individual glycosylation sites of Sindbis virion envelope glycoproteins, *J. Biol. Chem.* **258**:2555–2561.

Hsu, D., Yamanaka, M., Miller, J., Dale, B., Grubman, M., and Yilma, T., 1988, Cloning of the fusion gene of rinderpest virus: Comparative sequence analysis with other morbilliviruses, *Virology* **166**:149–153.

Hsu, M.-C., Scheid, A., and Choppin, P. W., 1981, Activation of the Sendai virus fusion protein (F) involves a conformational change with exposure of a new hydrophobic region, *J. Biol. Chem.* **256**:3557–3563.

Hsu, M-C., Scheid, A., and Choppin, P. W., 1983, Fusion of Sendai virus with liposomes: Dependence on the viral fusion protein (F) and the lipid composition of liposomes, *Virology* **126**:361–369.

Hull, J. D., Krah, D. L., and Choppin, P. W., 1987, Resistance of a measles virus mutant to fusion inhibiting oligopeptides is not associated with mutations in the fusion peptide, *Virology* **159**:368–372.

Iorio, R. M., and Bratt, M. A., 1984, Monoclonal antibodies as functional probes of the HN glycoprotein of Newcastle disease virus: Antigenic separation of the hemagglutinating and neuraminidase sites, *J. Immunol.* **133**:2215–2219.

Jambou, R. C., Elango, N., and Venkatesan, S., 1985, Proteins associated with human parainfluenza virus type 3, *J. Virol.* **56**:298–302.

Jimenez-Montano, M., and Zamora-Cortina, L., 1981, Evolutionary model for the generation of amino acid sequences and its application to the study of mammal alpha hemoglobin chains, *in* "Proceedings VII International Biophysics Congress," Mexico City.

Jorgensen, E. D., Collins, P. L., and Lomedico, P. T., 1987, Cloning and nucleotide sequence of Newcastle disease virus hemagglutinin-neuraminidase mRNA: Identification of a putative sialic acid binding site, *Virology* **156**:12–24.

Kohama, T., Garten, W., and Klenk, H.-D., 1981, Changes in conformation and charge paralleling proteolytic activation of Newcastle disease virus glycoproteins, *Virology* **111**:364–376.

Kornfeld, R., and Kornfeld, S., 1985, Assembly of asparagine-linked oligosaccharides, *Annu. Rev. Biochem.* **54:**631–664.

Kozutsumi, Y., Segal, M., Normington, K., Gething, M.-J., and Sambrook, J., 1988, The presence of malfolded protein in the endoplasmic reticulum signals induction of glucose regulated protein, *Nature (London)* **332:**462–464.

Lamb, R. A., Mahy, B. W. J., and Choppin, P. W., 1976, The synthesis of Sendai virus polypeptides in infected cells, *Virology* **69:**116–131.

Lambert, D., 1988, Role of oligosaccharides in the structure and function of respiratory syncytial virus glycoprotein, *Virology* **164:**458–466.

Leavitt, R., Schlesinger, S., and Kornfeld, S., 1977, Impaired intracellular migration and altered solubility of nonglycosylated glycoproteins of vesicular stomatitis virus and Sindbis virus, *J. Biol. Chem.* **252:**9018–9025.

Markwell, M. K., and Fox, C. F., 1980, Protein–protein interactions within paramyxoviruses identified by native disulfide bonding or reversible chemical crosslinking, *J. Virol.* **33:**152–166.

Matthews, R. E. F., 1982, Classification and nomenclature of viruses, *Intervirology* **17:**1–181.

McGinnes, L. W., and Morrison, T. G., 1986, Nucelotide sequence of the gene encoding the Newcastle disease virus fusion protein and comparisons of paramyxovirus fusion protein sequence, *Virus Res.* **5:**343–356.

McGinnes, L. W., Semerjian, A., and Morrison, T. G., 1985, Conformational changes in Newcastle disease virus fusion glycoprotein during intracellular transport, *J. Virol.* **56:**341–348.

McGinnes, L. W., Wilde, A., and Morrison, T. G., 1987, Nucleotide sequence of the gene encoding the Newcastle disease virus hemagglutinin-neuraminidase protein and comparisons of paramyxovirus hemagglutinin-neuraminidase protein sequence, *Virus Res.* **7:**187–202.

Merz, D. C., and Wolinsky, J. S., 1983, Conversion of non-fusing mumps virus infections to fusing infections by selective proteolysis of the HN glycoprotein, *Virology* **131:**328–340.

Millar, N. S., Chambers, P., and Emmerson, P. T., 1986, Nucleotide sequence analysis of the hemagglutinin-neuraminidase gene of Newcastle disease virus, *J. Gen. Virol.* **67:**1917–1927.

Miura, N., Uchida, T., and Okada, Y., 1982, HVJ (Sendai virus)-induced envelope fusion and cell fusion are blocked by monoclonal anti-HN protein antibody that does not inhibit hemagglutination activity of HVJ, *Exp. Cell Res.* **141:**409–420.

Miura, N., Nakatani, Y., Ishiura, M., Uchida, T., and Okada, Y., 1985, Molecular cloning of a full-length cDNA encoding the hemagglutinin-neuraminidase glycoprotein of Sendai virus, *FEBS Lett.* **188:**112–116.

Moore, N. F., and Burke, D. C., 1974, Characterization of the structural proteins of different strains of Newcastle disease virus, *J. Gen. Virol.* **25:**275–289.

Morrison, T. G., 1988, Structure, function and intracellular processing of paramyxovirus membrane proteins, *Virol. Res.* **10:**113–136.

Morrison, T. G., and McGinnes, L. W., 1989, Avian cells expressing the NDV hemagglutinin-neuraminidase protein are resistant to NDV infection, *Virology* **171:**10–17.

Morrison, T. G., and Simpson, D., 1980, Synthesis, stability, and cleavage of Newcastle disease virus glycoproteins in the absence of glycosylation, *J. Virol.* **36:**171–180.

Morrison, T. G., and Ward, L., 1984, Intracellular processing of the vesicular stomatitis virus glycoprotein and the Newcastle disease virus hemagglutinin-neuraminidase glycoprotein, *Virus Res.* **1:**225–239.

Morrison, T., McQuain, C., and Simpson, D., 1978, Assembly of viral membranes: Maturation of the vesicular stomatitis virus glycoprotein in the presence of tunicamycin, *J. Virol.* **28:**368–374.

Morrison, T. G., Chatis, P. A., and Simpson, D., 1981, Conformation and activity of the Newcastle disease virus HN protein in the absence of glycosylation, in "The Replication of Negative Strand Virus" (D. H. L. Bishop and R. H. Compans, eds.), pp. 471–477, Elsevier/North-Holland, Amsterdam.

Morrison, T., Ward, L., and Semerjian, A., 1985, Intracellular processing of the Newcastle disease virus fusion glycoprotein, *J. Virol.* **53:**851–857.

Morrison, T. G., Peeples, M. E., and McGinnes, L. W., 1987, Conformational changes in a viral glycoprotein during maturation due to disulfide bond disruption, *Proc. Natl. Acad. Sci. USA* **84:**1020–1024.

Morrison, T. G., McQuain, C., O'Connell, K. F., and McGinnes, L. W., 1990, Mature cell associated

HN protein of NDV exists in two forms differentiated by posttranslational modification, *Virus Res.* **5**:113–134.

Mottet, G., Portner, A., and Roux, L., 1986, Drastic immunoreactivity changes between the immature and mature forms of the Sendai virus HN and F_0 glycoproteins, *J. Virol.* **59**:132–141.

Munro, S., and Pelham, H. R. B., 1986, An Hsp-70-like protein in the ER: Identity with the 78kd glucose regulated protein and immunoglobulin heavy chain binding protein, *Cell* **46**:291–300.

Nagai, Y., and Klenk, H.-D., 1977, Activation of precursors to both glycoproteins of Newcastle disease virus by proteolytic cleavage, *Virology* **77**:125–134.

Nagai, Y., Klenk, H.-D., and Rott, R., 1976, Proteolytic cleavage of the viral glycoproteins and its significance for the virulence of Newcastle disease virus, *Virology* **72**:494–508.

Nakamura, K., Homma, M., and Compans, R. W., 1982, Effect of tunicamycin on the replication of Sendai virus, *Virology* **119**:474–487.

Needleman, S. B., and Wunsch, C. D., 1979, A general method applicable to the search for similarities in the amino acid sequence of two proteins, *J. Mol. Biol.* **48**:443–453.

Neyt, C., Geliebter, J., Slaoui, M., Morales, D., Meulemans, G., and Burny, A., 1989, Mutations located on both F1 and F2 subunits of the Newcastle disease virus fusion protein confer resistance to neutralization with monoclonal antibodies, *J. Virol.* **63**:952–954.

Nishikawa, K., Isomura, S., Suzuki, S., Watanabe, E., Hamaguchi, M., Yoshida, T., and Nagai, Y., 1983, Monoclonal antibodies to the HN glycoprotein of Newcastle disease virus, Biological characterization and use for strain comparisons, *Virology* **130**:318–330.

Nishikawa, K., Morishima, T., Toyoda, T., Miyadai, T., Yokochi, T., Yoshida, T., and Nagai, Y., 1986, Topological and operational delineation of antigenic sites on the HN glycoprotein of Newcastle disease virus and their structural requirements, *J. Virol.* **60**:987–993.

Novick, N. L., and Hoekstra, D., 1988, Membrane penetration of Sendai virus glycoprotein during the early stages of fusion with liposomes as determined by hydrophobic photoaffinity labeling, *Proc. Natl. Acad. Sci. USA* **85**:7433–7437.

Örvell, C., and Grandien, M., 1982, The effects of monoclonal antibodies on biologic activities of structural proteins of Sendai virus, *J. Immunol.* **129**:2779–2787.

Paterson, R. G., and Lamb, R. A., 1987, Ability of the hydrophobic fusion-related external domain of a paramyxovirus F protein to act as a membrane anchor, *Cell* **48**:441–452.

Paterson, R. G., Harris, T. J. R., and Lamb, R. A., 1984a, Analysis and gene assignment of mRNAs of a paramyxovirus, simian virus 5, *Virology* **138**:310–323.

Paterson, R. G., Harris, T. J. R., and Lamb, R. A., 1984b, Fusion protein of the paramyxovirus simian virus 5: Nucleotide sequence of mRNA predicts a highly hydrophobic glycoprotein, *Proc. Natl. Acad. Sci. USA* **81**:6706–6710.

Paterson, R. G., Shaughnessy, M. A., and Lamb, R. A., 1989, Analysis of the relationship between cleavability of paramyxovirus fusion protein and length of the connecting peptide, *J. Virol.* **63**:1293–1301.

Peeples, M. E., 1988, Newcastle disease virus replication, in "Developments in Veterinary Virology: Newcastle Disease Virus" (D. Alexander, ed.), pp. 45–78, Kluwer Academic Publishers, Boston, Massachusetts.

Peluso, R. W., Lamb, R. A., and Choppin, P. W., 1978, Infection with paramyxoviruses stimulates synthesis of cellular polypeptides that are also stimulated in cells transformed by Rous sarcoma virus or deprived of glucose, *Proc. Natl. Acad. Sci. USA* **75**:6120–6124.

Portner, A., 1981, Evidence for two different sites on the HN glycoprotein involved in neuraminidase and hemagglutinating activities, in "The Replication of Negative-Strand Viruses" (D. H. L. Bishop and R. W. Compans, eds.), pp. 465–470, Elsevier/North-Holland, New York.

Portner, A., 1984, Monoclonal antibodies as probes of the antigenic structure and functions of Sendai virus glycoproteins, in "Nonsegmented Negative Strand Viruses" (D. H. L. Bishop and R. W. Compans, eds.), pp. 345–350, Academic Press, Orlando, Florida.

Portner, A., and Kingsbury, D. W., 1976, Regulatory events in the synthesis of Sendai virus polypeptides and their assembly into virions, *Virology* **73**:79–88.

Portner, A., Scroggs, R. A., and Naeve, C. W., 1987a, The fusion glycoprotein of Sendai virus: Sequence analysis of an epitope involved in fusion and virus neutralization, *Virology* **157**:556–559.

Portner, A., Scroggs, R. A., and Metzger, D. W., 1987b, Distinct functions of antigenic sites of the HN glycoprotein of Sendai virus, *Virology* **158**:61–68.

Prehm, R., Scheid, A., and Choppin, P. W., 1979, The carbohydrate structure of the glycoproteins of the paramyxovirus SV5 grown in bovine kidney cells, *J. Biol. Chem.* **254:**9669–9677.

Ray, R., and Compans, R. W., 1986, Monoclonal antibodies reveal extensive antigenic differences between the hemagglutinin-neuraminidase glycoproteins of human and bovine parainfluenza 3 viruses, *Virology* **148:**232–236.

Richardson, C. D., and Choppin, P. W., 1983, Oligopeptides that specifically inhibit membrane fusion by paramyxoviruses: Studies on the site of action, *Virology* **131:**518–532.

Richardson, C. D., Scheid, A., and Choppin, P. W., 1980, Specific inhibition of paramyxovirus and myxovirus replication by oligopeptides with amino acid sequences similar to these at the N-terminal of the F_1 or HA_2 viral polypeptides, *Virology* **105:**205–222.

Richardson, C., Hull, D., Greer, P., Hasel, K., Berkovich, A., Englund, G., Bellini, W., Rima, B., and Lazzarini, R., 1986, The nucleotide sequence of the mRNA encoding the fusion protein of measles virus (Edmonston strain). A comparison of fusion proteins from several different paramyxoviruses, *Virology* **155:**508–523.

Rose, J. K., and Doms, R. W., 1988, Regulation of protein export from the endoplasmic reticulum, *Annu. Rev. Cell Biol.* **4:**258–288.

Rose, J. K., Adams, G. A., and Gallione, C. J., 1984, The presence of cysteine in the cytoplasmic domain of the vesicular stomatitis virus glycoprotein is required for palmitate addition, *Proc. Natl. Acad. Sci. USA* **81:**2050–2059.

Sakai, Y., and Shibuta, H., 1989, Syncytium formation by recombinant vaccinia viruses carrying bovine parainfluenza 3 virus envelope protein genes, *J. Virol.* **63:**3661–3668.

Satake, M., Coligan, J. E., Elango, N., Norrby, E., and Venkatesan, S., 1985, Respiratory syncytial virus envelope glycoprotein (G) has a novel structure, *Nucleic Acid Res.* **13:**7795–7812.

Sato, H., Hattori, S., Ishida, N., Imamura, Y., and Kawakita, M., 1987a, Nucleotide sequence of the hemagglutinin-neuraminidase gene of Newcastle disease virus avirulent strain D26: Evidence for a longer coding region with a carboxyl terminal extension as compared to virulent strains, *Virus Res.* **8:**217–232.

Sato, H., Oh-hira, M., Ishida, N., Imamura, Y., Hattori, S., and Kawahita, M., 1987b, Molecular cloning and nucleotide sequence of the *P*, *M*, and *F* genes of Newcastle disease virus avirulent strain D26, *Virus Res.* **7:**241–255.

Scheid, A., and Choppin, P. W., 1973, Isolation and purification of the envelope proteins of Newcastle disease virus, *J. Virol.* **11:**263–271.

Scheid, A., and Choppin, P. W., 1974, Identification of biological activities of paramyxovirus glycoproteins. Activation of cell fusion, hemolysis, and infectivity by proteolytic cleavage of an inactive precursor protein of Sendai virus, *Virology* **57:**470–490.

Scheid, A., and Choppin, P. W., 1976, Protease activation mutants of Sendai virus. Activation of biological properties by specific proteases, *Virology* **69:**265–277.

Scheid, A., and Choppin, P. W., 1977, Two disulfide-linked polypeptide chains constitute the active F protein of paramyxoviruses, *Virology* **80:**54–66.

Scheid, A., Caliguiri, L., Compans, R. W., and Choppin, P. W., 1972, Isolation of paramyxovirus glycoproteins. Association of both hemagglutinating and neuraminidase activities with the larger SV5 glycoprotein, *Virology* **50:**640–652.

Scheid, A., Graves, M., Silver, S., and Choppin, P. W., 1978, Studies on the structure and function of paramyxovirus glycoproteins, in "Negative Strand Viruses and the Host Cell" (B. W. J. Mahy and R. D. Barry, eds.), pp. 181–193, Academic Press, New York.

Schmidt, M. F. G., and Schlesinger, M. J., 1979a, Evidence for covalent attachment of fatty acids to Sindbis virus glycoproteins, *Proc. Natl. Acad. Sci. USA* **76:**1687–1691.

Schmidt, M. F. G., and Schlesinger, M. J., 1979b, Fatty acid binding to vesicular stomatitis virus glycoprotein: A new type of posttranslational modification of the viral glycoprotein, *Cell* **17:**813–819.

Schmidt, M. F. G., and Schlesinger, M. J., 1980, Relation of fatty acid attachment to the translation and maturation of vesicular stomatitis and Sindbis virus membrane glycoproteins, *J. Biol. Chem.* **255:**3334–3339.

Schulz, G. E., and Schirmer, R. H., 1979, "Principles of Protein Structure," Springer-Verlag, New York.

Schuy, W., Garten, W., Linder, D., and Klenk, H.-D., 1984, The carboxyterminus of the hemagglutinin-neuraminidase of Newcastle disease virus is exposed at the surface of the viral envelope, *Virus Res.* **1:**415–426.

Schwalbe, J. C., and Hightower, L. E., 1982, Maturation of the envelope glycoprotiens of Newcastle disease virus on cellular membranes, *J. Virol.* **41:**947–957.

Sechoy, O., Philippot, J. R., and Bienvenue, A., 1987, F Protein–F protein interaction within the Sendai virus identified by native bonding or chemical crosslinking, *J. Biol. Chem.* **262:**11519–11523.

Server, A. C., Smith, J. A., Waxham, M. N., Wolinsky, J. S., and Goodman, H. M., 1985, Purification and amino-terminal protein sequence analysis of the mumps virus fusion protein, *Virology* **144:**373–383.

Sheehan, J. P., Iorio, R. M., Syddall, R. J., Glickman, R. L., and Bratt, M. A., 1987, Reducing agent sensitive dimerization of the hemagglutinin–neuraminidase glycoprotein of Newcastle disease virus correlates with the presence of cysteine at residue 123, *Virology* **161:**603–606.

Shioda, T., Iwasaki, K., and Shibuta, H., 1986, Determination of the complete nucleotide sequence of the Sendai virus genome RNA and the predicted amino acid sequence of the F, HN and L proteins, *Nucleic Acids Res.* **14:**1545–1563.

Smith, G. W., and Hightower, L. E., 1980, Uncoupling of the hemagglutinating and neuraminidase activities of Newcastle disease virus (NDV), in "Animal Virus Genetics" (B. Fields, R. Jaenish, and C. F. Fox, eds.), pp. 623–632, Academic Press, New York.

Smith, G. W., and Hightower, L. E., 1981, Identification of the P proteins and other disulfide linked and phosphorylated proteins of Newcastle disease virus, *J. Virol.* **37:**256–267.

Smith, G. W., and Hightower, L. E., 1982, Revertant analysis of a temperature-sensitive mutant of Newcastle disease virus with defective glycoprotein: Implications of the fusion glycoprotein in cell killing and isolation of a neuraminidase-deficient hemagglutinating virus, *J. Virol.* **42:**659–668.

Spriggs, M. K., Olmsted, R. A., Venkatesan, S., Coligan, J. E., and Collins, P. L., 1986, Fusion glycoprotein of human parainfluenza virus type 3: Nucleotide sequence of the gene, direct identification of the cleavage-activation site, and comparison with other paramyxoviruses, *Virology* **152:**241–251.

Stallcup, K. C., and Fields, B. N., 1981, The replication of measles virus in the presence of tunicamycin, *Virology* **108:**391–404.

Steiner, D., Quinn, P., Chan, S., Marsh, J., and Tager, D., 1980, Processing mechanisms of the biosynthesis of proteins, *Ann. N.Y. Acad. Sci.* **343:**1–16.

Storey, D. G., Dimock, K., and Kang, C. Y., 1984, Structural characterization of virion proteins and genomic RNA of human parainfluenza virus 3, *J. Virol.* **52:**761–766.

Sugawara, K.-E., Tashiro, M., and Homma, M., 1982, Intermolecular association of HANA glycoprotein of Sendai virus in relation to the expression of biological activities, *Virology* **117:**444–455.

Thompson, S. D., and Portner, A., 1987, Localization of functional sites on the hemagglutinin-neuraminidase glycoprotein of Sendai virus by sequence analysis of antigenic and temperature sensitive mutants, *Virology* **160:**1–8.

Thompson, S. D., Laver, W. G., Murti, K., and Portner, A., 1988, Isolation of a biologically active soluble form of the hemagglutinin-neuraminidase protein of Sendai virus, *J. Virol.* **62:**4653–4660.

Toyoda, T., Sakaguchi, T., Imai, K., Inocencio, N. M., Gotoh, B., Hamaguchi, M., and Nagai, Y., 1987, Structural comparison of the cleavage-activation site of the fusion glycoprotein between virulent and avirulent strains of Newcastle disease virus, *Virology* **158:**242–247.

Toyoda, T., Gotoh, B., Sakaguchi, T., Kida, H., and Nagai, Y., 1988, Identification of amino acids relevant to three antigenic determinants on the fusion protein of Newcastle disease virus that are involved in fusion inhibition and neutralization, *J. Virol.* **62:**4427–4430.

Tozawa, H., Watanabe, M., and Ishida, N., 1973, Structural components of Sendai virus: Serological and physicochemical characterization of hemagglutinin subunit associated with neuraminidase activity, *Virology* **55:**242–253.

Tsukiyama, K., Sugiyama, M., Yoshikawa, Y., and Yamanouchi, K., 1987, Molecular cloning and sequence analysis of the rinderpest virus mRNA encoding the hemagglutinin protein, *Virology* **160:**48–54.

Tsukiyama, K., Yoshikawa, Y., and Yamanouchi, K., 1988, Fusion glycoprotein (F) of rinderpest virus; Entire nucleotide sequence of the F mRNA, and several features of the F protein, *Virology* **164**:523–530.

Undenfriend, S., and Kilpatrick, D., 1983, Biochemistry of the enkephalins and enkephalin-containing peptides, *Arch. Biochem. Biophys.* **221**:309–323.

Varghese, J. N., Laver, W. G., and Colman, P. M., 1983, Structure of the influenza virus glycoprotein antigen neuraminidase at 2.9 A resolution, *Nature* **303**:35–40.

Varsanyi, T. M., Jornvall, H., and Norrby, E., 1985, Isolation and characterization of the measles virus F1 polypeptide: Comparison with other paramyxovirus fusion proteins, *Virology* **147**: 110–117.

Veit, M., Schmidt, M. F. G., and Rott, R., 1989, Different palmitoylation of paramyxovirus glycoproteins, *Virology* **168**:173–176.

Vidal, S., Mottet, G., Kolakofsky, D., and Roux, L., 1989, Addition of high-mannose sugars must precede disulfide bond formation for proper folding of Sendai virus glycoproteins, *J. Virol.* **63**:892–900.

Von Heijne, G., 1983, Patterns of amino acids near signal sequence cleavage sites, *Eur. J. Biochem.* **133**:17–21.

Waxham, M. N., and Aronowski, J., 1988, Identification of amino acid involved in the sialidase activity of the mumps virus hemagglutinin-neuraminidase protein, *Virology* **167**:226–232.

Waxham, M. N ., Merz, D. C., and Wolinsky, J. S., 1986, Intracellular maturation of mumps virus hemagglutinin-neuraminidase glycoprotein: Conformation changes detected with monoclonal antibodies, *J. Virol.* **59**:392–400.

Waxham, M. N., Server, A. C., Goodman, H. M., and Wolinsky, J. S., 1987, Cloning and sequencing of the mumps virus fusion protein gene, *Virology* **159**:381–388.

Waxham, M. N., Aronowski, J., Server, A. C., Wolinsky, J. S., Smith, J. A., and Goodman, H. M., 1988, Sequence determination of the mumps virus *HN gene*, *Virology* **164**:318–325.

Webster, R. G., and Laver, W. G., 1975, Antigenic variation of influenza viruses, In: *The Influenza Viruses and Influenza* (E. D. Kilbourne, ed.), p. 209, Academic Press, New York.

Webster, R. G., Laver, W. G., Air, G. M., and Schild, G. C., 1982, Molecular mechanisms of variation in influenza viruses, *Nature (London)* **296**:115–121.

Wertz, G. W., Collins, P. L., Huang, Y., Gruber, C., Levine, S., and Ball, L. A., 1985, Nucleotide sequence of the G protein gene of human respiratory syncytial virus reveals an unusual type of membrane protein, *Proc. Natl. Acad. Sci. USA* **82**:4075–4079.

Wickner, W. T., and Lodish, H. F., 1985, Multiple mechanisms of protein insertion into and across membranes, *Science* **230**:400–407.

Wilson, C., Gilmore, R., and Morrison, T., 1987, Translation and membrane insertion of the hemagglutinin-neuraminidase glycoprotein of Newcastle disease virus, *Mol. Cell. Biol.* **7**: 1386–1392.

Wilson, C., Connolly, T., Morrison, T., and Gilmore, R., 1988, Integration of membrane proteins into the endoplasmic reticulum requires GTP, *J. Cell Biol.* **107**:69–77.

Wilson, C., Gilmore, R., and Morrison, T., 1990, Aberrant membrane insertion of a cytoplasmic tail deletion mutant of the hemagglutinin-neuraminidase glycoprotein of Newcastle disease virus, *Mol. Cell. Biol.* **10**:449–457.

Yamada, A., Takeuchi, K., and Hishiyama, M., 1988, Intracellular processing of mumps glycoproteins, *Virology* **165**:268–273

Yoshima, H., Nakanishi, M., Okada, Y., and Kobata, A., 1981, Carbohydrate structures of HVJ (Sendai virus) glycoproteins, *J. Biol. Chem.* **256**:5355–5361.

Yusoff, K., Nesbit, M., McCartney, H., Emmerson, P. T., and Samson, A. C. R., 1988, Mapping of three antigenic sites on the haemagglutinin-neuraminidase protein of Newcastle disease virus, *Virus Res.* **11**:319–334.

CHAPTER 14

The Unusual Attachment Glycoprotein of the Respiratory Syncytial Viruses
Structure, Maturation, and Role in Immunity

WAYNE M. SULLENDER AND GAIL W. WERTZ

I. INTRODUCTION

Human respiratory syncytial (RS) virus, a pneumovirus in the paramyxovirus family, differs from other members of this family in several ways: (1) it has additional genes, which include two independent genes for nonstructural proteins and a gene for a second matrixlike protein, (2) the organization of the genome differs from that of other paramyxoviruses, and (3) the major glycoprotein, G, is unique among paramyxovirus proteins (Collins and Wertz, 1986; Wertz *et al.*, 1985; see also Chapter 4). The G protein is characterized as the attachment protein of RS virus (Levine *et al.*, 1987). However, whereas the attachment proteins of the other paramyxoviruses possess hemagglutinin (HA) or hemagglutinin-neuraminidase (HN) activities, respectively, the G protein of RS virus lacks such activities (Richman *et al.*, 1971). Furthermore, analysis of the sequence of the G gene shows no similarity between the G protein of RS virus and the attachment proteins of the other paramyxoviruses, or indeed with any known RNA virus protein described to date (Wertz *et al.*, 1985). Instead, the structural features of G resemble those of a class of cellular proteins called the mucinous proteins. The purpose of this chapter is to describe the structure and function of the G protein and to discuss its role in infection and immunity to RS virus disease (see also Chapters 18 and 19).

WAYNE M. SULLENDER AND GAIL W. WERTZ • Department of Microbiology, University of Alabama School of Medicine, Birmingham, Alabama 35294.

II. THE STRUCTURE OF G

A. Primary Sequence

The sequence of complete cDNA clones of the G mRNA of the prototype A2 strain of human RS virus was the first to be described (see Appendix). A general overview of the structural features of the *G* gene based on that sequence will be described here. A discussion of the differences between the sequences of the G protein of human RS virus subgroups A and B and the bovine RS virus G protein will be presented below in the section on subgroup differences.

The G mRNA has 918 nucleotides and contains a single major open reading frame that encodes a polypeptide of 298 amino acids that has a predicted M_r of 32,600 (Wertz *et al.*, 1985; Satake *et al.*, 1985). This finding was surprising, since the mature G protein from the membranes of virions or virus-infected cells has a mobility on reducing SDS–polyacrylamide gels consistent with an M_r of approximately 88,000–90,000 (Dubovi, 1982; Gruber and Levine, 1983, 1985a; Lambert and Pons, 1983; Wertz *et al.*, 1985). Positive identification of the G-protein mRNA was provided by *in vitro* translation of G mRNA that had been hybrid-selected with the corresponding cDNA clones. A polypeptide with a mobility indicating an M_r of about 36,000 is the only product of the G mRNA and this polypeptide is specifically immunoprecipitated by antiserum prepared to the purified M_r 90,000 G protein isolated from RS virus-infected cells (Wertz *et al.*, 1985).

It is known that the G protein is extensively glycosylated (Gruber and Levine, 1985a; Lambert and Pons, 1983; Fernie *et al.*, 1985; Lambert, 1988). Initial data suggested that as much as 50–60% of the molecular weight of the G protein might be contributed by oligosaccharide addition to the M_r 32,600 polypeptide precursor to generate the mature M_r 90,000 G protein (Gruber and Levine, 1985a; Wertz *et al.*, 1985). Analysis of the deduced amino acid sequence of G, however, shows only four potential sites for N-linked oligosaccharide addition. Use of tunicamycin to inhibit N-linked sugar addition results in a reduction in the M_r of G by approximately 8000–10,000 (Gruber and Levine, 1985a; Fernie *et al.*, 1985; Wertz *et al.*, 1985). These data confirm the presence of some N-linked sugars in the structure of G, but not a contribution sufficient to account for the size increase from the M_r 32,600 polypeptide to the M_r 90,000 mature G. In addition, G is extensively glycosylated even in the presence of tunicamycin, a fact which indicates the presence of another type of oligosaccharide side chain, the most likely being O-linked oligosaccharides (Gruber and Levine, 1985a; Wertz *et al.*, 1985). This hypothesis is strengthened by analysis of the amino acid composition of G, which shows that G contains 30% serine plus threonine and 10% proline (Wertz *et al.*, 1985; Satake *et al.*, 1985). Serines and threonines are the attachment sites for O-linked oligosaccharides.

In summary, the sequence of the human RS virus G protein shows no relationships to the sequences of other RNA virus glycoproteins. Instead, the high content of serine, threonine, and proline in G and its extensive O-glycosylation are features characteristic of a class of cellular proteins termed the mucinous proteins (Wertz *et al.*, 1985; Kornfeld and Kornfeld, 1980). These observations suggest an evolutionary origin for the RS virus G protein distinct from the glycoproteins of other RNA viruses.

B. Structural Features of the G Polypeptide

Hydrophobicity analysis of the deduced amino acid sequence of G reveals two distinctive features. The protein contains neither a hydrophobic signal sequence at its N terminus nor a hydrophobic transmembrane anchor region near its C terminus (Wertz *et al.*, 1985). Instead, the N terminus is distinctly hydrophilic and the most hydrophobic region is between residues 38 and 66, which is postulated to serve both as a signal region and as the transmembrane anchor. Based on this analysis of the G protein sequence, it was proposed to be a type II integral membrane protein having a N-terminal anchor (residues 38–66) with the N terminus on the cytoplasmic side of the membrane and C terminus on the exoplasmic side. Consistent with this hypothesis, 77 of the 91 potential sites for the attachment of O-linked sugars and all of the potential N-linked attachment sites are located in the proposed external domain. A proposed model of the G protein structure is shown in Fig. 1. Present in the C-terminal 231-residue portion of the molecule are the 30 proline residues which contribute to the unusually high (10.1%) proline content of the protein and four cysteine residues, which are clustered in a 14-residue stretch (positions 173–186) (Wertz *et al.*, 1985; Satake *et al.*, 1985). As will be described below, these cysteines are highly conserved in different antigenic subgroups of RS virus (Johnson *et al.*, 1987a). Secondary structure predictions indicate that this region may contain a turn between two beta sheets (Garnier *et al.*, 1978). Norrby and colleagues (see Chapter 18) have generated synthetic peptides that include this cysteine-rich region and have analyzed immunological reactivity of a nested set of deleted peptides in reducing or nonreducing conditions. Their data indicate that the antigenic activity of this site is dependent on intrapeptide disulfide bond formation between Cys-176 and Cys-186.

Support for the proposed model for the orientation of G was provided by Vijaya *et al.* (1988), who constructed a series of C-terminally truncated mu-

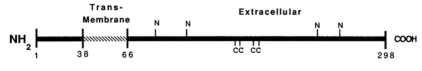

77   POTENTIAL O-LINKED SITES

FIGURE 1. Model for the structure of the RS virus glycoprotein G. The model is based on analysis of the deduced amino acid sequence and biochemical data discussed in the text. The major hydrophobic region between residues 38 and 66 (cross-hatched) is thought to serve as both signal and transmembrane anchor. The protein is oriented in the membrane with the N terminus in the cytoplasm and the C-terminal three-fourths of the molecule in the extracellular environment. Potential acceptor sites for N-linked carbohydrate attachment (N); 77 potential sites for O-linked carbohydrate attachment (Ser/Thr) located in the extracellular region are not indicated; conserved cysteine residues in the extracellular domain (C).

tants of the G protein linked to a reporter peptide and examined the ability of these mutants to be transported to and expressed on the cell surface. Mutants were constructed in which the G sequence was mutagenized at proline residues 71, 180, or 230 by inserting an oligonucleotide containing a restriction site for insertion of a repeating "reporter" peptide and a termination site. These truncated forms of G were placed in recombinant vaccinia virus vectors and intracellular transport and cell surface expression were analyzed. Similar studies using the same truncated mutants were subsequently carried out by Olmsted et al. (1989) using SV40 vectors. Removal of approximately three-quarters of the C terminus of G protein did not block membrane insertion and cell surface expression, as measured by indirect cell surface immunofluorescence. These data indicate that the N-terminal one-fourth of the 298-amino acid G protein containing the major hydrophobic region between residues 38 and 66 is sufficient to signal membrane insertion and cell surface expression. Further, detection of the C-terminal reporter peptide on the exterior surface of cells infected with the truncated mutants is consistent with the proposal that the C terminus of G is oriented extracellularly.

C. Carbohydrate of G

Peeples and Levine (1979), Gruber and Levine (1983, 1985a), Dubovi (1982), Bernstein and Hruska (1981), and Huang et al. (1985) characterized G as a glycoprotein. The sugars galactose, fucose, and glucosamine, but very little mannose, were reported in G (Gruber and Levine, 1985a; Levine et al., 1985). Fernie et al. (1985), Gruber and Levine (1985a), and Wertz et al. (1985) presented evidence using metabolic inhibitors that G contains both N- and O-linked oligosaccharides.

1. Analysis by Enzymes and Inhibitors

Direct evidence for the presence of O-linked as well as N-linked sugars on the G protein was provided by Lambert (1988), who used endo-β-N-acetylgalactosaminidase to confirm that sugar linkages on G are sensitive to this O-glycanase. Digestion with O-glycanase reduced the M_r 90,000 G to a population of species ranging from M_r 55,000 to 85,000, indicating a significant contribution of O-linked sugars to the structure of G. The size of G was reduced to approximately M_r 80,000 by treatment with endoglycosidase F and N-glycanase, enzymes which remove N-linked oligosaccharides.

In studies carried out in Chinese hamster ovary (CHO) cells using a recombinant vaccinia virus expression vector for the G protein (Wertz et al., 1989), three major forms of G protein were observed: (1) the nonglycosylated precursor G_p (predicted M_r 32,600, (2) an M_r 45,000 doublet intermediate, and (3) the M_r 88,000–90,000 mature form of G. Treatment of cells with tunicamycin to inhibit the addition of N-linked oligosaccharides results in synthesis of a fourth species of G having an electrophoretic mobility indicating that its M_r is approximately 8000–9000 less than the mature form. These forms of G have been observed in a variety of host cells infected with RS virus (Fernie *et al.*, 1985; Gruber and Levine, 1985a,b; Lambert, 1988; Wertz *et al.*, 1985). Previous studies of G in virus-infected cells using inhibitors and enzymatic digestion indicate that the M_r 45,000 form contains predominantly N-linked oligosaccharides (Gruber and Levine, 1985a; Fernie *et al.*, 1985; Lambert, 1988). The suggestion has been made that the mature G may consist of a dimer of two M_r 45,000 subunits (Fernie *et al.*, 1985). However, it is known that a glycosylated form of G with an apparent M_r of 82,000 is synthesized in the absence of N-linked sugar addition, and digestion with O-glycanase indicates that this form contains O-linked sugars (Lambert, 1988), making it unlikely that the mature form of G consists of dimers of the M_r 45,000 form.

2. Analysis of the Carbohydrate of G Using a Cell Line Deficient in O-Glycosylation

In other work, Wertz *et al.* (1989) used a CHO cell line that has a mutation in the ability to add O-linked sugars to demonstrate directly the contribution of O-linked sugars to the maturation of G. This cell line, termed ldlD, is defective both in the ability to add N-acetylgalactosamine (GalNAc), the first sugar in mucin-type O-linked oligosaccharide chain synthesis, to serine/threonine residues and in the ability to add galactose (Gal), which is required for the completion of both N- and O-linked chains (Kingsley *et al.*, 1986). Thus, these cells cannot add any O-linked oligosaccharides, although high mannose forms of N-linked sugars can be added. The defects in sugar addition are fully reversible by addition of GalNAc or Gal to culture media (Kingsley *et al.*, 1986). Thus, these cells, used in conjunction with an inhibitor of N-linked oligosaccharide synthesis such as tunicamycin, allow the establishment of conditions in which no carbohydrate addition occurs or in which either N-linked or O-linked carbohydrate addition occurs exclusively.

The contributions of N- or O-linked or both types of sugar addition to the synthesis of the human RS virus G protein were examined by comparing the synthesis of G protein in wild-type CHO cells with that in the mutant ldlD cells in the absence of any added sugars or in the presence of exogenously added Gal or GalNAc (Wertz *et al.*, 1989). The form of G synthesized in the ldlD cells in the presence of tunicamycin to prevent any N-linked sugar addition was also examined.

In the ldlD cells, a total of seven forms of the G protein can be discriminated under various conditions of sugar addition. These forms and the sugars they contain are summarized in Table I. The forms of G range from the nonglycosylated precursor G_p (M_r 32,600) to the M_r 45,000 intermediate containing only immature N-linked sugars, on to intermediates containing exclusively immature or mature O-linked sugars, termed G_{68} and G_{82}, respectively, or forms where addition of GalNAc allowed synthesis of both immature N-linked and immature O-linked sugars, termed G_{77-78}, up to the mature form termed G_{90}, containing both N- and O-linked sugars. These data show that the high-molecular-weight forms of G are observed only when O-glycosylation is permitted.

TABLE I. RS Virus G-Protein Intermediates Identified in ldlD Cells[a]

Media additions		Oligosaccharide attached	G protein product
Sugar[b]	Inhibitor		
0	Tunicamycin	None	G_p
0	0	N-linked, high mannose, immature complex	G_{45K}
Gal	0	N-linked mature complex	G_{45k_+}
GalNAc	Tunicamycin	O-linked (incomplete)	G_{68K}
GalNAc	0	N-linked immature complex, O-linked (incomplete)	G_{74-78K}
Gal + GalNAc	Tunicamycin	O-linked (complete)	G_{82K}
Gal + GalNAc	0	N-linked mature complex, O-linked (complete)	G_{90K}

[a]Summary of intermediates in G protein synthesis; for details, see Section II.C.2.
[b]Abbreviations: Gal, galactose; GalNAc, N-acetylgalactosamine.

III. MATURATION OF G

A. Intermediates in the Synthesis of G

Metabolic pulse-chase studies carried out in wild-type CHO cells or in the mutant ldlD cells under various conditions of sugar addition indicate that the pathway of maturation for the G protein proceeds from synthesis of the M_r 32,600 polypeptide accompanied by cotranslational attachment of high mannose forms of N-linked sugars in the rough endoplasmic reticulum to form an intermediate with M_r of 45,000. This step is followed by the Golgi-associated conversion of the N-linked sugars to the complex form. Addition of O-linked sugars results in the appearance of the mature form of G. Maturation of the M_r 45,000 N-linked form of G to the mature M_r 90,000 form occurs only in the presence of GalNAc, most probably because of the requirement of GalNAc for synthesis of O-linked sugars.

In the presence of the N-glycosylation inhibitor tunicamycin, G protein can mature directly from a nonglycosylated protein (precursor) to an M_r 82,000 form (addition of GalNAc plus galactose), due to GalNAc-dependent O-linked glycosylation. These data indicate that the mature G protein is not a dimer of two M_r 45,000, exclusively N-glycosylated species and that most of the shift in electrophoretic mobility is a consequence of GalNAc-dependent O-glycosylation. This O-glycosylation-dependent maturation can occur in the absence of N-glycosylation.

In other work, pulse-chase studies of the maturation of G in RS virus-infected cells also show that the M_r 45,000 form is converted to the mature form and that this conversion is inhibited by monensin, an ionophore that inhibits transport from the medial to the *trans* compartment of the Golgi complex (Gruber and Levine, 1985a,b; Fernie et al., 1985). These findings suggest that conversion to the mature form occurs in or beyond the *trans* Golgi. A preliminary report (Paradiso et al., 1987) suggested the M_r 45,000 form of G

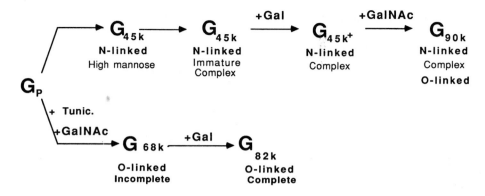

FIGURE 2. Proposed pathways of maturation for the G protein. For details see text, Section III.A.

contained N- and O-linked sugars, a finding which, if confirmed, would suggest that addition of O-linked sugars to G might occur before the *trans*-Golgi compartment, as observed for some cellular proteins (Elhammer and Kornfeld, 1984; Spielman *et al.*, 1987). However, there was no evidence for the presence of O-linked sugars in the M_r 45,000 intermediate in the ldlD cell studies (Wertz *et al.*, 1989), findings that are consistent with the results obtained by Gruber and Levine (1985a,b).

A tentative scheme for the sequence of events in maturation of the G protein is outlined in Fig. 2. It should be emphasized, however, that direct evidence, that is, evidence not involving inhibitors, for the time and place of O-glycosylation is not available.

The exact extent of O-glycosylation of G has not been determined. A preliminary report (Paradiso *et al.*, 1987) indicated that only 35% of the mass of G may be contributed by carbohydrate. For this to be the case, the mature molecular weight of G would have to be only 47,000–48,000. This appears unlikely in view of the other results reviewed here. We emphasize, however, that the molecular weight figures quoted throughout this chapter are approximations based on electrophoretic mobility. They are employed for ease of identification. The extensive glycosylation of G and the high content of proline (10%) may significantly alter the mobility of these proteins in gel electrophoresis and the resultant molecular weight determinations should be considered approximations at best.

B. Role of Carbohydrate in Cell Surface Maturation

The ldlD cells described above also provided the opportunity to examine the role of N- and O-linked sugar addition in transport to the cell surface. Expression of G on the surface of cells under each condition of sugar addition was monitored by indirect immunofluorescence and flow cytometry (Wertz *et al.*, 1989). These data show that transport of newly synthesized G to the surface of ldlD cells is severely inhibited when both N- and O-linked glycosylation are simultaneously blocked. However, the presence of either O-linked or N-linked (even immature N-linked) sugars allows cell surface expression. The mean level of expression of G on the surface in the absence of any O-glycosylation is approximately 50% of that observed under conditions of full sugar addition (Wertz et al., 1989).

C. Shed G

A soluble form of the attachment protein, G_s, is shed from infected cells (Hendricks et al., 1987, 1988). The G_s proteins have apparent M_r values 6000–9000 smaller than the virus-associated forms of G. The G_s protein is seen in culture fluids as early as 6 hr after infection, whereas virions were not released until 12 hr after infection, indicating that G_s is shed from intact, infected cells prior to the release of mature virus. Sequence analysis of purified G_s protein revealed two different N termini, indicating cleavage of the full–length G between amino acids 65 and 66 and 74 and 75 (Hendricks et al., 1988). These results suggest that two forms of G protein lacking the proposed intra-cytoplasmic and transmembrane domains are released from infected cells. The role of the shed G in infection and immunity is unknown.

IV. THE IMPORTANCE OF G IN THE ANTIGENIC DIVERSITY OF RS VIRUS

A. Recognition of Subgroups

It was not until 1985 that the existence of distinct antigenic subgroups of human RS virus was documented. Anderson et al. (1985) defined three sub-groups, designated groups 1, 2, and 3, while Mufson et al. (1985) described two distinct subtypes, which were called A and B. Prototype strains common to both publications are similarly categorized, with Anderson's group 1 and group 2 corresponding to Mufson's subtype A and subtype B, respectively. The excep-tion is the A2 strain, placed in group 3 by Anderson et al. and in subtype A by Mufson et al. Differentiation of viral isolates is based on their reactivity with monoclonal antibodies directed against several viral proteins; the most pro-nounced differences between strains are in the reactivity to the G protein (Anderson et al., 1985; Mufson et al., 1985).

Bovine RS virus, an important pathogen of cattle, is antigenically distinct from both of the human RS virus subtypes (Örvell et al., 1987). Neither poly-clonal nor monoclonal antibodies to the bovine RS virus G protein react with the human RS virus G, and antibodies directed against human RS virus G do not recognize the bovine virus G. Within the bovine viruses, variation in reactivity to the G protein occurs and distinct subgroups of bovine RS viruses may exist (Lerch et al., 1989).

1. Epidemiology of Subgroups

Documentation of the antigenic variability of RS strains and the existence of at least two major antigenic groups led to the analysis of collections of clinical isolates in an effort to understand the epidemiology and the relative importance of such variation. Hendry et al. (1986) examined isolates from infants hospitalized in Boston over two epidemic periods and found concurrent circulation of antigenically distinct strains of RS virus. In 1983–1984, 56% of the RS virus isolates were group 1, 41% were group 2, and 3% were intermediate. Isolates from 1981–1982 were 91% group 1 and 9% group 2. This establishes that year-to-year variation occurs in the incidence of isolation of the different groups. Among community-acquired isolates, both temporal and geographic clusterings of the two groups are observed in a single epidemic period (Hendry et al., 1986).

The different subgroups of RS virus play a role in infections in both hemispheres and around the world (Gimenez et al., 1986; Akerlind and Norrby, 1986; Storch and Park, 1987; Morgan et al. 1987; Mufson et al., 1988; Tsutsumi et al., 1988; Garcia-Barreno et al., 1989; Russi et al., 1989). Characterization of RS virus isolates from 12 university virology laboratories in the United States and Canada collected over two RS virus seasons, 1984–1986, revealed the presence of four antigenically distinct strains within the two subgroups. The authors noted the existence of multiple strains within each group, the cocirculation of multiple strains during outbreaks, and that outbreak strains follow local or regional, but not national patterns (Anderson et al., 1988).

2. Heterogeneity of the G Protein within Subgroups

Heterogeneity of the G protein among the subgroup B RS virus strains is reported, based on antigenic analysis of isolates with monoclonal antibodies directed against a subgroup B strain of RS virus (WV4843) (Örvell et al., 1987; Akerlind et al., 1988). Of 43 subgroup B isolates, 27 fail to react with four anti-G monoclonal antibodies representing a single epitope, G2. This nonreactive group is designated B1. The other 16 B strains which react with all four monoclonal antibodies directed to epitope G2 are designated B2. The B1 subgroup shows variability in the size of the G and P proteins, while the B2 subgroup has a larger G protein and a less variable P protein. The subgroup A proteins are more homogeneous than those of subgroup B. The B1 subgroup RS virus strains are seen annually, while the B2 strains were isolated in one season only (Alerlind et al., 1988).

Garcia-Barreno et al. (1989) analyzed 12 RS viruses collected over a 30-year period in a variety of geographic locations. Using monoclonal antibodies prepared against a subgroup A virus (Long), they detect more extensive antigenic variation in the G protein among strains of the same subgroup than has been previously described. Such variation occurs among members of both subgroups.

3. Conclusions

Recent studies thus define at least two major antigenic subgroups, A and B, of human RS virus. The greatest antigenic differences between strains are found in the reactivity of the G proteins (Mufson et al., 1985; Örvell et al., 1987; Storch and Park, 1987; Akerlind et al., 1988; Garcia-Barreno et al., 1989; Tsutsumi et al., 1989). These differences in immunological reactivity are re-

flected in the extensive sequence variability of the G protein between subgroups, as will be detailed in the next section. Within the subgroups, additional
antigenic heterogeneity of the G protein occurs (Anderson et al., 1985, 1988;
Akerlind et al., 1988; Garcia-Barreno et al., 1989; Tsutsumi et al., 1989). In
addition, structural differences between individual strains and/or between
members of the different subtypes have been noted, based on variability of
migration of the proteins N (Ward et al., 1983), P (Gimenez et al., 1986; Morgan
et al., 1987; Norrby et al., 1986; Akerlind et al., 1988), F, 22K (Norrby et al.,
1986), and G (Norrby et al., 1986; Walsh et al., 1987a; Storch and Park, 1987;
Akerlind et al., 1988).

B. Molecular Analysis of Subgroup Differences

1. Sequence Analysis of Human and Bovine RS Viruses

Additional G-gene nucleotide sequence information for another subgroup
B (8/60) and a bovine (391-2) RS virus allow us to confirm and extend the
earlier observations of Johnson et al. (1987a) with regard to molecular differences between subgroup A (A2, Long) and subgroups B (18537) RS viruses
(W. M. Sullender, K. Anderson, and G. W. Wertz, in press; R. A. Lerch, K.
Anderson and G. W. Wertz, in press). A general description of major features of
the G-protein amino acid sequence for the A2 RS virus is presented in a preceding section. These features, such as potential glycosylation sites, elevated proline contents, the presence of four cysteine residues in the extracellular domain, and plots of hydrophilicity, are similar for all four human (A2, Long,
18537, and 8/60) and the one bovine RS virus (391-2) G protein for which
sequence information is available (Table II; Fig. 4). Comparisons of the amino
acid sequences reveal fairly extensive divergence, particularly between sub-

TABLE II. Amino Acid Composition and Glycosylation Sites
of the Glycoprotein G of RS Viruses

RS virus (G length in amino acids)	Serine/threonine (%)	Proline (%)	N-linked sites (number)
A2 (298)	30	10	4
Long (298)	30	9	8
18537 (292)	28	9	3
8/60 (292)	30	9	3
Bovine (257)	26	8	4

[a]Deduced amino acid sequences were derived from published nucleotide sequences for the G proteins of A2 (Wertz et al., 1985), Long and 18537 (Johnson
et al., 1987a), 8/60 (W. M. Sullender, K. Anderson, and G. W. Wertz, unpublished data), and bovine RS viruses (R. A. Lerch, K. Anderson, and G. W.
Wertz, unpublished data). Protein lengths were deduced from apparent open
reading frames. Serine and threonine residues represent potential sites for
O-linked glycosylation. Numbers of potential N-linked glycosylation sites in
the extracellular domain are based on the sequence Asn–X–Ser/Thr with
suppression of X = Pro.

VIRUS STRAINS	CYTO-PLASMIC	TRANS-MEMBRANE	EXTRACELLULAR	TOTAL
Long vs A2	95%	93%	93%	93%
8/60 vs 18537	97%	100%	97%	98%
A2 vs 8/60	84%	86%	46%	55%
Bovine vs A2	43%	59%	21%	29%
Bovine vs 8/60	43%	59%	22%	30%

FIGURE 3. Amino acid identity comparisons between individual respiratory syncytial viruses for different domains of the glycoprotein G. Comparisons are of human viruses in the same subgroups (Long and A2 in subgroup A, 8/60 and 18537 in subgroup B), between subgroups (A2 and 8/60), and between the bovine virus (391-2) and members of both human virus subgroups. Amino acid sequences are deduced from nucleotide sequences reported for the G of A2 (Wertz et al., 1985), Long and 18537 (Johnson et al., 1987a), 8/60 (W. M. Sullender, K. Anderson, and G. W. Wertz, in press), and bovine RS virus (R. A. Lerch, K. Anderson, and G. W. Wertz, in press). The bar at the top of the figure indicates presumed domains and orientation of the protein in the membrane.

groups and between human and bovine viruses (Fig. 3). Examination of the various domains of the proteins (Fig. 3) shows the extracellular domain to be the most divergent, while the cytoplasmic and transmembrane regions have more shared amino acids. The bovine sequence is quite distinct from the human RS viruses and it lacks the 13-amino acid sequence in the extracellular domain which is shared by both A and B subgroup RS viruses (Johnson et al., 1987a; R. A. Lerch, K. Anderson and G. W. Wertz, in press).

Thus, the amino acid sequences for the G proteins of both human RS virus subgroups and bovine RS virus reveal extensive sequence divergence. Most similar are the amino acid sequences of the G proteins of viruses within the same subgroup, while greater differences exist between subgroups. The bovine RS virus G-protein amino acid sequence differs from that of human RS viruses to an even greater extent than do those of the human viruses between subgroups. Features which may be important for the secondary and tertiary structure of the proteins appear to be relatively conserved between the human and bovine viruses even though there is a wide divergence in terms of amino acid sequences.

The divergence of the G proteins between subgroups is in contrast to the similarities found between several other proteins. The deduced amino acid sequences are relatively highly conserved between subgroup B (18537) and subgroup A (A2) for the F (89%), N (96%), 1A (76%), 1B (92%), and 1C (87%) proteins (Johnson et al., 1987a; Johnson and Collins, 1988; Johnson and Collins, 1989).

2. Use of Nucleic Acid Hybridization in Subgroup Differentiation

Recently, we developed a nucleic acid hybridization assay which allows the determination of the subgroup classification of RS virus isolates. Replicate

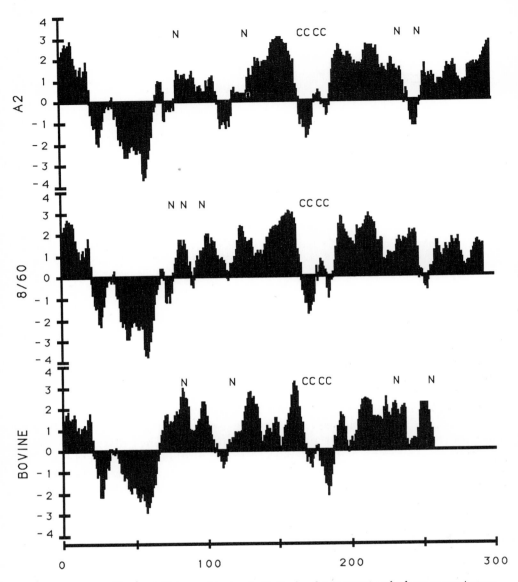

FIGURE 4. Plots of hydrophilicity and hydrophobicity for the G protein of a human respiratory syncytial virus subgroup A (A2) and subgroup B (8/60) and a bovine RS virus (391-2). The ordinates indicate hydrophilicity in positive and hydrophobicity in negative numbers. The abscissa scale extends from 0 to 300 amino acids with hash marks at 50-amino acid intervals. A window of nine amino acids was used to calculate the local hydrophobicity of each position by the procedure of Kyte and Doolittle (1982). C represents cysteine residues, N indicates potential N-linked glycosyl-ation sites (Asn–X–Ser/Thr with X = Pro suppressed) present in the extracellular domain. The 8/60 virus N-linked sites are identical to those of 18537 (Johnson et al, 1987a). Amino acid sequences were deduced from nucleotide sequences reported for the G of A2 (Wertz et al., 1985), Long and 18537 (Johnson et al., 1987a), 8/60 (W. M. Sullender, K. Anderson, and G. W. Wertz, unpublished data), and bovine RS virus (R. A. Lerch, K. Anderson, and G. W. Wertz, unpublished data).

templates were made from infected-cell suspensions prepared for RNA filter hybridization, as described by Paeratakul *et al.* (1988), and hybridized with isotope-labeled cDNA fragments corresponding to a major portion of the extra-cellular domain of either a subgroup A (A2) or a subgroup B (8/60) G protein. Using isolates previously grouped by analysis with monoclonal antibodies, all of 11 subgroup B and 15 subgroup A isolates were correctly classified (W. M. Sullender, L. J. Anderson, and G. W. Wertz, in press; Sullender *et al.*, 1989b). This technique offered an alternative to the use of monoclonal antibodies in the analysis of RS viruses. Preliminary evidence suggested that synthetic oligonucleotide probes may also be used to differentiate virus subgroups in this manner. The ability to obtain satisfactory hybridization to a variety of different isolates within each subgroup implied a relative conservation of the *G* gene within the subgroups. The lack of significant hybridization between subgroups confirms the extent of nucleic acid sequence divergence between subgroups noted above. This provides additional evidence for the divergence of the *G* gene between subgroups and relative conservation within the subgroups.

C. Immunological Differences

Recognition of the existence of at least two major subgroups of RS virus coincided with the availability of newer molecular and immunological tools for dissecting the immune response to RS virus. However, it has been shown recently that distinct subgroups of RS virus exist and that there is a preferen-tial pattern of antibody reactivity to the G protein of a homologous infecting strain compared to the reactivity to the G protein of a heterologous subgroup virus (described below). This finding indicates that any studies done to exam-ine the specific reactivity of antiserum to the G protein that have no documen-tation of the subgroup designation of the infecting strain or a measuring of responses to the G proteins of viral strains from both subgroups may not accurately describe such responses. We will focus on studies carried out since the identification of the existence of subgroups which meet these criteria. Additionally, until more information is available about the extent of hetero-geneity of the G protein of the subgroup B (Akerlind *et al.*, 1988) and the subgroup A (Garcia-Barreno *et al.*, 1989) viruses, caution should be exercised in evaluating the antibody response to the G protein of RS viruses.

1. Subgroup Dependence of Antibody Response to G

Differences in antibody reactivity between some RS viruses were recog-nized by early investigators. Coates *et al.* (1966) made estimations from the Archetti–Horsfall formula of only a 15% relatedness between the Long and 18537 RS viruses using ferret antisera in reciprocal plaque-reduction tests. Murphy *et al.* (1986) commented that Long strain glycoproteins are less effi-cient at detecting antibody rises in children infected with 18537-like strains. The degree of antigenic relatedness between subgroups estimated from the antibody responses of cotton rats using the Archetti–Horsfall formula reveals the F proteins to be more closely related (50% related), while the G proteins are much more different (5% related) between A and B subgroup RS viruses. Heterologous antibody responses to the G proteins are thus quite distinct (Johnson et al., 1987b).

Walsh et al. (1987a) prepared immunoaffinity-purified F and G proteins from both Long and 18537 strains and used them to produce antisera in rabbits. The Long and 18537 fusion proteins are similar in size and immunological reactivity; however, the Long and 18537 G proteins differ both in electrophoretic mobility, with 18537 G migrating at an apparent M_r of 78,000 and Long G at 84,000, and in immunological reactivity. Rabbit antiserum prepared after a subgroup A or subgroup B infection reacted preferentially with the homologous G protein by immune blotting and enzyme immunoassay. Furthermore, these monospecific antisera to G neutralized the homologous virus strain more efficiently than the heterologous strain (Walsh et al., 1987a). Synthetic peptides derived from corresponding regions on A and B subgroup virus G proteins also reacted preferentially with antisera produced after infection by the homologous subgroup virus (see Chapter 18).

2. Antibody Responses to G in Children

Hendry et al. (1988) reported the antibody responses after primary RS virus infections in a group of children in which the subgroup designation of the infecting strains was established. Antibody levels were measured by ELISA to immunoaffinity-purified F protein of Long strain RS virus and both Long and 18537 G proteins, and by neutralization assay. The response to the F protein (derived from Long, a subgroup A strain) was similar whether the infecting strain was Long-like or 18537-like (subgroup B). In contrast, ELISA antibody titers were significantly higher to the homologous rather than the heterologous strain G protein. Neutralizing antibody titers were also higher to the homologous type strain (Hendry et al., 1988).

Wagner et al. (1989) used purified Long F and G proteins from both Long and 18537 strains to analyze the IgG subclass responses of a group of children followed through their first three RS virus infections. The response to G was predominantly virus subgroup specific, and the F protein appeared to be more immunogenic than the G protein. Immunoglobulin subclass responses will be discussed below (Wagner et al., 1989).

V. ROLE OF CARBOHYDRATE IN THE IMMUNE RESPONSE TO THE G PROTEIN

The extensive glycosylation of the G glycoprotein has been postulated to affect its recognition by the immune system (Ward et al., 1983). The impaired ability of young infants to respond to carbohydrate antigens (Klein, 1982) led to questions about the ability of children to respond to this RS virus glycoprotein both after natural infection and if it were presented to them as a vaccine immunogen.

A. Immunoglobulin Subclass Responses

In humans, the antibody response to polysaccharide antigens is at least partially restricted to the IgG_2 subclass, while IgG_1 and IgG_3 are the predominant antiviral antibodies (Skavril, 1986). One approach to analyzing the ability of infants to respond to glycoprotein G has been to determine the IgG subclass composition of their response to the G protein after natural infection. Since responses to glycoprotein G show a relative RS virus subgroup specificity, studies which failed to define the subgroup of the infecting strain and/or measure antibody titers to the G protein of both A and B subgroups are difficult to interpret, as described in a previous section. Wagner et al. (1989), as mentioned above, reported an analysis of the responses of children experiencing two or three RS virus infections in the first 4 years of life, in which the subgroup of the initial infecting virus was either documented or inferred. Responses by ELISA to the G protein of both Long (an A subgroup virus) and 18537 (a B subgroup virus) were measured for IgG_1, IgG_2, and IgG_3 subclasses. IgG_1 and IgG_2 antibody levels were higher for the F than for the G protein after each infection, but the authors noted that substantial titers of IgG_2 antibody to both F and G proteins were detected after the first infection. The F protein appears to be a more potent immunogen than the G protein, particularly of an IgG_1 response. However, even after the first RS virus infection in children with an average age of less than 11 months, 90% or more had a fourfold rise in IgG_1 antibody titers to both the F and G glycoproteins. The IgG_1/IgG_2 ratio for the response to the G protein showed little change after each infection and is similar to that published for adults (Wagner et al., 1987). Thus, young children are able to mount a significant antibody response to G even though it is extensively glycosylated. The authors commented that the IgG subclass response to the RS virus glycoprotein G differs from that to polysaccharide antigens, and appears to be unique for a viral glycoprotein (Wagner et al., 1989).

B. Xid Mice

Another approach taken to help define the influence the extensive glycosylation of G might have on the immune response it stimulates used CBA/N mice with X chromosome-linked immune deficiency (Xid). Xid mice have a defect in their ability to respond to polysaccharide antigens, and barely detectable levels of circulating IgG_3 are present (Wicker and Scher, 1986). IgG_3 isotype antibodies account for most of the IgG anti-carbohydrate antibodies produced by mice (Briles, 1986). In CBA/N mice with the Xid defect, the females are immunologically normal and the males are affected. The ability of males and females to respond to the G and F proteins of RS virus expressed from vaccinia virus vectors was evaluated by measuring lung viral titers when mice were challenged with RS virus after immunization with these vectors. Equivalent protection was seen against RS replication in both the Xid mice and female mice, suggesting that in these mice the immune response to G is more complex than a simple humoral recognition of G as a polysaccharide antigen, as might have been predicted due to its extensive glycosylation (G. W. Wertz, W. M. Sullender, and D. E. Briles, unpublished observations).

VI. CHARACTERIZATION OF G AS A PROTECTIVE IMMUNOGEN

The G glycoprotein is of interest as a potential immunogen in vaccine development strategies because it is the viral attachment protein (Levine et al., 1987) and one of the major virus membrane proteins. Thus, it might be predicted to play an important role in interactions with the host immune response. The first evidence that G played an important role in the defense against RS virus was the report by Taylor et al. (1984) that passively administered monoclonal antibodies to G are capable of clearing an RS virus infection in mice. Analysis of responses to G have been performed in humans after RS virus infection and in animals after experimental infection and after exposure to immunoaffinity-purified G or to G expressed from recombinant vectors.

A. Recombinant Vaccinia Virus Vectors

1. Identification of G as a Protective Immunogen

Further evidence of the importance of G in providing a protective immune response is shown by analysis of the roles of individual RS viral gene products in providing protection, using recombinant vaccinia virus expression vectors. When mice (Stott et al., 1986) or cotton rats (Olmsted et al., 1986) were immunized with recombinant vaccinia virus vectors bearing the G gene of A2 RS virus, protection in mice was complete against A2 RS virus challenge and no virus was found in the lungs of the immunized mice, whereas the unimmunized control mice had $10^{4.5}$ plaque-forming units of virus per gram of lung tissue. The titer of virus in the lungs of immunized cotton rats was reduced by 2.7 $\log_{10}$ plaque-forming units of virus per gram of lung tissue. In the mouse experiments, vaccination with vaccinia virus vectors expressing G provided protection against homologous A2 RS virus as effectively as did vectors expressing F; in the cotton rat model, however, although G provided significant levels of protection, F consistently provided a higher level of protection (Stott et al., 1986, 1987; Johnson et al., 1987b; Wertz et al., 1987).

Olmsted et al. (1989) analyzed truncated cDNA mutants of the A2 virus G protein. These mutants contain cDNAs for the N-terminal 71 (G71), 180 (G180), and 230 (G230) amino acids of the G protein. When expressed from vaccinia virus recombinants and used to immunize cotton rats, only G230 afforded significant resistance to RS virus challenge, comparable to that of the complete G protein of 298 amino acids. These data suggest that the C-terminal 68 amino acids do not contribute to a major protective epitope of G.

2. Homotypic versus Heterotypic Protection

The ability of G to provide protection after immunization of cotton rats also has been evaluated against heterologous RS virus challenge. In the work of

Stott *et al.* (1987), animals immunized with vaccinia virus recombinants expressing the G of A2 RS virus (Ball *et al.*, 1986) manifest no statistically significant decrease in lung viral titers upon challenge with the B subgroup 8/60 virus, while Johnson *et al.* (1987b) reported a low but significant degree of protection after heterologous challenge with the subgroup B 18537 virus.

To examine the reciprocal situation, an 8/60 virus G cDNA clone was recombined into vaccinia virus and shown to express an authentic G protein. Cotton rats immunized with this virus showed a significant reduction in lung titers upon homologus (8/60 virus) challenge, but no significant reduction in lung titers was found after heterologous (A2) challenge (Sullender *et al.*, 1989a; W. M. Sullender, K. Anderson, and G. W. Wertz, in press). These differences in protection were in agreement with the relative subgroup specificity of the antibody response to the G protein after subgroup A or B infection described above (Wagner *et al.*, 1989; Walsh *et al.*, 1987a; Hendry *et al.*, 1988).

B. Immunoaffinity-Purified G

Walsh *et al.* (1987b) immunized cotton rats with immunoaffinity-purified glycoprotein G or F and demonstrated almost complete protection of the lungs against subsequent RS virus challenge. Viral replication in the nasal passages is not prevented, similar to the findings of Elango *et al.* (1986) of limited nasal resistance to RS virus challenge after immunization with vaccinia virus recombinants expressing the G protein. Routledge *et al.* (1988) immunized mice with purified RS virus G protein and observed lung protection similar to that in cotton rats.

C. Identification of Immunologically Important Epitopes of G

Examination of *in vitro* neutralization patterns with a panel of monoclonal antibodies to the F and G proteins demonstrated complete neutralization only with monoclonal antibodies to the F protein, and not with monoclonal antibodies to the G protein. The monoclonal antibodies to G reacted at ten epitopes corresponding to at least four antigenic sites on G and gave only partial or enhanced neutralization (Anderson *et al.*, 1988). Other reports of monoclonal antibodies to G show a similar lack of ability to neutralize virus (Mufson *et al.*, 1985; Örvell *et al.*, 1987). Walsh and Hruska (1983) described a single monoclonal antibody to G which is capable of neutralizing virus only in the presence of complement, and complement also enhanced the neutralizing ability of anti-G monoclonal antibodies reported by Garcia-Barreno *et al.* (1989). Tsutsumi *et al.* (1987) described three monoclonal antibodies which neutralize virus in the presence or absence of complement. In contrast, monospecific polyclonal antisera to G do neutralize virus in the absence of complement (Stott *et al.*, 1986; Olmsted *et al.*, 1986).

Variant viruses selected in the presence of complement with a monoclonal antibody directed against G no longer bind most of the remainder of a group of anti-G monoclonal antibodies, and polyclonal antisera raised against whole virus do not recognize the G protein of these variants. Competitive ELISAs using monoclonal antibodies suggest that the G protein includes a number of structurally overlapping individual epitopes which vary extensively. Garcia-Barreno et al. (1989) comment that even a few amino acid changes can produce significant antigenic alterations in the G protein. They observe that the Long and A2 viruses, which share a 94% G-protein amino acid identity, have only a 42% antigenic relatedness.

Norrby et al. (1987) synthesized a nested set of peptides 15 amino acids long, overlapping by five amino acids, that represented the extramembranous part of the G protein of a subgroup A RS virus. These peptides were analyzed by ELISA with human and animal antisera and murine anti-G monoclonal antibodies, and peptide 12 (amino acids 174–188 in the G of A2 RS virus) was the only peptide that reacted with all of these reagents.

In subsequent work (see Chapter 18), this epitope was demonstrated to be subgroup specific. The G protein of a subgroup B virus was found to have a highly antigenic site analogous to the site represented by peptide 12 of the subgroup A virus. The two sites showed subgroup-specific antigenic properties and evidence was found for the role of an intramolecular disulfide loop in the antigenic activity of the subgroup B G protein.

D. Role of Humoral versus Cell-Mediated Immunity

The importance of humoral immunity in the immune response to the G protein is shown by the following: (1) Passively administered monoclonal antibodies to G will clear an established RS virus infection in mice (Taylor et al., 1984). (2) High levels of antibody are produced in mice after immunization with vaccinia virus recombinants expressing the G gene and these mice have complete pulmonary protection against RS virus challenge (Stott et al., 1987).

Respiratory syncytial virus-specific cytotoxic T cells do not recognize G protein expressed by vaccinia virus recombinants (Bangham et al., 1986) and these same recombinants used as immunogens in mice are only capable of marginal priming of cytotoxic T cells (Pemberton et al., 1987). Thus, available data do not identify G as an important target for a cytotoxic T-cell response. Antibodies appear to play the major role in the response to the G protein. The mechanisms of RS virus neutralization by antibody have not been defined and are probably multiple (Anderson et al., 1989).

VII. SUMMARY AND PROSPECTS

The major glycoprotein G of respiratory syncytial virus clearly plays an important role in viral infection and in the immune response to infection. There has been significant progress in elucidating the structure, pathway of synthesis, and cell surface maturation of the G protein. Implicated as the attachment protein, G has a structure distinct from that of other paramyxovirus surface proteins and is remarkable for its high content of serine, threonine, and proline and its extensive modification by O-linked oligosaccharides, features which give it an overall resemblance to the cellular mucinous proteins.

The importance of both the N- and O-linked glycosylation of G in its structure and cell surface maturation has been documented. These studies have described new tools, such as the O-glycosylation-defective ldlD cells, which have allowed discrimination of the intermediates in glycosylation of G and will allow future investigations to determine the role of the carbohydrate of G in its biological function as the attachment protein and, importantly, in its recognition by the immune system.

The recognition of antigenically distinct subgroups of human RS virus and the antigenic and molecular variability of the G protein between subgroups documented to date have been reviewed here. In future work, molecular characterization of the extent of diversity within and between subgroups must be expanded. Also in the future, the role of subgroups and variation within subgroups on the epidemiology, clinical severity, and pattern of RS virus reinfection need to be documented.

Although both antigenic and amino acid sequence differences occur between subgroups, there are aspects of the sequences which suggest a conservation of structural features. The G proteins of all four human RS viruses and of bovine RS virus have high contents of serine and threonine, which are potential O-linked glycosylation sites. Evidence indicates that the extensive O-glycosylation is common to all RS virus G proteins. All five RS viruses share the location of four cysteine residues in the extracellular domain, which may be important in intramolecular bridging. In this same region, the four human RS viruses have a conserved 13-amino acid sequence. This conserved sequence is not shared by the bovine RS virus. All five viruses have elevated concentrations of proline, which have a strong effect on protein structure. Plots of overall hydrophobic/hydrophilic regions are similar, with a hydrophilic N-terminal cytoplasmic tail, a hydrophobic region believed to function as both transmembrane anchor and signal sequence, followed by the relatively hydrophilic extracellular domain. Thus, despite significant sequence divergence, there appear to be aspects common to all the RS virus G proteins which indicate that major structural features are conserved. These features may provide information useful for developing cross-subgroup-reactive immunogens or, for example, for constructing synthetic peptide immunogens that may be tested as potential vaccine candidates.

The evidence for conserved structural features of G in the presence of significant sequence diversity points to the importance of determining the three-dimensional structure of the G protein. An understanding of the structure of G at this level is a prerequisite for meaningful structure–function analyses of the G protein. Studies based on the known structure of G would allow identification of interacting sites in the mature molecule and, coupled with characterization of neutralization escape mutants and maturation-defective mutants, would allow identification of immunologically and biologically important regions of the protein. Such studies would open powerful new approaches for the design of antiviral substances and vaccines.

Efforts also need to be turned to the definition of the process of attachment and the mechanism of entry of RS virus into the cell. Characterization of the cellular receptor molecule for RS virus will provide insight into the tissue tropism of this respiratory tract pathogen and generate new opportunities for the preparation of reagents with which to investigate the pathogenesis of RS virus disease and, potentially, provide exciting new options for vaccine development, diagnosis, and therapy.

ACKNOWLEDGMENTS. The authors' work described herein was supported by Public Health Service grants AI 20181 and R37 AI 12464 to G.W.W. from the NIAID. This work also received support from the World Health Organization Program for Vaccine Development. W.S. received support from PHS grant F32 AI 07864 from NIAID. We thank Robert Lerch for making available data prior to publication.

VIII. REFERENCES

Akerlind, B., and Norrby, E., 1986, Occurrence of respiratory syncytial virus subtypes A and B strains in Sweden, *J. Med. Virol.* **19**:241–247.

Akerlind, B., Norrby, E., Orvell, C., and Mufson, M. A., 1988, Respiratory syncytial virus: Heterogeneity of subgroup B strains, *J. Gen. Virol.* **69**:2145–2154.

Anderson, L. J., Hierholzer, J. C., Tsou, C. Hendry, R. M., Fernie, B. F., Stone, Y., and McIntosh, K., 1985, Antigenic characterization of respiratory syncytial virus strains with monoclonal antibodies, *J. Infect. Dis.* **151**:626–633.

Anderson, L. J., Hendry, R. M., Pierik, L. T., and McIntosh, K., 1988, Multi-center study of strains of respiratory syncytial virus, *in* "Program and Abstracts of the Twenty-Eighth Interscience Conference on Antimicrobial Agents and Chemotherapy," 146.

Anderson, L. J., Bingham, P., and Hierholzer, J. C., 1989, Neutralization of respiratory syncytial virus by individual and mixtures of F and G protein monoclonal antibodies, *J. Virol.* **62**:4232–4238.

Ball, L. A., Young, K. Y., Anderson, K., Collins, P. L., and Wertz, G. W., 1986, Expression of the major glycoprotein G of human respiratory syncytial virus from recombinant vaccinia virus vectors, *Proc. Natl. Acad. Sci. USA* **83**:246–250.

Bangham, C. R. M., Openshaw, P. J. M., Ball, L. A., King, A. M. Q., Wertz, G. W., and Askonas, B. A., 1986, Human and murine cytotoxic T cells specific to respiratory syncytial virus recognize the viral nucleoprotein (N), but not the major glycoprotein (G), expressed by vaccinia virus recombinants, *J. Immunol.* **137**:3973–3977.

Bernstein, J. M., and Hruska, J. F., 1981, Respiratory syncytial virus proteins: Identification by immunoprecipitation, *J. Virol.* **38**:278–285.

Briles, D. E., Horowitz, J., McDaniel, L. S., Benjamin, W. H., Jr., Claflin, J. L., Booker, C. L., Scott, G., and Forman, C., 1986, Genetic control of the susceptibility to pneumococcal infection, *Curr. Top. Microbiol. Immunol.* **124**:103–120.

Coates, H. V., Alling, D. W., and Chanock, R. M., 1966, An antigenic analysis of respiratory syncytial virus isolates by a plaque reduction neutralization test, *Am. J. Epidemiol.* **83**:299–313.

Collins, P., and Wertz, G. W., 1986, Human respiratory syncytial virus genome and gene products, in "Concepts in Viral Pathogenesis II" (A. Notkins and M. Oldstone, eds.), pp. 40–46, Springer-Verlag, New York.

Dubovi, E. J., 1982, Analysis of proteins synthesized in respiratory syncytial virus-infected cells, *J. Virol.* **42**:372–378.

Elango, N., Prince, G. A., Murphy, B. R., Venkatesan, S., Chanock, R. M., and Moss, B., 1986, Resistance to human respiratory syncytial virus (RSV) infection induced by immunization of cotton rats with a recombinant vaccinia virus expressing the RSV G glycoprotein, *Proc. Natl. Acad. Sci. USA* **83**:1906–1910.

Elhammer, A., and Kornfeld, S., 1984, Two enzymes involved in the synthesis of O-linked oligosaccharides are localized on membranes of different densities in mouse lymphoma BW5147 cells, *J. Cell Biol.* **98:**327–331.

Fernie, B., Dapolito, G., Cote, P., and Gerin, J., 1985, Kinetics of synthesis of respiratory syncytial virus glycoproteins, *J. Gen. Virol.* **66:**1983–1990.

Garcia-Barreno, B., Palomo, C., Penas, C., Delgado, T., Perez-Brena, P., and Melero, J. A., 1989, Marked differences in the antigenic structure of human respiratory syncytial virus F and G glycoproteins, *J. Virol.* **63:**925–932.

Garnier, J., Osguthorpe, D., and Robson, B., 1978, Analysis of the accuracy for simple methods for predicting secondary structure of globular proteins, *J. Mol. Biol.* **120:**97–120.

Gimenez, H. B., Hardman, N., Keir, H. M., and Cash, P., 1986, Antigenic variation between human respiratory syncytial virus isolates, *J. Gen. Virol.* **67:**863–870.

Gruber, C., and Levine, S., 1983, Respiratory syncytial virus polypeptides, III. The envelope-associated proteins, *J. Gen. Virol.* **64:**825.

Gruber, C., and Levine, S., 1985a, Respiratory syncytial virus polypeptides. IV. The oligosaccharides of the glycoproteins, *J. Gen. Virol.* **66:**417–432.

Gruber, C., and Levine, S., 1985b, Respiratory syncytial virus polypeptides. V. The kinetics of glycoprotein synthesis, *J. Gen. Virol.* **66:**1241.

Hendricks, D. A., Baradaran, K., McIntosh, K., and Patterson, J. L., 1987, Appearance of a soluble form of the G protein of respiratory syncytial virus in fluids of infected cells, *J. Gen. Virol.* **68:**1705–1714.

Hendricks, D. A., McIntosh, K., and Patterson, J. L., 1988, Further characterization of the soluble form of the G glycoprotein of respiratory syncytial virus, *J. Virol.* **62:**2228–2233.

Hendry, R. M., Talis, A. L., Godfrey, E., Anderson, L. J., Fernie, B. F., and McIntosh, K., 1986, Concurrent circulation of antigenically distinct strains of respiratory syncytial virus during community outbreaks, *J. Inf. Dis.* **153:**291–297.

Hendry, R. M., Burns, J. C., Walsh, E. E., Graham, B. S., Wright, P. F., Hemming, V. G., Rodriguez, W. J., Kim, H. W., Gregory, A. P., McIntosh, K., Chanock, R. M., and Murphy, B. R., 1988, Strain-specific serum antibody responses in infants undergoing primary infection with respiratory syncytial virus, *J. Infect. Dis.* **157:**649–647.

Huang, Y. T., Collins, P. L., and Wertz, G. W., 1985, Characterization of the 10 proteins of human respiratory syncytial virus: Identification of a fourth envelope-associated protein, *Virus Res.* **2:**157–173.

Johnson, P. R., and Collins, P. L., 1988, The fusion glycoproteins of human respiratory syncytial virus of subgroups A and B: Sequence conservation provides a structural basis for antigenic relatedness, *J. Gen. Virol.* **69:**2623–2628.

Johnson, P. R., and Collins, P. L., 1989, The 1B (NS2), 1C (NS1) and N proteins of human respiratory syncytial virus (RSV) of antigenic subgroups A and B: Sequence conservation and divergence with RSV genomic RNA, *J. Gen. Virol.* **70:**1539–1547.

Johnson, P. R., Sprigs, M. K., Olmsted, R. A., and Collins, P. L., 1987a, The G glycoprotein of human respiratory syncytial virus of subgroups A and B: Extensive sequence divergence between antigenically related proteins, *Proc. Natl. Acad. Sci. USA* **84:**5625–5629.

Johnson, P. R., Olmsted, R. A., Prince, G. A., Murphy, B. R., Alling, D. W., Walsh, E. E., and Collins, P. L., 1987b, Antigenic relatedness between glycoproteins of human respiratory syncytial virus subgroups A and B: Evaluation of the contributions of F and G glycoproteins to immunity, *J. Virol.* **61:**3163–3166.

Kingsley, D. M., Kozarsky, K. F., Hobbie, L., and Krieger, M., 1986, Reversible defects in O-linked glycosylation and LDL receptor expression in a UDP-Gal/UDP-GalNAc 4-epimerase deficient mutant, *Cell* **44:**749–759.

Klein, J. O., Teele, D. W., Sloyer, J. L., Jr., Ploussard, J. H., Howie, V., Makela, P. H., and Karma, P., 1982, Use of pneumococcal vaccine for prevention of recurrent episodes of otitis media, *in* "Seminars in Infectious Diseases, Vol. 4, Bacterial Vaccines," (J. B. Robbins, J. C. Hill, and J. C. Sadoff, eds.), pp. 305–310, Thieme-Stratton, New York.

Klein, J. O., 1982, *in* "Seminars in Infectious Diseases, Vol. IV, Bacterial Vaccines" (J. B. Robbins, J. C. Hill, and J. C. Sadoff, eds.), p. 305.

Kornfeld, R., and Kornfeld, S., 1980, Structure of glycoprotein and their oligosaccharide units, *in* "The Biochemistry of Glycoproteins and Proteoglycans," (W. J. Lennarz, ed.), pp. 1–34, Plenum Press, New York.

Kyte, J., and Doolittle, R. F., 1982, A simple method for displaying the hydropathic character of a protein, *J. Mol. Biol.* **157:**105–132.

Lambert, D. M., 1988, Role of oligosaccharides in the structure and function of respiratory syncytial virus glycoprotein, *Virology* **164:**458–466.

Lambert, D. M., and Pons, M. W., 1983, Respiratory syncytial virus glycoproteins, *Virology* **130:**204–214.

Lerch, R. A., Stott, E. J., and Wertz, G. W., 1989, Characterization of bovine respiratory syncytial virus proteins and mRNAs and generation of cDNA clones to the viral mRNAs, *J. Virol.* **63:**833–840.

Levine, S. R., Klaiber-Franco, R., and Paradiso, P. R., 1987, Demonstration that glycoprotein G is the attachment protein of respiratory syncytial virus, *J. Gen. Virol.* **68:**2521–2524.

Morgan, L. A., Routledge, F. G., Willcocks, M. M., Samson, A. C. R., Scott, R., and Toms, G. L., 1987, Strain variation of respiratory syncytial virus, *J. Gen. Virol.* **68:**2781–2788.

Mufson, M. A., Orvel, C., Rafnar, B., and Norrby, E., 1985, Two distinct subtypes of human respiratory syncytial virus, *J. Gen. Virol.* **66:**2111–2124.

Mufson, M. A., Belshe, R. B., Orvell, C., and Norrby, E., 1988, Respiratory syncytial virus epidemics: Variable dominance of subgroups A and B strains among children, 1981–1986, *J. Infect. Dis.* **57:**143–148.

Murphy, B. R., Alling, D. W., Snyder, M. H., Walsh, E. E., Prince, G. A., Chanock, R. M., Hemming, V. G., Rodriguez, W. J., Kim, H. W., Graham, B. S., and Wright, P. F., 1986, Effect of age and preexisting antibody on serum antibody response of infants and children to the F and G glycoproteins during respiratory syncytial virus infection, *J. Clin. Microsc.* **24:**894–898.

Norrby, E., Mufson, M. A., and Sheshberadaran, H., 1986, Structural differences between subtype A and B strains of respiratory syncytial virus, *J. Gen. Virol.* **67:**2721–2729.

Norrby, E., Mufson, M. A., Alexander, H., Houghten, R. A., and Lerner, R. A., 1987, Site-directed serology with synthetic peptides representing the large glycoprotein G of respiratory syncytial virus, *Proc. Natl. Acad. Sci. USA* **84:**6572–6576.

Olmsted, R. A., Elango, N., Prince, G. A., Murphy, B. R., Johnson, P. R., Moss, B., Chanock, R. M., and Collins, P. L., 1986, Expression of the F glycoprotein of respiratory syncytial virus by a recombinant vaccinia virus: Comparison of the individual contributions of the F and G glycoproteins to host immunity, *Proc. Natl. Acad. Sci. USA* **83:**7462–7466.

Olmsted, R. A., Murphy, B. R., Lawrence, L. A., Elango, N., Moss, B., and Collins, P., 1989, Processing, surface expression, and immunogenicity of carboxy-terminally truncated mutants of G protein of human respiratory syncytial virus, *J. Virol.* **63:**411–420.

Örvell, C., Norrby, E., and Mufson, M. A., 1987, Preparation and characterization of monoclonal antibodies directed against five structural components of human respiratory syncytial virus subgroup B, *J. Gen. Virol.* **68:**1–11.

Paeratakul, U., DeStusio, P. R., and Taylor, W. M., 1988, A fast and sensitive method for detecting specific viral RNA in mammalian cells, *J. Virol.* **62:**1132–1135.

Paradiso, P., Hu, B., and Hildreth, S., 1987, Structure of the respiratory syncytial virus glycoprotein G, *in* "Abstracts of the Seventh International Congress of Virology, p. 189.

Peeples, M., and Levine, S., 1979, Respiratory syncytial virus polypeptides: Their location in the virion, *Virology* **95:**137–145.

Pemberton, R. M., Cannon, M. J., Openshaw, P. J. M., Ball, L. A., Wertz, G. W., and Askonas, B. A., 1987, Cytotoxic T cell specificity for respiratory syncytial virus proteins: Fusion protein is an important target antigen, *J. Gen. Virol.* **68:**2177–2182.

Richman, A. V., Pedreira, F. A., and Tauraso, N. M., 1971, Attempts to demonstrate hemagglutination and hemadsorption by respiratory syncytial virus, *Appl. Microsc.* **21:**1099–1100.

Routledge, E., Willcocks, M., Samson, A., Morgan, L., Scott, R., Anderson, J., and Toms, G., 1988, The purification of four RS virus proteins and their evaluation as protective agents against experimental infection in BALB/c mice, *J. Gen. Virol.* **69:**293–303.

Russi, J. C., Delfraro, A., Arbiza, J. R., Chiparelli, H., Orvell, C., Grandien, M., and Hortal, M., 1989, Antigenic characterization of respiratory syncytial virus associated with acute respiratory infections in Uruguayan children from 1985 to 1987, *J. Clin. Microsc.* **27:**1464–1466.

Satake, M., Coligan, J. E., Elango, N., Norrby, E., and Venkatesan, S., 1985, Respiratory syncytial virus envelope glycoprotein (G) has a novel structure, *Nucleic Acids Res.* **13:**7795–7812.

Skvaril, F., 1986, IgG subclasses in viral infections, *Monogr. Allergy* **19:**134–143.

Spielman, J., Rockley, N. L., and Carraway, K. L., 1987, Temporal aspects of O-glycosylation and cell surface expression of ascites sialoglycoprotein-1, the major cell surface sialomucin of 13762 mammary ascites tumor cells, *J. Biol. Chem.* **262:**269–275.

Storch, G. A., and Park, C. S., 1987, Monoclonal antibodies demonstrate heterogeneity in the G glycoprotein of prototype strains and clinical isolates of respiratory syncytial virus, *J. Med. Virol.*, **22:**345–356.

Stott, E., J., Ball, L. A., Young, K. K., Furze, J., and Wertz, G. W., 1986, Human respiratory syncytial virus glycoprotein G expressed from a recombinant vaccinia virus vector protects mice against live-virus challenge, *J. Virol.* **60:**607–613.

Stott, E. J., Ball, L. A., Anderson, K., Young, K. K., King. A. M. Q., and Wertz, G. W., 1987, Immune and histopathological responses in animals vaccinated with recombinant vaccinia viruses that express individual genes of human respiratory syncytial virus, *J. Virol.* **61:**3855–3861.

Sullender, W. M., Anderson, K., and Wertz, G. W., 1989a, Examination of cross-subgroup protection for respiratory syncytial virus, in "Program and Abstracts of the Twenty-Ninth Interscience Conference on Antimicrobial Agents and Chemotherapy."

Sullender, W. M., Anderson, L. J., and Wertz, G. W., 1989b, Differentiation of respiratory syncytial virus subgroups using nucleic acid hybridization , in "Program and Abstracts of the Twenty-Ninth Interscience Conference on Antimicrobial Agents and Chemotherapy."

Taylor, G., Stott, E. J., Bew, M., Fernie, B. F., Cote, P. J., Collins, A. P., Hughes, M., and Jebbett, J., 1984, Monoclonal antibodies protect against respiratory syncytial virus infection in mice, *Immunology* **52:**137–142.

Tsutsumi, H., Flanagan, T. D., and Ogra, P. L., 1987, Monoclonal antibodies to large glycoproteins of respiratory syncytial virus: Possible evidence of several functioning antigenic sites, *J. Gen. Virol.* **68:**2161–2167.

Tsutsumi, H., Onuma, M., Suga, K., Honjo, T., Chiba, Y., Chiba, S., and Ogra, P. L., 1988, Occurrence of respiratory syncytial virus subgroup A and B strains in Japan, 1980 to 1987, *J. Clin. Microsc.* **26:**1171–1174.

Tsutsumi, H., Nagai, K., Suga, K., Chiba, Y., Chiga, S., Tsugawa, S., and Ogra, P. L., 1989, Antigenic variation of human RSV strains isolated in Japan, *J. Med. Virol.* **27:**124–130.

Vijaya, S., Elango, N., Zavala, F., and Moss, B., 1988, Transport to the cell surface of a peptide sequence attached to the truncated C terminus of an N-terminally anchored integral membrane protein, *Mol. Cell. Biol.* **8:**1709–1714.

Wagner, D. K., Nelson, D. L., Walsh, E. E., Reimer, C. B., Henderson, F. W., and Murphy, B. R., 1987, Differential immunoglobulin G subclass antibody titers to respiratory syncytial virus F and G glycoproteins in adults, *J. Clin. Microsc.* **25:**748–750.

Wagner, D. K., Muelenaer, P., Henderson, F. W., Snyder, M. H., Reimer, C. B., Walsh, E. E., Anderson, L. J., Nelson, D. L., and Murphy, B. R., 1989, Serum immunoglobulin G antibody subclass response to respiratory syncytial virus F and G glycoproteins after first, second and third infections, *J. Clin. Microsc.* **27:**589–592.

Walsh, E. E., and Hruska, J., 1984, Monoclonal antibodies to respiratory syncytial virus proteins: Identification of the fusion proteins, *J. Virol.* **47:**171–177.

Walsh, E. E., Brandriss, M. W., and Schlesinger, J. J., 1987a, Immunological differences between the envelope glycoproteins of two strains of human respiratory syncytial virus, *J. Gen. Virol.* **68:**2169–2176.

Walsh, E. E., Hall, C. B., Briselli, M., Brandriss, M. W., and Schlesinger, J. J., 1987b, Immunization with glycoprotein subunits of respiratory syncytial virus to protect cotton rats against viral infection, *J. Infect. Dis.* **155:**1198–1204.

Ward, K. A., Lambden, P. R., Ogilvie, M. M., and Watt, P. J., 1983, Antibodies to respiratory syncytial virus polypeptides and their significance in human infection, *J. Gen. Virol.* **64:**1867–1876.

Wertz, G. W., Collins, P. L., Huang, Y., Gruber, C., Levine, S., and Ball, L. A., 1985, Nucleotide sequence of the G protein gene of human respiratory syncytial virus reveals an unusual type of viral membrane protein, *Proc. Natl. Acad. Sci. USA* **82:**4075–4079.

Wertz, G. W., Stott, E. J., Young, K. K. Y., Anderson, K., and Ball, L. A., 1987, Expression of the fusion protein of human respiratory syncytial virus from recombinant vaccinia virus vectors and protection of vaccinated mice, *J. Virol.* **61:**293–301.

Wertz, G. W. Krieger, M., and Ball, L. A., 1989, Structure and cell surface maturation of the attachment glycoprotein of human respiratory syncytial virus in a cell line deficient in O-glycosylation, *J. Virol.* **63:**4767–4776.

Wicker, L. S., and Scher, I., 1986, X-linked immune deficiency (xid) of CBA/N mice, *Curr. Top. Microbiol. Immunol.* **124:**87–101.

CHAPTER 15

New Frontiers Opened by the Exploration of Host Cell Receptors

MARY ANN K. MARKWELL

I. INTRODUCTION: DEFINITION OF A BIOLOGICAL RECEPTOR

The term virus receptor as used in this chapter is defined as a macromolecule or complex of macromolecules naturally occurring on the host cell surface which specifically binds the virus and through this binding facilitates the subsequent events of infection (Markwell *et al.*, 1984b). The specificity of the recognition between the viral binding protein and its cellular receptor and the involvement of the biological process of infection are the two key elements of the definition.

Although not intrinsic to the definition, certain assumptions about virus receptors have been commonly accepted. These include:

1. That the word "receptor" is synonymous with the word "protein."
2. That in any given cell type there is one and only one element that functions as the receptor.
3. That if a receptor is present in sufficient quantity on the cell surface, it will be recognized by its ligand.

As studies progressed on the receptors for the paramyxoviruses, it became evident at an early stage that new conceptual frontiers were ready for exploration by those willing to leave such preconceived ideas behind.

MARY ANN K. MARKWELL • Section on Molecular Pharmacology, Clinical Neuroscience Branch, National Institute of Mental Health, Bethesda, Maryland 20892.

II. CLUES FROM MODEL SYSTEMS: ERYTHROCYTES, LIPOSOMES, AND IMMOBILIZED SUBSTRATA

For more than four decades it has been accepted as dogma that molecules containing sialic acid (sialoglycoconjugates) serve as the cell surface receptors for paramyxoviruses. This assumption was based on the ability of *Vibrio cholerae* sialidase (formerly called neuraminidase) to destroy the receptors for these viruses and thus protect the host cell from infection (Stone, 1948; Marcus, 1959; Haff and Stewart, 1964).

Sialic acid is the group term for an acyl derivative of neuraminic acid. Approximately 20 different types of sialic acid have been discovered, but in humans the predominant, if not exclusive form, is N-acetylneuraminic acid (NeuAc) (Schauer, 1982). This form exists in humans both in proteins (sialoglycoproteins) and in lipids (sialoglycolipids or gangliosides).

Serum glycoproteins which contain NeuAc, such as fetuin, α_1-acid glycoprotein and β_2-macroglobulin, have been shown to cause the attachment of paramyxoviruses to sialidase-treated erythrocytes when coated on the surface of the membrane (Huang et al., 1973). In fact, from this knowledge, an elegant method for affinity purifying the virus attachment protein HN was devised using an immobilized fetuin substratum (Scheid and Choppin, 1974).

In a similar manner, gangliosides containing NeuAc incorporated into artificial membranes have been shown to cause the attachment of paramyxoviruses and to compete with erythrocytes for the virus, as measured by hemagglutination inhibition (Haywood, 1974; Sharom et al., 1976). Gangliosides containing more than one NeuAc group per molecule (di-, tri-, and tetrasialogangliosides) appeared to be more effective in hemagglutination inhibition (Haywood, 1975) and in promoting attachment of the virus in an ELISA-type assay (Holmgren et al., 1980a,b) than those containing only one NeuAc group (monosialogangliosides).

The consensus from a number of different laboratories working with different model systems, i.e., erythrocytes, liposomes, and artificial substrata, was that both sialoglycoproteins and gangliosides containing NeuAc could act as *attachment factors* for paramyxoviruses. The next step was to investigate whether one or both types of sialoglycoconjugates were actually involved in the infectious process as *receptors*.

III. RECEPTORS ON HOST CELLS

Enveloped viruses can penetrate host cells by at least two different routes: adsorptive endocytosis and direct membrane fusion (Fig. 1). The togaviruses, rhabdoviruses, and myxoviruses appear to enter cells via adsorptive endocytosis, a process which delivers the entire virus particle to the intracellular endosomal compartments (Fan and Sefton, 1978; Matlin et al., 1981; Yoshimura et al., 1982; Marsh, 1984). There, these viruses are uncoated by a membrane fusion event at acidic pH.

In contrast, for paramyxoviruses, the normal route of infection appears to be via a membrane fusion event at the neutral pH found at the cell surface

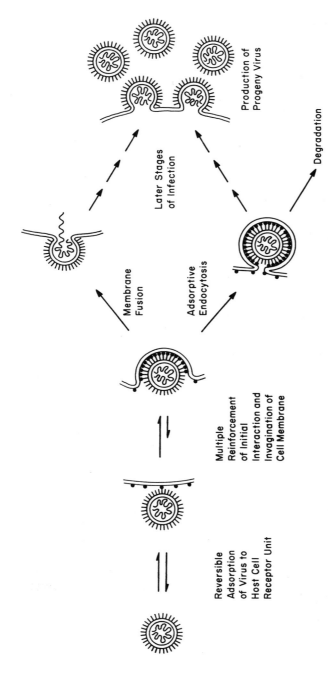

FIGURE 1. The role of the cell surface in infection of enveloped viruses. For Sendai virus, the route of entry that produces infection involves fusion of the viral membrane with the cell surface and a direct release of the viral nucleocapsid into the cytoplasm. Many other types of viruses, however, use adsorptive endocytosis to achieve infection. [Reprinted from Markwell *et al.* (1985) with permission.]

(Yasuda *et al.*, 1981). Fusion between the viral coat and the plasma membrane of the host cell directly releases the viral nucleocapsid into the cytoplasm to continue the subsequent events of infection (Homma and Ohuchi, 1973; Choppin and Scheid, 1980). Incompletely processed paramyxovirus particles which lack the ability to fuse at neutral pH still enter the cell by adsorptive endocytosis, but they are subsequently degraded.

Therefore, the receptor for paramyxoviruses must be an attachment factor which facilitates direct membrane fusion rather than adsorptive endocytosis. For membrane fusion to occur, it is estimated that the viral envelope and cell surface membrane must approach within 10–15 Å of each other (Gingell and Ginsberg, 1978), a situation favoring attachment to the proximal oligosaccharides of glycolipids over the more distal one of proteins (Yamakawa and Nagai, 1978). A study of Sendai virus fusing with liposomes at neutral pH (Haywood and Boyer, 1982) demonstrated that gangliosides caused both specific binding and fusion of the virus in the absence of any accompanying protein. This indicated that these glycolipids had the inherent capacity to act not only as attachment factors but also as receptors for this virus.

A. A Commonly Occurring Receptor Suggested by Fusion of Somatic Cells and by Systemic Infection by a Variant

A well-known corollary of viral envelope–cell fusion is the cell–cell fusion which occurs when paramyxoviruses are added at high multiplicities (Fig. 2). Paramyxoviruses, in particular Sendai virus, have been widely utilized as natural fusogens to create hybrid cells such as those which produce monoclonal antibodies. The observation that this virus not only can fuse a wide variety of somatic mammalian cells in culture, but can also grow readily in embryonated chicken eggs suggested that its receptor was a macromolecule conserved during evolution of these two classes of vertebrates.

In its natural host, the mouse, receptors for Sendai virus have been found in a number of organs, including the brain, lungs, heart, stomach, intestines, and kidneys (Ito *et al.*, 1983). However, when mice are infected intranasally or intravenously, the site of infection is localized in the lungs (Tashiro and Homma, 1983a). It has been demonstrated that the presence of specific proteases in the lung essential for the activation of the wild-type virus is the cause of this restriction of infection and is responsible for the development of pneumonia (Tashiro and Homma, 1983a,b). A pantropic variant (F1-R) of Sendai virus, which is cleaved by proteases present in many tissues, was found to cause a systemic infection of its natural host (Tashiro *et al.*, 1988). Viral antigens were detected in the brain, lung, pancreas, colon, kidney, and the testis, again emphasizing the wide distribution of the receptor.

B. Functional Receptor Assay in Host Cells

A functional receptor assay system based on the natural amplification of virus particles which occurs during infection was used to discriminate be-

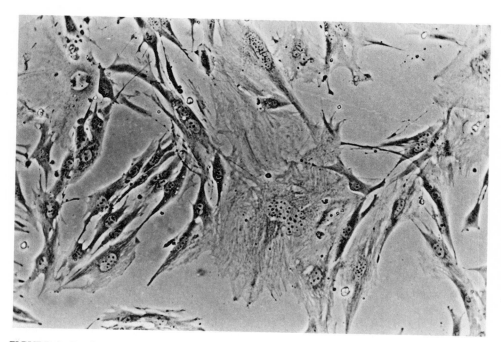

FIGURE 2. Sendai virus as a fusogen. The addition of high multiplicities of Sendai virus to fibroblastlike cells in culture results in the formation of polykaryons within 2–3 hr (M. A. K. Markwell, unpublished data).

tween attachment factors and receptors on the host cell surface and to quantify the latter (Markwell and Paulson, 1980). It was observed that the treatment of host cells with *Vibrio cholerae* sialidase which hydrolyzed endogenous cell surface sialoglycoconjugates rendered the cells resistant to infection by Sendai virus without altering their viability. Full susceptibility to infection could be restored to these cells by using a specific sialyltransferase which elaborates the NeuAc α2,3Gal (galactose) β1, 3GAlNAc (N-acetylgalactosamine) sequence on glycoproteins and glycolipids.

Treatment of host cells with proteases such as trypsin and Pronase had no effect on their susceptibility to infection (Markwell *et al.*, 1984a). It was also observed that sialoglycoproteins such as fetuin which are effective inhibitors of hemagglutination did not compete with host cell receptors during the process of infection. Thus, the focus on the search for the natural receptors for para-myxoviruses shifted from cell surface glycoproteins to gangliosides.

IV. GLYCOLIPIDS COME OF AGE AS RECEPTORS

The gangliosides containing the NeuAcα2,3Galβ1,3GalNAc sequence be-long to the gangliotetraose series, so named because all of its members contain the same neutral core of four (tetra) sugars (ose) and differ only in the number and attachment points of sialic acid residues to that core (Fig. 3). Members of

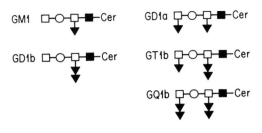

FIGURE 3. Chemical structure of gangliosides. Gangliosides on the right are isoreceptors for Sendai virus. Gangliosides on
the left belong to the same family as the isoreceptors, but lack the sialic acid on the terminal galactose and do not function as receptors for the virus. [Reprinted from Markwell
et al. (1981) with permission.]

this series have been shown to insert spontaneously into preformed living membranes when incubated with cells in culture (Fishman et al., 1976).

The simplest ganglioside of this series, the monosialoganglioside GM1, was first demonstrated to be a receptor for cholera toxin by Cuatrecasas (1973). The more complex gangliosides in the 1b subseries GD1b, GT1b, and GQ1b have been shown to be receptors for tetanus toxin (van Heyningen, 1976; Dimpfel et al., 1977; Holmgren et al., 1980a). Recently, the receptor structure of Clostridium botulinum neurotoxin type A was analyzed by thin-layer chromatography immunostaining (Takamizawa et al., 1986). GQ1b was found to be a high-affinity receptor and GT1b and GD1a moderate-affinity receptors for the neurotoxin.

A. Specificity and the Concept of Isoreceptors

The ability of members of the gangliotetraose series of gangliosides to function as receptors for Sendai virus was investigated in host cells which had been made receptor deficient by treatment with Vibrio cholerae sialidase. Individual, homogeneous preparations of each ganglioside were incubated with the cells before the addition of virus in the functional receptor assay system. It was observed that the gangliosides GM1 and GD1b, which lack the sialic acid on the terminal galactose (left, Fig. 3), do not function as receptors for the virus, but the gangliosides containing this sialic acid, GD1a, GT1b, and GQ1b (right, Fig. 3), did function as receptors for the virus. In addition, it was noted that the concentration of GQ1b needed for receptor response was 100-fold lower than that of GD1a and GT1b (Markwell et al., 1981). This is in good agreement with the results from ELISA-type binding of Sendai virus to gangliosides immobilized on plastic (Holmgren et al., 1980b), which suggested that GQ1b would be a high-affinity receptor for the virus and that GD1a and GT1b would have a lower binding capacity.

The fact that the virus binds to several different macromolecules in a receptor-type interaction does not imply a lack of specificity in the interaction. The manner in which gangliosides are synthesized, the addition of each new carbohydrate producing a different but structurally related member of the same series, creates a situation in which the same receptor determinant can be present on more than one macromolecule. Hence, isoreceptors are more the rule than the exception with gangliosides. And because each ganglioside iso-receptor serves as a biological precursor for the next most complex one in the series, these biosynthetically related isoreceptors exist within the same cell, unlike typical protein isoreceptors, which tend to be tissue or cell specific.

B. Identification in Host Cells

GD1a, GT1b, and GQ1b were shown to function as receptors when ex-ogenously added to receptor-deficient cells. The endogenous ganglioside con-tent of three cultured cell lines commonly used as hosts for paramyxoviruses, MDBK, HeLa, and MDCK cells, were analyzed to determine the amount and type of receptor ganglioside present. The presence of GM1, GD1a, and the

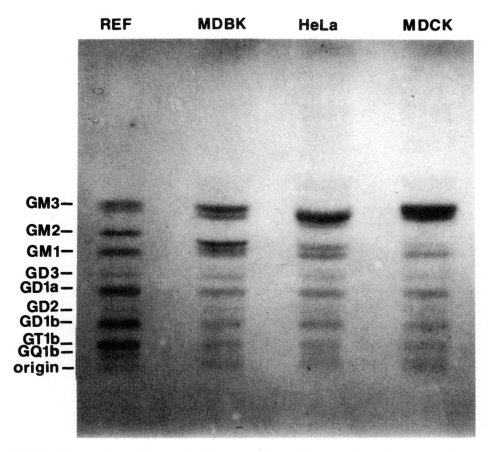

FIGURE 4. Comparison of the ganglioside content of typical host cells in culture. A total gang-lioside fraction was prepared from extracts of MDBK, HeLa, and MDCK cells and the individual species of gangliosides were resolved by chromatography on high-performance thin-layer chro-matographic plates developed in chloroform/methanol/0.25% KCl (aqueous) (50:40:10, v/v). The reference gangliosides were prepared from adult human brain with the addition of GM2 and GM3. Gangliosides were visualized by spraying with resorcinol reagent. [Reprinted from Markwell *et al.* (1984b) with permission.]

more complex homologs of the gangliotetraose series was established for each line (Fig. 4). In addition, the endogenous oligosialogangliosides present in each line of susceptible cells were shown to function as receptors for Sendai virus.

The ganglioside content of each cell line was shown not to be immutable, but instead to depend on the state of differentiation, passage number, and surface on which the cells were grown (Markwell et al., 1984b). Changes in conditions which decreased the receptor ganglioside content of the cells resulted in a corresponding decrease in susceptibility to infection.

C. Identification in Target Tissue

The primary site of infection for Sendai virus in the whole animal is the respiratory tract, which culminates in the lung. Therefore, the ganglioside content of this target organ in its natural host was analyzed to determine whether the same receptor gangliosides are present in lung as in host cells in culture.

The ganglioside pattern of mouse lung resembled that of the epithelial-like cultured MDBK, HeLa, and MDCK cells in its complexity. Most, if not all, of the sialic acid was found in the NeuAc form (Markwell and Sato, 1987). At least seven major gangliosides were observed. The presence of GD1a, the simplest of the isoreceptors for Sendai virus, was postulated from the migration of one of the major bands (Fig. 5). Confirmation of its presence and the presence of the more complex receptors GT1b and GQ1b was obtained by a thin-layer chromatogram overlay technique using radiolabeled cholera and tetanus toxins and an antibody to the gangliotetraose core structure in combination with hydrolysis in situ by bacterial sialidases as specific probes for these glycolipids. It was concluded that the endogenous receptor population available to Sendai virus in the lung was essentially the same as previously observed in cells in culture (Markwell and Sato, 1987).

D. Parallels between the Fusogenic Gradient and Receptor Function

The infectious mode of entry for Sendai virus is direct membrane fusion. The functional receptor assay pinpoints the effect of the gangliosides in host cells at the adsorption–fusion stage of viral infection. The close correlation between receptor specificities observed in host cells (Markwell et al., 1981) and the binding of virus to immobilized gangliosides (Holmgren et al., 1980b) demonstrates the involvement of gangliosides in the initial adsorption of the virus to the cell surface. The ability of Sendai virus to not only adsorb to, but also fuse with, liposomes containing receptor ganglioside in the complete absence of protein (Haywood and Boyer, 1982) opens the door to an intriguing question. Do these receptors facilitate not only the adsorption of the virus, but also its entry by fusion into the cell?

The structure of GD1a, the simplest of the isoreceptors, is shown in Fig. 6. Because gangliosides are amphiphilic lipids, they are natural detergents. Oligosialogangliosides, in particular, have the native ability to form bilayer-

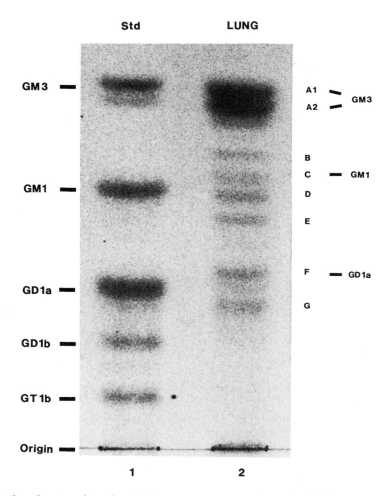

FIGURE 5. Identification of gangliosides in target tissue. A total ganglioside fraction was prepared from extracts of mouse lung and the individual species of gangliosides were resolved by chromatography on high-performance thin-layer chromatographic plates developed in chloroform/methanol/0.25% KCl (aqueous) (50:40:10, v/v). Gangliosides were visualized by spraying with resorcinol reagent. Individual gangliosides were identified by their migration and by the use of specific glycolipid probes, including bacterial toxins and an antibody to the gangliotetraose core structure. [Reprinted from Markwell and Sato, (1987) with permission.]

disrupting hexagonal mesophases that are conducive to membrane fusion when their local concentration exceeds 1.5% of the phospholipids (Sharom and Grant, 1978) or when Ca^{2+} is present as a cross-linking agent (Jaques *et al.*, 1977). It has been hypothesized (Markwell *et al.*, 1981) that the binding of a highly multivalent ligand such as Sendai virus to the cell surface may temporarily stabilize a clustering of its receptors, the oligosialogangliosides GD1a, GT1b, and GQ1b, in the very region of the host membrane where fusion must occur if infection is to proceed.

The hypothesis that oligosialogangliosides may be natural fusogens is sup-

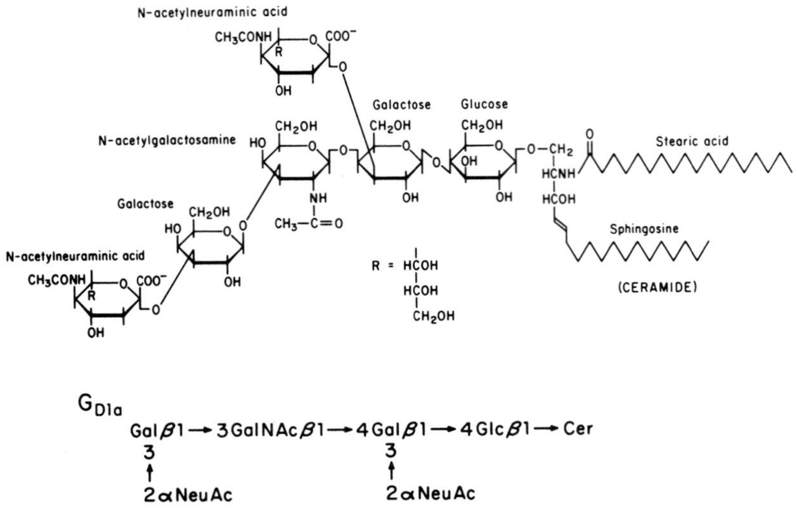

FIGURE 6. The structure of GDla. The amphiphilic nature of the simplest of the isoreceptor gangliosides GD1a is seen in the combination of the hydrophobic ceramide portion, which normally serves to anchor the ganglioside in the lipid bilayer of the membrane, and the hydrophilic carbohydrate portion, which provides the recognition site for the viral lectin, the HN protein. [M. A. K. Markwell (unpublished data).]

ported by several independent lines of evidence. First, in myogenesis, the concentration of oligosialogangliosides increases three-fold just prior to the fusion of myoblasts and returns to basal levels in myotubes (Whatley et al., 1976). Mutant myoblasts, unable to fuse into myotubes, do not synthesize the oligosialogangliosides. When these are added to the culture medium under conditions which would allow their insertion into the cell surface membrane, nonfusing subclones become fusogenic in the presence of Ca^{2+}.

The work of Maggio's group in Argentina has systematically shown that the oligosialogangliosides specifically interact with phosphatidylcholine in a mixed monolayer in the manner expected of lipids able to induce membrane fusion (Maggio and Lucy, 1976). They have also demonstrated that these potential fusogens, when added to chicken erythrocytes in the presence of Ca^{2+}, selectively cause fusion (Maggio et al., 1978). Polysialogangliosides are normal constituents of nerve endings, a site of repeated fusion between the synaptic vesicles and the synaptosomal plasma membrane. The addition of polysialogangliosides to an isolated synaptosomal preparation in the presence of Ca^{2+} mimicked this exocytotic event (Cumar et al., 1978). The fusogenic potential, as determined from these assays with cells and synaptosomes, i.e., that monosialogangliosides were relatively ineffective, disialogangliosides were moderately effective, and the polysialogangliosides were the most effective fusogens, matches the virus receptor potential for these same gangliosides.

Finally, there is the direct comparison of Sendai virus and the complex gangliosides as potential fusogens for the same cell line. The ability of the paramyxovirus to induce membrane fusion when incubated for 2–3 hr with fibroblast-type cells in culture is shown in Fig. 1. The ability of oligosialogan-

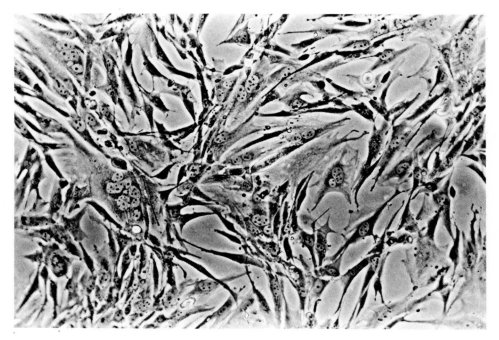

FIGURE 7. Polysialogangliosides as fusogens. The addition of microgram amounts of poly-sialogangliosides to the serum-free culture medium of fibroblastlike cells in culture results in the formation of polykaryons within 2–3 hr [M. A. K. Markwell (unpublished data).]

gliosides to induce membrane fusion in the same cell line when added to the culture media for 2–3 hr is shown in Figure 7 (M. A. K. Markwell, unpublished results). It is interesting to note that the susceptibility of four cell types (MK, MDBK, HaK, and BHK21-F) to fusion from within by SV5, a parainfluenza virus related to Sendai virus, corresponds directly to their endogenous ganglioside concentration (Choppin *et al.*, 1971).

V. EXPRESSION CONTROLS INFECTION

In their native state, NCTC 2071 cells are phenotypically receptor deficient for Sendai virus, i.e., they can neither be infected nor agglutinated by the virus unless receptor gangliosides are provided in the culture medium (Markwell *et al.*, 1986). However, a brief (10–15 sec) treatment of the cell surface with trypsin rendered more than 99% of the monolayer susceptible to infection. The concomitant increase in the binding of cholera toxin, a ganglioside-specific probe, was consistent with the hypothesis that the cells were genotypically positive, that the receptor gangliosides were at the cell surface, and that treatment with a protease increased the accessibility of the gangliosides at the cell surface to the virus.

Analyses of the sialoglycoconjugate content of NCTC 2071 cells indicated that (1) they had an unusually high ratio of sialoglycoproteins to gangliosides and (2) that the receptor ganglioside content of the cells was sufficient to render them susceptible to infection if they were accessible to the virus. The word "crypticity" was introduced in the literature in 1968 to explain the seeming inaccessibility to surface probes of glycolipids on the outer leaflet of the plasma membrane under certain conditions (Koscielak *et al.*, 1968). The reports that the phenotypic expression of globoside antigen on human erythrocytes (Koscielak *et al.*, 1968) and of GM1 on murine lymphocytes (Stein *et al.*, 1978) is enhanced after trypsin treatment indicate that membrane proteins can modulate the phenotypic expression of glycolipids. The studies in model systems by Peters *et al.* (1983) indicate that this is not due to an actual physical shielding of the glycolipids by these macromolecules, but rather is consistent with specific associations between these integral membrane components and the glycolipids which alter the organization of the membrane.

The ability of the receptor gangliosides to form clusters in the host cell membrane may be critical for the effective binding and entry of Sendai virus for two reasons. First, as a lectin, Sendai virus is unusually high in multivalency (Markwell, 1986). Each virus particle contains hundreds of HN (binding protein) spikes and each of these is tetravalent in itself (Markwell and Fox, 1980). This may allow the virus to make use of the more abundant but moderate-affinity receptors such as GQ1a and GT1b in addition to the rarer but high-affinity receptors such as GQ1b in binding to the targeted cell. Second, as previously mentioned, clustering of gangliosides induced by the binding of the virus may facilitate its entry into the cell by fusion.

In summary, three gangliosides belonging to the same family, specifically GD1a, GT1b, and GQ1b, have been identified as isoreceptors for Sendai virus both in cells in culture and in the target tissue. [Studies with model systems suggest that other gangliosides, such as the sialosylparagloboside, the major ganglioside of human erythrocytes, may also act as attachment factors for the virus (Umeda *et al.*, 1984), but their role as a receptor in infection remains to be demonstrated.] Cells such as NCTC 2071 cells, which do not express these glycolipids at their surface in an organizational manner recognized by viruses or toxins, are resistant to infection by Sendai virus.

VI. CURRENT STATUS OF RECEPTORS FOR PARAMYXOVIRUSES OTHER THAN SENDAI VIRUS

A. Human Parainfluenza Viruses, Respiratory Syncytial Virus, NDV, and Mumps Virus

Sendai virus (murine parainfluenza 1) is closely related antigenically to the human parainfluenza virus types 1–4. Of these, human parainfluenza 3 virus is the most prevalent and causes serious lower respiratory tract infections, particularly in children (Glezen *et al.*, 1984). Comparison of the amino acid sequence of the viral attachment protein HN of this virus to that of Sendai virus reveals some striking similarities (Storey *et al.*, 1987). There is a 43% overall similarity and up to 78% similarity in a number of highly conserved regions, including the 30 amino acids of the C terminus. These regions include sequences thought to be involved in the receptor-binding and sialidase activities of the virus. Little is known about the receptors for the human parainfluenza viruses beyond their classification as sialoglycoconjugates, based on their destruction by sialidases (Wassilewa, 1977; M. A. K. Markwell, unpublished results).

The epidemiology of human respiratory syncytial virus (RSV) resembles that of parainfluenza 3. RSV is the major cause of severe lower respiratory tract infection in infants and children throughout the world. Because of the age group it attacks, the lack of an effective vaccine, and the fact that infection can take place in the presence of circulating antibody (Chanock et al., 1982), annual epidemics which are a mix of new infections and reinfections occur each fall extending into early winter. Again, little is known about its receptor except its classification as a sialoglycoconjugate. Like Sendai virus, RSV mediates cell fusion (Walsh and Hruska, 1983), resulting in the formation of syncytia that constitute the characteristic cytopathology of its infection.

Newcastle disease virus (NDV) has been long known as one of the most diverse and deadly of the avian pathogens. The wide range of avian and mammalian cell lines it can infect or fuse in culture plus the isolation of the velogenic (highly virulent) strains from many different tissues, including the central nervous system in its native host, suggest that, like Sendai virus, its receptor is an entity both commonly found and highly conserved through evolution [see Peeples (1988) for recent review]. Although, like Sendai virus, NDV is considered nonpathogenic for humans, cases of conjuctivitis from researchers working with the virus have been reported and it readily infects human cell lines in culture. Thus, the availability of specific receptors for the viral attachment protein HN appears to be a less important determinant of host range than the interplay of the other products of the viral genome, antiviral substances, and the immunological defenses that are more species specific, e.g., the number of resident pulmonary alveolar macrophages which are of primary importance in the nonspecific defense of the lower respiratory tract of mammals [see Jakab (1984) for review]. This is in contrast to related viruses, such as influenza C virus and neurotropic strains of influenze A virus, which bind to other types of gangliosides (Herrler and Klenk, 1987a; M. A. K. Markwell, unpublished data) and whose tropism appears to be restricted by the more limited occurrence of these receptors (Herrler and Klenk, 1987b; M. A. K. Markwell, unpublished data).

Although the host cell receptor has not been defined for NDV, preliminary examinations of model systems, cells in culture, and tissue sections indicate that it is a sialoglycoconjugate in the NeuAc form, probably one or more gangliosides related to those which act as receptors for Sendai virus. Thus, treatment of host cells in culture with V. cholerae sialidase reduces the amount of virus attached (Wassilewa, 1977) and their susceptibility to infection (Haff and Stewart, 1964; M. A. K. Markwell, unpublished results). In a similar manner, treatment of tissue sections containing the basal layer of the epithelium of the chicken esophagus with this enzyme destroyed the ability of NDV to bind to it (Ito et al., 1982). Horse erythrocytes which contain the N-glycol form of sialic acid (NeuGly) rather than NeuAc are not agglutinated or lysed by Newcastle disease virus (Burnet and Lind, 1950). Suzuki et al. (1985) reported that asialoerythrocytes coated with various gangliosides interact with

Sendai virus and NDV, but their studies indicate that NDV recognizes both the NeuAc and NeuGly forms of sialoglycoconjugates, which seemingly is in contradiction to previous studies with horse erythrocytes.

As with the previously mentioned paramyxoviruses, infection by mumps virus appears to be initiated by droplet spread, with primary replication in the nasal mucosa or upper respiratory tract mucosal epithelium. Although the first clinical symptoms of swelling and inflammation usually relate to infection of the parotid gland, parotitis is not an obligate step in mumps infection. The virus also frequently disseminates to the kidneys and to the central nervous system (CNS). It has been estimated that half of all mumps infections are associated with replication of the virus within the CNS (Feldman, 1976). Therefore, the receptor for mumps virus, although not yet defined, is expected to be a sialoglycoconjugate present in both extraneural and neural tissues.

In summary, much less is known about the receptors for these other paramyxoviruses than for Sendai virus. Yet the similar epidemiology of these viruses, the sharing of the respiratory tract as a common primary site of infection, antigenic and structural similarities among their proteins, especially the attachment proteins, their relationship through evolution, and the emphasis on fusion as the mode of initial entry into the cell and in the cell-to-cell spread of infection suggest that the receptors for other paramyxoviruses will be structurally and functionally related to the isoreceptors defined for Sendai virus.

B. The Morbilliviruses

Measles, canine distemper, and rinderpest viruses are members of the genus *Morbillivirus* in the family *Paramyxoviridae* (Kingsbury *et al.*, 1978). Although infections by these viruses are also initiated in the respiratory tract, the members of this genus differ from the members of the genus *Paramyxovirus* in several features which directly bear on the identity of their receptors. They do not possess a receptor-destroying sialidase activity and their cellular receptors appear to be insensitive to the action of bacterial and viral sialidases (Howe and Lee, 1972). In keeping with these observed differences, the first reports of a receptor for measles virus focus on one or more proteins found in Vero cells (Krah and Choppin, 1988). Using the antiidiotypic antibody approach, five protein bands ranging in apparent molecular weight from 12,500 to 140,000 have been observed on SDS–PAGE gels after immunoprecipitation of Vero cell extract with mouse sera containing the antibody. These putative receptors, which are protease sensitive, have recently been solubilized from host cells in a form which inhibits measles virus attachment, infectivity, and hemagglutination activities (Krah, 1989).

VII. QUESTIONS REMAINING

A. A Second Receptor for the F Protein?

There are tantalizing clues in the literature that in addition to the specific interaction between the viral attachment protein and its cell surface receptor, the fusion protein of the virus may also interact with specific elements of the host-cell surface membrane. With regard to the paramyxoviruses, there is the suggestion that this may be the lipid phosphatidylcholine. Evidence for this comes from the report that the fusion protein F of Sendai virus possesses phophatidylcholine transport activity and that there are structural similarities between the viral fusion protein and a phosphatidylcholine transport protein isolated from beef liver (Demel *et al.*, 1987). It is tempting to speculate about a coordinate involvement between the receptor gangliosides for Sendai virus and phosphatidylcholine in the fusion process because it has been shown that these gangliosides form specific associations with phophatidylcholine through interactions with their polar head groups (Cumar *et al.*, 1978).

With regard to the morbilliviruses, especially measles virus, another membrane element may be involved in a specific interaction with its fusion protein. A synthetic hydrophobic peptide carbobenzoxy-D-Phe-L-Phe-Gly which mimics the highly conserved N-terminal region of the fusion protein of measles virus specifically inhibits membrane fusion by that virus (Richardson *et al.*, 1981). This peptide is homologous to the neuropeptide, substance P. The additon of substance P has been observed to reduce the amount of measles virus produced when measured after the first round of replication (Schroeder, 1986). Very recently it has been postulated that measles virus uses substance P receptors during infection of human IM-9 lymphocytes, because measles virus and substance P reciprocally compete in binding to an M_r 52,000–58,000 protein previously shown to comprise the substance P receptor (Harrowe *et al.*, 1990). The involvement of this protein in the fusion process is indicated by observations that anti-substance P antiserum inhibits the cell-to-cell spread of measles virus and that substance P itself blocks the fusion of measles virus with target cells.

B. Proteins As Receptors or Modulators?

With regard to the morbilliviruses, it is probable that one or more proteins will eventually be identified as receptors for these viruses. However, the possible involvement of proteins as receptors for the other paramyxoviruses is open to speculation. In the best-characterized receptor system, i.e., for Sendai virus, the cumulative evidence does not support the involvement of proteins as receptors in addition to gangliosides. Specifically, as already mentioned in Section IV.D, it has been shown that gangliosides in and of themselves not only directly bind the virus, but also facilitate the ensuing fusion event. Treatment of host cells with proteases either has no effect or enhances infection by the virus. Hemagglutination inhibitors such as glycophorin and fetuin do not compete with cellular receptors for the virus in infection. Cells such as the NCTC 2071 cells which are phenotypically negative for receptors involved in binding and infection by Sendai virus do not express the ganglioside isoreceptors, but do contain the normal amount of sialoglycoprotein. Finally, although some glycolipids and glycoproteins do share carbohydrate sequences in common, the sequences found on the ganglioside isoreceptors have not been identified on any typical sialoglycoprotein. However, as previously discussed in Section V, proteins may still prove to play an important role in infection as modulators, i.e., by their ability to mask cell surface gangliosides and thus prevent their interaction with the infecting virus.

C. Therapeutic Potential of Receptor Analogs?

With the identification of isoreceptors for one paramyxovirus, the question arises as to possible receptor analogs which might be used to prevent or treat viral infections. These would be especially valuable for viruses for which an effective vaccine is still not available and for those individuals who, because of age (infants, children, and the elderly) or because of other factors, cannot mount an appropriate protective immunological response. The initial outlook looks promising in this regard.

First, the site of primary infection, the respiratory tract, is one which readily lends itself to the introduction of compounds in the form of aerosols such as nasal sprays. Second, nature has already provided us with receptor analogs in the form of mucins lining the respiratory tract. These could be used as models for building more efficient, biologically compatible receptor analogs. Third, oligosaccharides and glycolipids are stable to storage and easy to administer. Fourth, in the limited clinical studies in which patients have received milligram amounts of gangliosides intramuscularly on a daily basis for treatment of nerve damage [see Haferkamp (1987) for review], there have been no negative side effects such as the development of antibodies to gangliosides. Thus, the study of receptors for paramyxoviruses may provide us with conceptually new approaches to limiting infection by these viruses.

ACKNOWLEDGMENTS. I would like to thank the many virologists who provided me with reprints and preprints of their work, but most especially Drs. David Krah and Donald Payan. Research cited from our own laboratory was supported in part by research grants from USPHS AI-15629-22817, the Kroc Foundation, and the National Multiple Sclerosis Society.

VIII. REFERENCES

Burnet, F. M., and Lind, P. E., 1950, Haemolysis by Newcastle disease virus II. General character of hemolytic action, *Aust. J. Exp. Biol. Med.* **28**:129–150.

Chanock, R. M., Kim, H. W., Brandt, C. D., and Parrott, R. H., 1982, *in* "Viral Infections of Humans: Epidemiology and Control," (A. S. Evans, ed.), pp. 471–489, Plenum Press, New York.

Choppin, P. W., and Scheid, A., 1980, The role of viral glycoproteins in adsorption, penetration, and pathogenicity of viruses, *Rev. Infect. Dis.* **2**:40–61.

Choppin, P. W., Klenk, H.-D., Compans, R. W., and Caliguiri, L. A., 1971, The parainfluenza virus SV5 and its relationship to the cell membrane, *Perspect. Virol.* **VII**:127–158.

Cuatrecasas, P., 1973, Interaction of *Vibrio cholerae* enterotoxin with cell membranes, *Biochemistry* **12**:3558–3566.

Cumar, F. A., Maggio, B., and Caputo, R., 1978, Dopamine release from nerve endings induced by polysialogangliosides, *Biochem. Biophys. Res. Commun.* **84**:65–69.

Demel, R. A., Sehgal, P. B., and Landsberger, F. R., 1987, A structural model for fusion of viral and cellular membranes, *in* "The Biology of Negative Strand Viruses" (B. Mahy and D. Kolakofsy, eds.), pp. 26–32, Elsevier Science Publishers, New York.

Dimpfel, W., Huang, R. T. C., and Habermann, E., 1977, Gangliosides in nervous tissue cultures and binding of ^{125}I-labeled tetanus toxin, a neuronal marker, *J. Neurochem.* **29**:329–334.

Fan, D. P., and Sefton, B. M., 1978, The entry into host cells of Sindbis virus, vesicular stomatitis virus and Sendai virus, *Cell* **15**:985–992.

Feldman, H. A., 1976, Mumps, in "Viral Infections of Humans: Epidemiology and Control" (A. S. Evans, ed.), pp. 317–336, Plenum Press, New York.

Fishman, P. H., Moss, J., and Vaughan, M., 1976, Uptake and metabolism of gangliosides in transformed mouse fibroblasts, *J. Biol. Chem.* **251**:4490–4494.

Gingell, D., and Ginsberg, L., 1978, Problems in the physical interpretation of membrane interaction and fusion, in "Membrane Fusion" (G. Poste and G. L. Nicolson, eds.), pp. 791–833, North Holland, New York.

Glezen, W. P., Frank, A. L., Taber, L. H., and Kasel, J. A., 1984, Parainfluenza virus type 3: Seasonality and risk of infection and reinfection in young children, *J. Infect. Dis.* **150**:851–857.

Haferkamp, G., 1987, Present state of clinical experience of ganglioside application in man, in "Gangliosides and Modulations of Neuronal Functions" (H. Rahmann, ed.), pp. 573–580, Springer-Verlag, Berlin.

Haff, R. F., and Stewart, R. C., 1964, Role of sialic acid receptors in adsorption of influenza virus to chick embryo cells, *J. Immunol.* **94**:842–851.

Harrowe, G., Mitsuhashi, M., and Payan, D. G., 1990, Measles virus–substance P receptor interactions: Possible novel mechanism of viral fusion, *J. Clin. Invest.* **85**:1324–1327.

Haywood, A. M., 1974, Characteristics of Sendai virus receptors in a model membrane, *J. Mol. Biol.* **83**:427–436.

Haywood, A. M., 1975, Model membranes and Sendai virus surface–surface interactions, in "Negative Strand Viruses," Vol. 2 (W. J. Mahy and R. D. Barry, eds.), pp. 923–928, Academic Press, New York.

Haywood, A. M., and Boyer, B. P., 1982, Sendai virus membrane fusion: Time course and effect of temperature, pH, calcium and receptor concentration, *Biochemistry* **21**:6041–6046.

Herrler, G., and Klenk, H.-D., 1987a, Restoration of receptors for influenza C virus on chicken erythorcytes by incorporation of bovine brain gangliosides, in "The Biology of Negative Strand Viruses" (B. Mahy and D. Kolakofsky, eds.), pp. 63–67, Elsevier, New York.

Herrler, G., and Klenk, H.-D., 1987b, The surface receptor is a major determinent of the cell tropism of influenza C virus, *Virology* **159**:102–108.

Holmgren, J., Elwing, H., Fredman, P., and Svennerholm, L., 1980a, Polystyrene-adsorbed gangliosides for investigation of the structure of the tetanus toxin receptor, *Eur. J. Biochem.* **106**:371–379.

Holmgren, J., Svennerholm, L., Elwing, H., Fredman, P., and Stannegard, O., 1980b, Sendai virus receptor: Proposed recognition based on binding to plastic-adsorbed gangliosides, *Proc. Natl. Acad. Sci. USA* **77**:1947–1950.

Homma, M., and Ohuchi, M., 1973, Trypsin action on the growth of Sendai virus in tissue culture cells. III. Structural difference of Sendai viruses grown in eggs and tissue culture cells, *J. Virol.* **12**:1457–1465.

Howe, C., and Lee, L. T., 1972, Virus–erythrocyte interactions, *Adv. Virus Res.* **17**:1–50.

Huang, R. T. C., Rott, R., and Klenk, H.-D., 1973, On the receptor of influenza viruses 1. Artificial receptor for influenza virus, *Z. Naturforsch* **28c**:342–345.

Ito, Y., Yamamoto, F., Takano, M., Maeno, K., Shimokata, K., Iinuma, M., Hara, K., and Iijima, S., 1982, Comparative studies on the distribution mode of orthomyxo-virus and paramyxo-virus receptor possessing cells in mice and birds, *Med. Microbiol. Immunol.* **171**:59–68.

Ito, Y., Yamamoto, F., Takano, M., Maeno, K., Shimokata, K., Iinuma, M., Hara, K., and Iijima, S., 1983, Detection of cellular receptors for Sendai virus in mouse tissue sections, *Arch. Virol.* **75**:103–113.

Jakab, G. J., 1984, Viral–bacterial interactions in respiratory infections: A review of the mechanisms of virus-induced suppression of pulmonary antibacterial defenses, in "Bovine Respiratory Disease" (R. W. Loan, ed.), pp. 223–286, Texas A&M University Press, College Station, Texas.

Jaques, L. W., Brown, E. B., Barrett, J. M., Wallace, Jr., S. B., and Weltner, Jr., W., 1977, Sialic acid: A calcium binding carbohydrate, *J. Biol. Chem.* **252**:4533–4538.

Kingsbury, D. W., Bratt, M. A., Choppin, P. W., Hanson, R. P., Hosaka, Y., ter Meulen, V., Norrby, E., Plowright, W., Rott, R., and Wunner, W. H., 1978, *Paramyxoviridae, Intervirology* **10**:137–152.

Koscielak, J., Hakomori, S.–I., and Jeanloz, R. W., 1968, Glycolipid antigen and its antibody, *Immunochemistry* **5**:441–455.

Krah, D. L., 1989, Characterization of octyl glucoside-solubilized cell membrane receptors for binding measles virus, *Virology* **172**:386–390.

Krah, D. L., and Choppin, P. W., 1988, Mice immunized with measles virus develop antibodies to a cell surface receptor for binding virus, *J. Virol.* **62**:1565–1572.

Maggio, B., and Lucy, J. A., 1976, Polar-group behavior in mixed monolayers of phospholipids and fusogenic lipids, *Biochem. J.* **155**:353–364.

Maggio, B., Cumar, F. A., and Capulto, R., 1978, Induction of membrane fusion by poly-sialogangliosides, *FEBS Lett.* **90**:149–152.

Marcus, P. I., 1959, Symposium on the biology of cells modified by viruses or antigens IV. Single-cell techniques in tracing virus–host interactions, *Bacteriol. Rev.* **23**:232–249.

Markwell, M. A. K., 1986, Viruses and hemagglutinins and lectins, in "Microbial Lectins and Agglutinins" (D. Mirelman, ed.), pp. 21–53, Wiley, New York.

Markwell, M. A. K., and Fox, C. F., 1980, Protein–protein interactions within paramyxoviruses identified by native disulfide bonding or reversible chemical cross-linking, *J. Virol.* **33**:152–166.

Markwell, M. A. K., and Paulson, J. C., 1980, Sendai virus utilizes specific sialyloligosaccharides as host cell receptor determinants, *Proc. Natl. Acad. Sci. USA* **77**:5693–5697.

Markwell, M. A. K., and Sato, E., 1987, Receptor gangliosides identified in target tissue for Sendai virus, in "Glycoconjugates: Proceedings of the 9th International Symposium" (J. Montreuil, A. Verbert, G. Spik, and B. Fournet, eds.), p. G33, A. Lerouge, Tourcolng, France.

Markwell, M. A. K., Svennerholm, L., and Paulson, J. C., 1981, Specific gangliosides function as host cell receptors for Sendai virus, *Proc. Natl. Acad. Sci. USA* **78**:5406–5410.

Markwell, M. A. K., Fredman, P., and Svennerholm, L., 1984a, Specific gangliosides are receptors for Sendai virus, *Adv. Exp. Biol.* **174**:369–379.

Markwell, M. A. K., Fredman, P., and Svennerholm, L., 1984b, Receptor ganglioside content of three hosts for Sendai virus: MDBK, HeLa, and MDCK Cells, *Biochim. Biophys. Acta* **775**:7–16.

Markwell, M. A. K., Portner, A., and Schwartz, A., 1985, An alternative route of infection for viruses: Entry by means of the asialoglycoprotein receptor of a Sendai virus mutant lacking its attachment protein, *Proc. Natl. Acad. Sci. USA* **82**:978–982.

Markwell, M. A. K., Moss, J., Hom, B. E., Fishman, P. H., and Svennerholm, L., 1986, Expression of gangliosides as receptors at the cell surface controls infection of NCTC 2071 cells by Sendai virus, *Virology* **155**:356–364.

Marsh, M., 1982, The entry of enveloped viruses into cells by endocytosis, *Biochem. J.* **218**:1–10.

Matlin, K. S., Reggio, H., Helenius, A., and Simons, K., 1981, Infectious entry pathways of influenza virus in a canine kidney line, *J. Cell Biol.* **91**:601–613.

Peeples, M. E., 1988, Newcastle disease virus replication, in "Newcastle Disease" (D. J. Alexander, ed.), pp. 45–78, Kluwer Academic Publishers, Boston.

Peters, M. W., Singleton, C., Barber, K. R., and Grant, C. W. M., 1983, Glycolipid crypticity in membranes—Not a simple shielding effect of macromolecules, *Biochim. Biophys. Acta* **731**:475–482.

Richardson, C. D., Scheid, A., and Choppin, P. W., 1981, Specific inhibition of paramyxovirus replication by hydrophobic oligopeptides, in "The Replication of Negative Strand Viruses" (D. H. L. Bishop and R. W. Compans, eds.), pp. 509–515, Elsevier/North-Holland, New York.

Schauer, R., 1982, Chemistry, metabolism, and biological functions of sialic acids, *Adv. Carbohydr. Chem. Biochem.* **40**:131–234.

Scheid, A., and Choppin, P. W., 1974, The hemagglutinating and neuraminidase protein of a paramyxovirus: Interaction with neuraminic acid in affinity chromatography, *Virology* **62**:125–133.

Schroeder, C., 1986, Substance P, a neuropeptide, inhibits measles virus replication in cell culture, *Acta Virol.* **30**:432–435.

Sharom, F. J., and Grant, C. W. M., 1978, A model for ganglioside behavior in cell membranes, *Biochim. Biophys. Acta* **507**:208–293.

Sharom, F. J., Barratt, D. G., Thede, A. E., and Grant, C. W. M., 1976, Glycolipids in model membranes: Spin-label and freeze-etch studies, *Biochim. Biophys. Acta* **455**:485–492.

Stein, K. E., Schwarting, G. A., and Marcus, D. M., 1978, Glycolipid markers of murine lymphocyte subpopulations, *J. Immunol.* **120**:676–679.

Stone, J. D., 1948, Prevention of virus infection with enzyme of *V. cholerae* I. Studies with viruses of mumps–influenza group in chick embryos, *Aust. J. Exp. Biol. Med. Sci.* **26**:49–64.

Storey, D. G., Coté, M.-J., Dimock, K., and Kang, C. Y., 1987, Nucleotide sequence of the coding and flanking regions of the human parainfluenza virus 3 hemagglutinin-neuraminidase gene: Comparison with other paramyxoviruses, *Intervirology* **27**:69–80.

Suzuki, Y., Suzuki, T., Matsunaga, M., and Matsumoto, M., 1985, Gangliosides as paramyxovirus receptor. Structural requirement of sialo-oligosaccharides in receptors for hemagglutinating virus of Japan (Sendai virus) and Newcastle disease virus, *J. Biochem.* **97**:1189–1199.

Takamizawa, K., Iwamori, M., Kozaki, S., Sakaguchi, G., Tanaka, R., Takayama, H., and Nagai, Y., 1986, TLC immunostaining of *Clostridium botulinum* type A neurotoxin binding to gangliosides and free fatty acids, *FEBS Lett.* **201**:229–232.

Tashiro, M., and Hommas, M., 1983a, Pneumotropism of Sendai virus in relation to protease-mediated activation in mouse lungs, *Infect. Immun.* **39**:879–888.

Tashiro, M., and Homma, M., 1983b, Evidence of proteolytic activation of Sendai virus in mouse lung, *Arch. Virol.* **77**:127–137.

Tashiro, M., Pritzer, E., Khoshnan, M. A., Yamakawa, M., Kuroda, K., Klenk, H.-D., Rott, R., and Seto, J. T., 1988, Characterization of a pantropic variant of Sendai virus derived from a host range mutant, *Virology* **165**:577–583.

Walsh, E. E., and Hruska, J., 1983, Monoclonal antibodies to respiratory syncytial virus proteins: Identification of the fusion protein, *J. Virol.* **47**:171–177.

Wassilewa, L., 1977, Cell receptors for paramyxoviruses, *Arch. Virol.* **54**:299–305.

Whatley, R., Ng, S. K.-C., Roger, J., McMurray, W. C., and Sanwal, B. D., 1976, Developmental changes in gangliosides during myogenesis of a rat myoblast cell line and its drug resistant variants, *Biochem. Biophys. Res. Commun.* **70**:180–185.

Umeda, M., Nojima, S., and Inoue, K., 1984, Activity of human erythrocyte gangliosides as a receptor to HVJ, *Virology* **133**:172–182.

Van Heyningen, S., 1976, Binding of ganglioside by the chains of tetanus toxin, *FEBS Lett.* **68**:5–7.

Yamakawa, T., and Nagai, Y., 1978, Glycolipids at the cell surface and their biological functions, *Trends Biochem. Sci.* **3**:128–131.

Yasuda, Y., Hosaka, Y., Fukami, Y., and Fukai, K., 1981, Immunoelectron microscopic study on interactions of noninfectious Sendai virus and murine cells, *J. Virol.* **39**:273–281.

Yoshimura, A., Kuroda, K., Kawasaki, K., Yamashina, S., Maeda, T., and Ohnishi, S., 1982, Infectious cell entry mechanism of influenza virus, *J. Virol.* **43**:284–293.

CHAPTER 16

Paramyxovirus M Proteins
Pulling It All Together and Taking It on the Road

MARK E. PEEPLES

I. INTRODUCTION

The paramyxovirus M protein has the job of gathering all of the virus components together at the plasma membrane in preparation for budding. It is a prototype for performing the assembly "bandleader" function: a function which all enveloped viruses must perform. While the paramyxovirus M protein is clearly involved in this orchestration, the mechanisms by which it functions are far from clear. In addition, assembly may not be the only role played by the M protein. The M protein may be involved in regulating another viral function, transcription. Furthermore, the striking localization of the M protein of one paramyxovirus, Newcastle disease virus (NDV), to the nucleus of infected cells may imply another, as yet unknown, function for the M protein.

The designation "M" originally came from virion fractionation studies in which viral proteins were separated into those associated with the nucleocapsid and those associated with the membrane (Scheid and Choppin, 1973). Detergent and high salt stripped three proteins from the virion: two glycoproteins and a nonglycosylated protein. The nonglycosylated protein was named the "membrane" protein because its only known characteristic was that it was removed from the virion along with the virion membrane. However, since the paramyxovirus M proteins have no hydrophobic sequences long enough to span a membrane and are not glycosylated, they are not intrinsic membrane proteins. In fact, Sendai virus particles treated with detergent under low-salt conditions lose their lipid envelope and glycoproteins while retaining the compact saclike structure of the particle. Only after these particles are treated with

MARK E. PEEPLES • Department of Immunology/Microbiology, Rush Medical College, Chicago, Illinois 60612.

additional salt is the M protein released, allowing the nucleocapsid structure to unfold (Shimizu and Ishida, 1975). It appears that the M protein is capable of maintaining the structure of the virion even in the absence of the virion membrane. For these reasons, the M protein is usually referred to as the "matrix" protein. This term is more descriptive of its position in the virion and its role as the assembly bandleader.

Evidence that the paramyxovirus M protein does perform the bandleader role during virion assembly has been most clearly derived from the measles virus mutants isolated from subacute sclerosing panencephalitis (SSPE) patients. Many of these viruses have a defect in their M protein such that it is undetectable in infected cells (Machamer et al., 1981). Infected cells produce very few infectious virus particles and therefore spread from cell to cell very poorly. SSPE measles virus is discussed at greater length in Chapter 12. Evidence from other paramyxovirus genetic systems also indicates that the M protein is involved in virion formation. Roux and Waldvogel (1982) demonstrated that cultured cells persistently infected with Sendai virus produced all of the viral proteins, but that the M protein was labile. These cells produced greatly reduced numbers of virus particles. The reduced M protein accumulation and the reduced virus production may be related to coinfection with defective-interfering particles (Roux et al., 1984; Tuffereau and Roux, 1988). Yoshida et al. (1979) analyzed a Sendai virus temperature-sensitive (ts) mutant isolated from a persistent infection which was blocked in a late assembly function at nonpermissive temperature. The M protein was not detectable by immunofluorescence at nonpermissive temperature, leading to the conclusion that the M protein is required for assembly. However, this virus also contained lesion(s) in its HN glycoprotein, making it difficult to conclude that the M protein mutation alone causes the functional defect in assembly.

Information from other negative-strand RNA viruses also indicates that their M proteins are required for final assembly. Lohmeyer et al. (1979) have examined the abortive infection of mouse L cells by influenza A (fowl plague) virus. Only the M protein was absent from these infected L cells, which were unable to produce infectious virus, whereas the M protein was readily detectable in infected chick embryo cells which produce infectious virus. Knipe et al. (1977) have demonstrated that ts mutations in the vesicular stomatitis virus M gene result in M protein molecules that are unstable at nonpermissive temperature. Under these conditions, very few virus particles are produced by cells infected with these mutants.

II. INTERACTIONS OF THE M PROTEIN

The evidence described above indicates that the M protein plays a central role in virion formation. It is a highly basic, somewhat hydrophobic protein that has been well conserved among the paramyxoviruses (Bellini et al., 1986; Galinski et al., 1987; Spriggs et al., 1987), indicating tight functional constraints on its structure. The M protein associates with the infected cell plasma membrane soon after it is synthesized (Bowen and Lyles, 1982; Nagai et al., 1976b) and is incorporated into virions after a short lag (Bowen and Lyles, 1982). Perhaps the lag represents the time required to draw the other viral components together.

Insights into the mechanisms by which the M protein is involved in the assembly of viral components come from many experiments which demonstrate that the M protein interacts with itself, with the viral glycoproteins, and with the nucleocapsid. The M protein also interacts with cell components, such as membrane lipids and cytoskeletal elements, which may also be involved in assembly or budding. In the following section, the evidence for each of these interactions is examined (see also Chapter 17).

A. M Protein Association with Lipid Membranes

Although the paramyxovirus M proteins contain no hydrophobic sequences long enough to span a membrane (see Section IV below), their general hydrophobic nature might allow them to associate with membranes. Li *et al.* (1980) found that the NDV M protein was an intrinsic protein, not associated with the hydrophobic lipid core in the virion envelope, since it could be eluted from these membranes with lithium diiodosalicylate. However, Bächi (1980) observed crystalline structures apparently composed of M protein which had penetrated into the hydrophobic interior of both the plasma membrane (Fig. 1B) and the virion membrane (Fig. 1C). These structures were found only in infected cells and are thought to be composed of M protein molecules, because somewhat similar structures are assembled *in vitro* by purified M protein (Figs. 2B–2E), as will be discussed below.

Perhaps the M protein is targeted to membranes by a direct association with lipid membranes either due to its charged nature or due to its hydrophobic character. Faaberg and Peeples (1988a) have demonstrated that the NDV M protein associates efficiently with lipid membranes presented in the form of liposomes. The forces responsible for this interaction do not appear to be electrostatic, for two reasons: (1) liposomes containing a net negative, positive, or neutral charge were all able to bind M protein; and (2) M protein was able to associate with these liposomes even in the presence of 0.5 M NaCl, which should shield electrostatic charges. It appears that the NDV M protein is able to interact with lipid membranes in a nonelectrostatic manner, presumably by a hydrophobic domain on the molecule. The influenza M protein has also been shown to bind to liposomes regardless of charge (Bucher *et al.*, 1980). Association of the VSV M protein with lipids appears to be more electrostatic, since it binds only to negatively-charged liposomes with forces which are inhibited by 0.5 M NaCl (Zakowski *et al.*, 1981).

It is clear from cell fractionation studies (Bowen and Lyles, 1982; Lamb and Choppin, 1977a; Nagai *et al.*, 1976) that the M protein is associated with the plasma membrane. Schwalbe and Hightower (1982) have shown that the NDV M protein associates preferentially with the plasma membrane. After a 2-min pulse of ^{35}S-methionine and a 3-min chase, 95% of the M was associated with the plasma membrane (Schwalbe and Hightower, 1982). (The nucleus was

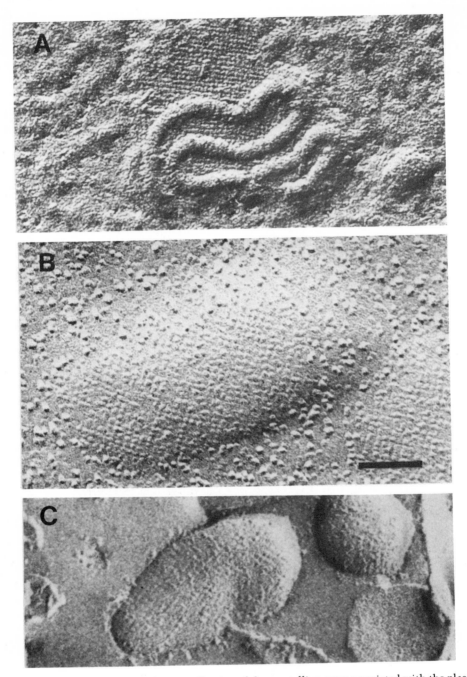

FIGURE 1. Electron micrographic visualization of the crystalline array associated with the plasma membrane of Sendai virus-infected EL-4 cells. (A) Cells were sheared to reveal the cytoplasmic surface of the plasma membrane, which was freeze-dried and freeze-etched (Büechi and Bächi, 1982). (B) Freeze-fractured plasma membrane showing the crystalline array at the inner leaflet in an area which is bulging, suggesting the budding of a virus particle (Bächi, 1980). (C) Freeze-fractured Sendai virus envelopes from virions harvested 24 hr (early) after infection of chicken eggs (Bächi, 1980). Bar = 100 nm.

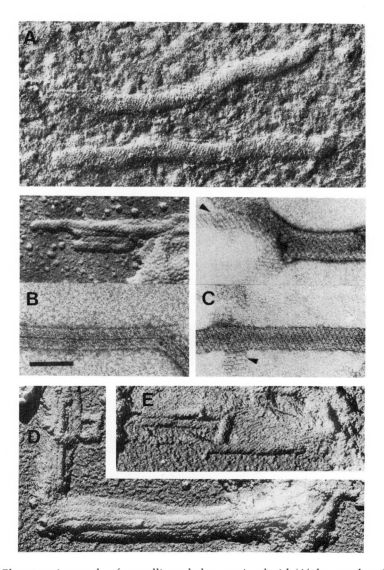

FIGURE 2. Electron micrographs of crystalline tubules associated with (A) the cytoplasmic face of an infected cell plasma membrane, and similar tubules of (C) large and (B,D,E) small diameter formed by purified M protein *in vitro*. (A) Sendai virus-infected EL-4 cells were sheared to reveal the cytoplasmic face of the infected cell plasma membrane, which was freeze-dried and freeze-etched (Büechi and Bächi, 1982). (B,D,E) Narrow tubes formed *in vitro* from purified M protein. M protein was isolated from detergent- and salt-disrupted virions and precipitated by dialysis against 10 mM phosphate buffer. The top picture in (B) is of freeze-dried, metal-shadowed M protein. The bottom picture in (B) is of negatively-stained M protein (Heggeness *et al.*, 1982). (D,E) Similarly prepared M protein aggregates clearly displaying a helical array (Hewitt and Nermut, 1977). (C) Wide tubes of M protein were visualized less frequently than the narrow tubes by Heggeness *et al.* (1982) after negative staining. Arrowheads indicate annuli of M protein which compose the tubes. Bar = 100 nm.

not analyzed in these experiments, a point that will be important below.) Cholesterol, which is found predominantly in the plasma membrane, does not appear to be the targeting signal, since liposomes prepared without cholesterol were able to bind M protein as well as those containing cholesterol (Faaberg and Peeples, 1988a). Alternatively, the M protein may be targeted to the plasma membrane by its interaction with the viral glycoproteins, both of which contain small cytoplasmic domains (Lyles, 1979).

B. M Protein Association with the Viral Glycoproteins

Evidence for such an interaction comes from several experiments. Yoshida *et al.* (1986) have demonstrated that monensin treatment of NDV-infected cells, which prevents transport of the viral glycoproteins from the Golgi to the plasma membrane, also prevents the M protein from reaching the plasma membrane. Interestingly, while the viral glycoproteins accumulated in the Golgi in these monensin-treated cells, the M protein did not. Perhaps the glycoprotein cytoplasmic domains need to be presented in the proper conformation or lipid context for the M protein to recognize them. Alternatively, one of the final steps in the maturation of one or both of the viral glycoproteins might involve a modification of the cytoplasmic tail of that glycoprotein making it receptive to interaction with the M protein. This modification might be independent or it could be coupled to another Golgi-based modification, such as carbohydrate processing. In the case of the F glycoprotein, the modification might be related to the attachment of palmitate (Chatis and Morrison, 1982; Schmidt, 1982) or to polypeptide chain cleavage. However, the fact that Sendai virus and many avirulent strains of NDV can produce virus particles containing uncleaved F_0 (Homma and Ohuchi, 1973; Nagai *et al.*, 1976a; Nagai and Klenk, 1977; Scheid and Choppin, 1974) argues against a requirement for cleavage in this proposed cytoplasmic domain alteration.

If the M protein is associated with a viral glycoprotein in the plasma membrane, redistribution of the glycoproteins should redistribute the M protein. Tyrrell and Ehrnst (1979) demonstrated that in a cell line chronically infected with measles virus, antibody-induced capping of the viral glycoproteins also resulted in capping of the viral M and NP proteins. Cytochalasin B, which dissociates actin filaments, decreased the NP protein cocapping without decreasing M protein cocapping (Tyrrell and Ehrnst, 1979), indicating that the cocapping is probably not ascribable to preformed virions and strengthening the argument that the M protein at the plasma membrane is associated with at least one of the viral glycoproteins. Yoshida *et al.* (1979) also examined antibody-induced capping of Sendai virus glycoproteins in cells infected with a *ts* mutant. At the nonpermissive temperature, the M protein was not detectable and the antibodies were able to efficiently cap the viral glycoproteins. However, at permissive temperature, the M protein was present and capping was slower and less complete. The glycoproteins therefore appear to be anchored to the M protein. It has not been possible to confirm this interaction in cross-linking experiments with virions of Sendai virus or NDV (Markwell and Fox, 1980), perhaps for technical reasons. The VSV G protein can be cross-linked to its M protein in virions (Dubovi and Wagner, 1977), indicating that in this virus the M protein physically contacts the viral glycoprotein.

Paramyxoviruses specifically defective in the interaction between their M protein and glycoproteins would be useful in exploring this interaction. One *ts* mutant of NDV might have these characteristics. Mutant D1 has a lesion in the gene encoding the M protein, yielding a protein that migrates more rapidly than the wild-type M protein on SDS–polyacrylamide gel electrophoresis (PAGE) (Peeples and Bratt, 1984a). At nonpermissive temperature, D1-infected cells produced virions lacking the F glycoprotein. Either the M protein mutation of D1 prevents its correct interaction with the F glycoprotein under these conditions or D1 contains a second lesion in its *F* gene. Using proteolytic methods, the lesion responsible for the migrational change of D1 was located to within 100 amino acids of the amino terminus of the protein (Peeples and Bratt, 1984b; M. Peeples, unpublished results). Both the *M* and *F* genes of D1 have recently been sequenced and two mutations have been found: A single mutation in the *M* gene which changes amino acid 72 from glutamic acid to glycine and a singe mutation in the *F* gene which changes amino acid 153 from Arg to Leu (C. Wang, G. Raghu, and M. Peeples, manuscript in preparation). Which lesion is responsible for the *ts* phenotype is now under investigation. Identification of the lesions in two other group D mutants and revertants of group D mutants may resolve this question and provide a genetic system either for studying the interaction of the M and F proteins or for studying F protein maturation and function.

It is interesting that two of the three SSPE measles viruses whose genes have been cloned and sequenced contain carboxyl-terminal deletions in their F glycoproteins: one is truncated by 24 amino acids and the other contains a frameshift resulting in 10 incorrect amino acids followed by the deletion of 15 amino acids (Cattaneo *et al.*, 1988). Both of these mutations destroy the majority of the 32-amino acid carboxyl-terminal domain of the F glycoprotein. The third cloned SSPE *M* gene, as well as an *M* gene cloned from a case of measles inclusion body encephalitis, encode grossly defective M proteins (Cattaneo *et al.*, 1986, 1988). While it is difficult to definitely assign the locations of the defects responsible for these persistent infections, the F glycoprotein cytoplasmic domain and the M protein are likely candidates in these cases. If the M protein must interact with the cytoplasmic domain of the F glycoprotein for virion formation, mutations in either component could prevent the assembly of infectious virions.

The M protein may also interact with the HN glycoprotein. Cells persistently infected with Sendai virus, as well as cells coinfected with wild-type and defective-interfering Sendai virus, produce diminished amounts of intracellular HN and M proteins (Roux *et al.*, 1984; Tuffereau and Roux, 1988). Apparently both proteins are unstable. If one of these proteins stabilizes the other, it is more likely that the M protein stabilizes HN than the other way around. This supposition is based on the finding that normal amounts of the M protein accumulate in Sendai virus *ts* 271 virus-infected cells at nonpermissive temperature, where no HN protein is present (Tuffereau *et al.*, 1985). The M protein is therefore stable in the absence of the HN protein.

One set of *in vitro* experiments also indicates that the M protein interacts with the viral glycoproteins. Yoshida *et al.* (1976) dissociated Sendai virus into nucleocapsid, M protein, and glycoprotein fractions by detergent and salt treatment. In reassociation experiments, the glycoproteins were unable to form a complex with the nucleocapsid unless the M protein fraction was added. The conclusion from this study is that the M protein interacts both with the glycoproteins and with the nucleocapsid.

Finally, while the M protein probably interacts with one or both of the viral glycoproteins at the plasma membrane, the specificity of this interaction has not been resolved. McSharry *et al.* (1971) demonstrated that cells coinfected with SV5 and VSV produce virus particles containing all of the VSV proteins and the two SV5 glycoproteins. Perhaps the cytoplasmic tails of viral glycoproteins contain sequences or conformations which broadly interact with a common feature of the M proteins of other viruses, allowing such pseudotypes to form. Alternatively, Zavada (1982) has suggested that the concentration of a viral M protein in one area of the plasma membrane may alter the membrane, resulting in a phase partitioning between viral and cellular glycoproteins. The viral glycoproteins may remain "soluble" in this membrane environment while cellular glycoproteins do not.

C. M Protein Association with M Protein

Cross-linking, electron microscopic, and biophysical studies all indicate that the M protein has a strong predilection for associating with itself. Membrane-permeable chemical cross-linkers readily form dimers, trimers, and perhaps larger multimers of M proteins in NDV or Sendai virus particles (Markwell and Fox, 1980; Nagai *et al.*, 1978). These studies indicate that at least some M protein molecules are located within 11 Å (the span of the cross-linkers used) of each other in the native virion.

Electron microscopic analysis of Sendai virus-infected cells by Büechi and Bächi (1982) has revealed a two-dimensional crystalline array of proteins on the cytoplasmic face of the plasma membrane. Virus nucleocapsids are associated with some of these areas, implying that they represent areas of virus assembly (Fig. 1A). Alternatively, large tubes with subunits forming a helical structure are found on the plasma membranes (Fig. 2A). It is interesting that the subunits in the tube structures appear to have helical symmetry (Fig. 2A), whereas the crystalline sheets of M protein associated with the plasma membrane appear to be assembled in a perpendicular format (Fig. 1A). The same perpendicular arrangement of the M protein appears in the freeze-fracture analyses of the infected cell plasma membrane and virion membrane (Fig. 1B and 1C), as discussed above. The difference in geometry between the tubes and the sheets of M protein may reflect a somewhat different interaction between M molecules. Perhaps the difference is due to an association with the membrane or with a glycoprotein before associating with another M protein.

The crystalline sheets of M protein were only present in 40% of the "early" virions (collected 24 hr after infection of embryonated eggs) observed by Bächi (1980). When these virions were "aged" (37°C, 48 hr) the crystalline structure disappeared (Bächi, 1980). Homma *et al.* (1976) and Shimizu *et al.* (1976) found that both "early" and "aged" virions were able to fuse with the erythrocyte, but that only aged virions were hemolytic. These authors also found that more of the "aged" virions were permeable to uranyl acetate during preparation for electron microscopy and suggested that they might contain holes which would be transferred to the erythrocyte membrane with fusion of the virion envelope. Interestingly, physical discontinuities in the virion membranes were not frequently found. Shimizu *et al.* (1976) suggested that fusion of the "aged" virus envelope might lead to an increased permeability of the erythrocyte membrane, swelling, and lysis. Dissociation of the crystalline array might be involved in this process if it results in altered virion membrane properties.

Russell and Almeida (1984) have reported that the virions of one avirulent strain of NDV have their glycoprotein spikes arranged in a very orderly two-dimensional crystalline array (periodicity of 8–9 nm) similar to the arrangement of the M proteins within the virion membrane of Sendai virus (Fig. 1C) (periodicity of 9 nm). Perhaps this glycoprotein array reflects a direct M protein–glycoprotein interaction. In the case of the avirulent NDV described, trypsin cleavage activated virion infectivity and resulted in a disordering of the glycoprotein array (Russell and Almeida, 1984), perhaps reflecting the dissociation of the M protein from the glycoproteins. This dissociation may be involved in the maturation of all paramyxovirus particles.

Interaction of the M protein with itself also occurs *in vitro* and has been used as a simple method to purify the protein. After NDV particles are treated with detergent and high salt to dissociate the envelope and the nucleocapsid is removed by centrifugation, dialysis against a low-salt buffer results in a precipitate of nearly pure M protein (Scheid and Choppin, 1973). Under these conditions, the Sendai virus M protein forms into narrow (Figs. 2B, 2D, and 2E) or wide (Fig. 2C) tubes consisting of M protein. These tubes are composed of a two-dimensional crystalline array of parallel strands wound in a helix. The width of each strand appears to be from 7.2 nm (Heggeness *et al.*, 1982) to 7.5 nm (Büechi and Bächi, 1982). The regular crystalline appearance of these tubes is reminiscent of the tubular forms of presumably M protein found on the cytoplasmic surface of infected cell plasma membranes (Fig. 2A). As mentioned above, the M protein packing in these tubule structures is quite regular, but its geometry appears somewhat different from the perpendicularly arranged membrane-associated (Fig. 1A) or membrane inner leaflet-penetrating structures thought to be the M protein (Figs. 1B and 1C).

While the *in vitro* precipitation of the M protein under low-salt conditions may indicate that the M protein's interaction with itself is a part of its function, it has made the design of *in vitro* association systems difficult. Aggregated M protein may not behave in the same way as individual molecules do, certainly not in the proper stoichiometry. Faaberg and Peeples (1988a) avoided the problem of M protein self-aggregation by isolating the NDV M protein on a monoclonal antibody affinity column and eluting and storing it in the presence of excess bovine serum albumin (BSA). While this approach may not be ideal, since interactions between the M protein and BSA may have been introduced, the M protein did remain in a small soluble complex under psychological conditions. This soluble M protein has been valuable in the association experiments with liposomes described above.

Hewitt (1977) has presented evidence that the M protein of Sendai virus is associated as a dimer. In that study, the M protein was isolated by repeated rounds of low-salt precipitation and high-salt resolubilization. M protein resolubilized in 1 M KCl appeared by both gel filtration and by velocity centrifugation to be approximately the size of a dimer. Interestingly, in electron micrographs of this soluble material, M proteins appear as annuli, 6 nm in diameter, with a 1.5-nm hole in the center (Hewitt and Nermut, 1977). Heggeness et al. (1982) also observed these M proteins annuli in the tubes produced by low-salt dialysis of the Sendai virus M protein (arrow in Fig. 2C). Varsanyi et al. (1984) isolated the measles virus M protein on a monoclonal antibody affinity column and found a similar disc, 8 nm in diameter, with a central hole. However, from the dimensions of side-on views of these complexes, these authors have suggested that the complex represents four to six M proteins.

D. M Protein Association with the Nucleocapsid

Yoshida et al. (1976) have demonstrated in in vitro association studies that purified Sendai virus nucleocapsids will only associate with purified viral glycoproteins in the presence of M protein, implying that the M protein associates with the nucleocapsid, as well as the glycoproteins. In the virion, the M protein appears to form a structure surrounding the nucleocapsid. As mentioned above, treatment of Sendai virions with detergent in low-salt buffer removes the virion membrane and glycoproteins without disrupting the saclike structure containing the nucleocapsid. The further addition of salt releases the M protein and allows the nucleocapsid structure to unfold (Shimizu and Ishida, 1975). The requirement for salt to disrupt this complex implies that the M protein interacts with the nucleocapsid via ionic bonds. The M protein of both Sendai virus and NDV could be cross-linked to the NP protein in fresh virions, indicating that the M protein is complexed with the major nucleocapsid protein (Markwell and Fox, 1980; Nagai et al., 1978).

The NP protein is highly basic at its amino terminus (Morgan et al., 1984), which is associated with the genomic RNA (Heggeness et al., 1981), but acidic in its carboxyl terminus (Morgan et al., 1984). The carboxyl terminus also contains most of the phosphorylated amino acids in the NP protein (Hsu and Kingsbury, 1982), giving it a highly negative charge. Since the M protein is a positively charged molecule, the ionic attraction between the M protein and the carboxyl terminus of the NP protein may be responsible for the M protein interaction with the nucleocapsid. Hamaguchi et al. (1983) have presented evidence that the Sendai virus M protein forms a complex with the L protein. Whether this association of the M and L proteins contributes to the M protein–nucleocapsid interaction is not clear.

In infected cells examined in cross section, nucleocapsids are found associated with the plasma membrane in areas which contain viral glycoprotein spikes (Compans et al., 1966). Viewed from the inside of the plasma membrane, nucleocapsids are found associated with the two-dimensional M protein crystalline structures on the plasma membrane (Fig. 1A). Caldwell and Lyles (1986) examined the M protein associated with the inner surface of infected cell plasma membranes after inverting the membranes on polycationic beads. They found two populations of Sendai virus M protein associated with the membranes: one population (65%) was removed by treatment with 4 M KCl, while the other was not. Interestingly, the nucleocapsids were not removed by this salt treatment either. Perhaps the salt-removable population of M protein is solely associated with membrane lipids or glycoproteins or both, but has not yet established contact with a nucleocapsid. Once the nucleocapsid binds to the M protein, it may change the association of M with the membranes (or with the cytoplasmic domain of the viral glycoproteins).

What role does the M protein association play in its interaction with the nucleocapsid? Is it simply involved with assembly, or might it regulate nucleocapsid functions? The nucleocapsid, which contains the viral genome, is capable of transcribing mRNA (Huang et al., 1971). Marx et al. (1974) demonstrated that M protein added to Sendai virus nucleocapsids decreased their ability to transcribe RNA by 90%. However, it was puzzling that the viral glycoproteins had a similar effect. The dampening effect of M protein on transcription has been clearly demonstrated in the rhabdovirus vesicular stomatitis virus (VSV). The transcriptive activity of VSV nucleocapsids is greatly diminished by addition of the VSV M protein (Carroll and Wagner, 1979; De et al., 1982). The M protein condenses the VSV nucleocapsid (De et al., 1982; Newcomb et al., 1982), presumably preventing the polymerase from traversing the nucleocapsid template. The portion of the VSV M protein which interacts with the nucleocapsid has been localized to its extreme amino-terminal region. This localization is based on: (1) monoclonal antibody and antipeptide serum reversal of transcription inhibition (Pal et al., 1985; Shipley et al., 1988); (2) inhibition of transcription by a highly basic peptide representing the 20 amino-terminal amino acids of the M protein (Shipley et al., 198); and (3) loss of transcription inhibition activity in the M protein of ts023 related to a lesion in amino acid 21 (Pal et al., 1985). Interestingly, the ts phenotype of this virus is not related to the substitution amino acid 21, but is related to one or both of the additional mutations in its M protein (Li et al., 1988; Morita et al., 1987).

E. M Protein Association with the Cytoskeleton

The M protein of NDV (Morrison and McGinnes, 1985) and measles virus (Bohn et al., 1986) as well as the other viral proteins appear to associate with the cell cytoskeleton, as demonstrated by detergent extraction procedures. Since actin is found as a component of many enveloped viruses, including paramyxoviruses (Lamb et al., 1976; Örvell, 1978; Tyrrell and Norrby, 1978), Giuffre et al. (1982) suggested that the M protein might associate directly with actin filaments. They presented evidence that the M protein associates with actin by affinity chromatography, cosedimentation, and circular dichroism.

III. M PROTEIN: ASSEMBLY BANDLEADER

Unlike the other viral proteins, M is produced in infected cells at half the level that it is represented in virions. In other words, the M protein may be the rate-limiting virion component, thereby ensuring that the pool of transcribing and replicating nucleocapsids will not be depleted (Choppin and Compans, 1975; Portner and Kingsbury, 1976; Lamb *et al.*, 1976). Taking into account the interactions described above, in conjunction with kinetic and video microscopic data, a scenario for the assembly process with more or less solid experimental confirmation as described above can be envisioned.

Soon after the M protein is synthesized on cytoplasmic ribosomes, it associates with the plasma membrane (Bowen and Lyles, 1982; Nagai *et al.*, 1976b; Schwalbe and Hightower, 1982), perhaps penetrating to the hydrophobic interior of the bilayer. Some, if not most, of the plasma membrane-associated M protein interacts with one or both of the viral glycoproteins via their cytoplasmic or transmembrane domains. An M protein, perhaps a dimer or larger, with a glycoprotein spike in tow may diffuse in the plane of the plasma membrane until it connects with another M protein and so on. A patch of M protein molecules forms on the cytoplasmic face of the plasma membrane, with its associated glycoprotein patch on the external plasma membrane. Shortly thereafter, free nucleocapsids associate with the plasma membrane-bound M protein. With all of the viral structural components assembled at the plasma membrane, the budding process begins.

The motivating force behind budding is not well defined. Bohn *et al.* (1986) have demonstrated by electron microscopy that many budding measles virus particles contain barbed actin filaments, many of which appear to be growing into the bud. These pictures suggest that the vectoral growth of the actin filaments may be the force behind budding. How the M protein interaction with actin might affect this process is not clear.

The budding process of respiratory syncytial virus occurs in localized areas of the cell and is rapid, as observed by video microscopy (Bächi, 1988). Filaments of RS virus bud to a final length of 5–10 nm with an average speed of 0.11–0.25 nm/sec, resulting in completed virions in 1–2 min (Bächi, 1988).

The stoichiometric relationship of viral proteins within the virion is not known. However, a comparison of the dimensions of the viral structures may be instructive. Sendai virus nucleocapsids isolated from virions display three different helical pitch conformations as determined by electron microscopy (Egelman *et al.*, 1989). The most likely conformation in the virion has a flexible 6.8-nm pitch. It is interesting that the periodicity of M proteins within the Sendai virus membrane is somewhat similar, 7.5 nm (Büechi and Bächi, 1982) or 9.0 nm (Bächi, 1980). Furthermore, the periodicity calculated for the orderly, pretrypsin NDV virions described above (Russell and Almeida, 1984) is 8–9 nm. Based on the similarity of each of these interprotein distances, a hypo-

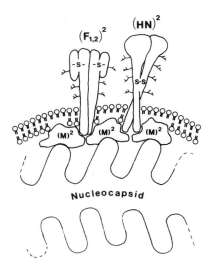

FIGURE 3. Hypothetical model of the stoichiometric relationship among the F and HN glycoproteins, the M protein, and the nucleocapsid. The rationale for this model is described in the text.

thetical model for paramyxovirus envelope stoichiometry is presented in Fig. 3. In this model, dimers of the F (Sechoy *et al.*, 1987; Walsh *et al.*, 1985) and dimers of the HN glycoproteins (Moore and Burke, 1974; Smith and Hightower, 1981) interact with the M protein, which interacts with a turn of the nucleocapsid in a 1:1:1 relationship.

Once the virion has formed, these interactions begin to be released: the M protein can no longer be chemically cross-linked to the NP protein (Markwell and Fox, 1980), the nucleocapsid association with the membrane is lost (Dubois-Dalcq *et al.*, 1984), the crystalline array of membrane-associated M protein disappears (Bächi, 1980), and the regular array of glycoproteins detected under some conditions is lost (Russell and Almeida, 1984). This intravirion dissociation process correlates temporally with the development of hemolytic virions (Homma *et al.*, 1976). In parallel, the virion shape changes from spherical or oval to pleomorphic (Kim *et al.*, 1979).

IV. STRUCTURE OF THE M PROTEIN

While several interactive functions of the paramyxovirus M protein have been described, the structural features of the M protein involved in these functions are not known. Circular dichroism studies indicate that the Sendai virus M protein contains little secondary structure: 11% α-helix, 19% β-pleated sheet, and 70% random coil (Giuffre *et al.*, 1982). It is possible that critical features may be reflected in conserved amino acids in the primary structure of the M proteins. Figure 4 presents the deduced amino acid sequences of the paramyxovirus M proteins. These M proteins are all approximately M_r 40,000. The respiratory syncytial virus M protein is much smaller, with an M_r of

Multiple sequence alignment (residues 1–150) for the protein sequences Pifm, Sndm, Mvm, Cdvm, Ndvm, Sv5m and Mumpm.

```
                    1                                              50
Pifm   M S I T N S A I Y R T F P E S S F S E N G H I E P L P L K V N E Q R K K A V P H I R V A K I G N P
Sndm   - - - M A D I Y R Y L D F P K F S Y E D N G T V S I A P L R T G S D D K K A L P Y P I R R I I K V G D P
Mvm    - - - M T E V Y Y D F D K S Q S S A W D I K G S I A P L Q T T Y S D G R L V P L P V R R V I I D
Cdvm   - - - M T E V Y D F D D Q S S W Y T K G S L A P L P T T Y P D G R L I P Q L P R L V I I D

Ndvm   M D S S R T I G L Y F D S A H S N L L A F F P I V L Q D T G D G . D G K K Q I A P Q Y R I Q R L D S W
Sv5m   M P S I S I P A D P T N P K A F I V F P V I N S D G G E . L R T T Y L N K L R T T Y L R K I L S G
Mumpm  M A G S Q I K V P L P K P P D S D S Q R L N A F P V I M A Q E G . D G E K G . K G R L L R Q I R L R K I L S G

                    51                                             100
Pifm   P K H G S R Y L D . . . V F L L G F F E M E R I K D K Y G S V N D L D S D P G Y K V C G S L P
Sndm   P K K H G S R Y L D . . . L L L G F F E E T P K Q T T N L G S V S D L T E P T S Y S I C G S L P
Mvm    P G L G D R K D E C F M Y I M F L M F F E . . . E E A T V G S . . . D D S D P L G G P P I G R A F G S L P
Cdvm   P G L G D R K D E C F M Y I F L M G I I E . . . . . . . . . D N D G L G P I G R T F G S L P

Ndvm   T D S K E D S . . . V F I T T Y G F I F Q V G N E E A T I V G E . M I N D N P K R E L L S A A M L C
Sv5m   D T H E P L V . . . T F I N T Y G F I H Y E Q D R G N T I V G S . D Q L G K K K R E A V T A A M V T
Mumpm  D P S D Q Q I . . . T F F V N T Y G I R A T P E T S E F I S S Q E E V T P V T A C M L S

                    101                                            150
Pifm   I G L A K Y T G N D D Q Q K E L L Q A A T K L D R D H H I E V R R T V R V K A K A G E M E M I V Y T V Q N I K P E L Y P W S S
Sndm   I G V A K Y Y G T A A R P E E L L L K A C T D E L L R D H V V R R T V R A G L N E M I V F F Y M V D S I G A P L L P W S G
Mvm    L G V G R S T A K P E E M K L L K E A T E L L Q D D R R T V R T A G V K E L V F F Y N N T P L T P W R K
Cdvm   L G V G R T T A R P E L K E A L L L K E A T T L D M E R M Q D I L F R K T A S D K E Q I L F Y N T P L H I T P W K K

Ndvm   I G S V P N T G D L V E L A R A C L T M V Q D T V D R T H T V Q D S I A P Q V L Q S C R V L
Sv5m   L G C G P N L P S L G N V L G Q L R E F Q P Q L F R G H T L
Mumpm  F G A G P V L E D P Q H M L K A L D Q T D H Q I
```

FIGURE 4. (continued)

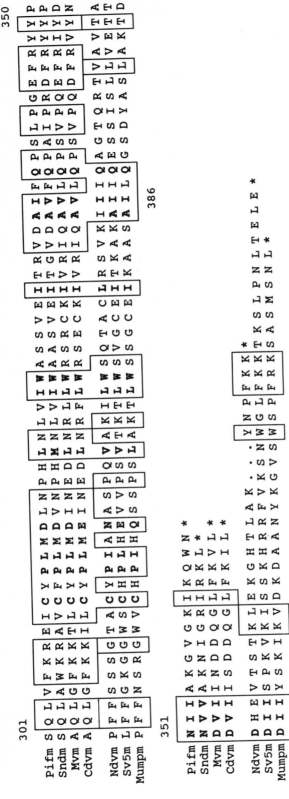

FIGURE 4. Comparison of paramyxovirus M-protein amino acid sequences. The GAP program from the University of Wisconsin Genetics Computer Group (Devereux *et al.*, 1984) was used to compare each pair of sequences. Parameters used were gap weight 5.0 and length weight 0.3. The sequences fall into two related "supergroups" as described in the text and in Table I. The M protein sequences of the four members of the Pif/MV supergroup are presented on the top four lines, separated from the analogous M protein sequences of the SV5/MV/NDV supergroup. The best fit of each sequence to the others was determined by eye, and manipulated in the LINEUP program and with a word processor. Boxes indicate positions at which all amino acids in one group are chemically similar (ILMV, KHR, DENQ, ST, AG, FWY). Boxes which span both groups indicate positions at which all residues or six out of seven residues are chemically similar. Within these boxes, the similar amino acids are in bold type. M-protein amino acid sequences (derived from nucleic acid sequences) used were human parainfluenza virus type 3 (Pifm) (Galinski *et al.*, 1987), Sendai virus (Sndm) (Blumberg *et al.*, 1984), measles virus (Mvm) and canine distemper virus (Cdvm) (Bellini *et al.*, 1986), Newcastle disease virus (Ndvm) (Chambers *et al.*, 1986), SV5 virus (SV5m) (H. Sheshberadaran and R. Lamb, personal communication), and mumps virus (Mumpm) (Elango, 1989). Shaded areas: amino acids 18–32 indicate regions of similarity with the VSV M protein which may represent the membrane-association region; amino acids 256–271 indicate a highly basic region of the NDV M protein hypothesized to contain the nuclear transport and nucleolar localization signals; and amino acids 283–284 through 293 indicate a highly conserved region described in the text.

29,000 (Satake and Venkatesan, 1984). The latter is omitted from the comparison in Fig. 4 and will be discussed separately because it bears little resemblance to the others.

The paramyxovirus M proteins fall into several groups (Table I). The "Pif group" contains parainfluenza virus type 3 and Sendai virus, another parainfluenza virus. The "MV group" contains measles and canine distemper viruses. The Pif and MV groups are more closely related to each other than they are to the remaining M proteins. Though not as closely related as the members of the Pif group or the MV group, the SV5 and mumps virus M proteins are related and will be called the "SV5/Mu group." The NDV M protein is more closely related to this SV5/Mu group than to the Pif or MV groups.

The M protein amino acid sequences of all seven of the viruses are displayed in Fig. 4, separated into two supergroups: the Pif and MV groups on the top; NDV and the SV5/Mu group on the bottom. The sequences were aligned using the pairwise comparisons as guides. Between 1 (SV5) and 4 (MV group) gaps were required to arrive at this alignment, many of which were inserted in the amino-terminal region of the sequences. The largest gap inserted was 12 amino acids, at position 72 of both members of the MV group. Amino acid positions described in the following discussion refer to the position in the

TABLE I. Amino Acid Sequence Relationships among the Paramyxovirus M Proteins[a]

	Pif	Snd	MV	CDV	NDV	SV5	Mump
Pif		63 (79)	36 (56)	35 (56)	21 (42)	18 (39)	19 (39)
Snd			35 (58)	34 (56)	17 (41)	16 (37)	19 (40)
MV				76 (87)	18 (42)	20 (38)	22 (42)
CDV					20 (42)	20 (40)	23 (43)
NDV						27 (48)	24 (47)
SV5							41 (61)
Mump							

[a]Pairwise alignments of each of the M protein sequences were made using the GAP program from the University of Wisconsin Genetics Computer Group (Devereux et al., 1984), using gap weight 5.0 and length weight 0.3. The percent of identical amino acids in the same positions in the two sequences being compared is presented, with the percent similarity in parentheses. Chemically similar amino acid groupings used were ILMV, KHR, DENQ, ST, AG, FWY. Boxes are drawn around groups of sequences which are related. Sequences used were human parainfluenza virus type 3 (Pif) (Galinski et al., 1987), Sendai virus (Snd) (Blumberg et al., 1984), measles virus (MV) and canine distemper virus (CDV) (Bellini et al., 1986), Newcastle disease virus (NDV) (Chambers et al., 1986), SV5 virus (SV5) (H. Sheshberadaran and R. Lamb, personal communication), and mumps virus (Mump) (Elango, 1989).

aligned sequences in Fig. 4. Interestingly, the NDV and SV5/Mu group M proteins all contain a carboxyl-terminal extension not found in the others. The extension contains a conserved sequence where four of six amino acids are similar.

In Fig. 4, boxes are drawn around identical or chemically similar amino acids occupying the same position in all of the sequences within a supergroup. The longest sequence of similar amino acids in the Pif/MV supergroup (amino acids 309–317) contains nine amino acids. Likewise, a sequence of eight similar amino acids is identified in the SV5/Mu and NDV supergroup (residues 283–291). Boxes are also drawn around identical or similar amino acids in at least six of the seven sequences to facilitate the identification of important amino acids. In 14 positions all seven amino acids were identical, indicating a strict requirement for that amino acid at that position. In 27 other positions, six of seven amino acids were similar, indicating a strong preference for that type of amino acid. Interestingly, in 15 of these 27 cases, the only virus which did not share a similar amino acid was NDV.

The cysteines of the M proteins are highly conserved among the members of each group: five out of six are shared within the Pif group and within the MV group. However, only two conserved cysteines are shared between these groups. Likewise, the two members of the SV5/Mu group share the position of seven of their nine cysteines. The position of one of these cysteines (residue 329) is shared with the MV group and is accompanied by a cysteine in the NDV sequence at the next position (residue 330). Interestingly, the position of one cysteine is shared by Sendai and NDV only (amino acid 117). There is only one cysteine whose position is shared by all of these paramyxoviruses: the cysteine at amino acid 310. Aside from this conserved cysteine, the two closest cysteines are at positions 266 and 270, where all but NDV have a cysteine, and at residues 329 and 330, where all but the Pif group have a cysteine. One of these cysteines may form a disulfide bond with the sole conserved cysteine, providing a similar loop in most of the M proteins. The unreduced NDV M protein does migrate more rapidly on SDS–PAGE than reduced M protein, indicating probable intrachain but not interchain disulfide bonds (M. Peeples, unpublished results).

Many of the prolines are conserved within each group: 17 in the Pif group and 14 in the MV group, ten of which are conserved between these two groups. There are ten prolines conserved within the SV5/Mu group and seven of these are shared with NDV. Interestingly, NDV shares two prolines with SV5, but not with mumps (amino acids 189 and 226). Five prolines are conserved to some extent between the two supergroups: NDV shares two prolines with the Pif/MV supergroup (residues 140 and 170), but not with the more closely related SV5/Mu group; one proline (amino acid 167) is shared by all but the MV group; one proline (position 197) is shared by all but the mumps virus M protein; and one proline (position 312) is common to all of these paramyxovirus M proteins. Interestingly, the sole conserved proline (residue 312) and the sole conserved cysteine (residue 310) are separated by a single amino acid. The proline might provide a bend in the amino acid chain necessary to position the conserved cysteine for disulfide bond formation with the proper cysteine partner.

All of the paramyxovirus M proteins are highly basic. Many of these basic residues occur in pairs: the Pif group shares four pairs of basic amino acids and the MV group shares five pairs, three of which are also shared with the Pif group. The M proteins of the SV5/Mu group share two pairs of basic amino acids. Interestingly, both of these pairs (residues 225 and 226 and residues 375 and 376) are also conserved in NDV. In addition, the SV5 and NDV M proteins share two other pairs of basic amino acids (positions 90 and 91 and positions 268 and 269) which are not found in mumps virus. Only one of the pairs of basic amino acids (positions 225 and 226) is conserved among all of the paramyxovirus M proteins shown in Fig. 4. This pair is embedded in the most highly conserved sequence among the paramyxovirus M proteins (amino acids 122–136). At either end of this short sequence, there are groups of similar amino acids: five of six (including the arginine and/or lysine basic pair) on the amino-terminal side and four of five (including the acidic glutamic acid residue) on the carboxyl-terminal side. Perhaps these amino acids of opposite charge interact with one another to form a critical structure.

McGinnes and Morrison (1987) have pointed out that the NDV M protein shares a region of striking similarity with the VSV M protein. This 15-amino acid region, which is near the amino terminus of the NDV M protein, is indicated by shading in Fig. 4. Compared to the VSV M protein (residues 113–127), the M protein of NDV has 10 identical and 11 similar (73%) amino acids. The related M proteins of SV5 and mumps virus also share 60% and 73% similarity with the VSV M protein. As McGinnes and Morrison (1987) have pointed out, this region of similarity in the VSV M protein is contained within a large fragment (81% of the molecule) which is involved in binding the M protein to the lipid bilayer (Ogden et al., 1986). This region of the M protein of the Pif/MV supergroup members does not share such a high degree of similarity with the VSV M protein: Pif (20%), Snd (33%), MV (40%), and CDV (47%).

As a possible alternative membrane-binding site, Galinski et al. (1987) pointed out that the carboxyl-terminal half of the M protein is particularly hydrophobic and these authors hypothesized that this part of the molecule might be involved in membrane attachment. Bellini et al. (1986) more specifically located a region of β-sheet, flanked by α-helices which have a hydrophobic character. The β-sheet (shaded in Fig. 4) initiates with a double glycine (positions 282 and 283) and ends with a single glycine (position 283). Interestingly, this region is somewhat similar among all of the paromyxoviruses. Instead of two glycines, the SV5/Mu and NDV supergroup have a glycine and a proline shifted one amino acid further (residues 283 and 284) and share the glycine at residue 293. Both glycine and proline promote turns in proteins (Chou and Fasman, 1974). Bellini et al. (1986) further speculate that the conserved cysteines which flank this region by approximately equal distances (amino acids 216 or 220, and 310) may maintain this region in a functional conformation. It is interesting, as noted above, that the M protein of NDV is the only one lacking the first of these cysteines (amino acid 216 or 220). This region in the NDV M protein coincides with the predicted nuclear/nucleolar signals of this protein, which will be discussed below.

The RSV M protein is, in general, similar to the other paramyxovirus M proteins, being somewhat hydrophobic, especially in its carboxyl terminus (Satake and Venkatesan, 1984). However, it has little similarity to the other paramyxovirus M proteins (Spriggs et al., 1987) and is much smaller. The RSV M protein is not as basic overall as the others and contains only one pair of basic amino acids, a motif that is common in the other M proteins. It is interesting that the RSV 22K protein has a more basic character and contains six pairs of basic amino acids (Collins and Wertz, 1985). This 22K protein is associated with the virus envelope and is similar to the M protein in its solubilization with nonionic detergent and high salt (Huang et al., 1984). Perhaps the M protein and the 22K protein divide the functions of the other paramyxovirus M proteins and interact at the plasma membrane to assemble the virus components.

At lease some of the paramyxovirus M proteins are modified by acetylation and phosphorylation. We (M. E. Peeples and M. A. Bratt, unpublished results) were unable to obtain an amino-terminal sequence of the NDV M protein by Edman degradation, presumably due to a blocked amino terminus. Blumberg et al. (1984) demonstrated by mass spectrometric analysis that the amino terminus of the Sendai virus M protein is modified in vivo by removal of the initiating methionine and acetylation of the following amino acid, alanine.

A proportion of the M protein molecules of Sendai virus, mumps virus, NDV, and SV5 also appear to be modified by the addition of phosphate. Virions produced by cultured cells in the presence of $^{32}PO_4$ contain phosphorylated M protein (Lamb, 1975; Hsu and Kingsbury, 1982; Naruse et al., 1981; Smith and Hightower, 1981; H. Sheshberadaran and R. A. Lamb, personal communication). Interestingly, a much larger proportion of the M proteins in Sendai-virus infected cells is phosphorylated (Lamb and Choppin, 1977b). The authors suggested several possible explanations: (1) phosphorylated and nonophosphorylated M protein may have different specific interactions with the plasma membranes, viral glycoproteins, or nucleocapsids; (2) the phosphorylated M protein may not be efficiently incorporated into virus particles; or (3) the M protein is dephosphorylated in the maturation process, perhaps driving the budding of new virions. The virion itself contains a kinase activity which will phosphorylate the M protein in vitro (Lamb, 1975; Roux and Kolakofsky, 1974). Whether this kinase activity is encoded by the virus is unknown.

The assumption that all paramyxovirus M proteins are phosphorylated should not be made. For instance, in the experiments which indicated that the NDV M protein is phosphorylated (Smith and Hightower, 1981), the M protein was identified by its migration on SDS–PAGE. However, an alternate form of the P protein (Collins et al., 1982) synthesized from the same reading frame (McGinnes et al., 1988) migrates very close to the M protein. Since the P protein is clearly phosphorylated, the phosphate in the M region of the gel may actually be due to the smaller form of the P protein. Until such experiments are repeated using monoclonal antibodies to precipitate the M protein, firm conclusions cannot be made.

Interestingly, there is no clear evidence for phosphorylation of the measles virus M protein. Multiple forms of the measles M protein have been observed (Fujinami and Oldstone, 1980; Graves, 1981), perhaps as a result of phosphorylation or proteolytic degradation. Attempts to detect phosphorylated M protein in infected cells (Robbins and Bussell, 1979; Wechsler and Fields, 1978) or in cells transfected with the M gene alone (Wong et al., 1987) have been unsuccessful.

The immunogenicity and antigenic structure of the paramyxovirus M protein have been studied with immune sera and cells and with monoclonal antibodies. Some studies have reported the detection of antibodies or lymphocyte responses to the M protein in paramyxovirus-infected patients (Giminez et al., 1987; Vainiopää, 1985; Rose et al., 1984), while others have reported a surprising lack of antibody to the M protein in the sera of such patients (Machamer et al., 1980; Norrby et al., 1981; Graves et al., 1984; Levine et al., 1988; Ward et al., 1983) or in infected mice (Levine et al., 1989). Likewise, hybridomas producing antibodies to the M protein have been isolated with reasonable frequency from mice immunized with paramyxoviruses in some studies (Nishikawa et al., 1987; Sheshberadaran et al., 1983; Rydbeck et al., 1986), but at a low frequency in other studies (Faaberg and Peeples, 1988b; A. Portner, personal communication). Whether these apparently conflicting results reflect differences in patient or mouse populations, methods of antibody detection, or methods of antigen presentation is not clear.

Using monoclonal antibodies in competition binding assays, several groups have detected between two and six discrete antigenic sites on paramyxovirus M proteins. Nishikawa et al. (1987) defined two distinct antigenic sites on the NDV M protein (one of which contains three distinct epitopes), while Faaberg and Peeples (1988b) defined three distinct antigenic sites. Sheshberadaran et al. (1985) and Sato et al. (1985) defined six and five distinct antigenic sites, respectively, on the measles virus M protein. Rydbeck et al. (1986) defined four distinct and two overlapping antigenic sites on the parainfluenza virus type 3 M protein. In addition to their usefulness in tracking strain variations and localizing the M protein in infected cells, these monoclonal antibodies may be useful in in vitro association systems as probes for functional domains of the M protein. If any of these antibodies block particular M protein interactions, it would be important to identify their antigenic sites on the three-dimensional structure of the M protein in order to connect structure with function.

V. M PROTEIN AND THE NUCLEUS

Paramyxoviruses are classic cytoplasmic viruses. Respiratory syncytial virus is able to replicate in enucleated cells (Follett et al., 1975). Polyvalent anti-NDV sera stain only the cytoplasm of uninfected cells (Traver et al., 1960; Wheelock and Tamm, 1959). However, monoclonal antibodies against the NDV M protein detect a large proportion of the M protein in the nuclei of infected cells (Faaberg and Peeples, 1988b; Hamaguchi, 1985; Peeples, 1988a).

The difference in these results might be explained by a low representation of antibodies to the M protein in the anti-NDV sera, especially since the M protein does not appear to be very immunogenic, as discussed above.

The ability of monoclonal antibodies against the M protein to stain the nucleus is not due to a cross-reacting host antigen, since these antibodies did not stain uninfected cells (Faaberg and Peeples, 1988b). Neither is the staining due to a host protein induced by virus infection, for two reasons. (1) Monoclonal antibodies to three discrete M-protein epitopes stain the nucleus of NDV (strain AV)-infected cells (Faaberg and Peeples, 1988b). A virus-induced host-cell protein sharing one epitope with the NDV M protein would be unlikely, but such a protein sharing three epitopes would be highly unlikely. (2) Two monoclonal antibodies, one of which does and the other of which does not recognize the M protein of NDV (strain L) virions, stain the nucleus of infected cells with the same specificity (Faaberg and Peeples, 1988b).

A second possible explanation for the apparent nuclear staining is that the M protein is associated with the outer nuclear envelope, since this envelope is directly accessible to cytoplasmic proteins. To allow the detecting antibodies access to the nuclear envelope, but not to the interior of the nucleus, fixed cells were treated with an intermediate concentration of Triton X-100 (0.02%) which disrupts the cytoplasmic membrane, but not the nuclear membrane (Peeples, 1988a). Under these conditions, the M protein was located in the cytoplasm, but not the nucleus (Fig. 5H) (Peeples, 1988a), indicating that the nuclear M protein is not on the outer nuclear envelope. Both the cytoplasmic and nuclear membranes of cells treated with a higher concentration of Triton X-100 (0.05%) are disrupted (Peeples, 1988a). Under these conditions, the M protein was located both in the cytoplasm and in the nucleus (Fig. 5I) (Peeples, 1988a).

In fact, a large proportion of the nuclear NDV M protein appears to localize to discrete regions of the nucleus, apparently the nucleolus (Hamaguchi et al., 1985; Peeples, 1988a). These discrete nuclear sites costain with antibodies to nucleoli (M. Peeples, unpublished results). A similar nucleolar concentration of M protein is found in at least six cell lines tested (M. Peeples, unpublished results), indicating that nuclear localization is controlled by the virus, not by the host cell.

The nuclear localization of the NDV M protein in an infected cell is not a late event. By 3.5 hr after infection, when the M protein is first detectable in the cell, it is found in the nucleus (M. Peeples, unpublished results). These results imply that the M protein is not entering the nucleus due to late-stage cytopathology. However, the mechanism by which the M protein reaches the nucleus is not known. It is possible that nuclear localization requires another of the viral proteins in the manner that the SV40 agnoprotein controls VP1 nuclear accumulation (Resnick and Shenk, 1986) or that the herpes simplex virus ICP4 protein controls nuclear transport of two other HSV proteins (Knipe and Smith, 1986). To test this possibility, we transiently expressed a cDNA copy of the M protein in cultured cells and found a very similar pattern to the infected cells: much of the M protein localized to the nucleus, and particularly to the nucleoli (N. Coleman-Fuller, K. Gupta, and M. Peeples, unpublished results).

Several proteins which accumulate in the nucleus contain specific amino

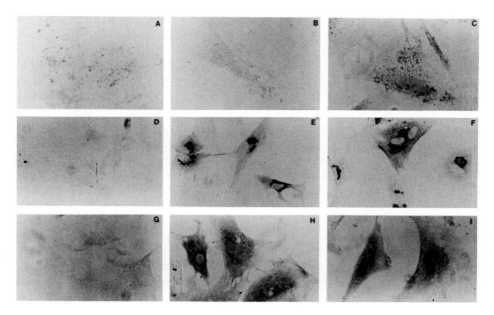

FIGURE 5. Immunoperoxidase staining of NDV-infected cells as a function of Triton X-100 concentration. Cells were fixed with 3% paraformaldehyde and treated with (B,E,H) 0.02% Triton, (C,F,I) 0.05% Triton, or (A,D,G) no Triton. Cells were subsequently stained with mouse monoclonal antibodies against NDV proteins (A,B,C) HN, (D,E,F) P, or (G,H,I) P, followed successively by goat anti-mouse Ig-biotin, streptavidin-peroxidase, and diaminobenzidine and H_2O_2. The details of this method are presented in Peeples (1988a). [Figure taken from Peeples (1988b).]

acid sequences which determine nuclear transport. Most of these sequences contain a core of highly basic amino acids, the most striking of which is shaded in Fig. 4 (amino acids 256–271). The NDV M protein contains clusters of basic amino acids (shaded area beginning at amino acid 256 in Fig. 4). In many proteins where a nuclear transport signal has been defined, a four-amino acid sequence of K-R/K-X-R/K has been found to be critical (Chelsky et al., 1989). The shaded area of the NDV M protein contains this potential consensus nuclear transport signal (amino acids 268–271). The basic amino acid-rich shaded area is also similar in its concentration of basic amino acids to the only nucleolar localization signal described, that of the pX protein of HTLV-I (Siomi et al., 1988). Perhaps this basic region is required for nuclear transport and/or nucleolar localization.

While the NDV M protein is the most flagrant in its nuclear localization, it may not be the only paramyxovirus M protein with this property. A small amount of the Sendai virus M protein has been detected in the nucleoli of infected cells (Yoshida et al., 1976). On the other hand, the M protein of measles virus (Tyrrell et al., 1980; Norrby et al., 1982) has not been detected in the nucleus of infected cells, although its NP protein has been (Norrby et al., 1982). While the members of the Pif/MV supergroup all have a stretch of three or four basic amino acids in a row near the region suggested for the NDV M-protein nuclear transport signal (Fig. 4), these sequences do not match the general consensus signal described above. However, since the transport signals of several nuclear proteins do not conform to this consensus (Greenspan *et al.*, 1988; Hall *et al.*, 1984; Moreland *et al.*, 1985; Siomi *et al.*, 1988), lack of this sequence does not necessarily indicate lack of nuclear transport. It is interesting that the influenza virus M protein (Patterson *et al.*, 1988) and the VSV M protein (Lyles *et al.*, 1988) have also been found in the nucleus of infected cells. Within the nucleus, the influenza M protein accumulates in the nucleolus (Bucher *et al.*, 1989) while the VSV M protein does not (Lyles *et al.*, 1988).

The role, if any, of the large amount of nuclear M protein in NDV-infected cells is unknown. There are at least several possibilities: (1) Perhaps the nucleolar M protein is modifying ribosomes as they assemble so that they will translate viral but not host mRNA. In fact, NDV is the most efficient of the paramyxoviruses in inhibiting host protein synthesis. (2) Perhaps all paramyxovirus M proteins migrate into the nucleus to obtain a modification and then migrate back into the cytoplasm. The NDV M protein might have a particularly strong entrance signal, but a weak exit signal compared to the other paramyxoviruses. (3) Perhaps the M protein is shuttling into and out of the nucleus, as do some of the cellular proteins localized in the nucleolus (Borer *et al.*, 1989). The M protein might be scavenging a nuclear factor to enhance replication or assembly.

VI. CONCLUSION AND OUTLOOK

The paramyxovirus M protein is the amphiphilic bandleader of virion assembly, similar in many respects to the M proteins of other enveloped RNA viruses [reviewed by Dubois-Dalcq *et al.* (1984)]. It interacts with viral components such as the nucleocapsid and probably the cytoplasmic domains of the viral glycoproteins, as well as with itself. In addition, the M protein interacts with cellular components such as actin and the plasma membrane. Most or all of these interactions are critical for its bandleading role in assembly. These interactions appear to be sequential and may even involve conformational changes subsequent to one coupling that exposes a site for the next coupling, though there is no evidence for such conformational changes as yet. After the virion has budded from the plasma membrane, many of these couplings are released. Conceptually, a virion in which the various components were freed from each other would seem best prepared to begin transcription upon fusion of the virion membrane with the plasma membrane of a target cell. Other possible M protein functions in the regulation of virus transcription and in the infected cell nucleus have not yet been well studied. These aspects of the M protein may be important in the virus replication cycle or in its pathogenic effect on host cells.

Our understanding of how the paramyxovirus M protein interacts with each of these viral and cellular components is in its infancy. Monoclonal antibodies may prove to be useful probes to locate functional areas of the molecule. In addition, the M protein genes of eight paramyxoviruses have now been sequenced and several highly conserved features have been identified. However, connecting these features with particular interactions is presently only speculation. Conserved features will be ideal targets in designing mutants to relate structure with function. Transient expression of mutant M proteins may help to locate areas of the M protein responsible for interacting with cellular structures, but identifying areas responsible for interacting with other viral proteins and structures will require complementation assays with viruses defective in their M protein.

ACKNOWLEDGMENTS. I thank Diana Huang, Laurent Roux, and Hooshmand Sheshberadaran for their critical reading of this manuscript, Kailash Gupta for help with the computer analyses, and Barbara Newton and Deborah Menchaca for their help with the artwork and manuscript preparation. This work was supported by Public Health Service Grant AI-29606 from the National Institute of Allergy and Infectious Diseaes.

VII. REFERENCES

Bächi, T., 1980, Intramembrane structural differentiation in Sendai virus maturation, *Virology* **106**:41–49.

Bächi, T., 1988, Direct observation of the budding and fusion of an enveloped virus by video microscopy of viable cells, *J. Cell Biol.* **107**:1689–1695.

Bellini, W. J., Englund, G., Richardson, C. D., Rozenblatt, S., and Lazzarini, R. A., 1986, Matrix genes of measles virus and canine distemper virus: Cloning, nucleotide sequences, and deduced amino acid sequences, *J. Virol.* **58**:408–416.

Blumberg, B. M., Rose, K., Simona, M. G., Roux, L., Giorgi, C., and Kolakofsky, D., 1984, Analysis of the Sendai virus *M* gene and protein, *J. Virol.* **52**:656–663.

Bohn, W., Rutter, G., Hohenberg, H., Mannweiler, K., and Nobis, P., 1986, Involvement of actin filaments in budding of measles virus: Studies on cytoskeletons of infected cells, *Virology* **149**:91–106.

Borer, R. A., Lehner, C. F., Eppenberger, H. M., and Nigg, E. A., 1989, Major nucleolar proteins shuttle between nucleus and cytoplasm, *Cell* **56**:379–390.

Bowen, H. A., and Lyles, D. S., 1982, Kinetics of incorporation of Sendai virus proteins into host plasma membranes and virions, *Virology* **121**:1–11.

Bucher, D. J., Kharitonenkov, I. G., Zakomirdin, J. A., Grigoriev, V. B., Klimenko, S. M., and Davis, J. F., 1980, Incorporation of influenza virus M-protein into liposomes, *J. Virol.* **36**:586–590.

Bucher, D., Popple, S., Baer, M., Mikhail, A., Gong, F.-F., Whitaker, C., Paoletti, E., and Judd, A., 1989, M protein (MI) of influenza virus: Antigenic analysis and intracellular localization with monoclonal antibodies, *J. Virol.* **63**:3622–3633.

Büechi, M., and Bächi, T., 1982, Microscopy of internal structures of Sendai virus associated with the cytoplasmic surface of host membranes, *Virology* **120**:349–359.

Caldwell, S. E., and Lyles, D. S., 1986, Dissociation of newly synthesized Sendai viral proteins from the cytoplasmic surface of isolated plasma membranes of infected cells, *J. Virol.* **57**:678–683.

Carroll, A. R., and Wagner, R. R., 1979, Role of the membrane (M) protein in endogenous inhibition of *in vitro* transcription by vesicular stomatitis virus, *J. Virol.* **29**:134–142.

Cattaneo, R., Schmid, A., Rebmann, G., Baczko, K., ter Meulen, V., Bellini, W. J., Rozenblatt, S., and Billeter, M. A., 1986, Accumulated measles virus mutations in a case of subacute sclerosing panencephalitis: Interrupted matrix protein reading frame and transcription alteration, *Virology* **154**:97–107.

Cattaneo, R., Schmid, A., Eschle, D., Baczko, K., ter Meulen, V., and Billeter, M. A., 1988, Biased hypermutation and other genetic changes in defective measles viruses in human brain infections, *Cell* **55**:255–265.

Chambers, P., Millar, N. S., Platt, S. G., and Emmerson, P. T., 1986, Nucleotide sequence of the gene encoding the matrix protein of Newcastle disease virus, *Nucleic Acids Res.* **14**:9051–9061.

Chatis, P. A., and Morrison, T. G., 1982, Fatty acid modification of Newcastle disease virus glycoproteins, *J. Virol.* **43**:342–347.

Chelsky, D., Ralph, R., and Jonak, G., 1989, Sequence requirements for synthetic peptide-mediated translocation to the nucleus, *Mol. Cell. Biol.* **9**:2487–2492.

Choppin, P. W., and Compans, R. W., 1975, Reproduction of paramyxoviruses, in "Comprehensive Virology," Vol. 4 (H. Fraenkel-Conrat and R. R. Wagner, eds., pp. 95–178, Plenum Press, New York.

Chou, P. Y., and Fasman, G. D., 1974, Conformational parameters for amino acids in helical, β sheet, and random coil regions calculated from proteins, *Biochemistry* **13**:211.

Collins, P. L., and Wertz, G. W., 1985, The envelope-associated 22K protein of human respiratory syncytial virus: Nucleotide sequence of the mRNA and a related polytranscript, *J. Virol.* **54**:65–71.

Collins, P. L., Wertz, G. W., Ball, L. A., and Hightower, L. E., 1982, Coding assignments of the five smaller mRNAs of Newcastle disease virus, *J. Virol.* **43**:1024–1031.

Compans, R. W., Holmes, K. V., Dales, S., and Choppin, P. W., 1966, An electron microscopic study of moderate and virulent virus–cell interactions of the parainfluenza virus SV5, *Virology* **30**:411–426.

De, B. P., Thornton, G. B., Luk, D., and Banerjee, A. K., 1982, Purified matrix protein of vesicular stomatitis virus blocks viral transcription *in vitro*, *Proc. Natl. Acad. Sci. USA* **79**:7137–7141.

Devereux, J., Haeberli, P., and Smithies, O., 1984, A comprehensive set of sequence analysis programs for the VAX, *Nucleic Acids Res.* **12**:387–395.

Dubois-Dalcq, M., Holmes, K. V., and Rentier, B., 1984, Assembly of *Paramyxoviridae*, in "Assembly of Enveloped RNA Viruses" (D. W. Kingsbury, ed.), pp. 44–65, Springer-Verlag, New York.

Dubovi, E. J., and Wagner, R. R., 1977, Spatial relationships of the proteins of vesicular stomatitis virus: Induction of reversible oligomers by cleavable protein cross-linkers and oxidation, *J. Virol.* **22**:500–509.

Egelman, E. H., Wu, S.-S., Amrein, M., Portner, A., and Murti, G., 1989, The Sendai virus nucleocapsid exists in at least four different helical states, *J. Virol.* **63**:2233–2243.

Elango, N., 1989, Complete nucleotide sequence of the matrix protein mRNA of mumps virus, *Virology* **168**:426–428.

Faaberg, K. S., and Peeples, M. E., 1988a, Association of soluble matrix protein of Newcastle disease virus with liposomes is independent of ionic conditions, *Virology* **166**:123–132.

Faaberg, K. S., and Peeples, M. E., 1988b, Strain variation and nuclear association of Newcastle disease virus matrix protein, *J. Virol.* **62**:586–593.

Follett, A. C., Pringle, C. R., and Pennington, T. H., 1975, Virus development in enucleate cells: Echovirus, poliovirus, pseudorabies virus, reovirus, respiratory syncytial virus and Semliki forest virus, *J. Gen. Virol.* **26**:183–196.

Fujinami, R. S., and Oldstone, M. B. A., 1980, Alterations in expression in measles virus peptides by antibodies: Molecular events in antibody-induced antigen modulation, *J. Immunol.* **125**:78–85.

Galinski, M. S., Mink, M. A., Lambert, D. M., Wechsler, S. L., and Pons, M. W., 1987, Molecular cloning and sequence analysis of the human parainfluenza 3 virus gene encoding the matrix protein, *Virology* **157**:24–30.

Gimenez, H. B., Keir, H. M., and Cash, P., 1987, Immunoblot analysis of the human antibody response to respiratory syncytial virus infection, *J. Gen. Virol.* **68**:1267–1275.

Giuffre, R. M., Tovell, D. R., Kay, C. M., and Tyrrell, D. L. J., 1982, Evidence for an interaction between the membrane protein of a paramyxovirus and actin, *J. Virol.* **42**:963–968.

Graves, M. C., 1981, Measles virus polypeptides in infected cells studied by immune precipitation and one-dimensional peptide mapping, *J. Virol.* **38**:224–230.

Graves, M., Griffin, D. E., Johnson, R. T., Hirsch, R. L., De Soriano, I. L., Roedenbeck, S., and Vaisberg, A., 1984, Development of antibody to measles virus polypeptides during complicated and uncomplicated measles virus infections, *J. Virol.* **49**:409–412.

Greenspan, D., Palese, P., and Krystal, M., 1988, Two nuclear localization signals in the influenza virus NS1 nonstructural protein, *J. Virol.* **62**:3020–3026.

Hall, M. N., Hereford, L., and Herskowitz, I., 1984, Targeting of *E. coli* β-galactosidase to the nucleus in yeast, *Cell* **36**:1057–1065.

Hamaguchi, M., Yoshida, T., Nishikawa, K., Naruse, H., and Nagai, Y., 1983, Transcriptive complex of Newcastle disease virus, I. Both L and P proteins are required to constitute an active complex, *Virology* **128**:105–117.

Hamaguchi, M., Nishikawa, K., Toyoda, T., Yoshida, T., Hanaichi, T., and Nagai, Y., 1985, Transcriptive complex of Newcastle disease virus. II. Structural and functional assembly associated with the cytoskeletal framework, *Virology* **147**:295–308.

Heggeness, M. H., Scheid, A., and Choppin, P. W., 1981, The relationship of conformational changes in the Sendai virus nucleocapsid to proteolytic cleavage of the NP polypeptide, *Virology* **114**:555–562.

Heggeness, M. H., Smith, P. R., and Choppin, P. W., 1982, *In vitro* assembly of the nonglycosylated membrane protein (M) of Sendai virus, *Proc. Natl. Acad. Sci. USA* **79**:6232–6236.

Hewitt, J. A., 1977, Studies on the subunit composition of the M-protein of Sendai virus, *FEBS Lett.* **81**:395–398.

Hewitt, J. A., and Nermut, M. V., 1977, A morphological study of the M-protein of Sendai virus, *J. Gen. Virol.* **34**:127–136.

Homma, M., and Ohuchi, M., 1973, Trypsin action on the growth of Sendai virus in tissue culture cells. III. Structural differences of Sendai viruses grown in eggs and tissue culture cells, *J. Virol.* **12**:1457–1465.

Homma, M., Shimizu, K., Shimizu, Y. K., and Ishida, N., 1976, On the study of Sendai virus hemolysis. I. Complete Sendai virus lacking in hemolytic activity, *Virology* **71**:41–47.

Hsu, C.-H., and Kingsbury, D. W., 1982, Topography of phosphate residues in Sendai virus proteins, *Virology* **120**:225–234.

Huang, A. S., Baltimore, D., and Bratt, M. A., 1971, Ribonucleic acid polymerase in virions of Newcastle disease virus: Comparison with the vesicular stomatitis virus polymerase, *J. Virol.* **7**:389–394.

Huang, Y. T., Collins, P. L., and Wertz, G. W., 1984, Identification of a new envelope-associated protein of human respiratory syncytial virus, *in* "Nonsegmented Negative Strand Viruses, Paramyxoviruses and Rhabdoviruses" (D. H. L. Bishop and R. W. Compans, eds.), pp. 365–368, Academic Press, New York.

Kim, J., Hama, K., Miyake, Y., and Okada, Y., 1979, Transformation of intramembrane particles of HVJ (Sendai virus) envelopes from an invisible to visible form on aging of virions, *Virology* **95**:523–535.

Knipe, D. M., and Smith, J. L., 1986, A mutant herpesvirus protein leads to a block in nuclear localization of other viral proteins, *Mol. Cell. Biol.* **6**:2371–2381.

Knipe, D. M., Baltimore, D., and Lodish, H. F., 1977, Maturation of viral proteins in cells infected with temperature-sensitive mutants of vesicular stomatitis virus, *J. Virol.* **21**:1149–1158.

Lamb, R. A., 1975, The phosphorylation of Sendai virus proteins by a virus particle-associated protein kinase, *J. Gen. Virol.* **26**:249–263.

Lamb, R. A., and Choppin, P. W., 1977a, The synthesis of Sendai virus polypeptides in infected cells, II. Intracellular distribution of polypeptides, *Virology* **81**:371–381.

Lamb, R. A., and Choppin, P. W., 1977b, The synthesis of Sendai virus polypeptides in infected cells, III. Phosphorylation of polypeptides, *Virology* **81**:382–397.

Lamb, R. A., Mahy, B. W. J., and Choppin, P. W., 1976, The synthesis of Sendai virus polypeptides in infected cells, *Virology* **69**:116–131.

Levine, S., Dajani, A., and Klaiber-Franco, R., 1988, The response of infants with bronchiolitis to the proteins of respiratory syncytial virus, *J. Gen. Virol.* **69**:1239.

Levine, S., Dillman, T. R., and Montgomery, P. C., 1989, The envelope proteins from purified respiratory syncytial virus protect mice from intranasal virus challenge, *Proc. Soc. Exp. Biol. Med.* **190**:349–356.

Li, J. K.-K., Miyakawa, T., and Fox, C. F., 1980, Protein organization in Newcastle disease virus as revealed by perturbant treatment, *J. Virol.* **34**:268–271.

Li, Y., Luo, L., Snyder, R. M., and Wagner, R. R., 1988, Site-specific mutations in vectors that express antigenic and temperature-sensitive phenotypes of the *M* gene of vesicular stomatitis virus, *J. Virol.* **62**:3729–3747.

Lohmeyer, J., Talens, L. T., and Klenk, H.-D., 1979, Biosynthesis of the influenza virus envelope in abortive infection, *J. Gen. Virol.* **42**:73–88.

Lyles, D. S., 1979, Glycoproteins of Sendai virus are transmembrane proteins, *Proc. Natl. Acad. Sci. USA* **76**:5621–5625.

Lyles, D. S., Puddington, L., and McCreedy, Jr., B. J., 1988, Vesicular stomatitis virus M protein in the nuclei of infected cells, *J. Virol.* **62**:4387–4392.

Machamer, C. E., Hayes, E. C., Gollobin, S. D., Westfall, L. K., and Zweerink, H. J., 1980, Antibodies against the measles matrix polypeptide after clinical infection and vaccination, *Infect. Immun.* **27**:817–825.

Machamer, C. E., Hayes, E. C., and Zweerink, H. J., 1981, Cells infected with a cell-associated subacute sclerosing panencephalitis virus do not express M protein, *Virology* **108**:515–520.

Markwell, M. A. K., and Fox, C. F., 1980, Protein–protein interactions within paramyxoviruses identified by native disulfide bonding or reversible chemical cross-linking, *J. Virol.* **33**:152–166.

Marx, P. A., Portner, A., and Kingsbury, D. W., 1974, Sendai virion transcriptase complex: Polypeptide composition and inhibition by virion envelope proteins, *J. Virol.* **13**:107–112.

McGinnes, L. W., and Morrison, T. G., 1987, The nucleotide sequence of the gene encoding the Newcastle disease virus membrane protein and comparisons of membrane protein sequences, *Virology* **156**:221–228.

McGinnes, L., McQuain, C., and Morrison, T., 1988, The P protein and the nonstructural 38K and 29K proteins of Newcastle disease virus are derived from the same open reading frame, *Virology* **164**:256–264.

McSharry, J. J., Compans, R. W., and Choppin, P. W., 1971, Proteins of vesicular stomatitis virus and of phenotypically mixed vesicular stomatitis virus–simian virus 5 virions, *J. Virol.* **8**:722–729.

Moore, N. V., and Burke, D. C., 1974, Characterization of the structural proteins of different strains of Newcastle disease virus, *J. Gen. Virol.* **25**:275–289.

Moreland, R. B., Nam, H. G., Hereford, L. M., and Fried, H. M., 1985, Identification of a nuclear localization signal of a yeast ribosomal protein, *Proc. Natl. Acad. Sci. USA* **82**:6561–6565.

Morgan, E. M., Re, G. G., and Kingsbury, D. W., 1984, Complete sequence of the Sendai virus *NP* gene from a cloned insert, *Virology* **135**:279–287.

Morita, K., Vanderoef, R., and Lenard, J., 1987, Phenotypic revertants of temperature-sensitive M protein mutants of vesicular stomatitis virus: Sequence analysis and functional characterization, *J. Virol.* **61**:256–263.

Morrison, T. G., and McGinnes, L. J., 1985, Cytochalasin D accelerates the release of Newcastle disease virus from infected cells, *Virus Res.* **4**:93–106.

Nagai, Y., and Klenk, H.-D., 1977, Activation of precursors to both glycoproteins of Newcastle disease virus by proteolytic cleavage, *Virology* **77**:125–134.

Nagai, Y., Klenk, H.-D., and Rott, R., 1976a, Proteolytic cleavage of the viral glycoproteins and its significance for the virulence of Newcastle disease virus, *Virology* **72**:494–508.

Nagai, Y., Ogura, H., and Klenk, H.-D., 1976b, Studies on the assembly of the envelope of Newcastle disease virus, *Virology* **69**:523–538.

Nagai, Y., Yoshida, T., Hamaguchi, M., Iinuma, M., Maeno, K., and Matsumoto, T., 1978, Cross-linking of Newcastle disease virus (NDV) proteins, *Arch. Virol.* **58**:15–28.

Naruse, H., Nagai, Y., Yoshida, T., Hamaguchi, M., Matsumoto, T., Isomura, S., and Suzuki, S., 1981, The polypeptides of mumps virus and their synthesis in infected chick embryo cells, *Virology* **112**:119–130.

Newcomb, W. W., Tobin, G. J., McGowan, J. J., and Brown, J. C., 1982, *In vitro* reassembly of vesicular stomatitis virus skeletons, *J. Virol.* **41**:1055–1062.

Nishikawa, K., Hanada, N., Morishima, T., Yoshida, T., Hamaguchi, M., Toyoda, T., and Nagai, Y., 1987, Antigenic characterization of the internal proteins of Newcastle disease virus by monoclonal antibodies, *Virus Res.* **7**:83–92.

Norrby, E., Örvell, C., Vandvik, B., and Cherry, J. D., 1981, Antibodies against measles virus polypeptides in different disease conditions, *Infect. Immun.* **34**:718–724.

Norrby, E., Chen, S.-N., Togashi, T., Shesberadaran, H., and Johnson, K. P., 1982, Five measles virus antigens demonstrated by use of mouse hybridoma antibodies in productively infected tissue culture cells, *Arch. Virol.* **71**:1–11.

Ogden, J., Pal, R., and Wagner, R., 1986, Mapping regions of the matrix protein of vesicular stomatitis virus which bind ribonucleocapsids, liposomes, and monoclonal antibodies, *J. Virol.* **58**:860–868.

Örvell, C., 1978, Structural polypeptides of mumps virus, *J. Gen. Virol.* **41**:527–539.

Pal, R., Grinnell, B. W., Snyder, R. M., and Wagner, R. R., 1985, Regulation of viral transcription by the matrix protein of vesicular stomatitis virus probed by monoclonal antibodies and temperature-sensitive mutants, *J. Virol.* **56**:386–394.

Patterson, S., Gross, J., and Oxford, J. S., 1988, The intracellular distribution of influenza virus matrix protein and nucleoprotein in infected cells and their relationship to haemagglutinin in the plasma membrane, *J. Gen. Virol.* **69**:1859–1872.

Peeples, M. E., 1988a, Differential detergent treatment allows immunofluorescent localization of the Newcastle disease virus matrix protein within the nucleus of infected cells, *Virology* **162**:255–259.

Peeples, M. E., 1988b, Newcastle disease virus replication, *in* "Newcastle Disease" (D. J. Alexander, ed.), pp. 45–78, Kluwer Academic Publishers, Boston, Massachusetts.

Peeples, M. E., and Bratt, M. A., 1984a, Mutation in the matrix protein of Newcastle disease virus can result in decreased fusion glycoprotein incorporation into particles and decreased infectivity, *J. Virol.* **51**:81–90.

Peeples, M. E., and Bratt, M. A., 1984b, Mapping mutant and wild-type M proteins of Newcastle disease virus (NDV) by repeated partial proteolysis, *in* "Nonsegmented Negative Strand Viruses," pp. 315–320, Academic Press, New York.

Picard, D., and Yamamoto, K., 1987, Two signals mediate hormone-dependent nuclear localization of the glucocorticoid receptor, *EMBO J.* **6**:3333–3340.

Portner, A., and Kingsbury, D. W., 1976, Regulatory events in the synthesis of Sendai virus polypeptides and their assembly into virions, *Virology* **73**:79–88.

Resnick, J., and Shenk, T., 1986, Simian virus 40 agnoprotein facilitates normal nuclear location of the major capsid polypeptide and cell-to-cell spread of virus, *J. Virol.* **60**:1098–1106.

Robbins, S. J., and Bussell, R. H., 1979, Structural phosphoproteins associated with purified measles virions and cytoplasmic nucleocapsids, *Intervirology* **12**:96.

Rose, J. W., Bellini, W. J., McFarlin, D. E., and McFarland, H. F., 1984, Human cellular immune response to measles virus polypeptides, *J. Virol.* **49**:988–991.

Roux, L., and Kolakofsky, D., 1974, Protein kinase associated with Sendai virions, *J. Virol.* **13**:545–547.

Roux, L., and Waldvogel, F. A., 1982, Instability of the viral M protein in BHK-21 vells persistently infected with Sendai virus, *Cell* **28**:293–302.

Roux, L., Beffy, P., and Portner, A., 1984, Restriction of cell surface expression of Sendai virus hemagglutinin-neuraminidase glycoprotein correlates with its higher instability in persistently and standard plus defective interfering virus infected BHK-21 cells, *Virology* **138**:118–128.

Russell, P. H., and Almeida, J. D., 1984, A regular subunit pattern seen on non-infectious Newcastle disease virus particles, *J. Gen. Virol.* **65**:1023–1031.

Rydbeck, R., Örvell, C., Löve, A., and Norrby, E., 1986, Characterization of four parainfluenza virus type 3 proteins by use of monoclonal antibodies, *J. Gen. Virol.* **67**:1531–1542.

Satake, M., and Venkatesan, S., 1984, Nucleotide sequence of the gene encoding respiratory syncytial virus matrix protein, *J. Virol.* **50**:92–99.

Sato, T. A., Fukuda, A., and Sugiura, A., 1985, Characterization of major structural proteins of measles virus with monoclonal antibodies, *J. Gen. Virol.* **66**:1397–1409.

Scheid, A., and Choppin, P. W., 1973, Isolation and purification of the envelope proteins of Newcastle disease virus, *J. Virol.* **11**:263–271.

Scheid, A., and Choppin, P. W., 1974, Identification of biological activities of paramyxovirus glycoproteins, Activation of cell fusion, hemolysis, and infectivity by proteolytic cleavage of an inactive precursor protein of Sendai virus, *Virology* **57**:475–490.

Schmidt, M. F. G., 1982, Acylation of viral spike glycoproteins: A feature of enveloped RNA viruses, *Virology* **116**:327–338.

Schwalbe, J. C., and Hightower, L. E., 1982, Maturation of the envelope glycoproteins of Newcastle disease virus on cellular membranes, *J. Virol.* **41**:947–957.

Sechoy, O., Philippot, J. R., and Bienvenue, A., 1987, F Protein–F protein interaction within the Sendai virus identified by native bonding or chemical cross-linking, *J. Biol. Chem.* **262**:11519–11523.

Sheshberadaran, H., and Lamb, R. A., 1990, Sequence characterization of the membrane protein gene of paramyxovirus simian virus 5, *Virology* **176**:234–243.

Sheshberadaran, H., Chen, S.-N., and Norrby, E., 1983, Monoclonal antibodies against five structural components of measles virus, *Virology* **128**:341–353.

Sheshberadaran, H., Norrby, E., and Rammohan, K. W., 1985, Monoclonal antibodies against five structural components of measles virus, *Arch. Virol.* **83**:251–268.

Shimizu, K., and Ishida, N., 1975, The smallest protein of Sendai virus: Its candidate function of binding nucleocapsid to envelope, *Virology* **67**:427–437.

Shimizu, Y. K., Shimizu, K., Ishida, N., and Homma, M., 1976, On the study of Sendai virus hemolysis, II. Morphological study of envelope fusion and hemolysis, *Virology* **71**:48–60.

Shipley, J. B., Pal, R., and Wagner, R. R., 1988, Antigenicity, function and conformation of synthetic oligopeptides corresponding to amino-terminal sequences of wild-type and mutant matrix proteins of vesicular stomatitis virus, *J. Virol.* **62**:2569–2577.

Siomi, H., Shida, H., Nam, S. H., Nosaka, T., Maki, M., and Hatanaka, M., 1988, Sequence requirements for nucleolar localization of human T cell leukemia virus type I pX protein, which regulates viral RNA processing, *Cell* **55**:197–209.

Smith, G. W., and Hightower, L. E., 1981, Identification of the P proteins and other disulfide-linked and phosphorylated proteins of Newcastle disease virus, *J. Virol.* **37**:256–267.

Spriggs, M. K., Johnson, P. R., and Collins, P. L., 1987, Sequence analysis of the matrix protein gene of human parainfluenza virus type 3: extensive sequence homology among paramyxoviruses, *J. Gen. Virol.* **68**:1491–1497.

Traver, M. I., Northrop, R. L., and Walker, D. L., 1960, Site of intracellular antigen production by myxoviruses, *Proc. Soc. Exp. Biol. Med.* **104**:268–273.

Tuffereau, C., and Roux, L., 1988, Direct adverse effects of Sendai virus DI particles on virus budding and on M protein fate and stability, *Virology* **162**:417–426.

Tuffereau, C., Portner, A., and Roux, L., 1985, The role of haemagglutinin-neuraminidase glycoprotein cell surface expression in the survival of Sendai virus-infected BHK-21 cells, *J. Gen. Virol.* **66**:2313–2318.

Tyrrell, D. L. J., and Ehrnst, A., 1979, Transmembrane communication in cells chronically infected with measles virus, *J. Cell Biol.* **81**:396–402.

Tyrrell, D. L. J., and Norrby, E., 1978, Structural polypeptides of measles virus, *J. Gen. Virol.* **39**:219–229.

Tyrrell, D. L. J., Rafter, D. J., Örvell, C., and Norrby, E., 1980, Isolation and immunological characterization of the nucleocapsid and membrane proteins of measles virus, *J. Gen. Virol.* **51**:307–315.

Vainionpää, R., Meurman, O., and Sarkkinen, H., 1985, Antibody response to respiratory syncytial structural proteins in children with acute respiratory synctial virus infection, *J. Virol.* **53**:976–979.

Varsanyi, T. M., Utter, G., and Norrby, E., 1984, Purification, morphology, and antigenic characterization of measles virus envelope components, *J. Gen. Virol.* **65**:355–366.

Walsh, E. E., Brandriss, M. W., and Schlesinger, J. J., 1985, Purification and characterization of the respiratory syncytial virus fusion protein, *J. Gen. Virol.* **66**:409–415.

Ward, K. A., Lambden, P. R., Ogilvie, M. M., and Watt, P. J., 1983, Antibodies to respiratory syncytial virus polypeptides and their significance in human infection, *J. Gen. Virol.* **64**:1867–1876.

Wechsler, S. L., and Fields, B. N., 1978, Intracellular synthesis of measles virus-specific polypeptides, *J. Virol.* **25**:285–297.

Wheelock, E. F., and Tamm, I., 1959, Mitosis and division in HeLa cells infected with influenza or Newcastle disease virus, *Virology* **8**:532–536.

Wong, T. C., Wipf, G., and Hirano, A., 1987, The measles virus matrix gene and gene product defined by *in vitro* and *in vivo* expression, *Virology* **157**:497–508.

Yoshida, T., Nagai, Y., Yoshii, S., Maeno, K., Matsumoto, T., and Hosino, M., 1976, Membrane (M) protein of HVJ (Sendai virus): Its role in virus assembly, *Virology* **71**:143–161.

Yoshida, T., Nagai, Y., Maeno, K., Iinuma, M., Hamaguchi, M., Matsumoto, T., Nagayoshi, S., and Hoshino, M., 1979, Studies on the role of M protein in virus assembly using a *ts* mutant of HVJ (Sendai virus), *Virology* **92**:139–154.

Yoshida, T., Nakayama, Y., Nagura, H., Toyoda, T., Nishikawa, K., Hamaguchi, M., and Nagai, Y., 1986, Inhibition of the assembly of Newcastle disease virus by monensin, *Virus Res.* **4**:179–195.

Zakowski, J. J., Petri, W. A., and Wagner, R. R., 1981, Role of matrix protein in assembling the membrane of vesicular stomatitis virus: Reconstitution of matrix protein with negatively charged phospholipid vesicles, *Biochemistry* **20**:3902–3907.

Zavada, J., 1982, The pseudotypic paradox, *J. Gen. Virol.* **63**:15–24.

CHAPTER 17

Intracellular Targeting and Assembly of Paramyxovirus Proteins

RANJIT RAY, LAURENT ROUX, AND
RICHARD W. COMPANS

I. INTRODUCTION

Paramyxoviruses are pleomorphic particles consisting of a spike-covered, lipid-containing envelope enclosing a helical nucleocapsid. The viral genome is a single-stranded, negative-sense RNA molecule. The helical nucleocapsid consists of RNA, nucleocapsid protein (NP), and two other proteins (P and L), and is covered by an envelope consisting of a lipid bilayer as well as two virus-coded glycoproteins, the hemagglutinin-neuraminidase (HN) and the fusion (F) proteins. Virus assembly occurs by a process of budding at the cellular plasma membrane. The viral glycoproteins are embedded as integral membrane proteins on the cell surface, and the internal viral components (M protein and nucleocapsid) associate with regions of the membrane which contain these glycoproteins. Virions are formed by a process of outfolding, or budding, at the cell surface, during which there is continuity between the plasma membrane of the host cell and that of the emerging virus particle (Compans et al., 1966). Stages in the assembly of a paramyxovirus are depicted in Fig. 1. The first recognizable change in the plasma membrane is the ability to detect viral antigens on the cell surface, as seen by immunoferritin labeling in Fig. 1A, in morphologically normal regions of the plasma membrane. A more striking change is seen in Fig. 1B, in which viral nucleocapsids are closely aligned under

RANJIT RAY • Secretech Inc., Birmingham, Alabama 35205.   LAURENT ROUX • Department of Microbiology, University of Geneva Medical School, C.M.U., 1211 Geneva 4, Switzerland.   RICHARD W. COMPANS • Department of Microbiology, University of Alabama at Birmingham, Birmingham, Alabama 35294.

457

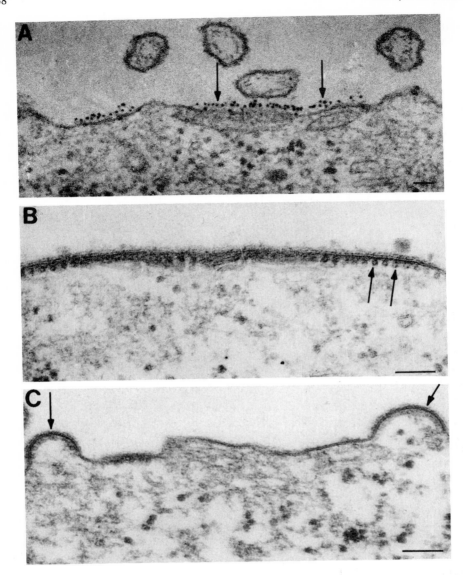

FIGURE 1. Stages in the assembly of a paramyxovirus. (A) Ferritin-conjugated antiviral antibody is seen to label areas of the surface of an SV5-infected MDBK cell which are of normal morphology, but presumably contain the viral glycoproteins ($\times 56,000$). [From Choppin and Compans (1975).] (B) SV5 nucleocapsids, many in cross section, are closely aligned under a long region of the plasma membrane of an infected monkey kidney cell. A layer of dense material corresponding to the viral glycoproteins is present on the outer surface of the membrane ($\times 105,000$). [From Compans *et al.* (1966).] (C) Budding SV5 virions containing helical nucleocapsids are seen as outfoldings of the plasma membrane (arrows) ($\times 105,000$). Magnification bar = 100 nm.

the plasma membrane. The external surface of the membrane exhibits an additional electron-dense layer in these regions, reflecting the concentration of viral glycoproteins on the external surface. Virions form by budding or outfolding of such modified areas of membrane, as seen in Fig. 1C. The virions acquire a lipid bilayer with a composition which closely reflects that of the plasma

membrane of the host cell (Klenk and Choppin, 1970). In contrast, all the
proteins of the virion are coded for by the viral genome, and host cell proteins
are effectively excluded from the virus during the budding process (Compans
and Klenk, 1979).

The process of virus assembly is highly specific, as illustrated in Fig. 2.
Budding virions are seen to be heavily labeled with ferritin-conjugated anti-

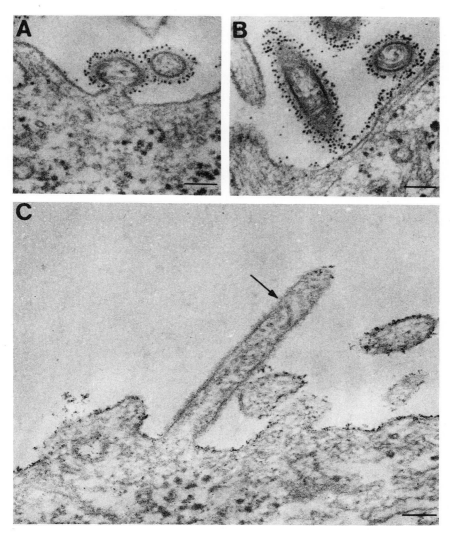

FIGURE 2. Specificity of the assembly process. (A) Two virions, one in the process of pinching off
from the cell surface, are tagged with ferritin-conjugated antibody, whereas the adjacent cell membranes are unlabeled (×85,000). (B) Two virions and the adjacent cell surface are tagged with
antibody (×85,000). [From Compans and Choppin (1971).] (C) Colloidal iron hydroxide staining of
neuraminic acid residues on the surface of SV5-infected cells. The surface of the plasma membrane
of the cell, including microvilli, is covered with electron-dense iron granules. The budding SV5
filament (arrow) is not stained, indicating that neuraminic acid residues are absent in the virions,
but are present in the adjacent cell membrane up to the base of the filament (×87,500). [From
Klenk et al. (1970).]

viral antibody, whereas the cell surface immediately adjacent to such budding particles is often devoid of such antibody tagging, as seen in Fig. 2A. Because of the presence of neuraminidase activity in paramyxoviruses, virions are devoid of neuraminic acid residues (Klenk *et al.*, 1970). However, as shown in Fig. 2C, the cell surface immediately adjacent to budding virions is found to contain neuraminic acid, as shown by colloidal iron hydroxide staining. Taken together, these observations demonstrate that there is an abrupt transition in the composition of the membrane at sites where virus budding occurs, with virus-specific components in the envelope of the emerging virion, and normal cellular membrane components in the cell surface just adjacent to budding particles.

The two viral glycoproteins HN and F form spikelike projections 8–12 nm in length on the external surface of the viral envelope (Chen *et al.*, 1971; Scheid *et al.*, 1972). Initiation of the infection process involves attachment of virus particles to neuraminic acid-containing receptors, which is mediated by the HN protein. The F protein is involved in the subsequent fusion of the viral membrane with the plasma membrane. A detailed review of the structure and function of these glycoproteins is presented elsewhere in this volume (Chapter 13). In the present chapter, we consider the pathways by which viral proteins are transported from their site of synthesis to the site of virus assembly and the interactions between proteins which are essential for assembly. We also review alteration of the normal assembly processes which occur under certain conditions. Finally, we discuss the use of paramyxoviruses as model systems for studies of membrane biogenesis and cell polarity.

II. SYNTHESIS AND TRANSPORT OF THE HEMAGGLUTININ-NEURAMINIDASE GLYCOPROTEIN

A. Site of Synthesis, Transport, and Kinetics

The envelope glycoproteins of Sendai virus have been shown to be transmembrane proteins (Lyles, 1979), since both glycoprotein species were found to have small segments exposed to proteolysis at the cytoplasmic surface. The deduced primary structure from the nucleotide sequence (Blumberg *et al.*, 1985b; Hiebert *et al.*, 1985; Alkatib and Briedis, 1986; Elango *et al.*, 1986; Millar *et al.*, 1986; Jorgensen *et al.*, 1987; McGinnes *et al.*, 1986; Tsukiyama *et al.*, 1987; Kovamees *et al.*, 1989) and the hydropathy plot demonstrate the presence of only one hydrophobic domain at the N terminus of HN, which is likely to represent the membrane anchor. Deletion mutations have been constructed in the cDNA derived from the NDV *HN* gene, and mRNAs derived from these DNAs were utilized to show that at the amino terminus a cytoplasmic domain of 26 amino acids and a hydrophobic region of 22 amino acids contain a signal sequence as well as a membrane anchoring sequence (Morrison, 1988). Moreover, it was suggested that the cytoplasmic domain may contain sequences responsible for the inverted orientation of the HN in the cell membrane.

Fractionation studies of cells infected with Sendai virus have provided evidence that the HN glycoprotein is synthesized in association with the rough endoplasmic reticulum and then slowly migrates via smooth cytoplasmic membranes to the plasma membrane (Nagai et al., 1976; Lamb and Choppin, 1977). The NDV HN protein has been shown to insert cotranslationally into microsomal membranes in a wheat germ cell-free system by a process which requires signal recognition particles (Wilson et al., 1987). A comparison of the rate of transport between the N-terminally anchored HN protein of NDV and the C-terminally anchored G protein of VSV showed that the G protein reaches the cell surface approximately three times faster than the HN protein, even though they have similar chain lengths (Morrison and Ward, 1984). Similarly, the Sendai virus N-terminally anchored HN protein was found to reach the cell surface three times more slowly than the C-terminally anchored Sendai virus F protein (Blumberg et al., 1985b). The rate differences are accounted for by different transit times of the proteins from the rough endoplasmic reticulum to the trans Golgi membranes. The reasons for such different rates of transport [which in the case of Sendai virus is reflected also in the rates at which the two proteins acquire their native conformation (Mottet et al., 1986)] are still uncertain. It is possible that the opposite orientations in which the proteins are anchored in the membrane may be involved here; the folding of the N-terminally anchored HN protein may be retarded until its synthesis is complete, providing a free C-terminal domain in the lumen of the ER.

Once the HN protein has reached the membrane, the fraction of molecules which is not incorporated into virus particles remains at the cell surface for a very long period of time (half-life much greater than 10 hr). During coinfection with defective-interfering Sendai virus particles, however, in which the cells survive the infection and where viral shedding is restricted, HN is removed from the membrane within 1 or 2 hr and is presumably reinternalized (Roux et al., 1985). Interestingly, this unstable expression (not observed for the F protein) at the surface correlates with high instability of the M protein. This suggests that interactions between M and HN are involved in maintaining HN at the surface (Tuffereau and Roux, 1988).

B. Glycosylation and Other Modifications

The HN protein is known to contain N-linked carbohydrates (Kohama et al., 1978; Yoshima et al., 1981; Nakamura et al., 1982; Herrler and Compans, 1983). Tunicamycin, an inhibitor of lipid-dolichol-mediated, N-linked glycosylation of asparagine residues in nascent proteins, has been widely used to follow the mechanism of intracellular processing and the biological role of the sugar moieties on viral glycoproteins. Treatment of Sendai virus-infected cells with tunicamycin inhibited the formation of virus particles (Nakamura et al., 1982). The synthesis of P, NP, and M polypeptides was partially inhibited and nonglycosylated forms of HN and F were found to be associated with the membrane fractions. Cell fractionation studies showed that migration of Sendai viral glycoproteins from rough to smooth membranes proceeded correctly in the absence of glycosylation. However, the lack of virion formation was suggested to be due to the failure of nonglycosylated forms of the viral glycoproteins to be transported to the plasma membrane. Indeed, after tunicamycin treatment, Sendai virus HN and F were shown to be aberrantly aggregated through disulfide bonds, and to be degraded before they reached the cell surface (Mottet et al., 1986; Vidal et al., 1989).

The requirement of glycosylation for virion formation in paramyxoviruses was shown to be different depending on the virus. Measles virus and NDV were shown to produce virus particles in the absence of glycosylation, in contrast to Sendai virus (Stallcup and Fields, 1981; Morrison et al., 1981). However, those particles were noninfectious. The nonglycosylated proteins L, P, NP, and M were synthesized in normal amounts and incorporated into released particles; however, glycosylated forms of the glycoproteins were not detected. Tunicamycin did not block intracellular transport of the nonglycosylated glycoproteins or particle formation of NDV, but eliminated hemagglutinating and neuraminidase activities (Morrison et al., 1981). These results indicate that glycosylation of paramyxoviruses glycoproteins is not a general requirement for their transport to the cell surface. It is likely that any requirement for glycosylation may be an indirect effect, in which nonglycosylated molecules are improperly folded and do not attain their native conformation (see Section II.C).

Another posttranslational modification which is observed for paramyxovirus glycoproteins is fatty acid acylation of cysteine residues in the transmembrane region. However, this is not consistently observed among all members of the paramyxoviruses, and there are examples of HN as well as F glycoproteins which are not acylated (Veit et al., 1989). Therefore, this modification is not essential for transport or function of the glycoproteins.

C. Oligomerization and Disulfide Bond Formation

Posttranslational modification accompanied by conformational changes has been recognized with both envelope glycoprotein species of paramyxoviruses (Hsu et al., 1981; McGinnes and Morrison, 1985; Morrison et al., 1987; Mottet et al., 1986; Waxham et al., 1986; Nishikawa et al., 1986; Yamada et al., 1988). The HN protein forms disulfide-linked multimers at the posttranslational level (Markwell and Fox, 1980; Schwalbe and Hightower, 1982). HN molecules of NDV and Sendai virus made in vivo in the presence of tunicamycin appear mainly as disulfide-linked aggregates, and lack of glycosylation probably allows aberrant disulfide bond formation. HN dimers of both viruses have been found in the rough endoplasmic reticulum. On the other hand, mumps virus HN oligomers appear to be formed later, before the protein reaches the trans Golgi cisternae (Yamada et al., 1988). Sendai and mumps virus HN were shown to display conformation-dependent antigenic epitopes (Mottet et al., 1986; Yamada et al., 1988). Reduction of Sendai virus HN largely eliminated its native immunoreactivity and thus implicated the importance of disulfide bonding. Correct disulfide bond formation in Sendai virus HN has been correlated with native structure formation of the glycosylated proteins and is preceded by glycosylation (Vidal et al., 1989). High-mannose sugar addition allows proper intramolecular disulfide bonding and proper folding, which, in turn, is presumably required for efficient transport to the cell surface. Internal disulfide bond formation is responsible for the appearance of HN dimers. However, Mottet *et al.* (1986) have shown that HN monomers are recognized by native antibodies to the native protein, and the kinetics of dimer formation does not correlate with antigenic maturation of the protein. Therefore, HN monomers probably acquire their native structure before intermolecular disulfide bond formation and the association of monomers into dimers does not seem to significantly perturb this mature structure.

III. SYNTHESIS AND TRANSPORT OF THE FUSION GLYCOPROTEIN

A. Site of Synthesis, Transport, and Kinetics

The F proteins of Sendai virus and NDV have been shown to be synthesized and transported to the cell surface by a similar pathway to that of HN (Sampson and Fox, 1973; Lamb and Choppin, 1977; Schwalbe and Hightower, 1982). However, the F protein, which is anchored to membranes by a C-terminal hydrophobic domain (Blumberg *et al.*, 1985a), was shown to reach the cell surface at a considerably faster rate than the oppositely oriented HN protein (Blumberg *et al.*, 1985b). On the other hand, studies with mumps virus indicated that both the HN and F glycoproteins reach the cell surface with similar kinetics, indicating that the proteins have similar migration rates on their pathway from the rough endoplasmic reticulum to the plasma membrane (Herrler and Compans, 1983). The fusion protein of NDV has been shown to undergo a conformational change during movement between the rough endoplasmic reticulum and the medial Golgi membranes (McGinnes and Morrison, 1985). The protein is inserted into the membrane of the rough endoplasmic reticulum in a signal recognition particle-dependent pathway (Morrison, 1988). The genes encoding the F proteins of several paramyxoviruses have been sequenced (Paterson *et al.*, 1984; Blumberg *et al.*, 1985a; Chambers *et al.*, 1986; McGinnes *et al.*, 1986; Richardson *et al.*, 1986; Spriggs *et al.*, 1986; Barrett *et al.*, 1987; Waxham *et al.*, 1987; Sato *et al.*, 1987; Merson *et al.*, 1988). The deduced amino acid sequences have shown a number of potential glycosylation sites at different positions of the protein backbone. The F glycoprotein of Sendai virus and NDV were shown to have asparagine-linked oligosaccharide chains (Morrison and Simpson, 1980; Schwalbe and Hightower, 1982; Nakamura *et al.*, 1982). Both complex and high-mannose-type oligosaccharides were found to be present in the F1 glycoprotein of mumps virus (Herrler and Compans, 1983). On the other hand, only high-mannose-type glycopeptides were detected in F2.

The F protein of paramyxoviruses is activated by proteolytic cleavage to an active fusion protein, with a new N-terminal hydrophobic domain exposed at the cleavage site (see Chapter 13). This hydrophobic domain [fusion-related external domain, FRED (Patterson and Lamb, 1987)] is thought to participate in virus-induced membrane fusion, by interacting with the lipid bilayer of an adjacent cell membrane. The ability of this hydrophobic domain to interact with lipid bilayers in inducing fusion raised questions about its ability to be

translocated across membranes of the endoplasmic reticulum during biosynthesis of the F protein. It was demonstrated by Patterson and Lamb (1987)
that the lack of a stop-transfer function was dependent on position; the FRED
sequence was sufficiently hydrophobic to function as a membrane anchor
when it was located at the terminus of a protein, and was capable of conferring
stable membrane anchorage upon a formerly soluble protein. However, when
located in an internal position, the FRED sequence was below the threshold of
hydrophobicity required to impede its translocation during biosynthesis.
Cleavage activation places the FRED sequence in a terminal position, in which
it is now able to interact stably with lipid bilayers, as would be expected during
membrane fusion.

B. Glycosylation, Oligomerization, and Disulfide Bond Formation

The spatial arrangement of the Sendai virus F protein was studied following reconstitution of the purified glycoprotein into lipid vesicles (Sechoy et al.,
1987). The components of the F protein spikes appeared as a structurally stable
complex, composed of a noncovalent association of four homo-oligomers, each
consisting of F1 and F2, linked by a disulfide bond. It was hypothesized that the
tetrameric form of the native F protein consists of two identical dimers that
can be chemically cross-linked in a stable complex.

The F protein of NDV has been shown to undergo a striking conformational change after its biosynthesis. The intramolecular disulfide bonds which
form on the nascent protein are subsequently either rearranged or disrupted
due to posttranslational modification, and the protein acquires new antigenic
determinants, as suggested from differences in reactivity patterns of monoclonal antibodies with the nascent and mature fusion proteins (McGinnes et
al., 1986; Morrison et al., 1987). Sendai virus F_0 maturation appears somewhat
different, since it was shown to acquire its native conformation as soon as the
protein was synthesized, and this correlated with immediate formation of
intramolecular disulfide bonds with no evidence of subsequent rearrangement
(Vidal et al., 1989; Mottet et al., 1986). Reduction of F_0 did not change its
reactivity with antibody to the native glycoprotein and thus the disulfide linkage did not appear to be required to maintain the native structure once it was
formed (Blumberg et al., 1985a; Mottet et al., 1986). As was observed for HN,
high-mannose sugar addition is thought to nevertheless precede formation of
correct disulfide bonds (Vidal et al., 1989).

IV. MATRIX (M) PROTEIN

A. Structural Organization in the Virion

For an extensive discussion of the M protein, see Chapter 16. Evidence that
the M protein is located internal to the lipid bilayer in parainfluenza virions
was obtained by demonstrating its resistance to protease treatment in intact
virus particles (Chen et al., 1971). The M protein is insoluble in water, and in

nonionic detergents in the absence of high salt. On the basis of these properties, a purification procedure was developed, and analysis revealed that the M protein of SV5 contains a high proportion of hydrophobic amino acids (McSharry et al., 1975). The M protein has high affinity for membranes, and is present in infected cells only in the membrane-containing fractions (Nagai et al., 1976; Lamb and Choppin, 1977).

Freeze-fracture electron microscopy of plasma membranes of Sendai virus-infected cells demonstrated patches of membrane-associated particles arranged in a crystalline array in the regions where spikes and the nucleoprotein strands were recruited for assembly of virus particles (Bächi, 1980). These arrays were associated with the inner lipid leaflet of the host membrane. After the process of viral budding, this structural modification of the membrane persists only for a short time in the viral envelope; newly-formed virions possess crystalline arrays of membrane particles which are not subsequently found in mature virus particles. The insertion of the spike glycoproteins into the lipid bilayer appeared to be insufficient for production of the membrane particles in crystalline arrays. It was therefore suggested that the M protein was involved in the formation of these patterns. The M protein is believed to act as a recognition site to anchor the viral glycoproteins at restricted areas of the plasma membrane to form a virus-specific membrane domain (Yoshida et al., 1979). These observations suggest interactions between the cytoplasmic tails of the glycoproteins and the M protein, since M only associates in a paracrystalline array with portions of the plasma membrane which contain F_0 and HN (Bächi, 1980; Büechi and Bächi, 1982).

B. Synthesis and Transport

The M protein is synthesized on free cytoplasmic polyribosomes and inserted rapidly and directly into the plasma membrane (Nagai et al., 1976; Lamb and Choppin, 1977). Cell fractionation studies (Nagai et al., 1976; Famulari et al., 1976; Lamb and Choppin, 1977) have suggested that the M protein is rapidly incorporated into the plasma membrane and into virions immediately after its synthesis. Furthermore, the M protein appears to be selectively inserted into the membrane regions which already contain viral glycoproteins. Interactions of M with Sendai virus HN have been suggested, since stable anchorage of HN in the plasma membrane appears to depend on the presence of M protein (Roux et al., 1985; Tuffereau and Roux, 1988). On the other hand, in NDV-infected cells, it is the efficient incorporation of F_0 into virus particles which appears to depend on M function (Peeples and Bratt, 1984). Sendai virus M also associates more effectively with microsomal vesicles containing HN in vitro compared with those containing F_0 (C. Bellocq and L. Roux, in preparation).

Interactions between M protein molecules probably occur by hydrophobic bonding (McSharry et al., 1975; Yoshida et al., 1976). Evidence for interaction between the M protein and nucleocapsid protein was also obtained in mixed infection with SV5 and a rhabdovirus, vesicular stomatitis virus (VSV) (McSharry et al., 1971). The progeny virions, although resembling VSV morphologically, contained SV5 glycoproteins, but no other SV5 proteins. This result suggested a specific interaction between the homologous nucleocapsid and M protein necessary for virus assembly. Yoshida *et al.* (1976) also demonstrated that the nucleocapsid protein subunits of Sendai virus do not combine with glycoproteins *in vitro* to form a coaggregate unless M protein is present in the reaction mixture. This suggested that the M protein may also be a mediator *in vivo* for the association of the nucleocapsid with the modified areas of plasma membrane in the final step of virus assembly. Cooperative interactions may mediate association of the nucleocapsid with an ordered array of M protein and glycoproteins at the membrane.

C. Domains of Interaction on the M Protein

During virus assembly, the M protein must interact with the two major viral components, the plasma membrane containing the glycoproteins, and the nucleocapsids. These interactions presumably involve different domains of the M protein. For instance, in the case of VSV, the amino-terminal 20% of M can be deleted without changing its ability to bind to liposomes, whereas the ability of this altered M protein to inhibit viral transcription (i.e., to interact with nucleocapsid) is decreased (Ogden *et al.*, 1986). No data which distinguish these two domains on paramyxovirus M proteins are available. The existence of such domains certainly cannot be deduced from sequence comparison studies among paramyxoviruses, which have revealed a "surprising lack of homology" (McGinnes and Morrison, 1987). The binding of M with nucleocapsids, but not with liposomes, is abolished by high-salt conditions (Faaberg and Peeples, 1988), suggesting that binding to nucleocapsids involves ionic interactions, while that with membranes depends on other interactions, presumably hydrophobic ones.

V. NUCLEOCAPSID-ASSOCIATED PROTEINS

A. Structural Organization in the Virion

The nucleocapsids of paramyxoviruses are single, left-handed helical structures (Compans *et al.*, 1972). The helical nucleocapsid is composed of a single-strand, negative-sense RNA, the nucleocapsid protein (NP), and two other proteins (P and L) which are involved in transcriptase activity (see also Chapters 5, 7, and 9). The NP protein is thought to determine the structure of the nucleocapsid and plays a role in regulation of transcription and replication (Mountcastle *et al.*, 1970; Stone *et al.*, 1972; McSharry *et al.*, 1975; Lamb *et al.*, 1976; Colonno and Stone, 1976). The conformations of the helical nucleocapsids of Sendai virus and SV5 have been shown to vary extensively with changes in salt concentration (Heggeness *et al.*, 1980). Flexibility of the nucleocapsids enables them to fit compactly within the viral envelope and may also provide a means to expose the RNA template to the RNA-synthesizing machinery without disassembly of the helix (Kingsbury, 1977).

The locations of NP, P, and M proteins on nucleocapsids derived from Sendai virus-infected cells and virions were analyzed by immunoelectron microscopy (Portner and Murti, 1986). The NP protein was shown to be distributed uniformly on nucleocapsids derived from both cells and virions. However, the distribution of M was different on cellular and virion-derived nucleocapsids. On nucleocapsids isolated from cells, the M molecules appeared as clusters and spanned the entire length of the nucleocapsid, while virion-derived nucleocapsids revealed little binding of M. The distribution of P molecules also showed differences between cell and virion nucleocapsids. P molecules were found in four to ten clusters on cell-derived nucleocapsids, while they were uniformly distributed on virion nucleocapsids.

The minor polypeptides L and P were efficiently cross-linked into large complexes, indicating that they possess abundant contacts with neighboring protein molecules in the helix (Raghow et al., 1979). About half of NP formed large cross-linked complexes, whereas most of the rest remained as monomers along with a small proportion of homodimers and low-order oligomers. Removal of the majority of P molecules from the nucleocapsids did not change the cross-linking pattern of NP, indicating that P is not required to maintain NP in a conformation favoring a high degree of polymerization.

B. Assembly of Nucleocapsid

Viral nucleocapsids are assembled in the cytoplasm. A segment of the nucleocapsids polypeptide chain is extremely susceptible to cleavage by proteolytic enzymes, and the remainder of the cleavage product is resistant (Mountcastle et al., 1970, 1974). The portion removed is probably exposed on the outer surface of the native subunit and may act as a specific recognition area during maturation. The extreme protease susceptibility of the nucleocapsid protein subunits could provide a possible mechanism by which nucleocapsids may accumulate, particularly in abortive or persistent virus infections.

The P protein is synthesized far in excess of its representation in the virion. Relative to the nucleocapsid structure unit NP, the rate of P synthesis has been found to be fourfold greater than its steady-state representation (Lamb and Mahy, 1975; Lamb et al., 1976; Portner and Kingsbury, 1976). Analysis of the kinetics of assembly of labeled P into virions during a chase with unlabeled precursors indicated that the overproduction of P has physiological significance. Early in the chase, labeled P was found in virions close to the proportion of its synthesis. Later, the proportion of P decreased toward its steady-state level, indicating that P has a greater probability of being nucleocapsid associated soon after its synthesis (Portner and Kingsbury, 1976). Kinetic studies have also shown that the P protein does not enter the nucleocapsid concurrently with the NP protein, but that NP first complexes with viral RNA to form helical nucleocapsids, and binding of the P protein then follows (Portner and Kingsbury, 1976; Kingsbury et al., 1978). Studies of deletion mutants of the Sendai virus P protein have provided evidence that carboxyl-terminal amino acid residues are required for binding of the P protein to the viral nucleocapsids (Ryan and Kingsbury, 1988).

The involvement of the host cell nucleus in measles virus infection has been reported, although its role is not clear. Both nuclei and cytoplasm of cells infected with measles virus were found to contain all of the viral structural proteins, with the exception of a surface glycoprotein. Immunofluorescence studies indicated that measles virus antigen appears in both acutely and persistently infected cell nuclei (Rapp *et al.*, 1960; Rustigian, 1966). Nucleocapsidlike viral inclusions in the nuclei of a variety of cells infected with measles virus have been observed and correlated with the detection in nuclei of nucleocapsid-related proteins NP, P2, and L (Wechsler and Fields, 1978).

VI. NONSTRUCTURAL PROTEINS

In addition to the proteins discussed above, which are incorporated into paramyxovirus particles during the assembly process, these viruses code for several nonstructural proteins. Several nonstructural proteins are encoded by the *P* gene; in some cases, their sequences are related to the *P* gene, whereas in other instances they are coded by different reading frames (see Chapter 6). Little information is available about the function of these proteins or their locations in infected cells. The C/C' proteins of Sendai virus have been localized by immunofluorescence (Portner *et al.*, 1986) and were found to differ in their staining pattern from that of nucleocapsids, indicating lack of association.

A second type of nonstructural protein, designated SH, of 44 amino acids, has been identified in SV5 (Hiebert *et al.*, 1985), and a coding sequence for a similar protein of 57 amino acids was found in mumps virus (Elango *et al.*, 1989). The coding sequences of these two proteins showed no similarity, but they were located in the same positions in the genomes of the two viruses, between the *F* and *HN* genes. A similar gene is not present in all paramyxoviruses, since the genomic sequence of Sendai virus (Blumberg *et al.*, 1985b) does not reveal such a gene. In both SV5 and mumps virus, a stretch of hydrophobic amino acids is present in these proteins, which could serve as a membrane-spanning domain. In the SH proteins of SV5, this domain is present at the C terminus, whereas the mumps sequence reveals a similar domain near the N terminus of the protein. Immunofluorescence and cell fractionation studies have provided evidence that SH is associated with the plasma membrane and that it follows the exocytic pathway for intracellular transport (Hiebert *et al.*, 1988). Proteolytic digestion experiments with microsomal vesicles indicated that the N-terminal hydrophobic domain is exposed on the cytoplasmic surface; thus, the protein is a type II membrane protein, with only five C-terminal amino acids apparently exposed on the cell surface (Hiebert *et al.*, 1988). The function of such small hydrophobic proteins in virus replication is unknown, but it is of interest that similar proteins have been described with other enveloped viruses, including the small membrane-spanning M2 protein of influenza A viruses (Zebedee *et al.*, 1985) and the NB protein of influenza B viruses (Shaw and Choppin, 1984; Williams and Lamb, 1986). It will be of interest to determine whether such proteins play a role in virus assembly.

VII. INVOLVEMENT OF HOST CELL COMPONENTS IN VIRUS ASSEMBLY

Although most cellular membrane proteins are effectively excluded from the envelopes of budding paramyxovirus particles, the possibility remains that some cellular components may play a role in virus maturation. Cellular actin has been identified in a variety of enveloped viruses, including paramyxoviruses (Wang *et al.*, 1976). This frequent association of cellular actin with virions may be attributed to its interaction with some other viral protein. An interaction of the M proteins of NDV and Sendai virus with cellular actin has been demonstrated (Guiffre *et al.*, 1982) and the possible role of actin filaments in the movement of nucleocapsids to the site of virus budding has been suggested (Tyrrell and Ehrnst, 1979). Measles virus release was shown to be inhibited in the presence of cytochalasin B (CB), a drug that disrupts actin microfilaments as well as inhibiting sugar transport and protein glycosylation (Stallcup *et al.*, 1983). Infected cells accumulated cell-associated virus in the presence of CB, whereas virus release decreased significantly. The inhibition occurred rapidly, even when the drug was added later in the replication cycle, whereas glycosylation inhibitors were only effective in inhibiting release if added early in the growth cycle. The alteration of microfilament structure was thus suggested to be responsible for the inhibition of the budding process of measles virus. A similar inhibition of the release of influenza virus and accumulation of cell-associated virus was observed in influenza virus-infected cells treated with CB (Griffin and Compans, 1979). However, it was found that this inhibition was due to partial inhibition of glycosylation in the presence of CB, which resulted in an inactive viral neuraminidase. Measles virus does not possess a neuraminidase activity; thus, the mechanism of inhibition of virus release by CB is unlikely to be the same. However, some similarities were observed in the two systems, including an alteration in the glycosylation of the measles virus hemagglutinin in the presence of CB, and the lack of comparable inhibition of release by cytochalasin D (CD), which inhibits microfilament function without affecting protein glycosylation (Griffin *et al.*, 1983; Stallcup *et al.*, 1983). It was suggested that, in the case of measles virus, the lack of marked inhibition of release in the presence of CD may be a result of a resident contractile function of actin in the presence of the inhibitor.

VIII. ALTERATIONS OF VIRUS ASSEMBLY IN PERSISTENTLY OR DI PARTICLE-INFECTED CELLS

Paramyxoviruses are well known to establish long-term persistent infections *in vitro* with the participation of temperature-sensitive viruses (*ts* mutants) or defective-interfering (DI) particles [for reviews, see Youngner and Preble (1980) and Holland *et al.* (1980)]. One common feature of these persistently-infected (pi) cells is their diminished ability to produce virus particles. These pi infections represent examples of perturbed virus assembly, and thus may provide information about the various interactions taking place in the assembly process.

The major alterations associated with defective budding are those involving the M protein. M protein was found to be absent or reduced in amount, changed in its apparent molecular weight, highly mutated, or unstable in persistent infections induced *in vitro* or naturally occurring long-term infections *in vivo* (Yoshida *et al.*, 1979; Hall and Choppin, 1979; Wechsler *et al.*, 1979; Roux and Waldvogel, 1982; Carter *et al.*, 1983; Sheppard *et al.*, 1985, 1986; Young *et al.*, 1985; Baczko *et al.*, 1986; Cattaneo *et al.*, 1987, 1988; Tuffereau and Roux, 1988). Most of these data speak for quantitative rather than qualitative defects, i.e., reduced presence of the protein rather than malfunction. Only one report suggested that malfunction (lack of recognition of M with an antibody raised against the "organized" form of M) was induced by DI particles, eventually leading to degradation (Tuffereau and Roux, 1988). A defect in expression of HN protein at the cell surface was associated with these M protein alterations and the reduction in viral budding (Yoshida *et al.*, 1979; Roux *et al.*, 1985). Moreover, cases with defects in the three envelope proteins HN, F, and M have been documented (Baczko *et al.*, 1984, 1986; Norrby *et al.*, 1985; Menna *et al.*, 1975). In rarer cases, modification of the NP protein has been reported, including its temperature-induced disappearance (Boriskin *et al.*, 1986) and the presence of two subsets of NP (Tuffereau and Roux, 1988). Finally, alterations of nucleocapsid conformation (Rozenblatt *et al.*, 1979) or organization under the plasma membrane (Dubois-Dalcq *et al.*, 1976) have been documented.

All these data point to an essential role of the M protein in virus assembly and viral particle budding. M appears necessary for the prebudding structure to form under the plasma membrane. In some instances, absence or alteration of the glycoproteins is associated with restriction of virus production. However, it is noteworthy that these alterations are observed in conjunction with the M protein defect, and therefore could result solely from the M protein defect. Efficient budding may take place in the absence of HN cell-surface expression, leading to virus particles devoid of HN (Portner *et al.*, 1975; Markwell *et al.*, 1985; Tuffereau *et al.*, 1985). Therefore, it appears that M can function independently of the glycoproteins (at least HN), but that these latter proteins need the presence of M to be efficiently incorporated into virus particles. The other component present in the prebudding structure is obviously the nucleocapsid. The nucleocapsid must interact with the virus-modified plasma membrane, and again the fact that budding may take place without the presence of glycoproteins (at least HN) may mean that the interaction of nucleocapsid is mainly with the M layer and not with the glycoproteins. Obviously, budding may be prevented by inability of the nucleocapsid to interact with the M at the plasma membrane. This inability might result from mutations affecting the contact points or from alterations in the organization of the nucleocapsids. It might also result from nucleocapsids which are too small, as in the case of DI nucleocapsids, which are known to bud less efficiently than nondefective ones (Re and Kingsbury, 1988; G. Mottet and L. Roux, 1989). In summary, it appears from these data dealing with defective budding that the minimum structure leading to bud formation contains nucleocapsids attached to the M protein. The M protein is probably required for efficient incorporation of the glycoproteins into virus particles because it presumably helps HN and F to anchor in a stable way in the plasma membrane (decreasing reinternalization) and/or it favors their condensation in prebudding patches, possibly by preventing further lateral movement once HN and F have made contact with M.

IX. VIRUS ASSEMBLY IN POLARIZED EPITHELIAL CELLS

A number of cell lines derived from epithelial tissues retain many of the differentiated properties of epithelial cell layers, including the presence of junctional complexes that separate the two plasma membrane domains, the apical and basolateral domains. It has been observed that when such cells are infected with enveloped viruses that are assembled by budding at the cell surface, the maturation of these viruses is restricted to either the apical or the basolateral membranes. Paramyxoviruses and orthomyxoviruses are assembled exclusively at apical membranes, whereas vesicular stomatitis virus (VSV) and C-type retroviruses are released from basolateral membranes (Rodriguez-Boulan and Sabatini, 1978; Roth et al., 1983a). The paramyxoviruses which have been studied in polarized cells include Sendai virus, SV5, and measles virus.

The system of virus-infected epithelial cells is an attractive one for studies of the mechanisms which determine the sorting of proteins to the distinct plasma membrane domains. Most studies of these aspects of protein sorting have been carried out with influenza virus and VSV; nevertheless, it is likely that the conclusions based on observations with these viruses will apply to paramyxoviruses as well. These observations include the following points: (1) The site of virus maturation corresponds to the site of surface expression of viral glycoproteins (Rodriguez-Boulan and Pendergast, 1980; Roth et al., 1983b), suggesting that the budding site is determined by the site of glycoprotein expression. (2) Glycoproteins of several viruses, when expressed from cloned cDNA using expression vectors, show polarized expression at the same membrane domains where virus budding occurs (Roth et al., 1983b; Jones et al., 1985; Stephens et al., 1986; Puddington et al., 1987). Thus, the information for protein sorting is contained in the glycoprotein molecule, and does not require other virus-coded proteins. (3) Polarized expression of glycoproteins occurs under conditions where glycosylation of proteins is completely inhibited (Roth et al., 1979; Green et al., 1981), indicating that the information required for sorting is contained in the amino acid sequence of the protein molecules. (4) The sorting of proteins occurs prior to their arrival at the cell surface, probably at the stage of exit from the trans-Golgi cisternae into vesicles that target the proteins to the correct plasma membrane domain (Misek et al., 1984; Matlin and Simons, 1984; Pfeiffer et al., 1985; Rindler et al., 1984). Current research efforts concerning protein traffic in polarized epithelial cells include determining the precise amino acid sequences which specify the sorting of the glycoprotein, as well as identifying the cellular components and mechanisms which are involved in the sorting process.

The polarized release of viruses from epithelial cells may also have consequences which are important for viral pathogenesis. Many viruses initially infect epithelial surfaces of the body; in some cases, infection is restricted to the cells of the initial epithelial surface, whereas in other cases the infection process spreads to cause a systemic disease. Paramyxoviruses include examples of both types of infections. Infections with parainfluenza viruses are restricted to the epithelial cells of the respiratory tract. These viruses are released by budding at the apical surfaces of epithelial cells, including cells in tracheal organ cultures, with little cytopathology (Hoorn and Tyrrell, 1969; Dourmashkin and Tyrrell, 1970). It is therefore readily apparent how infection by such viruses may be restricted to cells lining the respiratory tract. In contrast, measles virus causes a generalized infection, despite the fact that in epithelial cell cultures, assembly of the virus also appears to be restricted to apical cell membranes (Compans *et al.*, 1982). The ability of measles virus to multiply in lymphoid tissues may be related to the ability of this virus to cross the epithelial barrier and cause a systemic infection. Further information is needed on the possible biological consequences of virus release at restricted domains of polarized epithelial cells.

X. CONCLUDING REMARKS

The structural organization of paramyxoviruses as well as many aspects of their replication and assembly are now well understood. A diagram of the proposed sequence of events in virus assembly at the plasma membrane is shown in Fig. 3. Nevertheless, there are many unanswered questions about the assembly process. What are the precise molecular interactions involved? What is the possible role of virus-coded nonstructural proteins in virus assembly? Why are the proteins of paramyxoviruses targeted to the cell surface, whereas budding of virions in some other families takes place at intracellular membranes?

As discussed previously (Stephens and Compans, 1988), a selection mechanism must exist for the incorporation of proteins into the viral envelope, because host-cell membrane proteins are effectively excluded from the virion even in instances where viral proteins constitute only a minor portion of the proteins of the plasma membranes, as in the case with some paramyxovirus infections. On the other hand, envelope glycoproteins of one virus may be readily incorporated into the envelopes of unrelated viruses, forming phenotypically mixed virions or pseudotypes (Choppin and Compans, 1970; Zavada, 1982). The glycoproteins of viruses known to undergo phenotypic mixing exhibit considerable variation in their amino acid sequences, which indicates a lack of any linear sequence that specifies a recognition signal for virus assembly. Taken together, these observations suggest that there may be a general difference between cellular and viral membrane glycoproteins that determines their ability to be assembled into virions. The nature of this difference remains to be determined. One possibility may be a similar feature, as yet unidentified, in the three-dimensional structures of the proteins. Another possibility could be a difference in mobility: cellular membrane glycoproteins may be prevented from incorporation into virions because of interactions with each other or with other sets of cellular proteins underlying the membrane. Another possible difference could be related to the oligomeric structure of the

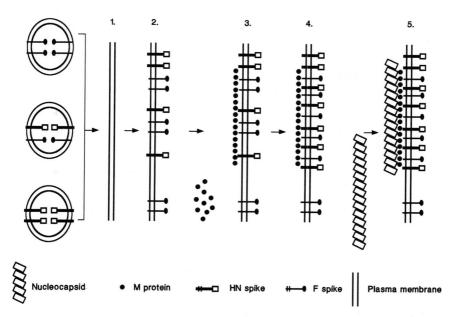

FIGURE 3. The sequence of events in the assembly of paramyxovirus proteins at the plasma membrane. (1) Transport vesicles carrying HN and F from the Golgi apparatus fuse with the plasma membrane. It is uncertain whether the two proteins are present in the same or different populations of vesicles. (2) The newly delivered proteins are probably distributed randomly on the cell surface, and are mobile and can diffuse laterally in the plane of the membrane. (3) The M protein molecules become associated with the plasma membrane, either individually or possibly as a preformed cluster. (4) Viral glycoproteins are localized in specific patches of membrane containing M proteins, from which host-cell membrane proteins are excluded. (5) Nucleocapsids associate with these modified areas of membrane, and presumably trigger the budding process.

viral proteins, which may have an important role in the lateral interactions needed to form a domain on the plasma membrane from which host-cell proteins are excluded.

ACKNOWLEDGMENTS. We thank Dr. Dan Kolakofsky for helpful comments and Joellynn Heaton and Betty Jeffrey for preparation of the manuscript. Research by Ranjit Ray was supported by contracts NOI AI 62604 and N44 AI 82504 from the National Institute of Allergy and Infectious Diseases. Research on Sendai virus glycoproteins by Laurent Roux was supported by the WHO and the Swiss National Foundation for Scientific Research. R. W. Compans thanks Dr. Dan Kolakofsky and the Fondation Louis Jeantet for support during the preparation of this review.

XI. REFERENCES

Alkhatib, G., and Briedis, D. J., 1986, The predicted primary structure of the measles virus hemagglutinin, *Virology* **150:**479–490.
Bächi, T., 1980, Intramembrane structural differentiation in Sendai virus maturation, *Virology* **106:**41–49.

Baczko, K., Carter, J. J., Billeter, M., and ter Meulen, V., 1984, Measles virus gene expression in subacute sclerosing panencephalitis, *Virus Res.* **1**:585–595.

Baczko, K., Liebert, U. G., Billeter, M., Cattaneo, R., Budka, H., and ter Meulen, V., 1986, Expression of defective measles virus genes in brain tissues of patients with subacute sclerosing panencephalitis, *J. Virol.* **59**:472–478.

Barrett, T., Clarke, D. K., Evans, S. A., and Rima, B. K., 1987, The nucleotide sequence of the gene encoding the F protein of canine distemper virus: A comparison of the deduced amino acid sequence with other paramyxoviruses, *Virus Res.* **8**:373–386.

Blumberg, B. M., Giorgi, C., Rose, K., and Kolakofsky, D., 1985a, Sequence determination of the Sendai virus fusion protein gene, *J. Gen. Virol.* **66**:317–331.

Blumberg, B., Giorgi, C., Roux, L., Raju, R., Dowling, P., Chollet, A., and Kolakofsky, D., 1985b, Sequence determination of the Sendai virus HN gene and its comparison to the influenza virus glycoproteins, *Cell* **41**:269–278.

Boriskin, Y. S., Bogomolova, N. N., Koptyaeva, I. B., Giraudon, P., and Wild, T. F., 1986, Measles virus persistent infection: Modification of the virus nucleocapsid protein, *J. Gen. Virol.* **67**:1979–1985.

Büechi, M., and Bächi, T., 1982, Microscopy of internal structures of Sendai virus associated with the cytoplasmic surface of the membranes, *Virology* **120**:349–359.

Carter, M. J., Willocks, M. M., and ter Meulen, V., 1983, Defective translation of measles virus protein in a subacute sclerosing panencephalitis cell line, *Nature* **305**:153–155.

Cattaneo, R., Rebmann, G., Scmid, A., Baczko, K., and ter Meulen, V., 1987, Altered transcription of a defective measles virus genome derived from a diseased human brain, *EMBO J.* **6**:681–688.

Cattaneo, R., Schmid, A., Billeter, M. A., Sheppard, R. D., and Udem, S. A., 1988, Multiple viral mutations rather than host factors cause defective measles virus gene expression in a subacute sclerosing panencephalitis cell line, *J. Virol.* **62**:1388–1397.

Chambers, P., Millar, N., Platt, S., and Emmerson, P., 1986, Nucleotide sequence of the gene encoding the matrix protein of Newcastle disease virus, *Nucleic Acids Res.* **14**:9051–9061.

Chen, C., Compans, R. W., and Choppin, P. W., 1971, Parainfluenza virus surface projections: Glycoproteins with hemagglutinin and neuraminidase activities, *J. Gen. Virol.* **11**:53–58.

Choppin, P. W., and Compans, R. W., 1970, Phenotypic mixing of envelope proteins of the parainfluenza virus SV5 and vesicular stomatitis virus, *J. Virol.* **5**:609–616.

Choppin, P. W., and Compans, R. W., 1975, Reproduction of paramyxoviruses, in "Comprehensive Virology," Vol. 4 (H. Fraenkel-Conrat and R. R. Wagner, eds.), pp. 95–178, Plenum Press, New York.

Colonno, R. J., and Stone, H. O., 1976, Isolation of a transcriptase complex from Newcastle virions, *J. Virol.* **19**:1080–1089.

Compans, R. W., and Choppin, P. W., 1971, The structure and assembly of influenza and parainfluenza viruses, in "Comparative Virology" (K. Maramorosch and F. Kurstak, eds.), pp. 407–432, Academic Press, New York.

Compans, R. W., and Klenk, H. D., 1979, Viral membranes, in "Comprehensive Virology," Vol. 13 (H. Fraenkel-Conrat and R. R. Wagner, eds.), pp. 293–407, Plenum Press, New York.

Compans, R. W., Holmes, K. V., Dales, S., and Choppin, P. W., 1966, An electron microscopic study of moderate and virulent virus–cell interactions of the parainfluenza virus SV5, *Virology* **30**:411–426.

Compans, R. W., Mountcastle, W. E., and Choppin, P. W., 1972, The sense of the helix of paramyxovirus nucleocapsids, *J. Mol. Biol.* **65**:167–169.

Compans, R. W., Roth, M. G., Alonso, F. V., Srinivas, R. V., Herrler, G., and Melsen, L. R., 1982, Do viral maturation sites influence disease processes?, in "Viral Diseases in South East Asia and the Western Pacific" (J. S. Mackenzie, ed.), pp. 328–332, Academic Press, Australia.

Dourmashkin, R. R., and Tyrrell, D. A. J., 1970, Attachment of two myxoviruses to ciliated epithelial cells, *J. Gen. Virol.* **9**:77–88.

Dubois-Dalcq, M., Reese, T. S., Murphy, M., and Fucillo, D., 1976, Defective bud formation in human cells chronically infected with subacute sclerosing panencephalitis virus, *J. Virol.* **19**:579–593.

Elango, N., Coligan, J. E., Jambou, R. C., and Venkatasan, S., 1986, Human parainfluenza type 3 virus hemagglutinin-neuraminidase glycoprotein: Nucleotide sequence of the mRNA and limited amino acid sequence of the purified protein, *J. Virol.* **57**:481–489.

Elango, N., Kovamees, J., Varsanyi, T. M., and Norrby, E., 1989, mRNA sequence and deduced amino acid sequence of the mumps virus small hydrophobic protein gene, *J. Virol.* **63:**1413–1415.

Faaberg, K. S., and Peeples, M. E., 1988, Association of soluble matrix protein of Newcastle disease virus with liposomes is independent of ionic conditions, *Virology* **166:**123–132.

Famulari, N. G., Buchhagen, D. L., Klenk, H. D., and Fleissner, E., 1976, Presence of murine leukemia virus envelope proteins gp70 and p15 (E) in a common polyprotein of infected cells, *J. Virol.* **20:**501–508.

Green, R. F., Meiss, H. K., and Rodriguez-Boulan, E. J., 1981, Glycosylation does not determine segregation of viral envelope proteins in plasma membranes of epithelial cells, *J. Cell. Biol.* **89:**230–239.

Griffin, J. A., and Compans, R. W., 1979, Effect of cytochalasin B on the maturation of envelope viruses, *J. Exp. Med.* **150:**379–391.

Griffin, J. A., Basak, S., and Compans, R. W., 1983, Effects of glucose starvation and the role of sialic acid in influenza virus release, *Virology* **125:**324–334.

Guiffre, R. M., Tovell, D. R., Kay, C. M., and Tyrrell, D. L. J., 1982, Evidence for an interaction between the membrane protein of a paramyxovirus and actin, *J. Virol.* **42:**963–968.

Hall, W. W., and Choppin, P. W., 1979, Evidence for lack of synthesis of the M polypeptide of measles virus in the brain from SSPE, *Virology* **99:**443–447.

Heggeness, M. H., Scheid, A., and Choppin, P. W., 1980, Conformation of the helical nucleocapsids of paramyxoviruses and vesicular stomatitis virus: Reversible coiling and uncoiling induced by changes in salt concentration, *Proc. Natl. Acad. Sci. USA* **77:**2631–2635.

Herrler, G., and Compans, R. W., 1983, Post translational modification and intracellular transport of mumps virus glycoproteins, *J. Virol.* **47:**354–362.

Hiebert, S. W., Paterson, R. G., and Lamb, R. A., 1985, The hemagglutinin-neuraminidase protein of the paramyxovirus SV5: Nucleotide sequence of the mRNA predicts an N-terminal membrane anchor, *J. Virol.* **54:**1–6.

Hiebert, S. W., Richardson, C. D., and Lamb, R. A., 1988, Cell surface expression and orientation in membranes of the 44-amino-acid SH protein of simian virus 5, *J. Virol.* **62:**2347–2357.

Holland, J. J., Kennedy, S. I. T., Semler, B. L., Jones, C. L., Roux, L., and Grabau, E. A., 1980, Defective interfering RNA viruses and the host-cell response, *in* "Comprehensive Virology," Vol. 16 (H. Fraenkel-Conrat and R. R. Wagner, eds.), pp. 137–192, Plenum Press, New York.

Hoorn, B., and Tyrrell, D. A. J., 1969, Organ cultures in virology, *Prog. Med. Virol.* **11:**408–450.

Hsu, M., Scheid, A., and Choppin, P. W., 1981, Activation of the Sendai virus fusion protein (F) involves a conformational change with exposure of a new hydrophobic region, *J. Biol. Chem.* **256:**3557–3563.

Jones, L. V., Compans, R. W., Davis, A. R., Bos, T. J., and Nayak, D. P., 1985, Surface expression of influenza virus neuraminidase, an aminoterminally anchored viral membrane glycoprotein, in polarized epithelial cells, *Mol. Cell. Biol.* **5:**2181–2189.

Jorgensen, E. D., Collins, P. L., and Lomedico, P. T., 1987, Cloning and nucleotide sequence of Newcastle disease virus hemagglutinin-neuraminidase mRNA: Identification of a putative sialic acid binding site, *Virology* **156:**12–24.

Kingsbury, D. W., 1977, Paramyxoviruses, *in* "The Molecular Biology of Animal Viruses," Vol. 1 (D. P. Nayak ed.), pp. 349–382, Marcel Dekker, New York.

Kingsbury, D. W., Hsu, C. H., and Murti, K. G., 1978, Intracellular metabolism of Sendai virus nucleocapsids, *Virology* **91:**86–94.

Klenk, H. D., and Choppin, P. W., 1970, Plasma membrane lipids and parainfluenza virus assembly, *Virology* **40:**939–947.

Klenk, H. D., Compans, R. W., and Choppin, P. W., 1970, An electron microscopic study of the presence or absence of neuraminic acid in enveloped viruses, *Virology* **42:**1158–1162.

Kohama, T., Shimizu, K., and Ishida, N., 1978, Carbohydrate composition of the envelope glycoproteins of Sendai virus, *Virology* **90:**226–234.

Kovamees, J., Norrby, E., and Elango, N., 1989, Complete nucleotide sequence of the hemagglutinin-neuraminidase (HN)mRNA of mumps virus and comparison of paramyxovirus HN proteins, *Virus Res.* **12:**87–96.

Lamb, R. A., and Choppin, P. W., 1977, The synthesis of Sendai virus polypeptides in infected cells II, Intracellular distribution of polypeptides, *Virology* **81:**371–381.

Lamb, R. A., and Mahy, B. W. J., 1975, The polypeptides and RNA of Sendai virus, *in* "Negative Strand Viruses" (B. W. J. Mahy and R. D. Barry, eds.), pp. 65–87, Academic Press, New York.

Lamb, R. A., Mahy, B. W. J., and Choppin, P. W., 1976, The synthesis of Sendai virus polypeptides in infected cells, *Virology* **69**:116–131.

Lyles, D. S., 1979, Glycoproteins of Sendai virus are transmembrane proteins, *Proc. Natl. Acad. Sci. USA* **76**:5621–5625.

Markwell, M. A. K., and Fox, C. F., 1980, Protein–protein interactions within paramyxoviruses identified by native disulfide bonding or reversible chemical cross-linking, *J. Virol.* **33**:152–166.

Markwell, M. A. K., Portner, A., and Schwartz, A. L., 1985, An alternative route of infection for viruses: Entry by means of the asialoglycoprotein receptor of a Sendai virus mutant lacking its attachment protein, *Proc. Natl. Acad. Sci. USA* **82**:978–982.

Matlin, K., and Simons, K., 1984, Sorting of an apical plasma membrane glycoprotein occurs before it reaches the cell surface in cultured epithelial cells, *J. Cell Biol.* **99**:2131–2139.

McGinnes, L. W., and Morrison, T. G., 1985, Conformational changes in Newcastle disease virus fusion glycoprotein during intracellular transport, *J. Virol.* **56**:341–348.

McGinnes, L. W., and Morrison, T. G., 1987, The nucleotide sequence of the gene encoding the Newcastle disease virus membrane protein and comparisons of membrane protein sequences, *Virology* **156**:221–228.

McGinnes, L. W., Wilde, A., and Morrison, T. G., 1986, Nucleotide sequence of the gene encoding the Newcastle disease virus fusion protein and comparisons of paramyxovirus fusion protein sequence, *Virus Res.* **5**:343–356.

McSharry, J. J., Compans, R. W., and Choppin, P. W., 1971, Proteins of vesicular stomatitis virus and phenotypically mixed vesicular stomatitis virus–simian virus 5 virions, *J. Virol.* **8**:722–729.

McSharry, J. J., Compans, R. W., Lackland, H., and Choppin, P. W., 1975, Isolation and characterization of the nonglycosylated membrane protein and a nucleocapsid complex from the paramyxovirus SV5, *Virology* **67**:365–374.

Menna, J. H., Collins, A. R., and Flanagan, T. D., 1975, Characterization of an *in vitro* persistent-state measles virus infection: Establishment and virological characterization of the BGM/MV cell line, *Infect. Immun.* **11**:152–158.

Merson, J. R., Hull, R. A., Estes, M. K., and Kasel, J. A., 1988, Molecular cloning and sequence determination of the fusion protein gene of human parainfluenza virus type 1, *Virology* **167**:97–105.

Millar, N. S., Chambers, P., and Emmerson, P. T., 1986, Nucleotide sequence analysis of the hemagglutinin-neuraminidase gene of Newcastle disease virus, *J. Gen. Virol.* **67**:1917–1927.

Misek, D. E., Bard, E., and Rodriguez-Boulan, E., 1984, Biogenesis of epithelial cells polarity: Intracellular sorting and vectorial exocytosis of an apical plasma membrane glycoprotein, *Cell* **39**:537–546.

Morrison, T. G., 1988, Structure, function, and intracellular processing of paramyxovirus membrane proteins, *Virus Res.* **10**:113–135.

Morrison, T. G., and Simpson, D., 1980, Synthesis, stability and cleavage of Newcastle disease virus glycoproteins in the absence of glycoproteins, *J. Virol.* **36**:171–180.

Morrison, T. G., and Ward, L. J., 1984, Intracellular processing of the vesicular stomatitis virus glycoprotein and the Newcastle disease virus hemagglutinin-neuraminidase glycoprotein, *Virus Res.* **1**:225–239.

Morrison, T. G., Chatis, P. A., and Simpson, D., 1981, Conformation and activity of the Newcastle disease virus HN protein in the absence of glycosylation, *in* "The Replication of Negative-Strand Viruses" (D. H. L. Bishop and R. W. Compans, eds.), pp. 471–477, Elsevier/North-Holland, Amsterdam.

Morrison, T. G., Peeples, M. E., and McGinnes, L. W., 1987, Conformational changes in a viral glycoprotein during maturation due to disulfide bond disruption, *Proc. Natl. Acad. Sci. USA* **84**:1020–1024.

Mottet, G., Portner, A., and Roux, L., 1986, Drastic immunoreactivity changes between the immature and mature form of the Sendai virus HN and F0 glycoproteins, *J. Virol.* **59**:132–141.

Mottet, G., and Roux, L., 1989, Budding efficiency of Sendai virus nucleocapsids: influence of size and ends of the RNA, *Virus Res.* **14**:175–188.

Mountcastle, W. E., Compans, R. W., Caliguiri, L. A., and Choppin, P. W., 1970, The nucleocapsid protein subunits of SV5, Newcastle disease virus and Sendai virus, *J. Virol.* **6:**677–684.

Mountcastle, W. E., Compans, R. W., Lackland, H., and Choppin, P. W., 1974, Proteolytic cleavage of subunits of the nucleocapsid of the paramyxovirus simian virus 5, *J. Virol.* **14:**1253–1261.

Nagai, Y., Ogura, H., and Klenk, H. D., 1976, Studies on the assembly of the envelope of Newcastle disease virus, *Virology* **69:**523–538.

Nakamura, K., Homma, M., and Compans, R. W., 1982, Effect of tunicamycin on the replication of Sendai virus, *Virology* **119:**474–487.

Nishikawa, K., Morishima, T., Tyroda, T., Miyadai, T., Yokochi, T., Yoshida, T., and Nagai, Y., 1986, Topological and operational delineation of antigenic sites and the HN glycoprotein of Newcastle disease virus and their structural requirements, *J. Virol.* **60:**987–993.

Norrby, E., Kristensson, K., Brzosko, W. J., and Kaspenberg, J. G., 1985, Measles virus matrix protein detected by immune fluorescence with monoclonal antibodies in the brain of patients with subacute sclerosing panencephalitis, *J. Virol.* **56:**337–340.

Ogden, J. R., Pal, R., and Wagner, R. R., 1986, Mapping regions of the matrix protein of vesicular stomatitis virus which bind to ribonucleocapsids, liposomes and monoclonal antibodies, *J. Virol.* **58:**860–868.

Paterson, R. G., and Lamb, R. A., 1987, Ability of the hydrophobic fusion-related external domain of a paramyxovirus F protein to act as a membrane anchor, *Cell* **48:**441–452.

Paterson, R. G., Harris, T. J. R., and Lamb, R. A., 1984, Analysis and gene assignment of mRNAs of a paramyxovirus, simian virus 5, *Virology* **138:**310–323.

Peeples, M. E., and Bratt, M. A., 1984, Mutation in the matrix protein of Newcastle disease virus can result in decreased fusion glycoprotein incorporation into particles and decreased infectivity, *J. Virol.* **51:**81–90.

Pfeiffer, S., Fuller, S. D., and Simons, K., 1985, Intracellular sorting and basolateral appearance of the G protein of vesicular stomatitis virus in MDCK cell, *J. Cell Biol.* **101:**470–476.

Portner, A., and Kingsbury, D. W., 1976, Regulatory events in the synthesis of Sendai virus polypeptides and their assembly into virions, *Virology* **73:**79–88.

Portner, A., and Murti, K. G., 1986, Localization of P, NP, and M proteins on Sendai virus nucleocapsids using immunogold labeling, *Virology* **150:**469–478.

Portner, A., Scroggs, R. A., Marx, P. A., and Kingsbury, D. W., 1975, A temperature-sensitive mutant of Sendai virus with an altered hemagglutinin-neuraminidase polypeptide: Consequences for virus assembly and cytopathology, *Virology* **67:**179–187.

Portner, A., Gupta, K. C., Seyer, J. M., Beachey, E. H., and Kingsbury, D. W., 1986, Localization and characterization of Sendai virus nonstructural C and C' proteins by antibodies against synthetic peptides, *Virus Res.* **6:**109–121.

Puddington, L., Woodgett, C., and Rose, J. K., 1987, Replacement of the cytoplasmic domain alters sorting of a viral glycoprotein in polarized cells, *Proc. Natl. Acad. Sci. USA* **84:**2756–2760.

Raghow, R., Kingsbury, D. W., Portner, A., and George, S., 1979, Topography of a flexible ribonucleoprotein helix: Protein–protein contacts in Sendai virus nucleocapsids, *J. Virol.* **30:**701–710.

Rapp, F., Gordon, I., and Baker, R. F., 1960, Observation of measles virus infection of cultured human cells. I. A study of development and spread of antigen by means of immunofluorescence, *J. Biophys. Biochem. Cytol.* **7:**43–48.

Re, G. G., and Kingsbury, D. W., 1988, Paradoxical effects of Sendai virus DI RNA size on survival: Inefficient envelopment of small nucleocapsids, *Virology* **165:**331–337.

Richardson, C., Hull, D., Greer, P., Hasel, K., Berkovich, A., Englund, G., Bellini, W., Rimia, B., and Lazzarini, R., 1986, The nucleotide sequence of the mRNA encoding the fusion protein of measles virus (Edmonston strain). A comparison of fusion proteins from several different paramyxoviruses, *Virology* **155:**508–523.

Rindler, M. J., Ivanov, I. E., Plesken, H., Rodriguez-Boulan, E. J., and Sabatini, D. D., 1984, Viral glycoproteins destined for apical or basolateral membrane domains traverse the same Golgi apparatus during their intracellular transport in Madin–Darby canine kidney cells, *J. Cell Biol.* **98:**1304–1319.

Rodriguez-Boulan, E., and Sabatini, D. D., 1978, Asymmetric budding of viruses in epithelial cell monolayers: A model system for study of epithelial polarity, *Proc. Natl. Acad. Sci. USA* **75:**5071–5075.

Rodriguez-Boulan, E., and Pendergast, M., 1980, Polarized distribution of viral envelope proteins in the plasma membrane of infected epithelial cells, *Cell* **20**:45–54.

Roth, M. G., Fitzpatrick, J. P., and Compans, R. W., 1979, Polarity of influenza and vesicular stomatitis virus maturation in MDCK cells: Lack of requirement of glycosylation of viral glycoproteins, *Proc. Natl. Acad. Sci. USA* **76**:6430–6434.

Roth, M. G., Srinivas, R. V., and Compans, R. W., 1983a, Basolateral maturation of retroviruses in polarized epithelial cells, *J. Virol.* **45**:1065–1073.

Roth, M. G., Compans, R. W., Giusti, L., Davis, A. R., Nayak, D. P., Gething, M.-J., and Sambrook, J., 1983b, Influenza hemagglutinin expression is polarized in cells infected with recombinant SV40 viruses carrying cloned hemagglutinin DNA, *Cell* **33**:435–443.

Roux, L., and Waldvogel, F. A., 1982. Instability of the viral M protein in BHK21 cells persistently infected with Sendai virus, *Cell* **28**:293–302.

Roux, L., Beffy, P., and Portner, A., 1985, Three variations in the cell surface expression of the hemagglutinin-neuraminidase glycoprotein of Sendai virus, *J. Gen. Virol.* **66**:987–1000.

Rozenblatt, S., Koch, T., Pinhasi, O., and Bratosin, S., 1979, Infective substructures of measles virus from acutely and persistently infected cells, *J. Virol.* **32**:329–333.

Rustigian, R., 1966, Persistence infection of cells in culture by measles virus, *J. Bacteriol.* **92**:1792–1804.

Ryan, K. W., and Kingsbury, D. W., 1988, Carboxyl-terminal region of Sendai virus P protein is required for binding to viral nucleocapsids, *Virology* **167**:106–112.

Sampson, A. C. R., and Fox, C. F., 1973, A precursor protein for Newcastle disease virus, *J. Virol.* **12**:579–587.

Sato, H., Ohihira, M., Ishida, N., Imamura, Y., Hattori, S., and Kawahita, M., 1987, Molecular cloning and nucleotide sequence of the *P, M,* and *F* genes of Newcastle disease virus avirulent straing D26, *Virus Res.* **7**:241–255.

Scheid, A., Caliguiri, L. A., Compans, R. W., and Choppin, P. W., 1972, Isolation of a paramyxoviruses glycoprotein. Association of both hemagglutinating and neuraminidase activities with the larger SV5 glycoprotein, *Virology* **50**:640–652.

Schwalbe, J. C., and Hightower, L. E., 1982, Maturation of the envelope glycoproteins of Newcastle disease virus on cellular membranes, *J. Virol.* **41**:947–957.

Sechoy, O., Philippot, J. R., and Bienvenue, A., 1987, F Protein–F protein interaction within the Sendai virus identified by native bonding or chemical cross-linking, *J. Biol. Chem.* **262**:11519–11523.

Shaw, M. W., and Choppin, P. W., 1984, Studies on the synthesis of the influenza B virus NB glycoprotein, *Virology* **139**:178–184.

Sheppard, R. D., Raine, C. S., Bornstein, M. B., and Udem, S. A., 1985, Measles virus matrix protein synthesized in a subacute sclerosing panencephalitis cell line, *Science* **228**:1219–1221.

Sheppard, R. D., Raine, C. S., Bornstein, M. B., and Udem, S. A., 1986, Rapid degradation restricts measles virus matrix protein expression in a subacute sclerosing panencephalitis cell line, *Proc. Natl. Acad. Sci. USA* **83**:7913–7917.

Spriggs, M. K., Olmsted, R. A., Venkatesan, S., Coligan, J. E., and Collins, P. L., 1986, Fusion glycoprotein of human parainfluenza virus type 3: Nucleotide sequence of the gene, direct identification of the cleavage activation site, and comparison with other paramyxoviruses, *Virology* **152**:241–251.

Stallcup, K. C., and Fields, B. N., 1981, The replication of measles virus in the presence of tunicamycin, *Virology* **108**:391–404.

Stallcup, K. C., Raine, C. S., and Fields, B. N., 1983, Cytochalasin B inhibits the maturation of measles virus, *Virology* **124**:59–74.

Stephens, E. B., and Compans, R. W., 1988, Assembly of animal viruses at cellular membranes, *Annu. Rev. Microbiol.* **42**:489–516.

Stephens, E. B., Compans, R. W., Earl, P., and Moss, B., 1986, Surface expression of viral glycoproteins is polarized in epithelial cells infected with recombinant vaccinia viral vectors, *EMBO J.* **5**:237–245.

Stone, H. O., Kingsbury, D. W., and Darlington, R. W., 1972, Sendai virus induced transcriptase from infected cells: Polypeptides in the transcriptive complex, *J. Virol.* **10**:1037–1043.

Tsukiyama, K., Sugiyama, M., Yoshikawa, Y., and Yamanouchi, K., 1987, Molecular cloning and sequence analysis of the rinderpest virus mRNA encoding the hemagglutinin protein, *Virology* **160**:48–54.

Tuffereau, C., and Roux, L., 1988, Direct adverse effects of Sendai virus DI particles on virus budding and on M protein fate and stability, *Virology* **162:**417–426.

Tuffereau, C., Portner, A., and Roux, L., 1985, The role of hemagglutinin-neuraminidase glycoprotein cell surface expression in the survival of Sendai virus-infected BHK-21 cells, *J. Gen. Virol.* **66:**2313–2318.

Tyrrell, D. L. J., and Ehrnst, A., 1979, Transmembrane communication in cells chronically infected with measles virus, *J. Cell Biol.* **81:**396–402.

Veit, M., Schmidt, M. F. G., and Rott, R., 1989, Different palmitoylation of paramyxovirus glycoproteins, *Virology* **168:**173–176.

Vidal, S., Mottet, G., Kolakofsky, D., and Roux, L., 1989, High mannose sugars must precede disulfide bond formation for proper folding of Sendai virus glycoproteins, *J. Virol.* **63:**892–900.

Wang, E., Walf, B. A., Lamb, R. A., Choppin, P., and Goldberg, R. A., 1976, The presence of actin in enveloped viruses, in "Cell Motility" (R. Goldman, T. Pollard, and J. Rosenbaum, eds.), pp. 589–599, Cold Spring Harbor Laboratory, Cold Spring Harbor, New York.

Waxham, M. N., Mertz, D. C., and Wolinsky, J. S., 1986, Intracellular maturation of mumps virus hemagglutinin-neuraminidase glycoprotein: Conformation changes detected with monoclonal antibodies, *J. Virol.* **59:**392–400.

Waxham, M. N., Sever, A., Goodman, H., and Wolinsky, J. S., 1987, Cloning and sequencing of the mumps virus fusion protein gene, *Virology* **159:**381–388.

Wechsler, S. L., and Fields, B. N., 1978, Intracellular synthesis of measles virus-specified polypeptides, *J. Virol.* **25:**285–297.

Wechsler, S. L., Rustigian, R., Stallcup, K. C., Byers, K. B., Winston, S. H., and Fields, B. N., 1979, Measles virus-specified polypeptide synthesis in two persistently infected HeLa cell lines, *J. Virol.* **31:**677–684.

Williams, M. A., and Lamb, R. A., 1986, Determination of the orientation of an integral membrane protein and sites of glycosylation by oligonucleotide-directed mutagenesis: Influenza B virus NB glycoprotein lacks a cleavable signal sequence and has an extracellular NH2-terminal region, *Mol. Cell. Biol.* **6:**4317–4328.

Wilson, C., Gilmore, R., and Morrison, T. G., 1987, Translation and membrane insertion of the hemagglutinin-neuraminidase glycoprotein of Newcastle disease virus, *Mol. Cell. Biol.* **7:**1386–1392.

Yamada, A., Takeuchi, K., and Hishigama, M., 1988, Intracellular processing of mumps virus glycoproteins, *Virology* **165:**268–273.

Yoshida, T., Nagai, Y., Yoshi, S., Maeno, K., Matsumoto, T., and Hoshino, M., 1976, Membrane (M) protein of HVJ (Sendai virus): Its role in virus assembly, *Virology* **71:**143–161.

Yoshida, T., Nagai, Y., Maeno, K., Iinuma, M., Hamaguchi, M., Matsumoto, T., Nagayoshi, S., and Hoshino, M., 1979, Studies on the role of M protein in virus assembly using a *ts* mutant of HVJ (Sendai virus), *Virology* **92:**139–154.

Yoshima, H., Nakanishi, M., Okada, Y. U., and Kobata, A., 1981, Carbohydrate structures of HVJ (Sendai virus) glycoproteins, *J. Biol. Chem.* **256:**5355–5361.

Young, K. K. Y., Heineke, B. E., and Wechsler, S. L., 1985, M Protein instability and lack of H protein processing associated with nonproductive persistent infection of HeLa cells by measles virus, *Virology* **143:**536–545.

Youngner, J. S., and Preble, O. T., 1980, Viral persistence: Evolution of viral populations, in "Comprehensive Virology," Vol. 16, (H. Fraankel-Conrat and R. R. Wanger eds.), pp. 73–135, Plenum Press, New York.

Zavada, J., 1982, The pseudotypic paradox, *J. Gen. Virol.* **63:**15–24.

Zebedee, S. L., Richardson, C. D., and Lamb, R. A., 1985, Characterization of the influenza virus M2 integral membrane protein and expression at infected cell surface from cloned cDNA, *J. Virol.* **56:**502–511.

CHAPTER 18

Immunobiology of Paramyxoviruses

ERLING NORRBY

I. INTRODUCTION

This chapter discusses the dissection of structural–functional properties of virus components by immunological methods. Monoclonal antibodies and antipeptide reagents represent remarkable tools for such ventures. Structural and nonstructural proteins of viruses can be identified and purified. Each protein species can be mapped with regard to the different antigenic sites it may expose. In cases when the protein, either isolated or participating in a more complex structure, carries a measurable biological activity, such as hemagglutination or a catalytic function, the involvement of each site of the folded protein in the activity can be evaluated. Sometimes it is possible to correlate variations in biological functions with substitution of a single amino acid.

When all of the proteins within a virus act in concert, infection of a susceptible cell will be achieved. It is of major importance to evaluate how specific immunological defense mechanisms can interfere with this process. These mechanisms include humoral and cell-associated phenomena targeting on mature virus particles and, in the case of cellular immunity, also on virus-infected cells. Knowledge pertaining to these issues assists the development of effective nonreplicating vaccines.

It will be assumed that the reader has knowledge of the types of viruses represented in the family of mammalian and avian paramyxoviruses with their three genera, *Paramyxovirus*, *Morbillivirus*, and *Pneumovirus*. Further, it is expected that some baseline acquaintance with the different structural proteins of paramyxoviruses is at hand. Distinguishing features of homologous components of different types of viruses will be emphasized. Functional units (antigenic sites) in different components identified by the use of immunologi-

ERLING NORRBY • Department of Virology, Karolinska Institute, School of Medicine, S-105 21 Stockholm, Sweden.

cal tools will be discussed and put into the perspective of evolutionary relationships, narrowing down to unique features of individual virus strains. The
section on immunoprotective responses in experimental systems and humans
will be selective. Predominantly nonreplicating components or components
synthesized in cells infected with a heterologous virus vector containing a
selected paramyxovirus gene will be discussed. The epidemiological consequences of use of the effective live measles and mumps virus vaccines are
presented in Chapter 19 and experimental live vaccine products are discussed
in Chapter 1. Finally, a separate chapter (Chapter 14) is devoted to the large
glycoprotein, G, of respiratory syncytial (RS) virus, because of the unique properties of this pneumovirus attachment protein. Thus, description of features of
this component will be deemphasized here.

Although paramyxoviruses predominantly cause acute infections, there
are examples of persistent paramyxovirus infections (see Chapter 11), preemi-
nently the noncontagious brain infection subacute sclerosing panencephalitis
caused by measles virus, as discussed by Billeter and Cattaneo (Chapter 12).
The acute infections can be either local or generalized and although the former
may have a repeated occurrence, the latter generally occur once in a lifetime.
Since paramyxoviruses survive in nature by sustaining a chain of acute infections, they have been a part of human infectious disease history only for a
relatively short time, 5000–6000 years. Infected animals were the source of
viruses which started to circulate among humans at the dawn of nonmigrating
extended civilizations. This may explain the relatively close relationships between certain human and animal viruses, as exemplified by three morbilliviruses, measles, canine distemper, and bovine rinderpest viruses, and also by
the relationships between human and bovine parainfluenza 3 viruses and between human and bovine RS viruses.

II. STRUCTURAL–FUNCTIONAL CHARACTERIZATION OF VIRUS COMPONENTS BY IMMUNOLOGICAL PROBING

A. Rationale

Identification of virus-specific products can be made in two different ways.
One way is to identify open reading frames in the virus genome and thereby
predict the primary structure of the purported gene product. The other way is
to directly identify a product after dissociation of purified virions or virus-
infected cells. In both cases, immunological techniques can be helpful in sin-
gling out individual components. To identify a protein predicted from an open
reading frame, synthetic peptides representing 10–20 amino acid segments of
the predicted polypeptide sequence can be used as immunogens. The resulting
serum is then employed in a search for the native protein. This technique has
been used for identification of, for example, the nonstructural C protein from
the phosphoprotein (P) gene in measles virus (Richardson et al., 1985), the 1A
protein of RS virus (Olmsted and Collins, 1989), and the large (L) protein of
measles virus (E. Norrby, to be published). In fact, by preparing antipeptide sera
specific for the N-terminal and C-terminal ends of a transmembranous protein,

it is possible to determine its orientation in the lipid bilayer. It is probable that antipeptide antibodies will be more extensively used in the future to follow structural–functional rearrangements of proteins as they mature by post-translational processing in cells or unfold in a modified environment to activate biological functions. So far, only monoclonal antibodies have been used for this purpose, as exemplified by a number of studies. Hidden sites in the measles virus glycoproteins were exposed by incubation at low pH (Sheshberadaran and Payne, 1988). During maturation, the availability of epitopes in Sendai virus (Mottet et al., 1986; Vidal et al., 1989), mumps virus (Waxham et al., 1986), and Newcastle disease virus (NDV) glycoproteins were found to vary (Long et al., 1986; Nishikawa et al., 1986). Monoclonal antibodies have also revealed alterations of neutralizing sites in selected non-neutralizable variants of measles virus (Sheshberadaran and Norrby, 1986) and Newcastle disease virus (Iorio et al., 1986). An example of topological and operational delineation of antigenic sites and their structures on the NDV hemagglutinin-neuraminidase (HN) protein is the study of Nishikawa et al., (1986). To define metabolism and cellular compartmentalization of specific viral proteins, the nascent and various processed forms of the molecules were identified by use of monoclonal antibodies in the presence of inhibitors of glycosylation and glycoprotein transport. Antigenic reactivity was also examined after the molecules had been deglycosylated by endoglycosidase F and reduced by 2-mercaptoethanol. Posttranslational organization was important for the emergence of all four antigenic sites examined.

Although a large amount of information on nucleotide sequences of paramyxovirus genomes has become available, the final word on the number of gene products involved in virus replication and maturation may not have been said. The number of separate consecutive genes identified in genomes characterized by nucleotide sequencing is ten in RS virus, seven in simian virus 5 (SV5) and mumps virus, and six in the remaining paramyxoviruses. However, there is ample evidence that a single gene can be the source of more than one protein by mechanisms operating both at the transcriptional and the translational level. Recently, a process of RNA "editing" during transcription as a means of changing the reading frame of the P gene in SV5 and measles virus was revealed (Thomas et al., 1988; Cattaneo et al., 1989). It is therefore likely that minor protein products of other paramyxovirus genes have been overlooked, and one wonders whether all paramyxoviruses may have a homologue of the small transmembranous RS virus 1A protein (Olmsted and Collins, 1989) or the SV5 and mumps virus small hydrophobic (SH) protein (Hiebert et al., 1988; Elango et al., 1989). Furthermore, one can also ask whether other paramyxoviruses have a second matrixlike protein corresponding to the M_r 22,000–24,000 protein of RS virus (Huang et al., 1985). To test these possibilities, antisera against synthetic peptides might be useful.

The advent of monoclonal antibodies made available reagents for identification of the existence of different paramyxovirus components and assisted in defining their biological functions. In a few studies, the peptide synthetic antigen and hybridoma technologies were combined in studies of paramyxovirus epitopes. A prerequisite for identification of the exact location of an epitope in a polypeptide chain by this approach is that it is linear or that a part of the chain is dominant in a discontinuous epitope. Sites for monoclonal antibody reactions were identified for measles virus antihemagglutinin (H) (Mäkela *et al.*, 1989) and RS virus anti-G monoclonal antibodies (Norrby *et al.*, 1987).

As mentioned, hybridoma technology has been extensively used for studies of paramyxovirus proteins. The results of these studies up to 1985 have been reviewed (Örvell and Norrby, 1985). The following comments pertain to the dissection of structural–functional characteristics of different components in subsequent studies.

B. Nonmembrane Components

Based on studies with NDV (Hamaguchi *et al.*, 1983), it has been proposed that in paramyxoviruses the complex of L, P, and nucleocapsid (N) proteins is responsible for viral transcription and replication. By use of monoclonal antibodies, several different antigenic sites harboring many epitopes have been identified in both paramyxovirus NP and P proteins, but almost no information is available on topological and operational allocation of antigenic sites. One exception is a study on the measles virus nucleocapsid protein (Buckland *et al.*, 1989). Deletion mutants of the *N* gene expressed in *Escherichia coli* were used to locate monoclonal antibody-specific antigenic sites.

C. Membrane Components

Studies with monoclonal antibodies have provided relatively more information on membrane components, which include one or two matrix (M) proteins and two or three transmembranous proteins. Considerable interest has been devoted to proteins included in the two dominating transmembrane glycoproteins, forming the peplomers responsible for the attachment of virions to cellular receptors and for their penetration into cells by membrane fusion. The properties of the attachment proteins differ markedly between members of different genera of paramyxoviruses. In paramyxoviruses, it has hemagglutinating as well as neuraminidase activity (HN protein), whereas in morbilliviruses it carries either an isolated hemagglutinating (H) activity, as exemplified by measles virus, or no corresponding activity, as in the remaining viruses representing this genus. As mentioned earlier, the RS virus attachment glycoprotein G has such unique features that it is the subject of Chapter 14 in this book. Suffice it to say in this context that G has a protein backbone which is much smaller than that in other paramyxoviruses, that it is heavily glycosylated to a large extent via O-linked sugars, and that it lacks demonstrable hemagglutinating or neuraminidase activity. However, the corresponding attachment protein of a murine pneumovirus, pneumonia virus of mice (PVM), carries hemagglutinating activity (Ling and Pringle, 1989b).

Groups of monoclonal antibodies reacting with the same components are characterized by competition assays. Studies of these kinds have been extensively performed with paramyxovirus HN and H proteins (Örvell and Grandien,

1982; Yewdell and Gerhard, 1982; Iorio and Bratt, 1983; Russell *et al.*, 1983; Örvell, 1984; Coelingh *et al.*, 1986; Samson *et al.*, 1985; Long *et al.*, 1986; Tozawa *et al.*, 1986; Portner *et al.*, 1987a; Thomson and Portner, 1987; Gotoh *et al.*, 1988; Komada *et al.*, 1989a). If a sufficiently large number of HN-specific monoclonal antibodies is employed, they can be divided into the following five principal categories with regard to their capacity to interfere with biological activities:

1. Blocks all biological activities: infectivity, hemagglutinating activity (HA), neuraminidase activity (NA), hemolysing activity (HLA).
2. Blocks infectivity to some extent and HA, but no other activity.
3. Blocks infectivity to some extent and NA, but no other activity.
4. Blocks infectivity to some extent and HLA, but no other activity.
5. Binds to the protein without blocking any activity.

These observations have a number of consequences. They are consistent with the view that HA and NA are located in antigenically distinct parts of the HN protein (Portner, 1981; Thomson and Portner, 1987). Furthermore, the involvement of the HN (or H) protein in the hemolysis and cell fusion reactions as evidenced by the effect of antibodies in group 4 above (Togashi *et al.*, 1981; Miura *et al.*, 1982; Örvell and Grandien, 1982; Nishikawa *et al.*, 1986; Tsurudome *et al.*, 1986), and effects of chymotrypsin treatment experiments linking a critical site in the mumps HN protein to cell fusion cytopathology (Merz and Wolinsky, 1983), indicate that additional functions separate from the HA and NA may reside in the HN protein. It remains to identify the proposed unique part of the attachment protein which acts in sequence with or in consort with the F protein to bring about membrane fusion.

In the special case of neutralizing antibodies it is possible to define the precise location of the epitopic site by use of nonneutralizing variants. Nucleotide sequencing of the parental virus and variant virus genes allows identification of amino acids involved in or influencing the contact surface of the epitope. In some cases, selection of nonneutralizable variants has led to recovery of strains which have lost one site for N-linked glycosylation (Löve *et al.*, 1985; Kövamees *et al.*, 1990; Gotoh *et al.*, 1988; van Wyke Coelingh and Tierney, 1989b). Such changes may indirectly affect the immunogenic properties of the protein. The altered protein has a reduced apparent M_r in SDS–polyacrylamide gel electrophoresis.

In contrast to the attachment protein, the fusion (F) proteins of different paramyxoviruses have comparatively uniform characteristics. They have a precursor form F_0 with an M_r of 60,000–70,000 which is cleaved into the disulfide-bonded F_1 (M_r 40,000–50,000) and F_2 (M_r 10,000–20,000) components, with N-linked glycosylation of both products or only the smaller one. The N-terminal part of the F_1 protein is uniquely conserved (Varsanyi *et al.*, 1985), although certain variations were found in the NDV protein (Toyoda *et al.*, 1987).

F protein-specific MAbs have been described which react with several different antigenic sites, but with regard to capacity to block biological activities they fall into two categories (Örvell and Grandien, 1982; Russell *et al.*, 1983; Tozawa *et al.*, 1986). They can show a certain capacity to block hemolysis, coupled with a variable potential to block infectivity, or with no demon-

strable ability to interfere with any biological activity. The technique of selecting nonneutralizing variants has a restricted application to F-protein antigenic
site mapping, since efficient neutralizing F-epitope-specific reagents are available only in some cases. In one study of Sendai virus (Portner et al., 1987b),
three F monoclonal antibody-resistant mutants were selected and the nucleotide sequence of the gene in each variant was determined. A single alteration of a proline to a glutamine residue was seen in position 399. This position
is localized at great distance from the N-terminal part of the F_1 protein deduced to be involved in fusion activity. In a recent extensive study (van Wyke
Coelingh and Tierney, 1989a), 26 monoclonal antibodies against the F protein
of parainfluenza virus type 3 were used to examine antigenic structure, biological properties, and natural variation of the protein. Twenty epitopes, 14 of
which participated in neutralization reactions, were clustered into seven antigenic sites. Three of these sites were involved in mediating fusion inhibition.
The amino acids recognized by the syncytium-inhibiting and neutralizing
monoclonal antibodies were identified (van Wyke Coelingh and Tierney,
1989b). Compared to other members of the paramyxovirus genus, NDV-specif-
ic anti-F monoclonal antibodies exhibit a more readily detectable neutralizing
activity (Abenes et al., 1986). By selection of nonneutralizable variants of NDV,
four distinct antigenic sites were defined. Also in this system the amino acids
relevant to three of the antigenic determinants have been determined (Toyoda
et al., 1988). Mutations located on both F_1 and F_2 subunits have been found
(Neyt et al., 1989). A more comprehensive picture has also been obtained in
studies of the RS virus F protein by use of monoclonal antibodies, reflecting
the fact that this component effectively induces production of both neutralizing and fusion-blocking antibodies (Walsh et al., 1985, 1986).

III. EVOLUTIONARY RELATIONSHIPS BETWEEN HOMOLOGOUS COMPONENTS REFLECTED IN IMMUNOLOGICAL CROSS-REACTIONS

The optimal way of evaluating evolutionary relationships between viruses
is to compare homologous nucleotide sequences. However, a mapping of immunological relationships may also provide evolutionary insight. In cases when
monoclonal antibodies are employed in the comparison, the outcome of the
investigation depends on the diversity of specificities represented by the reagents employed. No cross-reactions have been observed between paramyxoviruses representing different genera, except in rare cases of fortuitous occurrence of antigenic mimicry (Norrby et al., 1986a). Theoretically, such cross-
reactions might occur between highly conserved structures such as the hydrophobic N-terminal part of the F_1 protein, proposed to be involved in membrane
fusion, and RNA polymerase active sites in the L protein. However, no paramyxovirus immunogenic determinants have been defined in these positions,
probably, at least in part, because of needs to evade immunological selective
pressure on structures which have a high order of functional constringency.
Within the Paramyxovirus genus, a segregation into two subgenera is indicated on basis of immunological cross-reactions (Ito et al., 1987). The first

subgenus includes parainfluenza virus types 1 and 3, which show close relationships, exemplified by the shared properties of their HN proteins (Örvell *et al.*, 1986). Bovine parainfluenza virus type 3 shares antigenic properties with the homotypic human virus and thus belongs in this subgroup (Coelingh *et al.*, 1986; Rydbeck *et al.*, 1987). Nucleotide sequence analysis of variants selected by neutralizing antibodies was performed for characterization of subtype conserved and variable neutralizing epitopes in human and bovine parainfluenza 3 strains. Six different antigenic sites were identified and three of these were conserved. One of the conserved sites was a target for neutralizing antibodies, but the other two were nonneutralizing. In the second subgenus there are distinct relationships among mumps, parainfluenza types 2, 4A, and 4B, and SV5 viruses. Monoclonal antibodies have demonstrated a certain number of cross-reactions between homologous components of virus types belonging to the same subgenus (Örvell *et al.*, 1986; Randall and Young, 1988; Komada *et al.*, 1989a,b), but occasional cross-reactions are also seen between components of types representing the two subgenera (Örvell *et al.*, 1986). It is to be expected that the availability of further nucleotide sequence data will help to clarify the distinctiveness of this subdivision of the *Paramyxovirus* genus. Possibly, identification of several unique characteristics, such as the occurrence of the *SH* gene hitherto found only in SV5 and mumps virus, may lead to the establishment of two separate genera.

In the *Morbillivirus* genus, interesting close relationships have been found. Somewhat unexpectedly, it was shown that whereas the H protein shows the highest degree of diversity among measles, canine distemper, and bovine rinderpest viruses (sufficient data are not available on a fourth member of the group, peste des petits ruminants virus, to allow a comparative evaluation), the other major surface glycoprotein, F, is highly conserved (Sheshberadaran *et al.*, 1986). The internal components N, P, and M show intermediate degrees of cross-reactivity. On the basis of the relationships among the H proteins of these viruses, it has been proposed that bovine rinderpest may be the achetypal virus of the group, from which canine distemper virus evolved first, followed by measles virus (Norrby *et al.*, 1985).

The genus *Pneumovirus* includes human and bovine RS viruses and PVM. Recently, it was proposed that turkey rhinotracheitis virus should be classified in this genus (Barrett and Cavanagh, 1988), but it is not yet clear how this virus relates to the other members of this genus. It has been apparent for a few years that human RS viruses can be separated into two distinct subgroups, A and B (Anderson *et al.*, 1985; Mufson *et al.*, 1985; Walsh *et al.*, 1987a). The G protein shows the most pronounced antigenic differences between the two subgroups, but all the other structural (and possibly nonstructural) proteins also display certain subgroup-specific characteristics. In fact, subgroup-specific size characteristics were found in the case of F_1 and F_2 proteins and to some extent also in the P protein (Giminez *et al.*, 1986; Norrby *et al.*, 1986b; Åkerlind *et al.*, 1988). Although bovine RS virus shows a number of cross-reactions, it is distinctly different from both the B and the A subgroups of the human viruses. In particular, the G protein of bovine strains appears to be unique (Örvell *et al.*, 1987). Only a limited amount of information is available on cross-reactions between PVM and other pneumoviruses. An antigenic relationship was found between two PVM proteins and the N and P proteins of human RS virus in tests with polyclonal and monoclonal reagents (Giminez et al., 1984; Ling and Pringle, 1989a).

IV. STABILITY OF ANTIGEN CHARACTERISTICS AND INTRATYPIC VARIATIONS

All paramyxoviruses are antigenically stable. Hyperimmune sera define the existence of only a single type of each virus. This may appear remarkable in view of the high mutation frequency of negative-strand RNA viruses (Portner et al., 1980). However, it appears that there are functional constraints inherent to the biology of paramyxoviruses which do not apply to other virus families, such as the Orthomyxoviruses. Only members of the latter family have managed to functionally separate immunodominant sites from sites involved in peplomere-associated attachment and fusion phenomena. Measles virus exemplifies well the overall antigenic stability of paramyxoviruses. This virus was consecutively passaged in the presence of eight neutralizing anti-H monoclonal antibodies of different epitope specificities (Sheshberadaran and Norrby, 1986). The resulting virus had lost its capacity to be neutralized by all these antibodies, although it still could bind one of them. Nevertheless, the highly modified virus could be neutralized with unaltered efficacy by polyclonal antisera.

The fact that paramyxoviruses are monotypic does not imply that minor epitopic variations do not occur. Such variations have been seen in all paramyxoviruses examined by use of monoclonal antibodies, and strain-specific epitopic variations were noted in all protein species. The relative stability of different structural components may vary in different genera depending upon the correlation between immunodominant sites to which monoclonal antibodies would be expected to be directed and functionally critical sites which need to be retained in spite of changes by spontaneous mutations. Among measles virus envelope components, epitope variations appear to occur more frequently in the H protein than in the F protein (Giraudon and Wild, 1981; Sheshberadaran et al., 1983; Sato et al., 1985). In spite of the absence of any humoral immune selective pressure on internal proteins, marked epitope variations were found in the M protein, but less in the N and P proteins of this virus (Sheshberadaran et al., 1983; Sato et al., 1985). Likewise, there were certain variations between strains of the monotypic NDV both in the HN protein, even to the extent that it was referred to as remarkable antigenic drift (Nishikawa et al., 1983; Iorio et al., 1984), and in internal proteins, in particular the P and M proteins (Nishikawa et al., 1987). In fact, compared to other paramyxoviruses, NDV may show a more pronounced variation of surface epitopes (Nishikawa et al., 1983; Russell and Alexander, 1983; Ishida et al., 1985) involving sites mediating neutralizing activity (Iorio et al., 1986). In a recent study, the HN gene sequences of 13 strains of NDV isolated over 50 years was determined (Sakaguchi et al., 1989). Antigenic variations were not interpreted to be cumulative or progressive. Instead, three distinct lineages were identified. These lineages were demonstrated to have been cocirculating for considerable periods. Minor variations between both envelope and internal proteins of virus strains belonging to the same type were also seen in mumps virus (Server et al., 1982; Rydbeck et al., 1986), parainfluenza type 3 virus (Rydbeck et al., 1987; Coelingh et al., 1989; van Wyke Coelingh and Tierney, 1989b), and parainfluenza type 4A and 4B viruses (Komada et al., 1989a). In one study, the nucleotide and deduced amino acid sequences of HN genes of six strains of human parainfluenza type 3 viruses isolated between 1957 and 1983 were determined (van Wyke Coelingh et al., 1988). Only a limited genetic heterogeneity was registered. The percentage of nucleotide sequence identity between strains varied from 94.2 to 98.8. There was no tendency for accumulation of mutational changes of the kind seen in connection with antigenic drift in influenza virus.

Sometimes, strain-specific variations in apparent size of structural components have been encountered (McCarthy and Johnson, 1980; Örvell, 1980; Rima, 1983), but these variations do not correlate in a systematic manner to the absence of specific antigenic sites or to the expression of multicomponent-dependent phenomena such as pathogenicity. However, there are examples of modification of virulence in some neutralization-resistant variants. Certain monoclonal antibodies against NDV glycoproteins were suggested to distinguish between virulent and nonvirulent strains (Meulemans et al., 1987) and strain-specific NDV anti-HN monoclonal antibodies have been described (Iorio et al., 1984). These observations may partly reflect some unique properties of the NDV HN protein. In contrast to the attachment proteins of other paramyxoviruses, the NDV HN protein of some strains is formed as a precursor HN_0, from which 45 amino acids are cleaved to form the mature product (Gotoh et al., 1988). Sensitivity to proteolytic cleavage of both envelope glycoproteins was demonstrated more than a decade ago to correlate with virulence of NDV strains (Nagai et al., 1976). The HN cleavage site was identified by nucleotide sequencing analysis (Gotoh et al., 1988) and a corresponding analysis of the F protein was also performed (Toyoda et al., 1987). Perhaps some of these cleavage phenomena could be examined by immunochemical tools, especially synthetic peptide immunogens. In a recent study containing sequence data for HN and F proteins of multiple NDV strains, it was found that different degrees of virulence were associated with different lineages of evolution (Sakaguchi et al., 1989; Toyoda et al., 1989). There was no evidence for any occurrence of recombination in the generation of virulent strains. The Jeryl–Lynn vaccine strain of mumps virus showed a unique absence of reactivity with an anti-HN antibody clone in a comparison of nine strains (Rydbeck et al., 1986). Although it is tempting to use this kind of special MAb reaction pattern as a marker, one should be cautious about such an application. The character of a single epitope could readily change, as exemplified by the selection of nonneutralizable variants. Therefore, either more than one unique epitope characteristic of a strain has to be defined, or, alternatively, a distinct linkage between a marker such as virulence and the character of a single epitope has to be established. As yet, there is no example of the latter situation among the paramyxoviruses. In one study of mumps virus (Löve et al., 1985), four neutralization-resistant variants were selected after propagation in the presence of one anti-HN monoclonal antibody. One variant had a reduced neurovirulence, two had increased neuraminidase activity, and one had an HN protein of smaller apparent size. The correlating amino acid changes in these variants were recently deduced (Kövamees *et al.*, 1990). Changes at more than one amino acid were seen in each variant, but some of these changes were interpreted to be more critical. The variant with altered virulence showed such a critical change at a cysteine potentially influencing possibilities for intramolecular disulfide bonding. The two variants showing increased neuraminidase activity had changes within a region proposed to be involved in this function (Waxham and Aronowski, 1988). The fourth variant, finally, had lost one glycosylation site, explaining the reduced apparent M_r of the HN protein. A similar phenomenon of elimination of a glycosylation site was also observed in studies of NDV (Gotoh *et al.*, 1988). It might be commented that genetic changes in paramyxoviruses would be expected to be restricted to mutations and should not include deletions or duplications. Still, the apparent M_r of a protein may vary due to mutations causing a loss (or potentially a gain) of glycosylation sites, but also due to mutations causing steric rearrangements in the protein with resulting modifications of the apparent M_r.

V. IMMUNOPROTECTIVE ROLES OF ISOLATED VIRUS COMPONENTS

Live vaccines giving a durable, probably life-long immunity against the generalized infections measles and mumps are available. During the early era of viral vaccinology, attempts were also made to develop inactivated vaccines against these diseases. The testing of these candidate vaccine products brought to light particular problems that may concern the future development of both inactivated and subunit virus vaccines. When a formalin-inactivated measles virus vaccine was tested for its immunogenicity and protective efficacy, it was found that protection against disease after exposure to wild virus was relatively short lived, although readily measurable neutralizing antibody titers developed after immunization. Furthermore, some of the vaccinees exposed to naturally circulating virus contracted an aberrant form of disease referred to as atypical measles. This observation was very puzzling, particularly in the perspective that even minute amounts of antibodies passively provided in the form of immunoglobulin could prevent the emergence of disease. However, this protection appears to be based on the simultaneous presence of antibodies against both the H and F proteins and a full representation of antibodies against the latter component was not seen after immunization with a formalin-treated product, due to a selective destruction of critical immunoprotective antigens in this protein (Norrby *et al.*, 1975). Consequently, exposure to wild-type virus resulted in its replication at the portal of entry and dissemination in the body. Sensitization to the H component, which did not suffice to restrict virus replication, provided a basis for the development of immunopathological complications resulting from the Arthus phenomenon. A corresponding defective immune response was observed after immunization with a formalin-inactivated mumps virus vaccine in Finnish military recruits (Norrby and Penttinen, 1978). Also in this case, the vaccinees developed antibodies against the HN protein, but not against critical sites in the F protein. In this particular situation, no consequential immunopathological complications have been identified, but this may be due to the fact that 85–90% of the vaccinees were immune at the time of immunization and that possible complications among those that were not immune before vaccination may have had such a diverse symptomatology that they were not recognized as having mumps virus infections. Immunization of originally seropositive subjects gave antibody responses which were higher in subjects vaccinated with an inactivated virus preparation than in individuals immunized with live virus (Ilonen et al., 1984).

A third situation in which immunization caused enhanced disease in humans was the formalin-inactivated RS virus vaccine used in children (Chin et al., 1969; Kim et al., 1969). The nature of the defective immunoprotection in the vaccinees is still a matter of conjecture and both humoral and cell-mediated hypersensitivity have been proposed (McIntosh and Fishout, 1980). Discussions of malfunctioning humoral immunity need to consider separately the immunobiological properties of the G and the F proteins. The relative immunoprotective role of these two components differs from the situation prevailing for other paramyxoviruses. Monoclonal antibodies against the F protein give direct neutralization (Walsh and Hruska, 1983; Walsh et al., 1984; Mufson et al., 1985), but a corresponding direct neutralization with anti-G monoclonal antibodies is rare, although addition of complement or anti-immunoglobulin furnish some anti-G antibodies with neutralizing activity (Fernie et al., 1982; Walsh and Hruska, 1983). Also, experiments with vaccinia vector-borne RS virus glycoprotein genes emphasize the relatively dominating role of the F component (see below). The imbalance of the immune response in hitherto tested nonreplicating vaccines could reflect selective immunodestruction of either the G or the F protein. In cotton rats, immunopathological accentuated inflammation was seen after immunization with a killed vaccine (Prince et al., 1986). In later studies, immunization with high doses of purified G and in particular F glycoprotein, and in fact also preceding live-virus infection, could result in amplified pathological reactions (G. A. Prince, personal communications). It should finally be bought into the picture that children who received the inactivated vaccine had a disproportionate fraction of nonneutralizing anti-glycoprotein antibodies (Murphy et al., 1986), that in vitro enhancement of RS virus infection in a macrophage cell line by human sera has been shown (Giminez et al., 1989), and that infants respond relatively less efficiently than adults to the G protein (a consequence of the heavy glycosylation?) compared to the F protein (Ward et al., 1983). It should be kept in mind that naturally developing immunity to RS virus infection in humans is not complete (Henderson et al., 1979). Not until the third consecutive infection does one see a distinct amelioration of the disease. This may reflect the observation that neutralizing antibodies appear to become more stable after the second infection. The role of RS virus strains representing different subgroups warrants further studies. It has been shown that there is a preferential homosubtype protection against repeated infections (Mufson et al., 1987).

Even though it may appear from the experiences with formalin-inactivated measles and mumps vaccines that both the attachment and the fusion protein should be included in a component vaccine, this concept may not be relevant to other paramyxovirus vaccines. Completely different conditions may prevail in local respiratory infections with, for example, parainfluenzavirus type 3 or RS virus, compared to the generalized measles virus infection. Thus, in summary, if a vaccine-induced immunity against a paramyxovirus is to be based on circulating antibodies, a separate evaluation of the immunoprotective efficacy of different surface components should be made.

Since component vaccines unaided by adjuvants would not be expected to induce efficient cell-mediated immunity, it might be appropriate at this stage to define the sacrifice made by omitting this arm of the defense system. Only limited studies of cell-mediated immunity against paramyxoviruses have been performed. For some time, a delayed hypersensitivity test for mumps was advocated, but it was later found not to give reliable results. Most studies of cell-mediated immunity have been performed with measles and RS viruses.

The cell-mediated immune response to measles virus can be measured in antigen-specific T-cell proliferation or cytotoxic T-lymphocyte assays. Lymphocytes of high-responder individuals could be stimulated to proliferate when cultured in the presence of purified H, F, N, P, and M proteins (Rose et al., 1984). Cytotoxic cells have been identified both directly in peripheral blood (Kreth et al., 1979) and after in vitro culturing (Wright and Levy, 1979). However, in general the reactivity is weak and short lived. Thus, the proliferative response of peripheral blood lymphocytes in seropositive normal late convalescent individuals is much lower after measles than after, for example, a mumps virus infection (McFarland et al., 1980). The class dependence of histocompatibility restriction of measles virus-specific cytotoxic T cells is a matter of controversy. In earlier studies it was concluded that the restriction was class I antigen specific (Kreth et al., 1979; Lucas et al., 1982), but in more extended later studies it was found somewhat unexpectedly that the reaction was class II antigen restricted (Jacobson et al., 1989; Richert et al., 1985). This was demonstrated both with separated peripheral blood lymphocytes and with T-cell clones. The cytotoxic T cells show a specificity not only for surface antigen, but also for internal antigens, especially the N protein component.

The RS antigen specificity of cytotoxic T cells has been evaluated by use of cells infected with vaccinia virus vectors containing individual genes (Pemberton et al., 1987; Cannon and Bangham, 1989). The results indicated that the viral N and particularly the F proteins were recognized by cytotoxic T cells, but not the large viral glycoprotein G or the structural protein 1A. Passive transfer of memory T cells has been shown to clear a persistent RS virus infection in immunodeficient mice (Cannon et al., 1987) and cytotoxic T cells have been implicated not only in the clearance of infection, but also in augmentation of lung pathology in RS virus-infected mice (Cannon et al., 1988).

Taken together, the studies of cytotoxic T cells performed so far have a major importance for our understanding of the pathogenesis of infection, but they may be of less value in the design of immunoprotective interventions. Cytotoxic T cells would be expected to play a major role in the clearance of a paramyxovirus infection, but they may have a lower efficiency in providing long-term protection against repeated infections. However, T cells by definition play an important role in providing helper functions for antibody-producing cells. Thus, to facilitate effective cellular cooperation in the mobilization of B-lymphocyte responses, class II-dependent T cells epitopes need to be considered.

Three approaches can be taken to defining the immunoprotective role of different structural components. These are (1) passive immunoprophylaxis with component-specific reagents (MAbs or hyperimmune sera), (2) immunization with purified virus components, and (3) infection with vaccinia virus vectors containing genes representing individual components. For obvious reasons, the main emphasis hitherto has been on the attachment and fusion proteins of paramyxoviruses, but in future studies it might become important to consider also additional minor envelope-associated components. Some studies have already been made with the 1A protein of RS virus, but additional studies might be made with the SH proteins of SV5 and measles virus.

In humans, immunoglobulin has been found to effectively block a measles virus infection even when given 4–5 days after exposure to virus. Similarly, an effect on mumps has been proposed (Gellis et al., 1945), but this observation is less well substantiated. In experimental systems, the role of antibodies against different structural components was evaluated. Antibodies against both HN (or H or G) and F are effective, but their relative protective capacity varies from one type of virus to another. Generally, antibodies to the attachment protein are more effective than those against the fusion protein (Örvell and Grandien, 1982; Wolinsky et al., 1985; Löve et al., 1986; Rydbeck et al., 1988; Umino et al., 1987), but in the case of RS virus, antibodies against the F protein are relatively more efficient (Taylor et al., 1984; Walsh et al., 1984). In one passive immune protection study in cotton rats, the role of the subgroup of the infecting RS virus was evaluated (Johnson et al., 1987). G-specific and F-specific antisera representing both subgroups A and B were employed. The F-specific reagents gave equal protection against homologous and heterologous virus challenge. However, in tests with G-specific sera, the protection against challenge with homologous virus was 13 times better than that against heterologous virus.

In the case of Sendai virus, mumps virus, and NDV, protection was seen even with monoclonal antibodies lacking detectable neutralizing, hemolysis-inhibiting, or hemagglutinating-inhibiting activities. Even though in vitro neutralization provides the best guidance for predictions of the capacity of antibodies to give in vivo protection, extrapolations from phenomena measured under in vitro conditions have to be made with caution. The properties of monoclonal antibodies are determined not only by their direct monoepitopic specificity, but also by their capacity to mediate secondary F_c-dependent interactions with complement factors, phagocytes, and other cells. The latter properties are complex and depend not only on antibody class and subclass, but also on antibody affinity, epitope spacing, protein flexibility, and other factors. Already, the relatively simplified phenomenon of in vitro neutralization can be amplified by addition of complement or anti-immunoglobulin and by combining different antibodies (Iorio and Bratt, 1984; Russell, 1986). It is further noteworthy that one neutralizing monoclonal antibody specific for the Sendai virus HN protein could protect against an infection with parainfluenza type 3 virus (Rydbeck et al., 1988). The protective effect of monoclonal antibodies against the H component of measles virus was found to be related to their hemagglutinating-inhibiting activity (Giraudon and Wild, 1985). It was also shown that such antibodies can change an acute lethal form of measles encephalitis in mice into a late subacute disease (Rammohan et al., 1983).

In the case of RS virus infections of experimental animals, immunoglobulin containing neutralizing antibodies has been found not only to have a prophylactic effect, but even a therapeutic effect (Hemming et al., 1985; Prince et al., 1985, 1987). The therapeutic effect was documented in cotton rats and monkeys. There was a 50- to 500-fold reduction of virus in pulmonary tissue (less in the nose) and, in the overall majority of animals, virus was cleared within 24 hr. The therapeutic effect was 160 times greater by the topical route than by parenteral inoculation.

Only limited studies have been performed in which purified (usually by immunoaffinity chromatography) HN (or the corresponding attachment protein) or F components were used. In studies of parainfluenza virus type 3 (Ray et al., 1988), immunization with each of the components did not provide protection, but when they were used in conjunction, a higher neutralizing activity was detected in bronchial lavages and a complete protection from challenge infection could be documented. Parainfluenza virus type 3 HN glycoprotein was also expressed by a recombinant baculovirus in insect cells (van Wyke Coelingh et al., 1987). Repeated injection of cell-associated material with or without Freund's complete adjuvant gave an immunity in cotton rats against virus challenge. Similar studies were also performed with the parainfluenza virus type 3 F protein (Ray et al., 1989).

The relative immunoprotective effects of H and F components have been evaluated in different morbillivirus systems. The dominating antigenic sites inducing in vitro neutralizing antibodies are associated with the H component. Although monoclonal antibodies and polyclonal hyperimmune sera against F protein generally have little or no in vitro neutralizing activity (Varsanyi et al., 1984), such antibodies appear to play an important role in in vivo protection. This was shown in rodents with separated components associated with immunostimulating complexes (Varsanyi et al., 1987; De Vries et al., 1988a). In dogs, both purified canine distemper virus H and F proteins induced a protection against disease after challenge with virus (Norrby et al., 1986c; De Vries et al., 1988b), and in the same species, measles virus also gave protection against canine distemper virus based on cross-reactive F antigenic sites (Norrby and Appel, 1980; Appel et al., 1984). An infection-permissive immunity was established in all cases, but the most efficient immunity was seen when the homologous H protein was used.

Finally, immunization with isolated RS virus proteins has been performed in cotton rats (Walsh et al., 1987b). Again, both components could provide protection, but in this case the F protein was the dominating immunogen. The pulmonary resistance against replication of challenge virus was complete, but the nasal resistance was only incomplete. Sera from G protein-immunized animals had neutralizing activity, but sera from F-immunized animals had both neutralizing and cell-fusion-inhibiting activity.

Vaccinia virus vectors including individual heterologous genes have been effectively used to examine the immunogenicity of supposedly native paramyxovirus proteins. The parainfluenza type 3 HN and F proteins were expressed in cotton rats by use of recombinant vaccinia viruses (Spriggs et al.,

1987). Both proteins gave almost complete protection against respiratory infection by challenge virus. The neutralizing antibody titer induced in the upper respiratory tract was 500 times higher with the HN than with the F recombinant, but in serum the corresponding difference was only threefold. In further studies in monkeys, the two recombinants were combined (Spriggs et al., 1988). An infection-permissive immunity was established and the titer of virus in the upper and lower respiratory tract, was reduced about 1000 and 100 times, respectively, and the time of virus shedding was shortened. The corresponding vaccinia virus vectors were also prepared with SV5 virus HN and F genes (Paterson et al., 1987). The F construct somewhat unexpectedly induced higher titers of circulating neutralizing antibodies than the H construct. However, with regard to protection against challenge, the H-specific immunity was superior. The immunity established after the simultaneous use of HN and F recombinants was almost equal to postinfection immunity in capacity to protect against virus challenge.

Vaccinia virus recombinants encoding measles virus H and F genes were tested with regard to their capacity to prevent encephalitis in mice after intracerebral challenge (Drillien et al., 1988). Animals vaccinated with either recombinant resisted an inoculation of a cell-associated measles virus subacute sclerosing panencephalitis virus strain which was lethal in control animals. Followup during 270 days after challenge showed development of delayed encephalitis leading to death in some animals which had developed anti-H immunity, but this was not observed in animals with anti-F immunity. Vaccinia virus constructs have also been used to express glycoprotein genes of another morbillivirus, rinderpest virus. Vaccination of rabbits with a construct containing the complete F gene of rinderpest virus protected the animals against challenge with a lethal dose of virus (Barrett et al., 1989). In parallel experiments, vaccinia virus recombinants expressing the H or F gene of rinderpest virus provided protection in cattle. A dose 1000 times lethal was resisted by animals vaccinated with either recombinant or with the combined recombinants (Yilma et al., 1988).

Extensive studies with vaccinia virus recombinants containing the G and F genes of RS virus, representing both subgroups A and B, have been performed (Stott et al., 1986; Olmstedt et al., 1986, 1989; Wertz et al., 1987). Immunity against the F protein in cotton rats gave effective protection against lower respiratory tract infection with viruses representing both subgroups. There was also effective protection against upper respiratory tract infections, but this immunity predominantly blocked replication of virus of the homologous subgroup. Immunity against the G protein was much less effective than anti-F immunity and it only provided protection against challenge with homologous virus (see Chapter 14).

VI. LIVE VACCINES BASED ON VIRUSES WITH A DIFFERENT HOST SPECIFICITY: THE JENNERIAN APPROACH

Effective live vaccines against systematic infections with mumps, measles, and canine distemper viruses have been developed. The technique employed has been to adapt the virus to growth in nonnatural host cells. So far, no live vaccine effective against local, mucosal paramyxovirus infections has been developed by the use of this approach. Still, a live vaccine carries the potential of possibly providing an efficient local immunity as a result of restricted replication at the portal of entry of locally replicating viruses. An alternative approach to heterologous cell adaptation for live-virus vaccine development would be to exploit a naturally occurring, closely related virus of a different host species origin. Since in evolutionary terms paramyxoviruses were transferred from acutely infected animals into humans relatively recently, several examples of such closely related subtypes can be found in this family.

Human parainfluenza type 3 virus is second only to RS virus as a causative agent of serious respiratory infections in infants and children and has therefore been considered as an important target for vaccine development. The possible use of bovine parainfluenza 3 virus for immunization of humans against human parainfluenza 3 virus has been considered (Chanock et al., 1988). Deduced amino acid sequence data on the two viruses show a considerable conservation. In the case of the F and HN proteins, the degree of identity is 80 and 77%, respectively (Suzu et al., 1987). This degree of relatedness is reflected in the already mentioned (Coelingh et al., 1986) occurrence of three shared epitopes in an antigenic site A (recognized by neutralizing monoclonal antibodies) and two epitopes in antigenic sites D and F (recognized by nonneutralizing monoclonal antibodies).

Replication of bovine parainfluenza type 3 virus in the respiratory tracts of both cotton rats and squirrel monkeys (Coelingh et al., 1987) has induced resistance to replication of human parainfluenza 3 virus. The duration and magnitude of replication of prototype human and bovine parainfluenza 3 virus strains in the upper and lower respiratory tracts of squirrel monkeys were found to be similar (Coelingh et al., 1988). The animals did not develop symptoms. Protection experiments in these animals were performed by intratracheal infection with either virus strain and the animals were then challenged 4 weeks later with the human strain by the same route. Only one of six animals initially infected with the homologous human virus shed virus after challenge. In contrast, all six animals with a prior infection with bovine parainfluenza 3 virus excreted virus after challenge. However, there was a marked reduction in the duration of viral shedding and the amount of virus shed compared to that in challenged nonimmunized animals. Evidence was obtained for restriction of replication of the bovine strain in chimpanzees and rhesus monkeys. Compared with the human strain, there was a 100- to 1000-fold reduction in the tracheal content of virus. It is likely that this reflects a property of attenuation for humans, but no definite conclusion on this matter can be drawn, since no illness was seen in the two monkey species studied. The degree of attenuation of bovine parainfluenza 3 virus in humans and its capacity to restrict heterologous virus replication at different times after immunization deserves further study. Reversing the approach and determining if human parainfluenza 3 virus has any capacity to protect cattle against infection with the bovine virus may also be warranted.

The Jennerian approach also readily applies to heterospecies immunization between the three closely related morbilliviruses measles, canine distemper, and rinderpest viruses. As in the parainfluenzavirus type 3 situation, a close amino acid sequence relationship between the F proteins was found, 71–81% identity (Barrett et al., 1987; Buckland et al., 1987; Richardson et al., 1986; Tsukiyama et al., 1988). However, the attachment proteins are relatively divergent in morbilliviruses, at least as evidenced by comparison of deduced measles and rinderpest virus H-protein amino acid sequence data (no sequence data on canine distemper virus H have been published), yielding 58% identity (Alkhatib and Briedis, 1986; Gerald et al., 1986; Yamanaka et al., 1988). The potential heterologous cross-protection by rinderpest virus in humans or dogs has not been exploited because of the risk of spread of virus. Nor has exploitation of the possible protective effect of measles virus against rinderpest come into consideration (Plowright, 1962), probably because of a superior effect of homologous strains of attenuated virus. However, live measles virus is used for immunization of dogs against distemper. An analysis of the antibody response in the immunized animals has revealed that the protection is mediated by an antibody response to the F protein (Appel et al., 1984). The protection appears in many cases not to be complete but only to limit development of symptoms in the presence of virus replication. Finally, canine distemper virus was tried in early experiments for the immunization of humans against measles (Adams et al., 1959), but, again, homologous attenuated virus was soon found to be the preparation of choice.

The genus *Pneumovirus* also includes viruses which represent two closely related groups of different host species (human and bovine) origin. The apparent evolutionary difference between the G proteins is larger than between the F proteins. This conclusion is based on immunobiological characterization by use of monoclonal antibodies. In one comprehensive study (Örvell et al., 1987) employing previously characterized monoclonal antibodies against a strain of subgroup A virus (Mufson et al., 1985) and an additional large collection of monoclonal antibodies against a subgroup B virus, it was found that none of the antibodies against the G protein reacted with bovine RS virus, but that a considerable fraction of F-specific monoclonal antibodies reacted. In fact, all of the latter antibodies which showed neutralizing activity blocked human strains of both subgroups and bovine strains equally well. Thus, dominating neutralizing epitopes must be shared between RS viruses in the two species. In view of this, the replicative capacity of bovine RS virus in humans might be evaluated. There already have been attempts to use human RS virus for the immunization of cattle. A bovine strain of RS virus and a human strain were adapted to *in vitro* growth in bovine cell cultures and then inoculated into gnotobiotic calves (Thomas et al., 1984). The viruses replicated without causing symptoms, and specific antibody responses were recorded. In further studies (Stott et al., 1984), an inactivated bovine virus vaccine was compared with two candidate live-virus vaccines. These live vaccines were a modified bovine strain and a temperature-sensitive mutant of a human strain. Only the inactivated vaccine gave complete protection, whereas the live immunogens gave an infection-permissive immunity. However, the mean peak virus titers and the mean duration of shedding were reduced significantly in the candidate live-vaccine virus-immunized animals compared to the nonimmunized control animals.

VII. PROSPECTS FOR ADVANCES IN DIAGNOSTIC VIROLOGICAL PROCEDURES

The refinement of available immunological methodology offers new possibilities both for detection of viral antigens and for identification of specific antibodies in clinical materials. Concerning antigen detection, one may decide to use intertype cross-reacting, type-specific, or strain-specific reagents, depending upon the goal of the investigation. Currently, such a distinctive diagnosis may be performed by the use of MAbs, but it is possible that in the future antipeptide reagents could serve a similar purpose. Cross-reactions may complicate a serological diagnosis in particular in the case of parainfluenza virus infection. An example is a study of cerebrospinal fluid and serum samples from patients with multiple sclerosis, which was interpreted to suggest a particular role of SV5 virus in the disease (Goswami et al., 1987). In a followup study (Vandvik and Norrby 1989), it was confirmed that patients with multiple sclerosis had a preferential intrathecal production of antibodies reacting with SV5. However, by use of imprint immunofixation, it was documented that the population of IgG carrying SV5-specific activity was a part of a larger population of oligoclonal IgG specific for parainfluenza virus type 2 or mumps virus. Thus, the SV5 antibodies were cross-reacting and derived from infection with human viruses. Their presence in the cerebrospinal fluid was due to the general polyclonal activation of intrathecal IgG synthesis, which includes a range of virus-specific antibodies, seen in patients with multiple sclerosis (see also Chapter 11).

In order to direct a serological test to be specific for intertype cross-reacting, type, subgroup, or strain reactions, fragments of proteins need to be used as antigen. Such fragments may be prepared by the use of recombinant DNA technology. An alternative, attractive approach may be to use synthetic peptide antigens. So far, this technique has only been of limited use in studies of paramyxoviruses. Antipeptide antibodies were used for the identification of the measles virus C protein in cells (Richardson et al., 1985) and to study the orientation of RS virus 1A antigen in membranes (Olmsted and Collins, 1989). Except for this use of peptides as immunogen, the only use of polypeptides as antigens was in studies of RS virus G protein (Norrby et al., 1987) and 1A protein (Nicholas et al., 1988). Since the G protein is considered in Chapter 14, it suffices to note here that further studies (B. Åkerlind-Stopner et al., 1990) have highlighted the possibility of developing a subgroup-specific, site-directed serological assay. It could also be added that the use of overlapping peptides representing the F protein of RS virus in a search for immunodominant linear epitopes has given less encouraging results (E. Norrby, unpublished). Thus, the usefulness of technology employing peptides as antigen has to be evaluated individually with regard to virus system and to virus components.

VIII. CONCLUDING REMARKS

During the last decade, we have witnessed a rapid development of knowledge of the immunobiology of paramyxoviruses. This new knowledge derives from applications of the whole arsenal of modern technology, including gene technology, monoclonal antibodies, and synthetic peptides. Gradually, a picture of the evolutionary relationships between the large number of paramyxoviruses of medical and veterinary importance has begun to emerge. An important task remaining is to map some potential structural and nonstructural proteins and thereby determine their role in immune protection. Some live vaccines have proven highly protective against generalized paramyxovirus infections, but an effective means of preventing respiratory tract infections with these viruses has not yet been found. Locally replicating attenuated virus strains offer an attractive approach for immune interventions, but it is also possible that structural components properly selected and presented not to give an immunity predisposing to immunopathological complications, perhaps by addition of a suitable adjuvant, may provide an effective and durable immunity.

IX. REFERENCES

Abenes, G., Kida, H., and Yanagawa, R., 1986, Antigenic mapping and functional analysis of the F protein of Newcastle disease virus using monoclonal antibodies, *Arch. Virol.* **90**:97–110.

Adams, J. M., Imagawa, D. T., Wright, S. W., and Tarjan, G., 1959, Measles immunisation with live avian distemper virus, *Virology* **7**:351–353.

Åkerlind, B., Norrby, E., Örvell, C., and Mufson, M. A., 1988, Respiratory syncytial virus: Heterogeneity of subgroup B strains, *J. Gen. Virol.* **69**:2145–2154.

Åkerlind-Stopner, B., Utter, G., Mufson, M. A., Örvell, C., Lerner, R. A., and Norrby, E., 1990, A subgroup-specific antigenic site in the G protein of respiratory syncitial virus forms a disulphide-bonded loop, *J. Virol.*, in press.

Alkhatib, G., and Briedis, D. J., 1986, The predicted primary structure of the measles virus hemagglutinin, *Virology* **150**:479–490.

Anderson, L. J., Hierholzer, J. C., Tsou, C., Hendry, R. M., Fernie, B. F., Stone, Y., and McIntosh, K., 1985, Antigenic characterization of respiratory syncytial virus strains with monoclonal antibodies, *J. Infect. Dis.* **151**:626–633.

Appel, M. J. G., Shek, W. R., Sheshberadaran, H., and Norrby, E., 1984, Measles virus and inactivated canine distemper virus induce incomplete immunity to canine distemper, *Arch. Virol.* **82**:73–82.

Barrett, T., and Cavanagh, D., 1988, Pneumovirus-like characteristics of the mRNA and proteins of turkey rhinotracheitis virus, *Virus Res.* **11**:241–246.

Barrett, T., Clarke, D. K., Evans, S. A., and Rima, B. K., 1987, The nucleotide sequence of the gene encoding the F protein of canine distemper virus: A comparison of the deduced amino acid sequence with other paramyxoviruses, *Virus Res.* **8**:373–386.

Barrett, T., Belsham, G. J., Subbarao, S. M., and Evans, S. A., 1989, Immunization with a vaccinia recombinant expressing the F protein protects rabbits from challenge with a lethal dose of rinderpest virus, *Virology* **170**:11–18.

Buckland, R., Gerald, C., Barker, R., and Wild, T. F., 1987, Fusion glycoprotein of measles virus: Nucleotide sequence of the gene and comparison with other paramyxoviruses, *J. Gen. Virol.* **68**:1695–1703.

Buckland, R., Giraudon, P., and Wild, F., 1989, Expression of measles virus nucleocapsid in *Escherichia coli:* Use of deletion mutants to locate antigenic sites, *J. Gen. Virol.* **70**:435–441.

Cannon, M. J., and Bangham, C. R. M., 1989, Recognition of respiratory syncytial virus fusion protein by mouse cytotoxic T cell clones and a human cytotoxic T cell line, *J. Gen. Virol.* **70**:79–87.

Cannon, M. J., Stott, E. J., Taylor, G., and Asconas, B. A., 1987, Clearance of persistent respiratory syncytial virus infection in immunodeficient mice following transfer of primed T cells, *Immunology* **62**:133–138.

Cannon, M. J., Openshaw, P. J. M., and Asconas, B. A., 1988, Cytotoxic T cells clear virus but augment lung pathology in mice infected with respiratory syncytial virus, *J. Exp. Med.* **168:**1163–1168.

Cattaneo, R., Kaelin, K., Baczko, K., and Billeter, M. A., 1989, Measles virus editing provides an additional cysteine-rich protein, *Cell* **56:**759–764.

Chanock, R. M., Murphy, B. R., Collins, P. L., Coelingh, K. L. W., Olmsted, R. A., Snyder, M. H., Spriggs, M. K., Prince, G. A., Moss, B., Flores, J., Gorziglia, M., and Kapikian, A. Z., 1988, Live vaccines for respiratory and enteric tract diseases, *Vaccine* **6:**129–133.

Chin, J., Magoffin, R. L., Shearer, L. A., Schieble, J. H., and Lennette, E. H., 1969, Field evaluation of a respiratory syncytial virus vaccine and a bivalent parainfluenza virus vaccine in a pediatric population, *Am. J. Epidemiol.* **89:**449–463.

Coelingh, K. L., Winter, C. C., Murphy, B. R., Rice, J. M., Kimball, P. C., Olmstead, R. A., and Collins, P. L., 1986, Conserved epitopes on the hemagglutinin-neuraminidase protein of human and bovine parainfluenza type 3 viruses: Nucleotide sequence analysis of variants selected with monoclonal antibodies, *J. Virol.* **60:**90–96.

Coelingh, K. L. W., Winter, C. C., and Murphy, B. R., 1987, Antigenic relationships between hemagglutinin-neuraminidase glycoproteins of human and bovine type 3 parainfluenza viruses, in "Vaccines '87: Modern Approaches to New Vaccines" (R. M. Chanock, R. A. Lerner, F. Brown, and H. Ginsberg, eds.), pp. 290–301, Cold Spring Harbor Laboratory, Cold Spring Harbor, New York.

Coelingh, K. L. W., Battey, J., Lebacq-Verheyden, A. L., Collins, P. L., and Murphy, B. R., 1988, Development of live virus and subunit vaccines for parainfluenza type 3 virus, in "Vaccines '88: Modern Approaches to New Vaccines" (R. M. Chanock, R. A. Lerner, F. Brown, and H. Ginsberg, eds.), pp. 171–177, Cold Spring Harbor Laboratory, Cold Spring Harbor, New York.

Coelingh, K. L., Winter, C. C., and Murphy, B. R., 1989, Antigenic variation in the hemagglutinin-neuraminidase protein of human parainfluenza type 3 virus, *Virology* **143:**569–582.

De Vries, P., Van Binnendijk, R. S., Van Der Marel, P., Van Wezel, A. L., Voorma, H. O., Sundquist, B., Uytdehaag, F. G. C. M., and Osterhaus, A. D. M. E., 1988a, Measles virus fusion protein presented in an immune stimulating complex (Iscom) induces haemolysis-inhibiting and fusion-inhibiting antibodies, virus-specific T cells and protection in mice, *J. Gen. Virol.* **69:**549–559.

De Vries, P., Uytdehaag, F. G. C. M., and Osterhaus, A. D. M. E., 1988b, Canine distemper virus (CDV) immune-stimulating complexes (Iscoms) protect dogs against CDV infection, *J. Gen. Virol.* **69:**2071–2083.

Drillien, R., Spehner, D., Kirn, A., Giraudon, P., Buckland, R., Wild, F., and Lecocq, J. P., 1988, Protection of mice from fatal measles encephalitis by vaccination with vaccinia virus recombinants encoding either the hemagglutinin or the fusion protein, *Proc. Natl. Acad. Sci. USA* **85:**1252–1256.

Elango, N., Kövamees, J., Varsanyi, T. M., and Norrby, E., 1989, mRNA sequence and deduced amino acid sequence of the mumps virus small hydrophobic protein gene, *J. Virol.* **63:**1413–1415.

Fernie, B. F., Cote, Jr., P., and Gerin, J. L., 1982, Classification of hybridomas to respiratory syncytial virus glycoproteins, *Proc. Soc. Exp. Biol. Med.* **171:**266–271.

Gellis, S. S., McGuiness, A. C., and Peters, M., 1945, A study of the prevention of mumps orchitis by gamma globulin, *Am. J. Med. Sci.* **210:**661–664.

Gerald, C., Buckland, R., Barker, R., Freeman, G., and Wild, T. F., 1986, Measles virus hemagglutinin gene: Cloning, complete nucleotide sequence analysis and expression in COS cells, *J. Gen. Virol.* **67:**2695–2703.

Giminez, H. B., Cash, P., and Melvin, W. T., 1984, Monoclonal antibodies to human respiratory syncytial virus and their use in comparison of different virus isolates, *J. Gen. Virol.* **65:**963–971.

Giminez, H. B., Hardman, N., Keir, H. M., and Cash, P., 1986, Antigenic variation between human respiratory syncytial virus isolates, *J. Gen. Virol.* **67:**863–870.

Giminez, H. B., Keir, H. M., and Cash, P., 1989, In vitro enhancement of respiratory syncytial virus infection of U937 cells by human sera, *J. Gen. Virol.* **70:**89–96.

Giraudon, P., and Wild, T. F., 1981, Differentiation of measles virus strains and a strain of canine distemper by monoclonal antibodies, *J. Gen. Virol.* **57:**179–183.

Giraudon, P., and Wild, T. F., 1985, Correlation between epitopes on hemagglutinin of measles virus and biological activities: Passive protection by monoclonal antibodies is related to their hemagglutination-inhibiting activity, *Virology* **144:**46–58.

Goswami, K. K. A., Randall, R. E., Lange, L. S., and Russell, W. C., 1987, Antibodies against the paramyxovirus SV5 in the cerebrospinal fluid of some multiple sclerosis patients, *Nature* **327**:244–247.

Gotoh, B., Sakaguchi, T., Nishikawa, K., Inocencio, N. M., Hamaguchi, M., Toyoda, T., and Nagai, Y., 1988, Structural features unique to each of the three antigenic sites on the hemagglutinin-neuraminidase protein of Newcastle disease virus, *Virology* **163**:174–182.

Hamaguchi, M., Yoshida, T., Nishikawa, K., Naruse, H., and Nagai, Y., 1983, Transcriptive complex of Newcastle disease virus. I. Both L and P proteins are required to constitute an active complex, *Virology* **128**:105–117.

Hemming, V. G., Prince, G. A., Horswood, R. L., London, W. T., Murphy, B. R., Walsh, E. E., Fischer, G. W., Weisman, L. E., Baron, P. A., and Chanock, R. M., 1985, Studies of passive immunity for infections of respiratory syncytial virus in the respiratory tract of a primate model, *J. Infect. Dis.* **152**:1083–1086.

Henderson, F. W., Collier, A. M., Clyde, W. A., and Denny, F. W., 1979, Respiratory-syncytial-virus infections, reinfections and immunity. A prospective, longitudinal study in young children, *N. Engl. J. Med.* **300**:530–534.

Hiebert, S. W., Richardson, C., and Lamb, R. A., 1988, Cell surface expression and orientation in membranes of the 44-amino-acid SH protein of simian virus 5, *J. Virol.* **62**:2347–2357.

Huang, Y. T., Collins, P. L., and Wertz, G. W., 1985, Characterization of the 10 proteins of human respiratory syncytial virus, identification of a fourth envelope associated protein, *Virus Res.* **2**:157–173.

Ilonen, J., Salmi, A., Tuokko, H., Herva, E., and Penttinen, K., 1984, Immune responses to live attenuated and inactivated mumps virus vaccines in seronegative and seropositive young adult males, *J. Med. Virol.* **13**:331–338.

Iorio, R. M., and Bratt, M. A., 1983, Monoclonal antibodies to Newcastle disease virus: Delineation of four epitopes on the HN glycoprotein, *J. Virol.* **48**:440–450.

Iorio, R. M., and Bratt, M. A., 1984, Neutralization of Newcastle disease virus by monoclonal antibodies to the hemagglutinin-neuraminidase glycoprotein: Requirement for antibodies to four sites for complete neutralization, *J. Virol.* **51**:445–451.

Iorio, R. M., Lawton, K. A., Nicholson, P., and Bratt, M. A., 1984, Monoclonal antibodies identify a strain-specific epitope on the HN glycoprotein of Newcastle disease virus strain Australia-Victoria, *Virus Res.* **1**:513–525.

Iorio, R. M., Bergman, J. B., Glickman, R. L., and Bratt, M. A., 1986, Genetic variation within a neutralizing domain on the hemagglutinin-neuraminidase glycoprotein of Newcastle disease virus, *J. Gen. Virol.* **67**:1393–1403.

Ishida, M., Nerome, K., Matsumoto, M., Mikami, T., and Oya, A., 1985, Characterization of reference strain of Newcastle disease virus (NDV) and NDV-like isolates by monoclonal antibodies to HN subunits, *Arch. Virol.* **85**:109–121.

Ito, Y., Tsurudome, M., Hishiyama, M., and Yamada, A., 1987, Immunological interrelationships among human and non-human paramyxoviruses revealed by immunoprecipitation, *J. Gen. Virol.* **68**:1289–1297.

Jacobson, S., Sekaly, R. P., Jacobson, C. L., McFarland, H. F., and Long, E. O., 1989, HLA class II-restricted presentation of cytoplasmic measles virus antigens to cytotoxic T cells, *J. Virol.* **63**:1756–1762.

Johnson, Jr., P. R., Olmsted, R. A., Prince, G. A., Murphy, B. R., Alling, D. W., Walsh, E. E., and Collins, P. L., 1987, Antigenic relatedness between glycoproteins of human respiratory syncytial virus subgroups A and B: Evaluation of the contributions of F and G glycoproteins to immunity, *J. Virol.* **61**:3163–3166.

Kim, H. W., Canchola, J. G., Brandt, C. D., Pyles, G., Chanock, R. M., Jensen, K., and Parrott, R. H., 1969, Respiratory syncytial virus disease in infants despite prior administration of antigenic inactivated vaccine, *Am. J. Epidemiol.* **89**:422–434.

Komada, H., Tsurudome, M., Ueda, M., Nishio, M., Bando, H., and Ito, Y., 1989a, Isolation and characterization of monoclonal antibodies to human parainfluenza virus type 4 and their use in revealing antigenic relation between subtypes 4A and 4B, *Virology* **171**:28–37.

Komada, H., Tsurudome, M., Bando, H., Nishio, M., Yamada, A., Hishiyama, M., and Ito, N., 1989b, Virus-specific polypeptides of human parainfluenza virus type 4 and their synthesis in infected cells, *Virology* **171**:254–259.

Kövamees, J., Rydbeck, R., Örvell, C., and Norrby, E., 1990, Hemagglutinin-neuraminidase (HN) amino acid alterations in neutralization escape mutants of Kilham mumps virus, *Virus Res.* in press.

Kreth, H. W., ter Meulen, V., and Eckert, G., 1979, Demonstration of HLA restricted killer cells in patients with acute measles, *Med. Microbiol. Immunol.* **165**:203–214.

Ling, R., and Pringle, C. R., 1989a, Polypeptides of pneumonia virus of mice. I Immunological cross-reactions and post-translation modifications, *J. Gen. Virol.* **70**:1427–1440.

Ling, R., and Pringle, C. R., 1989b, Polypeptides of pneumonia virus of mice. II Characterization of the glycoproteins, *J. Gen. Virol.* **70**:1441–1452.

Long, L., Portetelle, D., Ghysdael, J., Gonze, M., Burny, A., and Meulemans, G., 1986, Monoclonal antibodies to hemagglutinin-neuraminidase and fusion glycoproteins of Newcastle disease virus: Relationship between glycosylation and reactivity, *J. Virol.* **57**:1198–1202.

Löve, A., Rydbeck, R., Kristensson, K., Örvell, C., and Norrby, E., 1985, Hemagglutinin-neuraminidase glycoprotein as a determinant of pathogenicity in mumps virus hamster encephalitis: Analysis of mutants selected with monoclonal antibodies, *J. Virol.* **53**:67–74.

Löve, A., Rydbeck, R., Utter, G., Örvell, C., Kristensson, K., and Norrby, E., 1986, Monoclonal antibodies against the fusion protein are protective in mumps meningoencephalitis, *J. Virol.* **58**:220–222.

Lucas, J. C., Biddison, E. W., Nelson, L. D., and Shaw, S., 1982, Killing of measles virus infected cells by human cytotoxic T cells, *Infect. Immun.* **38**:226–232.

Mäkelä, M. J., Lund, G. A., and Salmi, A. A., 1989, Antigenicity of the measles virus hemagglutinin studied by using synthetic peptides, *J. Gen. Virol.* **70**:603–614.

McCarthy, M., and Johnson, R. T., 1980, A comparison of the structural polypeptides of five strains of measles virus, *J. Gen. Virol.* **46**:15–27.

McFarland, H. F., Pedone, C. A., Mingioli, E. S., and McFarlin, D. E., 1980, The response of human lymphocyte subpopulation to measles, mumps and vaccinia viral antigens, *J. Immunol.* **125**:221–225.

McIntosh, K., and Fishaut, J. M., 1980, Immunopathologic mechanisms in lower respiratory tract disease of infants due to respiratory syncytial virus, *Prog. Med. Virol.* **26**:94–118.

Merz, D. C., and Wolinsky, J. S., 1983, Conversion of nonfusing mumps virus infections to fusing infections by selective proteolysis of the HN glycoprotein, *Virology* **131**:328–340.

Meulemans, G., Genze, M., Cartier, M. C., Petit, P., Burny, A., and Long, L., 1987, Evaluation of the use of monoclonal antibodies to hemagglutinin and fusion glycoproteins of Newcastle disease virus for *in vitro* identification and strain differentiation purposes, *Arch. Virol.* **92**:55–62.

Miura, N., Uchida, T., and Okada, Y., 1982, HVJ (Sendai virus)-induced envelope fusion and cell fusion are blocked by monoclonal anti-HN protein antibody that does not inhibit hemagglutination activity of HVJ, *Exp. Cell. Res.* **141**:409–420.

Mottet, G., Portner, A., and Roux, L., 1986, Drastic immunoreactivity changes between the immature and mature forms of the Sendai virus HN and F_0 glycoproteins, *J. Virol.* **59**:132–141.

Mufson, M. A., Örvell, C., Rafnar, B., and Norrby, E., 1985, Two distinct subtypes of human respiratory syncytial virus, *J. Gen. Virol.* **66**:2111–2124.

Mufson, M. A., Belshe, R. B., Örvell, C., and Norrby, E., 1987, Subgroup characteristics of respiratory syncytial virus strains recovered from children with two consecutive infections, *J. Clin. Microbiol.* **25**:1535–1539.

Murphy, B. R., Prince, G. A., Walsh, E. E., Kim, H. W., Parrott, R. H., Hemming, V. G., Rodriguez, W. J., and Chanock, R. M., 1986, Dissociation between serum neutralizing and glycoprotein antibody responses of infants and children who received inactivated respiratory syncytial virus vaccine, *J. Clin. Microbiol.* **24**:197–202.

Nagai, Y., Klenk, H. D., and Rott, R., 1976, Proteolytic cleavage of the viral glycoproteins and its significance for the virulence of Newcastle disease virus, *Virology* **72**:494–508.

Neyt, C., Geliebter, J., Slaoui, M., Morales, D., Meulemans, G., and Burney, A., 1989, Mutations located on both F_1 and F_2 subunits of Newcastle disease virus fusion protein confer resistance to neutralization with monoclonal antibodies, *J. Virol.* **63**:952–954.

Nicholas, J. A., Mitchell, M. A., Levely, M. E., Rubino, K. L., Kinner, J. H., Ham, N. K., and Smith, C. W., 1988, Mapping an antibody-binding site and a T-cell-stimulating site on the 1A protein of respiratory syncytial virus, *J. Virol.* **62**:4465–4473.

Nishikawa, K., Isomura, S., Suzuki, S., Watanabe, E., Hamaguchi, M., Yoshida, T., and Nagai, Y.,

1983, Monoclonal antibodies to the HN glycoprotein of Newcastle disease virus. Biological characterization and use for strain comparison, *Virology* **130**:318–330.

Nishikawa, K., Morishima, T., Toyoda, T., Miyadai, T., Yokochi, T., Yoshida, T., and Nagai, Y., 1986, Topological and operational delineation of antigenic sites on the HN glycoprotein of Newcastle disease virus and their structural requirement, *J. Virol.* **60**:987–993.

Nishikawa, K., Hanada, N., Morishima, T., Yoshida, T., Hamaguchi, M., Toyoda, T., and Nagai, Y., 1987, Antigenic characterization of the internal proteins of Newcastle disease virus by monoclonal antibodies, *Virus Res.* **7**:83–92.

Norrby, E., and Appel, M. J., 1980, Immunity to canine distemper after immunization of dogs with inactivated and live measles virus, *Arch. Virol.* **66**:169–177.

Norrby, E., and Penttinen, K., 1978, Differences in antibodies to surface components of mumps after immunization with formalin-inactivated and live virus vaccines, *J. Infect. Dis.* **139**:672–676.

Norrby, E., Enders-Ruckle, G., and ter Meulen, V., 1975, Differences in the appearance of antibodies to structural components of measles virus after immunization with inactivated and live virus, *J. Infect. Dis.* **132**:262–269.

Norrby, E., Sheshberadaran, H., McCullough, K. C., Carpenter, W. C., and Örvell, C., 1985, Is rinderpest virus the archevirus of the morbillivirus genus?, *Intervirology* **23**:228–232.

Norrby, E., Sheshberadaran, H., and Rafnar, B., 1986a, Antigen mimicry involving measles virus hemagglutinin and human respiratory syncytial virus nucleoprotein, *J. Virol.* **57**:394–396.

Norrby, E., Mufson, M. A., and Sheshberadaran, H., 1986b, Structural differences between subtype A and B strains of respiratory syncytial virus *J. Gen. Virol.* **67**:2721–2729.

Norrby, E., Utter, G., Örvell, C., and Appel, M. J. G., 1986c, Protection against canine distemper virus in dogs after immunization with isolated fusion protein, *J. Virol.* **58**:536–541.

Norrby, E., Mufson, M. A., Alexander, H., Houghten, R. A., and Lerner, R. A., 1987, Site-directed serology with synthetic peptides representing the large glycoprotein G of respiratory syncytial virus, *Proc. Natl. Acad. Sci. USA* **84**:6572–6576.

Olmsted, R. A., and Collins, P. L., 1989, The 1A protein of respiratory syncytial virus is an integral protein present as multiple, structurally distinct species, *J. Virol.* **63**:2019–2029.

Olmsted, R. A., Elango, N., Prince, G. A., Murphy, B. R., Johnson, P. H., Moss, B., Chanock, R. M., and Collins, P. L., 1986, Expression of the F glycoprotein of respiratory syncytial virus by a recombinant vaccinia virus: Comparison of the individual contributions of the F and G glycoproteins to host immunity, *Proc. Natl. Acad. Sci. USA* **83**:7462–7466.

Olmsted, R. A., Murphy, B. R., Lawrence, L. A., Elango, N., Moss, B., and Collins, P. L., 1989, Processing, surface expression, and immunogenicity of carboxy-terminally truncated mutants of G protein of human respiratory syncytial virus, *J. Virol.* **63**:411–420.

Örvell, C., 1980, Structural polypeptides of canine distemper virus, *Arch. Virol.* **66**:193–206.

Örvell, C., 1984, The reactions of monoclonal antibodies with structural proteins of mumps virus, *J. Immunol.* **132**:2622–2629.

Örvell, C., and Grandien, M., 1982, The effects of monoclonal antibodies on biologic activities of structural proteins of Sendai virus, *J. Immunol.* **129**:2779–2787.

Örvell, C., and Norrby, E., 1985, Antigenic structure of paramyxoviruses, in "Immunochemistry of Viruses—The Basis for Serodiagnosis and Vaccines" (A. R. Neurath and M. H. V. van Regenmortel, eds.), pp. 241–264, Elsevier, Amsterdam.

Örvell, C., Rydbeck, R., and Löve, A., 1986, Immunological relationships between mumps virus and parainfluenza viruses studied with monoclonal antibodies, *J. Gen. Virol.* **67**:1929–1939.

Örvell, C., Norrby, E., and Mufson, M. A., 1987, Preparation and characterization of monoclonal antibodies directed against five structural components of human respiratory syncytial virus subgroup B, *J. Gen. Virol.* **68**:3125–3135.

Paterson, R. G., Lamb, R. A., Moss, B., and Murphy, B. R., 1987, Comparison of the relative roles of the F and HN surface glycoproteins of the paramyxovirus simian virus 5 in inducing protective immunity, *J. Virol.* **61**:1972–1977.

Pemberton, R. M., Cannon, M. J., Openshaw, P. J. M., Ball, L. A., Wertz, G. W., and Askonas, B. A., 1987, Cytotoxic T cell specificity for respiratory syncytial virus proteins: Fusion protein is an important target antigen, *J. Gen. Virol.* **68**:2177–2182.

Plowright, W., 1962, Rinderpest virus, *Ann. N. Y. Acad. Sci.* **101**:548–563.

Portner, A., 1981, The HN glycoprotein of Sendai virus: Analysis of site(s) involved in hemag-glutinating and neuraminidase activities, *Virology* **115**:375–384.

Portner, A., Webster, R. G., and Bean, W. J., 1980, Similar frequencies of antigenic variants in Sendai, vesicular stomatitis and influenza A viruses, *Virology* **104**:235–238.

Portner, A., Scroggs, R. A., and Metzger, P. W., 1987a, Distinct functions of antigenic sites of the HN glycoprotein of Sendai virus, *Virology* **158**:61–68.

Portner, A., Scroggs, R. A., and Naeve, C. W., 1987b, The fusion protein of Sendai virus: Sequence analysis of an epitope involved in fusion and virus neutralization, *Virology* **157**:556–559.

Prince, G. A., Horswood, R. L., and Chanock, R. M., 1985, Quantitative aspects of passive immu-nity to respiratory syncytial virus infection in infant cotton rats, *J. Virol.* **55**:517–520.

Prince, G. A., Jenson, A. B., Hemming, V. G., Murphy, B. R., Walsh, E. E., Horswood, R. L., and Chanock, R. M., 1986, Enhancement of respiratory syncytial virus pulmonary pathology in cotton rats by prior intramuscular inoculation of formalin inactivated virus, *J. Virol.* **57**:721–728.

Prince, G. A., Hemming, V. G., Horswood, R. L., Baron, P. A., and Chanock, R. M., 1987, Effective-ness of topically administered neutralizing antibodies in experimental immunotherapy of respiratory syncytial virus infection in cotton rates, *J. Virol.* **61**:1851–1854.

Rammohan, K. W., McFarland, H. F., Bellini, W. J., Gheuens, J., and McFarlin, D. E., 1983, Antibody mediated modification of encephalitis induced by hamster neurotropic measles virus, *J. Infect. Dis.* **147**:546–550.

Randall, R. E., and Young, D. F., 1988, Comparison between parainfluenza virus type 2 and simian virus 5: Monoclonal antibodies reveal major antigenic differences, *J. Gen. Virol.* **69**:2051–2060.

Ray, R., Glaze, B. J., and Compans, R. W., 1988, Role of individual glycoproteins of human para-influenza virus type 3 in the induction of a protective immune response, *J. Virol.* **62**:783–787.

Ray, R., Galinski, M. S., and Compans, R. W., 1989, Expression of the fusion glycoprotein of human parainfluenza type 3 virus in insect cells by a recombinant baculovirus and analysis of its immunogenic property, *Virus Res.* **12**:169–180.

Richardson, C. D., Berkovich, A., Rozenblatt, S., and Bellini, W. J., 1985, Use of antibodies directed against synthetic peptides for identifying cDNA clones, establishing reading frames and de-ducing the gene order of measles virus, *J. Virol.* **54**:186–193.

Richardson, C., Hull, D., Greer, P., Hasel, K., Berkovich, A., Englund, G., Bellini, W., Rima, B., and Lazzarini, R., 1986, The nucleotide sequence of the mRNA encoding the fusion protein of measles virus (Edmonston strain): A comparison of fusion protein from several different para-myxoviruses, *Virology* **155**:508–523.

Richert, J. R., McFarland, H. F., McFarlin, D. E., Johnson, A. J., Woody, J. N., and Harzman, R. J., 1985, Measles-specific T cell clones derived from a twin with multiple sclerosis; genetic restriction studies, *J. Immunol.* **134**:1561–1566.

Rima, B. K., 1983, The proteins of morbilliviruses, *J. Gen. Virol.* **64**:1205–1219.

Rose, J. W., Bellini, W. J., McFarlin, D. B., and McFarland, H., 1984, Human cellular immune response to measles virus polypeptides, *J. Virol.* **49**:988–991.

Russell, P. H., 1986, The synergistic neutralization of Newcastle disease virus by two monoclonal antibodies to its hemagglutinin-neuraminidase protein, *Arch. Virol.* **90**:135–144.

Russell, P. H., and Alexander, D. J., 1983, Antigenic variation of Newcastle disease virus strain, detected by monoclonal antibodies, *Arch. Virol.* **75**:243–253.

Russell, P. H., Griffiths, P. C., Goswami, K. K. A., Alexander, D. J., Cannon, M. J., and Russell, W. C., 1983, The characterization of monoclonal antibodies to Newcastle disease virus *J. Gen. Virol.* **64**:2069–2072.

Rydbeck, R., Löve, A., Örvell, C., and Norrby, E., 1986, Antigenic variation of envelope and internal proteins of measles virus strains detected with monoclonal antibodies, *J. Gen. Virol.* **67**:281–287.

Rydbeck, R., Löve, A., Örvell, C., and Norrby, E., 1987, Antigenic analysis of human and bovine parainfluenza virus type 3 strains with monoclonal antibodies, *J. Gen. Virol.* **68**:2153–2160.

Rydbeck, R., Löve, A., and Norrby, E., 1988, Protective effects of monoclonal antibodies against parainfluenza virus type 3-induced brain infection in hamsters, *J. Gen. Virol.* **69**:1019–1024.

Sakaguchi, T., Toyoda, T., Gotoh, B., Inocencia, N. M., Kuma, K., Miyata, T., and Nagai, Y., 1989,

Newcastle disease virus evolution. I. Multiple lineages defined by sequence variability of the hemagglutinin-neuraminidase gene, *Virology* **169**:260–272.

Samson, A. C. R., Russell, P. H., and Hallum, S. E., 1985, Isolation and characterization of monoclonal antibody-resistent mutants of Newcastle disease virus, *J. Gen. Virol.* **66**:357–361.

Sato, T. A., Fukuda, A., and Sugiura, A., 1985, Characterization of major structural proteins of measles virus with monoclonal antibodies, *J. Gen. Virol.* **66**:1397–1409.

Server, A. C., Merz, D. C., Waxham, M. N., and Wolinsky, J. S., 1982, Differentiation of mumps virus strains with monoclonal antibody to the HN glycoprotein, *Infect. Immun.* **35**:179–186.

Sheshberadaran, H., and Norrby, E., 1986, Characterization of epitopes on the measles virus hemagglutinin, *Virology* **152**:58–69.

Sheshberadaran, H., and Payne, L. G., 1988, Protein antigen–monoclonal antibody contact sites investigated by limited proteolysis of monoclonal antibody-bound antigen: Protein "footprinting," *Proc. Natl. Acad. Sci. USA* **85**:1–5.

Sheshberadaran, H., Chen, S.-N., and Norrby, E., 1983, Monoclonal antibodies against five structural components of measles virus. I. Characterization of antigenic determinants on nine strains of measles virus, *Virology* **128**:341–358.

Sheshberadaran H., Norrby, E., McCullough, K. C., Carpenter, W. C., and Örvell, C., 1986, The antigenic relationship between measles, canine distemper and rinderpest viruses studied with monoclonal antibodies, *J. Gen. Virol.* **67**:1381–1392.

Spriggs, M. K., Murphy, B. R., Prince, G. A., Olmsted, R. A., and Collins, P., 1987, Expression of the F and HN glycoproteins of human parainfluenza-virus type 3 by recombinant vaccinia viruses: Contribution of the individual proteins to host immunity, *J. Virol.* **61**:3416–3423.

Spriggs, M. K., Collins, P. L., Tiesney, E., Landa, W. T., and Murphy, B. R., 1988, Immunization with vaccinia virus recombinants that express the surface glycoproteins of human parainfluenza virus type 3 (PIV3) protects monkeys against PIV3 infection, *J. Virol.* **62**:1293–1296.

Stott, E. J., Thomas, L. H., Taylor, G., Collins, A. P., Jebbett, J., and Crouch, S., 1984, A comparison of three vaccines against respiratory syncytial virus in calves, *J. Hyg.* (Camb.) **93**:251–261.

Stott, E. J., Ball, A., Young, K. K., Furze, J., and Wertz, G. W., 1986, Human respiratory syncytial virus glycoprotein G expressed from a recombinant vaccinia virus vector protects mice against live-virus challenge, *J. Virol.* **60**:607–613.

Suzu, S., Sakai, Y., Shioda, T., and Shibuta, H., 1987, Nucleotide sequence of the bovine parainfluenza 3 virus genome: The genes of the F and HN glycoproteins, *Nucleic Acids Res.* **15**:2945–2958.

Taylor, G., Stott, E. J., Bew, M., Fernie, B. F., Cote, P. J., Collins, A. P., Hughes, M., and Jebbett, J., 1984, Monoclonal antibodies protect against respiratory syncytial virus infection in mice, *Immunology* **52**:137–142.

Thomas, L. H., Stott, E. J., Collins, A. P., Crouch, S., and Jebbett, J., 1984, Infection of gnotobiotic calves with a bovine and human isolate of respiratory syncytial virus. Modification of the response by dexamethasone, *Arch. Virol.* **79**:67–77.

Thomas, S. M., Lamb, R. A., and Paterson, R. G., 1988, Two mRNAs that differ by two nontemplated nucleotides encode the amino coterminal proteins P and V of the paramyxovirus SV5, *Cell* **54**:891–902.

Thomson, S. D., and Portner, A., 1987, Localization of functional sites on the hemagglutinin-neuraminidase glycoprotein of Sendai virus by sequence analysis of antigenic and temperature-sensitive mutants, *Virology* **160**:1–8.

Togashi, T., Örvell, C., Vartdal, F., and Norrby, E., 1981, Production of antibodies against measles virions by use of the mouse hybridoma technique, *Arch. Virol.* **67**:149–157.

Toyoda, T., Sakaguchi, T., Imai, K., Inocencio, N. M., Gotoh, B., Hamaguchi, M., and Nagai, Y., 1987, Structural comparison of the cleavage-activation site of the fusion glycoprotein between virulent and avirulent strains of Newcastle disease virus, *Virology* **158**:242–247.

Toyoda, T. B., Gotoh, T., Sakaguchi, T., Kida, H., and Nagai, Y., 1988, Identification of amino acids relevant to three antigenic determinants on the fusion protein of Newcastle disease virus that are involved in fusion inhibition and neutralization, *J. Virol.* **62**:4427–4430.

Toyoda, T., Sakaguchi, T., Hiroda, H., Gotoh, B., Kuma, K., Miyata, T., and Nagai, Y., 1989, Newcastle disease virus evolution. II Lack of gene recombination in generating virulent and avirulent strains, *Virology* **169**:273–282.

Tozawa, H., Komatsu, H., Ohkata, K., Nakajima, T., Watanabe, M., Tanaka, Y., and Arifuku, M., 1986, Neutralizing activity of the antibodies against two kinds of envelope glycoproteins of Sendai virus, *Arch. Virol.* **91:**145–161.

Tsukiyama, K., Yoshikawa, Y., and Yamanouchi, K., 1988, Fusion glycoprotein (F) of rinderpest virus: Entire nucleotide sequence of the F mRNA and several features of the F protein, *Virology* **164:**523–530.

Tsurudome, M., Yamada, A., Hishiyama, M., and Ito, Y., 1986, Monoclonal antibodies against the glycoproteins of mumps virus: Fusion inhibition by anti-HN monoclonal antibody, *J. Gen. Virol.* **67:**2259–2265.

Umino, Y., Kohama, T., Kohase, M., Sugiura, A., Klenk, H.-D., and Rott, R., 1987, Protective effect of antibodies to two viral envelope glycoproteins on lethal infection with Newcastle disease virus, *Arch. Virol.* **94:**97–107.

Vandvik, B., and Norrby, E., 1989, Paramyxovirus SV5 and multiple sclerosis, *Nature* **338:**769–770.

Van Wyke Coelingh, K., Murphy, B. R., Collins, P. L., Lebacq-Verheyden, A. M., and Battey, J. F., 1987, Expression of biologically active and antigenically authentic parainfluenza type 3 virus hemagglutinin-neuraminidase glycoprotein by a recombinant baculovirus, *Virology* **160:**465–472.

Van Wyke Coelingh, K., Winter, C. C., and Murphy, B. R., 1988, Nucleotide and deduced amino acid sequence of hemagglutinin-neuraminidase genes of human parainfluenza viruses isolated from 1957–1983, *Virology* **162:**137–143.

Van Wyke Coelingh, K., and Tierney, E. L., 1989a, Antigenic and functional organization of human parainfluenza virus type 3 fusion glycoprotein, *J. Virol.* **63:**375–382.

Van Wyke Coelingh, K., and Tierney, E. L., 1989b, Identification of amino acids recognized by syncytium-inhibiting and neutralizing monoclonal antibodies to the human parainfluenza type 3 virus protein, *J. Virol.* **63:**3755–3760.

Varsanyi, T. M., Utter, G., and Norrby, E., 1984, Purification, morphology and antigenic characterization of measles virus envelope components, *J. Gen. Virol.* **65:**355–366.

Varsanyi, T. M., Jörnvall, H., and Norrby, E., 1985, Isolation and characterization of the measles virus F1 polypeptide: Comparison with other paramyxovirus fusion proteins, *Virology* **147:**110–117.

Varsanyi, T. M., Morein, B., Löve, A., and Norrby, E., 1987, Protection against lethal measles virus infection in mice by immune stimulating complexes containing the hemagglutinin or fusion protein, *J. Virol.* **61:**3896–3901.

Vidal, S., Mottet, G., Kolakofsky, D., and Roux, L., 1989, Addition of high-mannose sugars must precede disulfide bond formation for proper folding of Sendai virus glycoproteins, *J. Virol.* **63:**892–900.

Walsh, E. E., and Hruska, J., 1983, Monoclonal antibodies to respiratory syncytial virus proteins: Identification of the fusion protein, *J. Virol.* **47:**171–177.

Walsh, E. E., Schlesinger, J. J., and Brandriss, M. W., 1984, Protection from respiratory syncytial virus infection in cotton rats by passive transfer of monoclonal antibodies, *Infect. Immun.* **43:**756–758.

Walsh, E. E., Brandriss, M. W., and Schlesinger, J. J., 1985, Purification and characterization of the respiratory syncytial virus fusion protein, *J. Gen. Virol.* **66:**409–415.

Walsh, E. E., Cote, P. J., Fernie, B. F., Schlesinger, J. J., and Brandriss, M. W., 1986, Analysis of the respiratory syncytial virus fusion protein using monoclonal and polyclonal antibodies, *J. Gen. Virol.* **67:**505–513.

Walsh, E. E., Brandriss, M. W., and Schlesinger, J. J., 1987a, Immunologic differences between the envelope glycoproteins of two strains of human respiratory syncytial virus, *J. Gen. Virol.* **68:**2169–2176.

Walsh, E. E., Hall, C. B., Brandriss, M. W., and Schlesinger, J. J., 1987b, Immunization with glycoprotein subunits of respiratory syncytial virus to protect cotton rats against infection, *J. Infect. Dis.* **155:**1198–1204.

Ward, K. A., Lambden, P. R., Ogilvie, M. M., and Watt, P. J., 1983, Antibodies to respiratory syncytial virus polypeptides and their significance in human infection, *J. Gen. Virol.* **64:**1867–1876.

Waxham, M. N., and Aronowski, J., 1988, Identification of amino acids involved in the sialidase activity of the mumps virus hemagglutinin-neuraminidase protein, *Virology* **167:**226–232.

Waxham, M. N., Merz, D. C., and Wolinsky, J. S., 1986, Intracellular maturation of mumps virus hemagglutinin-neuraminidase glycoprotein: Conformational changes detected with monoclonal antibodies, *J. Virol.* **59**:392–400.

Wertz, G. W., Stott, E. J., Young, K. K. Y., Anderson, K., and Ball, L. A., 1987, Expression of the fusion protein of human respiratory syncytial virus from recombinant vaccinia virus vectors and protection of vaccinated mice, *J. Virol.* **61**:293–301.

Wolinsky, J. S., Waxham, M. N., and Server, A. C., 1985, Protective effects of glycoprotein-specific monoclonal antibodies on the cause of experimental mumps virus meningoencephalitis, *J. Virol.* **53**:727–734.

Wright, L. L., and Levy, N. L., 1979, Generation of infected fibroblasts of human T and non-T lymphocytes with specific cytotoxicity influenced by histocompatibility, against measles virus-infected cells, *J. Immunol.* **122**:2379–2387.

Yamanaka, M., Hsu, P., Crisp, T., Dale, B., Grubman, M., and Yilma, T., 1988, Cloning and sequence analysis of the hemagglutinin gene of the virulent strain of rinderpest virus, *Virology* **166**:251–253.

Yewdell, J., and Gerhard, W., 1982, Delineation of four antigenic sites on a paramyxovirus glycoprotein via which monoclonal antibodies mediate distinct antiviral activities, *J. Immunol.* **128**:2670–2675.

Yilma, T., Hsu, D., Jones, L., Owens, S., Grubman, M., Mebus, C., Yamanaka, M., and Dale, B., 1988, Protection of cattle against rinderpest with vaccinia virus recombinants expressing the HA or F gene, *Science* **242**:1058–1061.

CHAPTER 19

Epidemiology of *Paramyxoviridae*

Francis L. Black

I. COMMON CHARACTERISTICS AND CONTRASTS

All *Paramyxoviridae* are extremely infectious, and there is little chance for anybody living in a cosmopolitan community to get through a full life without being infected by wild or attenuated forms of all the paramyxoviruses adapted to humans. Their common structure is labile and all viruses of this family are dependent on transmission by close association of hosts. They do not survive drying on a solid surface, nor are they sufficiently stable in water to be transmitted by this vehicle with regularity. They are inactivated by stomach acid and intestinal enzymes, and do not infect via the gut. Very efficient infection occurs through aerosolized virus, and direct physical contact between hosts is not required. Drying in an aerosol does not inactivate them and it is probably the surface tension forces involved in drying on a surface, not drying per se, that inactivates. Urine may be an important source of the aerosols (Gresser and Katz, 1960). These viruses infect through mucous membranes (Papp, 1956; Black and Sheridan, 1960), but measles virus, at least, infects most efficiently when introduced in droplets small enough to reach into the lungs (McCrumb *et al.*, 1962).

For all these similarities, members of the paramyxovirus family have immunological differences of great epidemiological consequence (see also Chapter 18). These probably derive from the different cell receptors used by different viruses and from the fact that some can replicate in cells that reside in membrane surfaces, whereas others are subject to the immunological defenses of deeper tissues. The diseases they cause follow diverse epidemiological patterns because the immune responses they induce differ in degree of specificity and in durability. Immunity to members of the *Morbillivirus* and to mumps virus is

FRANCIS L. BLACK • Department of Epidemiology and Public Health, Yale University School of Medicine, New Haven, Connecticut 06510.

sufficiently strong and durable that individuals who have once been infected play no further role in spreading these diseases, but rather dampen the rate of virus spread in a population. On the other hand, repeated infections play a major role in dissemination of the parainfluenza viruses and of respiratory syncytial virus (RSV). The *Morbillivirus* and *Pneumovirus* genera both include several antigenically related species that are separated by distinct host ranges. The spread of measles virus is independent of the distribution of all other viruses, but in the *Parainfluenza* and *Pneumovirus* genera there are immunologically distinct strains infecting the same species with ill-defined cross-effects. Specific host-range analogs of mumps virus are not known, but paramyxoviruses that infect hosts other than humans are common. These contrasts determine the structure of the discussion that follows.

The epidemiology of some of these diseases has literally been studied for centuries, and a voluminous literature has accumulated. In this review, no attempt has been made to be comprehensive in listing this literature, and little preference is given to reports with chronological priority. Rather the references have been selected for their ability to be understood individually and on the basis of their usefulness as leads to earlier papers.

II. *MORBILLIVIRUS*

A. Characteristics of the Genus As a Whole

The three main members of *Morbillivirus*, measles virus, canine distemper virus (CDV), and rinderpest virus, affect different species, and therefore the distribution of one does not influence the distribution of the others. Peste des petits ruminants virus (PPRV) the fourth recognized *Morbillivirus*, is closely related to rinderpest virus but its degree of epidemiological interaction with that agent has not been well defined. Strong serological reactions among all members of the group can be demonstrated in laboratory systems (Sheshberadaran *et al.*, 1986), and partial cross-immunity to CDV and to rinderpest virus can be elicited by injection of measles virus into dogs or cattle (Plowright, 1968; Appel *et al.*, 1984) or CDV into humans (Adams *et al.*, 1959). However, natural infection does not occur across these species lines and the cross-reactions have no epidemiological significance.

A very useful epidemic parameter R_0, the transmission potential, is defined as the number of secondary cases to be expected from one index case introduced into a totally inexperienced population. Disease endemicity in an infinitely large population will continue as long as $R_0 > 1$. To eliminate a disease, it is only necessary to hold R_0 below 1. All the members of *Morbillivirus* have extremely high R_0 values, that is, they are all very efficient at infecting susceptible hosts and one index case can lead to very many secondary cases. Because of this, epidemics in populations that have not had prior experience are explosive. CDV and rinderpest virus often, and measles virus occasionally, kill most of the individuals of an infected population. Their epidemics have devastated wild species and caused human famine by the destruction of livestock. All morbilliviruses, however, induce solid, lasting immunity in surviving hosts after rather short acute infections, and the outbreaks terminate as abruptly as they begin. Unless the initial population is very large, or there is a reservoir for the virus in some species or organ, the virus will then die out, leaving subsequent generations fully susceptible and ripe for another major epidemic. Only for CDV has a reservoir been identified.

CDV has a natural host range that is limited to the order Carnivora, but includes most families of this order (Appel and Gillespie, 1972; Montali *et al.*, 1983). The recent CDV epidemic that decimated the seals of the North Sea and Baltic (Dickson, 1988; Osterhaus and Vedder, 1988) is typical of the devastating effects of all morbilliviruses when they enter populations where virus spread is not dampened by a substantial proportion of previously immunized individuals. Except for the Pinnipediae, members of Carnivora characteristically live in small population groups, and a virus that persists only a short time in any one host would have difficulty maintaining itself in them. So many species are subject to CDV infection that one cannot say how many might experience persistent infection or be subject to repeated infection and thus serve as reservoirs for the virus. The dog is the only identified reservoir in that it is subject to a relatively rare persistent neurological infection, old dog distemper (Tobler and Imagawa, 1984). In many ways, old dog distemper is reminiscent of subacute sclerosing panencephalitis (SSPE) in humans, but, unlike SSPE, it is accompanied by reactivation of fully infectious virus. Norrby *et al.* (1985) have presented molecular evidence that rinderpest is the primordial member of the *Morbillivirus* genus, but old dog distemper would seem to give CDV the greatest epidemiological stability.

Rinderpest and PPRV are also naturally confined to one mammalian order, this time to the Artiodactyla. Families affected include the Bovidae, including the antelopes, giraffes, pigs, and deer. Sheep and goats may be infected with PPRV. PPRV is not easily transmitted to cattle, but otherwise it has not been clearly distinguished from rinderpest virus. The fact that Indian cattle are not as susceptible to rinderpest as African or European breeds has been interpreted to imply that the disease has a longer history in Asia than elsewhere (Plowright, 1968). Massive epidemics of rinderpest in African game animals have followed the introduction of infected cattle (Plowright, 1968; Shanthikumar *et al.*, 1985) and the disease has been quiescent in wild animals where, as in South Africa, it has been controlled in domestic cattle. This has led to the hypothesis that the only reservoir for this virus is in cattle and that it would be eradicated if it could be eliminated from them. An African continent-wide rinderpest extermination campaign has recently been launched on this premise (Walsh, 1987). Even without an attempt to immunize wild animals, this campaign faces tremendous logistic problems. These are compounded by the fact that the only available vaccine must be kept cold all the way to the farmers' herds. Nevertheless, the fact that rinderpest seems to have died out from several African countries in the natural course of events has led proponents to believe that even a limited effort might turn the tide against the virus.

The host range of measles virus is limited to the primates. In this instance, relatively few species, only the family Hominoidiae, maintain the virus, although many monkey species are sometimes naturally infected. The monkeys do not maintain viral endemicity without human contact (Bhatt *et al.*, 1966).

Because the apes are now too few to play a significant role, discussion of measles virus epidemiology will be confined to infection in humans. Measles virus is remarkably uniform (Black 1989a) and, being free of immune reactions related to other viruses, it offers a particularly useful model for study of the basic factors in the epidemiology of an infectious agent.

B. Measles Virus

1. Prevaccine Era

a. Isolated Populations

i. Virgin Soil Epidemics. The simplest manifestations of the epidemiology of measles come from virgin soil epidemics in isolated populations. The first, and perhaps most important, of these studies was that carried out by Panum in the Faroe islands in 1846 (Panum, 1940). The absence of previous immunization, except in some of the eldest members of this population, permitted demonstration of the universality of susceptibility, and the involvement of all age groups allowed comparison of effects across the whole age spectrum. It was in the Faroe Islands that it was first convincingly demonstrated that measles is always caused by contagion from an active case, and never by some vague miasma. The incubation period of 14 days for measles was firmly established, although it may be longer in adults. The Faroe Islands were not truly virgin in 1846 and Panum was able to show that immunity persisted in those who had been infected in the most recently preceding epidemic 65 years earlier.

Another virgin soil epidemic in Danish lands that contributed much to our knowledge was the 1951 outbreak in southwest Greenland (Christensen *et al.*, 1953). The initial patient in this epidemic, Manasse, a Greenlander returning from Copenhagen, escaped quarantine because 3 weeks elapsed between the time when he visited a sick friend and arrived in Julianahaab. Manasse went to a dance on his first night home and 250 cases of measles occurred 2 weeks later. It is seldom possible to measure R_0 directly, but here we can. Judging from this rare instance of direct measurement, the R_0 for measles had extraordinarily high value of 250. Doubtless, the fact that Manasse went to a large gathering just before developing overt disease affected this value, but it is not unusual for adults to congregate in large groups and this value may not be unique. Southwest Greenland had a total population of 4320 persons at the time of the epidemic and Christensen *et al.* were able to record 4257 cases, or an overall attack rate of 98.5%. Most of the unaffected persons had been treated with gamma globulin, or had a history of past measles, and the attack rate in exposed persons without immunity was 99.9%.

ii. Mortality. Measles-related mortality in different age groups has been most effectively studied in virgin soil populations (Christensen, 1953; Peart and Nagler, 1954). The highest rates are in the youngest children; the rate is lowest from about 5 to 15 years of age, and then increases gradually throughout adulthood.

Mortality in these epidemics can be very high, up to 50% (Black *et al.*, 1971; Cliff and Haggett, 1985), and measles played a major part in the extermination of several newly contacted human populations. A major reason for the high mortality is simply the breakdown of routine nursing care when all ages are sick at once (Neel *et al.*, 1970), but even where nursing care is provided, rates have remained unusually high. Aaby (1988) has suggested that the reason is the large inoculum that is commonly received from close contact with a sick person. It is plausible, but unproven, that inexperienced populations may be genetically more susceptible. If measles is a relatively new disease, as suggested below, then populations that were isolated until modern times have had no prior experience with the virus, and, like the African cattle, have not be selected for resistance. This, however, presumes either preexisting polymorphism that affects susceptibility or that experienced populations have selected new mutations that confer resistance. It has been shown that inexperienced populations may develop higher fevers when given attenuated measles virus, but it has not been possible to demonstrate linkage between this characteristic and known genetic markers (Black *et al.*, 1977).

    iii. Critical Population Size. A basic theory of epidemics was worked out by Reed and Frost in the 1930s, but not published until much later (Frost, 1976). Frost showed that the number of new cases in each wave I_{t+1} is dependent on the number in the previous wave I_t, times the number of persons susceptible S_t, times an infectivity factor, lambda. Lambda includes such variables as the frequency and closeness of contact between persons, the amount of virus excreted by infectious persons, and the efficiency of the virus in establishing infection when it contacts a new host. It differs from R_0 in that it reflects transmission rates in the presence of acquired immunity, not the idealized virgin soil situation. Although it can be shown that changes in any of the components of lambda affect the epidemic curve in a predictable manner, it has not been possible to measure their separate effects. The effect of changes in proportion susceptible and proportion infectious, has, however, been well worked out (Black and Singer, 1987). If one starts with a susceptible population and one infectious individual, the number of cases will increase until S becomes small enough that $I_{k+1} < I_k$. Even then, because I_k has reached a high level, the number of new cases will continue high for several disease generations. By the time the incidence wanes, the number remaining susceptible will be reduced to a very low value. It was through this mechanism that the number of susceptible persons in Greenland was reduced to 0.1%, although 10% susceptibility is a usual equilibrium value in cosmopolitan communities (Hedrich, 1933). Even with a steady input of new susceptible individuals through births, the virus will frequently fail to perpetuate itself in such a situation, and it will die out. Based on both the urban studies of Bartlett (1957) and studies in isolated populations by Black (1965), a figure of 300,000 is usually taken as the minimum number that will support endemicity of measles virus.

    Actually, this minimal population for endemicity is influenced by population dispersion and by seasonality. Measles might persist in a slightly smaller population if it were widely dispersed yet interconnected, but it may die out even in a population of one million when the people are concentrated in a small area, as in Hawaii. The virus is more likely to die out, and hence the critical population will be smaller, where seasonal changes in lambda caused by climate or in season changes of human activity are superimposed on the epidemic cycles (Yorke *et al.*, 1979). Because populations of 300,000 persons did not arise until the development of the Middle-Eastern river valley civilizations about 4300 years ago, measles, as we know it, cannot have existed earlier.

b. Developed Countries

i. Incidence. Before vaccine became available, essentially everybody experienced measles infection in the more developed countries of the world, and most had recognizable symptoms. The efficiency of reporting was often low, but if both mother and proband were questioned, a history of clinical measles could be elicited in about 80% of children of primary school age (Kress *et al.*, 1961; Kogon *et al.*, 1968). When sensitive, specific serological tests became available, it could be shown that at least 99% of young adults had serological evidence of past infection (Black, 1989a). A countermeasure to measles commonly used at this time was "prophylactic" gamma globlulin. Immune serum, or concentrated gamma globulin, was administered after infection had occurred. The goal was to give just enough antibody to attenuate the disease, but often symptoms of infection were completely suppressed (Black and Yannet, 1960). Many persons with measles antibody, but no history of disease, have had this treatment. The epidemiology of measles in the prevaccine era is, then, a question of, "When did infection occur?", not, "Who was infected?"

ii. Seasonality and Age at Infection. The question has two parts, "When?" in terms of age, and "When?" in terms of time of year. The two parts are closely interrelated, because schools have played a major role since data collection has been systematized. A close association between rate of epidemic growth and times when school was in session in Britain was demonstrated by Fine and Clarkson (1982). The effect is generally to build cumulatively increasing attack rates through much of the school year, with a peak in the late spring, and rapidly diminishing rates in the summer. This could be attributed to changes in climate, but Fine and Clarkson were able to show minor changes associated with shorter vacations. Climate does play an independent, synergistic role: in mid-nineteenth century Europe, before schooling was sufficiently general to have had much effect, Hirsch (1883) showed that spring was the season when most epidemics peaked. Seventy-seven (36%) of 213 epidemics which he studied peaked in this season. It seems likely that indoor association of persons has always been greater in winter in temperate climates and that measles, because of its relatively long incubation period, continues to increase its prevalence throughout the colder months.

The age at infection is also strongly related to schooling in developed countries. Case reporting was doubtless biased by greater efficiency where school nurses were involved, but serological studies confirm the fact that the highest incidence was in the primary school grades (Black, 1959). The probability of a child being infected before entering school depended on the number of children in the family and several indicators of socioeconomic status, the lower-status children being more likely to get infected (Black and Davis, 1968). Before automobiles were generally available, persons living in rural areas were less frequently exposed and a proportion of the rural population reached adulthood without immunity to measles (Babbitt and Gordon, 1954). This is reflected by measles in the military. The incidence of measles in the U.S. Armed forces declined steadily from 32 per 1000 man-years in the Civil War to 0.9 per 1000 in the Viet Nam war (Black, 1989b). In 1962, 98.8% of the U.S. population had demonstrable measles virus antibody at the age of induction into the military (Black, 1964) and modern methods could probably detect significant antibody in some of the remaining 1.2%. As noted earlier, R_0 cannot be measured directly in experienced populations, but it is approximately 1 + L/A, where L is the life expectancy and A the average at infection, slightly above the age at which 50% of the population has antibody (R. M. Anderson and May, 1985). Using this formula and data from England and Wales, R. M. Anderson (1982) estimated the R_0 for measles as 16. It will become evident that this is an average figure for the situation at the time of Anderson's study. R_0 varies in different population subgroups and trends upward with age until the early 20's, as persons associate in larger and larger groups. Where most young children have been vaccinated, the residual pool of susceptible persons is concentrated in these upper age groups and the effective R_0 can exceed 100. To reduce it below 1.0, more than 99% of these age groups must be immunized.

iii. Epidemic Cycles. The cyclic nature of measles epidemics in large populations has drawn the interest of mathematical modelers at least since the work of Hamer (1906). Measles virus persists in most large populations, but its prevalence varies from year to year as well as from season to season. Each spring, in any large temperate-land population, there is an increase in the number of cases, but they only reach epidemic proportions in years spaced with regular periodicity. For many years in New York City there were epidemics in the even-numbered years, but not in the intervening years (Yorke and London, 1973). The pattern in London was similar (Fine and Clarkson, 1982). In smaller cities, the epidemics were more widely spaced. The original Reed–Frost formula suggested that the extreme swings in disease prevalence observed when the virus is first introduced would gradually dampen and periodicity would ultimately be lost. Bartlett (1956) showed that, in a real community, stochastic effects would maintain the kind of amplitude that is commonly seen, and Yorke *et al.* (1979) showed seasonal changes in transmissibility have the same effect.

iv. Mortality and Late Sequelae. Measles mortality used to be high in European countries, possibly as high as in modern Africa (Panum, 1940; Chalmers, 1930; Aaby *et al.*, 1986), but as these countries developed economically, their measles mortality rates declined steadily (Babbitt and Gordon, 1954). Schutz (1925) found a sevenfold decline in measles mortality in several German cities between 1875 and 1920. There was a 45-fold decline in the U.S. measles mortality rate between 1912 and 1965 (Communicable Disease Center, 1966). By the 1950s the mortality rate in the England and Wales was 2 per 10,000 cases (Babbitt *et al.*, 1963), and in the United States there were an average of 525 deaths per year among 3.8 million cases (estimated from the birth rate and serological survey), for a case fatality rate of 1.4 per 10,000. This decline in measles mortality was so gradual and so broad that it cannot be related to any single change in treatment, but seems to parallel the general improvement in standard of living. One wishes for a more specific explanation for this decline of measles mortality in developed countries, as we seek to improve the lot of those living the less developed lands.

Although much reduced, measles mortality in the 1950s remained at about one-third the mortality associated with poliomyelitis. This seems inconsistent with the fact that poliomyelitis was viewed with great fear, while measles was considered a laughing matter. The explanation was partly due to the fact that nonfatal sequelae of poliomyelitis were more obvious, but also probably because poliomyelitis seemed more lethal, in that most nonlethal infections were entirely inapparent. The low level of concern induced by measles is illustrated by Nevil Shute's (1957) choice of measles, in the novel "On the Beach," as a ridiculously minor problem that delayed the search for a way out from atomic annihilation. Even the medical community was reluctant to use a vaccine for measles, justifying this by concern that a poor vaccine might postpone the age at infection, and that the most serious complication of the disease, encephalitis, increased in incidence with age (Greenberg et al., 1955). Generally, measles seemed to be one of life's minor irritants; this acceptance of a bad, but tolerable situation persists to the present time in some countries, making global control efforts difficult.

A question that was never fully answered was whether measles quite commonly caused minor neurological damage. Neurological effects could be regularly observed in acute disease (Gibbs et al., 1959; Pampiglione, 1964), but it was difficult to get an uninfected control group in whom late sequelae could be studied (Fox et al., 1968).

c. Less Developed Areas

i. Mortality. In less developed countries also, measles virus infection is almost universal. This is evident from the nearly 100% seropositivity rates found in young women of many countries (Black et al., 1987). The most striking difference in less developed countries is that measles is still associated with high mortality. Katz (1985) recently estimated that 1.5 million children die of measles every year. The vast majority of these deaths occur in the less developed countries. In these countries, the case fatality rate must be about 1.3%, or nearly 100 times greater than the current rates in the developed world. The difference cannot be attributed to differences in the virus, because a large part of the measles now occurring in the United States derives from importation, and these outbreaks are not associated with greater severity than the endogenous cases (Centers for Disease Control, 1983). Not only is the average mortality rate high in developing countries, but in some outbreaks the rate may be extraordinary—over 10% (Morley, 1967; Aaby et al., 1983). The reasons for these very high rates are threefold.

The most obvious problem is malnutrition. In a series of studies of hospitalized cases compiled by Aaby (1988), the mortality rate in children who were 80% or more of the normal weight for age was 12%, but it was 28% in children below 65% of the norm. Intermediate nutrition levels had intermediate mortality rates. This correlation has been difficult to confirm in community studies, however. Scrimshaw et al. (1966) carried out an experimental study with food supplements for the test group, but the mortality in the control group decreased nearly as much as in the supplemented group. A prospective study carried out in Bangladesh (Chen *et al.*, 1980) showed a higher risk of dying in children below 65% of the weight for age norm, but no effect of lesser degrees of malnutrition. Two complications may explain these inconsistencies. Most studies measure malnutrition caused by current protein-calorie deficiency, but the factor leading to increased risk from measles may be more specific. Collins (1987) found that the amount of measles passive antibody that a child receives at birth is reduced in proportion to indicators of long-term maternal malnutrition, but not by short-term factors. Attention has recently been given to the possibility that vitamin A nutrition may be particularly important (WHO/UNICEF, 1987). The effect of vitamin A deficiency seems to be particularly strong in the association of measles with blindness, but when an attempt was made to see if the blindness persisted, it was found that few of the affected children had survived even a few months (Pepping *et al.*, 1988). This exemplifies the second complication encountered in relating nutrition to measles mortality—much of the associated mortality is delayed. The delay may reflect a synergistic effect of measles and other diseases in progressively pushing a child's development farther and farther down. This effect was first described in prospective studies of individual cases by Mata (1978), but has been extended on a population basis by Koster *et al.* (1981).

The second factor leading to high measles mortality in less developed countries is the early age at infection. As shown in the virgin soil studies above, and, in a narrower frame, in vital statistics from many developed and little developed countries (Carrada Bravo and Velazquez Diaz, 1980; Aaby, 1988), the younger the child when infected, the more likely he or she will die. A minor deviation in this pattern appears in children less than 1 year old who retain sufficient maternal IgG to modify, but not prevent, disease. This narrow band of relatively mild disease, however, shifts with the mean age at which a population becomes susceptible, and does not change the relationship between mean age at infection and mortality rate. Three factors contribute to the lower age-specific attack rate. The early age at infection is a reflection of the broad base of the age pyramid. The high birth and death rates in these countries means that if 10% of the population is susceptible in an equilibrium endemic state, as in Baltimore (Hedrich, 1933), the mean age at infection would be much lower (Black, 1982). Second, interpersonal space is commonly much less than in modern developed countries. It was shown years ago by Chalmers (1930) that the rate of spread of measles in Glasgow was related to the number of persons per room, because measles spreads more easily where there is crowding. Finally, children in less developed countries lose passive immunity at an earlier age (Christie *et al.*, 1990) and hence measles infection can occur in younger infants. The early loss of passive immunity is apparently due to the rapid acquisition of a variety of antibodies as a result of early exposure to many infectious agents and a homeostatic feedback mechanism that accelerates the degradation of all IgG species to maintain physiological balance. When all these factors coalesce, the mean age at infection with measles may be as low as 24 months (Black, 1962; Aaby *et al.*, 1983).

The third factor that increases measles mortality in the less developed world appears to be an increased frequency of large infecting doses and more adverse outcomes when the virus is thus given a head start. This thesis has been expounded by Aaby (1988). Children of large families and children in-fected in the home, have poorer prognoses than the average child infected on the street. The prevalence of large, often polygamous families and generally small house size in underdeveloped areas would increase the chance of trans-ferring large amounts of virus. This effect is clearly confounded with circum-stances that lead to malnutrition and early age at infection. Quantitative data that would permit proportionment of the relative roles of these three factors through analysis of variance have not been available, but the case for a signifi-cant effect of dose seems strong.

 ii. Seasonality. In the less developed areas of the world, school schedules play a minor role, partly because schooling is less common, but also because a large proportion of the children are infected before they reach school age. Even so, seasonality persists in the prevalence of measles. Factors that lead to indoor congregation and factors that lead to intercommunity spread generally work in opposition to one another, and it is difficult to predict the season of maximal incidence. In Brazil, most measles occurs in July and August, their winter, but the peak is less marked than in the United States (Fig. 1). In south India, strong seasonality is associated with the end of the monsoon, which permits people to move more freely between villages (John *et al.*, 1980). In Sahelian Senegal, epi-demics may occur in either the wet or the dry season (Aaby, 1988).

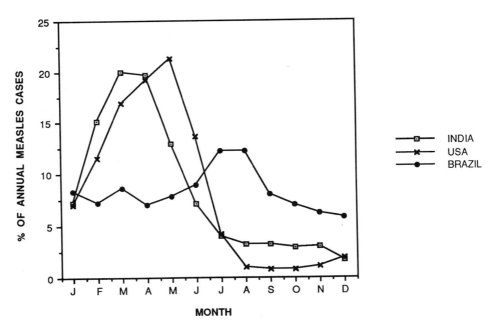

FIGURE 1. Distribution of reported measles cases by month of the year in the United States and south India for 1977 and in Brazil for 1979. U.S. data are from Center for Disease Control (1978), Indian data are from John *et al.* (1980), and Brazilian data are from unpublished information from the Divisão de Epidemiologia, FSESP, Ministerio de Saude, Brasilia, Brazil.

2. Vaccine Era

a. Developed Countries

Widespread use of measles vaccine, which began in the United States in 1967 and more gradually in other developed countries, has had an epochal effect on the epidemiology of measles. Incidence in the United States, which must have run near 4 million cases per year before that time, has dropped to an average of 4500 a year during the 1980s, nearly three orders of magnitude. A measles death is now national news.

In spite of this success, the vaccination program has left disappointment because measles has not been eliminated. Twice (Sencer *et al.*, 1967; Hinman *et al.*, 1979) the Centers for Disease Control have called for an eradication, and twice the campaign has failed. It is not just that importations have continued, but most cases in the United States have continued to be acquired from endogenous sources (Centers for Disease Control, 1989a). There is no evidence that waning of the immunity derived from early vaccinations is a factor, but the disease continues in that small part of the population that has not been vaccinated, or in whom the vaccine failed to immunize initially. Through the 1970s, measles in college populations was a troublesome problem. A cohort too old to have been vaccinated and too young to have been thoroughly immunized by natural infections has passed through the age at which congregation in very large groups is most frequent. In 1988 the proportion of cases in persons over 20 was holding stable at 13% and two other age groups, 0–15 months and 10–19 years, were the biggest problems (Centers for Disease Control, 1989b). The 10- to 19-year stratum represents a continuation of the problem of a cohort with a high level of herd immunity that is still inadequate when the cohort associates in large groups. The American Academy of Pediatrics has recommended that they be revaccinated, not to boost waning titers, but to provide a second opportunity for primary immunization.

The problem in infants is new and growing. Seventeen percent of all cases now occur in infants younger than 16 months, by far the highest age-specific attack rate. Present policy dictates that these children are too young to be vaccinated because too many have residual maternal antibody and will fail to respond. Part of the trend in young babies is probably due to increased use of child care centers, but part is due to lower levels of passive immunity. A progressively greater proportion of new mothers derive their immunity from vaccination. In the last 10 years in New Haven, Connecticut, the proportion of women giving birth who derive their measles immunity from vaccination has gone from 10 to 65%. The antibody titers of the vaccinated women are only one-third of the titers of mothers who were naturally infected, and their children lose protection nearly 3 months sooner (Lennon and Black, 1986). An obvious remedy is to lower the age for vaccination, and that is now recommended in high-risk areas (Advisory Committee on Immunization Practices, 1989). However, we presently have a heterogeneous population and it is not possible to make one recommendation that will optimally serve all children.

These problems with specific age groups are too small to have foiled the eradication plans if the assumptions on which those plans were based had held true. The estimate by R. M. Anderson (1982) of the R_0 for measles, 16, implies that if the proportion of the population susceptible to measles could be reduced to 6%, the disease could be eliminated. Other contemporary studies reached similar conclusions. In fact, this would only work if the efficiency of transmission were homogeneous. With reduction of transmission in the preschool and elementary school years, those who were not immunized progressed into high school, college, and the military without acquiring immunity. There, the efficiency of transmission is much greater and the R_0 much higher (Black, 1989a). It seems likely that a population immunity rate of 99% may be needed to stop measles transmission. That is difficult to attain.

b. Less Developed Countries

If the failure to eradicate measles from developed countries has proved a disappointment, a failure to make even a dent in measles attack rates in many less developed countries is tragic. The problem starts with case reporting. Where increased attention is paid to a previously poorly reported disease, an almost inevitable effect will be an increase in reporting efficiency and a greater number of reported cases. If substantial impact were made by vaccination, it should nevertheless be possible to see a shift in the age distribution of the cases. If the number of susceptible were reduced by half, there would be a proportional reduction in new cases in the next disease cycle; in the third cycle, there would be both half as many susceptible and half as many infectious and therefore one-quarter the number of new contacts. Thus, the average time between loss of passive immunity and infection will be quadrupled. If this interval was initially 15 months and the average child 24 months old when infected, it would now be 60 months old and the average case would be in a 5-year-old. This change will be somewhat reduced by the circumstances that the susceptible pool is no longer being reduced at the same pace by natural disease, and that these older children get around more than the infants, but the effect should still be obvious. In fact, the age shift has been minimal in well-studied campaigns in various African cities (Guyer and McBean, 1982; Dabis et al., 1988; Taylor et al., 1988).

In less developed countries, the window of opportunity between the time of loss of maternal antibody and the time of infection with measles virus during which vaccine might be protective is very narrow (Black, 1989c). A good public health system such as those of Hong Kong (Chang, 1982), Singapore (Doraisingham, 1982), or Cuba (Pan American Health Organization, 1988) can get the vaccine to a sufficient part of the population to slow virus circulation and widen this window, but less developed countries have commonly failed. Too often, children are not bought for vaccination before they are infected; too often, health care workers, fearing that they will not see the child again, give the vaccine too early; too often, the vaccine has been mishandled and inactivated; and too many neighborhoods are not adequately covered by any health program.

3. Subacute Sclerosing Panencephalitis

SSPE is a universally fatal delayed sequel of measles virus infection that has distinctive epidemiological characteristics (Detels et al., 1973; Halsey et *al.*, 1980). Viruses isolated from SSPE differ from standard measles strains (Schleuderberg *et al.*, 1974), but not in a consistent manner (see Chapter 12). The differences must be the result of mutations occurring in the affected individual, probably during the long latent period in the brain. An epidemiologically distinct virus fully subject to immune defenses raised by measles virus could not persist in the same ecological niche as the common agent. It would have to propagate in a narrow stream, and yet be sufficiently prevalent that it commonly gets to children at a very early age. Rather, SSPE occurs when measles virus establishes an infection of glial tissue that releases few if any mature virus particles, and is relatively inaccessible to the immune defenses but propagates slowly by spread to adjacent cells. The epidemiological question, then is, "Under what conditions is chronic glial cell measles virus infection established?"

SSPE most frequently ensues when measles infection occurs before the second birthday in boys living in a rural environment in contact with birds (Halsey *et al.*, 1980). No nontrivial explanation for the rural and avian associations has been found, but the age association is important. Most measles in developing countries occurs at this early age and hence SSPE is likely to be particularly common there. The data are generally too sparse to confirm this, but 14 cases of SSPE have been identified in Jamaica between 1977 and 1986 (Grey *et al.*, 1986). This is a rate of 3.4 cases per 100,000 children at risk, about ten times the rate seen in the United States before the vaccine era (Sutton, 1979).

The other consequence of susceptibility to SSPE at an early age is that if vaccine virus strains retained the neurological tropism of wild virus, programs to vaccinate children before they are naturally infected, the only useful time, would increase the incidence of encephalitis. It was early claimed that vaccine virus does not give rise to the neurological signs common in natural measles (St. Geme, 1968), but the first case–control studies suggested that vaccine virus might sometimes cause SSPE (Modlin *et al.*, 1977). Subsequent experience has essentially proven otherwise; the incidence of SSPE declined more slowly than the incidence of measles, as would be expected of a late sequel, but after measles incidence in the United States had been held at a low level for several years, SSPE practically disappeared (Centers for Disease Control, 1982).

III. *PARAMYXOVIRUS*

A. Mumps Virus

1. Prevaccine Era

a. Measurement of Attack Rates

In contrast to measles virus, antigenic differences between mumps virus isolates have been reported (Server *et al.*, 1982), but these seem not to have epidemiological significance because, once acquired, immunity to mumps provides protection wherever one may travel. A more relevant epidemiological difference between mumps and measles infections is the large proportion of mumps cases that go unrecognized. Whereas surveys reveal that parent or proband know that about 80% of young North American adults have had measles, only about 60% know they have been infected with mumps (Center for Disease Control, 1973). Determination of the true number infected by mumps virus is made difficult by the fact that mumps serologic tests must compromise between sensitivity and specificity, because of cross-reactions with the parainfluenza viruses. The neutralization test (Buynak et al., 1967), especially when plaques are counted (Wagonvoort et al., 1988), is probably the most specific. The complement-fixation test was used in many early studies (Gordon and Kilham, 1949), the hemagglutination-inhibition test in others (Black and Houghton, 1967). The allowance that needs to be made for in-complete detection depends on which test is used. This uncertainty is in-creased by the fact that, as measured by any test, mumps antibody titers decline with time after infection, even though susceptibility to disease does not recur (Black and Houghton, 1967). A study of Dutch families using the plaque neutralization test detected antibody in more than 95% of young adults, but even the remaining 5% included persons who proved to be immune when exposed (Wagonvoort et al., 1988). As in the case of measles, it seems that nearly all adults living in cosmopolitan communities have immunity to mumps.

b. Isolated Populations

As with measles, virgin soil epidemics have been very helpful in defining some basic parameters of mumps epidemiology. All the well-studied epidemics occurred in Alaska, on islands in the Bering Sea (Philip et al., 1959; Reed et al., 1967; Maynard et al., 1970). Similar outbreaks must have occurred elsewhere in the arctic and in the tropics, but they did not draw the attention given to measles epidemics. The 1957 epidemic on St. Lawrence Island was the largest and cleanest, in terms of absence of prior immunity, of the well-studied epi-demics. Even this outbreak showed evidence of prior immunity in the oldest people, so percentages used here are based on the 499 persons less than 50 years old. Ninety-three percent developed mumps complement-fixing antibody, but even with an active search for signs of disease at the appropriate time, only 70% were recognizably sick. There was a strong age relationship, particularly in the group under 5 years of age, in the proportion with apparent disease (Fig. 2).

The less frequent recognition of mumps epidemics is presumably due to the fact that uncomplicated mumps is not as severe as uncomplicated measles; the concern caused by mumps relates to its complications, particularly or-chitis and encephalitis. Again, the Bering Sea epidemics are helpful in defining the frequency of these effects. Orchitis was seen in all ages of males from 3 to 51 years, but was concentrated in those past puberty. Twenty-five percent of all males had this sign and, on St. Lawrence Island, 38% of men more than 15 years old had it. The orchitis was frequently bilateral. There were four cases of encephalitis and 40 cases with stiff neck in the St. Lawrence population, for a 10.5% prevalence of neurological signs.

It is not possible to set a lower limit on the incubation period of mumps

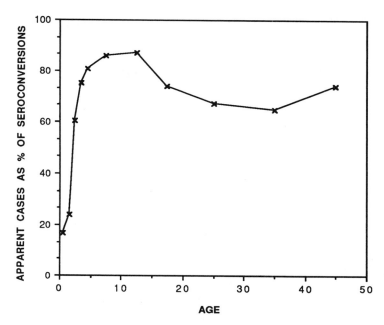

FIGURE 2. Age-specific attack rate of clinically recognizable mumps in the population of St. Lawrence Island, Alaska during an epidemic in which essentially the whole population developed antibody. [From Philip *et al.* (1959).]

from these data, but cases within a household were spaced by up to 50 days, demonstrating the potential for a long incubation.

c. Developed Countries

R. M. Anderson (1982) estimated the R_0 for mumps in North America as 5. This suggests that mumps virus is one-third as infectious as measles virus, and that the average age at infection would be 15 instead of 5 years. Anderson later gave the mean age of infection with mumps in Britain as 6–7 years (R. M. Anderson *et al.*, 1987). While such a large difference in infectiousness is patently wrong, the idea that mumps virus is less infectious than measles virus is widely held. This impression doubtless derives from the fact that many adults gave a negative history and lacked antibody detectable by the early tests (Black, 1964; Vicens *et al.*, 1966; Niederman *et al.*, 1967). It was often assumed that persons without detectable mumps antibody and a negative history had not been infected. However, as noted above, modern methods show specific immunity in nearly all adults living in Holland or on islands affected by the Bering Sea outbreaks. If one takes the clinical history data from the 1967 U.S. Epidemic Intelligence Service survey (Center for Disease Control, 1973) and adds the inapparent attack rate for each age as it was seen on St. Lawrence Island (Philip *et al.*, 1959), one finds that 95% of the U.S. population is probably infected by 15 years of age. This is not significantly different from the proportion that have been infected by measles virus.

Family studies carried out in Baltimore by Meyer (1962) confirmed the

TABLE I. Mumps-Associated Encephalitis or Aseptic Meningitis
in 1972 by Age[a]

Age (years)	Reported cases of mumps	With encephalitis or aseptic meningitis	CNS disease per 1000 cases mumps
0–4	37,100	39	1.05
5–9	26,000	103	3.96
10–14	4,800	40	8.33
15–19	3,000	13	4.33
>20	3,000	39	13.00

[a]Cases per 1000 reported cases of mumps. Data from Center for Disease Control (1975).

occasionally long incubation period observed on St. Lawrence Island, but showed that the interval in 95% of the cases was between 14 and 24 days. Meyer also confirmed a higher frequency of inapparent cases in young children.

A question that has never been fully answered is whether cross-reactive parainfluenza antibodies play a role in determining the severity of symptoms in mumps infection. Mumps and parainfluenza, especially type 3, hemagglutination-inhibition titers are strongly correlated (Black and Houghton, 1967). If, however, cross-reactive antibody were a major protective element, one would expect to find more severe disease in very young children and in the Bering Island populations. The observed patterns were the reverse of this prediction.

Mumps outbreaks show a seasonality similar to, although somewhat broader than, that of measles (Center for Disease Control, 1975). However, in New York at least, there was little evidence of interyear periodicity (Yorke and London, 1973). The longer and less uniform incubation period of mumps is a plausible reason for this difference from measles.

Mumps used to be the commonest cause of hospital admissions for encephalitis in the United States. The prognosis of mumps encephalitis is good, but it is nevertheless a major cause of worry and expense. Like measles, the incidence of encephalitis goes up with age (Table I). Besides encephalitis and orchitis, a variety of other complications have been associated with mumps. Hearing impairment and juvenile onset diabetes are two of the more important, but mumps is not the major cause of either of these problems, and it has been difficult to determine its exact role (Feldman, 1989).

d. Less Developed Countries

The epidemiology of mumps in the less developed countries has received very little attention. Mumps infection is as prevalent there as in the more developed world; M. Whitehurst (unpublished) found neutralizing antibody in sera from 97% of 100 Jamaican women of childbearing age. Presumably, like measles, it hits at an earlier age in the less developed areas. If this is so, then the incidence of apparent disease may be quite low and the incidence of encephalitis and other complications low, even in relation to the number of apparent cases.

2. Vaccine Era

A live mumps vaccine was licensed in the United States in 1967 and somewhat earlier in the USSR, but it was not extensively used until it was combined with measles and rubella vaccines several years later. Mumps vaccine use has proceeded in the United States with very little specific planning or control, but since the early 1970s almost every child getting measles vaccine has also received mumps vaccine, and there has been a dramatic decrease in mumps incidence. Prevaccine-era mumps case reporting was very incomplete, but there must, like measles, have been about 3,800,000 infections per year to account for the observed immunity. Through most of the 1980s the number of reported cases has been about 5000 per year, although there was an increase in the 1986–1987 season (Cochi *et al.*, 1988; Centers for Disease Control, 1988).

Center for Disease Control (1973) data suggest that the highest reported attack rate was in the 10- to 14-year-old group and it continued high in the next five-year age stratum (Centers for Disease Control, 1988). This is a repetition of the measles pattern, some 10 years later. Like the cohort who retained a high susceptibility to measles because they were too old to be included in vaccination programs but too young all to have been infected, the mumps-susceptible group seem to be passing through the period of their greatest risk. During this transition, mumps outbreaks in the workplace have drawn popular attention (Kaplan *et al.*, 1988; Centers for Disease Control, 1987). As with measles, there are calls for revaccinating this age group (Pachman, 1988), but as yet there is no evidence that the vaccine-induced immunity is less permanent than the naturally-induced immunity.

There is cause for worry that partial control of mumps might defer the age at infection in many persons until adolescence or adulthood, and lead to an actual increase in the number of serious complications. The mathematics of this phenomenon has been worked out by R. M. Anderson *et al.* (1987), and appears alarming. Anderson *et al.*, however, used their very low estimate of the mumps R_0 and their conclusions need revision. A comparison of reported cases in the United States in 1970 and 1987 is shown in Table II. The proportion of cases that occur in persons past puberty is up fivefold, but only in the 15- to 19-

TABLE II. Age-Specific Incidence of Mumps in the United States in 1970 and 1987[a]

Age (years)	1970 reported cases		1987 reported cases	
	Percent of all cases	Cases per million at risk	Percent of all cases	Cases per million at risk
0–1	3	918	2.9	110
1–4	47	3591	7.2	68
5–9	33	1739	17.8	140
10–14	9	453	39.7	311
15–19	3	163	26.8	201
20–24	} 3	} 102	4.6	32
25–29			1.1	7

[a]Data from Center for Disease Control (1973) and Centers for Disease Control (1988).

year-old age stratum has there been an absolute increase in number of reported cases. When the probably great increase in efficiency of reporting is taken into account, it seems most unlikely that there has been an actual increase in number of mumps cases in any age group. So far, the problem has been avoided in the United States.

B. Parainfluenza Viruses

The parainfluenza viruses, together with Newcastle disease virus (NDV), comprise a spectrum of antigenically related agents infecting a wide variety of vertebrate species. There are parainfluenza viruses of monkeys (Hull, 1956), dogs (McCandish et al., 1978), cattle (Rossi and Keisel, 1972), laboratory rodents (Parker and Richter, 1986), chickens (Andrewes, 1972), and many other species, including five types that infect humans (Hsiung et al., 1963). Host species specificity is not strong. Sendai virus, now generally agreed to be a murine pathogen, is sufficiently similar to human parainfluenza type 1 that it was long regarded as a strain of the human virus (Parker and Richter, 1986). NDV, the chicken agent, causes a localized infection in humans (Andrewes, 1972), but seems not to be serially propagated in humans. Bovine parainfluenza 3 virus is very similar to, but distinct from, human parainfluenza 3 virus (Rydbeck et al., 1987). Some of the veterinary members of the group, especially bovine type 3 and NDV, cause economically important diseases and successful vaccines against these agents have been developed (Thomson et al., 1986; American Veterinary Medical Association, 1986).

All members of the genus, including mumps virus, cross-react extensively in serological tests, but each of the human viruses maintains its own distinctive identity in diverse human populations. Some idea of the cause of an infection can be gained from serology, but this may be misleading, because anamnestic cross reactions are often stronger than the reaction to the infecting type (Tyeryar et al., 1978). Isolation is difficult because of the lability of these viruses, but new techniques for detection of viral antigens in nasal secretions offer promise of increased specificity as well as speed (Schopfer et al., 1986).

Spread of the parainfluenza viruses is by aerosol (Karim et al., 1985), and there are data to indicate that human type 3, at least, is effectively preserved by rapid drying in aerosol (W. S. Miller and Artenstein, 1967). There was no loss of its infectivity in 60 min at 20% relative humidity, but there was one $\log_{10}$ decline in titer at 80% humidity.

The predominant diseases caused by these viruses in all species involve the respiratory tract. The incubation period is characteristically 3–6 days (Glezen et al., 1989). It is characteristic that, while immunity acquired during a primary infection protects against repetition of the most severe disease, it does not prevent reinfection. Unlike the situation with measles and mumps, where persons who have been infected act like heavy water in a nuclear reactor to slow the flux, those who have been infected by the parainfluenza viruses return to the susceptible pool after just a few months of immunity. For this reason, and because persistent infections with continuous virus release may occur (Gross et al., 1973), these viruses circulate much more commonly than their R_0

would indicate, and human parainfluenza virus type 3 is more prevalent than either measles or mumps viruses. Annual epidemics of parainfluenza 3 are the rule, but types 1 and 2 commonly occur in alternate years (Brandt *et al.*, 1974; Glezen *et al.*, 1984). The type 1 and type 2 epidemics build quickly to a peak in the fall or early winter, but there may be evidence of type 3 in circulation through much of the year. The pattern of high type 3 prevalence is found in diverse human communities, and children are commonly infected with type 3 at a younger age than with the other viruses [Prague, Pecenkova' *et al.* (1972); rural Mexico, Golubjatnikov *et al.* (1975); Seattle, Cooney *et al.* (1975); Michigan, Monto *et al.* (1986)]. Extreme isolation may protect against these viruses temporarily (Black *et al.*, 1974), but they have penetrated even to the Alaskan Eskimo (Maynard *et al.*, 1967), and epidemiologists have never been able to study a virgin soil epidemic.

Maternally derived antibodies give substantial protection for the first 4 months (Glezen *et al.*, 1984). The attack rate rises rapidly after this and is highest during the next 18 months (Bisno, 1970). In a typical cosmopolitan population, 50% of children have experienced type 3 infection by the end of the second year of life, and types 1 and 2 before 5 years of age (Pecenkova' *et al.*, 1972). Type 1 causes more overt disease than type 2 (Glezen *et al.*, 1984), and type 4 is by far the least prevalent (Gardner, 1969). There are differences between the types in their relative predilection for different parts of the respiratory system (Denny and Clyde, 1986), but these may be due as much to the different ages at infection as to differences in the agents.

The human viruses cause serious disease in young children (Joosting *et al.*, 1979) and annoying illness in adults (Cooney *et al.*, 1975). These infections spread readily in a hospital setting and they are a significant added burden to already sick children (Mufson *et al.*, 1973). The successful veterinary vaccines make it seem probable that vaccination could reduce this problem. Intensive efforts to develop killed-virus vaccine were initiated early (Vella *et al.*, 1969; Chin *et al.*, 1969). Although these vaccines did not clearly exacerbate subsequent infections, like the RSV vaccine, there were sometimes more cases recorded in the vaccinated group than in the controls. Live vaccines have also been tested, but when given by injection, these, too, have failed (Potash *et al.*, 1970), and the work has lagged. The titer of IgG antibody in the serum correlates poorly with immunity (Smith *et al.*, 1966; Tremonti *et al.*, 1968), although it must be the mediator of maternally derived protection. The IgA titer in nasal secretions correlates better, and more recent attempts to develop a vaccine have focused on boosting this antibody (Wigley *et al.*, 1970; Belshe and Hissom, 1982; Ray *et al.*, 1985).

IV. *PNEUMOVIRUS:* RESPIRATORY SYNCYTIAL VIRUS

Although veterinary viruses provide useful models for measles and two veterinary analogues of RSV, pneumonia virus of mice (PVM) and bovine RSV, have been extensively studied, they are not particularly useful models for human RSV disease (see Chapter 14). PVM is not known as a naturally occurring disease, but is known only in laboratory animals (Richter, 1986). It is not demonstrably related to RSV by immunology and may have quite different patterns of transmission. Bovine RSV seems to be a more useful model. Like RSV, it causes the most severe disease in young, but not the youngest hosts (Kimman et al., 1988). The fact that the most severely affected calves do not make IgG antibody in response to the infection finds a parallel in the uneven immune response to RSV in humans (McIntosh and Fishaut, 1980). Maternally derived antibody gives much shorter protection against RSV than against the other Paramyxoviridae, and the more serious effects of infection may be due either to low levels of this antibody or the unbalanced immune response that is mounted as when this antibody wanes.

RSV remains the commonest viral cause of serious disease in infancy (Hendry et al., 1988), although efforts have been underway to develop a vaccine for more than 20 years (Kapikian et al., 1969). Serological studies are not as helpful in defining the epidemiology of RSV as with viruses of the other para-myxovirus genera. The primary problem is that antibody titer measured by traditional methods is not stable (Henderson et al., 1979) and, even if present, is not necessarily protective (Parrott et al., 1973). Only recently was it discovered that two quite distinct subtypes of RSV exist and the two may circulate contemporaneously in the same community (L. J. Anderson et al., 1985; Mufson et al., 1988). These two types, variously referred to as A and B or 1 and 2, are distinguishable only by the immune reaction with the 90K glycoprotein, and even then infection with one will give an immune response to the other in a substantial proportion of infected children. Generally A seems to be more commonly associated with disease than B, but data are sparse.

With the proviso that much of the older data need to be reevaluated for the effects of the two subtypes, there are certain things that can be said about the epidemiology of RSV. Virus circulation is strongly seasonal, occurring somewhat earlier in the winter than does measles or mumps (Brandt et al., 1973; L. J. Anderson et al., 1985); and infection commonly occurs at a very early age so that most children have been infected at the end of two winter seasons (Henderson et al., 1979). This does not mean that the virus is more infectious than measles or mumps viruses, but, like parainfluenza virus, more RSV circulates because virus-experienced persons can be readily reinfected to resume virus excretion (Beem, 1967). The susceptibility to reinfection is more than successive infection with the other subtype, because in the child care setting, even children who have already been twice infected may have a 65% reinfection rate (Henderson et al., 1979). Adults in an exposed family may have a 45% infection rate (Hall et al., 1976).

The most persistent question regards the reasons why RSV infection is so severe in the age span 1–6 months (Channock et al., 1989). The immune system is not fully developed at this age, particularly with regard to complement and phagocytic and NK T-cells (M. E. Miller and Steihm, 1983). It is clear from the unfortunate consequences of the killed-vaccine trials (Kapikian et al., 1969; Kim et al., 1969) that certain components of the immune response can cause the disease to be more serious and can extend the time period in which severe disease may occur beyond the normal high-risk period. It was at first thought that marginal levels of IgG, whether passively derived or vaccine induced, might be the cause of this effect, but the effect does not apply to all anti-

RSV IgG, and various other ideas have been investigated (McIntosh and Fishaut, 1980; Channock *et al.*, 1989). Examination of the immune responses to individual virus proteins has shown that antibody to the receptor-binding 90K protein is not made until relatively late in the first year of life (Ward *et al.*, 1983). Antibody to this protein is believed necessary for immunity (L. J. Anderson *et al.*, 1985). This delay is consistent with the facts that the protein has 60% carbohydrate content (Hendry *et al.*, 1988) and that the capacity to produce IgG_2, which includes much of the body's anticarbohydrate response, is delayed in infants (Watt *et al.*, 1986).

The problem of the early age of peak susceptibility to serious disease is particularly troublesome in the search for control by vaccination. Any effective immunization must occur early in life, yet circulating maternally derived antibodies interfere with immunization by attenuated live virus administered by injection (Buynak *et al.*, 1978), and, as already noted, a killed vaccine worsened the situation. The possibility of administering a live vaccine intranasally (Wright *et al.*, 1976) is the chief current hope. An approach used in cattle against bovine RSV (American Veterinary Medical Association, 1986) and in humans against tetanus (Newell *et al.*, 1966) that has not been tried is that of increasing passive immunity in the child so as to carry it through the critical phase by boosting immunity in the mother. This approach was side-tracked by early reports that passive antibody titer does not correlate with protection (Parrott *et al.*, 1973). More recent studies by Glezen *et al.* (1981) and Ogilvie *et al.* (1981) have reversed this conclusion and maternal immunization has again been proposed. Such an approach still faces the legal impediment that applies in this country to all treatment of pregnant women.

V. REFERENCES

Aaby, P., 1988, Malnutrition and overcrowding/intensive exposure in severe measles infection: Review of community studies, *Rev. Infect. Dis.* **10**:478–491.

Aaby, P., Bukh, J., Lisse, I. M., and Smits, A. J., 1983, Measles mortality, state of nutrition and family structure: A community study from Guinea Bissau, *J. Infect. Dis.* **147**:693–701.

Aaby, P., Bukh, J., Lisse, I. M., and Smits, A. J., 1986, Severe measles in Sunderland, 1885; A European–African comparison of causes of severe infection, *Int. J. Epidemiol.* **15**:101–107.

Adams, J. M., Imagawa, D. T., Wright, S. W., and Tarjan, G., 1959, Measles immunization with live avian distemper virus, *Virology* **7**:353–354.

Advisory Committee on Immunization Practices, 1989, Measles prevention. Supplemental statement, *Morbid. Mortal. Weekly Rep.* **38**:11–14.

American Veterinary Medical Association, 1986, *J. Am. Vet. Med. Assoc.* **1986**:281–284.

Anderson, L. J., Hierholzer, J. C., Tsou, C., Hendry, R. M., Fernie, B. F., Stone, Y., and McIntosh, K., 1985, Antigenic characterization of repiratory syncytial strains with monoclonal antibodies, *J. Infect. Dis.* **151**:633.

Anderson, R. M., 1982, Transmission dynamics and control of infectious agents, *in* "Population Biology of Infectious Disease" (R. M. Anderson and R. M. May, eds.), pp. 149–176, Springer-Verlag, New York.

Anderson, R. M., and May, R. M., 1985, Vaccination and herd immunity to infectious diseases, *Nature* **318**:323–329.

Anderson, R. M., Crombie, J. A., and Grenfell, B. T., 1987, The epidemiology of mumps in the U.K.: A preliminary study of virus transmission, herd immunity and the potential impact of immunization, *Epidemiol. Infect.* **99**:65–84.

Andrewes, C. H., 1972, "Viruses of Vertebrates," Williams and Wilkins, Baltimore, Maryland.

Appel, M. J. G., and Gillespie, J. H., 1972, "Canine Distemper Virus," Springer-Verlag, Berlin.

Appel, M. J. G., Shek, W. R., Sheshberadaran, H., and Norrby, E., 1984, Measles virus and inactivated canine distemper virus induced incomplete immunity to canine distemper, *Arch. Virol.* **82:**73–82.

Babbitt, F. L., Jr., and Gordon, J. E., 1954, Modern measles, *Am. J. Med. Sci.* **228:**334–361.

Babbitt, F. L., Galbraith, N. S., McDonald, J. C., Shaw, A., and Zuckerman, A. J., 1963, Deaths from measles in England and Wales in 1961, *Mon. Bull. Minist. Health Pub. Health Lab. Serv.* **22:**167–175.

Bartlett, M. S., 1956, Deterministic and stochastic models for recurrent epidemics, *in* "Third Berkeley Symposium on Mathematical Statistics and Probability," Vol. 4, pp. 81–109.

Bartlett, M. S., 1957, Measles periodicity and community size, *J. R. Stat. Soc. A* **120:**40–70.

Beem, M., 1967, Repeated infections with respiratory syncytial virus, *J. Immunol.* **98:**1115–1122.

Belshe, R. B., and Hissom, F. K., 1982, Cold-adaptation of parainfluenza virus type 3: Induction of three phenotype markers, *J. Med. Virol.* **10:**235–242.

Bhatt, P. N., Brandt, C. D., Weiss, R. A., Fox, J. P., and Shaffer, M. F., 1966, Viral infections of monkeys in the natural habitat in southern India, *Am. J. Trop. Med. Hyg.* **15:**561–566.

Bisno, A. L., Barratt, N. P., Swanston, W. H., and Spence, L. P., 1970, An outbreak of acute respiratory disease in Trinidad associated with parainfluenza viruses, *Am. J. Epidemiol.* **91:**68–77.

Black, F. L., 1959, Measles antibodies in the population of New Haven, Conn, 1959, *J. Immunol.* **83:**74–82.

Black, F. L., 1962, Measles antibody prevalence in diverse populations, *Am. J. Dis. Child.* **103:**242–249.

Black, F. L., 1964, A nationwide serum survey of United States military recruits, 1962. III. Measles and mumps antibodies, *Am. J. Hyg.* **80:**304–307.

Black, F. L., 1965, Measles endemicity ininsular populations: Critical community size and its evolutionary implication, *Theor. Biol.* **11:**207–211.

Black, F. L., 1982, The role of herd immunity in the control of measles, *Yale J. Biol. Med.* **55:**351–360.

Black, F. L., 1989a, Measles active and passive immunity in a worldwide perspective, *Prog. Med. Virol.* **36:**1–33.

Black, F. L., 1989b, Measles, *in* "Viral Infections of Humans," 3rd ed. (A. S. Evans, ed.), pp. 450–470, Plenum Press, New York.

Black, F. L., 1989c, The age when the window of opportunity for vaccination against measles opens and closes in diverse countries, *in* "Proceedings of the Symposium on Infectious Diseases in Developing Countries," pp. 93–100, South African Medical Research Council, Johannesberg, South Africa.

Black, F. L., and Davis, D. E. M., 1968, Measles and readiness for reading and learning. II. New Haven study, *Am. J. Epidemiol.* **88:**337–344.

Black, F. L., and Houghton, W. J., 1967, The significance of mumps hemagglutination inhibition titers in normal populations, *Am. J. Epidemiol.* **85:**101–107.

Black, F. L., and Sheridan, S. R., 1960, Studies on attenuated measles vaccine. IV. Administration of vaccine by several routes, *N. Engl. J. Med.* **263:**166–170.

Black, F. L., and Singer, B., 1987, Elaboration versus simplification in refining mathematical models of infectious disease, *Annu. Rev. Microbiol.* **41:**677–701.

Black, F. L., and Yannet, H., 1960, Inapparent measles after gamma globulin administration, *J. Am. Med. Assoc.* **173:**1183–1188.

Black, F. L., Hierholzer, W. H., Woodall, J. P., and Pinheiro, F. P., 1971, Intensified reactions to measles vaccine in unexposed populations of American Indians, *J. Infect. Dis.* **124:**306–317.

Black, F. L., Hierholzer, W. H., Pinheiro, F. P., Evans, A. S., Woodall, J. P., Opton, E. M., Emmons, J. E., West, B. S., Edsall, G., Downs, W. G., and Wallace, G. D., 1974, Evidence for persistence of infectious agents in isolated populations, *Am. J. Epidemiol.* **100:**230–250.

Black, F. L., Pinheiro, F. P., Hierholzer, W. H., and Lee, R. V., 1977, Epidemiology of infectious disease: The example of measles, *in* "Health and Disease in Tribal Societies," pp. 115–143, Ciba Foundation.

Black, F. L., Berman, L. L., Borgono, J. M., Capper, R. A., Carvalho, A. A., Collins, C., Glover, O.,

Hijazi, Z., Jacobson, F. L., Lee, Y.-L., Libel, M., Linhares, A. C., Mendizabal-Morris, C. A., Simoes, E., Siqueira-Campos, E., Stevenson, J., and Vecchi, N., 1987, Geographic variation in infant loss of maternal antibody and in prevalence of rubella antibody, *Am. J. Epidemiol.* **124**:442–452.

Brandt, C. D., Kim, H. W., Arrobio, J. O., Wood, S. C., Channock, R. M., and Parrott, R. H., 1973, Epidemiology of respiratory syncytial virus infection in Washington, D.C. III. Composite analysis of eleven consecutive yearly epidemics, *Am. J. Epidemiol.* **98**:355–364.

Brandt, C. D., Kim, H. W., Channock, R. M., and Parrott, R. S., 1974, Parainfluenza epidemiology, *Pediatr. Res.* **8**:422.

Buynak, E. B., Whitman, J. E., Roehm, R. R., Morton, D. H., Lampson, G. P., and Hilleman, M. R., 1967, Comparison of neutralization and hemagglutination-inhibition techniques for measuring mumps antibody, *Proc. Soc. Exp. Biol. Med.* **125**:1068–1071.

Buynak, E. B., Weibel, R. E., McLean, A. A., and Hilleman, M. R., 1978, Live respiratory syncytial virus vaccine administered parenterally, *Proc. Soc. Exp. Biol. Med.* **157**:636–642.

Carrada Bravo, T., and Velazquez Diaz, G., 1980, El impacto del sarampion en Mexico, *Salud Pub. Mex.* **22**:359–405.

Communicable Disease Center, 1966, "Measles Surveillance, Report No. 3," Department of Health Education and Welfare Publication.

Center for Disease Control, 1973, Mumps Surveillance, DHEW Publication 73-8178, Washington, D.C.

Center for Disease Control, 1975, Mumps Surveillance January 1972–June 1974, DHEW Publication 75-8178, Washington, D.C.

Center for Disease Control, 1978, Annual summary, 1977, *Morbid. Mortal. Weekly Rep.* **26**(53):56.

Centers for Disease Control, 1982, Subacute sclerosing panencephalitis surveillance—United States, *Morbid. Mortal. Weekly Rep.* **31**:385–388.

Centers for Disease Control, 1983, Chains of measles transmission—United States, *Morbid. Mortal. Weekly Rep.* **32**:282–284.

Centers for Disease Control, 1987, Mumps outbreaks on university campuses—Illinois, Wisconsin, South Dakota, *Morbid. Mortal. Weekly Rep.* **36**:496–498, 503–505.

Centers for Disease Control, 1988, Summary of notifiable diseases. United States, 1987, *Morbid. Mortal. Weekly Rep.* **36**:(54):30.

Centers for Disease Control, 1989a, Measles—Chicago, *Morbid. Mortal. Weekly Rep.* **38**:591–592.

Centers for Disease Control, 1989b, Measles—United States, 1988, *Morbid. Mortal. Weekly Rep.* **38**:601–605.

Chalmers, A. K., 1930, "The Health of Glasgow 1818–1925," Bell and Blain, Glasgow.

Chang, W. K., 1982, A review of viral diseases in Honk Kong, *in* "Viral Diseases in South-East Asia and the Western Pacific" (J. S. Mackenzie, ed.), pp. 229–235, Academic Press, Sydney, Australia.

Channock, R. M., Kim, H. W., Brandt, C. D., and Parrott, R. H., 1989, Respiratory syncytial virus, in "Viral Infections of Humans," 3rd ed., (A. S. Evans ed.), pp. 525–544, Plenum Press, New York.

Chen, L. C., Chowdhury, A. K. M. A., and Huffman, S. L., 1980, Anthropomorphic assessment of energy-protein malnutrition and subsequent risk of mortality among preschool aged children, *Am. J. Clin. Nutr.* **33**:1836–1845.

Chin, J., Magoffin, R. L., Shearer, L. A., Schieble, J. H., and Lennette, E. H., 1969, Field evaluation of a respiratory syncytial virus vaccine and a trivalent parainfluenza virus vaccine in a pediatric population, *Am. J. Epidemiol.* **89**:449–463.

Christensen, P. E., Schmidt, H., Bang, H. O., Andersen, V., Jordal, B., and Jensen, O., 1953, An epidemic of measles in southern Greenland, 1951. Measles in virgin soil. II. The epidemic proper, *Acta Med. Scand.* **144**:430–449.

Christie, C. D., Lee-Hirsh, J., Rogall, B., Merrill, S., Ramlal, A., Karian, V., and Black, F. L., 1990, Durability of passive measles antibody in Jamaican Children, *Int. J. Epidemiol.* **19**:in press.

Cliff, A. D., and Haggett, P., 1985, The Spread of Measles in Fiji and the Pacific, Department of Human Geography, Publication 18, Australian National University, Canberra, Australia.

Cochi, S. L., Preblud, S. R., and Orenstein, W. A., 1988, Perspectives on the relative resurgence of mumps in the United States, *Am. J. Dis. Child.* **142**:499–507.

Collins, C. M., 1987, Two studies which evaluate measles control in South Africa: I. Status of maternal–fetal immunoglobulin transfer in a South African "Homeland." II. Premature vaccination in Black children and the effect on revaccination, Thesis, Yale University School of Medicine, New Haven, Connecticut.

Communicable Disease Center, 1966, Measles Surveillance, Report No. 3, DHEW, Washington, D.C.

Cooney, M. K., Fox, J. P., and Hall, C. E., 1975, The Seattle virus watch. VI. Observation of infections with and illness due to parainfluenza, mumps, and respiratory syncytial viruses and *Mycoplasma pneumoniae*, *Am. J. Epidemiol.* **101**:532–542.

Dabis, F., Sow, A., Waldman, R. J., Bikakouri, P., Senga, J., Madzou, G., and Jones, T. S., 1988, Loss of maternal antibody during infancy in an African city, *Am. J. Epidemiol.* **127**:171–178.

Denny, F. W., and Clyde, W. A., 1986, Acute lower respiratory tract infections in non-hospitalized children, *J. Pediatr.* **108**:635–646.

Detels, R., Brody, J. A., McNew, J., and Edgar, A. H., 1973, Further epidemiologic studies of subacute sclerosing panencephalitis, *Lancet* **2**:11–14.

Dickson, D., 1988, Canine distemper may be killing North Sea seals, *Science* **241**:1284.

Doraisingham, S., 1982, Viral diseases in Singapore—A national overview, in "Viral diseases in South-East Asia and the Western Pacific" (J. S. Mackenzie, ed.), pp. 229–235, Academic Press, Sydney, Australia.

Feldman, H. A., 1989, Mumps, in "Viral Infections in Humans," 3rd ed. (A. S. Evans, ed.), pp. 471–491, Plenum Press, New York.

Fine, P. E. M., and Clarkson, J. A., 1982, Measles in England and Wales—I: An analysis of factors underlying seasonal patterns, *Int. J. Epidemiol.* **11**:5–14.

Fox, J. P., Black, F. L., and Kogon, S. A., 1968, Measles and readiness for reading and learning. V. Evaluative comparison of the studies and overall conclusions, *Am. J. Epidemiol.* **88**:359–367.

Frost, W. H., 1976, Some conceptions of epidemics in general, *Am. J. Epidemiol.* **94**:179–89.

Gardner, S. D., 1969, The isolation of parainfluenza 4 subtypes A and B in England and serological studies of their prevalence, *J. Hyg.* (Camb.) **67**:545–550.

Gibbs, F. A., Gibbs, E. L., Carpenter, P. R., and Spies, H. W., 1959, Electroencephalographic abnormality in "uncomplicated" measles, *J. Am. Med. Assoc.* **171**:1050–1055.

Glezen, W. P., Paredes, A., Allison, J. E., Tabor, L. H., and Frank, A. L., 1981, Risk of respiratory syncytial virus infection for infants from low-income families in relationship to age, sex, ethnic group and maternal antibody titer, *J. Pediatr.* **98**:708–715.

Glezen, W. P., Frank, A. L., Taber, L. H., and Kasel, J. A., 1984, Parainfluenza virus type 3: Seasonality and risk of infection and reinfection in young children, *J. Infect. Dis.* **150**:851–857.

Glezen, W. P., Loda, F. A., and Denny, F. W., 1989, Parainfluenza viruses, in "Viral Infections of Humans," 3rd ed. (A. S. Evans, ed.), pp. 493–507, Plenum Press, New York.

Golubjatnikov, R., Allen, V. D., Olmos-Blancarte, M. P., and Inhorn, S. L., 1975, Serologic profile of children in a Mexican highland community: Prevalence of complement-fixing antibody to *Mycoplasma pneumoniae*, respiratory syncytial virus and parainfluenza viruses, *Am. J. Epidemiol.* **101**:458–464.

Gordon, J. E., and Kilham, L., 1949, Ten years in the epidemiology of mumps, *Am. J. Med. Sci.* **218**:338–359.

Greenberg, M., Pelliteri, O., and Eisenstein, D. T., 1955, Measles encephalitis. I. Prophylactic effect of gamma globulin, *J. Pediatr.* **46**:642–647.

Gresser, I., and Katz, S. L., 1960, Isolation of measles virus from urine, *N. Engl. J. Med.* **263**:452–454.

Grey, R. H., Char, G., Prabhakar, R., Bainbridge, R., and Johnson, B., 1986, Subacute sclerosing panencephalitis in Jamaica, *W. Indies Med. J.* **35**:27–34.

Gross, P. A., Green, R. H., and McCrea-Curnen, M. G., 1973, Persistent infection with parainfluenza type 3 virus in man, *Am. Rev. Respir. Dis.* **108**:894–898.

Guyer, B., and McBean, A. M., 1981, The epidemiology and control of measles in Yaounda, Cameroun, 1968–1975, *Int. J. Epidemiol.* **10**:263–269.

Hall, C. B., Geiman, J. M., Biggar, R., Kotok, D. I., Hogan, P. M., and Douglas, R. G., 1976, Respiratory syncyial virus infections within families, *N. Engl. J. Med.* **294**:414–419.

Halsey, N. A., Modlin, J. F., Jabbour, J. T., Dubey, L., Eddins, D. L., and Ludwig, D. L., 1980, Risk factors in SSPE: A case control study, *Am. J. Epidemiol.* **111**:415–431.

Hamer, W. H., 1906, The Milroy lectures on epidemic disease in England—The evidence of variability and persistency of type, *Lancet* **1**:733–739.

Hedrich, A. W., 1933, Monthly estimates of the child population susceptible to measles 1900–1931, *Am. J. Hyg.* **17**:613–636.

Henderson, F. W., Collier, A. M., Clyde, W. A., and Denny, F. W., 1979, Respiratory syncytial virus infection, reinfection and immunity: A prospective longitudinal study in young children, *N. Engl. Med. J.* **300**:530–535.

Hendry, R. M., Burns, J. C., Walsh, E. E., Graham, B. S., Wright, P. F., Hemming, V. G., Rodriguez, W. J., Kim, H. W., Prince, G. A., McIntosh, K., Channock, R. M., and Murphy, B. R., 1988, Strain-specific antibody responses in infants undergoing primary infection with respiratory syncytial virus, *J. Infect. Dis.* **157**:640–647.

Hinman, A. R., Brandlin-Bennett, A. D., and Nieberg, P. I., 1979, The opportunity and obligation to eliminate measles from the United States, *J. Am. Med. Assoc.* **242**:1157–1161.

Hirsh, A., 1883, "Handbook of Geographical and Historical Pathology," Vol. I, p. 163, New Sydenham Society, London.

Hsiung, G. D., Isacson, P., and Tucker, G., 1963, Studies of parainfluenza viruses. II. Serologic interrelationships in humans, *Yale J. Biol. Med.* **35**:534–544.

Hull, R. N., Minner, J. R., and Smith, J. W., 1956, New agents recovered from tissue cultures of monkey kidney cells, *Am. J. Hyg.* **63**:204–215.

John, T. J., Joseph, A., George, T. I., Radakrishnan, J., Singh, R. D. P., and George, K. G., 1980, Epidemiology and prevention of measles in rural South India, *Ind. J. Med. Res.* **72**:153–158.

Joosting, A. C. C., Harwin, R. M., Orchard, M., Martin, E., and Gear, J. H. S., 1979, Respiratory viruses in hospital patients on the Witwatersrand, *S. Afr. Med. J.* **55**:403–409.

Kapikian, A. Z., Mitchell, R. H., Channock, R. M., Shvedoff, R. A., and Stewart, C. E., 1969, An epidemiologic study of altered clinical reactivity to respiratory syncytial (RS) virus infection in children previously vaccinated with an inactivated RS virus vaccine, *Am. J. Epidemiol.* **89**:405–421.

Kaplan, K. M., Marder, D. C., Cochi, S. L., and Preblud, S. R., 1988, Mumps in the workplace. Further evidence of the changing epidemiology of a childhood vaccine-preventable disease, *J. Am. Med. Assoc.* **260**:1434–1438.

Karim, A., Ghoneim, N. H., Wood, G. T., Marsfield, M. E., and Brown, J. R., 1985, A serologic epidemiological study of parainfluenza 1, 2 and 3 viruses in beef cattle and calves in Illinois, *Res. Commun. Chem. Pathol. Pharmacol.* **47**:437–440.

Katz, S. L., 1985, Measles—Forgotten but not gone, *N. Engl. J. Med.* **313**:577–578.

Kim, H. W., Canchola, J. G., Brandt, C. D., Pyles, G., Channock, R. M., Jensen, K., and Parrott, R. H., 1969, Respiratory syncytial disease in infants despite prior administration of antigenic inactivated vaccine, *Am. J. Epidemiol.* **89**:422–434.

Kimman, T. G., Zimmer, G. M., Westrenbrink, F., Mars, J., and van Leuwen, E., 1988, Epidemiologic study of bovine respiratory syncytial virus infection in calves: Influence of maternal antibodies on the outcome of disease, *Vet. Rec.* **123**:104–109.

Kogon, A., Hall, C. E., Cooney, M. K., and Fox, J. P., 1968, Measles and readiness for reading and learning. IV Shoreline school district study, *Am. J. Epidemiol.* **88**:351–358.

Koster, F. T., Curlin, G. C., Aziz, K. M. A., and Haque, A., 1981, Synergistic impact of measles and diarrhoea on nutrition and mortality in Bangladesh, *Bull. World Health Org.* **59**:901–908.

Kress, S., Schluederberg, A. E., Hornick, R. B., Morse, L. J., Cole, J. L., Slater, E. A., and McCrumb, F. R., 1961, Studies with live attenuated measles vaccine, *Am. J. Dis. Child.* **101**:701–707.

Lennon, J. L., and Black, F. L., 1986, Maternally derived measles immunity in the era of vaccine-protected mothers, *J. Pediat.* **108**:671–676.

Mata, L. J., 1978, "The children of Santa Maria Cauque," MIT Press, Cambridge, Massachusetts.

Maynard, J. E., Felz, E. T., Wulff, H., Fortuine, R., Poland, J. D., and Chin, T. D. Y., 1967, Surveillance of respiratory infection among Alaskan Eskimo children, *J. Am. Med. Assoc.* **200**:927–931.

Maynard, J. E., Shramek, G., Noble, G. R., Deinhardt, F., and Clark, P., 1970, Use of attenuated live mumps vaccine during a "virgin soil" epidemic of mumps on St. Paul Island, Alaska, *Am. J. Epidemiol.* **92**:301–306.

McCandish, I. E., Thompson, H., Cornwell, H. G., and Wright, N. G., 1978, A study of dogs with kennel cough, *Vet. Rec.* **102**:293–301.

McCrumb, F. R., Jr., Kress, S., and Saunders, E., 1962, Studies with live attenuated measles-virus vaccine. I. Clinical and immunologic responses in institutionalized children, *Am. J. Dis. Child.* **101**:687–707.

McIntosh, K., and Fishaut, M., 1980, Immunopathologic mechanisms in lower respiratory tract disease of infants due to respiratory syncytial virus, *Prog. Med. Virol.* **26**:94–118.

Meyer, M. B., 1962, An epidemiologic study of mumps; Its spread in schools and families, *Am. J. Hyg.* **75**:259–281.

Miller, M. E., and Stiehm, E. R., 1983, Immunology and resistance to infection, *in* "Infectious Diseases of the Fetus and Newborn Infant" (J. S. Remington and J. O. Klein, eds.), Saunders, Philadelphia.

Miller, W. S., and Artenstein, M. S., 1967, Aerosol stability of three acute respiratory disease viruses, *Proc. Soc. Exp. Biol. Med.* **127**:222–227.

Modlin, J. F., Jabbour, J. T., Witte, J. J., and Halsey, N. A., 1977, Epidemiologic studies of measles, measles vaccine and subacute sclerosing panencephalitis, *Pediatrics* **59**:505–512.

Montali, R. J., Bartz, C. R., Teare, J. A., Allen, J. T., Appel, M. J. G., and Bush, M., 1983, Clinical trials with canine distemper vaccines in exotic carnivores, *J. Am. Vet. Med. Assoc.* **183**:1163–1167.

Monto, A. S., Koopman, J. S., and Bryan, E. R., 1986, The Tecumseh study of illness. XIV. Occurrence of respiratory viruses, 1976–1981, *Am. J. Epidemiol.* **124**:359–367.

Morely, D. C., Martin, W. J., and Allen, I., 1967, Measles in West Africa, *W. Afr. Med. J.* **16**:24–31.

Mufson, M. A., Mocega, H. E., and Krause, H. E., 1973, Acquisition of parainfluenza 3 virus infection by hospitalized children. I. Frequencies, rates, and temporal data, *J. Infect. Dis.* **128**:141–147.

Mufson, M. A., Belshe, R. B., Örvell, C., and Norrby, E., 1988, Respiratory syncytial virus epidemic: Variable dominance of subgroups A and B strains among children 1981–1986, *J. Infect. Dis.* **157**:143–148.

Neel, J. V., Centerwall, W. R., Chagnon, N. A., and Casey, H. L., 1970, Notes on the effect of measles and measles vaccine in a virgin-soil population of South American Indians, *Am. J. Epidemiol.* **91**:418–429.

Newell, K. W., Lehmann, A. D., Leblanc, D. R., and Osorio, N. G., 1966, The use of toxoid for the prevention of tetanus neonatorum. Final report of a double-blind controlled field trial, *Bull. World Health Org.* **35**:863–871.

Niederman, J. C., Henderson, J. R., Opton, E. M., Black, F. L., and Skrnova, K., 1967, A nationwide survey of Brazilian military recruits, 1964. II, Antibody patterns with arboviruses, polioviruses, measles and mumps, *Am. J. Epidemiol.* **86**:319–329.

Norrby, E., Sheshberadaran, H., and McCollogh, K. C., 1985, Is rinderpest the archetype of the *Morbillivirus* genus?, *Intervirology* **23**:228–232.

Ogilvie, M. M., Santhire Vathenen, A., Radford, M., Codd, J., and Key, S., 1981, Maternal antibody and respiratory syncytial virus infection in infancy, *J. Med. Virol.* **7**:263–272.

Osterhause, A. D. M. E., and Vedder, E. J., 1988, Identification of virus causing recent seal deaths, *Nature* **335**:20.

Pachman, D. J., 1988, Mumps occurring in previously vaccinated adolescents *Am. J. Dis. Child.* **142**:478–479.

Pan American Health Organization, 1988, Reported cases of EPI diseases, *EPI Newsl.* **10**:7.

Pampiglione, G., 1964, Prodromal phase of measles: Some neurophysiological studies, *Br. Med. J.* **ii**:1296–1300.

Panum, P. L., 1940, "Observations Made during the Epidemic of Measles on the Faroe Islands in the Year 1846," American Publishing Association, New York.

Papp, K., 1956, Experiences prouvant que la voie d'infection de la rougeole est la contamination de la muguese conjontivale, *Rev. Immunol.* **20**:27–36.

Parker, J. C., and Richter, C. D., 1986, Viral diseases of the respiratory system, *in* "The Mouse in Biomedical Research" (H. L. Foster, J. D. Small, and J. C. Foxe, eds.), pp. 109–158, Academic Press, New York.

Parrott, R. H., Kim, H. W., Arrobio, J. O., Hodes, D. S., Murphy, B. R., Brandt, C. D., Camargo, E., and Channock, R. M., 1973, Epidemiology of respiratory syncytial virus infection in Washington, D.C. II. Infection and disease with respect to age, immunologic status, race and sex, *Am. J. Epidemiol.* **98**:289–300.

Peart, A. F. W., and Nagler, F. P., 1954, Measles in the Canadian Arctic 1952, *Can. J. Pub. Health* **45**:146–157.

Pecenková, I., Maixnerová, M., Fedová, D., and Tumová, B., 1972, Study of the distribution of *M. parainfluenzae* in the child population by serological survey methods, *J. Hyg. Epidemiol. Microbiol. Immunol.* **16**:194–201.

Pepping, F., Hackenitz, E. A., West, C. E., Duggan, M. B., and Franken, S., 1988, Relationship between measles, malnutrition and blindness, *Am. J. Clin. Nutr.* **47**:341–343.

Philip, R. N., Reinhard, K. R., and Lackman, D. B., 1959, Observations on a mumps epidemic in a "virgin population," *Am. J. Hyg.* **69**:91–111.

Plowright, W., 1968, "Rinderpest," Springer-Verlag, Vienna.

Potash, L., Lees, R., Greenberger, J. L., Hoyrup, A., Denny, L. D., and Channock, R. M., 1970, A mutant parainfluenza type 1 virus with decreased capacity for growth at 38°C and 39°C, *J. Infect. Dis.* **121**:640–647.

Reed, D., Brown, G., Merrick, R., Sever, J., and Felz, E., 1967, A mumps epidemic on St. George Island, Alaska, *J. Am. Med. Assoc.* **199**:967–971.

Ray, R., Brown, V. E., and Compans, R. W., 1985, Glycoproteins of parainfluenza type 3: Characterization and evaluation of a subunit vaccine, *J. Infect. Dis.* **152**:1219–1230.

Richter, C. B., 1986, Mouse adenovirus, K virus, and pneumonia virus of mice, in "Viral and Mycoplasmal Infections of Laboratory Rodents" (P. N. Bhatt, R. O. Jacoby, H. C. Morse, and A. E. New, eds.), Academic Press, Orlando, Florida.

Rossi, C. R., and Keisel, G. K., 1972, Epizootiologic study of parainfluenza-3 viral infection in calves in two Alabama herds, *Am. J. Vet. Res.* **33**:2341–2349.

Rydbeck, R., Löve, A., Örvell, C., and Norrby, E., 1987, Antigenic analysis of human and bovine parainfluenza virus type 3 strain with monoclonal antibodies, *J. Gen. Virol.* **68**:2153–2160.

St. Geme, J. W., Jr., Wright, F. S., Halbertt, F., and Anderson, J. A., 1959, Failure to detect subtle neurotropism of live, attenuated measles virus vaccine, *J. Pediatr.* **70**:36–45.

Schleuderberg, A. E., Chavanich, S., Litman, M. B., and Carter, C., 1974, Comparative molecular weight estimates of measles and subacute sclerosing panencephatitis virus structural polypeptides by simultaneous electrophoresis in acrylamide gel slabs, *Biochem. Biophys. Res. Commun.* **58**:547–551.

Schopfer, K., Germann, D., Eggenburger, K., Bachler, A., and Wunderli, W., 1986, Virale respiratorishe Infectionen bei Kindern: Neue diagnostische Methode zur Fruherfassung, *Schweitzer Med. Woch.* **116**:502–507.

Schutz, F., 1925, "Die Epidemiologie der Masern," Gustav Fischer, Jena.

Scrimshaw, N. S., Solomon, J. B., Bruch, H. A., and Gordon, J. E., 1966, Studies of diarrheal disease in Central America. VIII. Measles, diarrhea and nutritional deficiency in Guatemala, *Am. J. Trop. Med. Hyg.* **15**:625–631.

Sencer, D., Dull, H. B., and Langmuir, A. D., 1967, Epidemiologic basis for eradication of measles in 1967, *Public Health Rep.* **82**:253–256.

Server, A. C., Merz, D. C., Waxham, M. N., and Wolinsky, J. S., 1982, Differentiation of mumps virus strains with monoclonal antibody, *Infect. Immun.* **35**:179–186.

Shanthikumar, S. R., Malachi, S. A., and Majiyagabe, K. A., 1985, Rinderpest outbreak in free living wildlife in Nigeria, *Vet. Rec.* **117**:469–470.

Sheshberadaran, S. R., Norrby, E., McCullogh, K. C., Carpenter, W. C., and Örvell, C., 1986, The antigenic relationship between measles, canine distemper and rinderpest viruses studied with monoclonal antibodies, *J. Gen. Virol.* **67**:1381–1392.

Shute, N., 1957, "On the Beach," Morrow, New York.

Smith, C. B., Purcell, R. H., Belanti, J. A., and Channock, R. M., 1966, Protective effect of antibody to parainfluenza virus type 1 virus, *N. Engl. J. Med.* **275**:1145–1152.

Sutton, R. N. P., 1979, Slow viruses and chronic disease of the nervous system, *Postgrad. Med.* **55**:143–149.

Taylor, W. R., Mambu, R. K., ma-Disu, M., and Weinman, J. M., 1988, Measles control in urban Africa complicated by high incidence of measles in the first year of life, *Am. J. Epidemiol.* **127**:788–794.

Thomson, J. R., Nettleton, P. F., Greig, A., and Barr, J., 1986, A bovine respiratory virus vaccination trial, *Vet. Rec.* **119:**450–453.

Tobler, L. H., and Imagawa, D. T., 1984, Mechanism of persistence with canine distemper virus: Difference between a laboratory strain and an isolate from a dog with chronic neurological disease, *Intervirology* **21:**77–86.

Tremonti, L. P., Lin, J. S. L., and Jackson, G. C., 1968, Neutralizing activity in nasal secretions and serum in volunteers to parainfluenza type 2, *J. Immunol.* **101:**572–577.

Tyeryar, F. J., Richards, L. S., and Belshe, R. B., 1978, Report of a workshop on respiratory synscytial virus and parainfluenza viruses, *J. Infect. Dis.* **137:**835–846.

Vella, P. P., Weibel, R. E., Woodhour, A. F., Mascoli, C. C., Leagus, M. B., Ittensohn, O. L., Stokes, J. J., Jr., and Hilleman, M. R., 1969, Respiratory virus vaccine. VIII. Field evaluation of trivalent parainfluenza virus vaccine among preschool children living in families 1967–68, *Am. Rev. Respir. Dis.* **99:**526–541.

Vicens, C. N., Nobrega, F. T., Joseph, J. M., and Meyer, M. M., 1966, Evaluation of tests for measurement of previous mumps infection and analysis of mumps exposure by blood group, *Am. J. Epidemiol.* **84:**371–381.

Wagonvoort, J. H. T., Harmsen, M., Khader, B. J., Kraaijveld, C. A., and Winkler, K. C., 1988, Epidemiology of mumps in the Netherlands, *J. Hyg.* **85:**313–326.

Walsh, J., 1987, War on cattle disease divides the troops, *Science* **237:**1289–1291.

Ward, K. A., Lambden, P. R., Ogilvie, M. M., and Watt, P. J., 1983, Antibodies to respiratory syncytial virus polypeptides and their significance in human infection, *J. Gen. Virol.* **64:**1867–1876.

Watt, P. J., Zardis, M., and Lambden, P. R., 1986, Age related IgG subclass response to respiratory syncytial virus fusion protein in infected infants, *Clin. Exptl. Immunol.* **64:**503–509.

WHO/UNICEF, 1987, Extended program on immunization and nutrition. Joint WHO/UNICEF statement on vitamin A for measles, *EPI Newsl. Pan Am. Health Org.* **9:**(5):6–8.

Wigley, F. M., Fructman, M. H., and Waldman, R. H., 1970, Aerosol immunization of humans with inactivated parainfluenza type 2 vaccine, *N. Engl. J. Med.* **283:**1250–1253.

Wright, P. F., Shinovaki, T., Fleet, W., Sell, S. H., Thompson, J., and Karzon, D. T., 1976, Evaluation of live, attenuated respiratory syncytial virus vaccine in infants, *J. Pediatr.* **88:**931–936.

Yorke, J. A., and London, W. P., 1973, Recurrent outbreaks of measles, chickenpox and mumps. II. Systematic differences in contact rates and stochastic effects, *Am. J. Epidemiol.* **98:**469–78.

Yorke, J. A., Nathanson, N., Pianigiani, G., and Martin, J., 1979, Seasonality and the requirements for perpetuation and eradication of viruses in populations, *Am. J. Epidemiol.* **109:**103–123.

Annotated Nucleotide and Protein Sequences for Selected *Paramyxoviridae*

Mark S. Galinski

I. INTRODUCTION

Recently, the molecular cloning and sequence analyses of a number of paramyxovirus mRNAs and genes have provided the complete nucleotide sequence and genomic organization for several different *Paramyxoviridae*, including Sendai virus (murine parainfluenza virus type 1), human parainfluenza virus type 3, measles virus, and human respiratory syncytial virus. In addition to these viruses, nucleotide sequence analyses of several Newcastle disease virus serotypes, mumps virus, simian virus 5, bovine parainfluenza virus 3, canine distemper virus, and rinderpest virus have contributed additional information concerning the molecular organization of the genomes, protein sequences and structures, and evolutionary relatedness of the various viruses in this family.

The four viral nucleotide sequences compiled here are representative of the three genera grouped within the *Paramyxoviridae* family. The genus *Paramyxovirus* is represented by Sendai virus (Blumberg *et al.*, 1984, 1985a,b; Giorgi *et al.*, 1983; Gupta and Kingsbury, 1984; Hidaka *et al.*, 1984; Morgan and Kingsbury, 1984; Shioda *et al.*, 1983, 1986) and human parainfluenza virus type 3 (Dimock *et al.*, 1986; Galinski *et al.*, 1986a,b, 1987a,b, 1988; Spriggs and Collins, 1986), *Morbillivirus* by measles virus (Alkhatib and Briedis, 1986; Bellini *et al.*, 1985, 1986; Blumberg *et al.*, 1988; Crowley *et al.*, 1988; Richardson *et al.*, 1986; Rozenblatt *et al.*, 1985), and *Pneumovirus* by the human

MARK S. GALINSKI • Department of Molecular Biology, Cleveland Clinic Foundation, Cleveland, Ohio 44195.

respiratory syncytial virus [Collins and Wertz, 1985a,b; Collins et al., 1984, 1985, 1986; Elango et al., 1985a,b; Satake et al., 1985; Wertz et al., 1985; D. Steck and P. L. Collins, unpublished data (on the L gene); M. A. Mink and P. L. Collins, unpublished data (on 3' viral RNA)]. Although all of the data presented here have been published, only some are available in the nucleic acid data banks (EMBL, GenBank, and DNA Data Bank of Japan).

The nucleotide sequences are given as cDNA in the positive-strand (5' → 3'), since almost all of the information was derived from recombinant molecular clones or by direct dideoxy sequence analysis of the viral genes. This orientation accommodates the inclusion of the amino acid sequences of the encoded proteins. The single-letter amino acid sequences of all the "known" viral proteins are shown centered below their respective codons. Selected landmarks have been emphasized in the nucleotide sequences, including the gene-end boundaries (transcription termination T_{stop}, intergenic I, and transcription initiation T_{start} sequences) and the extracistronic sequences located at the ends of the genomes. These latter sequences are believed to be analogous to the positive- and negative-strand leader sequences found in vesicular stomatitis virus-infected cells (Banerjee, 1987), although their synthesis as unique subgenomic RNAs is uncertain in some of the paramyxoviruses.

A peculiar feature of the phosphoprotein gene of some members of the *Paramyxoviridae* is the presence of overlapping and discontinuous cistrons (see Chapters 6 and 7). The accession of overlapping cistrons occurs during protein translation through initiation at alternate methionine codons. Interestingly, initiation at a nonstandard codon (leucine) occurs during synthesis of the C' protein of Sendai virus (Gupta and Padwardhan, 1988; Patwardhan and Gupta, 1988; Curran et al., 1986). The accession of discontinuous cistrons occurs through "aberrant" transcription termination, resulting in the synthesis of mRNAs containing nontemplated nucleotides. This results in specific frameshifts and fusion of the overlapping coding sequences in some of the mRNA transcripts (Thomas et al., 1988, Cattaneo et al., 1989). The synthesis of novel proteins following the fusion of two cistrons is a process which has only recently been recognized. The positions of the nucleotides which are believed to function in this "aberrant" transcription termination are underlined. In addition, the amino acids encoded in the accessed cistron have been provided.

A number of landmarks in the protein sequences of the surface glycoproteins have been indicated (boxed amino acids). These include the hydrophobic domains representing the signal peptide and transmembranal anchor of the uncleaved fusion precursor protein F_0, the amino-terminal fusion domain of the F_1 subunit polypeptide which is generated following posttranslational cleavage of F_0, and the hydrophobic domain found near the amino terminus of the receptor-binding proteins (HN, hemagglutinin-neuraminidase; H, hemagglutinin; and G, glycoprotein).

The nucleotide sequences and predicted amino acid sequences compiled here have been submitted to GenBank.

SENDAI VIRUS

```
         10        20        30        40        50        60        70        80        90       100       110       120
ACCAAACAAGAGAAAAAACATGTATGGGATATGTAATGAAGTTATACAGGATTTTAGGGTCAAAGTATCCACCCTGAGGAGCAGGTTCCAGACCCTTTGCTTTGCTGCCAAAGTTCACGA
                    EXTRACISTRONIC SEQUENCES                      T start                      NUCLEOCAPSID PROTEIN ▶

        130       140       150       160       170       180       190       200       210       220       230       240
TGGCCGGGTTGTTGAGCACCTTCGATACATTTAGCTCTAGGAGGAGCGAAAGTATTAATAAGTCGGGAAGAGGTGCTGTTATCCCCGGCCAGAGGAGCACAGTCTCAGTGTTCGTACTAG
 M  A  G  L  L  S  T  F  D  T  F  S  S  R  R  S  E  S  I  N  K  S  G  R  G  A  V  I  P  G  Q  R  S  T  V  S  V  F  V  L

        250       260       270       280       290       300       310       320       330       340       350       360
GCTTAAGTGTGACTGATGATGCAGACAAGTTATTCATTGCAACTACCTTCCTAGCTCACTCATTGGACACAGATAAGCGGCACTCTCAGAGAGGGGGGTTCCTCGTCTCTCTGCTTGCCA
 G  L  S  V  T  D  D  A  D  K  L  F  I  A  T  T  F  L  A  H  S  L  D  T  D  K  R  H  S  Q  R  G  G  F  L  V  S  L  L  A

        370       380       390       400       410       420       430       440       450       460       470       480
TGGCTTACAGTAGTCCAGAATTGTACTTGACAACAAACGGAGTAAACGCCGATGTCAAATATGTGATCTACAACATAGAGAAAGACCCTAAGAGGACGAAGACAGACGGATTCATTGTGA
 M  A  Y  S  S  P  E  L  Y  L  T  T  N  G  V  N  A  D  V  K  Y  V  I  Y  N  I  E  K  D  P  K  R  T  K  T  D  G  F  I  V

        490       500       510       520       530       540       550       560       570       580       590       600
AGACGAGAGATATGGAATATGAGAGGACCCACAGAATGGCTGTTTGGACCTATGGTCAACAAGAGCCCACTCTTCCAGGGTCAACGGGATGCTGCAGACCCTGACACACTCCTTCAAATCT
 K  T  R  D  M  E  Y  E  R  T  T  E  W  L  F  G  P  M  V  N  K  S  P  L  F  Q  G  Q  R  D  A  A  D  P  D  T  L  L  Q  I

        610       620       630       640       650       660       670       680       690       700       710       720
ATGGGTATCCTGCATGCCTAGGAGCAATAATTGTCCAAGTCTGGATTGTGCTGGTGAAGGCCATCACAAGCAGCGCCGGCTTAAGGAAAGGGTTCTTCAACAGGTTAGAGGCGTTCAGAC
 Y  G  Y  P  A  C  L  G  A  I  I  V  Q  V  W  I  V  L  V  K  A  I  T  S  S  A  G  L  R  K  G  F  F  N  R  L  E  A  F  R

        730       740       750       760       770       780       790       800       810       820       830       840
AAGACGGCACCGTGAAAGGTGCCTTAGTTTTCACTGGGGAGACAGTTGAGGGGTAGGCTCGGTTATGAGATCTCAGCCAAAGCCTTGTATCTCTCATGGTTGAGACCCTTGTGACTATGA
 Q  D  G  T  V  K  G  A  L  V  F  T  G  E  T  V  E  G  I  G  S  V  M  R  S  Q  Q  S  L  V  S  L  M  V  E  T  L  V  T  M

        850       860       870       880       890       900       910       920       930       940       950       960
ATACTGCAAGATCTGATCTCACCACATTAGAGAAGAACATCCAGATCGTTGGGAACTACATCCGAGATGCAGGGCTGGCTTCCTTCATGAACACTATTAAATATGGGGTGGAAACAAAGA
 N  T  A  R  S  D  L  T  T  L  E  K  N  I  Q  I  V  G  N  Y  I  R  D  A  G  L  A  S  F  M  N  T  I  K  Y  G  V  E  T  K

        970       980       990      1000      1010      1020      1030      1040      1050      1060      1070      1080
TGGCAGCTCTAACGTTGTCAAACCTGAGGCCCGATATTAATAAGCTTAGAAGCCTCATAGACACCTACCTGTCAAAAGGCCCCAGAGCTCCCTTTATCTGTATCCTCAAGGACCCTGTTC
 M  A  A  L  T  L  S  N  L  R  P  D  I  N  K  L  R  S  L  I  D  T  Y  L  S  K  G  P  R  A  P  F  I  C  I  L  K  D  P  V

       1090      1100      1110      1120      1130      1140      1150      1160      1170      1180      1190      1200
ATGGTGAATTTGCTCCAGGCAATTATCCTGCACTATGGAGTTACGCCATGGGAGTCGCCGTCGTACAGAACAAGGCAATGCAGCAGTACGTCACAGGGAGGACATACCTTGATATGGAAA
 H  G  E  F  A  P  G  N  Y  P  A  L  W  S  Y  A  M  G  V  A  V  V  Q  N  K  A  M  Q  Q  Y  V  T  G  R  T  Y  L  D  M  E

       1210      1220      1230      1240      1250      1260      1270      1280      1290      1300      1310      1320
TGTTCTTACTAGGACAAGCCGTGGCAAAGGACGCTGAATCGAAGATTAGCAGTGCCTTGGAAGATGAGTTAGGAGTGACGGAAGCAGCCAAGGGGAGGCTCAGACATCATCTGGCGAGCT
 M  F  L  L  G  Q  A  V  A  K  D  A  E  S  K  I  S  S  A  L  E  D  E  L  G  V  T  E  A  A  K  G  R  L  R  H  H  L  A  S

       1330      1340      1350      1360      1370      1380      1390      1400      1410      1420      1430      1440
TGTCCGGTGGAAATGGTGCTTACCGCAAACCAACAGCCGGTGGTGCAATTGAGGTAGCTCTAGACAATGCCGACATCGACCTAGAAACAAAAGCCCATGCGGACCAGGACGCTAGGGGTT
 L  S  G  G  N  G  A  Y  R  K  P  T  G  G  G  A  I  E  V  A  L  D  N  A  D  I  D  L  E  T  K  A  H  A  D  Q  D  A  R  G

       1450      1460      1470      1480      1490      1500      1510      1520      1530      1540      1550      1560
GGGGTGGAGATAGTGGTGAAAGATGGGCACGTCAGGTGAGTGGTGGCCACTTTGTCACACTACATGGGGCTGAACGGTTAGAGGAGGAAACCAATGATGAGGATGTATCAGACATAGAGA
 W  G  G  D  S  G  E  R  W  A  R  Q  V  S  G  G  H  F  V  T  L  H  G  A  E  R  L  E  E  E  T  N  D  E  D  V  S  D  I  E

       1570      1580      1590      1600      1610      1620      1630      1640      1650      1660      1670      1680
GAAGAATAGCCATGAGACTGGCAGAGAGACGGCAAGAGATTCTGCAACCCATGGAGATGAAGGCCGCAATAACGGTGTCGATTATGACGAAGATGACGATACCGCAGCAGTAGCTGGGGGT
 R  R  I  A  M  R  L  A  E  R  R  Q  E  I  L  Q  P  M  E  M  K  A  A  I  T  V  S  I  M  T  K  M  T  I  P  Q  Q  *

       1690      1700      1710      1720      1730      1740      1750      1760      1770      1780      1790      1800
AGGAGGAATCTAGGATCATACGAGGCTTCAAGGTACTTGATCCGTAGTAAGAAAAACTTAGGGTGAAAGTTCATCCACCGATCGGCTCAGGCAAGGCCACACCCAACCCCACCGACCACA
                                   T stop    I   T start

       1810      1820      1830      1840      1850      1860      1870      1880      1890      1900      1910      1920
CCCAGCAGTCGAGACAGCCACGGCTTCGGCTACACTTACCGCATGGATCAAGATGCCTTCATTCTTAAAGAAGATTCTGAAGTTGAGAGGGAGGCGCCAGGAGGACGAGAGTCGCTCTCG
PHOSPHOPROTEIN ▶                     M  D  Q  D  A  F  I  L  K  E  D  S  E  V  E  R  E  A  P  G  G  R  E  S  L  S
                          T  A  S  A  T  L  T  A  W  I  K  M  P  S  F  L  K  K  I  L  K  L  R  G  R  R  Q  E  D  E  S  R  S  R
C & C' PROTEINS ▶
       1930      1940      1950      1960      1970      1980      1990      2000      2010      2020      2030      2040
GATGTTATCGGATTCCTCGATGCTGTCCTGTCGAGTGAACCAACTGACATCGGAGGGGACAGAAGCTGGCTCCACAACACCATCAACACTCCCCAAGGACCAGGCTCTGCTCATAGAGCC
 D  V  I  G  F  L  D  A  V  L  S  S  E  P  T  D  I  G  G  D  R  S  W  L  H  N  T  I  N  T  P  Q  G  P  G  S  A  H  R  A
 M  L  S  D  S  S  M  L  S  C  R  V  N  Q  L  T  S  E  G  T  E  A  G  S  T  T  P  S  T  L  P  K  D  Q  A  L  L  I  E  P

       2050      2060      2070      2080      2090      2100      2110      2120      2130      2140      2150      2160
AAAAGTGAGGGCGAAGGAGAAGTCTCAACACCGTCGACCCAAGATAATCGATCAGGTGAGGAGAGTAGAGTCTCTGGGGAACAAGCAAGCCGAGGCAGAAGCACATGCTGGAAACCTT
 K  S  E  G  E  G  E  V  S  T  P  S  T  Q  D  N  R  S  G  E  E  S  R  V  S  G  R  T  S  K  P  E  A  E  A  H  A  G  N  L
 K  V  R  A  K  E  K  S  Q  H  R  R  P  K  I  I  D  Q  V  R  R  V  E  S  L  G  E  Q  A  S  Q  R  Q  K  H  M  L  E  T  L
```

```
        2170      2180      2190      2200      2210      2220      2230      2240      2250      2260      2270      2280
GATAAACAAAATATACACCGGGCCTTTGGGGGAAGAACTGGTACAAACTCTGTATCTCAGGATCTGGGCGATGGAGGAGACTCCGGAATCCTTGAAAATCCTCCAAATGAGAGAGGATAT
D  K  Q  N  I  H  R  A  F  G  G  R  T  G  T  N  S  V  S  Q  D  L  G  D  G  G  D  S  G  I  L  E  N  P  P  N  E  R  G  Y
 I  N  K  I  Y  T  G  P  L  G  E  E  L  V  Q  T  L  Y  L  R  I  W  A  M  E  E  T  P  E  S  L  K  I  L  Q  M  R  E  D  I

        2290      2300      2310      2320      2330      2340      2350      2360      2370      2380      2390      2400
CCGAGATCAGGTATTGAAGATGAAAACAGAGAGATGGCTGCGCACCCTGATAAGAGGGGAGAAGACCAAGCTGAAGGACTTCCAGAAGAGGTACGAGGAAGTACATCCCTACCTGATGAA
P  R  S  G  I  E  D  E  N  R  E  M  A  A  H  P  D  K  R  G  E  D  Q  A  E  G  L  P  E  E  V  R  G  S  T  S  L  P  D  E
 R  D  Q  V  L  K  M  K  T  E  R  W  L  R  T  L  I  R  G  E  K  T  K  L  K  D  F  Q  K  R  Y  E  E  V  H  P  Y  L  M  K

        2410      2420      2430      2440      2450      2460      2470      2480      2490      2500      2510      2520
GGAGAAGGTGGAGCAAGTAATAATGGAAGAAGCATGGAGCCTGGCAGCTCACATAGTGCAAGAGTAACTGGGGTCCTGGTGATTCCTAGCCCCGAACTTGAAGAGGCTGTGCTACGGAGG
G  E  G  G  A  S  N  N  G  R  S  M  E  P  G  S  S  H  S  A  R  V  T  G  V  L  V  I  P  S  P  E  L  E  E  A  V  L  R  R
 E  K  V  E  Q  V  I  M  E  E  A  W  S  L  A  A  H  I  V  Q  E  *

        2530      2540      2550      2560      2570      2580      2590      2600      2610      2620      2630      2640
AACAAAAGAAGACCTACCAACAGTGGGTCCAAACCTCTTACTCCAGCAACCGTGCCTGGCACCCGGTCCCCACCGCTGAATCGTTACAACAGCACAGGGTCACCACCAGGAAAACCCCCA
N  K  R  R  P  T  N  S  G  S  K  P  L  T  P  A  T  V  P  G  T  R  S  P  P  L  N  R  Y  N  S  T  G  S  P  P  G  K  P  P
 E  Q  K  K  T  Y  Q  Q  W  V  Q  T  S  Y  S  S  N  R  A  W  H  P  V  P  T  A  E  S  L  Q  Q  H  R  V  T  T  R  K  T  P

        2650      2660      2670      2680      2690      2700      2710      2720      2730      2740      2750      2760
TCTACACAGGATGAGCACATCAACTCTGGGGACACCCCCGCCGTCAGGGTCAAAGACCGGAAACCACCAATAGGGACCCGCTCTGTCTCAGATTGTCCAGCCAACGGCCGTTCAATCCAC
S  T  Q  D  E  H  I  N  S  G  D  T  P  A  V  R  V  K  D  R  K  P  P  I  G  T  R  S  V  S  D  C  P  A  N  G  R  S  I  H

        2770      2780      2790      2800      2810      2820      2830      2840      2850      2860      2870      2880
CCGGGTCTAGAGACCGACTCAACAAAAAAGGGCATAGGAGAGAACACATCATCTATGAAAGAGATGGCTACATTGTTGACGAGTCTTGGTGTAATCCAGTCTGCTCAAGAATTCGAATCA
P  G  L  E  T  D  S  T  K  K  G  I  G  E  N  T  S  S  M  K  E  M  A  T  L  L  T  S  L  G  V  I  Q  S  A  Q  E  F  E  S
    V CISTRON  ▶    G  H  R  R  E  H  I  I  Y  E  R  D  G  Y  I  V  D  E  S  W  C  N  P  V  C  S  R  I  R  I
```

V CISTRON ▶

```
        2890      2900      2910      2920      2930      2940      2950      2960      2970      2980      2990      3000
TCCCGAGACGCGAGTTATGTGTTTGCAAGACGTGCCCTAAAGTCTGCAAACTATGCAGAGATGACATTCAATGTATGCGGCCTGATCCTTTCTGCCGAGAAATCTTCCGCTCGTAAGGTA
S  R  D  A  S  Y  V  F  A  R  R  A  L  K  S  A  N  Y  A  E  M  T  F  N  V  C  G  L  I  L  S  A  E  K  S  S  A  R  K  V
 I  P  R  R  E  L  C  V  C  K  T  C  P  K  V  C  K  L  C  R  D  D  I  Q  C  M  R  P  D  P  F  C  R  E  I  F  R  S  *

        3010      3020      3030      3040      3050      3060      3070      3080      3090      3100      3110      3120
GATGAGAACAAACAACTGCTCAAACAGATCCAAGAGAGCGTGGAGTCATTCCGGGACACCTATAAGAGATTCTCTGAGTATCAGAAAGAACAGAACTCATTGCTGATGTCCAACCTATCT
D  E  N  K  Q  L  L  K  Q  I  Q  E  S  V  E  S  F  R  D  T  Y  K  R  F  S  E  Y  Q  K  E  Q  N  S  L  L  M  S  N  L  S

        3130      3140      3150      3160      3170      3180      3190      3200      3210      3220      3230      3240
ACACTTCATATCATCACAGATAGAGGTGGCAAGACTGACAACACAGACTCCCTTACAAGGTCCCCCTCCGTTTTTGCAAAATCAAAAGAGAACAAGACTAAGGCTACCAGGTTTGACCCA
T  L  H  I  I  T  D  R  G  G  K  T  D  N  T  D  S  L  T  R  S  P  S  V  F  A  K  S  K  E  N  K  T  K  A  T  R  F  D  P

        3250      3260      3270      3280      3290      3300      3310      3320      3330      3340      3350      3360
TCTATGGAGACCCTAGAAGATATGAAGTACAAACCGGACCTAATCCGAGAGGATGAATTTAGAGATGAGATCCGCAACCCGGTGTACCAAGAGAGGGACACAGAACCTAGGGCCTCAAAC
S  M  E  T  L  E  D  M  K  Y  K  P  D  L  I  R  E  D  E  F  R  D  E  I  R  N  P  V  Y  Q  E  R  D  T  E  P  R  A  S  N

        3370      3380      3390      3400      3410      3420      3430      3440      3450      3460      3470      3480
GCATCCACGTCTCCTCCCCCTCCAAAGAGAAGCCCACAATGCACTCTCTCAGGCTCGTCATAGAGAGCAGTCCCCTAAGCAGAGCTGAGAAAGCAGCATATGTGAAATCATTATCCAAGTGC
A  S  R  L  L  P  S  K  E  K  P  T  M  H  S  L  R  L  V  I  E  S  S  P  L  S  R  A  E  K  A  A  Y  V  K  S  L  S  K  C

        3490      3500      3510      3520      3530      3540      3550      3560      3570      3580      3590      3600
AAGACAGACCAAGAGGTTAAGGCAGTCATGGAACTCGTAGAAGAGGACATAGAGTCACTGACCAACTAGATCCCGGGTGAGGCATCCTACCATCCTCAGTCATAGAGAGATCCAATCTAC
K  T  D  Q  E  V  K  A  V  M  E  L  V  E  E  D  I  E  S  L  T  N  *

        3610      3620      3630      3640      3650      3660      3670      3680      3690      3700      3710      3720
CATCAGCATCAGCCAGTAAAGATTAAGAAAAACTTAGGGTGAAAGAAATTTCACCTAACACGGCGCAATGGCAGATATCTATAGATTCCCTAAGTTCTCATATGGGGATAACGGTACTGT
                                                                                        M  A  D  I  Y  R  F  P  K  F  S  Y  E  D  N  G  T  V
```

```
          ⌐─┐⌐⌐┐⌐┐
           T_stop  I  T_start          MATRIX PROTEIN ▶
```

MATRIX PROTEIN ▶

```
        3730      3740      3750      3760      3770      3780      3790      3800      3810      3820      3830      3840
GGAGCCCCTGCCTCTGAGGACTGGTCCGGATAAGAAAGCCATCCCCCACATCAGGATTGTCAAGGTAGGAGACCCTCCTAAACATGGAGTGAGATACCTAGATTTATTGCTCTTGGGTTT
E  P  L  P  L  R  T  G  P  D  K  K  A  I  P  H  I  R  I  V  K  V  G  D  P  P  K  H  G  V  R  Y  L  D  L  L  L  L  G  F

        3850      3860      3870      3880      3890      3900      3910      3920      3930      3940      3950      3960
CTTTGAGACACCGAAACAAACAACCAATCTAGGGAGCGTATCTGACTTGACAGAGCCGACCAGCTACTCAATATGCGGCTCCGGGTCGTTACCCATAGGTGTGGCCAAATACTACGGGAC
F  E  T  P  K  Q  T  T  N  L  G  S  V  S  D  L  T  E  P  T  S  Y  S  I  C  G  S  G  S  L  P  I  G  V  A  K  Y  Y  G  T

        3970      3980      3990      4000      4010      4020      4030      4040      4050      4060      4070      4080
TGATCAGGAACTCTTAAAGGCCTGCACCGATCTCAGAATTACGGTGAGGAGGACTGTTCGAGCAGGAGAGGATGATCGTATACATGGTGGATTCGATTGGTGCTCCACTCCTACCAGTGTC
D  Q  E  L  L  K  A  C  T  D  L  R  I  T  V  R  R  T  V  R  A  G  E  M  I  V  Y  M  V  D  S  I  G  A  P  L  L  P  W  S

        4090      4100      4110      4120      4130      4140      4150      4160      4170      4180      4190      4200
AGGCAGGCTGAGACAGGGAATGATATTTAATGCAAACAAGGTCGCACTAGCTCCCCAATGCCTCCCTGTGGACAAGGACTAAGACTCAGAGTGGTGTTTGTCAATGGGACATCTCTAGG
G  R  L  R  Q  G  M  I  F  N  A  N  K  V  A  L  A  P  Q  C  L  P  V  D  K  D  I  R  L  R  V  V  F  V  N  G  T  S  L  G
```

```
      4210      4220      4230      4240      4250      4260      4270      4280      4290      4300      4310      4320
GGCAATCACCATAGCCAAGATCCCAAAGACCCTTGCAGACCTTGCATTGCCCAACTCTATATCCGTTAATTTACTGGTGACACTCAAGACCGGGATCTCCACAGAACAAAAGGGGGTACT
  A  I  T  I  A  K  I  P  K  T  L  A  D  L  A  L  P  N  S  I  S  V  N  L  L  V  T  L  K  T  G  I  S  T  E  Q  K  G  V  L

      4330      4340      4350      4360      4370      4380      4390      4400      4410      4420      4430      4440
CCCAGTACTTGATGATCAAGGGGAGAAAAAGCTCAATTTTATGGTGCACCTCGGGTTGATCAGGAGAAAGGTCGGGAAGATATACTCTGTTGAGTACTGCAAGAGCAAGATTGAGAGAAT
  P  V  L  D  D  Q  G  E  K  K  L  N  F  M  V  H  L  G  L  I  R  R  K  V  G  K  I  Y  S  V  E  Y  C  K  S  K  I  E  R  M

      4450      4460      4470      4480      4490      4500      4510      4520      4530      4540      4550      4560
GCGGCTGATTTTCTCACTTGGGTTAATCGGCGGTATAAGCTTCCATGTTCAGGTTAATGGGACACTATCTAAGACATTCATGAGTCAGCTCGCATGGAAGAGGGCAGTCTGCTTCCCATT
  R  L  I  F  S  L  G  L  I  G  G  I  S  F  H  V  Q  V  N  G  T  L  S  K  T  F  M  S  Q  L  A  W  K  R  A  V  C  F  P  L

      4570      4580      4590      4600      4610      4620      4630      4640      4650      4660      4670      4680
AATGGATGTGAATCCCCATATGAACATGGTGATTTGGGCGGCATCTGTAGAAATCACAGGCGTCGATGCGGTGTTCCAACCGGCCATCCCTCGTGATTTCCGCTACTACCCTAATGTTGT
  M  D  V  N  P  H  M  N  M  V  I  W  A  A  S  V  E  I  T  G  V  D  A  V  F  Q  P  A  I  P  R  D  F  R  Y  Y  P  N  V  V

      4690      4700      4710      4720      4730      4740      4750      4760      4770      4780      4790      4800
GGCTAAGAACATCGGAAGGATCAGAAAGCTGTAAATGTGCACCCATCAGAGACCTGCGACAATGCCCCAAGCAGACACCACCTGGCAGTCGGAGCCACCGGGTCACTCCTTGTCTTAAAT
  A  K  N  I  G  R  I  R  K  L  *
```

```
      4810      4820      4830      4840      4850      4860      4870      4880      4890      4900      4910      4920
AAGAAAAACTTAGGGATAAAGTCCCTTGTGAGTGCTTGGTTGCAAAACTCTCCCCTTGGGAAACATGACAGCATATATCCAGAGATCACAGTGCATCTCAACATCACTACTGGTTGTTCT
```

T stop   I   T start   **FUSION PROTEIN ▶**

```
                                                             M  T  A  V  I  Q  R  S  Q  C  I  S  T  S  L  L  V  V  L
```

```
      4930      4940      4950      4960      4970      4980      4990      5000      5010      5020      5030      5040
CACCACATTGGTCTCGTGTCAGATTCCCAGGGATAGGCTCTCTCAACATAGGGGTCATAGTCGATGAAGGGAAATCACTGAAGATAGCTGGATCCCACGAATCGAGGTACATAGTACTGAG
  T  T  L  V  S  C  Q  I  P  R  D  R  L  S  N  I  G  V  I  V  D  E  G  K  S  L  K  I  A  G  S  H  E  S  R  Y  I  V  L  S

      5050      5060      5070      5080      5090      5100      5110      5120      5130      5140      5150      5160
TCTAGTTCCGGGGGTAGACTTTGAGAATGGGTGCGGAACAGCCCAGGTTATCCAGTACAAGAGCCTACTGAACAGGCTGTTAATCCCATTGAGGGATGCCTTAGATCTTCAGGAGGCTCT
  L  V  P  G  V  D  F  E  N  G  C  G  T  A  Q  V  I  Q  Y  K  S  L  L  N  R  L  L  I  P  L  R  D  A  L  D  L  Q  E  A  L

      5170      5180      5190      5200      5210      5220      5230      5240      5250      5260      5270      5280
GATAACTGTCACCAATGATACGACACAAAATGCCGGTGCTCCCCAGTCGAGATTCTTCGGTGCTGTGATTGGTACTATCGCACTTGGAGTGGCGACATCAGCACAAATCACCGCAGGGAT
  I  T  V  T  N  D  T  T  Q  N  A  G  A  P  Q  S  R  F  F  G  A  V  I  G  T  I  A  L  G  V  A  T  S  A  Q  I  T  A  G  I

      5290      5300      5310      5320      5330      5340      5350      5360      5370      5380      5390      5400
TGCACTAGCCGAAGCGAGGGAGGCCAAAAGAGACATAGCGCTCATCAAAGAATCGATGACAAAAACACACAAGTCTATAGAACTGCTGCAAAACGTGTGGGGGAACAAATTCTTGCTCT
  A  L  A  E  A  R  E  A  K  R  D  I  A  L  I  K  E  S  M  T  K  T  H  K  S  I  E  L  L  Q  N  A  V  G  E  Q  I  L  A  L

      5410      5420      5430      5440      5450      5460      5470      5480      5490      5500      5510      5520
AAAGACACTCCAGGATTTCGTGAATGATGAGATCAAACCCGCAATAAGCGAATTAGGCTGTGAGACTGCTGCCTTAAGACTGGGTATAAAATTGACACAGCATTACTCCGAGCTGTTAAC
  K  T  L  Q  D  F  V  N  D  E  I  K  P  A  I  S  E  L  G  C  E  T  A  A  L  R  L  G  I  K  L  T  Q  H  Y  S  E  L  L  T

      5530      5540      5550      5560      5570      5580      5590      5600      5610      5620      5630      5640
TGCCGTTCGGCTCGAATTTCGGAACCATCGGAGAGAAGAGCCTCACGCTGCAGGCGCTGTCTTCACTTTACTCTGCTAACATTACTGAGATTATGACCACAATCAGGACAGGGCAGTCTAA
  A  F  G  S  N  F  G  T  I  G  E  K  S  L  T  L  Q  A  L  S  S  L  Y  S  A  N  I  T  E  I  M  T  T  I  R  T  G  Q  S  N

      5650      5660      5670      5680      5690      5700      5710      5720      5730      5740      5750      5760
CATCTCTGATGTCATTTATACAGAACAGATCAAAGGAACGGTGATAGATGTGGATCTAGAGAGATACATGGTCACCCTGTCTGTGAAGATCCCTATTCTTTCTGAAGTCCCAGGTGTGCT
  I  S  D  V  I  Y  T  E  Q  I  K  G  T  V  I  D  V  D  L  E  R  Y  M  V  T  L  S  V  K  I  P  I  L  S  E  V  P  G  V  L

      5770      5780      5790      5800      5810      5820      5830      5840      5850      5860      5870      5880
CATACACAAGGCATCATCTATTTCTTACAACATAGACGGGGAGGAATGGTATGTGACTGCCCCAAGCCATATACTCAGTCGTGCTTCTTTCTTAGGGGGTGCAGACATAACCGATTGTGT
  I  H  K  A  S  S  I  S  Y  N  I  D  G  E  E  W  Y  V  T  A  P  S  H  I  L  S  R  A  S  F  L  G  G  A  D  I  T  D  C  V

      5890      5900      5910      5920      5930      5940      5950      5960      5970      5980      5990      6000
TGAGTCCAGATTGACCTATATATGCCCCAGGGATCCCGCACAACTGATACCTGACAGCCAGCAAAAGTGTATCCTGGGGGACACAACAAGGTGTCCTGTCACAAAAGTTGTGGACAGCCT
  E  S  R  L  T  Y  I  C  P  R  D  P  A  Q  L  I  P  D  S  Q  Q  K  C  I  L  G  D  T  T  R  C  P  V  T  K  V  V  D  S  L

      6010      6020      6030      6040      6050      6060      6070      6080      6090      6100      6110      6120
TATCCCCAAGTTTGCTTTTGTGAATGGGGGCGTTGTTGCTAACCGCATAGCATCCACATGTACCTGCGGGACAGGCCGAGACCAATCAGTCAGGATCGCTCTAAAGGTGTAGCATTCCT
  I  P  K  F  A  F  V  N  G  G  V  V  A  N  R  I  A  S  T  C  T  C  G  T  G  R  R  P  I  S  Q  D  R  S  K  G  V  A  F  L

      6130      6140      6150      6160      6170      6180      6190      6200      6210      6220      6230      6240
AACCCATGACAACTGTGGTCTTATAGGTGTCAATGGGGTAGAATTGTATGCTAACCGGAGAGGGCACGATGCCACTTGGGGGGTCCAGAACTTGACAGTCGGTCCTGCAATTGCTATCAG
  T  H  D  N  C  G  L  I  G  V  N  G  V  E  L  Y  A  N  R  R  G  H  D  A  T  W  G  V  Q  N  L  T  V  G  P  A  I  A  I  R

      6250      6260      6270      6280      6290      6300      6310      6320      6330      6340      6350      6360
ACCCGTTGATATTTCTCTCAACCTTGCTGATGCTACAAATTTCCTGCAAGACTCTAAGGCTGAGCTTGAGAAAGCACGGAAAATCCTCTCTGAGGTAGGTAGATGGTACAACTCAAGAGA
  P  V  D  I  S  L  N  L  A  D  A  T  N  F  L  Q  D  S  K  A  E  L  E  K  A  R  K  I  L  S  E  V  G  R  W  Y  N  S  R  E

      6370      6380      6390      6400      6410      6420      6430      6440      6450      6460      6470      6480
GACTGTGATTACGATCCATAGTAGTTTATGGTCGTAATATTGGTGGTCATTATAGTGATCGTCATCGTGCTTTATAGACTCAAAAGGTCAATGCTAATGGGTAATCCAGATGAGCGTATACC
  T  V  I  T  I  I  V  V  M  V  V  I  L  V  V  I  I  V  I  V  L  Y  R  L  K  R  S  M  L  M  G  N  P  D  E  R  I  P
```

```
      6490      6500      6510      6520      6530      6540      6550      6560      6570      6580      6590      6600
GAGGGACACATATACATTAGAGCCGAAGATCAGACATATGTACACAAACGGTGGGTTTGATGCGATGGCTGAGAAACGATGATCACGACCTATTCAGATGTCTTGTAAAGCAGGCATGG
 R  D  T  Y  T  L  E  P  K  I  R  H  M  Y  T  N  G  G  F  D  A  M  A  E  K  R  *  S  R  P  L  S  D  V  L  *
```

```
      6610      6620      6630      6640      6650      6660      6670      6680      6690      6700      6710      6720
TATCCGTTGAGATCTGTATATAATAAGAAAAACTTAGGTGAAAGTGAGGTCGCGCGGTACTTTAGCTTTCACCCCAAACTTGCACAGATCATGGATGGTGATAGGGGCAAACGTGACTCG
                                                                                          M  D  G  D  R  G  K  R  D  S
```

T_{stop} I T_{start} **HEMAGGLUTININ-NEURAMINIDASE PROTEIN ▶**

```
      6730      6740      6750      6760      6770      6780      6790      6800      6810      6820      6830      6840
TACTGGTCTACCTCTCCTAGTGGTAGCACTACAAAATTAGCATCAGGTTGGGAGAGGTCAAGTAAAGTTGACACCTGGTTGCTGATTCTCTCATTCACCCAGTGGGCTTTGTCAATTGCC
 Y  W  S  T  S  P  S  G  S  T  T  K  L  A  S  G  W  E  R  S  S  K  V  D  T  W  L  L  I  L  S  F  T  Q  W  A  L  S  I  A
```

```
      6850      6860      6870      6880      6890      6900      6910      6920      6930      6940      6950      6960
ACGGTGATCATCTGTATCATAATTTCTGCTAGACAAGGGTATAGTACGAAAGAGTACTCAATGACTGTAGAGGCATTGAACATGAGCAGCAGGGAGGTGAAAGAGTCACTTACCAGTCTA
 T  V  I  I  C  I  I  I  S  A  R  Q  G  Y  S  T  K  E  Y  S  M  T  V  E  A  L  N  M  S  S  R  E  V  K  E  S  L  T  S  L
```

```
      6970      6980      6990      7000      7010      7020      7030      7040      7050      7060      7070      7080
ATAAGGCAAGAGGTTATCGCAAGGGCTGTCAACATTCAGAGCTCTGTGCAAACCGGAATCCCAGTCTTGTTGAACAAAAACAGCAGGGATGTCATCCAGATGATTGATAAGTCGTGCAGC
 I  R  Q  E  V  I  A  R  A  V  N  I  Q  S  S  V  Q  T  G  I  P  V  L  L  N  K  N  S  R  D  V  I  Q  M  I  D  K  S  C  S
```

```
      7090      7100      7110      7120      7130      7140      7150      7160      7170      7180      7190      7200
AGACAAGAGCTCACTCAGCTCTGTGAGAGTACGATCGCAGTCCAACCATCCCGAGGGAATTGCCCCACTTGAGCCACATAGTTTCTGGGATGCCCTGTCGGAGAACCGTATCTTAGCTCA
 R  Q  E  L  T  Q  L  C  E  S  T  I  A  V  H  H  P  E  G  I  A  P  L  E  P  H  S  F  W  R  C  P  V  G  E  P  Y  L  S  S
```

```
      7210      7220      7230      7240      7250      7260      7270      7280      7290      7300      7310      7320
GATCCTAAAATCTCATTGCTGCTTGGTCCGAGCTTGTTATCTGGTTCTACAACGATCTCTGGATGTGTTAGGCTCCCTTCACTCTCAATTGGCGAGGCAATCTATGCCTATTCATCAAAT
 D  P  K  I  S  L  L  L  G  P  S  L  L  S  G  S  T  T  I  S  G  C  V  R  L  P  S  L  S  I  G  E  A  I  Y  A  Y  S  S  N
```

```
      7330      7340      7350      7360      7370      7380      7390      7400      7410      7420      7430      7440
CTCATTACACAAGGTTGTGCTGACATAGGGAAATCATATCAGGTCCTGCGACTAGGGTACATATCACTCAATTCAGATATGTTCCCTGATCTTAACCCCGTAGTGTCCCACACTTATGAC
 L  I  T  Q  G  C  A  D  I  G  K  S  Y  Q  V  L  Q  L  G  Y  I  S  L  N  S  D  M  F  P  D  L  N  P  V  V  S  H  T  Y  D
```

```
      7450      7460      7470      7480      7490      7500      7510      7520      7530      7540      7550      7560
ATCAACGACAATCGGAAATCATGCTCTGTGGTGGCAACCGGGACTAGGGGTTTATCAGCTTTGCTCCATGCCGACTGTAGACGAAAGAACCGACTACTCTAGTGATGGTATCGAGGATCTG
 I  N  D  N  R  K  S  C  S  V  V  A  T  G  T  R  G  Y  Q  L  C  S  M  P  T  V  D  E  R  T  D  Y  S  S  D  G  I  E  D  L
```

```
      7570      7580      7590      7600      7610      7620      7630      7640      7650      7660      7670      7680
GTCCTTGATGTCCTGGATCTCAAAGGGAGCACTAAGTCTCACCGGTATCGCAACAGCGAGGTAGATCTTGATCACCCGTTCTCTGCACTATACCCCAGTGTAGGCAACGGCATTGCAACA
 V  L  D  V  L  D  L  K  G  S  T  K  S  H  R  Y  R  N  S  E  V  D  L  D  H  P  F  S  A  L  Y  P  S  V  G  N  G  I  A  T
```

```
      7690      7700      7710      7720      7730      7740      7750      7760      7770      7780      7790      7800
GAAGGCTCATTGATATTTCTTGGGTATGGTGGGCTAACCACCCCTCTACAGGGTGATACAAAATGTAGGACCCAAGGATGCCAACAGGTGTCGCAAGCACACATGCAATGAGGCTCTGAAA
 E  G  S  L  I  F  L  G  Y  G  G  L  T  T  P  L  Q  G  D  T  K  C  R  T  Q  G  C  Q  Q  V  S  Q  D  T  C  N  E  A  L  K
```

```
      7810      7820      7830      7840      7850      7860      7870      7880      7890      7900      7910      7920
ATTACATGGCTAGGAGGGAAACAGGTGGTCAACGTGATCATCCGGGTCAATGACTATCTCTCAGAGAGGCCAAAGATAAGAGTCACAACCATTCCAATCACTCAAAACTATCTCGGGGCG
 I  T  W  L  G  G  K  Q  V  V  N  V  I  I  R  V  N  D  Y  L  S  E  R  P  K  I  R  V  T  T  I  P  I  T  Q  N  Y  L  G  A
```

```
      7930      7940      7950      7960      7970      7980      7990      8000      8010      8020      8030      8040
GAAGGTAGATTATTAAAATTGGGTGATCGGGTGTACATCTATACAAGATCATCAGGCTGGCACTCTCAATGCAGATAGGAGTACTTGATGTCAGCCACCCTTTGACTATCAACTGGACA
 E  G  R  L  L  K  L  G  D  R  V  Y  I  Y  T  R  S  S  G  W  H  S  Q  L  Q  I  G  V  L  D  V  S  H  P  L  T  I  N  W  T
```

```
      8050      8060      8070      8080      8090      8100      8110      8120      8130      8140      8150      8160
CCTCATGAAGCCTTGTCTGACCAGGAGAAATAAAGAGTGCAATTGGTACAATACGTGTCCGAAGGAATGCATATCAGGCGTATACACTGATGCTTATCCATTGTCCCCTGATGCAGCTAAC
 P  H  E  A  L  S  R  P  G  N  K  E  C  N  W  Y  N  T  C  P  K  E  C  I  S  G  V  Y  T  D  A  Y  P  L  S  P  D  A  A  N
```

```
      8170      8180      8190      8200      8210      8220      8230      8240      8250      8260      8270      8280
GTCGCTACCGTCACGCTATATGCCAATACATCGCGTGTCAACCCAACAATCATGTATTCTAACACTACTAACATTATAAATATGTTAAGGATAAAGGATGTTCAATTAGAGGTTGCATAT
 V  A  T  V  T  L  Y  A  N  T  S  R  V  N  P  T  I  M  Y  S  N  T  T  N  I  I  N  M  L  R  I  K  D  V  Q  L  E  V  A  Y
```

```
      8290      8300      8310      8320      8330      8340      8350      8360      8370      8380      8390      8400
ACCACGACGATTTCATCGTGTATCACGCATTTTGGTAAAGGCTACTGCTTTCACATCATCGAGATCAATCGAGAGAGCCTGAATACCTTACAGCCGATGCTCTTTAAGACTAGCATCCCTAAA
 T  T  I  S  S  C  I  T  H  F  G  K  G  Y  C  F  H  I  I  E  I  N  Q  K  S  L  N  T  L  Q  P  M  L  F  K  T  S  I  P  K
```

```
      8410      8420      8430      8440      8450      8460      8470      8480      8490      8500      8510      8520
TTATGCAAGGCCGAGTCTTAAATTTAACTGACTAGCAGGCTTGTCGGCCTTGCTGACACTAAAGTCATCTCCGAACATCCACAATATCTCTCAGTCTCTTACGTCTCTCACAGTATTAAG
 L  C  K  A  E  S  *
```
                                                                                                            T_{stop}

```
      8530      8540      8550      8560      8570      8580      8590      8600      8610      8620      8630      8640
AAAAAACCCAGGGTGAATGGGAAGCTTGCCATAGGTCATGGATGGGCAGGAGTCCTCCCAAAACCCTTCTGCACATACTCTATCCAGAATGCCACCTGAACTCTCCCATAGTCAGGGGGAAG
                                                            M  D  G  Q  E  S  S  Q  N  P  S  D  I  L  Y  P  E  C  H  L  N  S  P  I  V  R  G  K
```

I T_{start} **L PROTEIN ▶**

```
      8650      8660      8670      8680      8690      8700      8710      8720      8730      8740      8750      8760
ATAGCACAGTTGCACGTCTTGTTGATGTGAACCAGCCCTACAGACTGAAGGACGACAGCATAATAAATATTACAAAGCACAAAATTAGGAACGGAGGATTGTCCCCCCGTCAAATTAAG
 I  A  Q  L  H  V  L  L  D  V  N  Q  P  Y  R  L  K  D  D  S  I  I  N  I  T  K  H  H  K  I  R  N  G  G  L  S  P  R  Q  I  K
```

SENDAI VIRUS

```
      3770      3780      8790      8800      8810      8820      8830      8840      8850      8860      8870      8880
ATCAGGTCTCTGGGTAAGGCTCTTCAACGCACAATAAAGGATTTAGACCGATACACGTTTGAACCGTACCCAACCTACTCTCAGGAATTACTTAGGCTTGATATACCAGAGATATGTGAC
  I  R  S  L  G  K  A  L  Q  R  T  I  K  D  L  D  R  Y  T  F  E  P  Y  P  T  Y  S  Q  E  L  L  R  L  D  I  P  E  I  C  D

      8890      8900      8910      8920      8930      8940      8950      8960      8970      8980      8990      9000
AAAATCCGATCCGTCTTCGCGGTCTCGGATCGGCTGACCAGGGAGTTATCTAGTGGGTTCCAGGATCTTTGGTTGAATATCTTCAAGCAACTAGGCAATATAGAAGGAAGAGAGGGGTAC
  K  I  R  S  V  F  A  V  S  D  R  L  T  R  E  L  S  S  G  F  Q  D  L  W  L  N  I  F  K  Q  L  G  N  I  E  G  R  E  G  Y

      9010      9020      9030      9040      9050      9060      9070      9080      9090      9100      9110      9120
GATCCGGTTGCAGGATATCGGCACCATCCCGGAGATAACTGATAAGTACAGCAGGAATAGATGGTATAGGCCCATTCCTAACTTGGTTCAGCATCAAATATGACATGCGGTGGATGCAGAAG
  D  P  L  Q  D  I  G  T  I  P  E  I  T  D  K  Y  S  R  N  R  W  Y  R  P  F  L  T  W  F  S  I  K  Y  D  M  R  W  M  Q  K

      9130      9140      9150      9160      9170      9180      9190      9200      9210      9220      9230      9240
ACCAGACCGGGGGGACCCCTTGATACCTCTAATTCACATAACCTCCTAGAATGCAAATCATACACTCTAGTAACATACGGAGATCTTGTCATGATACTGAACAAGTTGACATTGACAGGG
  T  R  P  G  G  P  L  D  T  S  N  S  H  N  L  L  E  C  K  S  Y  T  L  V  T  Y  G  D  L  V  M  I  L  N  K  L  T  L  T  G

      9250      9260      9270      9280      9290      9300      9310      9320      9330      9340      9350      9360
TATATCCTAACCCCTGAGCTGGTCTTGATGTATTGTGATGTTGTAGAAGGAAGGTGGAATATGTCTGCTGCAGGGCATCTAGATAAGAAGTCCATTGGGATAACAAGCAAAGGTGAGGAA
  Y  I  L  T  P  E  L  V  L  M  Y  C  D  V  V  E  G  R  W  N  M  S  A  A  G  H  L  D  K  K  S  I  G  I  T  S  K  G  E  E

      9370      9380      9390      9400      9410      9420      9430      9440      9450      9460      9470      9480
TTATGGGAACTAGTGGATTCCCTCTTCTCCAAGTCTTGGAGAGGAAATATACAATGTCATCGCACTATTGGAGCCCCTATCTCTTGCTCTCCATACAACTAAATGATCCTGTTATACCTCTA
  L  W  E  L  V  D  S  L  F  S  S  L  G  E  E  I  Y  N  V  I  A  L  L  E  P  L  S  L  A  L  I  Q  L  N  D  P  V  I  P  L

      9490      9500      9510      9520      9530      9540      9550      9560      9570      9580      9590      9600
CGTGGGGCATTTATGAGGCATGTGTTGACAGAGCTACAGACTGTTTTAACAAGTAGAGACGTGTACACAGATGCTGAAGCAGACACTATTGTGGAGTCGTTACTCGCCATTTTCCATGGA
  R  G  A  F  M  R  H  V  L  T  E  L  Q  T  V  L  T  S  R  D  V  Y  T  D  A  E  A  D  T  I  V  E  S  L  L  A  I  F  H  G

      9610      9620      9630      9640      9650      9660      9670      9680      9690      9700      9710      9720
ACCTCTATTGATGAGAAAGCAGAGATCTTTTCCTTCTTTAGGACATTTGGCCACCCCAGCTTAGAGGCTGTCACTGCCGCCGACAAGGTAAGGGCCCATATGTATGCACAAAAGGCAATA
  T  S  I  D  E  K  A  E  I  F  S  F  F  R  T  F  G  H  P  S  L  E  A  V  T  A  A  D  K  V  R  A  H  M  Y  A  Q  K  A  I

      9730      9740      9750      9760      9770      9780      9790      9800      9810      9820      9830      9840
AAGCTTAAGACCCTATACGAGTGTCATGCAGTTTTTTGCACTATCATCATAAATGGGTATAGAGAGAGGCATGGCGGACAGTGGCCCCCCTGTGACTTCCCTGATCACGTGTGTCTAGAA
  K  L  K  T  L  Y  E  C  H  A  V  F  C  T  I  I  I  N  G  Y  R  E  R  H  G  G  Q  W  P  P  C  D  F  P  D  H  V  C  L  E

      9850      9860      9870      9880      9890      9900      9910      9920      9930      9940      9950      9960
CTAAGGAACGCTCAAGGGTCCAATACGGCACTCTCTTATGAATGTGCTGTAGACAACTATACAAGTTTCATAGGCTTCAAGTTTCGGAAGTTTATAGAACCACAACTAGATGAAGATCTC
  L  R  N  A  Q  G  S  N  T  A  L  S  Y  E  C  A  V  D  N  Y  T  S  F  I  G  F  K  F  R  K  F  I  E  P  Q  L  D  E  D  L

      9970      9980      9990      10000     10010     10020     10030     10040     10050     10060     10070     10080
ACAATATATATGAAAGACAAAGCACTATCCCCCAGGAAGGAGGCATGGGACTCTGTATACCCGGATAGTAATCTGTACTATAAAGCCCCAGAGTCTGAAGAGACCCGGCGGCTTATTGAA
  T  I  Y  M  K  D  K  A  L  S  P  R  K  E  A  W  D  S  V  Y  P  D  S  N  L  Y  Y  K  A  P  E  S  E  E  T  R  R  L  I  E

      10090     10100     10110     10120     10130     10140     10150     10160     10170     10180     10190     10200
GTGTTCATAAATGATGAGAATTTCAACCCAGAAGAAATTATCAATTATGTGGAGTCAGGAGATTGGTTGAAAGACGAGGAGTTCAACATCTCGTACAGTCTCAAAGAGAAAGAGATCAAG
  V  F  I  N  D  E  N  F  N  P  E  E  I  I  N  Y  V  E  S  G  D  W  L  K  D  E  E  F  N  I  S  Y  S  L  K  E  K  E  I  K

      10210     10220     10230     10240     10250     10260     10270     10280     10290     10300     10310     10320
CAAGAGGGTCGTCTATTCGCAAAAATGACTTATAAGATGCGAGCCGTACAGGTGCTGGCAGAGACACTACTGGCTAAAGGAATAGGAGAGCTATTCAGGGAAAATGGGATGGTTAAGGGA
  Q  E  G  R  L  F  A  K  M  T  Y  K  M  R  A  V  Q  V  L  A  E  T  L  L  A  K  G  I  G  E  L  F  R  E  N  G  M  V  K  G

      10330     10340     10350     10360     10370     10380     10390     10400     10410     10420     10430     10440
GAGATAGACCTACTTAAAAGATTGACTACTCTTTCTGTCTCAGGCGTCCCCAGGACTGATTCAGTGTACAATAACTCTAAATCATCAGAGAAGAGAAACGAAGGCATGGAAAATAAGAAC
  E  I  D  L  L  K  R  L  T  T  L  S  V  S  G  V  P  R  T  D  S  V  Y  N  N  S  K  S  S  E  K  R  N  E  G  M  E  N  K  N

      10450     10460     10470     10480     10490     10500     10510     10520     10530     10540     10550     10560
TCTGGGGGGTACTGGGACGAAAAGAAGAGGTCCAGACATGAATTCAAGGCAACAGATTCATCAACAGACGGCTATGAAACGTTAAGTTGCTTCCTCACAACAGACCTCAAGAAATACTGC
  S  G  G  Y  W  D  E  K  K  R  S  R  H  E  F  K  A  T  D  S  S  T  D  G  Y  E  T  L  S  C  F  L  T  T  D  L  K  K  Y  C

      10570     10580     10590     10600     10610     10620     10630     10640     10650     10660     10670     10680
TTAAACTGGAGATTTGAGAGTACTGCATTGTTTGGTCAGAGATGCAACGAGATATTTGGCTTCAAGACCTTCTTTAACTGGATGCATCCAGTCCTTGAAAGGTGTACAATATATGTTGGA
  L  N  W  R  F  E  S  T  A  L  F  G  Q  R  C  N  E  I  F  G  F  K  T  F  F  N  W  M  H  P  V  L  E  R  C  T  I  Y  V  G

      10690     10700     10710     10720     10730     10740     10750     10760     10770     10780     10790     10800
GATCCTTACTGTCCAGTCGCCGACCGGATGCATCGACAACTCCAGGATCATGCAGACTCTGGCATTTTCATACATAATCCTAGGGGGGCATAGAAGGTTACTGCCAGAAGCTGTGGACC
  D  P  Y  C  P  V  A  D  R  M  H  R  Q  L  Q  D  H  A  D  S  G  I  F  I  H  N  P  R  G  G  I  E  G  Y  C  Q  K  L  W  T

      10810     10820     10830     10840     10850     10860     10870     10880     10890     10900     10910     10920
TTAATCTCAATCAGTGCAATCCACCTAGCAGCTGTGAGAGTGGGTGTCAGGGTCTCTGCAATAGTTCAGGGTGACAATCAAGCTATAGCCGTGACATCAAGAGTACCTGTAGCTCAGACT
  L  I  S  I  S  A  I  H  L  A  A  V  R  V  G  V  R  V  S  A  M  V  Q  G  D  N  Q  A  I  A  V  T  S  R  V  P  V  A  Q  T

      10930     10940     10950     10960     10970     10980     10990     11000     11010     11020     11030     11040
TACAAGCAGAAGAAAAATCATGTCTATGAGGAGATCACCAAATATTTCGGTGCTCTAAGACACGTCATGTTTGATGTAGGGCACGAGCTAAAATTGAACGAGACCATCATTAGTAGCAAG
  Y  K  Q  K  K  N  H  V  Y  E  E  I  T  K  Y  F  G  A  L  R  H  V  M  F  D  V  G  H  E  L  K  L  N  E  T  I  I  S  S  K
```

```
      11050      11060      11070      11080      11090      11100      11110      11120      11130      11140      11150      11160
ATGTTTGTCTATAGTAAAAGGATATACTATGATGGGAAGATTTTACCACAGTGCCTGAAAGCCTTGACCAAGTGTGTATTCTGGTCCGAGACACTGGTAGATGAAAACAGATCTGCTTGT
 M  F  V  Y  S  K  R  I  Y  Y  D  G  K  I  L  P  Q  C  L  K  A  L  T  K  C  V  F  W  S  E  T  L  V  D  E  N  R  S  A  C

      11170      11180      11190      11200      11210      11220      11230      11240      11250      11260      11270      11280
TCGAACATCTCAACATCCATAGCAAAAGCTATCGAAAATGGGTATTCTCCTATACTAGGCTACTGCATTGCGTTGTATAAGACCTGTCAGCAGGTGTGCATATCACTAGGGATGACTATA
 S  N  I  S  T  S  I  A  K  A  I  E  N  G  Y  S  P  I  L  G  Y  C  I  A  L  Y  K  T  C  Q  Q  V  C  I  S  L  G  M  T  I

      11290      11300      11310      11320      11330      11340      11350      11360      11370      11380      11390      11400
AATCCAACTATCAGCCCGACCGTAAGAGATCAATACTTTAAGGGTAAGAATTGGCTGAGATGTGCAGTGTTGATTCCAGCAAATGTTGGAGGATTCAACTACATGTCTACATCTAGATGC
 N  P  T  I  S  P  T  V  R  D  Q  Y  F  K  G  K  N  W  L  R  C  A  V  L  I  P  A  N  V  G  G  F  N  Y  M  S  T  S  R  C

      11410      11420      11430      11440      11450      11460      11470      11480      11490      11500      11510      11520
TTTGTTAGAAATATTGGAGACCCCGCAGTAGCAGCCCTAGCTGATCTCAAAAGATTCATCAGAGCGGATCTGTTAGACAAGCAGGTATTATACAGGGTCATGAATCAAGAACCCGGTGAC
 F  V  R  N  I  G  D  P  A  V  A  A  L  A  D  L  K  R  F  I  R  A  D  L  L  D  K  Q  V  L  Y  R  V  M  N  Q  E  P  G  D

      11530      11540      11550      11560      11570      11580      11590      11600      11610      11620      11630      11640
TCTAGTTTTCTAGATTGGGCTTCAGACCCTTATTCGTGTAACCTCCCGCATTCTCAGAGTATAACTACGATTATAAAGAATATCACTGCTAGATCTGTGCTGCAGGAATCCCCGAATCCT
 S  S  F  L  D  W  A  S  D  P  Y  S  C  N  L  P  H  S  Q  S  I  T  T  I  I  K  N  I  T  A  R  S  V  L  Q  E  S  P  N  P

      11650      11660      11670      11680      11690      11700      11710      11720      11730      11740      11750      11760
CTACTGTCTGGTCTCTTCACCGAGACTAGTGGAGAAGAGGATCTCAACCTGGCCTCGTTCCTTATGGACCGGAAAGTCATCCTGCCGAGAGTGGCTCATGAGATCCTGGGTAATTCCTTA
 L  L  S  G  L  F  T  E  T  S  G  E  E  D  L  N  L  A  S  F  L  M  D  R  K  V  I  L  P  R  V  A  H  E  I  L  G  N  S  L

      11770      11780      11790      11800      11810      11820      11830      11840      11850      11860      11870      11880
ACTGGAGTTAGGGAGGCGGATTGCAGGGATGCTTGATACGACCAAGTCTCTAGTGAGAGCCAGCGTTAGGAAAGGAGGATTATCATATGGGATATTGAGGAGGCTTGTCAATTATGATCTA
 T  G  V  R  E  A  I  A  G  M  L  D  T  T  K  S  L  V  R  A  S  V  R  K  G  G  L  S  Y  G  I  L  R  R  L  V  N  Y  D  L

      11890      11900      11910      11920      11930      11940      11950      11960      11970      11980      11990      12000
TTGCAGTACGAGACACTGACTAGAACTCTCAGGAAACCGGTGAAAGACAACATCGAATATGAGTATATGTGTTCAGTTGAGCTAGCTGTCGGTCTAAGGCAGAAAATGTGGATCCACCTG
 L  Q  Y  E  T  L  T  R  T  L  R  K  P  V  K  D  N  I  E  Y  E  Y  M  C  S  V  E  L  A  V  G  L  R  Q  K  M  W  I  H  L

      12010      12020      12030      12040      12050      12060      12070      12080      12090      12100      12110      12120
ACTTACGGGAGACCCATACATGGGCTAGAAACACCGACCCCTTTAGAGCTCTTGAGGGGAATATTTATCGAAGGTTCAGAGGTGTGCAAGCTTTGCAGGTCTGAAGGAGCAGACCCCATC
 T  Y  G  R  P  I  H  G  L  E  T  P  D  P  L  E  L  L  R  G  I  F  I  E  G  S  E  V  C  K  L  C  R  S  E  G  A  D  P  I

      12130      12140      12150      12160      12170      12180      12190      12200      12210      12220      12230      12240
TATACATGGTTCTATCTTCCTGACAATATAGACCTGGACACGCTTACAAACGGATGTCCGGCTATAAGAATCCCCTATTTTGGATCAGCCACTGATGAAAGGTCGGAAGCCCAACTCGGG
 Y  T  W  F  Y  L  P  D  N  I  D  L  D  T  L  T  N  G  C  P  A  I  R  I  P  Y  F  G  S  A  T  D  E  R  S  E  A  Q  L  G

      12250      12260      12270      12280      12290      12300      12310      12320      12330      12340      12350      12360
TATGTAAGAAATCTAAGCAAACCCGCAAAGGCGGCCATCCGGATAGCTATGGTGTATACGTGGGCCTACGGGACTGATGAGATATCGTGGATGGAAGCCGCTCTTATAGCCCAAACAAGA
 Y  V  R  N  L  S  K  P  A  K  A  A  I  R  I  A  M  V  Y  T  W  A  Y  G  T  D  E  I  S  W  M  E  A  A  L  I  A  Q  T  R

      12370      12380      12390      12400      12410      12420      12430      12440      12450      12460      12470      12480
GCTAATCTGAGCTTAGAGAATCTAAAGCTGCTGACTCCTGTTTCAACCTCCACTAATCTATCTCATAGGTTGAAAGATACGGCAACCCAGATGAAGTTCTCTAGTGCAACACTAGTCCGT
 A  N  L  S  L  E  N  L  K  L  L  T  P  V  S  T  S  T  N  L  S  H  R  L  K  D  T  A  T  Q  M  K  F  S  S  A  T  L  V  R

      12490      12500      12510      12520      12530      12540      12550      12560      12570      12580      12590      12600
GCAAGTCGGTTCATAACAATATCAAATGATAACATGGCACTCAAAGAAGCAGGGGAGTCGAAGGATACTAATCTCGTGTATCAGCGATTATGCTAACTGGGCTAAGCTTGTTCGAGTTC
 A  S  R  F  I  T  I  S  N  D  N  M  A  L  K  E  A  G  E  S  K  D  T  N  L  V  Y  Q  Q  I  M  L  T  G  L  S  L  F  E  F

      12610      12620      12630      12640      12650      12660      12670      12680      12690      12700      12710      12720
AATATGAGATATAAGAAAGGTTCCTTAGGGAAGCCACTGATATTGCACTTACATCTTAATAACGGGTGCTGTATAATGGAGTCCCCACAGGAGGCGAATATCCCCCCAAGGTCCACATTA
 N  M  R  Y  K  K  G  S  L  G  K  P  L  I  L  H  L  H  L  N  N  G  C  C  I  M  E  S  P  Q  E  A  N  I  P  P  R  S  T  L

      12730      12740      12750      12760      12770      12780      12790      12800      12810      12820      12830      12840
GATTTAGAGATTACACAGAGAACAATAAATTGATCTATGATCCTGATCCACTCAAGGATGTGGACCTTGAGCTATTTAGCAAGGTCAGAGATGTTGTACACACAGTTGACATGACTTAT
 D  L  E  I  T  Q  E  N  N  K  L  I  Y  D  P  D  P  L  K  D  V  D  L  E  L  F  S  K  V  R  D  V  V  H  T  V  D  M  T  Y

      12850      12860      12870      12880      12890      12900      12910      12920      12930      12940      12950      12960
TGGTCAGATGATGAAGTTATCAGAGCAACCAGTATCTGTACTGCAATGACGATAGCTGATACAATGTCTCAATTAGATAGAGACAACTTAAAAGAGATGATCGCACTAGTAAATGACGAT
 W  S  D  D  E  V  I  R  A  T  S  I  C  T  A  M  T  I  A  D  T  M  S  Q  L  D  R  D  N  L  K  E  M  I  A  L  V  N  D  D

      12970      12980      12990      13000      13010      13020      13030      13040      13050      13060      13070      13080
GATGTCAACAGCTTGATTACTGAGTTTATGGTGATTGATGTTCCTTTATTTTGCTCAACGTTCGGGGGTATTCTAGTCAATCAGTTTGCATACTCACTCTACGGCTTAAACCTCAGAGGA
 D  V  N  S  L  I  T  E  F  M  V  I  D  V  P  L  F  C  S  T  F  G  G  I  L  V  N  Q  F  A  Y  S  L  Y  G  L  N  L  R  G

      13090      13100      13110      13120      13130      13140      13150      13160      13170      13180      13190      13200
AGGGAAGAAATATGGGGACATGTAGTCCGGATTCTTAAAGATACCTCCCACGCAGTTTTAAAAGTCTTATCTAATGCTCTATCTCATCCCAAAATCTTCAAACGATTCTGGAATGCAGGT
 R  E  E  I  W  G  H  V  V  R  I  L  K  D  T  S  H  A  V  L  K  V  L  S  N  A  L  S  H  P  K  I  F  K  R  F  W  N  A  G

      13210      13220      13230      13240      13250      13260      13270      13280      13290      13300      13310      13320
GTCGTGGAACCTGTGTATGGGCCTAACCTCTCAAATCAGGATAAGATACTCTTGGCCCTCTCTGTCTGTGAATATTCTGTGGATCTATTCATGCACGATTGGCAAGGGGGTGTACCGCTT
 V  V  E  P  V  Y  G  P  N  L  S  N  Q  D  K  I  L  L  A  L  S  V  C  E  Y  S  V  D  L  F  M  H  D  W  Q  G  G  V  P  L
```

SENDAI VIRUS

```
        13330     13340     13350     13360     13370     13380     13390     13400     13410     13420     13430     13440
GAGATCTTTATCTGTGACAATGACCCAGATGTGGCCGACATGAGGAGGTCCTCTTTCTTGGCAAGACATCTTGCATACCTATGCAGCTTGGCAGAGATATCTAGGGATGGGCCAAGATTA
 E  I  F  I  C  D  N  D  P  D  V  A  D  M  R  R  S  S  F  L  A  R  H  L  A  Y  L  C  S  L  A  E  I  S  R  D  G  P  R  L

        13450     13460     13470     13480     13490     13500     13510     13520     13530     13540     13550     13560
GAATCAATGAACTCTCTAGAGAGGCTGGAGTCACTAAAGAGTTACCTGGAACTCACATTTCTTGATGACCCGGTACTGAGGTACAGTCAGTTGACTGGCCTAGTCATCAAAGTATTCCCA
 E  S  M  N  S  L  E  R  L  E  S  L  K  S  Y  L  E  L  T  F  L  D  D  P  V  L  R  Y  S  Q  L  T  G  L  V  I  K  V  F  P

        13570     13580     13590     13600     13610     13620     13630     13640     13650     13660     13670     13680
TCTACTTTGACCTATATCCGGAAGTCATCTATAAAAGTGTTAAGGACAAGAGGTATAGGAGTCCCTGAAGTCTTAGAAGATTGGGATCCCGAGGCAGATAATGCACTGTTAGATGGTATC
 S  T  L  T  Y  I  R  K  S  S  I  K  V  L  R  T  R  G  I  G  V  P  E  V  L  E  D  W  D  P  E  A  D  N  A  L  L  D  G  I

        13690     13700     13710     13720     13730     13740     13750     13760     13770     13780     13790     13800
GCGGCAGAAATACAACGAATATTCCTTTGGGACATCAGACTAGAGCCCCTTTTTTGGGGGTTGAGAGTATCCAAGTCACAGGTACTGCGTCTCCGGGGGTACAAGGAGATCACAAGAGGT
 A  A  E  I  Q  Q  N  I  P  L  G  H  Q  T  R  A  P  F  W  G  L  R  V  S  K  S  Q  V  L  R  L  R  G  Y  K  E  I  T  R  G

        13810     13820     13830     13840     13850     13860     13870     13880     13890     13900     13910     13920
GAGATAGGCAGATCAGGTGTTGGTCTGACGTTACCATTCGATGGAAGATATCTATCTCACCAGCTGAGGCTCTTTGGCATCAACAGTACTAGCTGCTTGAAAGCACTTGAACTTACCTAC
 E  I  G  R  S  G  V  G  L  T  L  P  F  D  G  R  Y  L  S  H  Q  L  R  L  F  G  I  N  S  T  S  C  L  K  A  L  E  L  T  Y

        13930     13940     13950     13960     13970     13980     13990     14000     14010     14020     14030     14040
CTATTGAGCCCCTTAGTTGACAAGGATAAAGATAGGCTATATTTAGGGGAAGGAGCTGGGGCCATGCTTTCCTGTTTATGACGCTACTCTTGGCCCATGCATCAACTATTATAACTCAGGG
 L  L  S  P  L  V  D  K  D  K  D  R  L  Y  L  G  E  G  A  G  A  M  L  S  C  Y  D  A  T  L  G  P  C  I  N  Y  Y  N  S  G

        14050     14060     14070     14080     14090     14100     14110     14120     14130     14140     14150     14160
GTATACTCTTGTGATGTCAATGGGCAGAGAGAGTTAAATATATATCCTGCTGAGGTGGCACTAGTGGGAAAGAAATTAAACAATGTTACTAGTCTGGGTCAAAGAGTTAAAGTGTTATTC
 V  Y  S  C  D  V  N  G  Q  R  E  L  N  I  Y  P  A  E  V  A  L  V  G  K  K  L  N  N  V  T  S  L  G  Q  R  V  K  V  L  F

        14170     14180     14190     14200     14210     14220     14230     14240     14250     14260     14270     14280
AACGGGAATCCTGGCTCGACATGGATTGGGAATGATGAGTGTGAGGCTTTGATTTGGAATGAATTACAGAATAGCTCGATAGGCCTAGTCCACTGTGACATGGAGGGAGGAGATCATAAG
 N  G  N  P  G  S  T  W  I  G  N  D  E  C  E  A  L  I  W  N  E  L  Q  N  S  S  I  G  L  V  H  C  D  M  E  G  G  D  H  K

        14290     14300     14310     14320     14330     14340     14350     14360     14370     14380     14390     14400
GATGATCAAGTTGTACTGCATGAGCATTACAGTGTAATCCGGATCGCGTATCTGGTGGGGGATCGAGACGTTGTGCTTATAAGCAAGATTGCTCCCAGGCTGGGCACGGATTGGACCAGG
 D  D  Q  V  L  H  E  H  Y  S  V  I  R  I  A  Y  L  V  G  D  R  D  V  V  L  I  S  K  I  A  P  R  L  G  T  D  W  T  R

        14410     14420     14430     14440     14450     14460     14470     14480     14490     14500     14510     14520
CAGCTCAGCCTATATCTGAGATACTGGGACGAGGTTAACCTAATAGTGCTTAAAACATCTAACCCTGCTTCCACAGAGATGTATCTCCTATCGAGGCACCCCAAATCTGACATTATAGAG
 Q  L  S  L  Y  L  R  Y  W  D  E  V  N  L  I  V  L  K  T  S  N  P  A  S  T  E  M  Y  L  L  S  R  H  P  K  S  D  I  I  E

        14530     14540     14550     14560     14570     14580     14590     14600     14610     14620     14630     14640
GACAGCAAGACAGTGTTAGCTAGTCTCCTCCCTTTGTCAAAAGAAGATAGCATCAAGATAGAAAAGTGGATCTTAATAGAGAAGGCAAAGGCTCACGAATGGGTTACTCGGGAATTGAGA
 D  S  K  T  V  L  A  S  L  L  P  L  S  K  E  D  S  I  K  I  E  K  W  I  L  I  E  K  A  K  A  H  E  W  V  T  R  E  L  R

        14650     14660     14670     14680     14690     14700     14710     14720     14730     14740     14750     14760
GAAGGAAGCTCTTCATCAGGGATGCTTAGACCTTACCATCAAGCACTGCAGACGTTTGGCTTTGAACCAAACTTGTATAAATTGAGCAGAGATTTCTTGTCCACCATGAACATAGCTGAT
 E  G  S  S  S  S  G  M  L  R  P  Y  H  Q  A  L  Q  T  F  G  F  E  P  N  L  Y  K  L  S  R  D  F  L  S  T  M  N  I  A  D

        14770     14780     14790     14800     14810     14820     14830     14840     14850     14860     14870     14880
ACACACAACTGCATGATAGCTTTCAACAGGGTTTTGAAGGATACAATCTTCGAATGGGCTAGAATAACTGAGTCAGATAAAAGGCTTAAACTAACTGGTAAGTATGACCTGTATCCTGTG
 T  H  N  C  M  I  A  F  N  R  V  L  K  D  T  I  F  E  W  A  R  I  T  E  S  D  K  R  L  K  L  T  G  K  Y  D  L  Y  P  V

        14890     14900     14910     14920     14930     14940     14950     14960     14970     14980     14990     15000
AGAGATTCAGGCAAGTTGAAGACAATTTCTAGAAGACTTGTGCTATCTTGGATATCTTTATCTATGTCCACAAGATTGGTAACTGGGTCATTCCCTGACCAGAAGTTTGAAGCAAGACTT
 R  D  S  G  K  L  K  T  I  S  R  R  L  V  L  S  W  I  S  L  S  M  S  T  R  L  V  T  G  S  F  P  D  Q  K  F  E  A  R  L

        15010     15020     15030     15040     15050     15060     15070     15080     15090     15100     15110     15120
CAATTGGGAATAGTTTCATTATCATCCCGTGAAATCAGGAACCTGAGGGTTATCACAAAAACTTTATTAGACAGGTTTGAGGATATTATACATAGTATAACGTATAGATTCCTCACCAAA
 Q  L  G  I  V  S  L  S  S  R  E  I  R  N  L  R  V  I  T  K  T  L  L  D  R  F  E  D  I  I  H  S  I  T  Y  R  F  L  T  K

        15130     15140     15150     15160     15170     15180     15190     15200     15210     15220     15230     15240
GAAATAAAGATTTTGATGAAGATTTTAGGGCAGTCAAGATGTTCGGGGCCAGGCAAAATGAATACAGCACCGTGATTGATGATGGATCACTAGGTGATATCGAGCCATATGACAGCTCG
 E  I  K  I  L  M  K  I  L  G  A  V  K  M  F  G  A  R  Q  N  E  Y  T  T  V  I  D  D  G  S  L  G  D  I  E  P  Y  D  S  S

        15250     15260     15270     15280     15290     15300     15310     15320     15330     15340     15350     15360
TAATAATTAGTCCCTATCGTGCAGAACGATCGAAGCTCCGCGGTACCTGGAAGTCTTGGACTTGTCCATATGACAATAGTAAGAAAAAACTTACAAGAAGACAAGAAAATTTAAAAGGATA
 *

        15370     15380
CATATCTCTTAAACTCTTGTCTGGT
```

T$_{stop}$ I EXTRACISTRONIC SEQUENCES

HUMAN PARAINFLUENZA VIRUS TYPE 3

```
          10        20        30        40        50        60        70        80        90       100       110       120
ACCAAACAAGAGAAGAAACTTGTTTGGAAATATAAATTTAAATTAAAATTAACTTAGGATTAAAGACATTGACTAGAAGGTCAAGAAAAGGGAACTCTATAATTTCAAAAATGTTGAGCC
                                                                                    NUCLEOCAPSID PROTEIN ▸   M  L  S
      EXTRACISTRONIC SEQUENCES                          T start

         130       140       150       160       170       180       190       200       210       220       230       240
TATTTGATACATTTAATGCACGTAGGCAAGAAAACATAACAAAATCAGCTGGTGGAGCTATCATTCCTGGACAGAAAAATACTGTCTCCATATTTGCCCTTGGACCGACAATAACTGATG
L  F  D  T  F  N  A  R  R  Q  E  N  I  T  K  S  A  G  G  A  I  I  P  G  Q  K  N  T  V  S  I  F  A  L  G  P  T  I  T  D

         250       260       270       280       290       300       310       320       330       340       350       360
ACGATGAGAAAATGACATTAGCTCTTCTATTTCTATCTCATTCACTAGATAATGAGAAACAACATGCACAAAGGGCAGGGTTCTTGGTGTCTTTATTGTCAATGGCTTATGCCAATCCAG
D  D  E  K  M  T  L  A  L  L  F  L  S  H  S  L  D  N  E  K  Q  H  A  Q  R  A  G  F  L  V  S  L  L  S  M  A  Y  A  N  P

         370       380       390       400       410       420       430       440       450       460       470       480
AGCTTTACCTGACAACAAATGGAAGTAATGCAGATGTTAAATATGTCATATATATGATTAGGAAAGATCTAAAACGGCAAAAGTATGGAGGATTTGTGGTTAAGACGAGAGAGATGATAT
E  L  Y  L  T  T  N  G  S  N  A  D  V  K  Y  V  I  Y  M  I  R  K  D  L  K  R  Q  K  Y  G  G  F  V  V  K  T  R  E  M  I

         490       500       510       520       530       540       550       560       570       580       590       600
ATGAAAAGACAACTGAGTGGATATTTGGAAGTGACCTGGATTATGACCAGGAAACTATGCTGCAGAACGGCAGAAACAATTCAACGATTGAAGATCTTGTTCACACATTTGGGTATCCAT
Y  E  K  T  T  E  W  I  F  G  S  D  L  D  Y  D  Q  E  T  M  L  Q  N  G  R  N  N  S  T  I  E  D  L  V  H  T  F  G  Y  P

         610       620       630       640       650       660       670       680       690       700       710       720
CATGTTTAGGAGCTCTTATAATACAGATCTGGATAGTTTTGGTCAAAGCCATCACTAGCATCTCAGGGTTAAGAAAAGGCTTTTTCACTGATTAGAGGCTTTCAGACAAGATGGAACGA
S  C  L  G  A  L  I  I  Q  I  W  I  V  L  V  K  A  I  T  S  I  S  G  L  R  K  G  F  F  T  R  L  E  A  F  R  Q  D  G  T

         730       740       750       760       770       780       790       800       810       820       830       840
TGCAAGCAGGGCTGGTATTGAGCGGTGACACAGTGGATCAGATTGGGTCAATCATGCGGTCTCAACAGAGCTTGGTAACTCTTATGGTTGAGACATTAATAACAATGACTAGCAGAA
V  Q  A  G  L  V  L  S  G  D  T  V  D  Q  I  G  S  I  M  R  S  Q  Q  S  L  V  T  L  M  V  E  T  L  I  T  M  N  T  S  R

         850       860       870       880       890       900       910       920       930       940       950       960
ATGACCTCACAACCATAGAAAAGAATATACAAATTGTTGGTAACTACATAAGAGATGCAGGTCTTGCTTCATTCTTCAATACAATCAGGTATGGAATTGAGACTAGAATGGCAGCTTTGA
N  D  L  T  T  I  E  K  N  I  Q  I  V  G  N  Y  I  R  D  A  G  L  A  S  F  F  N  T  I  R  Y  G  I  E  T  R  M  A  A  L

         970       980       990      1000      1010      1020      1030      1040      1050      1060      1070      1080
GTCTATCTACTCTCAGACCAGATATCAATGAGTTAAAAGCTCTGATGGAATTGTATTTATCAAAGGGACCACGCGCTCCTTTTATCTGTATCCTCAGAGATCCTATACATGGTGAGTTCG
S  L  S  T  L  R  P  D  I  N  R  L  K  A  L  M  E  L  Y  L  S  K  G  P  R  A  P  F  I  C  I  L  R  D  P  I  H  G  E  F

        1090      1100      1110      1120      1130      1140      1150      1160      1170      1180      1190      1200
CACCAGGCAACTATCCTGCCATATGGAGTTATGCAATGGGGGTGGCAGTTGTACAAAACAGAGCCATGCAACAGTATGTGACGGGAAGATCATATCTAGATATTGATATGTTCCAGCTGG
A  P  G  N  Y  P  A  I  W  S  Y  A  M  G  V  A  V  V  Q  N  R  A  M  Q  Q  Y  V  T  G  R  S  Y  L  D  I  D  M  F  Q  L

        1210      1220      1230      1240      1250      1260      1270      1280      1290      1300      1310      1320
GACAAGCAGTAGCACGTGATGCTGAAGCTCAGATGAGCTCAACACTGGAAGATGAACTTGGAGTGACACACGAAGCCAAAGAAAGCTTGAAAAGACATATAAGGAACATAAACAGTTCAG
G  Q  A  V  A  R  D  A  E  A  Q  M  S  S  T  L  E  D  E  L  G  V  T  H  E  A  K  E  S  L  K  R  H  I  R  N  I  N  S  S

        1330      1340      1350      1360      1370      1380      1390      1400      1410      1420      1430      1440
AGACATCTTTCCACAAACCAACAGGCGGATCAGCCATAGAGATGGCAATAGATGAAGAGCCAGAACAATTTGAACACGAGCAGATCAAGAACAAGATGGAGAACCTCAATCATCTATAA
E  T  S  F  H  K  P  T  G  G  S  A  I  E  M  A  I  D  E  E  P  E  Q  F  E  H  R  A  D  Q  E  Q  D  G  E  P  Q  S  S  I

        1450      1460      1470      1480      1490      1500      1510      1520      1530      1540      1550      1560
TCCAATATGCTTGGGCAGAAGGAAACAGAAGTGATGATCGGACCGGAGCAAGCTACAGAATCCGACAATATCAAGACTGAACAACAAAACATCAGAGACAGACTAAACAAGAGACTCAACG
I  Q  Y  A  W  A  E  G  N  R  S  D  D  R  T  E  Q  A  T  E  S  D  N  I  K  T  E  Q  Q  N  I  R  D  R  L  N  K  R  L  N

        1570      1580      1590      1600      1610      1620      1630      1640      1650      1660      1670      1680
ACAAGAAGAAACAAGGCAGTCAACCATCCACCAATCCACAAACAGAACGAACCAGGACGAAATAGACGATCTGTTCAATGCATTTGGAAGCAACTAACTGAGTCAACATTTTGATCTAA
D  K  K  K  Q  G  S  Q  P  S  T  N  P  T  N  R  T  N  Q  D  E  I  D  D  L  F  N  A  F  G  S  N  *

        1690      1700      1710      1720      1730      1740      1750      1760      1770      1780      1790      1800
ATCAATAATAAATAAGAAAAACTTAGGATTAAAGAATCCTATCATACCAGAACATAGAGTGGTAAATTTAGAGTCTGCTTGCAACTCAATCAATAGAGAGTTGATGGAAAGCGATGCTAA
                     T stop    I    T start                              PHOSPHOPROTEIN ▸    M  E  S  D  A  K
                                                                         C PROTEIN ▸       M  L

        1810      1820      1830      1840      1850      1860      1870      1880      1890      1900      1910      1920
AAACTATCAAATCATGGATTCTTGGGAAGAGGAACCAAGAGATAAATCAACTAATATCTCCTCGGCCCTCAACATCATTGAATTCATACTCAGCACCGACCCCCAAGAAGACCTATCGGA
                 N  Y  Q  I  M  D  S  W  E  E  E  P  R  D  K  S  T  N  I  S  S  A  L  N  I  I  E  F  I  L  S  T  D  P  Q  E  D  L  S  E
K  T  I  K  S  W  I  L  G  K  R  N  Q  E  I  N  Q  L  I  S  P  R  P  S  T  S  L  N  S  Y  S  A  P  T  P  K  K  T  Y  R

        1930      1940      1950      1960      1970      1980      1990      2000      2010      2020      2030      2040
AAACGACACAATCAACACAAGAACCCAACTCAGCGCCCACCATCTGTCAACCAGAAATCAAACCAACAGAAACAAGTGAAAAAGTTAGTGGATCAACTGACAAAAATAGACAGTCTGG
N  D  T  I  N  T  R  T  Q  Q  L  S  A  T  I  C  Q  P  E  I  K  P  T  E  T  S  E  K  V  S  G  S  T  D  K  N  R  Q  S  G
K  T  T  Q  S  T  Q  E  P  S  N  S  A  P  P  S  V  N  Q  K  S  N  Q  Q  K  Q  V  K  K  L  V  D  Q  L  T  K  I  D  S  L

        2050      2060      2070      2080      2090      2100      2110      2120      2130      2140      2150      2160
GTCATCACACGAATGTACAACAGAACAAAAGATAGAAATATTGATCAGGAAACTGTACAGGGAGGATCTGGGAAGAAGCAGCTCAGATAGTAGAGCTGAGCTGTGGTCTCTGGAGG
S  S  H  E  C  T  T  E  A  K  D  R  N  I  D  Q  E  T  V  Q  G  G  S  G  R  R  S  S  S  D  S  R  A  E  T  V  V  S  G  G
G  H  H  T  N  V  Q  Q  K  Q  K  I  E  I  L  I  R  K  L  Y  R  E  D  L  G  E  E  A  A  Q  I  V  E  L  R  L  W  S  L  E
```

HUMAN PARAINFLUENZA VIRUS TYPE 3

```
      2170      2180      2190      2200      2210      2220      2230      2240      2250      2260      2270      2280
AATCTCTGGAAGCATCACAGATTCTAAAAATGGAACCCAAACACGGAGAATATTGATCTCAATGAAATTAGAAAGATGGATAAGGACTCTATTGAGAGGAAAATGCGACAATCTGCAGA
    I  S  G  S  I  T  D  S  K  N  G  T  Q  N  T  E  N  I  D  L  N  E  I  R  K  M  D  K  D  S  I  E  R  K  M  R  Q  S  A  D
  E  S  L  E  A  S  Q  I  L  K  M  E  P  K  T  R  R  I  L  I  S  M  K  L  E  R  W  I  R  T  L  L  R  G  K  C  D  N  L  Q

      2290      2300      2310      2320      2330      2340      2350      2360      2370      2380      2390      2400
TGTTCCAAGCGAGATATCAGGAAGTGATGTCATATTTACAACAGAACAAAGTAGAAACAGTGATCATGGAAGAAGCTTGGAACCTATCAGTACACCTGATACAAGATCAATGAGTGTTGT
    V  P  S  E  I  S  G  S  D  V  I  F  T  T  E  Q  S  R  N  S  D  H  G  R  S  L  E  P  I  S  T  P  D  T  R  S  M  S  V  V
  M  F  Q  A  R  Y  Q  E  V  M  S  Y  L  Q  Q  N  K  V  E  T  V  I  M  E  E  A  W  N  L  S  V  H  L  I  Q  D  Q  *

      2410      2420      2430      2440      2450      2460      2470      2480      2490      2500      2510      2520
TACTGCTGCGACACCAGATGATGAAGAAGAAATACTAATGAAAAATAGTAGGATGAAGAAAAGTTCTTCAACACACCAAGAAGATGACAAAAGAATTAAAAAAGGGGAGAAAGGGAAAGA
                                                                                                    G  K  R  E  R
    T  A  A  T  P  D  D  E  E  E  I  L  M  K  N  S  R  M  K  K  S  S  S  T  H  Q  E  D  D  K  R  I  K  K  G  G  K  G  K  D
                                                         D CISTRON ▸  M  T  K  E  L  K  K  G  E  K  G  K

      2530      2540      2550      2560      2570      2580      2590      2600      2610      2620      2630      2640
CTGGTTTAAGAAATCAAGAGATACTGACAACCAGACATCAACATCAGATCACAAACCCACATCAAAAGGGCAAAAGAAAATCTCAAAAACAACAACCACCAACACCGACACAAAGGGGCA
  L  V  *
    W  F  K  K  S  R  D  T  D  N  Q  T  S  T  S  D  H  K  P  T  S  K  G  Q  K  K  I  S  K  T  T  T  T  N  T  D  T  K  G  Q
  T  G  L  R  N  Q  E  I  L  T  T  R  H  Q  H  Q  I  T  N  P  H  Q  K  G  K  R  K  S  Q  K  Q  Q  P  P  T  P  T  Q  R  G

      2650      2660      2670      2680      2690      2700      2710      2720      2730      2740      2750      2760
AACAGAACACAGACAGAATCATCAGAAACACAATCCCCATCATGGAATCCCATTATCGACAACAACACTGACCGAACGGACAAGCACAACCCCCCCAACAACAACTCCCAGATC
    T  E  T  Q  T  E  S  S  E  T  Q  S  P  S  W  N  P  I  I  D  N  N  T  D  R  T  E  R  T  S  T  T  P  P  T  T  P  R  S
  K  Q  K  H  R  Q  N  H  Q  K  H  N  P  H  H  G  I  P  L  S  T  T  T  L  T  E  P  N  G  Q  A  Q  P  P  Q  Q  Q  L  P  D

      2770      2780      2790      2800      2810      2820      2830      2840      2850      2860      2870      2880
AACTCGTACAAAAGAATCAATCCGAACAAACTCTGAATCCAAACCCAAGACACAAAAGACAATTGGAAAGGAAAGGAAGGATACAGAAGAGAGCAATCGATTTACAGAGAGGGCAATTAC
                                                                    V CISTRON ▸      G  K  E  G  Y  R  R  E  Q  S  I  Y  R  E  G  N  Y
    T  R  T  K  E  S  I  R  T  N  S  E  S  K  P  K  T  Q  K  T  I  G  K  E  R  K  D  T  E  E  S  N  R  F  T  E  R  A  I  T
  Q  L  V  Q  K  N  Q  S  E  Q  T  L  N  P  N  P  R  H  K  R  Q  L  E  R  K  G  R  I  Q  K  R  A  I  D  L  Q  R  G  Q  L

      2890      2900      2910      2920      2930      2940      2950      2960      2970      2980      2990      3000
TCTATTGCAGAATCTTGGTGTAATTCAATCTACATCAAAACTAGATTTATATCAGACAAACGAGTTGTATGTGTAGCAAATGTACTAAACAATGTAGATACTGCATCAAAGATAGACTT
    S  I  A  E  S  W  C  N  S  I  Y  I  K  T  R  F  I  S  R  Q  T  S  C  M  C  S  K  C  T  K  Q  C  R  Y  C  I  K  D  R  L
      L  L  Q  N  L  G  V  I  Q  S  T  S  K  L  D  L  Y  Q  D  K  R  V  V  C  V  A  N  V  L  N  N  V  D  T  A  S  K  I  D  F
  L  Y  C  R  I  L  V  *

      3010      3020      3030      3040      3050      3060      3070      3080      3090      3100      3110      3120
CCTAGCAGGATTAGTCATAGGGGTTTCAATGGACAATGACACAAAATTAATACAGATACAAAATGAAATGTTAAACCTCAAAGCAGATCTAAAGAGAATGGACGAATCACATAGAAGATT
  P  S  R  I  S  H  R  G  F  N  G  Q  *
    L  A  G  L  V  I  G  V  S  M  D  N  D  T  K  L  I  Q  I  Q  N  E  M  L  N  L  K  A  D  L  K  R  M  D  E  S  H  R  R  L

      3130      3140      3150      3160      3170      3180      3190      3200      3210      3220      3230      3240
GATAGAAAATCAAAGAGAACAACTGTCATTGATCACATCGTTAATTTCAAATCTTAAAATTATGACTGAGAGAGGAGGAAAGAAAGACCAAAATGAATCCAATGAGGAGTATCTATGAT
    I  E  N  Q  R  E  Q  L  S  L  I  T  S  L  I  S  N  L  K  I  M  T  E  R  G  G  K  K  D  Q  N  E  S  N  E  R  V  S  M  I

      3250      3260      3270      3280      3290      3300      3310      3320      3330      3340      3350      3360
CAAGACAAAATTGAAGAAGAAAAGATCAAGAAAACCAGGTTTGACCCACTTATGGAGGCACAAGGTATTGACAAGAATATACCTGATCTATATCGACATGCAGGAAATACGTTAGAGAA
    K  T  K  L  K  E  E  K  I  K  K  T  R  F  D  P  L  M  E  A  Q  G  I  D  K  N  I  P  D  L  Y  R  H  A  G  N  T  L  E  N

      3370      3380      3390      3400      3410      3420      3430      3440      3450      3460      3470      3480
CGACGTACAAGTTAAATCAGAGATATTAAGTTCATACAACGAGTCAAATGCAACAAGACTAATACCCAGAAAAGTGAGCAGTACAATGAGACTACTAGTTGCAGTCATCAACAACAGCAA
    D  V  Q  V  K  S  E  I  L  S  S  Y  N  E  S  N  A  T  R  L  I  P  R  K  V  S  S  T  M  R  S  L  V  A  V  I  N  N  S  N

      3490      3500      3510      3520      3530      3540      3550      3560      3570      3580      3590      3600
TCTCCCACAAAGCACAAAACAATCATATATAAACGAACTCAAAACATTGCAAAAGTGATGAAAGATATCTGAATTGATGGACATGTTCAATGAAGATGTTAACAATTGCTAAAGATCAAA
    L  P  Q  S  T  K  Q  S  Y  I  N  E  L  K  H  C  K  S  D  E  E  V  S  E  L  M  D  M  F  N  E  D  V  N  N  C  *

      3610      3620      3630      3640      3650      3660      3670      3680      3690      3700      3710      3720
TAAAAAAAAACAACACCGAATAAATAGACAAGAAACAACAGTAGATCAAAACCTATCAACACACACAAAATCAAGCAGAGTGAAACAATAGACATCAATCAATATACAAATAAGAAAACT
                                                                                                    └┴──┴──┴┴──┴┘
                                                                                                     T stop      I

      3730      3740      3750      3760      3770      3780      3790      3800      3810      3820      3830      3840
TAGGATTAAAGAATAAATTAATCCTTGTCCAAAATGAGTATAACTAACTCTGCAATATACACATTCCCGGAGTCATCATTCTCTGAGAATGGTCATATAGAACCATTACCACTCAAAGTC
└┴──┴┘                                           M  S  I  T  N  S  A  I  Y  T  F  P  E  S  S  F  S  E  N  G  H  I  E  P  L  P  L  K  V
 T start    MATRIX PROTEIN ▸

      3850      3860      3870      3880      3890      3900      3910      3920      3930      3940      3950      3960
AATGAACGAGAAAAGCAGTACCTCACATTAGAGTTGCCAAAATCGGAAATCCACCAAAACATGGATCCCGGTATTGGATGTCTTCTTACTCGGCTTCTTCGAGATGGAACGAATCAAA
  N  E  Q  R  K  A  V  P  H  I  R  V  A  K  I  G  N  P  P  K  H  G  S  R  Y  L  D  V  F  L  L  G  F  F  E  M  E  R  I  K

      3970      3980      3990      4000      4010      4020      4030      4040      4050      4060      4070      4080
GACAAATACGGGAGTGTGAATGATCTTGACAGTGACCCGGGTTACAAAGTTTGTGGCTCTGGATCATTACCAATCGGATTAGCCAAATACACTGGGAATGACCAGGAATTATTACAGGCT
  D  K  Y  G  S  V  N  D  L  D  S  D  P  G  Y  K  V  C  G  S  G  S  L  P  I  G  L  A  K  Y  T  G  N  D  Q  E  L  L  Q  A
```

HUMAN PARAINFLUENZA VIRUS TYPE 3

```
      4090      4100      4110      4120      4130      4140      4150      4160      4170      4180      4190      4200
GCAACTAAACTGGACATAGAAGTGAGAAGAACAGTCAAAGCGAAAGAAATGATTGTTTATACGGTACAAAATATAAAACCAGAACTGTACCCATGGTCCAGTAGACTAAGAAAAGGAATG
 A  T  K  L  D  I  E  V  R  R  T  V  K  A  K  E  M  I  V  Y  T  V  Q  N  I  K  P  E  L  Y  P  W  S  S  R  L  R  K  G  M

      4210      4220      4230      4240      4250      4260      4270      4280      4290      4300      4310      4320
TTGTTCGATGCCAACAAAGTTGCTCTTGCTCCTCAATGTCTTCCACTAGATAGGAGCATAAAATTCAGAGTAATCTTCGTTAATTGTACGGCAATTGGATCAATAACCTTGTTTAAAATT
 L  F  D  A  N  K  V  A  L  A  P  Q  C  L  P  L  D  R  S  I  K  F  R  V  I  F  V  N  C  T  A  I  G  S  I  T  L  F  K  I

      4330      4340      4350      4360      4370      4380      4390      4400      4410      4420      4430      4440
CCCAAGTCAATGGCACTACTATCTCTACCCAGCACAATATCAATCAATCTGCAGGTACACATCAAAACAGGGGTTCAGACTGATTCTAAAGGGATAGTTCAAATTTTGGATGAGAAGGGT
 P  K  S  M  A  S  L  S  L  P  S  T  I  S  I  N  L  Q  V  H  I  K  T  G  V  Q  T  D  S  K  G  I  V  Q  I  L  D  E  K  G

      4450      4460      4470      4480      4490      4500      4510      4520      4530      4540      4550      4560
GAAAAATCACTGAATTTCATGGTCCATCTCGGATTGATCAAAAGAAAAGTAGGCAGAATGTACTCTGTCGAGTACTGTAAACAGAAAATCGAGAAAATGAGATTGATATTTTCTTTGGGA
 E  K  S  L  N  F  M  V  H  L  G  L  I  K  R  K  V  G  R  M  Y  S  V  E  Y  C  K  Q  K  I  E  K  M  R  L  I  F  S  L  G

      4570      4580      4590      4600      4610      4620      4630      4640      4650      4660      4670      4680
TTAGTTGGAGGAATCAGTCTTCATGTCAATGCAACTGGATCTATATCAAAAACACTAGCAAGTCAGCTGGTATTCAAAAGGGAGATTTGTTATCCCTTAATGGATCTAAATCCACATCTC
 L  V  G  G  I  S  L  H  V  N  A  T  G  S  I  S  K  T  L  A  S  Q  L  V  F  K  R  E  I  C  Y  P  L  M  D  L  N  P  H  L

      4690      4700      4710      4720      4730      4740      4750      4760      4770      4780      4790      4800
AATCTAGTTATCTGGGCTTCATCAGTAGAGATTACAAGAGTGGATGCAATTTTCCAACCTTCTTTACCTGGCGAGTTCAGATACTATCCTAACATTATTGCAAAAGGAGTTGGGAAAATC
 N  L  V  I  W  A  S  S  V  E  I  T  R  V  D  A  I  F  Q  P  S  L  P  G  E  F  R  Y  Y  P  N  I  I  A  K  G  V  G  K  I

      4810      4820      4830      4840      4850      4860      4870      4880      4890      4900      4910      4920
AAACAATGGAACTAGTAATCTCTATTTTGATCTGGATATATCTATTAAGCCAAAGCAAATAAGAGTAATCAAAAACTTAGGACAAAAGAAGTCAATACCAACAACTATTAGCAGCCACA
 K  Q  W  N  *
```
⌐T____⌐ T____⌐ ⌐‖‖‖‖‖⌐ ‖T‖‖T‖
T $_{stop}$ I   T $_{start}$ **FUSION PROTEIN** ▶

```
      4930      4940      4950      4960      4970      4980      4990      5000      5010      5020      5030      5040
CTCGCTGGAACAAGAAAGAAGGGATAAAAAAAGTTTAACAGAAGAAACAAAAACAAAAAGCACAGAACACCAGAACAACAAGATCAAAACACCCAACCCACTCAAAACGAAAATCTCAAA
```

```
      5050      5060      5070      5080      5090      5100      5110      5120      5130      5140      5150      5160
AGAGATTGGCAACACAACAAACACTGAACATCATGCCAACCTCAATACTGCTAATTATTACAACCATGATTATGGCATCTTTCTGCCAAATAGATATCACAAAACTACAGCATGTAGGTG
                                        M  P  T  S  I  L  L  I  I  T  T  M  I  M  A  S  F  C  Q  I  D  I  T  K  L  Q  H  V  G
```

```
      5170      5180      5190      5200      5210      5220      5230      5240      5250      5260      5270      5280
TATTGGTTAACAGTCCCAAAGGGATGAAGATATCACAAAACTTTGAAACAAGATATCTAATTTTGAGCCTCATACCAAAAATAGAAGATTCTAACTCTTGTGGTGACCAACAGATCAAGC
 V  L  V  N  S  P  K  G  M  K  I  S  Q  N  F  E  T  R  Y  L  I  L  S  L  I  P  K  I  E  D  S  N  S  C  G  D  Q  Q  I  K
```

```
      5290      5300      5310      5320      5330      5340      5350      5360      5370      5380      5390      5400
AATACAAGAGGTTATTGGATAGACTGATCATTCCTTTATATGATGGATTAAGATTACAGAAGGATGTGATAGTGTCCAATCAAGAATCCAATGAAAACACTGACCCCAGAACAAAACGAT
 Q  Y  K  R  L  L  D  R  L  I  I  P  L  Y  D  G  L  R  L  Q  K  D  V  I  V  S  N  Q  E  S  N  E  N  T  D  P  R  T  K  R
```

```
      5410      5420      5430      5440      5450      5460      5470      5480      5490      5500      5510      5520
TCTTTGGAGGGGTAATTGGAACTATTGCTCTGGGAGTGGCCAACCTCAGCACACAAATTACAGCGGCAGTTGCTCTGGTTGAAGCCAAGCAGGCAAGATCAGACATTGAAAAACTCAAGGAAG
 F  F  G  G  V  I  G  T  I  A  L  G  V  A  T  S  A  Q  I  T  A  A  V  A  L  V  E  A  K  Q  A  R  S  D  I  E  K  L  K  E
```

```
      5530      5540      5550      5560      5570      5580      5590      5600      5610      5620      5630      5640
CAATCAGGGACACAAACAAAGCAGTGCAGTCAGTCCAGAGCTCCATAGGAAATTTGATAGTAGCAATTAAATCGGTCCAGGATTATGTCAACAAAGAAATCGTGCCATCAATTGCGAGAT
 A  I  R  D  T  N  K  A  V  Q  S  V  Q  S  S  I  G  N  L  I  V  A  I  K  S  V  Q  D  Y  V  N  K  E  I  V  P  S  I  A  R
```

```
      5650      5660      5670      5680      5690      5700      5710      5720      5730      5740      5750      5760
TAGGTTGTGAAGCAGCAGGACTTCAGTTAGGAATTGCATTAACACAGCATTACTCAGAATTAACAAACATATTCGGTGATAACATAGGATCATTACAAGAAAAAGGGATAAAATTACAAG
 L  G  C  E  A  A  G  L  Q  L  G  I  A  L  T  Q  H  Y  S  E  L  T  N  I  F  G  D  N  I  G  S  L  Q  E  K  G  I  K  L  Q
```

```
      5770      5780      5790      5800      5810      5820      5830      5840      5850      5860      5870      5880
GTATAGCATCATTATACCGCACAAATATCACAGAGATATTCACAACATCCACAGTTGATAAATATGATATTTATGATCTTATTATTTACGAATCAATAAAGGTGAGAGTTATAGATGTTG
 G  I  A  S  L  Y  R  T  N  I  T  E  I  F  T  T  S  T  V  D  K  Y  D  I  Y  D  L  L  F  T  E  S  I  K  V  R  V  I  D  V
```

```
      5890      5900      5910      5920      5930      5940      5950      5960      5970      5980      5990      6000
ACTTGAATGATTACTCAATCACCCTCCAAGTCAGACTCCCTTTATTAACTAGACTGCTGAACACCCAGATTTACAGAGTAGATTCCATATCATATAACATCCAAAACAGAGAATGGTATA
 D  L  N  D  Y  S  I  T  L  Q  V  R  L  P  L  L  T  R  L  L  N  T  Q  I  Y  R  V  D  S  I  S  Y  N  I  Q  N  R  E  W  Y
```

```
      6010      6020      6030      6040      6050      6060      6070      6080      6090      6100      6110      6120
TCCCTCTTCCCAGCCACATCATGACAAAAGGGGCATTTCTAGGTGGAGCAGATGTCAAAGAATGTATAGAAGCATTCAGCAGTTATATATGCCCTTCTGATCCAGGATTTGTACTAAACC
 I  P  L  P  S  H  I  M  T  K  G  A  F  L  G  G  A  D  V  K  E  C  I  E  A  F  S  S  Y  I  C  P  S  D  P  G  F  V  L  N
```

```
      6130      6140      6150      6160      6170      6180      6190      6200      6210      6220      6230      6240
ATGAAATGGAGAGCTGTTTATCAGGAAACATATCCCAATGTCCAAGAACCGTGGTTAAATCAGACATTGTTCCAAGATATGCATTTGTCAATGGAGGAGTGGTTGCAAATTGTATAACAA
 H  E  M  E  S  C  L  S  G  N  I  S  Q  C  P  R  T  V  V  K  S  D  I  V  P  R  Y  A  F  V  N  G  G  V  V  A  N  C  I  T
```

```
      6250      6260      6270      6280      6290      6300      6310      6320      6330      6340      6350      6360
CCACATGTACATGCAACGGTATCGGTAATAGAATCAATCAACCACCTGATCAAGGAGTGAAAAATTATAACACATAAAGAATGTAATCAATAGGTATCAACGGAATGCTGTTCAATACAA
 T  T  C  T  C  N  G  I  G  N  R  I  N  Q  P  P  D  Q  G  V  K  I  I  T  H  K  E  C  N  T  I  G  I  N  G  M  L  F  N  T
```

HUMAN PARAINFLUENZA VIRUS TYPE 3

```
      6370      6380      6390      6400      6410      6420      6430      6440      6450      6460      6470      6480
ATAAAGAGGAACTCTTGCATTTTACACACCAAATGATATAACATTAAACAATTCTGTTGCACTTGATCCCAATTGACATATCAATCGAGCTCAATAAGGCCAAATCAGATCTAGAAGAGT
 N  K  E  G  T  L  A  F  Y  T  P  N  D  I  T  L  N  N  S  V  A  L  D  P  I  D  I  S  I  E  L  N  K  A  K  S  D  L  E  E

      6490      6500      6510      6520      6530      6540      6550      6560      6570      6580      6590      6600
CAAAAGAATGGATAAGAAGGTCAAATCAAAAACTAGAATTCCATTGGAAATTGGCATCAATCTAGCACCACAATCATAATTGTTTTGATAATGATAATTATATTGTTTATAATTAATGTAA
 S  K  E  W  I  R  R  S  N  Q  K  L  D  S  I  G  N  W  H  Q  S  S  T  T │I  I  I  V  L  I  M  I  I  I  L  F  I  I  N  V

      6610      6620      6630      6640      6650      6660      6670      6680      6690      6700      6710      6720
CGATAATTATAATTGCAGTTAAGTATTACAGAATTCAAAAGAGAAATCGAGTGGATCAAAATGATAAACCATATGTATTAACAAACAAATGACAGATCTATAGATCATTAGATATTAAAA
 T  I  I  I  I  A  V │K  Y  Y  R  I  Q  K  R  N  R  V  D  Q  N  D  K  P  Y  V  L  T  N  K  *

      6730      6740      6750      6760      6770      6780      6790      6800      6810      6820      6830      6840
TTATAAAAAACTTAGGAGTAAAGTTACGCAATTCAACTCTTACTCATATAATTGAGAAAGAACCCAACAGACAAATCCAAATCCGAGATGGAATACTGGAAGCACACCAATCACGGGAAAG
T stop   I   T start      HEMAGGLUTININ-NEURAMINIDASE PROTEIN ▶         M  E  Y  W  K  H  T  N  H  G  K

      6850      6860      6870      6880      6890      6900      6910      6920      6930      6940      6950      6960
ATGCTGGTAATGAGTGGAAACATCCATGGCTACTCATGGCAACAAGATCACCAACAAGATAACATATATATTATGGACAATAATCCTGGTGTTATTATCAATAGTCTTCATCATAGTGC
 D  A  G  N  E  L  E  T  S  M  A  T  H  G  N  K  I  T  N  K │I  T  Y  I  L  W  T  I  I  L  V  L  L  S  I  V  F  I  I  V

      6970      6980      6990      7000      7010      7020      7030      7040      7050      7060      7070      7080
TAATTAATTCCATCAAAAGTGAAAAAGCCCATGAATCATTGCTACAAGACGTAAACAATGAGTTTATGGAAGTTACAGAAAAGATCCAAATGGCATCGGATAATATTAATGATCTAATAC
 L  I │N  S  I  K  S  E  K  A  H  E  S  L  L  Q  D  V  N  N  E  F  M  E  V  T  E  K  I  Q  M  A  S  D  N  I  N  D  L  I

      7090      7100      7110      7120      7130      7140      7150      7160      7170      7180      7190      7200
AGTCAGGAGTGAATACAAGGCTTCTTACAATTCAGAGTCATGTCCAGAATTATATACCGATATCATTGACACAACAAATGTCGGATCTTAGGAAATTCATTAGTGAAATTACAATTAGGA
 Q  S  G  V  N  T  R  L  L  T  I  Q  S  H  V  Q  N  Y  I  P  I  S  L  T  Q  Q  M  S  D  L  R  K  F  I  S  E  I  T  I  R

      7210      7220      7230      7240      7250      7260      7270      7280      7290      7300      7310      7320
ATGATAATCGAGAAGTGCCTCCACAAAGAATAACACATGATGCGGGCATAAAACCTTTAAATCCAGATGATTTTTGGAGATGCACGTCTGGTCTTCCATCTTTAATGAAAACTCCAAAAA
 N  D  N  R  E  V  P  P  Q  R  I  T  H  D  A  G  I  K  P  L  N  P  D  D  F  W  R  C  T  S  G  L  P  S  L  M  K  T  P  K

      7330      7340      7350      7360      7370      7380      7390      7400      7410      7420      7430      7440
TAAGGTTAATGCCGGGCCGGGATTATTAGCTATGCCAACGACTGTTGATGGCTGTGTTGTAAGAACTCCGTCCTTAGTTATAAATGATCTGATTTATGCTTATACCTCAAATCTAATTACTC
 I  R  L  M  P  G  P  G  L  L  A  M  P  T  T  V  D  G  C  V  R  T  P  S  L  V  I  N  D  L  I  Y  A  Y  T  S  N  L  I  T

      7450      7460      7470      7480      7490      7500      7510      7520      7530      7540      7550      7560
GAGGTTGCCAGGATATAGGAAAATCATATCAAGTATTACAGATAGGGATAATAACTGTAAACTCAGACTTGGTACCTGACTTAAATCCTAGGATCTCTCATACTTTCAACATAAATGACA
 R  G  C  Q  D  I  G  K  S  Y  Q  V  L  Q  I  G  I  I  T  V  N  S  D  L  V  P  D  L  N  P  R  I  S  H  T  F  N  I  N  D

      7570      7580      7590      7600      7610      7620      7630      7640      7650      7660      7670      7680
ATAGAAAGTCATGTTCTCTAGCACTCCTAAACACAGATGTATATCAACTGTGTTCGACTCCCAAAGTTGATGAAAGATCAGATTATGCATCATCAGGCATAGAAGATATTGTACTTGATA
 N  R  K  S  C  S  L  A  L  L  N  T  D  V  Y  Q  L  C  S  T  P  K  V  D  E  R  S  D  Y  A  S  S  G  I  E  D  I  V  L  D

      7690      7700      7710      7720      7730      7740      7750      7760      7770      7780      7790      7800
TCGTCAATCATGATGGTTCAATCTCAACAACAAGATTTAAGAACAATAATATAAGTTTTGATCAACCATATGCGGCATTATACCCATCTGTTGGACCAGGGATATACTACAAAGGCAAAA
 I  V  N  H  D  G  S  I  S  T  T  R  F  K  N  N  N  I  S  F  D  Q  P  Y  A  A  L  Y  P  S  V  G  P  G  I  Y  Y  K  G  K

      7810      7820      7830      7840      7850      7860      7870      7880      7890      7900      7910      7920
TAATATTTCTCGGGTATGGAGGTCTTGAACATCCAATAAATGAGAATGCAATCTGCAACACAACTGGGTGTCCCGGGAAAACGCAGAGAGACTGCAATCAGGCATCTCATAGTCCTTGGT
 I  I  F  L  G  Y  G  G  L  E  H  P  I  N  E  N  A  I  C  N  T  T  G  C  P  G  K  T  Q  R  D  C  N  Q  A  S  H  S  P  W

      7930      7940      7950      7960      7970      7980      7990      8000      8010      8020      8030      8040
TTTCAGACAGAAGGATGGTCAACTCCATTATTGTTGTTGACAAGGGCTTAAACTCAATTCCAAAGCTGAAGGTATGGACGATATCCATGAGACAAAATTACTGGGGGTCAGAAGGAAGGC
 F  S  D  R  R  M  V  N  S  I  I  V  V  D  K  G  L  N  S  I  P  K  L  K  V  W  T  I  S  M  R  Q  N  Y  W  G  S  E  G  R

      8050      8060      8070      8080      8090      8100      8110      8120      8130      8140      8150      8160
TACTTCTACTAGGTAACAAGATCTATATATATCAAGATCTACAAGTTGGCATAGCAAGTTACAATTAGGAATAATTGATATTACTGATTACAGTGATATAAGAATAAAATGACATGGC
 L  L  L  L  G  N  K  I  Y  I  Y  T  R  S  T  S  W  H  S  K  L  Q  L  G  I  I  D  I  T  D  Y  S  D  I  R  I  K  W  T  W

      8170      8180      8190      8200      8210      8220      8230      8240      8250      8260      8270      8280
ATAAATGTGCTATCAAGACCAGGAAACAATGAATGTCCATGGGGACATTCATGCCCAGATGGATGTATAACAGGAGTATATACTGATGCATATCCACTCAATCCCACAGGGAGCATTGTGT
 H  N  V  L  S  R  P  G  N  N  E  C  P  W  G  H  S  C  P  D  G  C  I  T  G  V  Y  T  D  A  Y  P  L  N  P  T  G  S  I  V

      8290      8300      8310      8320      8330      8340      8350      8360      8370      8380      8390      8400
CATCTCGTCATATTAGACTCGCAAAAATCGAGAGTAAACCCAGTCATAACTTACTCAACATCAACTGAAAGGGTAAACGAGCTGGCCATCCGAAACAAAACACTCTCAGCTGGATATACAA
 S  S  V  I  L  D  S  Q  K  S  R  V  N  P  V  I  T  Y  S  T  S  T  E  R  V  N  E  L  A  I  R  N  K  T  L  S  A  G  Y  T

      8410      8420      8430      8440      8450      8460      8470      8480      8490      8500      8510      8520
CAACGAGCTGCATTACACACTATAACAAAGGATATTGTTTTCATATAGTAGAAATAAATCATAAAAGCTTAGACACATTCCAACCTATGTTGTTCAAAACAGAGATTCCAAAAAGCTGCA
 T  T  S  C  I  T  H  Y  N  K  G  Y  C  F  H  I  V  E  I  N  H  K  S  L  D  T  F  Q  P  M  L  F  K  T  E  I  P  K  S  C

      8530      8540      8550      8560      8570      8580      8590      8600      8610      8620      8630      8640
GTTAACATAATTAACCATAATATGTATTAACCTATCTATAATACAAGTATATGATAAGTAATCAGCAATCAGACAATAGATAAAAGAGAAAATATAAAAAACTTAGGAGCAAAGCATGCT
 S  *                                                                    T stop   I   T start                       L
```

```
      8650      8660      8670      8680      8690      8700      8710      8720      8730      8740      8750      8760
CGAAAAATGGACACTGAATCTAACAATGGCACTGTATCTGACATACTCTATCCTGAGTGTCACCTTAATTCTCCTATCGTTAAGGGTAAAATAGCACAATTACACACTATTATGAGTCGTSL
      M  D  T  E  S  N  N  G  T  V  S  D  I  L  Y  P  E  C  H  L  N  S  P  I  V  K  G  K  I  A  Q  L  H  T  I  M  S  L
PROTEIN ▸
      8770      8780      8790      8800      8810      8820      8830      8840      8850      8860      8870      8880
CCACAGCCTTACGATATGGATGACGACTCAATACTAGTTATCACTAGACAGAAAATAAAACTCAATAAATTAGATAAAAGACAAGATCTATTAGAAGATTAAAATTAATATTAACTGAG
   P  Q  P  Y  D  M  D  D  D  S  I  L  V  I  T  R  Q  K  I  K  L  N  K  L  D  K  R  Q  R  S  I  R  R  L  K  L  I  L  T  E
      8890      8900      8910      8920      8930      8940      8950      8960      8970      8980      8990      9000
AAAGTGAATGACTTAGGAAAAATACACATTTATTAGATATCCAGAAATGTCAAAAGAAATGTTCAAATTACATATACCTGGTATTAACAGTAAAGTGACTGAATTATTACTTAAAGCAGAT
   K  V  N  D  L  G  K  Y  T  F  I  R  Y  P  E  M  S  K  E  M  F  K  L  H  I  P  G  I  N  S  K  V  T  E  L  L  L  K  A  D
      9010      9020      9030      9040      9050      9060      9070      9080      9090      9100      9110      9120
AGAACATATAGTCAAATGACTGATGGATTAAGAGATCTATGGATTAATGTGCTATCGAAATTAGCCTCAAAAAATGATGGAAGCAATTATGATCTTAATGAAGAAATTAATAATATATCA
   R  T  Y  S  Q  M  T  D  G  L  R  D  L  W  I  N  V  L  S  K  L  A  S  K  N  D  G  S  N  Y  D  L  N  E  E  I  N  N  I  S
      9130      9140      9150      9160      9170      9180      9190      9200      9210      9220      9230      9240
AAAGTTCACACAACCTATAAATCAGATAAATGGTATAATCCATTCAAAACATGGTTCACTATCAAGTATGATATGAGAAGATTGCAAAAAGCTCGAAATGAGGTCACTTTTAATATGGGG
   K  V  H  T  T  Y  K  S  D  K  W  Y  N  P  F  K  T  W  F  T  I  K  Y  D  M  R  R  L  Q  K  A  R  N  E  V  T  F  N  M  G
      9250      9260      9270      9280      9290      9300      9310      9320      9330      9340      9350      9360
AAAGATTATAACTTGTTAGAAGACCAGAAGAATTCTTATTGATACATCCGAATTGGTTTTAATATTAGATAAACAAAACTATAATGGTTATCTAATTACTCCTGAATTAGTATTGCCTG
   K  D  Y  N  L  L  E  D  Q  K  N  F  L  L  I  H  P  E  L  V  L  I  L  D  K  Q  N  Y  N  G  Y  L  I  T  P  E  L  V  L  P
      9370      9380      9390      9400      9410      9420      9430      9440      9450      9460      9470      9480
TATTGTGACGTAGTTGAAGGCCGATGGAATATAAGTGCATGTGCTAAGTTAGATCCAAAATTACAATCTATGTATCAGAAAGGCAATAATCTGTGGGAAGTGATAGATAAATTGTTTCCA
   Y  C  D  V  V  E  G  R  W  N  I  S  A  C  A  K  L  D  P  K  L  Q  S  M  Y  Q  K  G  N  N  L  W  E  V  I  D  K  L  F  P
      9490      9500      9510      9520      9530      9540      9550      9560      9570      9580      9590      9600
ATTATGGGAGAAAAGACATTTGATGTGATATCATTATTAGAACCACTTGCATTATCTCTAATTCAAACTCATGATCCTGTTAAACAATTAAGGGGAGCTTTTTTAAATCATGTGTTATCC
   I  M  G  E  K  T  F  D  V  I  S  L  L  E  P  L  A  L  S  L  I  Q  T  H  D  P  V  K  Q  L  R  G  A  F  L  N  H  V  L  S
      9610      9620      9630      9640      9650      9660      9670      9680      9690      9700      9710      9720
GAGATGGAATTGATATTTGAATCTAGAGAATCGATTAAAGAATTTCTGAGTGTAGATTACATTGATAAAATCTTAGATATATTTAATAAATCTACAATAGATGAAATAGCAGAGATTTTC
   E  M  E  L  I  F  E  S  R  E  S  I  K  E  F  L  S  V  D  Y  I  D  K  I  L  D  I  F  N  K  S  T  I  D  E  I  A  E  I  F
      9730      9740      9750      9760      9770      9780      9790      9800      9810      9820      9830      9840
TCTTTTTTTAGAACATTTGGGCATCCTCCATTAGAGGCTAGTATTGCAGCAGAAAAAGTTAGAAAATATATGTATATTGGGAAACAATTAAAATTTGACACTATTAATAAATGTCATGCT
   S  F  F  R  T  F  G  H  P  P  L  E  A  S  I  A  A  E  K  V  R  K  Y  M  Y  I  G  K  Q  L  K  F  D  T  I  N  K  C  H  A
      9850      9860      9870      9880      9890      9900      9910      9920      9930      9940      9950      9960
ATCTTCTGTACAATAATAATTAACGGATATAGAGAAAGGCATGGTGGACAGTGGCCTCCTGTGACATTACCTGATCATGCACACGAATTCATCATAAATGCTTACGGTTCAAATTCTGCG
   I  F  C  T  I  I  I  N  G  Y  R  E  R  H  G  G  Q  W  P  P  V  T  L  P  D  H  A  H  E  F  I  I  N  A  Y  G  S  N  S  A
      9970      9980      9990     10000     10010     10020     10030     10040     10050     10060     10070     10080
ATATCATATGAAAACGCTGTTGATTATTACCAGAGCTTTATAGGAATAAAATTTAATAAATTCATAGAACCTCAGTTAGATGAAGATTTGACAATTTATATGAAAGATAAAGCATTGTCT
   I  S  Y  E  N  A  V  D  Y  Y  Q  S  F  I  G  I  K  F  N  K  F  I  E  P  Q  L  D  E  D  L  T  I  Y  M  K  D  K  A  L  S
     10090     10100     10110     10120     10130     10140     10150     10160     10170     10180     10190     10200
CCAAAAAAATCAAACTGGGACACAGTTTCTCCTGCCTCATAATTTACTGTACCGTACTAACGCATCCAACGAATCACGAAGGATTAGTTGAAAAATTTATAGCAGATAGTAAATTTGATCCT
   P  K  K  S  N  W  D  T  V  S  P  A  S  N  L  L  Y  R  T  N  A  S  N  E  S  R  R  L  V  E  K  F  I  A  D  S  K  F  D  P
     10210     10220     10230     10240     10250     10260     10270     10280     10290     10300     10310     10320
AATCAGATATTAGATTATGTAGAATCTGGGGACTGGTTAGATGATCCAGAATTTAATATTTCTTATAGTCTTAAAGAAAATGAGATCAAACAAGAAGGTAGACTCTTTGCAAAAATGACA
   N  Q  I  L  D  Y  V  E  S  G  D  W  L  D  D  P  E  F  N  I  S  Y  S  L  K  E  N  E  I  K  Q  E  G  R  L  F  A  K  M  T
     10330     10340     10350     10360     10370     10380     10390     10400     10410     10420     10430     10440
TATAAAATGAGAGCTACACAAGTTTTATCAGAGACACTACTTGCAAATAATATAGGGGAAATTCTTTCAAGAAAATGGGATGGTGTAAAAGGAGAGATTGAATTACTTAAGAGATTAACAACC
   Y  K  M  R  A  T  Q  V  L  S  E  T  L  L  A  N  N  I  G  K  F  F  Q  E  N  G  M  V  K  G  E  I  E  L  L  K  R  L  T  T
     10450     10460     10470     10480     10490     10500     10510     10520     10530     10540     10550     10560
ATATCAATATCAGGAGTTCCACGGTATAATGAAGTATACAATAATTCTAAAAGTCATACAGATGATCTTAAAACCTACAATAAAATAAGTAATCTCAATTTGTCTTCTAATCAGAAATCA
   I  S  I  S  G  V  P  R  Y  N  E  V  Y  N  N  S  K  S  H  T  D  D  L  K  T  Y  N  K  I  S  N  L  N  L  S  S  N  Q  K  S
     10570     10580     10590     10600     10610     10620     10630     10640     10650     10660     10670     10680
AAGAAATTTGAATTCAAGTCAACGGATATTTACAATGATGGATACGAGACTGTGAGCTGTTTTCTAACAACAGATCTCAAAAAATACTGTCTTAATTGGAGATATGAATCAACAGCTCTA
   K  K  F  E  F  K  S  T  D  I  Y  N  D  G  Y  E  T  V  S  C  F  L  T  T  D  L  K  K  Y  C  L  N  W  R  Y  E  S  T  A  L
     10690     10700     10710     10720     10730     10740     10750     10760     10770     10780     10790     10800
TTTGGAGAAACTTGCAACCAAATATTTGGATTAAATAAATTGTTTAATTGGTTACACCCTCGTCTTGAAGGAAGTACAATCTATGTAGGTGATCCCTATTGTCCTCCATCAGATAAGGAA
   F  G  E  T  C  N  Q  I  F  G  L  N  K  L  F  N  W  L  H  P  R  L  E  G  S  T  I  Y  V  G  D  P  Y  C  P  P  S  D  K  E
     10810     10820     10830     10840     10850     10860     10870     10880     10890     10900     10910     10920
CATATATATCATTAGAGGATCACCCTGATTCTGGATTTTATGTTCATAACCCAAGAGGGGGTATAGAAGGATTTTGTCAAAAATTGTGGACACTCATATCTATAAGTGCAATACATCTAGCA
   H  I  S  L  E  D  H  P  D  S  G  F  Y  V  H  N  P  R  G  G  I  E  G  F  C  Q  K  L  W  T  L  I  S  I  S  A  I  H  L  A
```

HUMAN PARAINFLUENZA VIRUS TYPE 3

```
        10930     10940     10950     10960     10970     10980     10990     11000     11010     11020     11030     11040
GCTGTTAGAATAGGCGTAAGGGTAACTGCAATGGTTCAAGGAGATAATCAAGCTATAGCTGTAACAACAAGAGTACCCAACAATTATGACTACAGAGTTAAGAAGGAGATAGTTTATAAA
 A  V  R  I  G  V  R  V  T  A  M  V  Q  G  D  N  Q  A  I  A  V  T  T  R  V  P  N  N  Y  D  Y  R  V  K  K  E  I  V  Y  K

        11050     11060     11070     11080     11090     11100     11110     11120     11130     11140     11150     11160
GATGTGGTGAGATTTTTTGATTCATTAAGAGAAGTAATGGATGATCTAGGTCATGAACTTAAATTAAATGAAACAATTATAAGTAGCAAGATGTTCATATATAGCAAAAGAATATATTAC
 D  V  V  R  F  F  D  S  L  R  E  V  M  D  D  L  G  H  E  L  K  L  N  E  T  I  I  S  S  K  M  F  I  Y  S  K  R  I  Y  Y

        11170     11180     11190     11200     11210     11220     11230     11240     11250     11260     11270     11280
GATGGGAGAATTCTTCCCCAAGCTCTGAAAGCATTATCTAGATGTGTCTTCTGGTCAGAGACAGTAATAGACGAAACAAGATCAGCATCTTCAAACTTGGCAACATCATTTGCAAAAGCA
 D  G  R  I  L  P  Q  A  L  K  A  L  S  R  C  V  F  W  S  E  T  V  I  D  E  T  R  S  A  S  S  N  L  A  T  S  F  A  K  A

        11290     11300     11310     11320     11330     11340     11350     11360     11370     11380     11390     11400
ATTGAGAATGGTTATTCACCTGTTCTAGGATATGCATGCTCAATTTTTAAGAACATTCAACAACTATATATTGCCCTTGGGATGAATATCAATCCAACTATAACACAGAATATCAAAGAT
 I  E  N  G  Y  S  P  V  L  G  Y  A  C  S  I  F  K  N  I  Q  Q  L  Y  I  A  L  G  M  N  I  N  P  T  I  T  Q  N  I  K  D

        11410     11420     11430     11440     11450     11460     11470     11480     11490     11500     11510     11520
TTATATTTTAGGAATCCAAATTGGATGCAATATGCATCTTTAATACCTGCTAGTGTTGGGGGATTCAATTACATGGCCATGTCAAGATGTTTTGTAAGGAATATTGGCGATCCATCAGTT
 L  Y  F  R  N  P  N  W  M  Q  Y  A  S  L  I  P  A  S  V  G  G  F  N  Y  M  A  M  S  R  C  F  V  R  N  I  G  D  P  S  V

        11530     11540     11550     11560     11570     11580     11590     11600     11610     11620     11630     11640
GCCGCATTAGCTGATATTAAAAGATTTATTAAGGCGAACCTATTAGACCGAAGTGTTCTTTATAGGATTATGAATCAAGAACCAGGTGAGTCATCTTTTTTTGGACTGGGCTTCAGACCCA
 A  A  L  A  D  I  K  R  F  I  K  A  N  L  L  D  R  S  V  L  Y  R  I  M  N  Q  E  P  G  E  S  S  F  L  D  W  A  S  D  P

        11650     11660     11670     11680     11690     11700     11710     11720     11730     11740     11750     11760
TATTCATGCAATTTACCACAATCTCAAAATATAACCACTATGATAAAAAATATAACAGCAAGAAATGTATTACAGGATTCACCGAATCCATTATTATCTGGATTATTCACAAATACAATG
 Y  S  C  N  L  P  Q  S  Q  N  I  T  T  M  I  K  N  I  T  A  R  N  V  L  Q  D  S  P  N  P  L  L  S  G  L  F  T  N  T  M

        11770     11780     11790     11800     11810     11820     11830     11840     11850     11860     11870     11880
ATAGAAGAAGATGAAGAATTAGCTGAGTTCTTGATGGATAGGAAGGTAATTCTCCCTAGAGTTGCACATGATATTCTAGATAATTCTCTCACAGGAATCAGAAATGCTATAGCTGGAATG
 I  E  E  D  E  E  L  A  E  F  L  M  D  R  K  V  I  L  P  R  V  A  H  D  I  L  D  N  S  L  T  G  I  R  N  A  I  A  G  M

        11890     11900     11910     11920     11930     11940     11950     11960     11970     11980     11990     12000
TTGCATACGACAAAATCTCTAATTCGGGTTGGCATAAATAGAGGAGGACTGACATACAGTTTGTTGAGGAAAATCAGTAATTACGATCGTAGTACAATATGAAACACTAAGTAGGACTTTG
 L  D  T  T  K  S  L  I  R  V  G  I  N  R  G  G  L  T  Y  S  L  L  R  K  I  S  N  Y  D  L  V  Q  Y  E  T  L  S  R  T  L

        12010     12020     12030     12040     12050     12060     12070     12080     12090     12100     12110     12120
CGACTAATTGTAAGCGATAAAATCAGGTATGAAGATATGTGTTCGGTAGACCTTGCTATAGCATTGCGTCAAAAGATGTGGATTCATTTATCAGGAGGAAGGATGATAAGTGGACTTGAA
 R  L  I  V  S  D  K  I  R  Y  E  D  M  C  S  V  D  L  A  I  A  L  R  Q  K  M  W  I  H  L  S  G  G  R  M  I  S  G  L  E

        12130     12140     12150     12160     12170     12180     12190     12200     12210     12220     12230     12240
ACACCTGATCCATTAGAATTACTATCTGGGGTGATAATAACAGGATCGGAACATTGTAAAATATGTTATTCTTCAGATGGCACAAACCCATATACTTGGATGTATTTACCGGGTAATATT
 T  P  D  P  L  E  L  L  S  G  V  I  I  T  G  S  E  H  C  K  I  C  Y  S  S  D  G  T  N  P  Y  T  W  M  Y  L  P  G  N  I

        12250     12260     12270     12280     12290     12300     12310     12320     12330     12340     12350     12360
AAAATAGGATCAGCAGAAACAGGTATATCATCATTGAGAGTTCCTTATTTTGGATCAGTCACTGGTGAGAGATCTGAGGCACAATTGGGATATATCAAGAATCTTAGTAAACCTGCAAAA
 K  I  G  S  A  E  T  G  I  S  S  L  R  V  P  Y  F  G  S  V  T  G  E  R  S  E  A  Q  L  G  Y  I  K  N  L  S  K  P  A  K

        12370     12380     12390     12400     12410     12420     12430     12440     12450     12460     12470     12480
GCCGCAATAAGAATAGCAATGATATATACATGGGCATTTGGTAATGATGAGATATCTTGGATGGAGGCCTTCACAAATAGCACAAACACGTGCAAATTTTACACTAGATAGTCTCAAAATT
 A  A  I  R  I  A  M  I  Y  T  W  A  F  C  N  D  E  I  S  W  M  E  A  S  Q  I  A  Q  T  R  A  N  F  T  L  D  S  L  K  I

        12490     12500     12510     12520     12530     12540     12550     12560     12570     12580     12590     12600
CTAACACCGGTAGCTACATCAACAAATTTATCACACAGATTAAAGGATACTGCAACCCAGATGAAGTTCTCCAGTACATCATTGATTAGGGTCAGCAGATTCATAACAATGTCCAATGAT
 L  T  P  V  A  T  S  T  N  L  S  H  R  L  K  D  T  A  T  Q  M  K  F  S  S  T  S  L  I  R  V  S  R  F  I  T  M  S  N  D

        12610     12620     12630     12640     12650     12660     12670     12680     12690     12700     12710     12720
AACATGTCTATCAAGGAAGCTAATGAGACCAAAGATACCAATCTTATTTATCAACAAATAATGTTAACAGGATTAAGTGTTTTCGAATATTTATTTAGATTAGAAGAAACCACAGGACAC
 N  M  S  I  K  E  A  N  E  T  K  D  T  N  L  I  Y  Q  Q  I  M  L  T  G  L  S  V  F  E  Y  L  F  R  L  E  E  T  T  G  H

        12730     12740     12750     12760     12770     12780     12790     12800     12810     12820     12830     12840
AACCCCATAGTTATGCATCTGCACATAGAAGATGAGTGTTGTATTAAAGAAAGTTTTAATGATGAGCATATTAATCCAAAGTCTACATTAGAATCAATTAGGCACCCTGAAAGTAATGAA
 N  P  I  V  M  H  L  H  I  E  D  E  C  C  I  K  E  S  F  N  D  E  H  I  N  P  K  S  T  L  E  S  I  R  H  P  E  S  N  E

        12850     12860     12870     12880     12890     12900     12910     12920     12930     12940     12950     12960
TTTATTTATGATAAAGACCCGCTCAAGGACGTGGACTTATCAAAACTTATGGTTATTAAAGATCTTTCTTACACAATTGATATGAATTATTGGGATGATACTGACATCATACATGCAATT
 F  I  Y  D  K  D  P  L  K  D  V  D  L  S  K  L  M  V  I  K  D  L  S  Y  T  I  D  M  N  Y  W  D  D  T  D  I  I  H  A  I

        12970     12980     12990     13000     13010     13020     13030     13040     13050     13060     13070     13080
TCAATATGTACTGCAATTACAATAGCAGACACTATGTCACAATTAGATCGAGATAACTTAAAAGAGATAATAGTCATTGCAAATGATGATGATATTAATAGCTTAATCACTGAATTTTTG
 S  I  C  T  A  I  T  I  A  D  T  M  S  Q  L  D  R  D  N  L  K  E  I  I  V  I  A  N  D  D  D  I  N  S  L  I  T  E  F  L

        13090     13100     13110     13120     13130     13140     13150     13160     13170     13180     13190     13200
ACTCTTGATATACTTGTATTTCTTAAGACATTTGGTGGATTATTAGTAAATCAATTGCATACACTCTTTATAGTTTAAAAACCGAAGGCAGGGACCTCATTTGGGATTATATAATGAGA
 T  L  D  I  L  V  F  L  K  T  F  G  G  L  L  V  N  Q  F  A  Y  T  L  Y  S  L  K  T  E  G  R  D  L  I  W  D  Y  I  M  R
```

HUMAN PARAINFLUENZA VIRUS TYPE 3

```
        13210     13220     13230     13240     13250     13260     13270     13280     13290     13300     13310     13320
ACACTGAGAGATACTTCCCATTCAATATTAAAAGTATTATCTAATGCATTATCTCATCCTAAAGTATTCAAGAGGTTCTGGGATTGTGGAGTCTTAAACCCTATTTATGGCCCTAATACT
  T  L  R  D  T  S  H  S  I  L  K  V  L  S  N  A  L  S  H  P  K  V  F  K  R  F  W  D  C  G  V  L  N  P  I  Y  G  P  N  T

        13330     13340     13350     13360     13370     13380     13390     13400     13410     13420     13430     13440
GCTAGTCAAGACCAGATAAAACTTGCCCTCTCTATATGTGAATATTCACTAGATCTATTTATGAGAGAATGGTTGAATGGTGTATCACTTGAAATATACATTTGTGACAGCGATATGGAA
  A  S  Q  D  Q  I  K  L  A  L  S  I  C  E  Y  S  L  D  L  F  M  R  E  W  L  N  G  V  S  L  E  I  Y  I  C  D  S  D  M  E

        13450     13460     13470     13480     13490     13500     13510     13520     13530     13540     13550     13560
GTTGCGAATGATAGGAAACAAGCCTTTATTTCTAGACACCTTTCATTTGTTTGTTGTTTAGCAGAAATTGCATCTTTTGGACCTAACCTGTTAAACTTAACATACTTAGAGAGACTTGAT
  V  A  N  D  R  K  Q  A  F  I  S  R  H  L  S  F  V  C  C  L  A  E  I  A  S  F  G  P  N  L  L  N  L  T  Y  L  E  R  L  D

        13570     13580     13590     13600     13610     13620     13630     13640     13650     13660     13670     13680
CTATTGAAACAATATCTTGAATTAAATATTAAAGACGACCCTACTCTTAAATATGTACAAATATCTGGATTATTAATTAAATCGCTCCCATCAACTGTAACATACGTAAGAAAGACTGCA
  L  L  K  Q  Y  L  E  L  N  I  K  D  D  P  T  L  K  Y  V  Q  I  S  G  L  L  I  K  S  L  P  S  T  V  T  Y  V  R  K  T  A

        13690     13700     13710     13720     13730     13740     13750     13760     13770     13780     13790     13800
ATCAAATATTTAAGGATTCGTGGTATTAGTCCACCTGAGGTAATTGATGATTGGGATCCGATAGAAGATGAAAATATGCTGGATAACATTGTCAAAACTATAAATGATAATTGTAATAAA
  I  K  Y  L  R  I  R  G  I  S  P  P  E  V  I  D  D  W  D  P  I  E  D  E  N  M  L  D  N  I  V  K  T  I  N  D  N  C  N  K

        13810     13820     13830     13840     13850     13860     13870     13880     13890     13900     13910     13920
GATAATAAAGGGAATAAAATTAACAATTTCTGGGGACTAGCGCTTAAGAACTATCAGGTCCTTAAAATCAGATCTATAACAAGTGATTCTGATAATAATGATAGATCAGATGCTAGTACC
  D  N  K  G  N  K  I  N  N  F  W  G  L  A  L  K  N  Y  Q  V  L  K  I  R  S  I  T  S  D  S  D  N  N  D  R  S  D  A  S  T

        13930     13940     13950     13960     13970     13980     13990     14000     14010     14020     14030     14040
GGTGGTTTGACACTTCCTCAAGGGGGGAATTATCTATCACATCAATTGAGATCATTCGGAATCAACAGCACTAGTTGTCTGAAAGCTCTTGAGTTATCACAAATTCCAACGAAGGAAGTT
  G  G  L  T  L  P  Q  G  G  N  Y  L  S  H  Q  L  R  S  F  G  I  N  S  T  S  C  L  K  A  L  E  L  S  Q  I  P  T  K  E  V

        14050     14060     14070     14080     14090     14100     14110     14120     14130     14140     14150     14160
AATAAAGACCAGGACAGGCTCTTCCTAGGAGAAGGAGCAGGAGCTATGCTAGCATGGTTATGATGCCACATTAGGACCTGCAGTTAATTATTATAATTCCGGTTTGAATATAACAGATGTA
  N  K  D  Q  D  R  L  F  L  G  E  G  A  G  A  M  L  A  C  Y  D  A  T  L  G  P  A  V  N  Y  Y  N  S  G  L  N  I  T  D  V

        14170     14180     14190     14200     14210     14220     14230     14240     14250     14260     14270     14280
ATTGGTCAACGAGAATTGAAAATATTCCCTTCAGAGGTATCATTAGTAGGTAAAAAATTAGGAAATGTGACACAGATCCTTAATAGGGTAAAAGTACTGTTCAATGGAAATCCAAATTCA
  I  G  Q  R  E  L  K  I  F  P  S  E  V  S  L  V  G  K  K  L  G  N  V  T  Q  I  L  N  R  V  K  V  L  F  N  G  N  P  N  S

        14290     14300     14310     14320     14330     14340     14350     14360     14370     14380     14390     14400
ACATGAGATAGGAAATATGGAATGCGAGACGTTAATATGGAGTGAATTGAATGATAAGTCTATTGGATTGGTACATTGTGATATGGAAGGAGCTATCGGTAAATCAGAAGAAACTGTTCTA
  T  W  I  G  N  M  E  C  E  T  L  I  W  S  E  L  N  D  K  S  I  G  L  V  H  C  D  M  E  G  A  I  G  K  S  E  E  T  V  L

        14410     14420     14430     14440     14450     14460     14470     14480     14490     14500     14510     14520
CATGAACACTATAGTGTTATAAGAATTACATACTTGATTGGGGATGATGATGTTGTTTTAATTTCCAAAATTATACCTCCAATCACTCCGAATTGGTCTAGAATACTTTATCTATATAAG
  H  E  H  Y  S  V  I  R  I  T  Y  L  I  G  D  D  D  V  V  L  I  S  K  I  I  P  P  I  T  P  N  W  S  R  I  L  Y  L  Y  K

        14530     14540     14550     14560     14570     14580     14590     14600     14610     14620     14630     14640
TTATATTGGAAAGATGTAAGTATAATATCACTTAAAACTTCTAATCCTGCATCAACAGAATTATATCTAATTTCAAAAGATGCGTATTGTACTATAATGGAACCTAGTGAAGTTGTTTTA
  L  Y  W  K  D  V  S  I  I  S  L  K  T  S  N  P  A  S  T  E  L  Y  L  I  S  K  D  A  Y  C  T  I  M  E  P  S  E  V  V  L

        14650     14660     14670     14680     14690     14700     14710     14720     14730     14740     14750     14760
TCAAAACTTAAAAGATTGTCACTCTTGGAAGAAAAATAATCTATTAAAATGGATCATTTTATCAAAGAAGAAAAATAATGAATGGTTACATCATGAAATCAAAGAAGGGGAAAGAGATTAT
  S  K  L  K  R  L  S  L  L  E  E  N  N  L  L  K  W  I  I  L  S  K  K  K  N  N  E  W  L  H  H  E  I  K  E  G  E  R  D  Y

        14770     14780     14790     14800     14810     14820     14830     14840     14850     14860     14870     14880
GGAGTTATGAGACCATATCATATGGCATTACAAATTTTTGGATTTCAAATCAATTTAAATCATCTGGCGAAAGAATTTTTTATCAACTCCAGATCTGACTAATATCAACAATATAATCCAA
  G  V  M  R  P  Y  H  M  A  L  Q  I  F  G  F  Q  I  N  L  N  H  L  A  K  E  F  L  S  T  P  D  L  T  N  I  N  N  I  I  Q

        14890     14900     14910     14920     14930     14940     14950     14960     14970     14980     14990     15000
AGTTTTCAGAGAACAATCAAGGATGTTTTGTTTGAATGGATTAATATAACTCATGATGGTAAGAGACATAAATTAGGCGGGAGATATAACATATTCCCACTGAAAAATAAGGGGAAATTA
  S  F  Q  R  T  I  K  D  V  L  F  E  W  I  N  I  T  H  D  G  K  R  H  K  L  G  G  R  Y  N  I  F  P  L  K  N  K  G  K  L

        15010     15020     15030     15040     15050     15060     15070     15080     15090     15100     15110     15120
AGACTGCTATCGAGAAGACTAGTATTAAGTTGGATTTCATTGTCATTATCGACTGATTACTTACAGGTCGTTTTCCTGATGAAAAATTTGAACATAGAGCACAGACTGGATATGTGTCA
  R  L  L  S  R  R  L  V  L  S  W  I  S  L  S  L  S  T  R  L  L  T  G  R  F  P  D  E  K  F  E  H  R  A  Q  T  G  Y  V  S

        15130     15140     15150     15160     15170     15180     15190     15200     15210     15220     15230     15240
TTAGCTGATACTGATTTAGAATCATTAAAGTTATTGTTGTCAAAAAACACCATTAAGAATTACAGAGAGTGTATAGGGTCAATATCATATTGGTTTCTAACCAAAGAAGTTAAAATACTTATG
  L  A  D  T  D  L  E  S  L  K  L  L  S  K  N  T  I  K  N  Y  R  E  C  I  G  S  I  S  Y  W  F  L  T  K  E  V  K  I  L  M

        15250     15260     15270     15280     15290     15300     15310     15320     15330     15340     15350     15360
AAATTGATTGGTGGTGCTAAATTATTAGGAATTCCCGGACAATAATAAGGAACCCGAAGAACAGTTATTAGAAGACTACAATCACATGATGAATTTGATATGAGATTAAAATACAAATACA
  K  L  I  G  G  A  K  L  L  G  I  P  G  Q  Y  K  E  P  E  E  Q  L  L  E  D  Y  N  Q  H  D  E  F  D  I  D  *

        15370     15380     15390     15400     15410     15420     15430     15440     15450     15460
ATAAAGATATATCCTAACCTTTATCATTAAGCCTAAAGATAGACAAAAAGTAAGAAAAACATGTAATATATATATACCAAACAGAGTTCTTCTCTTGTTTGGT
```

T_{stop}   I   EXTRACISTRONIC SEQUENCES

MEASLES VIRUS

```
       10        20        30        40        50        60        70        80        90       100       110       120
ACCAAACAAAGTTGGGTAAGGATAGTTCAATCAATGATCATCTTCTAGTGCACTTAGGATTCAAGATCCTATTATCAGGGACAAGAGCAGGATTAGGGATATCCGAGATGGCCACACTTT
                EXTRACISTRONIC SEQUENCES                               T start            NUCLEOCAPSID PROTEIN ▸ M  A  T  L

      130       140       150       160       170       180       190       200       210       220       230       240
TAAGGAGCTTAGCATTGTTCAAAAGAAACAAGGACAAACCACCCATTACATCAGGATCCGGTGGAGCCATCAGAGGAATCAAACACATTATTATAGTACCAATCCCTGGAGATTCCTCAA
L  R  S  L  A  L  F  K  R  N  K  D  K  P  P  I  T  S  G  S  G  G  A  I  R  G  I  K  H  I  I  I  V  P  I  P  G  D  S  S

      250       260       270       280       290       300       310       320       330       340       350       360
TTACCACTCGATCCAGACTTCTGGACACGGTTGGTCAGGTTAATTGGAAACCCGGATGTGAGCGGGCCCAAACTAACAGGGGCACTAATAGGTATATTATCCTTATTTGTGGAGTCTCCAG
I  T  T  R  S  R  L  L  D  R  L  V  R  L  I  G  N  P  D  V  S  G  P  K  L  T  G  A  L  I  G  I  L  S  L  F  V  E  S  P

      370       380       390       400       410       420       430       440       450       460       470       480
GTCAATTGATTCAGAGGATCACCGATGACCCTGACGTTAGCATAAGGCTGTTAGAGGTTGTCCAGAGTGACCAGTCACAATCTGGCCTTACCTTCGCATCAAGAGGTACCAACATGGAGG
G  Q  L  I  Q  R  I  T  D  D  P  D  V  S  I  R  L  L  E  V  V  Q  S  D  Q  S  Q  S  G  L  T  F  A  S  R  G  T  N  M  E

      490       500       510       520       530       540       550       560       570       580       590       600
ATGAGGCGGACCAATACTTTTCACATGATGATCCAATTAGTAGTGATCAATCCAGGTTCGGATGGTTCGAGAACAAGGAAATCTCAGATATTGAAGTGCAAGACCCTGAGGGATTCAACA
D  E  A  D  Q  Y  F  S  H  D  D  P  I  S  S  D  Q  S  R  F  G  W  F  E  N  K  E  I  S  D  I  E  V  Q  D  P  E  G  F  N

      610       620       630       640       650       660       670       680       690       700       710       720
TGATTCTGGGTACCATCCTAGCCCAAATTTGGGTCTTGCTCGCAAAGGCGGTTACGGCCCCAGACACGGCAGCTGATTGGGAGCTAAGAAGGTGGATAAAGTACACCCAACAAAGAAGGG
M  I  L  G  T  I  L  A  Q  I  W  V  L  L  A  K  A  V  T  A  P  D  T  A  A  D  W  E  L  R  R  W  I  K  Y  T  Q  Q  R  R

      730       740       750       760       770       780       790       800       810       820       830       840
TAGTTGGTGAATTAGATTGGAGAGAAAGATGGTTGGATGTGGTGAGGAACATTATTGCCGAGGACCTCTCCTTACGCCGATTCATGGTCGCTCTAATCCTGGATATCAAGAGAACACCCG
V  V  G  E  F  R  L  E  R  K  W  L  D  V  V  R  N  I  I  A  E  D  L  S  L  R  R  F  M  V  A  L  I  L  D  I  K  R  T  P

      850       860       870       880       890       900       910       920       930       940       950       960
GAAACAAACCCAGGATTGCTGAAATGATATGTGACATTGATACATATATCGTAGAGGCAGGATTAGCCAGTTTTATCCTGACTATTAAGTTTGGGATAGAAACTATGTATCCTGCTCTTG
G  N  K  P  R  I  A  E  M  I  C  D  I  D  T  Y  I  V  E  A  G  L  A  S  F  I  L  T  I  K  F  G  I  E  T  M  Y  P  A  L

      970       980       990      1000      1010      1020      1030      1040      1050      1060      1070      1080
GACTGCATGAATTTGCTGGTGAGTTATCCACACTTGAGTCCTTGATGAACCTTTACCAGCAAATGGGGAAACCTGCACCCTACATGGTAAACCTGGAGAACTCAATTCAGAACAAGTTCA
G  L  H  E  F  A  G  E  L  S  T  L  E  S  L  M  N  L  Y  Q  Q  M  G  K  P  A  P  Y  M  V  N  L  E  N  S  I  Q  N  K  F

     1090      1100      1110      1120      1130      1140      1150      1160      1170      1180      1190      1200
GTGCAGGATCATACCCTCTGCTCTGGAGCTATGCCATGGGAGTAGGAGTGGAACTTGAAAACTCCATGGGAGGTTTGAACTTTGGCCGATCTTACTTTGATCCAGCATATTTTAGATTAG
S  A  G  S  Y  P  L  L  W  S  Y  A  M  G  V  G  V  E  L  E  N  S  M  G  G  L  N  F  G  R  S  Y  F  D  P  A  Y  F  R  L

     1210      1220      1230      1240      1250      1260      1270      1280      1290      1300      1310      1320
GGCAAGAGATGGTAAGGAGGTCAGCTGGAAAGGTCAGTTCCACATTAGCATCTGAACTCGGTATCACTGCCGAGGATGCAAGGCTTGTTTCAGAGATTGCAATGCATACTACTGAGGACA
G  Q  E  M  V  R  R  S  A  G  K  V  S  S  T  L  A  S  E  L  G  I  T  A  E  D  A  R  L  V  S  E  I  A  M  H  T  T  E  D

     1330      1340      1350      1360      1370      1380      1390      1400      1410      1420      1430      1440
AGATCAGTAGAGCGGTTGGACCCAGACAAGCCCAAGTATCCATTTCTACAGGGTGATCAAAGTGAGAATGAGCTACCGCGATTGGGGGGCAAGGAAGATAGGGGGTCAAACAGAGTCGAG
K  I  S  R  A  V  G  P  R  Q  A  Q  V  S  F  L  Q  G  D  Q  S  E  N  E  L  P  R  L  G  G  K  E  D  R  R  V  K  Q  S  R

     1450      1460      1470      1480      1490      1500      1510      1520      1530      1540      1550      1560
GAGAAGCCAGGGAGAGCTACAGAGAAACCGGGCCCAGCAGAGCAAGTGATGCGAGAGCTGCCCATCTTCCAACCGGCACACCCCTAGACATTGACACTGCATCGGAGTCCAGCCAAGATC
G  E  A  R  E  S  Y  R  E  T  G  P  S  R  A  S  D  A  R  A  A  H  L  P  T  G  T  P  L  D  I  D  T  A  S  E  S  S  Q  D

     1570      1580      1590      1600      1610      1620      1630      1640      1650      1660      1670      1680
CGCAGGACAGTCGAAGGTCAGCTGAGCCCCTGCTTAGCTGCAAGCCATGGCAGGAATCTCGGAAGAACAAGGCTCAGACACGGACACCCCTACAGTGTACAATGACAGAAATCTTCTAGA
P  Q  D  S  R  R  S  A  E  P  L  L  S  C  K  P  W  Q  E  S  R  K  N  K  A  Q  T  R  T  P  L  Q  C  T  M  T  E  I  F  *

     1690      1700      1710      1720      1730      1740      1750      1760      1770      1780      1790      1800
CTAGGTGCGAGAGGCCGAGGGCCAGAACAACATCCGCCTACCCTCCATCATTGTTATAAAAAACTTAGGAACCAGGTCCACACAGCCGCCAGCCCATCAACCATCCACTCCCACGATTGG
                                    T stop          I  T start            PHOSPHOPROTEIN ▸

     1810      1820      1830      1840      1850      1860      1870      1880      1890      1900      1910      1920
AGCCGATGGCAGAAGAGCAGGCACGCCATGTCAAAAACGGACTGGAATGCATCCGGGCTCTCAAGGCCGAGCCCATCGGCTCACTGGCCATCGAGGAAGCTATGGCAGCATGGTCAGAAA
C PROTEIN ▸        M  S  K  T  D  W  N  A  S  G  L  S  R  P  S  P  S  A  H  W  P  S  R  K  L  W  Q  H  G  Q  K
         M  A  E  E  Q  A  R  H  V  K  N  G  L  E  C  I  R  A  L  K  A  E  P  I  G  S  L  A  I  E  E  A  M  A  A  W  S  E

     1930      1940      1950      1960      1970      1980      1990      2000      2010      2020      2030      2040
TATCAGACAACCCAGGACAGGAGCGAGCCACCTGCAGGGAAGAGAAGGCAGGCAGTTCGGGTCTCAGCAAACCATGCCTCTCAGCAATTGGATCAACTGAAGGCGGTGCACCTCGCATCC
Y  Q  T  T  Q  D  R  S  E  P  P  A  G  K  R  R  Q  A  V  R  V  S  A  N  H  A  S  Q  Q  L  D  Q  L  K  A  V  H  L  A  S
I  S  D  N  P  G  Q  E  R  A  T  C  R  E  E  K  A  G  S  S  G  L  S  K  P  C  L  S  A  I  G  S  T  E  G  G  A  P  R  I

     2050      2060      2070      2080      2090      2100      2110      2120      2130      2140      2150      2160
GCGGTCAGGGACCTGGAGAGAGCGATGACGACGCTGAAACTTTGGGAATCCCCCCAAGAAATCTCCAGGCATCAAGCACTGGGTTACAGTGTTATTATGTTTTATGCACAGCGGTGAAG
A  V  R  D  L  E  R  A  M  T  T  L  K  L  W  E  S  P  Q  E  I  S  R  H  Q  A  L  G  Y  S  V  I  M  F  M  I  T  A  V  K
R  G  Q  G  P  G  E  S  D  D  D  A  E  T  L  G  I  P  P  R  N  L  Q  A  S  S  T  G  L  Q  C  Y  Y  V  V  Y  D  H  S  G  E
```

```
        2170      2180      2190      2200      2210      2220      2230      2240      2250      2260      2270      2280
CGGTTAAGGGAATCCAAGATGCTGACTCTATCATGGTTCAATCAGGCCTTGATGGTGATAGCACCCTCTCAGGAGGAGACAATGAATCTGAAAACAGCCGATGTGGATATTGGCGAACCTG
 R  L  R  E  S  K  M  L  T  L  S  W  F  N  Q  A  L  M  V  I  A  P  S  Q  E  E  T  M  N  L  K  T  A  M  W  I  L  A  N  L
 A  V  K  G  I  Q  D  A  D  S  I  M  V  Q  S  G  L  D  G  D  S  T  L  S  G  G  D  N  E  S  E  N  S  D  V  D  I  G  E  P

        2290      2300      2310      2320      2330      2340      2350      2360      2370      2380      2390      2400
CCGAGGGATATGCTATCACTGACCGGGGATCTGCTCCCATCTCTATGGGGTTCAGGGCTTCTGATGTTGAAACTGCAGAAGGAGGGGAGATCCACGAGCTCCTGAGACTCCAATCCA
 I  P  R  D  M  L  S  L  T  G  D  L  L  P  S  L  W  G  S  G  L  L  M  L  K  L  Q  K  E  G  R  S  T  S  S  *
 D  T  E  G  Y  A  I  T  D  R  G  S  A  P  I  S  M  G  F  R  A  S  D  V  E  T  A  E  G  G  E  I  H  E  L  L  R  L  Q  S

        2410      2420      2430      2440      2450      2460      2470      2480      2490      2500      2510      2520
GAGGCAACAACTTTCCGAGGCTTGGGAAAACTCTCAATGTTCCTCCGCCCCCGGACCCCGGTAGGGCCAGCACTTCGGGACACCCATTAAAAAGGGCACAGAGCGCAGATTAGCCTCAT
                                                                             V CISTRON ▶ G  H  R  A  Q  I  S  L  I
 R  G  N  N  F  P  R  L  G  K  T  L  N  V  P  P  P  P  D  P  G  R  A  S  T  S  G  T  P  I  K  K  G  T  E  R  R  L  A  S

        2530      2540      2550      2560      2570      2580      2590      2600      2610      2620      2630      2640
TTGGAACGGAGATCGCGTCTTTATTGACAGGTGGTGCAACCCAATGTGCTCGAAAGTCACCCTCGGAACCATCAGGGCCAGGTGCACCTGCGGGGAATGTCCCCGAGTGTGTGAGCAATG
  W  N  G  D  R  V  F  I  D  R  W  C  N  P  M  C  S  K  V  T  L  G  T  I  R  A  R  C  T  C  G  E  C  P  R  V  C  E  Q  C
 F  G  T  E  I  A  S  L  L  T  G  G  A  T  Q  C  A  R  K  S  P  S  E  P  S  G  P  G  A  P  A  G  N  V  P  E  C  V  S  N

        2650      2660      2670      2680      2690      2700      2710      2720      2730      2740      2750      2760
CCGCACTGATACAGGAGTGGACACCCGAATCTGGTACCACAATCTCCCCGAGATCCCAGAATAATGAAGAAGGGGGAGACTATTATGATGATGAGCTGTTCTCTGATGTCCAAGATATTA
  R  T  D  T  G  V  D  T  R  I  W  Y  H  N  L  P  E  I  P  E  *
 A  A  L  I  Q  E  W  T  P  E  S  G  T  T  I  S  P  R  S  Q  N  N  E  E  G  G  D  Y  Y  D  D  E  L  F  S  D  V  Q  D  I

        2770      2780      2790      2800      2810      2820      2830      2840      2850      2860      2870      2880
AAACAGCCTTGGCCAAAATACACGAGGATAATCAGAAGATAATCTCCAAGCTAGAATCACTGCTGTTATTGAAGGGAGAAGTTGAGTCAATTAAGAAGCAGATCAACAGGCAAAATATCA
 K  T  A  L  A  K  I  H  E  D  N  Q  K  I  I  S  K  L  E  S  L  L  L  L  K  G  E  V  E  S  I  K  K  Q  I  N  R  Q  N  I

        2890      2900      2910      2920      2930      2940      2950      2960      2970      2980      2990      3000
GCATATCCACCCTGGAAGGACACCTCTCAAGCATCATGATCGCCATTCCTGGACTTGGGAAGGATCCCAACGACCCCACTGCAGATGTCGAAATCAATCCCGACTTGAAACCCATCATAG
 S  I  S  T  L  E  G  H  L  S  S  I  M  I  A  I  P  G  L  G  K  D  P  N  D  P  T  A  D  V  E  I  N  P  D  L  K  P  I  I

        3010      3020      3030      3040      3050      3060      3070      3080      3090      3100      3110      3120
GCAGAGATTCAGGCCGACTGGCCGAAGTTCTCAGGAAACCCGTTGCCAGCCGACAACTCCAAGGAATGACAAATGGACGGACCAGTTCCAGAGGACAGCTGCTGAAGGAATTTCAGC
 G  R  D  S  G  R  A  L  A  E  V  L  R  K  P  V  A  S  R  Q  L  Q  G  M  T  N  G  R  T  S  S  R  G  Q  L  L  K  E  F  Q

        3130      3140      3150      3160      3170      3180      3190      3200      3210      3220      3230      3240
TAAAGCCGATCGGGAAAAAGTGAGCTCAGCCGTCGGGTTTGTTCCTGACACCGGCCCTGCATCACGCAGTGTAATCCGCTCCATTATAAATCCAGCCGGCTAGAGGAGGATCGGAAGC
 L  K  P  I  G  K  K  M  S  S  A  V  G  F  V  P  D  T  G  P  A  S  R  S  V  I  R  S  I  I  K  S  S  R  L  E  E  D  R  K

        3250      3260      3270      3280      3290      3300      3310      3320      3330      3340      3350      3360
GTTACCTGATGACTCTCCTTGATGATATCAAAGGAGCCAATGATCTTGCCAAGTTCCACCAGATGCTGATGAAGATAATAATGAAGTAGCTACAGCTCAACTTACCTGCCAACCCCATGC
 R  Y  L  M  T  L  L  D  D  I  K  G  A  N  D  L  A  K  F  H  Q  M  L  M  K  I  I  M  K  *

        3370      3380      3390      3400      3410      3420      3430      3440      3450      3460      3470      3480
CAGTCGACCAACTAGTACAACCTAAATCCATTATAAAAAACTTAGGAGCAAAGTGATTGCCTCCCAAGTTCCACAATGACAGAGATCTACGACTTCGGCAAGTCGGCATGGGACATCAAA
                                                                             M  T  E  I  Y  D  F  D  K  S  A  W  D  I  K
        3490      3500      3510      3520      3530      3540      3550      3560      3570      3580      3590      3600
        T_stop        I  T_start   MATRIX PROTEIN ▶
GGGTCGATCGCTCCGATACAACCCACCACCTACAGTGATGGCAGGCTGGTGCCCCAGGTCAGAGTCATAGATCCTGGTCTAGGCGACAGGAAGGATGAATGCTTTATGTACATGTTTCTG
 G  S  I  A  P  I  Q  P  T  T  Y  S  D  G  R  L  V  P  Q  V  R  V  I  D  P  G  L  G  D  R  K  D  E  C  F  M  Y  M  F  L

        3610      3620      3630      3640      3650      3660      3670      3680      3690      3700      3710      3720
CTGGGGGTTGTTGAGGACAGCGATCCCCTAGGGCCTCCAATCGGGCGAGCATTTGGGTCCCTGCCCTTAGGTGTTGGCAGATCCACAGCAAAGCCCGAAAAACTCCTCAAAGAGGCCACT
 L  G  V  V  E  D  S  D  P  L  G  P  P  I  G  R  A  F  G  S  L  P  L  G  V  G  R  S  T  A  K  P  E  K  L  L  K  E  A  T

        3730      3740      3750      3760      3770      3780      3790      3800      3810      3820      3830      3840
GAGCTTGACATAGTTGTTAGACGTACAGCAGGGCTCAATGAAAAACTGGTGTTCTACAACAACACCCCACTAACTCTCCTCACACCTTGGAGAAAGGTCCTAACAACAGGGAGTGTCTTC
 E  L  D  I  V  V  R  R  T  A  G  L  N  E  K  L  V  F  Y  N  N  T  P  L  T  L  L  L  T  P  W  R  K  V  L  T  T  G  S  V  F

        3850      3860      3870      3880      3890      3900      3910      3920      3930      3940      3950      3960
AACGCAAACCAAGTGTGCAGTGCGGTTAATCTGATACCGCTCGATACCCCGCAGAGGTTCCGTGTTGTTTATATGAGCATCACCCGTCTTTCGGATAACGGGTATTACACCGTTCCTAGA
 N  A  N  Q  V  C  S  A  V  N  L  I  P  L  D  T  P  Q  R  F  R  V  V  Y  M  S  I  T  R  L  S  D  N  G  Y  Y  T  V  P  R

        3970      3980      3990      4000      4010      4020      4030      4040      4050      4060      4070      4080
AGAATGCTGGAATTCAGATCGGTCAATGCAGTGGCCTTCAACCTGCTGGTGACCCTTAGGATTGACAAGGCCGATAGGCCCTGGGAAGATCATCGACAATACAGAGCAACTTCCTGAGGCA
 R  M  L  E  F  R  S  V  N  A  V  A  F  N  L  L  V  T  L  R  I  D  K  A  I  G  P  G  K  I  I  D  N  T  E  Q  L  P  E  A

        4090      4100      4110      4120      4130      4140      4150      4160      4170      4180      4190      4200
ACATTTATGGTCCACATCGGGAACTTCAGGAGAAAGAAGAGTGAAGTCTACTCTGCCGATTATTGCAAAATGAAAATCGAAAAGATGGGCCTGGTTTTTGCACTTGGTGGGATAGGGGGC
 T  F  M  V  H  I  G  N  F  R  R  K  K  S  E  V  Y  S  A  D  Y  C  K  M  K  I  E  K  M  G  L  V  F  A  L  G  G  I  G  G

        4210      4220      4230      4240      4250      4260      4270      4280      4290      4300      4310      4320
ACCAGTCTTCACATTAGAAGCACAGGCAAAATGAGCAAGACTCTCCATGCACAACTCGGGTTCAAGAAGACCTTATGTTACCCGCTGATGGATATCAATGAAGACCTTAATCGATTACTC
 T  S  L  H  I  R  S  T  G  K  M  S  K  T  L  H  A  Q  L  G  F  K  K  T  L  C  Y  P  L  M  D  I  N  E  D  L  N  R  L  L
```

MEASLES VIRUS

```
     4330      4340      4350      4360      4370      4380      4390      4400      4410      4420      4430      4440
TGGAGGAGCAGATGCAAGATAGTAAGAATCCAGGCAGTTTTGCAGCCATCAGTTCCTCAAGAATTCCGCATTTACGACGACGTGATCATAAATGATGACCAAGGACTATTCAAAGTTCTG
  W  R  S  R  C  K  I  V  R  I  Q  A  V  L  Q  P  S  V  P  Q  E  F  R  I  Y  D  D  V  I  I  N  D  D  Q  G  L  F  K  V  L

     4450      4460      4470      4480      4490      4500      4510      4520      4530      4540      4550      4560
TAGACCGTAGTGCCCAGCAATGCCCGAAAACGACCCCCCTCACAATGACAGCCAGAAGGCCCGGACAAAAAAGCCCCCTCCGAAAGACTCCACGGACCAAGCGAGAGGCCAGCCAGCAGC
  *

     4570      4580      4590      4600      4610      4620      4630      4640      4650      4660      4670      4680
CGACGGCAAGCGCGAACACCAGGCGGCCCCAGCACAGAACAGCCCTGACACAAGGCCACCACCAGCCACCCCAATCTGCATCCTCCTCGTGGGACCCCGAGGACCAACCCCCAAGGCTG

     4690      4700      4710      4720      4730      4740      4750      4760      4770      4780      4790      4800
CCCCCGATCCAAACCACCAACCGCACCCCCACCACCCCCGGGAAAGAAACCCCCAGCAATTGGAAGGCCCCTCCCCCTCTTCCTCAACACAAGAACTTCCACAACCGAACCGCACAAGCGA

     4810      4820      4830      4840      4850      4860      4870      4880      4890      4900      4910      4920
CCGAGGTGACCCAACCGCGCGGCATCCGACTCCCTAGACAGATCCTCTCTCCCGGCAAACTAAACAAAACTTAGGGCCAAGGNNNNATACACACCCAACAGAACCCAGACCCCGGCCCAC
                                                                                                      FUSION PROTEIN ▶
     4930      4940      4950      4960      4970      4980      4990      5000      5010      5020      5030      5040
                                                    stop      start
GGCGCCGCGCCCCCAACCCCCGACAACCAGAGGGAGCCCCCAACCAATCCCGCCGGCTCCCCCGGTGCCCACAGGCAGGGACACCAACCCCCGAACAGACCCAGCACCCAACCATCGACA

     5050      5060      5070      5080      5090      5100      5110      5120      5130      5140      5150      5160
ATCCAAGACGGGGGGGCCCCCCAAAAAAAGGCCCCCAGGGGCCGACAGCCAGCACCGCGAGGAAGCCCACCCACCCCACACACGACCACGGCAACCAAACCAGAACCCAGACCACCCTG

     5170      5180      5190      5200      5210      5220      5230      5240      5250      5260      5270      5280
GCCACCAGCTCCCAGACTCGGCCATCACCCCGCAGAAAGGAAAGGCCACAACCCGCGCACCCCAGCCCCGATCCGGCGGGGAGCCACCCAACCCGAACCAGCACCCAAGAGCGATCCCCG

     5290      5300      5310      5320      5330      5340      5350      5360      5370      5380      5390      5400
AAGGACCCCGAACCGCAAAGGACATCAGTATCCCACAGCCTCTCCAAGTCCCCCGGTCTCCTCCTCTTCGAGGACCAAAAGATCAATCCACCACACCCGACGACACTCAACTCCC

     5410      5420      5430      5440      5450      5460      5470      5480      5490      5500      5510      5520
CACCCCTAAAGGAGACACCGGAATCCCAGAATCAAGACTCATCCAATGTCCATCATGGGTCTCAAGGTGAACGTCTCTCTGCCATATTCATGGCAGTACTGTTAACTCTCCAAACACCCAC
                                              M  S  I  M  G  L  K  V  N  V  S  A  I  F  M  A  V  L  L  T  L  Q  T  P  T

     5530      5540      5550      5560      5570      5580      5590      5600      5610      5620      5630      5640
CGGTCAAATCCATTGGGGCAATCTCTCTAAGATAGGGGTGGTAGGAATAGGAAGTGCAAGCTACAAAGTTATGACTCGTTCCAGCCATCAATCATTAGTCATAAAATTAATGCCCAATAT
  G  Q  I  H  W  G  N  L  S  K  I  G  V  V  G  I  G  S  A  S  Y  K  V  M  T  R  S  S  H  Q  S  L  V  I  K  L  M  P  N  I

     5650      5660      5670      5680      5690      5700      5710      5720      5730      5740      5750      5760
AACTCTCCTCAATAACTGCACGAGGGTAGAGATTGCAGAATACAGGAGACTACTGAGAACTGTTTTGGAACCACTTAGAGATGCACTTAATGCAATGACCCAGAATATAAGACCGGTTCA
  T  L  L  N  N  C  T  R  V  E  I  A  E  Y  R  R  L  L  R  T  V  L  E  P  L  R  D  A  L  N  A  M  T  Q  N  I  R  P  V  Q

     5770      5780      5790      5800      5810      5820      5830      5840      5850      5860      5870      5880
GAGTGTAGCTTCAAGTAGGAGACAAAGAGATTTGCTGGAGTAGTCCTGGCAGGTGCGGCCCTAGGCGTTGCCACAGCTGCTCAGATAACAGCCGGCATTGCACTTCACCAGTCCATGCT
  S  V  A  S  S  R  R  H  K  R  F  A  G  V  V  L  A  G  A  A  L  G  V  A  T  A  A  Q  I  T  A  G  I  A  L  H  Q  S  M  L

     5890      5900      5910      5920      5930      5940      5950      5960      5970      5980      5990      6000
GAACTCTCAAGCCATCGACAATCTGAGAGCGAGCCTGGAAACTACTAACTCAGGCAATTGAGGCAATCAGACAAGCAGGGCAGGAGATGATATTGGCTGTTCAGGGTGTCCAAGACTACAT
  N  S  Q  A  I  D  N  L  R  A  S  L  E  T  T  N  Q  A  I  E  A  I  R  Q  A  G  Q  E  M  I  L  A  V  Q  G  V  Q  D  Y  I

     6010      6020      6030      6040      6050      6060      6070      6080      6090      6100      6110      6120
CAATAATGAGCTGATACCGTCTATGAACCAACTATCTTGTGATTTAATCGGCCAGAAGCTCGGGCTCAAATTGCTCAGATACTATACAGAAATCCTGTCATTATTTGGCCCCAGCTTACG
  N  N  E  L  I  P  S  M  N  Q  L  S  C  D  L  I  G  Q  K  L  G  L  K  L  L  R  Y  Y  T  E  I  L  S  L  F  G  P  S  L  R

     6130      6140      6150      6160      6170      6180      6190      6200      6210      6220      6230      6240
GGACCCCATATCTGCGGAGATATCTATCCAGGCTTTGAGCTATGCGCTTGGAGGAGACATCAATAAGGTGTTAGAAAAGCTCGGATACAGTGGAGGTGATTTACTGGGCATCTTAGAGAG
  D  P  I  S  A  E  I  S  I  Q  A  L  S  Y  A  L  G  G  D  I  N  K  V  L  E  K  L  G  Y  S  G  G  D  L  L  G  I  L  E  S

     6250      6260      6270      6280      6290      6300      6310      6320      6330      6340      6350      6360
CAGAGGAATAAAGGCCCGGATAACTCACGTCGACACAGAGTCCTACTTCATTGTCCTCAGTATAGCCTATCCGACGCTGTCCGAGATTAAGGGGGTGATTGTCCACCGGCTAGAGGGGGT
  R  G  I  K  A  R  I  T  H  V  D  T  E  S  Y  F  I  V  L  S  I  A  Y  P  T  L  S  E  I  K  G  V  I  V  H  R  L  E  G  V

     6370      6380      6390      6400      6410      6420      6430      6440      6450      6460      6470      6480
CTCGTACAACATAGGCTCTCAAGAGTGGTATACCACTGTGCCCAAGTATGTCGCAACCCAAGGGTACCTTATCTCGAATTTTGATGAGTCATCGTGTACTTTCATGCCAGAGGGGACTGT
  S  Y  N  I  G  S  Q  E  W  Y  T  T  V  P  K  Y  V  A  T  Q  G  Y  L  I  S  N  F  D  E  S  S  C  T  F  M  P  E  G  T  V

     6490      6500      6510      6520      6530      6540      6550      6560      6570      6580      6590      6600
GTGCAGCCAAAATCGGTTGTACCCGATGAGTCCTCTGCTCCAAGAATGCCTCCGGGGGTCCACTAAGTCCTGTGCTCGTACACTCGTATCCGGGTCTTTTGGGAACCGGTTCATTTTATC
  C  S  Q  N  R  L  Y  P  M  S  P  L  L  Q  E  C  L  R  G  S  T  K  S  C  A  R  T  L  V  S  G  S  F  G  N  R  F  I  L  S

     6610      6620      6630      6640      6650      6660      6670      6680      6690      6700      6710      6720
ACAAGGGAACCTAATAGCCAATTGTGCATCAATCCTTTGCAAGTGTTACACAACAGGAACGATCATTAATCAAGACCCTGACAAGATCCTAACATACATTGCTGCCGATCACTGCCCGGT
  Q  G  N  L  I  A  N  C  A  S  I  L  C  K  C  Y  T  T  G  T  I  I  N  Q  D  P  D  K  I  L  T  Y  I  A  A  D  H  C  P  V

     6730      6740      6750      6760      6770      6780      6790      6800      6810      6820      6830      6840
GTCGAGGTGAACGGCGTGACCATCCAAGTCGGGAGCAGGAGGTATCCAGACGCTGTGTACTTGCACAGAATTGACCTCGGTCCTCCCATATCATTGGAGAGGTTGGACGTAGGGACAAA
  V  E  V  N  G  V  T  I  Q  V  G  S  R  R  Y  P  D  A  V  Y  L  H  R  I  D  L  G  P  P  I  S  L  E  R  L  D  V  G  T  N
```

```
      6850      6860      6870      6880      6890      6900      6910      6920      6930      6940      6950      6960
TCTGGGGAATGCAATTGCTAAGTTGGAGGATGCCAAGGAATTGTTGGAGTCATCGGACCAGATATTGAGGAGTATGAAAGGTTTATCGAGCACTAGCAGTCTACATCCTGATTGCAGT
 L  G  N  A  I  A  K  L  E  D  A  K  E  L  L  E  S  S  D  Q  I  L  R  S  M  K |G  L  S  S  T  S  I  V  Y  I  L  I  A  V

      6970      6980      6990      7000      7010      7020      7030      7040      7050      7060      7070      7080
GTGTCTTGGAGGGTTGATAGGGATCCCCGCTTTAATATGTTGCTGCAGGGGGCGTTGTAATAAAAAGGGAGAACAAGTTGGTATGTCAAGACCAGGCCTAAAGCCTGATCTTACGGGAAC
  C  L  G  G  L  I  G  I  P  A  L  I  C  C  C |R  G  R  C  N  K  K  G  E  Q  V  G  M  S  R  P  G  L  K  P  D  L  T  G  T

      7090      7100      7110      7120      7130      7140      7150      7160      7170      7180      7190      7200
ATCAAAATCCTATGTAAGGTCGCTCTGATCCTCTACAACTCTTGAAACACAAATGTCCCACAAGTCTCCTCTTCGTCATCAAGCAACCACCGCACCCAGCATCAAGCCCACCTGAAATTA
 S  K  S  Y  V  R  S  L  *

      7210      7220      7230      7240      7250      7260      7270      7280      7290      7300      7310      7320
TCTCCGGCTTCCCTCTGGCCGAACAATATCGGTAGTTAATTAAAACTTAGGGTGCAAGATCATCCACAATGTCACCACAACGAGACCGGATAAATGCCTTCTACAAAGATAACCCCCATC
                        ┌──┐  ┌┐ ┌┐                                        M  S  P  Q  R  D  R  I  N  A  F  Y  K  D  N  P  H
                        └──┴──┴┴─┴┴────────                                HEMAGGLUTININ PROTEIN ►
                           T_stop    I  T_start
      7330      7340      7350      7360      7370      7380      7390      7400      7410      7420      7430      7440
CCAAGGGAAGTAGGATAGTCATTAACAGAGAACATCTTATGATTGATAGACCCTTATGTTTTGCTGGCTGTTCTGTTTGTCATGTTTCTGAGCTTGATCGGGTTGCTAGCCATTGCAGGCA
 P  K  G  S  R  I  V  I  N  R  E  H  L  M  I  D  R |P  Y  V  L  L  A  V  L  F  V  M  F  L  S  L  I  G  L  L  A  I  A  G

      7450      7460      7470      7480      7490      7500      7510      7520      7530      7540      7550      7560
TTAGACTTCATCGGGCAGCCATCTACACCGCAGAGATCCATAAAAGCCTCAGCACCAATCTAGATGTAACTAACTCAATCGAGCATCAGGTCAAGGACGTGCTGACACCACTCTTCAAAA
 I| R  L  H  R  A  A  I  Y  T  A  E  I  H  K  S  L  S  T  N  L  D  V  T  N  S  I  E  H  Q  V  K  D  V  L  T  P  L  F  K

      7570      7580      7590      7600      7610      7620      7630      7640      7650      7660      7670      7680
TCATCGGTGATGAAGTGGGCCTGAGGACACCTCAGAGATTCACTGACCTAGTGAAATTCATCTCTGACAAGATTAAATTCCTTAATCCGGATAGGGAGTACGACTTCAGAGATCTCACTT
 I  I  G  D  E  V  G  L  R  T  P  Q  R  F  T  D  L  V  K  F  I  S  D  K  I  K  F  L  N  P  D  R  E  Y  D  F  R  D  L  T

      7690      7700      7710      7720      7730      7740      7750      7760      7770      7780      7790      7800
GGTGTATCAACCCGCCAGAGAGAATCAAATTGGATTGATGATCAATACTGTGCAGATGTGGCTGCTGAAGAGCTCATGAATGCATTGGTGAACTCAACTCTACTGGAGACCAGAACAACCA
 W  C  I  N  P  P  E  R  I  K  L  D  Y  D  Q  Y  C  A  D  V  A  A  E  E  L  M  N  A  L  V  N  S  T  L  L  E  T  R  T  T

      7810      7820      7830      7840      7850      7860      7870      7880      7890      7900      7910      7920
ATCAGTTCCTAGCTGTCTCAAAGGGAAACTGCTCAGGGCCCACTACAATCAGAGGTCAATTCTCAAACATGTCGCTGTCCCTGTTAGACTTGTATTTAGGTCGAGGTTACAATGTGTCAT
 N  Q  F  L  A  V  S  K  G  N  C  S  G  P  T  T  I  R  G  Q  F  S  N  M  S  L  S  L  L  D  L  Y  L  G  R  G  Y  N  V  S

      7930      7940      7950      7960      7970      7980      7990      8000      8010      8020      8030      8040
CTATAGTCACTATGACATCCCAGGGAATGTATGGGGGAACTTACCTAGTGGAAAAGCCTAATCTGAGCAGCAAAAGGTCAGAGTTGTCACAACTGAGCATGTACCGAGTGTTTGAAGTAG
 S  I  V  T  M  T  S  Q  G  M  Y  G  G  T  Y  L  V  E  K  P  N  L  S  S  K  R  S  E  L  S  Q  L  S  M  Y  R  V  F  E  V

      8050      8060      8070      8080      8090      8100      8110      8120      8130      8140      8150      8160
GTGTTATCAGAAATCCGGGTTTGGGGGCTCCGGTGTTCCATATGACAAACTATCTTGAGCAACCAGTCAGTAATGATCTCAGCAACTGTATGGTGGCTTTGGGGGAGCTCAAACTCGCAG
 G  V  I  R  N  P  G  L  G  A  P  V  F  H  M  T  N  Y  L  E  Q  P  V  S  N  D  L  S  N  C  M  V  A  L  G  E  L  K  L  A

      8170      8180      8190      8200      8210      8220      8230      8240      8250      8260      8270      8280
CCCTTTGTCACGGGGAAGATTCTATCACAATTCCCTATCAGGGATCAGGGAAAGGTGTCAGCTTCCAGCTCGTCAAGCTAGGTGTCTGGAAATCCCCAACCGACATGCAATCCTGGGTCC
 A  L  C  H  G  E  D  S  I  T  I  P  Y  Q  G  S  G  K  G  V  S  F  Q  L  V  K  L  G  V  W  K  S  P  T  D  M  Q  S  W  V

      8290      8300      8310      8320      8330      8340      8350      8360      8370      8380      8390      8400
CCTTATCAACGGATGATCCCAGTGATAGACAGGCTTTACCTCTCATCTCACAGAGGTGTTATCGCTGATAATCAAGCAAAATGGGCTGTCCCGACAACACGAGATGCAAGTTGCGAA
 P  L  S  T  D  D  P  V  I  D  R  L  Y  L  S  S  H  R  G  V  I  A  D  N  Q  A  K  W  A  V  P  T  T  R  T  D  D  K  L  R

      8410      8420      8430      8440      8450      8460      8470      8480      8490      8500      8510      8520
TGGAGACATGCTTCCAACAGGCGTGTAAGGGTAAAATCCAAGCACTCTGCGAGAATCCCGAGTGGGCACCATTGAAGCATAACAGGATTCCTTCATCAGGGGTCTTGTCTGTTGATCTGA
 M  E  T  C  F  Q  Q  A  C  K  G  K  I  Q  A  L  C  E  N  P  E  W  A  P  L  K  H  N  R  I  P  S  Y  G  V  L  S  V  D  L

      8530      8540      8550      8560      8570      8580      8590      8600      8610      8620      8630      8640
GTCTGACAGTTGAGCTTAAAATCAAAATTGCTTCGGGATTCGGGCCATTGATCACACACGGTTCAGGGATGGACCTATACAAATCCAACCACAACAATGTGTATTGGCTGACTATCCCGC
 S  L  T  V  E  L  K  I  K  I  A  S  G  F  G  P  L  I  T  H  G  S  G  M  D  L  Y  K  S  N  H  N  N  V  Y  W  L  T  I  P

      8650      8660      8670      8680      8690      8700      8710      8720      8730      8740      8750      8760
CAATGAAGAACCTAGCCTTAGGTGTAATCAACACATTGGAGTGGATACCGAGATTCAAGGTTAGTCCCTACCTCTTCAATGTCCCAATTAAGGAAGCAGGCGAAGACTGCCATGCCCCAA
 P  M  K  N  L  A  L  G  V  I  N  T  L  E  W  I  P  R  F  K  V  S  P  Y  L  F  N  V  P  I  K  E  A  G  E  D  C  H  A  P

      8770      8780      8790      8800      8810      8820      8830      8840      8850      8860      8870      8880
CATACCTACCTGCGGAGGTGGATGGTGATGTCAAACTCAGTTCCAATCTGGTGATTCTACCTGGTCAAGATCTCCAATATGTTTTGGCAACCTACGATACTTCCAGGGTTGAACATGCTG
 T  Y  L  P  A  E  V  D  G  D  V  K  L  S  S  N  L  V  I  L  P  G  Q  D  L  Q  Y  V  L  A  T  Y  D  T  S  R  V  E  H  A

      8890      8900      8910      8920      8930      8940      8950      8960      8970      8980      8990      9000
TGGTTTATTACGTTTACAGCCCAAGCCGCTCATTTTCTTACTTTTATCCTTTTAGGTTGCCTATAAAGGGGGTCCCCATCGAATTACAAGTGGAATGCTTCACATGGGACCAAAAACTCT
 V  V  Y  Y  V  Y  S  P  S  R  S  F  S  Y  F  Y  P  F  R  L  P  I  K  G  V  P  I  E  L  Q  V  E  C  F  T  W  D  Q  K  L

      9010      9020      9030      9040      9050      9060      9070      9080      9090      9100      9110      9120
GGTGCCGTCACTTCTGTGTGCTTGCGGACTCAGAATCTGGTGGACATATCACTCACTCTGGGATGGAGGGCATGGGAGTCAGCTGCACAGTCACCCGGGAAGATGGAACCAATCGCAGAT
 W  C  R  H  F  C  V  L  A  D  S  E  S  G  G  H  I  T  H  S  G  M  E  G  M  G  V  S  C  T  V  T  R  E  D  G  T  N  R  R
```

MEASLES VIRUS

```
      9130      9140      9150      9160      9170      9180      9190      9200      9210      9220      9230      9240
AGGGCTGCTAGTGAACCAATCTCATGATGTCACCCAGACATCAGGCATACCCACTAGTGTGAAATAGACATCAGAATTAAGAAAAACGTAGGGTCCAAGTGGTTCCCCGTTATGGACTCG
*                                                                                                       M  D  S
                                                                          T_stop        I   T_start    L PROTEIN ▶
      9250      9260      9270      9280      9290      9300      9310      9320      9330      9340      9350      9360
CTATCTGTCAACCAGATCTTATACCCTGAAGTTCACCTAGATAGCCCGATAGTTACCAATAAGATAGTAGCCATCTGGAGTATGCTCGAGTTCCTCACGCTTACAGCCTGGAGGACCCT
 L  S  V  N  Q  I  L  Y  P  E  V  H  L  D  S  P  I  V  T  N  K  I  V  A  I  L  E  Y  A  R  V  P  H  A  Y  S  L  E  D  P

      9370      9380      9390      9400      9410      9420      9430      9440      9450      9460      9470      9480
ACACTGTGTCAGAACATCAAGCACCGCCTAAAAAACGGATTTTCCAACCAAATGATTATAAACAATGTGGAAGTTGGGAATGTCATCAAGTCCAAGCTTAGGAGTTATCCGGCCCACTCT
 T  L  C  Q  N  I  K  H  R  L  K  N  G  F  S  N  Q  M  I  I  N  N  V  E  V  G  N  V  I  K  S  K  L  R  S  Y  P  A  H  S

      9490      9500      9510      9520      9530      9540      9550      9560      9570      9580      9590      9600
CATATTCCATATCCAAATTGTAATCAGGATTTATTTAACATAGAAGACAAAGAGTCAACGAGGAAGATCCGTGAACTCCTCAAAAAGGGGAATTCGCTGTACTCCAAAGTCAGTGATAAG
 H  I  P  Y  P  N  C  N  Q  D  L  F  N  I  E  D  K  E  S  T  R  K  I  R  E  L  L  K  K  G  N  S  L  Y  S  K  V  S  D  K

      9610      9620      9630      9640      9650      9660      9670      9680      9690      9700      9710      9720
GTTTTCCAATGCTTAAGGGACACTAACTCACGGCTTGGCCTAGGCTCCGAATTGAGGGAGGACATCAAGGAGAAAGTTATTAACTTGGGAGTTTACATGCACAGCTCCCAGTGGTTTGAG
 V  F  Q  C  L  R  D  T  N  S  R  L  G  L  G  S  E  L  R  E  D  I  K  E  K  V  I  N  L  G  V  Y  M  H  S  S  Q  W  F  E

      9730      9740      9750      9760      9770      9780      9790      9800      9810      9820      9830      9840
CCCTTTCTGTTTTGGTTTACAGTCAAGACTGAGATGAGGTCAGTGATTAAATCACAAACCCATACTTGCCATAGGAGGAGGACACACACCTGTATTCTTCACTGGTAGTTCAGTTGAGTTG
 P  F  L  F  W  F  T  V  K  T  E  M  R  S  V  I  K  S  Q  T  H  T  C  H  R  R  R  H  T  P  V  F  F  T  G  S  S  V  E  L

      9850      9860      9870      9880      9890      9900      9910      9920      9930      9940      9950      9960
CTAATCTCTCGTGACCTTGTTGCTATAATCAGTAAAGAGTCTCAACATGTATATTACCTGACATTTGAACTGGTTTTGATGTATTGTGATGTCATAGAGGGGAGGTTAATGACAGAGACC
 L  I  S  R  D  L  V  A  I  I  S  K  E  S  Q  H  V  Y  Y  L  T  F  E  L  V  L  M  Y  C  D  V  I  E  G  R  L  M  T  E  T

      9970      9980      9990     10000     10010     10020     10030     10040     10050     10060     10070     10080
GCTATGACTATTGATGCTAGGTATACAGAGCTTCTAGGAAGAGTCAGATACATGTGGAAACTGATAGATGGTTTCTTCCCTGCACTCGGGAATCCAACTTATCAAATTGTAGCCATGCTG
 A  M  T  I  D  A  R  Y  T  E  L  L  G  R  V  R  Y  M  W  K  L  I  D  G  F  F  P  A  L  G  N  P  T  Y  Q  I  V  A  M  L

     10090     10100     10110     10120     10130     10140     10150     10160     10170     10180     10190     10200
GAGCCTCTTTCACTTGCTTACCTGCAGCTGAGGGATATAACAGTAGAACTCAGAGGTGCTTTCCTTAACCACTGCTTTACTGAAATACATGATGTTCTTGACCAAACGGGTTTTCTGAT
 E  P  L  S  L  A  Y  L  Q  L  R  D  I  T  V  E  L  R  G  A  F  L  N  H  C  F  T  E  I  H  D  V  L  D  Q  N  G  F  S  D

     10210     10220     10230     10240     10250     10260     10270     10280     10290     10300     10310     10320
GAAGGTACTTATCATGAGTTAATTGAAGCTCTAGATTACATTTTCATAACTGATGACATACATCTGACAGGGGAGATTTTCTCATTTTTCAGAGTTTCGGCCACCCCAGACTTGAAGCA
 E  G  T  Y  H  E  L  I  E  A  L  D  Y  I  F  I  T  D  D  I  H  L  T  G  E  I  F  S  F  F  R  S  F  G  H  P  R  L  E  A

     10330     10340     10350     10360     10370     10380     10390     10400     10410     10420     10430     10440
GTAACGGCTGCTGAAAATGTTAGGAAATACATGAATCAGCCTAAAGTCATTGTGTATGAGACTCTGATGAAAGGTCATGCCATATTTTGTGGAATCATAATCAACGGCTATCGTGACAGG
 V  T  A  A  E  N  V  R  K  Y  M  N  Q  P  K  V  I  V  Y  E  T  L  M  K  G  H  A  I  F  C  G  I  I  I  N  G  Y  R  D  R

     10450     10460     10470     10480     10490     10500     10510     10520     10530     10540     10550     10560
CACGGAGGCAGTTGGCCACCGCTGACCCTCCCCCTGCATGCTGCAGACACAATCCGGAATGCTCAAGCTTCAGGTGAAGGGTTAACACATGAGCAGTGCGTTGATAACTGGAAATCTTTT
 H  G  G  S  W  P  P  L  T  L  P  L  H  A  A  D  T  I  R  N  A  Q  A  S  G  E  G  L  T  H  E  Q  C  V  D  N  W  K  S  F

     10570     10580     10590     10600     10610     10620     10630     10640     10650     10660     10670     10680
GCTGGAGTGAAATTTGGCTGCTTTATGCCTCTTAGCCTGGATAGTGATCTGACAATGTACCTAAAGGACAAGGCACTTGCTGCTCTCCAAAGGGAATGGGATTCAGTTTACCCGAAAGAG
 A  G  V  K  F  G  C  F  M  P  L  S  L  D  S  D  L  T  M  Y  L  K  D  K  A  L  A  A  L  Q  R  E  W  D  S  V  Y  P  K  E

     10690     10700     10710     10720     10730     10740     10750     10760     10770     10780     10790     10800
TTCCTGCGTTACGACCCTCCCAAGGGAACCGGGTCACGGAGGCTTGTAGATGTTTTCCTTAATGATTCGAGCTTTGACCCATATGATGTGATAATGTATGTTGTAAGTGGAGCTTACCTC
 F  L  R  Y  D  P  P  K  G  T  G  S  R  R  L  V  D  V  F  L  N  D  S  F  D  P  Y  D  V  I  M  Y  V  V  S  G  A  Y  L

     10810     10820     10830     10840     10850     10860     10870     10880     10890     10900     10910     10920
CATGACCCTGAGTTCAACCTGTCTTACAGCCTGCAAGAAAAGGAGATCAAGGAAACAGGTAGACTTTTTGCTAAAATGACTTACAAAATGAGGGCATGCCAAGTGATTGCTGAAAATCTA
 H  D  P  E  F  N  L  S  Y  S  L  Q  E  K  E  I  K  E  T  G  R  L  F  A  K  M  T  Y  K  M  R  A  C  Q  V  I  A  E  N  L

     10930     10940     10950     10960     10970     10980     10990     11000     11010     11020     11030     11040
ATCTCAAACGGGATTGGCAAATATTTTAAGGACAATGGGATGGCCAAGGATGAGCAAGATTTGACTAAGGCACTCCACACTCTAGCTGTCTCAGGAGTCCCCAAAGATCTCAAAGAAAGT
 I  S  N  G  I  G  K  Y  F  K  D  N  G  M  A  K  D  E  Q  D  L  T  K  A  L  H  T  L  A  V  S  G  V  P  K  D  L  K  E  S

     11050     11060     11070     11080     11090     11100     11110     11120     11130     11140     11150     11160
CACAGGGGGGGGCCAGTCTTAAAAACCTACTCCCGAAGCCCAGTCCACACAAGTACCAGGAACGTGAGAGCACAAAAGGGTTTATAGGGTTCCCTCAAGTAATTCGGCAGGACCAAGAC
 H  R  G  G  P  V  L  K  T  Y  S  R  S  P  V  H  T  S  T  R  N  V  R  A  A  K  G  F  I  G  F  P  Q  V  I  R  Q  D  Q  D

     11170     11180     11190     11200     11210     11220     11230     11240     11250     11260     11270     11280
ACTGATCATCCGGAGAATATGGAAGCTTACGAGACAGTCAGTGCATTTATCACGACTGATCTCAAGAAGTACTGCCTTAATTGGAGATATGAGACCATCAGCTTGTTTGCACAGAGGCTA
 T  D  H  P  E  N  M  E  A  Y  E  T  V  S  A  F  I  T  T  D  L  K  K  Y  C  L  N  W  R  Y  E  T  I  S  L  F  A  Q  R  L

     11290     11300     11310     11320     11330     11340     11350     11360     11370     11380     11390     11400
AATGAGATTTACGGATTGCCCTCATTTTTCCAGTGGCTGCATCCGCGGCTTGAGACCTCTGTCCTGTATGTAAGTGACCCTCATTGCCCCCCCGACCTTGACGCCCATATCCCGTTATAT
 N  E  I  Y  G  L  P  S  F  F  Q  W  L  H  P  R  L  E  T  S  V  L  Y  V  S  D  P  H  C  P  P  D  L  D  A  H  I  P  L  Y
```

MEASLES VIRUS

```
      11410     11420     11430     11440     11450     11460     11470     11480     11490     11500     11510     11520
AAAGTCCCCAATGATCAAATCTTCATTAAGTACCCTATGGGAGGTATAGAAGGGTATTGTCAGAAGCTGTGGACCATCAGCACCATTCCCTATCTATACCTGGCTGCTTATGAGAGCGGA
  K  V  P  N  D  Q  I  F  I  K  Y  P  M  G  G  I  E  G  Y  C  Q  K  L  W  T  I  S  T  I  P  Y  L  Y  L  A  A  Y  E  S  G
```

```
      11530     11540     11550     11560     11570     11580     11590     11600     11610     11620     11630     11640
GTCCGGATTGCTTCGTTAGTGCAAGGGGACAATCAGACCATAGCCGTAACAAAAAGGGTACCCAGCACATGGCCCTACAACCTTAAGAAACGGGAAGCTGCTAGAGTAACTAGAGATTAC
  V  R  I  A  S  L  V  Q  G  D  N  Q  T  I  A  V  T  K  R  V  P  S  T  W  P  Y  N  L  K  K  R  E  A  A  R  V  T  R  D  Y
```

```
      11650     11660     11670     11680     11690     11700     11710     11720     11730     11740     11750     11760
TTTGTAATTCTTAGGCAAAGGCTACATGATATTGGCCATCACCTCAAGGCAAATGAGACAATTGTTTCATCACATTTTTTTGTCTATTCAAAAGGAATATATTATGATGGGCTACTTGTG
  F  V  I  L  R  Q  R  L  H  D  I  G  H  H  L  K  A  N  E  T  I  V  S  S  H  F  F  V  Y  S  K  G  I  Y  Y  D  G  L  L  V
```

```
      11770     11780     11790     11800     11810     11820     11830     11840     11850     11860     11870     11880
TCCCAATCACTCAAGAGCATCGCAAGATGTGTATTCTGGTCAGAGACTATAGTTGATGAAACAAGGGCAGCATGCAGTAATATTGCTACAACAATGGCTAAAAGCATCGAGAGAGGTTAT
  S  Q  S  L  K  S  I  A  R  C  V  F  W  S  E  T  I  V  D  E  T  R  A  A  C  S  N  I  A  T  T  M  A  K  S  I  E  R  G  Y
```

```
      11890     11900     11910     11920     11930     11940     11950     11960     11970     11980     11990     12000
GACCGTTACCTTGCATATTCCCTGAACTTCCTAAAAGTGATCCAGCAAATTCTGATCTCTCTTGGCTTCACAATCAATTCAACCATGACCCGGGATGTAGTCATACCCCTCCTCACAAAC
  D  R  Y  L  A  Y  S  L  N  F  L  K  V  I  Q  Q  I  L  I  S  L  G  F  T  I  N  S  T  M  T  R  D  V  V  I  P  L  L  T  N
```

```
      12010     12020     12030     12040     12050     12060     12070     12080     12090     12100     12110     12120
AACGACCTCTTAATAAGGATGGCACTGTTGCCCGCTCCTATTGGGGGGATGAATTATCTGAATATGAGCAGGCTGTTTGTCAGAAACATCGGTGATCCAGTAACATCATCAATTGCTGAT
  N  D  L  L  I  R  M  A  L  L  P  A  P  I  G  G  M  N  Y  L  N  M  S  R  L  F  V  R  N  I  G  D  P  V  T  S  S  I  A  D
```

```
      12130     12140     12150     12160     12170     12180     12190     12200     12210     12220     12230     12240
CTCAAGAGAATGATTCTCGCCTCACTAATGCCTGAAGAGACCCTCCATCAGGTAATGACACAACAACCGGGGGACTCTTCATTCCTAGACTGGGCTAGCGACCCCTTACTCAGCAAATCTT
  L  K  R  M  I  L  A  S  L  M  P  E  E  T  L  H  Q  V  M  T  Q  Q  P  G  D  S  S  F  L  D  W  A  S  D  P  Y  S  A  N  L
```

```
      12250     12260     12270     12280     12290     12300     12310     12320     12330     12340     12350     12360
GTATGTGTCCAGAGCATCACTAGACTCCTCAAGAACATAACTGCAAGGTTTGTCCTGATCCATAGTCCAAACCCAATGTTAAAAGGATTATTCCATGATGACAGTAAAGAAGAGGACGAG
  V  C  V  Q  S  I  T  R  L  L  K  N  I  T  A  R  F  V  L  I  H  S  P  N  P  M  L  K  G  L  F  H  D  D  S  K  E  E  D  E
```

```
      12370     12380     12390     12400     12410     12420     12430     12440     12450     12460     12470     12480
GGACTGGCGGCATTCCTCATGGACAGGCATATTATAGTACCTCGGGCAGCTCATGAAATCCTGGATCATAGTGTCACAGGGGCAAGAGAGTCTATTGCAGGCATGCTGGATACCACAAAA
  G  L  A  A  F  L  M  D  R  H  I  I  V  P  R  A  A  H  E  I  L  D  H  S  V  T  G  A  R  E  S  I  A  G  M  L  D  T  T  K
```

```
      12490     12500     12510     12520     12530     12540     12550     12560     12570     12580     12590     12600
GGCTTGATTCGAGCCAGCATGAGGAAGGGGGGTTTAACCTCTCGAGTGATAACCAGATTGTCCAATTATGACTATGAACAATTCAGAGCAGGGATGGTGCTATTGACAGGAAGAAAGAGA
  G  L  I  R  A  S  M  R  K  G  G  L  T  S  R  V  I  T  R  L  S  N  Y  D  Y  E  Q  F  R  A  G  M  V  L  L  T  G  R  K  R
```

```
      12610     12620     12630     12640     12650     12660     12670     12680     12690     12700     12710     12720
AATGTCCTCATTGACAAAGAGTCATGTTCAGTGCAGCTGGCGAGAGCTCTAAGAAGCCATATGTGGGCGAGGCTAGCTCGAGGACGGCCTATTTACGGCCTTGAGGTCCCTGATGTACTA
  N  V  L  I  D  K  E  S  C  S  V  Q  L  A  R  A  L  R  S  H  M  W  A  R  L  A  R  G  R  P  I  Y  G  L  E  V  P  D  V  L
```

```
      12730     12740     12750     12760     12770     12780     12790     12800     12810     12820     12830     12840
GAATCTATGCGAGGCCACCTTATTCGGCGTCATGAGACATGTGTCATCTGCGAGTGTGGATCAGTCAACTACGGATGGTTTTTTGTCCCCTCGGGTTGCCAACTGGATGATATTGACAAG
  E  S  M  R  G  H  L  I  R  R  H  E  T  C  V  I  C  E  C  G  S  V  N  Y  G  W  F  F  V  P  S  G  C  Q  L  D  D  I  D  K
```

```
      12850     12860     12870     12880     12890     12900     12910     12920     12930     12940     12950     12960
GAAACATCATCCTTGAGAGTCCCATATATTGGTTCTACCACTGATGAGAGAACAGACATGAAGCTTGCCTTCGTAAGAGCCCCAAGTCGATCCTTGCGATCTGCTGTTAGAATAGCAACA
  E  T  S  S  L  R  V  P  Y  I  G  S  T  T  D  E  R  T  D  M  K  L  A  F  V  R  A  P  S  R  S  L  R  S  A  V  R  I  A  T
```

```
      12970     12980     12990     13000     13010     13020     13030     13040     13050     13060     13070     13080
GTGTACTCATGGGCTTACGGTGATGATGATAGCTCTTGGAACGAAGCCTGGTTGTTGGCTAGGCAAAGGGCCAATGTGAGCCTGGAGGAGCTAAGGGTGATCACTCCCATCTCAACTTCG
  V  Y  S  W  A  Y  G  D  D  D  S  S  W  N  E  A  W  L  L  A  R  Q  R  A  N  V  S  L  E  E  L  R  V  I  T  P  I  S  T  S
```

```
      13090     13100     13110     13120     13130     13140     13150     13160     13170     13180     13190     13200
ACTAATTTAGCGCATAGGTTGAGGGATCGTAGCACTCAAGTGAAATACTCAGGTACATCCCTTGTCCGAGTGGCGAGGTATACCACAATCTCCAACGACAATCTCTCATTTGTCATATCA
  T  N  L  A  H  R  L  R  D  R  S  T  Q  V  K  Y  S  G  T  S  L  V  R  V  A  R  Y  T  T  I  S  N  D  N  L  S  F  V  I  S
```

```
      13210     13220     13230     13240     13250     13260     13270     13280     13290     13300     13310     13320
GATAAGAAGGTTGATACTAACTTTATATACCAACAAGGAATGCTTCTAGGGTTGGGTGTTTTAGAAACATTGTTTCGACTCGAGAAAGATACCGGATCATCTAACACGGTATTACATCTT
  D  K  K  V  D  T  N  F  I  Y  Q  Q  G  M  L  L  G  L  G  V  L  E  T  L  F  R  L  E  K  D  T  G  S  S  N  T  V  L  H  L
```

```
      13330     13340     13350     13360     13370     13380     13390     13400     13410     13420     13430     13440
CACGTCGAAACAGATTGTTGCGTGATCCCGATGATAGATCATCCCAGGATACCCAGCTCCCGCAAGCTAGAGCTGAGGGCAGAGCTATGTACCAACCCATTGATATATGATAATGCACCT
  H  V  E  T  D  C  C  V  I  P  M  I  D  H  P  R  I  P  S  S  R  K  L  E  L  R  A  E  L  C  T  N  P  L  I  Y  D  N  A  P
```

```
      13450     13460     13470     13480     13490     13500     13510     13520     13530     13540     13550     13560
TTAATTGACAGAGATGCAACAAGGCTATACACCCAGAGCCATAGGAGGCACCTTGTGGAATTTGTTCACATGGTCCACACCCCAACTATATCACATTTTAGCTAAGTCCACAGCACTATCT
  L  I  D  R  D  A  T  R  L  Y  T  Q  S  H  R  R  H  L  V  E  F  V  T  W  S  T  P  Q  L  Y  H  I  L  A  K  S  T  A  L  S
```

```
      13570     13580     13590     13600     13610     13620     13630     13640     13650     13660     13670     13680
ATGATTGACCTGGTAACAAAATTTGAGAAGGACCATATGAATGAAATTTCAGCTCTCATAGGGGATGACGATATCAATAGTTTCATAACTGAGTTTCTGCTCATAGAGCCAAGATTATTC
  M  I  D  L  V  T  K  F  E  K  D  H  M  N  E  I  S  A  L  I  G  D  D  D  I  N  S  F  I  T  E  F  L  L  I  E  P  R  L  F
```

```
        13690      13700      13710      13720      13730      13740      13750      13760      13770      13780      13790      13800
ACTATCTACTTGGGCCAGTGTGCGGCCATCAATTGGGCATTTGATGTACATTATCATAGACCATCAGGGAAATATCAGGATGGGTGAGCTGTTGTCATCGTTCCTTTCTAGAATGAGCAAA
 T  I  Y  L  G  Q  C  A  A  I  N  W  A  F  D  V  H  Y  H  R  P  S  G  K  Y  Q  M  G  E  L  L  S  S  F  L  S  R  M  S  K

        13810      13820      13830      13840      13850      13860      13870      13880      13890      13900      13910      13920
GGAGTGTTTAAGGTGCTTGTCAATGCTCTAAGCCACCCCAAAGATCTACAAGAAATTCTGGCATTGTGGTATTATAGAGCCTATCCATGGTCCTTCACTTGATGCTCAAAACTTGCACACA
 G  V  F  K  V  L  V  N  A  L  S  H  P  K  I  Y  K  K  F  W  H  C  G  I  I  E  P  I  H  G  P  S  L  D  A  Q  N  L  H  T

        13930      13940      13950      13960      13970      13980      13990      14000      14010      14020      14030      14040
ACTGTGTGCAACATGGTTTACACATGCTATATGACCTACCTCGACCTGTTGTTGAATGAAGAGTTAGAAGAGTTCACATTTCTCTTGTGTGTGAAAGCGACGAGGATGTAGTACCGGACAGA
 T  V  C  N  M  V  Y  T  C  Y  M  T  Y  L  D  L  L  L  N  E  E  L  E  E  F  T  F  L  L  C  E  S  D  E  D  V  P  D  R

        14050      14060      14070      14080      14090      14100      14110      14120      14130      14140      14150      14160
TTCGACAACATCCAGGCAAAACACTTATGTGTTCTGGCAGATTTGTACTGTCAACCAGGGACCTGCCCACCAATTCAAGGTCTAAGACCGGTAGAGAAATGTGCAGTTCTAACCGACCAT
 F  D  N  I  Q  A  K  H  L  C  V  L  A  D  L  Y  C  Q  P  G  T  C  P  P  I  Q  G  L  R  P  V  E  K  C  A  V  L  T  D  H

        14170      14180      14190      14200      14210      14220      14230      14240      14250      14260      14270      14280
ATCAAGGCAGAGGCTATGTTATCTCCAGCAGGATCTTCGTGGAACATAAATCCAATTATTGTAGACCATTACTCATGCTCCCTGACTTATCTCCGGCGAGGATCGATCAAACAGATAAGA
 I  K  A  E  A  M  L  S  P  A  G  S  S  W  N  I  N  P  I  I  V  D  H  Y  S  C  S  L  T  Y  L  R  R  G  S  I  K  Q  I  R

        14290      14300      14310      14320      14330      14340      14350      14360      14370      14380      14390      14400
TTGAGAGTTGATCCAGGATTCATTTTCGACGCCCTCGCTGAGGTAAATGTCAGTCAGCCAAAGATCGGCAGCAACATCTCAAATATGAGCATCAAGGCTTTCAGACCCCCACACGAT
 L  R  V  D  P  G  F  I  F  D  A  L  A  E  V  N  V  S  Q  P  K  I  G  S  N  N  I  S  N  M  S  I  K  A  F  R  P  P  H  D

        14410      14420      14430      14440      14450      14460      14470      14480      14490      14500      14510      14520
GATGTTGCAAAATTGCTCAAAGATATCAACACAAGCAAGCACAATCTTCCCATTTCAGGGGGCAATCTCGCCAATTATGAAATCCATGCTTTCCGCAGAATCGGGTTGAACTCATCTGCT
 D  V  A  K  L  L  K  D  I  N  T  S  K  H  N  L  P  I  S  G  G  N  L  A  N  Y  E  I  H  A  F  R  R  I  G  L  N  S  S  A

        14530      14540      14550      14560      14570      14580      14590      14600      14610      14620      14630      14640
TGCTACAAAGCTGTTGAGATATCAACATTAATTAGGAGATGCCTTGAGCCAGGGGAGGACGGCTTGTTCTTGGGTGAGGGATCGGGTTCTATGTTGATCACTTATAAGGAGATACTTAAA
 C  Y  K  A  V  E  I  S  T  L  I  R  R  C  L  E  P  G  E  D  G  L  F  L  G  E  G  S  G  S  M  L  I  T  Y  K  E  I  L  K

        14650      14660      14670      14680      14690      14700      14710      14720      14730      14740      14750      14760
CTAAGCAAGTGCTTCTATAATAGTGGGGTTTCCGCCAATTCTAGATCTGGTCAAAGGGAATTAGCACCCTATCCCTCCGAAGTTGGCCTTGTCGAACACAGAATGGGAGTAGGTAATATT
 L  S  K  C  F  Y  N  S  G  V  S  A  N  S  R  S  G  Q  R  E  L  A  P  Y  P  S  E  V  G  L  V  E  H  R  M  G  V  G  N  I

        14770      14780      14790      14800      14810      14820      14830      14840      14850      14860      14870      14880
GTCAAAGTGCTTCTTTAACGGGAGGCCCGAAGTCACGTGGGTAGGCAGTGTAGATTGCTTCAATTTCATAGTTAGTAATATCCCTACCTCTAGTGTGGGGTTTATCCATTCAGATATAGAG
 V  K  V  L  F  N  G  R  P  E  V  T  W  V  G  S  V  D  C  F  N  F  I  V  S  N  I  P  T  S  S  V  G  F  I  H  S  D  I  E

        14890      14900      14910      14920      14930      14940      14950      14960      14970      14980      14990      15000
ACCTTGCCTGACAAAGATACTATAGAGAAGCTAGAGGAATTGGCAGCCATCTTATCGATGGCTCTGCTCCTGGGCAAAATAGGATCAATACTGGTGATTAAGCTTATGCCTTTCAGCGGG
 T  L  P  D  K  D  T  I  E  K  L  E  E  L  A  A  I  L  S  M  A  L  L  L  G  K  I  G  S  I  L  V  I  K  L  M  P  F  S  G

        15010      15020      15030      15040      15050      15060      15070      15080      15090      15100      15110      15120
GATTTTGTTCAGGGATTTATAAGTTATGTAGGGTCTCATTATAGAGAAGTGAACCTTGTATACCCTAGATACAGCAACTTCATATCTACTGAATCTTATTTGGTTATGACAGATCTCAAG
 D  F  V  Q  G  F  I  S  Y  V  G  S  H  Y  R  E  V  N  L  V  Y  P  R  Y  S  N  F  I  S  T  E  S  Y  L  V  M  T  D  L  K

        15130      15140      15150      15160      15170      15180      15190      15200      15210      15220      15230      15240
GCTAACCGGCTAATGAATCCTGAAAAGATTAAGCAGCAGATAATTGAATCATCTGTGAGGACTTCACCTGGACTTATAGGTCACATCCTATCCATTAAGCAACTAAGCTGCATACAAGCA
 A  N  R  L  M  N  P  E  K  I  K  Q  Q  I  I  E  S  S  V  R  T  S  P  G  L  I  G  H  I  L  S  I  K  Q  L  S  C  I  Q  A

        15250      15260      15270      15280      15290      15300      15310      15320      15330      15340      15350      15360
ATTGTGGGAGACGCAGTTGTAGAGGTGATATCAATCCTACTCTGAAAAAACTTACACCTATAGAGCAGGTGCTGATCAATTGCGGGTTGGCAATTAACGGACCTAAGCTGTGCAAAGAA
 I  V  G  D  A  V  S  R  G  D  I  N  P  T  L  K  K  L  T  P  I  E  Q  V  L  I  N  C  G  L  A  I  N  G  P  K  L  C  K  E

        15370      15380      15390      15400      15410      15420      15430      15440      15450      15460      15470      15480
TTGATCCACCATGATGTTGCCTCAGGGCAAGATGGATTGCTTAATTCTATACTCATCCTCTACAGGGAGTTGGCAAGATTCAAAGACAACCAAAGAAGTCAACAAGGGATGTTCCACGCT
 L  I  H  H  D  V  A  S  G  Q  D  G  L  L  N  S  I  L  I  L  Y  R  E  L  A  R  F  K  D  N  Q  R  S  Q  Q  G  M  F  H  A

        15490      15500      15510      15520      15530      15540      15550      15560      15570      15580      15590      15600
TACCCCGTATTGGTAAGTAGCAGGCAACGAGAACTTATATCTAGGATCACCCGCAAATTTTGGGGGCACATTCTTCTTTACTCCGGGAACAGAAAGTTGATAAATAAGTTTATCCAGAAT
 Y  P  V  L  V  S  S  R  Q  R  E  L  I  S  R  I  T  R  K  F  W  G  H  I  L  L  Y  S  G  N  R  K  L  I  N  K  F  I  Q  N

        15610      15620      15630      15640      15650      15660      15670      15680      15690      15700      15710      15720
CTCAAGTCCGGCTATCTGATACTAGACTTACACCAGAATATCTTCGTTAAGAATCTATCCAAGTCAGAGAAACAGATTATTATGACGGGGGTTTGAAACGTGAGTGGGTTTTTAAGGTA
 L  K  S  G  Y  L  I  L  D  L  H  Q  N  I  F  V  K  N  L  S  K  S  E  K  Q  I  I  M  T  G  G  L  K  R  E  W  V  F  K  V

        15730      15740      15750      15760      15770      15780      15790      15800      15810      15820      15830      15840
ACAGTCAAGGAGACCAAAGAATGGTATAAAGTTAGTCGGATACAGTGCCCTGATTAAGGACTAATTGGTTGAACTCCGGAACCCTAATCCTGCCCTAGGTGGTTAGGCATTATTTGCAATA
 T  V  K  E  T  K  E  W  Y  K  L  V  G  Y  S  A  L  I  K  D  *
```

HUMAN RESPIRATORY SYNCYTIAL VIRUS

```
        10        20        30        40        50        60        70        80        90       100       110       120
NGGAAAAAAATGCGTACAACAAACTTGCATAAACCAAAANNNNGGGGCAAATAAGAATTTGATAAGTACCACTTAAATTTAACTCCCTTGGTTAGAGATGGGCAGCAATTCATTGAGTAT
                                                                                 1C PROTEIN ►M  G  S  N  S  L  S  M
     EXTRACISTRONIC SEQUENCES                    T start
       130       140       150       160       170       180       190       200       210       220       230       240
GATAAAAGTTAGATTACAAAATTTGTTTGACAATGATGAAGTAGCATTGTTAAAAATAACATGCTATACTGATAAATTAATACATTTAACTAACGCTTTGGCTAAGGCAGTGATACATAC
  I  K  V  R  L  Q  N  L  F  D  N  D  E  V  A  L  L  K  I  T  C  Y  T  D  K  L  I  H  L  T  N  A  L  A  K  A  V  I  H  T
       250       260       270       280       290       300       310       320       330       340       350       360
AATCAAATTGAATGGCATTGTGTTTGTGCATGTTATTACAAGTAGTGATATTTGCCCTAATAATAATATTGTAGTAAAATCCAATTTCACAACAATGCCAGTACTACAAAATGGAGGTTA
  I  K  L  N  G  I  V  F  V  H  V  I  T  S  S  D  I  C  P  N  N  N  I  V  V  K  S  N  F  T  T  M  P  V  L  Q  N  G  G  Y
       370       380       390       400       410       420       430       440       450       460       470       480
TATATGGGAAATGATGGAATTAACACATTGCTCTCAACCTAATGGTCTACTAGATGACAATTGTGAAATTAAATTCTCCAAAAAACTAAGTGATTCAACAATGACCAATTATATGAATCA
  I  W  E  M  M  E  L  T  H  C  S  Q  P  N  G  L  L  D  D  N  C  E  I  K  F  S  K  K  L  S  D  S  T  M  T  N  Y  M  N  Q
       490       500       510       520       530       540       550       560       570       580       590       600
ATTATCTGAATTACTTGGATTTGATCTTAATCCATAAATTATAATTAATATCAACTAGCAAATCAATGTCACTAACACCATTAGTTAATATAAAACTTAACAGAAGACAAAAATGGGGCA
  L  S  E  L  L  G  F  D  L  N  P  *                                                  T stop          I          T start
       610       620       630       640       650       660       670       680       690       700       710       720
AATAAATCAATTCAGCCAACCCAACCATGGACACAACCCACAATGATAATACACCAAAAGACTGATGATCACAGACATGAGACCGTTGTCACTTGAGACCATAATAACATCACTAACCA
 ◄  ║ 1B PROTEIN ►M  D  T  T  H  N  D  N  T  P  Q  R  L  M  I  T  D  M  R  P  L  S  L  E  T  I  I  T  S  L  T
       730       740       750       760       770       780       790       800       810       820       830       840
GAGACATCATAACACACAAATTTATATACTTGATAAATCATGAATGCATAGTGAGAAAACTTGATGAAAAACAGGCCACATTTACATTCCTGGTCAACTATGAAATGAAACTATTACACA
  R  D  I  I  T  H  K  F  I  Y  L  I  N  H  E  C  I  V  R  K  L  D  E  K  Q  A  T  F  T  F  L  V  N  Y  E  M  K  L  L  H
       850       860       870       880       890       900       910       920       930       940       950       960
AAGTAGGAAGCACTAAATATAAAAAATATACTGAATACAACACAAAATGGCACTTTCCCTATGCCAATATTCATCAATCATGATGGGTTCTTAGAATGCATTGGCATTAAGCCTACAA
  K  V  G  S  T  K  Y  K  K  Y  T  E  Y  N  T  K  Y  G  T  F  P  M  P  I  F  I  N  H  D  G  F  L  E  C  I  G  I  K  P  T
       970       980       990      1000      1010      1020      1030      1040      1050      1060      1070      1080
AGCATACTCCCATAATATACAAGTATGATCTCAATCCATAAATTTCAACACAATATTCACACAATCTAAAACAACAACTCTATGCATAACTATACTCCATAGTCCAGATGGAGCCTGAAA
  K  H  T  P  I  I  Y  K  Y  D  L  N  P  *
      1090      1100      1110      1120      1130      1140      1150      1160      1170      1180      1190      1200
ATTATAGTAATTTAAAATTAAGGAGAGATATAAGATAGAAGATGGGGCAAATACAAAGATGGCTCTTAGCAAAGTCAAGTTGAATGATACACTCAACAAAGATCAACTTCTGTCATCCAG
                                                             MALSKVKLNDTLNKDQLLSSS
   T stop          I                    T start    NUCLEOCAPSID PROTEIN ►
      1210      1220      1230      1240      1250      1260      1270      1280      1290      1300      1310      1320
CAAATACACCATCCAACGGAGCACAGGAGGATAGTATTGATACTCCTAATTATGATGTGCAGAAACACATCAATAAGTTATGTGGCATGTTATTAATCACAGAAGATGCTAATCATAAATT
  K  Y  T  I  Q  R  S  T  G  D  S  I  D  T  P  N  Y  D  V  Q  K  H  I  N  K  L  C  G  M  L  L  I  T  E  D  A  N  H  K  F
      1330      1340      1350      1360      1370      1380      1390      1400      1410      1420      1430      1440
CACTGGGTTAATAGGTATGTTATATGCGATGTCTAGGTTAGGAAGAGAAGACACCATAAAAATACTCAGAGATGCGGGATATCATGTAAAAGCAAATGGAGTAGATGTAACAACACATCG
  T  G  L  I  G  M  L  Y  A  M  S  R  L  G  R  E  D  T  I  K  I  L  R  D  A  G  Y  H  V  K  A  N  G  V  D  V  T  T  H  R
      1450      1460      1470      1480      1490      1500      1510      1520      1530      1540      1550      1560
TCAAGACATTAATGGAAAAGAAATGAAATTTGAAGTGTTAACATTGGCAAGCTTAACAACTGAAATTCAAATCAACATTGAGATAGAATCTAGAAAATCCTACAAAAAAATGCTAAAAGA
  Q  D  I  N  G  K  E  M  K  F  E  V  L  T  L  A  S  L  T  T  E  I  Q  I  N  I  E  I  E  S  R  K  S  Y  K  K  M  L  K  E
      1570      1580      1590      1600      1610      1620      1630      1640      1650      1660      1670      1680
AATGGGAGAGGTAGCTCCAGAATACAGGCATGACTCTCCTGATTGTGGGATGATAATATTATGTATAGCAGCATTAGTAATAACTAAATTAGCAGCAGGGGACAGATCTGGTCTTACAGC
  M  G  E  V  A  P  E  Y  R  H  D  S  P  D  C  G  M  I  I  L  C  I  A  A  L  V  I  T  K  L  A  A  G  D  R  S  G  L  T  A
      1690      1700      1710      1720      1730      1740      1750      1760      1770      1780      1790      1800
CGTGATTAGGAGGCTAATAATGTCCTAAAAAATGAAACGTTACAAAGGCTTACTACCCAAGGACATAGCCAACAGCTTCTATGAAGTGTTTGAAAAACATCCCCACTTTATAGA
  V  I  R  R  A  N  N  V  L  K  N  E  M  K  R  Y  K  G  L  L  P  K  D  I  A  N  S  F  Y  E  V  F  E  K  H  P  H  F  I  D
      1810      1820      1830      1840      1850      1860      1870      1880      1890      1900      1910      1920
TGTTTTTGTTCATTTTGGTATAGCACAATCTTCTACCAGAGGTGGCAGTAGAGTTGAAGGGATTTTTGCAGGATTGTTTATGAATGCCTATGGTGCAGGGCAAGTGATGTTACGGTGGGG
  V  F  V  H  F  G  I  A  Q  S  S  T  R  G  G  S  R  V  E  G  I  F  A  G  L  F  M  N  A  Y  G  A  G  Q  V  M  L  R  W  G
      1930      1940      1950      1960      1970      1980      1990      2000      2010      2020      2030      2040
AGTCTTAGCAAAATCAGTTAAAAATATTATGTTAGGACATGCTAGTGTGCAAGCAGAAATGGAACAAGTGTTGAGGTTTATGAATATGCCCAAAAATTGGGTGGTGAAGCAGGATTCTA
  V  L  A  K  S  V  K  N  I  M  L  G  H  A  S  V  Q  A  E  M  E  Q  V  V  E  V  Y  E  Y  A  Q  K  L  G  G  E  A  G  F  Y
      2050      2060      2070      2080      2090      2100      2110      2120      2130      2140      2150      2160
CCCATATATTGAACAACCCAAAAGCATCATTATTATCTTTGACTCAATTTCCTCACTTCTCCAGTGTAGTATTAGGCAATGCTGCTGGCCTAGGCATAATGGGAGAGTACAGAGGTACACC
  H  I  L  N  N  P  K  A  S  L  L  S  L  T  Q  F  P  H  F  S  S  V  V  L  G  N  A  A  G  L  G  I  M  G  E  Y  R  G  T  P
      2170      2180      2190      2200      2210      2220      2230      2240      2250      2260      2270      2280
GAGGAATCAAGATCTATATGATGCAGCAAAGGCATATGCTGAACAACTCAAAGAAAATGGTGTGATTAACTACAGTGTACTAGACTTGACAGCAGAAGAACTAGAGGCTATCAAACATCA
  R  N  Q  D  L  Y  D  A  A  K  A  Y  A  E  Q  L  K  E  N  G  V  I  N  Y  S  V  L  D  L  T  A  E  E  L  E  A  I  K  H  Q
```

```
      2290      2300      2310      2320      2330      2340      2350      2360      2370      2380      2390      2400
GCTTAATCCAAAAGATAATGATGTAGAGCTTTGAGTTAATAAAAAATGGGGCAAATAAATCATCATGAAAAGTTTGCTCCTGAATTCCATGGAGAAGATGCAAACAACAGGGCTACTAA
 L  N  P  K  D  N  D  V  E  L  *                      M  E  K  F  A  P  E  F  H  G  E  D  A  N  N  R  A  T  K
                                      T_stop    I  T_start    PHOSPHOPROTEIN ▶

      2410      2420      2430      2440      2450      2460      2470      2480      2490      2500      2510      2520
ATTCCTAGAATCAATAAAGGGCAAATTCACATCACCCAAAGATCCCAAGAAAAAAGATAGTATCATATCTGTCAACTCAATAGATATAGAAGTAACCAAAGAAAGCCCTATAACATCAAA
 F  L  E  S  I  K  G  K  F  T  S  P  K  D  P  K  K  K  D  S  I  I  S  V  N  S  I  D  I  E  V  T  K  E  S  P  I  T  S  N

      2530      2540      2550      2560      2570      2580      2590      2600      2610      2620      2630      2640
TTCAACTATTATCAACCCAACAAATGAGACAGATGATACTGCAGGGAACAAGCCCAATTATCAAAGAAAACCTCTAGTAAGTTTCAAAGAAGACCCTACACCAAGTGATAATCCCTTTTC
 S  T  I  I  N  P  T  N  E  T  D  D  T  A  G  N  K  P  N  Y  Q  R  K  P  L  V  S  F  K  E  D  P  T  P  S  D  N  P  F  S

      2650      2660      2670      2680      2690      2700      2710      2720      2730      2740      2750      2760
TAAACTATACAAAGAAACCATAGAAACATTTGATAACAATGAAGAAGAATCCAGCTATTCATACGAAGAAATAAATGATCAGACAAACGATAATATAACAGCAAGATTAGATAGGATTGA
 K  L  Y  K  E  T  I  E  T  F  D  N  N  E  E  E  S  S  Y  S  Y  E  E  I  N  D  Q  T  N  D  N  I  T  A  R  L  D  R  I  D

      2770      2780      2790      2800      2810      2820      2830      2840      2850      2860      2870      2880
TGAAAAATTAAGTGAAATACTAGGAATGCTTCACACATTAGTAGTGGCAAGTGCAGGACCTACATCTGCTCGGGATGGTATAAGAGATGCCATGATTGGTTTAAGAGAGAAATGATAGA
 E  K  L  S  E  I  L  G  M  L  H  T  L  V  V  A  S  A  G  P  T  S  A  R  D  G  I  R  D  A  M  I  G  L  R  E  E  M  I  E

      2890      2900      2910      2920      2930      2940      2950      2960      2970      2980      2990      3000
AAAAATCAGAACTGAAGCATTAATGACCAATGACAGATTAGAAGCTATGGCAAGACTCAGGAATGAGGAAAGTGAAAAGATGGCAAAAGACACATCAGATGAAGTGTCTCTCAATCCAAC
 K  I  R  T  E  A  L  M  T  N  D  R  L  E  A  M  A  R  L  R  N  E  E  S  E  K  M  A  K  D  T  S  D  E  V  S  L  N  P  T

      3010      3020      3030      3040      3050      3060      3070      3080      3090      3100      3110      3120
ATCAGAGAAATTGAACAACCTATTGGAAGGGAATGATAGTGACAATGATCTATCACTTGAAGATTTCTGATTAGTTACCACTCTTCACATCAACACACAATACCAACAGAAGCACAACAA
 S  E  K  L  N  N  L  L  E  G  N  D  S  D  N  D  L  S  L  E  D  F  *

      3130      3140      3150      3160      3170      3180      3190      3200      3210      3220      3230      3240
ACTAACCAACCCAATCATCCAACCAAACATCCATCCGCCAATCAGCCAACAGCCAACAAAACAACCAGCCAATCCAAAACTAACCACCCGGAAAAAATCTATAATATAGTTACAAAAAA
                                                                                          T_stop

      3250      3260      3270      3280      3290      3300      3310      3320      3330      3340      3350      3360
AGGAAAGGGTGGGGCAAATATGGAAACATACGTGAACAAGCTTCACGAAGGCTCCACATACACAGCTGCTGTTCAATACAATGTCTTAGAAAAAGACGATGACCCTGCATCACTTACAAT
               M  E  T  Y  V  N  K  L  H  E  G  S  T  Y  T  A  A  V  Q  Y  N  V  L  E  K  D  D  D  P  A  S  L  T  I
        I     T_start  MATRIX PROTEIN ▶

      3370      3380      3390      3400      3410      3420      3430      3440      3450      3460      3470      3480
ATGGGTGCCCATGTTCCAATCATCTATGCCGCAGATTTACTTATAAAAGAACTAGCTAATGTCAACATACTAGTGAAACAAATATCCACACCCAAGGGACCTTCACTAAGAGTCATGAT
 W  V  P  M  F  Q  S  S  M  P  A  D  L  L  I  K  E  L  A  N  V  N  I  L  V  K  Q  I  S  T  P  K  G  P  S  L  R  V  M  I

      3490      3500      3510      3520      3530      3540      3550      3560      3570      3580      3590      3600
AAACTCAAGAAGTGCAGTGCTAGCACAAATGCCCAGCAAATTTACCATATGCGCTAATGTGTCCTTGGATGAAAGAAGCAAACTAGCATATGATGTAACCACACCCTGTGAAATCAAGGC
 N  S  R  S  A  V  L  A  Q  M  P  S  K  F  T  I  C  A  N  V  S  L  D  E  R  S  K  L  A  Y  D  V  T  T  P  C  E  I  K  A

      3610      3620      3630      3640      3650      3660      3670      3680      3690      3700      3710      3720
ATGTAGTCTAACATGCCTAAAATCAAAAAAATATGTTGACTACAGTTAAAGATCTCACTATGAAGACACTCAACCCTACACATGATATTATTGCTTTATGTGAATTTGAAAACATAGTAAC
 C  S  L  T  C  L  K  S  K  N  M  L  T  T  V  K  D  L  T  M  K  T  L  N  P  T  H  D  I  I  A  L  C  E  F  N  I  V  T

      3730      3740      3750      3760      3770      3780      3790      3800      3810      3820      3830      3840
ATCAAAAAAAGTCATAATACCAACATACCTAAGATCCATCAGTGTCAGAATAAAGATCTGAACACACTTGAAAATATAACAACCACTGAATTCAAAAATGCTATCACAAATGCAAAAAT
 S  K  K  V  I  I  P  T  Y  L  R  S  I  S  V  R  N  K  D  L  N  T  L  E  N  I  T  T  T  E  F  K  N  A  I  T  N  A  K  I

      3850      3860      3870      3880      3890      3900      3910      3920      3930      3940      3950      3960
CATCCCTTACTCAGGATTACTATTAGTCATCACAGTGACTGACAACAAAGGAGCATTCAAATACATAAAGCCACAAAGTCAATTCATAGTAGATCTTGGAGCTTACCTAGAAAAAGAAAG
 I  P  Y  S  G  L  L  L  V  I  T  V  T  D  N  K  G  A  F  K  Y  I  K  P  Q  S  Q  F  I  V  D  L  G  A  Y  L  E  K  E  S

      3970      3980      3990      4000      4010      4020      4030      4040      4050      4060      4070      4080
TATATATTATGTTACCACAAATTGGAAGCACACAGCTACACGATTTGCAATCAAACCCATGGAAGATTAACCTTTTTCCTCTACATCAGTGTGTTAATTCATACAAACTTTCTACCTACA
 I  Y  Y  V  T  T  N  W  K  H  T  A  T  R  F  A  I  K  P  M  E  D  *

      4090      4100      4110      4120      4130      4140      4150      4160      4170      4180      4190      4200
TTCTTCACTTCACCATCACAATCACAAACACTCTGTGGTTCAACCAATCAAACAAAACTTATCTGAAGTCCCAGATCATCCCAAGTCATTGTTTATCGATCTAGTACTCAAATAAGTTA
                                                                                                          T_stop

      4210      4220      4230      4240      4250      4260      4270      4280      4290      4300      4310      4320
ATAAAAAATATACACATGGGGCAAATAATCATTGGAGGAAATCCAACTAATCACAATATCTGTTAACATAGACAAGTCCACACACCATACAGAATCAACCAATGGAAAATACATCCATAA
                                                                            1B PROTEIN ▶   M  E  N  T  S  I
T_stop        I     T_start

      4330      4340      4350      4360      4370      4380      4390      4400      4410      4420      4430      4440
CAATAGAATTCTCAAGCAAATTCTGGCCTTACTTTACACTAATACACATGCACAACAATAATCTCTTTGCTAATCATAATCTCCATCATGATTGCAATACTAAACAAACTTTGTGAATT
 T  I  E  F  S  S  K  F  W  P  Y  F  T  L  I  H  M  I  T  T  I  I  S  L  L  I  I  I  S  I  M  I  A  I  L  N  K  L  C  E
```

RESPIRATORY SYNCYTIAL VIRUS

```
      4450      4460      4470      4480      4490      4500      4510      4520      4530      4540      4550      4560
ATAACGTATTCCATAACAAAACCTTTGAGTTACCAAGAGCTCGAGTCAACACATAGCATTCATCAATCCAACAGCCCAAAACAGTAACCTTGCATTTAAAAATGAACAACCCCTACCTCT
Y  N  V  F  H  N  K  T  F  E  L  P  R  A  R  V  N  T  *

      4570      4580      4590      4600      4610      4620      4630      4640      4650      4660      4670      4680
TTACAACACCTCATTAACATCCCACCATGCAAACCACTATCCATACTATAAAGTAGTTAATTAAAAATAGTCATAACAATGAACTAGGATATCAAGACTAACAATAACATTGGGGCAAAT
                                       T_stop                                    I                          T_start

      4690      4700      4710      4720      4730      4740      4750      4760      4770      4780      4790      4800
GCAAACATGTCCAAAAACAAGGACCAACGCACCGCTAAGACATTAGAAAGGACCTGGGACACTCTCAATCATTTATTATTCATATCATCGTGCTTATATAAGTTAAATCTTAAATCTGTA
             M  S  K  N  K  D  Q  R  T  A  K  T  L  E  R  T  W  D  T  L  N  H  L  L  F  I  S  S  C  L  Y  K  L  N  L  K  S  V
GLYCOPROTEIN ▶
      4810      4820      4830      4840      4850      4860      4870      4880      4890      4900      4910      4920
GCACAAAATCACATTATCCATTCTGGCAATGATAATCTCAACTTCACTTATAATTGCAGCCATCATATTCATAGCCTCGGCAAACCACAAAGTCACACCAACAACTGCAATCATACAAGAT
A  Q  I  T  L  S  I  L  A  M  I  I  S  T  S  L  I  I  A  A  I  I  F  I  A  S  A  N  H  K  V  T  P  T  T  A  I  I  Q  D

      4930      4940      4950      4960      4970      4980      4990      5000      5010      5020      5030      5040
GCAACAAGCCAGATCAAGAACACAACCCCAACATACCTCACCCAGAATCCTCAGCTTGGAATCAGTCCCTCTAATCCGTCTGAAATTACATCACAAATCACCACCATACTAGCTTCAACA
A  T  S  Q  I  K  N  T  T  P  T  Y  L  T  Q  N  P  Q  L  G  I  S  P  S  N  P  S  E  I  T  S  Q  I  T  T  I  L  A  S  T

      5050      5060      5070      5080      5090      5100      5110      5120      5130      5140      5150      5160
ACACCAGGAGTCAAGTCAACCCTGCAATCCAACAGTCAAGACCAAAAACACAACAACAACTCAAACACAACCCAGCAAGCCCACCACAAAACAACGCCAAAACAAACCACCAAGCAAA
T  P  G  V  K  S  T  L  Q  S  T  T  V  K  T  K  N  T  T  T  T  Q  T  Q  P  S  K  P  T  T  K  Q  R  Q  N  K  P  P  S  K

      5170      5180      5190      5200      5210      5220      5230      5240      5250      5260      5270      5280
CCCAATAATGATTTTCACTTTGAAGTGTTCAACTTTGTACCCTGCAGCATATGCAGCAACAATCCAACCTGCTGGGCTATCTGCAAAAGAATACCAAACAAAAAACCAGGAAAGAAAACC
P  N  N  D  F  H  F  E  V  F  N  F  V  P  C  S  I  C  S  N  N  P  T  C  W  A  I  C  K  R  I  P  N  K  K  P  G  K  K  T

      5290      5300      5310      5320      5330      5340      5350      5360      5370      5380      5390      5400
ACTACCAAGCCCACAAAAAAACCAACCCTCAAGACAACCAAAAAAGATCCCAAACCTCAAACCACTAAATCAAAGGAAGTACCCACCACCAAGCCCACAGAAGAGCCAACCATCAACACC
T  T  K  P  T  K  K  P  T  L  K  T  T  K  K  D  P  K  P  Q  T  T  K  S  K  E  V  P  T  T  K  P  T  E  E  P  T  I  N  T

      5410      5420      5430      5440      5450      5460      5470      5480      5490      5500      5510      5520
ACCAAAACAAACATCATAACTACACTACTCACCTCCAACACCACAGGAAATCCAGAACTCACAAGTCAAATGGAAACCTTCCACTCAACTTCCTCCGAAGGCAATCCAAGCCCTTCTCAA
T  K  T  N  I  I  T  T  L  L  T  S  N  T  T  G  N  P  E  L  T  S  Q  M  E  T  F  H  S  T  S  S  E  G  N  P  S  P  S  Q

      5530      5540      5550      5560      5570      5580      5590      5600      5610      5620      5630      5640
GTCTCTACAACATCCGAGTACCCATCACAACCTTCATCTCCACCCAACACCACCACGCCAGTAGTTACTTAAAAACATATTATCACAAAAAGCCATGACCAACTTAAACAGAATCAAATA
V  S  T  T  S  E  Y  P  S  Q  P  S  S  P  P  N  T  P  R  Q                T_stop                        I

      5650      5660      5670      5680      5690      5700      5710      5720      5730      5740      5750      5760
AACTCTGGGGCAAATAACAATGGAGTTGCTAATCCTCAAAGCAAATGCAATTACCACAATCCTCACTGCAGTCACATTTTGTTTTGCTTCTGGTCAAACATCACTGAAGAATTTTATCA
                   T_start  FUSION PROTEIN ▶   M  E  L  L  I  L  K  A  N  A  I  T  T  I  L  T  A  V  T  F  C  F  A  S  G  Q  N  I  T  E  E  F  Y  Q

      5770      5780      5790      5800      5810      5820      5830      5840      5850      5860      5870      5880
ATCAACATGCAGTGCAGTTAGCAAAGGCTATCTTAGTGCTCTGAGAACTGGTTGGTATACCAGTGTTATAACTATAGAATTAAGTAATATCAAGGAAAATAAGTGTAATGGAACAGATGC
S  T  C  S  A  V  S  K  G  Y  L  S  A  L  R  T  G  W  Y  T  S  V  I  T  I  E  L  S  N  I  K  E  N  K  C  N  G  T  D  A

      5890      5900      5910      5920      5930      5940      5950      5960      5970      5980      5990      6000
TAAGGTAAAATTGATAAAACAAGAATTAGATAAATATAAAAATGCTGTAACAGAATTGCAGTTGCTCATGCAAAGCACACCACCAACAAACAATCGAGCCAGAAGAGAACTACCAAGGTT
K  V  K  L  I  K  Q  E  L  D  K  Y  K  N  A  V  T  E  L  Q  L  L  M  Q  S  T  P  P  T  N  N  R  A  R  R  E  L  P  R  F

      6010      6020      6030      6040      6050      6060      6070      6080      6090      6100      6110      6120
TATGAATTATACACTCAACAATGCCAAAAAAACCAATGTAACATTAAGCAAGAAAAGGAAAAGAAGATTTCTTGGTTTTTTGTTAGGTGTTGGATCTGCAATCGCCAGTGGCGTTGCTGT
M  N  Y  T  L  N  N  A  K  K  T  N  V  T  L  S  K  K  R  K  R  R  F  L  G  F  L  L  G  V  G  S  A  I  A  S  G  V  A  V

      6130      6140      6150      6160      6170      6180      6190      6200      6210      6220      6230      6240
ATCTAAGGTCCTGCACCTAGAAGGGGAAGTGAACAAGATCAAAAGTGCTCTACTATCCACAAACAAGGCTGTAGTCAGCTTATCAAATGGAGTTAGTGTCTTAACCAGCAAAGTGTTAGA
S  K  V  L  H  L  E  G  E  V  N  K  I  K  S  A  L  L  S  T  N  K  A  V  V  S  L  S  N  G  V  S  V  L  T  S  K  V  L  D

      6250      6260      6270      6280      6290      6300      6310      6320      6330      6340      6350      6360
CCTCAAAAACTATATAGATAAACAATTGTTACCTATTGTGAACAAGCAAAGCTGCAGCATATCAAATATAGAAACTGTGATAGAGTTCCAACAAAAGAACAACAGACTACTAGAGATTAC
L  K  N  Y  I  D  K  Q  L  L  P  I  V  N  K  Q  S  C  S  I  S  N  I  E  T  V  I  E  F  Q  Q  K  N  N  R  L  L  E  I  T

      6370      6380      6390      6400      6410      6420      6430      6440      6450      6460      6470      6480
CAGGGAATTTAGTGTTAATGCAGGTGTAACTACACCTGTAAGCACTTACATGTTAACTAATAGTGAATTATTGTCATTAATCAATGATATGCCTATAACAAATGATCAGAAAAAGTTAAT
R  E  F  S  V  N  A  G  V  T  T  P  V  S  T  Y  M  L  T  N  S  E  L  L  S  L  I  N  D  M  P  I  T  N  D  Q  K  K  L  M

      6490      6500      6510      6520      6530      6540      6550      6560      6570      6580      6590      6600
GTCCAACAATGTTCAAAATAGTTAGACAGCAAAGTTACTCTATCATGTCCATAATAAAAGAGGAAGTCTTAGCATATGTAGTACAATTACCACTATATGGTGTTATAGATACACCCTGTTG
S  N  N  V  Q  I  V  R  Q  Q  S  Y  S  I  M  S  I  I  K  E  E  V  L  A  Y  V  V  Q  L  P  L  Y  G  V  I  D  T  P  C  W

      6610      6620      6630      6640      6650      6660      6670      6680      6690      6700      6710      6720
GAAACTACACACATCCCCTCTATGTACAACCAACACAAAAGAAGGGTCCAACATCTGTTTAACAAGAACTGACAGAGGATGGTACTGTGACAATGCAGGATCAGTATCCTTCTTCCCACA
K  L  H  T  S  P  L  C  T  T  N  T  K  E  G  S  N  I  C  L  T  R  T  D  R  G  W  Y  C  D  N  A  G  S  V  S  F  F  P  Q
```

RESPIRATORY SYNCYTIAL VIRUS

```
      6730      6740      6750      6760      6770      6780      6790      6800      6810      6820      6830      6840
AGCTGAAACATGTAAAGTTCAATCAAATCGAGTATTTTGTGACACAATGAACAGTTTAACATTACCAAGTGAAATAAATCTCTGCAATGTTGACATATTCAACCCCAAATATGATTGTAA
  A  E  T  C  K  V  Q  S  N  R  V  F  C  D  T  M  N  S  L  T  L  P  S  E  I  N  L  C  N  V  D  I  F  N  P  K  Y  D  C  K

      6850      6860      6870      6880      6890      6900      6910      6920      6930      6940      6950      6960
AATTATGACTTCAAAAACAGATGTAAGCAGCTCCGTTATCACATCTCTAGGAGCCATTGTGTCATGCTATGGCAAAACTAAATGTACAGCATCCAATAAAAATCGTGGAATCATAAAGAC
  I  M  T  S  K  T  D  V  S  S  S  V  I  T  S  L  G  A  I  V  S  C  Y  G  K  T  K  C  T  A  S  N  K  N  R  G  I  I  K  T

      6970      6980      6990      7000      7010      7020      7030      7040      7050      7060      7070      7080
ATTTTCTAACGGGTGCGATTATGTATCAAATAAAGGGATGGACACTGTGTCTGTAGGTAACACATTATATTATGTAAATAAGCAAGAAGGTAAAGTCTCTATGTAAAGGTGAACCAAT
  F  S  N  G  C  D  Y  V  S  N  K  G  M  D  T  V  S  V  G  N  T  L  Y  Y  V  N  K  Q  E  G  K  S  L  Y  V  K  G  E  P  I

      7090      7100      7110      7120      7130      7140      7150      7160      7170      7180      7190      7200
AATAAATTTCTATGACCCATTAGTTATTCCCCTCTGATGAATTTGATGCATCAATATCTCAAGTCAACGAGAAGATTAACCAGAGCCTAGCATTTATTCGTAAATCCGATGAATTATTACA
  I  N  F  Y  D  P  L  V  F  P  S  D  E  F  D  A  S  I  S  Q  V  N  E  K  I  N  Q  S  L  A  F  I  R  K  S  D  E  L  L  H

      7210      7220      7230      7240      7250      7260      7270      7280      7290      7300      7310      7320
TAATGTAAATGCTGGTAAATCCACCACAAATATCATGATAACTACTAATAATTATAGTGATTATAGTAATATTGTTATCATTAATTGCTGTTGGACTGCTCTTATACTGTAAGGCCAGAAG
  N  V  N  A  G  K  S  T  T  N |I  M  I  T  T  I  I  I  V  I  I  V  I  L  L  S  L  I  A  V  G  L  L  L  Y  C| K  A  R  S

      7330      7340      7350      7360      7370      7380      7390      7400      7410      7420      7430      7440
CACACCAGTCACACTAAGCAAAGATCAACTGAGTGGTATAAATAATTGCATTTAGTAACTAAATAAAAATAGCACCTAATCATGTTCTTACAATGGTTTACTATCTGCTCATAGACAA
  T  P  V  T  L  S  K  D  Q  L  S  G  I  N  N  I  A  F  S  N  *

      7450      7460      7470      7480      7490      7500      7510      7520      7530      7540      7550      7560
CCCATCTGTCATTGGATTTTCTTAAAATCTGAACTTCATCGAAACTCTCATCTATAAACCATCTCACTTACACTATTTAAGTAGATTCCTAGTTTATAGTTATATAAAACACAATTGAAT
                                                                                            ▔▔▔▔▔▔▔▔▔▔▔▔▔▔▔▔
                                                                                                   ᵀstop
      7570      7580      7590      7600      7610      7620      7630      7640      7650      7660      7670      7680
GCCAGATTAACTTACCATCTGTAAAAATGAAAACTGGGCAAATATGTCACGAAGGAATCCTTGCAAATTTGAAATTCGAGGTCATTGCTTAAATGGTAAGAGGTGTCATTTTAGTCATA
════════════════════════════════════════════════════════════════  M  S  R  R  N  P  C  K  F  E  I  R  G  H  C  L  N  G  K  R  C  H  F  S  H
             I                               ᵀstart      22K PROTEIN ▶
      7690      7700      7710      7720      7730      7740      7750      7760      7770      7780      7790      7800
ATTATTTTGAATGGCCACCCCATGCACTGCTTGTAAGACAAAACTTTATGTTTAAACAGAATACTTAAGTCTATGGATAAAAGTATAGATACCTTATCAGAAATAAGTGGAGCTGCAGAGT
  N  Y  F  E  W  P  P  H  A  L  L  V  R  Q  N  F  M  L  N  R  I  L  K  S  M  D  K  S  I  D  T  L  S  E  I  S  G  A  A  E

      7810      7820      7830      7840      7850      7860      7870      7880      7890      7900      7910      7920
TGGACAGAACAGAAGAGTATGCTCTTGGTGTAGTTGGAGTGCTAGAGAGTTATATAGGATCAATAAACAATATAACTAAACAATCAGCATGTGTTGCCATGAGCAAACTCCTCACTGAAC
  L  D  R  T  E  E  Y  A  L  G  V  V  G  V  L  E  S  Y  I  G  S  I  N  N  I  T  K  Q  S  A  C  V  A  M  S  K  L  L  T  E

      7930      7940      7950      7960      7970      7980      7990      8000      8010      8020      8030      8040
TCAATAGTGATGATATCAAAAAGCTGAGGGACAATGAAGAGCTAAATTCACCCAAGATAAGAGTGTACAATACTGTCATATCATATATTGAAAGCAACAGGAAAAACAATAAACAAACTA
  L  N  S  D  D  I  K  K  L  R  D  N  E  E  L  N  S  P  K  I  R  V  Y  N  T  V  I  S  Y  I  E  S  N  R  K  N  N  K  Q  T

      8050      8060      8070      8080      8090      8100      8110      8120      8130      8140      8150      8160
TCCATCTGTTAAAAAGATTGCCAGCAGACGTATTGAAGAAAACCATCAAAAACACATTGGATATCCATAAGAGCATAACCATCAACAACCCAAAAGAATCAACTGTTAGTGATACAAATG
  I  H  L  L  K  R  L  P  A  D  V  L  K  K  T  I  K  N  T  L  D  I  H  K  S  I  T  I  N  N  P  K  E  S  T  V  S  D  T  N

      8170      8180      8190      8200      8210      8220      8230      8240      8250      8260      8270      8280
ACCATGCCAAAAATAATGATACTACCTGACAAATATCCTTGTAGTATAACTTCCATACTAATAACAAGTAGATGTGGAGTTACTATGTATAATCAAAAGAACACACTATATTTCAATCAA
  D  H  A  K  N  N  D  T  T  *

      8290      8300      8310      8320      8330      8340      8350      8360      8370      8380      8390      8400
AACAACCCAAAATAACCATATGTACTCACCGAATCAAACATTCAATGAAATCCATTGGACCTCTCAAGAATTGATTGACACAATTCAAAATTTTCTACAACATCTAGGTATTATTGAGGAT

      8410      8420      8430      8440      8450      8460      8470      8480      8490      8500      8510      8520
ATATATACAATATATATATTAGTGTCATAACACTCAATTCTAACACTCACCACATCGTTACATTATTAATTCAAACAATTCAAGTTGTGGGACAAAATGGATCCCATTATTAATGGAAAT
                                                                                      ▔▔▔▔▔▔▔▔▔▔▔▔
                                                                                            ᵀstart    M  D  P  I  I  N  G  N  ▶.
                                                                                                   L PROTEIN ▶.
      8530      8540       8550      8560      8570      8580      8590      8600      8610      8620      8630      8640
               ᵀstop    22K PROTEIN
TCTGCTAATGTTTATCTAACCGATAGTTATTTAAAAGGTGTTATCTCTTTCTCAGAGTGTAATGCTTTAGGAAGTTACATATTCAATGGTCCTTATCTCAAAAATGATTATACCAACTTA
  S  A  N  V  Y  L  T  D  S  Y  L  K  G  V  I  S  F  S  E  C  N  A  L  G  S  Y  I  F  N  G  P  Y  L  K  N  D  Y  T  N  L

      8650      8660      8670      8680      8690      8700      8710      8720      8730      8740      8750      8760
ATTAGTAGACAAAATCCATTAATAGAACACATGAATCTAAAGAAACTAAATATAACACAGTCCTTAATATCTAAGTATCATAAAGGTGAAATAAAATTAGAAGAACCTACTTATTTTCAG
  I  S  R  Q  N  P  L  I  E  H  M  N  L  K  K  L  N  I  T  Q  S  L  I  S  K  Y  H  K  G  E  I  K  L  E  E  P  T  Y  F  Q

      8770      8780      8790      8800      8810      8820      8830      8840      8850      8860      8870      8880
TCATTACTTATGACATACAAGAGTGACCTCGTCAGAACAGATTGCTACCACTAATTTACTTAAAAGATAATAAGAGAGCTATAGAAATAAGTGATGTCAAAGTCTATGCTATATTG
  S  L  L  M  T  Y  K  S  M  T  S  S  E  Q  I  A  T  T  N  L  L  K  K  I  I  R  R  A  I  E  I  S  D  V  K  V  Y  A  I  L

      8890      8900      8910      8920      8930      8940      8950      8960      8970      8980      8990      9000
AATAAACTAGGGCTTAAAGAAAAAGGACAAGATTAAATCCAACAATGGACAAGATGAAGACAACTCAGTTATTACGACCATAATCAAAGATGATATACTTTCAGCTGTTAAAGATAATCAA
  N  K  L  G  L  K  E  K  D  K  I  K  S  N  N  G  Q  D  E  D  N  S  V  I  T  T  I  I  K  D  D  I  L  S  A  V  K  D  N  Q
```

```
      9010      9020      9030      9040      9050      9060      9070      9080      9090      9100      9110      9120
TCTCATCTTAAAGCAGACAAAAATCACTCTACAAAACAAAAAGACACAATCAAAACAACACTCTTGAAGAAATTGATGTGTTCAATGCAACATCCTCCATCATGGTTAATACATTGGTTT
 S  H  L  K  A  D  K  N  H  S  T  K  Q  K  D  T  I  K  T  T  L  L  K  K  L  M  C  S  M  Q  H  P  P  S  W  L  I  H  W  F
```

```
      9130      9140      9150      9160      9170      9180      9190      9200      9210      9220      9230      9240
AACTTATACACAAAATTAAACAACATATTAACACAGTATCGATCAAATGAGGTAAAAAACCATGGGTTTACATTGATAGATAATCAAACTCTTAGTGGATTTCAATTTATTTTGAACCAA
 N  L  Y  T  K  L  N  N  I  L  T  Q  Y  R  S  N  E  V  K  N  H  G  F  T  L  I  D  N  Q  T  L  S  G  F  Q  F  I  L  N  Q
```

```
      9250      9260      9270      9280      9290      9300      9310      9320      9330      9340      9350      9360
TATGGTTGTATAGTTTATCATAAGGAACTCAAAAGAATTACTGTGACAACCTATAATCAATTCTTGACATGGAAAGATATTAGCCTTAGTAGATTAAATGTTTGTTTAATTACATGGATT
 Y  G  C  I  V  Y  H  K  E  L  K  R  I  T  V  T  T  Y  N  Q  F  L  T  W  K  D  I  S  L  S  R  L  N  V  C  L  I  T  W  I
```

```
      9370      9380      9390      9400      9410      9420      9430      9440      9450      9460      9470      9480
AGTAACTGCTTGAACACATTAAATAAAAGCTTAGGCTTAAGATGCGGATTCAATAATGTTATCTTGACACAACTATTCCTTTATGGAGATTGTATACTAAAGCTATTTCACAATGAGGGG
 S  N  C  L  N  T  L  N  K  S  L  G  L  R  C  G  F  N  N  V  I  L  T  Q  L  F  L  Y  G  D  C  I  L  K  L  F  H  N  E  G
```

```
      9490      9500      9510      9520      9530      9540      9550      9560      9570      9580      9590      9600
TTCTACATAATAAAAGAGGTAGAGGGATTTATTATGTCTCTAATTTTAAATATAACAGAAGAAGATCAATTCAGAAAACGATTTTATAATAGTATGCTCAACAACATCCACAGATGCTGCT
 F  Y  I  I  K  E  V  E  G  F  I  M  S  L  I  L  N  I  T  E  E  D  Q  F  R  K  R  F  Y  N  S  M  L  N  N  I  T  D  A  A
```

```
      9610      9620      9630      9640      9650      9660      9670      9680      9690      9700      9710      9720
AATAAAGCTCAGAAAAATCTGCTATCAAGAGTATGTCATACATTATTAGATAAGACAGTGTCCGATAATAATAAATGGCAGATGGATAATTCTATTAAGTAAGTTCCTTAAATTAATT
 N  K  A  Q  K  N  L  L  S  R  V  C  H  T  L  L  D  K  T  V  S  D  N  I  I  N  G  R  W  I  I  L  L  S  K  F  L  K  L  I
```

```
      9730      9740      9750      9760      9770      9780      9790      9800      9810      9820      9830      9840
AAGCTTGCAGGTGACAATAACCTTAACAATCTGAGTGAACTATATTTTTTGTTCAGAATATTTGGACACCCAATGGTAGATGAAAGACAAGCCATGGATGCTGTTAAAATTAATTGCAAT
 K  L  A  G  D  N  N  L  N  N  L  S  E  L  Y  F  L  F  R  I  F  G  H  P  M  V  D  E  R  Q  A  M  D  A  V  K  I  N  C  N
```

```
      9850      9860      9870      9880      9890      9900      9910      9920      9930      9940      9950      9960
GAGACCAAATTTTACTTGTTAAGCAGTCTGAGTATGTTAAGAGGTGCCTTTATATATAGAATTATAAAAGGGTTTGTAAATAATTACAACAGATGGCCTACTTTAAGAAATGCTATTGTT
 E  T  K  F  Y  L  L  S  S  L  S  M  L  R  G  A  F  I  Y  R  I  I  K  G  F  V  N  N  Y  N  R  W  P  T  L  R  N  A  I  V
```

```
      9970      9980      9990     10000     10010     10020     10030     10040     10050     10060     10070     10080
TTACCCTTAAGATGGTTAACTTACTATAAACTAAACACTTATCCTTCTTTGTTGGAACTTACAGAAAGAGATTTGATTGTGTTATCAGGACTACGTTTCTATCGTGAGTTTCGGTTGCCT
 L  P  L  R  W  L  T  Y  Y  K  L  N  T  Y  P  S  L  L  E  L  T  E  R  D  L  I  V  L  S  G  L  R  F  Y  R  E  F  R  L  P
```

```
     10090     10100     10110     10120     10130     10140     10150     10160     10170     10180     10190     10200
AAAAAAGTGGATCTTGAAATGATTATAAATGATAAAGCTATATCACCTCCTAAAAATTTGATATGGACTAGTTTCCCTAGAAATTACATGCCATCACACATACAAAACTATATAGAACAT
 K  K  V  D  L  E  M  I  I  N  D  K  A  I  S  P  P  K  N  L  I  W  T  S  F  P  R  N  Y  M  P  S  H  I  Q  N  Y  I  E  H
```

```
     10210     10220     10230     10240     10250     10260     10270     10280     10290     10300     10310     10320
GAAAAATTAAAATTTTCCGAGAGTGATAAATCAAGAAGAGTATTAGAGTATTATTTAAGAGATAACAAATTCAATGAATGTGATTTATACAACTGTGTAGTTAATCAAAGTTATCTCAAC
 E  K  L  K  F  S  E  S  D  K  S  R  R  V  L  E  Y  Y  L  R  D  N  K  F  N  E  C  D  L  Y  N  C  V  V  N  Q  S  Y  L  N
```

```
     10330     10340     10350     10360     10370     10380     10390     10400     10410     10420     10430     10440
AACCCCTAATCATGTGGTATCATTGACAGGCAAAGAAAGAGAACTCAGTGTAGGTAGAATGTTTGCAATGCAACCGGGAATGTTCAGACAGGTTCAAATATTGCAGAGAAAATGATAGCT
 N  P  N  H  V  V  S  L  T  G  K  E  R  E  L  S  V  G  R  M  F  A  M  Q  P  G  M  F  R  Q  V  Q  I  L  A  E  K  M  I  A
```

```
     10450     10460     10470     10480     10490     10500     10510     10520     10530     10540     10550     10560
GAAAAACATTTTACAATTCTTTCCTGAAAGTCTTACAAGATATGGTGATCTAGAACTACAAAAAATATTAGAATTGAAAGCAGGAATAAGTAACAAATCAAATCGCTACAATGATAATTAC
 E  N  I  L  Q  F  F  P  E  S  L  T  R  Y  G  D  L  E  L  Q  K  I  L  E  L  K  A  G  I  S  N  K  S  N  R  Y  N  D  N  Y
```

```
     10570     10580     10590     10600     10610     10620     10630     10640     10650     10660     10670     10680
AACAATTACATTAGTAAGTGCTCTATCATCACAGATCTCAGCAAATTCAATCAAGCATTTCGATATGAAACGTCATGTATTTGTAGTGATGTGCTGGATGAACTGCATGGTGTACAATCT
 N  N  Y  I  S  K  C  S  I  I  T  D  L  S  K  F  N  Q  A  F  R  Y  E  T  S  C  I  C  S  D  V  L  D  E  L  H  G  V  Q  S
```

```
     10690     10700     10710     10720     10730     10740     10750     10760     10770     10780     10790     10800
CTATTTTCCTGGTTACATTTAACTATTCCTCATGTCACAATAATATGCACATATAGGCATGCACCCCCCTATATAGGAGATCATATTGTAGATCTTAACAATGTAGATGAACAAAGTGGA
 L  F  S  W  L  H  L  T  I  P  H  V  T  I  I  C  T  Y  R  H  A  P  P  Y  I  G  D  H  I  V  D  L  N  N  V  D  E  Q  S  G
```

```
     10810     10820     10830     10840     10850     10860     10870     10880     10890     10900     10910     10920
TTATATAGATATACACATGGGTGGCATCGAAGGGTGGTGTCAAAAACTGTGGACCATAGAAGCTATATCACTATTGGATCTAATATCTCTCAAAGGGAAATTCTCAATTACTGCTTTAATT
 L  Y  R  Y  H  M  G  G  I  E  G  W  C  Q  K  L  W  T  I  E  A  I  S  L  L  D  L  I  S  L  K  G  K  F  S  I  T  A  L  I
```

```
     10930     10940     10950     10960     10970     10980     10990     11000     11010     11020     11030     11040
AATGGTGACAATCAATCAATAGATATAAGCAAACCAATCAGACTCATGGAAGGTCAAACTCATGCTCAAGCAGATTATTTGCTAGCATTAAATAGCCTTAAATTACTGTATAAAGAGTAT
 N  G  D  N  Q  S  I  D  I  S  K  P  I  R  L  M  E  G  Q  T  H  A  Q  A  D  Y  L  L  A  L  N  S  L  K  L  L  Y  K  E  Y
```

```
     11050     11060     11070     11080     11090     11100     11110     11120     11130     11140     11150     11160
GCAGGCATAGGCCACAAATTAAAAGGAACTGAGACTTATATCACGAGATATGCAATTTATGAGTAAAACAATCAACATAACGGTGTATATTACCCAGCTAGTATAAAGAAAGTCCTA
 A  G  I  G  H  K  L  K  G  T  E  T  Y  I  S  R  D  M  Q  F  M  S  K  T  I  Q  H  N  G  V  Y  Y  P  A  S  I  K  K  V  L
```

```
     11170     11180     11190     11200     11210     11220     11230     11240     11250     11260     11270     11280
AGAGTGGGACCGTGGATAAACACTATACTTGATGATTTCAAAGTGAGTCTAGAATCTATAGGTAGTTTGCACACAAGAATTAGAATATAGAGGTGAAAGTCTATTATGCAGTTTAATATTT
 R  V  G  P  W  I  N  T  I  L  D  D  F  K  V  S  L  E  S  I  G  S  L  T  Q  E  L  E  Y  R  G  E  S  L  L  C  S  L  I  F
```

RESPIRATORY SYNCYTIAL VIRUS

```
      11290      11300      11310      11320      11330      11340      11350      11360      11370      11380      11390      11400
AGAAATGTATGGTTATATAATCAGATTGCTCTACAATTAAAAAATCATGCATTATGTAACAATAAACTATATTTGGACATATTAAAGGTTCTGAAACACTTAAAAACCTTTTTTAATCTT
  R   N   V   W   L   Y   N   Q   I   A   L   Q   L   K   N   H   A   L   C   N   N   K   L   Y   L   D   I   L   K   V   L   K   H   L   K   T   F   F   N   L

      11410      11420      11430      11440      11450      11460      11470      11480      11490      11500      11510      11520
GATAATATTGATACAGCATTAACATTGTATATGAATTTACCCATGTTATTTGGTGGTGGTGATCCCAACTTGTTATATCGAAGTTTCTATAGAAGAACTCCTGACTTCCTCACAGAGGCT
  D   N   I   D   T   A   L   T   L   Y   M   N   L   P   M   L   F   G   G   G   D   P   N   L   L   Y   R   S   F   Y   R   R   T   P   D   F   L   T   E   A

      11530      11540      11550      11560      11570      11580      11590      11600      11610      11620      11630      11640
ATAGTTCACTCTGTGTTCATACTTAGTTATTATACAAACCATGACTTAAAAGATAAACTTCAAGATCTGTCAGATGATAGATTGAATAAGTTCTTAACATGCATAATCACGTTTGACAAA
  I   V   H   S   V   F   I   L   S   Y   Y   T   N   H   D   L   K   D   K   L   Q   D   L   S   D   D   R   L   N   K   F   L   T   C   I   I   T   F   D   K

      11650      11660      11670      11680      11690      11700      11710      11720      11730      11740      11750      11760
AACCCTAATGCTGAATTCGTAACATTGATGAGAGATCCTCAAGCTTTAGGGTCTGAGAGACAAGCTAAAATTACTAGCGAAATCAATAGACTGGCAGTTACAGAGGTTTTGAGTACAGCT
  N   P   N   A   E   F   V   T   L   M   R   D   P   Q   A   L   G   S   E   R   Q   A   K   I   T   S   E   I   N   R   L   A   V   T   E   V   L   S   T   A

      11770      11780      11790      11800      11810      11820      11830      11840      11850      11860      11870      11880
CCAAACAAAATATTCTCCAAAAGTGCACAACATTATACTACTACAGAGATAGATCTAAATGATATTATGCAAAATATAGAACCTACATATCCTCATGGGCTAAGAGTTGTTTATGAAAGT
  P   N   K   I   F   S   K   S   A   Q   H   Y   T   T   T   E   I   D   L   N   D   I   M   Q   N   I   E   P   T   Y   P   H   G   L   R   V   V   Y   E   S

      11890      11900      11910      11920      11930      11940      11950      11960      11970      11980      11990      12000
TTACCCTTTTATAAAGCAGAGAAAATAGTAAATCTTATATCAGGTACAAATCTATAACTAACATACTGGAAAAAACTTCTGCCATAGACTTAACAGATATTGATAGAGCCACTGAGATG
  L   P   F   Y   K   A   E   K   I   V   N   L   I   S   G   T   K   S   I   T   N   I   L   E   K   T   S   A   I   D   L   T   D   I   D   R   A   T   E   M

      12010      12020      12030      12040      12050      12060      12070      12080      12090      12100      12110      12120
ATGAGGAAAAACATAACTTTGCTTATAAGGATACTTCCATTGGATTGTAACAGAGATAAAAGAGAGATATTGAGTATGGAAAACCTAAGTATTACTGAATTAAGCAAATATGTTAGGGAA
  M   R   K   N   I   T   L   L   I   R   I   L   P   L   D   C   N   R   D   K   R   E   I   L   S   M   E   N   L   S   I   T   E   L   S   K   Y   V   R   E

      12130      12140      12150      12160      12170      12180      12190      12200      12210      12220      12230      12240
AGATCTTGGTCTTTATCCAATATAGTTGGTGTTACATCACCCAGTATCATGTATACAATGGACATCAAATATACTACAAGCACTATATCTAGTGGCATAATTATAGAGAAATATAATGTT
  R   S   W   S   L   S   N   I   V   G   V   T   S   P   S   I   M   Y   T   M   D   I   K   Y   T   T   S   T   I   S   S   G   I   I   I   E   K   Y   N   V

      12250      12260      12270      12280      12290      12300      12310      12320      12330      12340      12350      12360
AACAGTTTAACACGTGGTGAGAGAGGACCCACTAAACCATGGGTTGGTTCATCTACACAAGAGAAAAAACAATGCCAGTTTATAATAGACAAGTCTTAACCAAAAAACAGAGAGATCAA
  N   S   L   T   R   G   E   R   G   P   T   K   P   W   V   G   S   S   T   Q   E   K   K   T   M   P   V   Y   N   R   Q   V   L   T   K   K   Q   R   D   Q

      12370      12380      12390      12400      12410      12420      12430      12440      12450      12460      12470      12480
ATAGATCTTATTAGCAAAATTGGATTGGGTGTATGCATCTATAGATAACAAGGATGAATTCATGGAAGAACTCAGCATAGGAACCCTTGGGTTAACATATGAAAAGGCCAAGAAATTATTT
  I   D   L   L   A   K   L   D   W   V   Y   A   S   I   D   N   K   D   E   F   M   E   E   L   S   I   G   T   L   G   L   T   Y   E   K   A   K   K   L   F

      12490      12500      12510      12520      12530      12540      12550      12560      12570      12580      12590      12600
CCACAATATTTAAGTGTCAATTATTTGCATCGCCTTACAGTCAGTAGTAGACCATGTGAATTCCCTGCATCAATACCAGCTTATAGAACAACAAATTATCACTTTGACACTAGCCCTATT
  P   Q   Y   L   S   V   N   Y   L   H   R   L   T   V   S   S   R   P   C   E   F   P   A   S   I   P   A   Y   R   T   T   N   Y   H   F   D   T   S   P   I

      12610      12620      12630      12640      12650      12660      12670      12680      12690      12700      12710      12720
AATCGCATATTAACAGAAAAGTATGGTGATGAAGATATTGACATAGTATTCCAAAACTGTATAAGCTTTGGCCTTAGTTTAATGTCAGTAGTAGAACAATTTACTAATGTATGTCCTAAC
  N   R   I   L   T   E   K   Y   G   D   E   D   I   D   I   V   F   Q   N   C   I   S   F   G   L   S   L   M   S   V   V   E   Q   F   T   N   V   C   P   N

      12730      12740      12750      12760      12770      12780      12790      12800      12810      12820      12830      12840
AGAATTATTCTCATACCTAAGCTTAATGAGATACATTTGATGAAACCTCCCATATTCACAGGTGATGTTGATATTCACAAGTTAAAACAAGTGATACAAAAACAGCATATGTTTTTACCA
  R   I   I   L   I   P   K   L   N   E   I   H   L   M   K   P   P   I   F   T   G   D   V   D   I   H   K   L   K   Q   V   I   Q   K   Q   H   M   F   L   P

      12850      12860      12870      12880      12890      12900      12910      12920      12930      12940      12950      12960
GACAAAATAAGTTTGACTCAATATGTGGAATTATTCTTAAGTAAACACTCAAATCTGGATCTCATGTTAATTCTAATTTAATATTGGCACATAAAATATCTGACTATTTTCATAAT
  D   K   I   S   L   T   Q   Y   V   E   L   F   L   S   N   K   T   L   K   S   G   S   H   V   N   S   N   L   I   L   A   H   K   I   S   D   Y   F   H   N

      12970      12980      12990      13000      13010      13020      13030      13040      13050      13060      13070      13080
ACTTACATTTTAAGTACTAATTTAGCTGGACATTGGATTCTGATTATACAACTTATGAAAGATTCTAAAGGTATTTTTGAAAAAGATTGGGGAGAGGGATATATAACTGATCATATGTTT
  T   Y   I   L   S   T   N   L   A   G   H   W   I   L   I   I   Q   L   M   K   D   S   K   G   I   F   E   K   D   W   G   E   G   Y   I   T   D   H   M   F

      13090      13100      13110      13120      13130      13140      13150      13160      13170      13180      13190      13200
ATTAATTTGAAAGTTTTCTTCAATGCTTATAAGACCTATCTCTTGTGTTTTCATAAAGGTTATGGCAAAGCAAAGCTGGAGTGTGATATGAACACTTCAGATCTTCTATGTGTATTGGAA
  I   N   L   K   V   F   F   N   A   Y   K   T   Y   L   L   C   F   H   K   G   Y   G   K   A   K   L   E   C   D   M   N   T   S   D   L   L   C   V   L   E

      13210      13220      13230      13240      13250      13260      13270      13280      13290      13300      13310      13320
TTAATAGACAGTAGTTATTGGAAGTCTATGTCTAAGGTATTTTTGAACAAAAAGTTATCAAATACATTCTTAGCCAAGATGCAAGTTTACATAGAGTAAAAGGATGTCATAGCTTCAAA
  L   I   D   S   S   Y   W   K   S   M   S   K   V   F   L   E   Q   K   V   I   K   Y   I   L   S   Q   D   A   S   L   H   R   V   K   G   C   H   S   F   K

      13330      13340      13350      13360      13370      13380      13390      13400      13410      13420      13430      13440
TTATGGTTTCTTAAACGTCTTAATGTAGCAGAATTCACAGTTTGCCCTTGGGTTGTTAACATAGATTATCATCCAACACATATGAAAGCAATATTAACTTATATAGATCTTGTTAGAATG
  L   W   F   L   K   R   L   N   V   A   E   F   T   V   C   P   W   V   V   N   I   D   Y   H   P   T   H   M   K   A   I   L   T   Y   I   D   L   V   R   M

      13450      13460      13470      13480      13490      13500      13510      13520      13530      13540      13550      13560
GGATTGATAAATATAGATAGGAATACACATTAAAAATAAACACAAATTCAATGATGAATTTTATACTTCTAATCTCTTCTACATTAATTATAACTTCTCAGATAATACTCATCTATTAACT
  G   L   I   N   I   D   R   I   H   I   K   N   K   H   K   F   N   D   E   F   Y   T   S   N   L   F   Y   I   N   Y   N   F   S   D   N   T   H   L   L   T
```

```
      13570     13580     13590     13600     13610     13620     13630     13640     13650     13660     13670     13680
AAACATATAAGGATTGCTAATTCTGAATTAGAAAATAATTACAACAAATTATATCATCCTACACCAGAAACCCTAGAGAATATACTAGCCAATCCGATTAAAAGTAATGACAAAAAGACA
  K  H  I  R  I  A  N  S  E  L  E  N  N  Y  N  K  L  Y  H  P  T  P  E  T  L  E  N  I  L  A  N  P  I  K  S  N  D  K  K  T

      13690     13700     13710     13720     13730     13740     13750     13760     13770     13780     13790     13800
CTGAATGACTATTGTATAGGTAAAAATGTTGACTCAATAATGTTACCATTGTTATCTAATAAGAAGCTTATTAAATCGTCTGCAATGATTAGAACCAATTACAGCAAACAAGATTTGTAT
  L  N  D  Y  C  I  G  K  N  V  D  S  I  M  L  P  L  L  S  N  K  K  L  I  K  S  S  A  M  I  R  T  N  Y  S  K  Q  D  L  Y

      13810     13820     13830     13840     13850     13860     13870     13880     13890     13900     13910     13920
AATTTATTCCCTATGGTTGTGATTGATAGAATTATAGATCATTCAGGCAATACAGCCAAATCCAACCAACTTTACACTACTACTTCCCACCAAATATCTTTAGTGCACAATAGCACATCA
  N  L  F  P  M  V  V  I  D  R  I  I  D  H  S  G  N  T  A  K  S  N  Q  L  Y  T  T  T  S  H  Q  I  S  L  V  H  N  S  T  S

      13930     13940     13950     13960     13970     13980     13990     14000     14010     14020     14030     14040
CTTTACTGCATGCTTCCTTGGCATCATATTAATAGATTCAATTTTGTATTTAGTGTTCTACAGGTTGTAAAATTAGTATAGAGTATATTTTAAAAGATCTTAAAATTAAAGATCCCAATTGT
  L  Y  C  M  L  P  W  H  H  I  N  R  F  N  F  V  F  S  S  T  G  C  K  I  S  I  E  Y  I  L  K  D  L  K  I  K  D  P  N  C

      14050     14060     14070     14080     14090     14100     14110     14120     14130     14140     14150     14160
ATAGCATTCATAGGTGAAGGAGCAGGGAATTTATTATTGCGTACAGTAGTGGAACTTCATCCTGACATAAGATATATTTACAGAAGTCTGAAAGATTGCAATGATCATAGTTTACCTATT
  I  A  F  I  G  E  G  A  G  N  L  L  L  R  T  V  V  E  L  H  P  D  I  R  Y  I  Y  R  S  L  K  D  C  N  D  H  S  L  P  I

      14170     14180     14190     14200     14210     14220     14230     14240     14250     14260     14270     14280
GAGTTTTTAAGGCTGTACAATGGACATATCAACATTGATTATGGTGAAAATTTGACCATTCCTGCTACAGATGCAACCAACAACATTCATTGGTCTTATTTACATATAAAAGTTTGCTGAA
  E  F  L  R  L  Y  N  G  H  I  N  I  D  Y  G  E  N  L  T  I  P  A  T  D  A  T  N  N  I  H  W  S  Y  L  H  I  K  F  A  E

      14290     14300     14310     14320     14330     14340     14350     14360     14370     14380     14390     14400
CCTATCAGTCTTTTTGTCTGTGATGCCGAATTGTCTGTAACAGTCAACTGGAGTAAAATTATAATAGAATGGAGCAAGCATGTAAGAAAGTGCAAGTACTGTTCCTCAGTTAATAAATGT
  P  I  S  L  F  V  C  D  A  E  L  S  V  T  V  N  W  S  K  I  I  I  E  W  S  K  H  V  R  K  C  K  Y  C  S  S  V  N  K  C

      14410     14420     14430     14440     14450     14460     14470     14480     14490     14500     14510     14520
ATGTTAATAGTAAAATATCATGCTCAAGATGATATTGATTTCAAATTAGACAATATAACTATATTAAAAACTTATGTATGCTTAGGCAGTAAGTTAAAGGGATCGGAGGTTTACTTAGTC
  M  L  I  V  K  Y  H  A  Q  D  D  I  D  F  K  L  D  N  I  T  I  L  K  T  Y  V  C  L  G  S  K  L  K  G  S  E  V  Y  L  V

      14530     14540     14550     14560     14570     14580     14590     14600     14610     14620     14630     14640
CTTTACAATAGGTCCTGCGAATATATTCCCAGTATTTAATGTAGTACAAAATGCTAAATTGATACTATCAAGAACCAAAAATTTCATCATGCCTAAGAAAGCTGATAAAGAGTCTATTGAT
  L  T  I  G  P  A  N  I  F  P  V  F  N  V  Q  N  A  K  L  I  L  S  R  T  K  N  F  I  M  P  K  K  A  D  K  E  S  I  D

      14650     14660     14670     14680     14690     14700     14710     14720     14730     14740     14750     14760
GCAAATATTAAAAGTTTGATACCCTTTCTTTGTTACCCTATAACAAAAAAGGAATTAATACTGCATTGTCAAAACTAAAGAGTGTTGTTAGTGGAGATATACTATCATATTCTATAGCT
  A  N  I  K  S  L  I  P  F  L  C  Y  P  I  T  K  K  G  I  N  T  A  L  S  K  L  K  S  V  V  S  G  D  I  L  S  Y  S  I  A

      14770     14780     14790     14800     14810     14820     14830     14840     14850     14860     14870     14880
GGACGTAATGAAGTTTTCAGCAATAAACTTATAAATCATAAGCATATGAACATCTTAAAATGGTTCAATCATGTTTTAAATTTCAGATCAACAGAACTAAACTATAACCATTTATATATG
  G  R  N  E  V  F  S  N  K  L  I  N  H  K  H  M  N  I  L  K  W  F  N  H  V  L  N  F  R  S  T  E  L  N  Y  N  H  L  Y  M

      14890     14900     14910     14920     14930     14940     14950     14960     14970     14980     14990     15000
GTAGAATCTACATATCCTTACCTAAGTGAATTGTTAAACAGCTTGACAACCAATGAACTTAAAAAACTGATTAAAATCACAGGTAGTCTGTTATACAACTTTCATAATGAATAATGAATA
  V  E  S  T  Y  P  Y  L  S  E  L  L  N  S  L  T  T  N  E  L  K  K  L  I  K  I  T  G  S  L  L  Y  N  F  H  N  E  *

      15010     15020     15030     15040     15050     15060     15070     15080     15090     15100     15110     15120
AAGATCTTTATAATAAAAAATTCCCATAGCTATACACTAACACTGTATTCAATTATAGTTGTTAAAAAATTAAAAATCATATAATTTTTTAAATAACTTTTAGTGAACTAATCCTAAAGTTAT
                                                                  T_stop                  EXTRACISTRONIC SEQUENCES
      15130     15140     15150     15160     15170     15180     15190     15200     15210     15220
CATTTTAATCTTGGAGGAATAAATTTAAACCCTAATCTAATTGGTTTATATGTGTATTAACTAAATTACGAGATATTAGTTTTTGACACTTTTTTTCTCGT
```

I. REFERENCES

Alkhatib, G., and Briedis, D. J., 1986, The predicted primary structure of the measles virus hemagglutinin, *Virology* **150**:479–490.

Banerjee, A. K., 1987, Transcription and replication of rhabdoviruses, *Microbiol. Rev.* **51**:66–87.

Bellini, W. J., Englund, G., Rozenblatt, S., Arnheiter, H., and Richardson, C. D., 1985, Measles virus P gene codes for two proteins, *J. Virol.* **53**:908–919.

Bellini, W. J., Englund, G. Richardson, C. D., Rozenblatt, S., and Lazzarini, R. A., 1986, Matrix genes of measles virus and canine distemper virus: Cloning, nucleotide sequences, and deduced amino acid sequences, *J. Virol.* **58**:408–416.

Blumberg, B. M., Rose, K., Simona, M. G., Roux, L., Giorgi, C., and Kolakofsky, D., 1984, Analysis of the Sendai virus M Gene and protein, *J. Virol.* **52:**656–663.

Blumberg, B. M., Giorgi, C., Rose, K., and Kolakofsky, D., 1985a, Sequence determination of the Sendai virus fusion protein gene, *J. Gen. Virol.* **66:**37–331.

Blumberg, B., Giorgi, C., Roux, L., Raju, R., Dowling, P., Chollet, A., and Kolakofshy, D., 1985b, Sequence determination of the Sendai virus HN gene and its comparison to the influenza virus glycoproteins, *Cell* **41:**269–278.

Blumberg, B. M., Crowley, J. C., Silverman, J. L., Menonna, J., Cook, S. D., and Dowling, P. E., 1988, Measles virus L protein evidences elements of ancestral RNA polymerase, *Virology* **164:**487–497.

Cattaneo, R., Kaelin, K., Baczko, K., and Billeter, M. A., 1989, Measles virus editing provides an additional cysteine-rich protein, *Cell* **56:**759–764.

Collins, P. L., and Wertz, G. W., 1985a, Nucleotide sequences of the 1B and 1C nonstructural protein mRNAs of human respiratory syncytial virus, *Virology* **143:**442–451.

Collins, P. L., and Wertz, G. W., 1985b, The envelope-associated 22K protein of human respiratory syncytial virus: Nucleotide sequence of the mRNA and a related polytranscript, *J. Virol.* **54:**65–71.

Collins, P. L., Huang, Y. T., and Wertz, G. W., 1984, Nucleotide sequence of the gene encoding the fusion (F) glycoprotein of human respiratory syncytial virus, *Proc. Natl. Acad. Sci. USA* **81:**7683–7687.

Collins, P. L., Anderson, K., Langer, S. L., and Wertz, G. W., 1985, Correct sequence for the major nucleocapsid protein mRNA of respiratory syncytial virus, *Virology* **146:**69–77.

Collins, P. L., Dickens, L. E., Buckler-White, A., Olmsted, R. A., Spriggs, M. K., Camargo, E., and Coelingh, K. V. W., 1986, Nucleotide sequences for the gene junctions of human respiratory syncytial virus reveal distinct features of intergenic structure and gene order, *Proc. Natl. Acad. Sci. USA* **83:**4594–4598.

Crowley, J. C., Dowling, P. C., Menonna, J., Silverman, J. I., Schuback, D., Cook, S. D., and Blumberg, B. M., 1988, Sequence variability and function of measles virus 3′ and 5′ ends and intercistronic regions, *Virology* **164:**498–506.

Curran, J. A., Richardson, C., and Kolakofsky, D., 1986, Robosomal initiation at alternate AUGs on the Sendai virus P/C mRNA, *J. Virol.* **57:**684–687.

Dimock, K., Rud, E. W., and Kang, C. Y., 1986, 3′-Terminal sequence of human parainfluenza virus 3 genomic RNA, *Nucleid Acids Res.* **14:**4694.

Elango, N., Satake, M., Coligan, J. E., Norrby, E., Camargo, E., and Venkatesan, S., 1985, Respiratory syncytial virus fusion glycoprotein: Nucleotide sequence of mRNA, identification of cleavage activation site and amino acid sequence of N-terminus of F1 subunit, *Nucleic Acids Res.* **13:**1559–1574.

Elango, N., Satake, M., and Venkatesan, S., 1985, mRNA sequence of three respiratory syncytial virus genes encoding two nonstructural proteins and a 22K structural protein, *J. Virol.* **55:**101–110.

Galinski, M. S., Mink, M. A., Lambert, D. M., Wechsler, S. L., and Pons, M. W., 1986a, Molecular cloning and sequence analysis of the human parainfluenza 3 virus RNA encoding the nucleocapsid protein, *Virology* **149:**139–151.

Galinski, M. S., Mink, M. A., Lambert, D. M., Wechsler, S. L., and Pons, M. W., 1986b, Molecular cloning and sequence analysis of the human parainfluenza 3 virus mRNA encoding the P and C proteins, *Virology* **154:**46–60.

Galinski, M. S., Mink, M. A., Lambert, D. M., Wechsler, S. L., and Pons, M. W., 1987a, Molecular cloning and sequence analysis of the human parainfluenza 3 virus gene encoding the matrix protein, *Virology* **152:**24–30.

Galinski, M. S., Mink, M. A., Lambert, D. M., Wechsler, S. L., and Pons, M. W., 1987b, Molecular cloning and sequence analysis of the human parainfluenza 3 virus genes encoding the surface glycoproteins, F and HN, *Virus Res.* **12:**169–180.

Galinski, M. S., Mink, M. A., and Pons, M. W., 1988, Molecular cloning and sequence analysis of the human parainfluenza 3 virus gene encoding the L protein, *Virology* **165:**499–510.

Giorgi, C., Blumberg, B. M., and Kolakofsky, D., 1983, Sendai virus contains overlapping genes expressed from a single mRNA, *Cell* **35:**829–836.

Gupta, K. C., and Kingsbury, D. W., 1984, Complete sequences of the intergenic and mRNA start signals in the Sendai virus genome: Homologies with the genome of vesicular stomatitis virus, *Nucleic Acids Res.* **12:**3829–3841.

Gupta, K. C., and Patwardhan, S., 1988, ACG, the initiator codon far a Sendai virus protein, *J. Biol. Chem.* **263:**8553–8556.

Hidaka, Y., Kanda, T., Iwasaki, K., Nomoto, A., Shioda, T., and Shibuta, H., 1984, Nucleotide sequence of a Sendai virus genome region covering the entire M gene and the 3′ proximal 1013 nucleotides of the F gene, *Nucleic Acids Res.* **12:**7965–7973.

Morgan, E. M., and Kingsbury, D. W., 1984, Complete sequence of the Sendai virus NP gene from a cloned insert, *Virology* **135:**279–287.

Patwardhan, S., and Gupta, K. C., 1988, Translation initiation potential of the 5′ proximal AUGs of the polycistronic P/C mRNA of Sendai virus, *J. Biol. Chem.* **263:**4907–4913.

Richardson, C., Hull, D., Greer, P., Hasel, K., Berkovich, A., Englund, G., Bellini, B., Rima, B., and Lazzarini, R., 1986, The nucleotide sequence of the mRNA encoding the fusion protein of measles virus (Edmonston strain): A comparison of fusion proteins from several different paramyxoviruses, *Virology* **155:**508–523.

Rozenblatt, S., Eizenberg, O., Ben-Levy, R., Lavie, V., and Bellini, W. J., 1985, Sequence homology within the morbilliviruses, *J. Virol.* **53:**684–690.

Satake, M., Coligan, J. E., Elango, N., Norrby, E., and Venkatesan, S., 1985, Respiratory syncytial virus envelope glycoprotein (G) has a novel structure, *Nucleic Acids Res.* **13:**7795–7812.

Shioda, T., Hidaka, Y., Kanda, T., Shibuta, H., Nomoto, A., and Iwasaki, K., 1983, Sequence of 3,687 nucleotides from the 3′ end of Sendai virus genome RNA and the predicted amino acid sequences of viral NP, P, and C proteins, *Nucleid Acids Res.* **11:**7217–7330.

Shioda, T., Iwasaki, K., and Shibuta, H., 1986, Determination of the complete nucleotide sequence of the Sendai virus genome RNA and the predicted acid sequences of the F, HN and L protein, *Nucleic Acids Res.* **14:**1545–1563.

Spriggs, M. K., and Collins, P. L., 1986a, Human parainfluenza virus type 3: Messenger RNAs, polypeptide coding assignments, intergenic sequences, and genetic map, *J. Virol.* **59:**646–654.

Thomas, S., Lamb, R. A., and Paterson, R. G., 1988, Two mRNAs that differ by two nontemplated nucleotides encode the amino coterminal proteins P and V of the paramyxovirus SV5, *Cell* **54:**891–902.

Wertz, G., Collins, P., Huang, Y., Gruber, C., Levine, S., and Ball, L., 1985, Nucleotide sequence of the G protein gene of human respiratory syncytial virus reveals an unusual type of membrane protein, *Proc. Natl. Acad. Sci. USA* **82:**4075–4079.

Index

A549 cells, 263–265
Acetylation, paramyxovirus M protein, 446
N-Acetylgalactosamine 64, 387
N-Acetylneuraminic acid (sialic acid)
  functional domains of attachment proteins,
    353–354
  ganglioside chemical structure, 411–412
  hemagglutinin-neuraminidase glycopro-
    teins, 53
  hemagglutinin protein gene, 89
  host cell receptors and, 408
    in Newcastle disease virus, 419–420
  in model systems, 408
  mumps virus, 25–26
  Newcastle disease virus, 419
  target tissue identification, 414
  viral replication, 63–64
  virion envelope proteins, 46
  virus assembly, 459–460
Acetylgalactosaminidase, 386
ACG, alternate initiation codon
  coding potential, 7
  C protein, 59, 184, 186
  ribosomal choice, 228–230
  nucleotide sequences, 328–329
α, 1-Acid glycoprotein, 408
Actin, interactions with virus components, 72
  cellular actin and virions, 469
  cocapping and protein association, 431
  M protein, 438, 450
  RNA synthesis, 266–268
  Sendai virus M protein, 241
  transcriptional choice, 226
  virus assembly, host cells and, 469
Actinomycin D, 11, 107, 264
Active site
  fusion protein, 370
  neuraminidase, 48, 53, 353–355, 370
  RNA polymerase, 6, 240
Acylation
  fatty acid, of viral glycoproteins, 45, 46,
    363, 374, 462

Acylation (cont.)
  neuraminic acid, 408
Adaptation, virus–host, 25, 89, 496
Adeno-associated virus, protein initiation at
  ACG, 184
Adenovirus
  antibody, in MS patients, 311
  cistron usage, 56
  C protein, 87, 189
  vaccine potential, 99
  V protein, 196
Adjuvants 492, 494, 499
Adsorption, virus-cell
  attachment G protein, 131
  fusion glycoprotein, 47
  mumps virus adaptation, 25–26
  receptor function, 414
  temperature-sensitive mutants, 20–21
  viral replication, 63–64
Adsorptive endocytosis, 408–410
Adsorption, paramyxovirus replication and,
  63–65
Aerosols
  receptor analogs, prophylactic, 422
  virus transmission, 509, 526
  autoimmune chronic active hepatitis, 309–
    310
  immunoglobulin subclass responses, 397
  immunologic response
    measles virus, 512–521
    mumps infection, 521–526
    parainfluenza virus, 526–527, 529
    pneumovirus, 527–529
    receptor analogs, 422
  measles virus
    developed countries, incidence in, 514–
      515
    mortality rates, less developed areas, 517–
      518
    virgin soil epidemics, 512–513
  mumps virus epidemiology, 522–523
    vaccine era, 525

569

DATE DUE

JAN 19 1992		
AUG 27 1992		
JAN 18 1994		
JAN 20 1998		
MAY 3 0 2006		